# Environmental Pollution
## Principles, Analysis and Control

# Environmental Pollution

## Principles, Analysis and Control

**P Narayanan**
MSc, PhD (X-ray Crystallography/Biophysics)
*Former* Professor, University of Mumbai, Mumbai

CBSPD

**CBS Publishers & Distributors** Pvt Ltd

New Delhi • Bengaluru • Chennai • Kochi • Kolkata • Lucknow • Mumbai
Gujarat • Hyderabad • Jharkhand • Nagpur • Patna • Pune • Uttarakhand

**Environmental Pollution**
Principles, Analysis and Control

**ISBN:** 978-81-239-1451-0



**First Edition:** 2007
Reprint: 2008, 2009, 2011, 2014, 2016, 2018, **2025**



Published by **Satish Kumar Jain** and produced by **Varun Jain** for
**CBS Publishers & Distributors** Pvt Ltd
4819/XI Prahlad Street, 24 Ansari Road, Daryaganj, New Delhi 110 002, India.
Ph: 011-23266838, 23289259 Website: www.cbspd.com
e-mail: delhi@cbspd.com

*Corporate Office:* 204 FIE, Industrial Area, Patparganj, Delhi 110 092
Ph: 011-4934 4934 Fax: 011-4934 4935
e-mail: publishing@cbspd.com; publicity@cbspd.com

*Branches*

- **Bengaluru:** Seema House 2975, 17th Cross, KR Road, Banasankari 2nd Stage, Bengaluru 560 070, Karnataka, India
  Ph: +91-80-26771678/79 Fax: +91-80-26771680 e-mail: bangalore@cbspd.com
- **Chennai:** 18/8B, Subbaraya Street, Shenoy Nagar, Chennai 600 030, Tamil Nadu, India
  Ph: +91-044-42032115, 044-26681266 e-mail: chennai@cbspd.com
- **Kochi:** 42/1325, 1326, Power House Road, Opp KSEB, Power House, Ernakulum Kochi 682 018, Kerala, India
  Ph: +91-484-4059061-65,67 Fax: +91-484-4059065 e-mail: kochi@cbspd.com
- **Kolkata:** 147, Hind Ceramics Compound, 1st Floor, Nilgunj Road, Belghoria, Kolkata-700056, West Bengal, India
  Ph: +033-25633055, 033-25633056 e-mail: kolkata@cbspd.com
- **Lucknow:** Basement, Khushnuma Complex, 7 Meerabai Marg (Behind Jawahar Bhawan), Lucknow-226001, UP, India
  Ph: +0522-4000032 e-mail: tiwari.lucknow@cbspd.com
- **Mumbai:** PWD Shed, Gala no 25/26, Ramchandra Bhatt Marg, Next to JJ Hospital Gate no. 2, Opp. Union Bank of India, Noorbaug, Mumbai-400009, Maharashtra, India
  Ph: 022-66661880/89 e-mail: mumbai@cbspd.com

*Representatives*

- Gujarat 0-9879558667
- Hyderabad 0-9885175004
- Jharkhand 0-9811541605
- Nagpur 0-8692091830
- Patna 0-9334159340
- Pune 0-9664372571
- Uttarakhand 0-9716462459

*Printed at* **Srk Graphics**

*Dedicated to my parents,*
*in their memory, and*
*to my wife and son, for*
*their constant support*
*and patience.*

# Preface

Environmental Science is a multi-disciplinary subject, with inputs from physics, chemistry, biology, geology, public health and other allied fields. However, there is no single book that covers all the core topics in Environmental Science, with emphasis on quantitative physical (and biophysical) principles and instrumental techniques in the analysis of environmental pollutants and their effects. The books that are available are methods and handbook texts that deal on air or water pollution and specific pollutants on particular systems that are of interest to experts and specific groups only. This book, "*Environmental Pollution: Principles, Analysis and Control,*" is intended to fill this gap—as an academic textbook to students on the subject, as a reference book for the environmental studies and as a source of information about environmental science. As such, it can be of use to physics, chemistry, biochemistry, biology, medicine, ecology (study of interrelationships between the forms of life and environment in the biosphere), pharmacology, and for anyone holding an environmentally-related interest.

The general outlay and the sequence of the topics are logically organized. Each section comprises interrelated chapters. An overview to each section is included to highlight the essential features of the chapters under it with a common theme, and also to enable interconnectivity between sections and chapters. Several sections (each section comprising several chapters) are devoted to individual systems (atmosphere, hydrosphere and lithosphere), collection and sampling and quality analysis methods. A conscious effort has been made in emphasizing the quantitative and analytical methods of analysis in environment pollution monitoring. Hence, a major portion of the book is devoted to these analytical chemical and physical (biophysical) techniques that include separation, electrolytic, spectroscopic, nuclear, diffraction (and other quantitative) methods. Attempt has also been made, though not explicitly stated, to bring out the underlying the biophysical principles and mechanisms while correlating the effects of environmental pollutants on ecology and living organisms.

A list of abbreviations, appendix and glossary terms and index would facilitate the reader with quick cross-reference to technical terms, phrases and concepts. With these efforts, it is hoped that the book would serve the purpose that it is intended for.

(Mumbai, 2006) **P. Narayanan**

# Acknowledgements

It is with pleasure to thank Mr. T. K. S. Murthy for his advice, guidance and his untiring efforts and help in the reorganization of subject material, proof reading and constant encouragement throughout the course of writing this book. Bibliographic help rendered by Ms. Swarna Murthy is gratefully acknowledged.

# Contents

*Acknowledgements* ..... vi
*Preface* ..... vii
*List of Tables* ..... xxvii
*List of Abbreviations* ..... xxxiii

**Chapter 1: Environment: Introduction** ..... **1**

1.1 Environmental Systems ..... 2
1.2 Environmental Cycles ..... 3
1.2.1 The Carbon Cycle ..... 3
1.2.2 The Nitrogen Cycle ..... 4
1.2.3 The Sulfur Cycle ..... 5
1.2.4 The Other Nutrient Cycles ..... 6
1.2.5 The Hydrologic Cycle ..... 6
1.3 Environmental Pollution ..... 6
1.3.1 Contaminants and Pollutants ..... 7
1.4 The Energy Problem ..... 9
1.4.1 Alternate Energy Sources ..... 9
1.5 Pollution Control Methods ..... 14
Bibliography ..... 15

**SECTION I: THE ATMOSPHERE**

**Chapter 2: The Atmosphere and its Characteristics** ..... **21**

2.1 Composition and Structure of the Atmosphere ..... 22
2.1.1 Troposphere ..... 24
2.1.2 Stratosphere ..... 24
2.1.3 Mesosphere and Thermosphere ..... 24

2.2 Energetics of the Earth-atmospheric System ........ 25
2.3 Weather and Climate ........ 29
2.3.1 Weather ........ 29
2.3.2 Climate ........ 29
2.4 Meterology ........ 31
2.4.1 Atmospheric Systems ........ 31
2.4.2 Atmospheric Dispersion ........ 32
2.5 Atmospheric Stability/Instability ........ 34
2.5.1 Temperature Inversion ........ 35
2.5.2 Lapse Rates and Atmospheric Stability ........ 35
2.5.3 Topological Influences ........ 40
2.6 Atmospheric Measurements ........ 40
2.6.1 Measurement of Wind Speed and Wind Direction ........ 41
2.6.2 Measurement of Pressure ........ 42
2.6.3 Measurement of Volume ........ 45
2.6.4 Measurement of Temperature ........ 45
Bibliography ........ 46

**Chapter 3: Air Pollution ........ 47**
3.1 Air Pollution Determinants ........ 48
3.1.1 Air Quality Impact Assessment (AQIA) ........ 49
3.2 Monitoring of Air Pollution ........ 49
3.2.1 Atmospheric Dispersion ........ 50
3.3 Air Pollution Dispersion Modeling ........ 52
3.3.1 Temperature Inversion and Atmospheric Stability ........ 53
3.3.2 Atmospheric and Topographic Effects ........ 53
3.3.3 Diurnal Variation of Air Pollution ........ 55
3.3.4 Plume Behavior ........ 56
3.4 Air Pollution Due to Reacting Constituents ........ 58
3.4.1 Principles of Thermodynamics ........ 59
3.4.2 Reaction Kinetics ........ 64
Bibliography ........ 69

**Chapter 4: Atmospheric Pollutants ........ 71**
4.1 Nature of Pollutants ........ 72
4.1.1 Physical Forms of Pollutants ........ 72
4.1.2 Origin of Air Pollutants ........ 72
4.1.3 Classification of Air Pollutants ........ 73

4.2 Role of Different Pollutants ..... 75

4.2.1 Particulates ..... 75

4.2.2 Gaseous Pollutants ..... 76

4.2.3 Radioactive Pollutants ..... 77

4.2.4 Organic Pollutants ..... 77

Bibliography ..... 78

**Chapter 5: Impact of Atmospheric Pollutants ..... 79**

5.1 Atmospheric Effects ..... 80

5.2 Role of Different Air Pollutants ..... 81

5.2.1 Particulates ..... 81

5.2.2 Aerosols ..... 82

5.3 Gaseous Air Pollutants ..... 82

5.3.1 Oxides of Sulfur ($SO_x$) ..... 83

5.3.2 Oxides of Nitrogen ($NO_x$) ..... 84

5.3.3 Ozone ($O_2$) ..... 88

5.3.4 Chlorofluorocarbons (CFCs) ..... 89

5.3.5 Oxides of Carbon ($CO_x$) ..... 90

5.4 Thermal Pollution ..... 93

5.4.1 Effects on Organisms ..... 93

5.4.2 Effects on Flora ..... 94

5.4.3 Global Impact of Thermal Pollution ..... 94

Bibliography ..... 100

**Chapter 6: Air Pollution Effects on Objects and Organisms ..... 101**

6.1 Effects of Air Pollutants on Inanimate Materials ..... 102

6.2 Effects of Air Pollutants on Ecosystems ..... 103

6.2.1 Global Effects ..... 103

6.2.2 Effects on Vegetation ..... 104

6.3 Effects of Air Pollution on Animal Kingdom ..... 106

6.3.1 Physicochemical Effects ..... 107

6.3.2 Physiological Systems ..... 109

6.3.3 Physiological Processes ..... 111

6.3.4 Toxic Effects of Particulate Matter ..... 112

6.3.5 Metal Toxicity ..... 112

6.3.6 Physiological-Effects of Gaseous Pollutants ..... 119

Bibliography ..... 121

## SECTION II: THE HYDROSPHERE

**Chapter 7: Water** .......... **127**

7.1 Distribution of Water and the Hydrologic Cycle .......... 128

7.1.1 The Hydrologic Cycle .......... 128

7.1.2 Other Effects of Water Cycle .......... 129

7.2 Characteristics of Water .......... 130

7.2.1 Properties of Water (General) .......... 130

7.2.2 Hydrogen Bond and its Influence .......... 130

7.3 Physical Properties of Water .......... 132

7.3.1 Density of Water .......... 132

7.3.2 Surface Tension of Water and Water Potential .......... 133

7.3.3 Thermal Properties of Water .......... 133

7.3.4 The Dipole Nature of Water .......... 134

7.3.5 Viscosity of Water .......... 135

7.4 Physicochemical Properties of Water .......... 135

7.4.1 Vapor Pressure .......... 137

7.4.2 Osmotic Pressure .......... 138

7.4.3 Humidity .......... 138

7.4.4 Dew Point .......... 139

7.4.5 Turbidity and Color .......... 139

7.5 Chemical Properties of Water .......... 139

7.5.1 Solubility (Solvent Action) .......... 139

7.5.2 Salinity .......... 140

7.5.3 Water Hardness .......... 141

7.5.4 Acidity and Alkalinity .......... 142

7.6 Biological Characteristics of Water .......... 142

Bibliography .......... 143

**Chapter 8: Water Pollution** .......... **145**

8.1 Fresh Water .......... 146

8.2 Factors that Affect Water Quality .......... 147

8.2.1 Water Quality Index (WQI) .......... 148

8.2.2 Parameters Influencing Water Quality .......... 148

8.3 Sources of Water Pollutants .......... 149

8.3.1 Point- and Non-point Sources .......... 150

8.3.2 Domestic Sewage .......... 151

8.3.3 Industrial Sewage .......... 151

8.3.4 Storm Sewage .......... 151

8.3.5 Agricultural Sources ... 151
8.3.6 Mining Wastes ... 152
8.3.7 Atmospheric Deposition ... 152
8.4 Ecological Impact ... 153
Bibliography ... 154

**Chapter 9: Ecological Effects of Water Pollution ... 155**
9.1 Ecosystems ... 156
9.1.1 Ecological Cycles (Food Chain) ... 157
9.1.2 Processes of Ecosystems ... 157
9.2 Plant Ecology ... 158
9.2.1 Standing Crop ... 158
9.3 Microbial Ecology ... 159
9.3.1 Microbial Transformation of Carbon ... 160
9.3.2 Microbial Transformation of Nitrogen (Nitrogen Fixation) ... 160
9.3.3 Other Microbial Activities ... 161
9.4 Ecology of the Water Bodies ... 162
9.4.1 Geographical Factors ... 162
9.4.2 Physical Factors ... 162
9.4.3 Chemical Factors ... 163
9.4.4 Biological Factors ... 164
Bibliography ... 166

**Chapter 10: Behavior of Pollutants in Water Bodies ... 167**
10.1 Water Bodies ... 168
10.2 Oxygen Demand in Water Bodies ... 168
10.2.1 Biochemical Oxygen Demand (BOD) ... 169
10.2.2 Chemical Oxygen Demand (COD) ... 171
10.3 Behavior of Pollutants in Streams ... 172
10.3.1 Flow Characteristics in Streams ... 173
10.4 Behavior of Pollutants in Stationary Water Bodies ... 173
10.4.1 Ecology of Stationary Water Bodies ... 174
10.5 Environmental Impact of Wastewater Discharge ... 177
10.5.1 Spatial Separation of the P-R Cycle ... 178
10.6 Environmental Risks ... 179
10.6.1 Environmental Risk Evaluation ... 179
10.6.2 Relative Risk ... 182
Bibliography ... 183

## SECTION III: THE LITHOSPHERE

**Chapter 11: Soil Pollution** ........ **189**
11.1 Soil Components ........ 190
11.1.1 Clays ........ 191
11.1.2 Humus Substances ........ 191
11.2 Role of the Soil ........ 191
11.3 Soil Pollutants ........ 192
11.4 Solid Waste Pollution ........ 194
11.5 Chemical Pollution ........ 196
11.5.1 Chemical Reactions in the Soil ........ 197
11.6 Industrial Wastes ........ 199
Bibliography ........ 200

**Chapter 12: Hazardous Pollutants** ........ **201**
12.1 Toxic Substances ........ 202
12.1.1 Physical Classification ........ 202
12.1.2 Physiological Classification ........ 203
12.2 Hazardous Chemicals ........ 203
12.3 Argochemicals ........ 205
12.3.1 Mode of Action of Pesticides ........ 206
12.4 Pharmaceuticals and Personal Care Products (PPCPs) ........ 211
Bibliography ........ 211

**Chapter 13: Radioactive Pollution** ........ **213**
13.1 Sources of Radioactivity ........ 214
13.1.1 Radionuclides ........ 215
13.2 Measurement of Radiation ........ 217
13.2.1 Half-life ........ 218
13.2.2 Absorbed Dose, D ........ 218
13.2.3 Dose Equivalent, H ........ 218
13.2.4 Linear Energy Transfer (LET) ........ 219
13.3 Radiation Exposure ........ 219
13.3.1 Natural Radioactive Hazards ........ 220
13.3.2 Man-made Radioactive Hazards ........ 221
13.3.3 Biological Effects of Radiation ........ 225
13.3.4 Unusual Exposure Situations ........ 226
Bibliography ........ 227

## SECTION IV: COLLECTION AND SAMPLING

**Chapter 14: Sample Collection** ........ **233**

14.1 Collection Methods ........ 234

14.2 Inertial Impactors ........ 236

14.2.1 High-volume (Hi-vol) Air Sampler ........ 237

14.2.2 Low-volume Air Sampler ........ 238

14.2.3 Anderson Multi-stage Sampler ........ 239

14.2.4 Quartz Crystal Microbalance Cascade Impactor (QCM) ........ 240

14.2.5 Radon Progeny Detectors ........ 240

14.3 Electrostatic Precipitators ........ 241

14.4 Thermal Precipitators ........ 242

14.4.1 Freeze-out Collection ........ 242

14.5 Gravitational Settling Chambers ........ 242

14.6 Filters ........ 242

14.7 Centrifugal Separators ........ 244

14.7.1 Cyclones ........ 244

14.7.2 Goetz Particle Spectrometer ........ 245

14.8 Scrubbers ........ 245

Bibliography ........ 249

**Chapter 15: Sampling Analysis Methods** ........ **251**

15.1 Sampling Process ........ 252

15.2 Sampling of Gaseous Pollutants ........ 253

15.2.1 Sampling of Particulate Matter ........ 253

15.2.2 Air Flow Measurement ........ 254

15.2.3 Sampling of Inorganic Gases ........ 255

15.2.4 Sampling of Gas-phase Organic Pollutants ........ 256

15.3 Sampling of Soil Pollutants ........ 256

15.3.1 Derivatization Method ........ 258

15.3.2 Soxhlet Extraction Method ........ 258

15.4 Sampling of Liquid-Phase Pollutants ........ 259

15.4.1 Equilibrium Dialysis Method ........ 259

15.4.2 Solvent Removal Methods ........ 259

15.5 Sample Preparation (Concentration) Methods ........ 260

15.5.1 Evaporation ........ 261

15.5.2 Sublimation ........ 261

15.5.3 Precipitation and Co-precipitation ........ 262

15.5.4 Filtration ........ 262

15.5.5 Floatation ........ 263
15.5.6 Extraction ........ 263
15.5.7 Sorption ........ 263
15.5.8 Electrochemical Methods ........ 264
15.5.9 Mineralization (Ashing) ........ 264
15.5.10 Zone Melting ........ 265
15.6 Environmental Quality Management (Eqm) ........ 266
15.6.1 Quality Assurance and Quality Control (QA/QC) ........ 266
15.6.2 Statistical Methods in Environmental Analysis ........ 268
Bibliography ........ 271

**SECTION V: QUALITY ANALYSIS**

**Chapter 16: General Techniques in Air Quality Monitoring ........ 279**
16.1 Qualitative Methods for Pollution Monitoring ........ 280
16.1.1 Indicator Methods ........ 281
16.1.2 Visibility Monitoring ........ 282
16.1.3 Odor ........ 283
16.2 Quantitative Methods for Pollution Monitoring ........ 283
16.2.1 Chemical Methods ........ 283
16.2.2 Electrochemical Methods ........ 286
16.2.3 Physical Methods ........ 286
16.2.4 Thermal Methods ........ 287
16.3 Other Methods ........ 288
16.3.1 Electrical Conductivity ........ 288
16.3.2 Radiation Monitoring Methods ........ 289
16.3.4 Nuclear Analytical Methods ........ 293
16.4 Analysis of Air Pollutants ........ 293
16.4.1 Analysis of Particulate Pollutants ........ 294
16.4.2 Surface Composition Analysis ........ 300
16.4.3 Analysis of Hazardous Air Pollutants (HAPs) ........ 301
16.4.4 Analysis of Radioactive Pollutants ........ 303
Bibliography ........ 305

**Chapter 17: General Techniques in Water Quality Monitoring ........ 307**
17.1 Factors Affecting Water Quality ........ 308
17.2 Physical Parameters in Water Quality Analysis ........ 308
17.2.1 Water flow Patterns ........ 308
17.2.2 Organoleptic Properties ........ 309

17.2.3 Temperature ..... 312
17.2.4 Electrical Conductivity ..... 314
17.3 Chemical Parameters in Water Quality Analysis ..... 315
17.3.1 Dissolved and Suspended Matter in Water ..... 315
17.3.2 Water Salinity and Hardness ..... 316
17.3.3 pH and Ion-sensitive Electrodes ..... 316
17.3.4 Determination of Inorganic Pollutants in Water ..... 319
17.3.5 Determination of Organic Pollutants in Water ..... 321
17.4 Biological Indicators of Water Quality ..... 325
17.4.1 Photosynthesis Measurement ..... 325
17.4.2 Bacterial Estimation of Pathogens ..... 329
Bibliography ..... 330

**SECTION VI: CHEMICAL METHODS IN ENVIRONMENTAL POLLUTION ANALYSIS**

**Chapter 18: Chemical Methods in Environmental Pollution Analysis—Gases and Aerosols ..... 337**
18.1 General Analysis Protocols for Gases and Aerosols ..... 338
18.1.1 Determination by the Ring-oven Technique ..... 338
18.1.2 Continuous Flow Chemical Analysis ..... 338
18.2 Oxidants and Radicals ..... 340
18.2.1 Oxygen ($O_2$) ..... 340
18.2.2 Ozone ($O_3$) ..... 341
18.2.3 Peroxides ..... 342
18.2.4 Radicals ..... 342
18.3 Carbon and its (Inorganic) Compounds ..... 343
18.3.1 Elemental Carbon ..... 343
18.3.2 Inorganic Carbon (IC) ..... 343
18.3.3 Total Organic Carbon (TOC) ..... 343
18.3.4 Carbon Monoxide (CO) ..... 345
18.3.5 Carbon Dioxide ($CO_2$) ..... 346
18.3.6 Methane ($CH_4$) ..... 347
18.3.7 Combustible Gases ..... 347
18.4 Nitrogen and Nitrogen Compounds ..... 347
18.4.1 Nitrogen ($N_2$) ..... 348
18.4.2 Dinitrogen Oxide (Nitrous Oxide, ($N_2O$) ..... 349
18.4.3 Nitrogen Monoxide (Nitric Oxide, NO) ..... 349
18.4.4 Nitrogen Dioxide ($NO_2$) and Nitrite ($NO_2^-$) ..... 350

18.4.5 Nitrous Acid ($HNO_2$) and Nitric Acid ($HNO_3$) ........ 352
18.4.6 Nitrate ($NO_3^-$) ........ 353
18.4.7 Ammonia ($NH_3$) ........ 354
18.4.8 Photochemical Smog ........ 355
18.4.9 Cyanide ($CN^-$) ........ 355
18.5 Sulfur and Sulfur Compounds ........ 356
18.5.1 Sulfide ($S_2^-$) ........ 356
18.5.2 Hydrogen Sulfide ($H_2S$) ........ 356
18.5.3 Sulfur Dioxide ($SO_2$) ........ 357
18.5.4 Sulfite $(SO_3)^{2-}$ ........ 360
18.5.5 Sulfate $(SO_4)^{2-}$ ........ 360
18.5.6 Sulfuric Acid ($H_2SO_4$) ........ 360
18.6 Halogens and Halogen Compounds ........ 360
18.6.1 Fluoride ($F^-$) ........ 361
18.6.2 Chloride ($Cl^-$) ........ 361
18.6.3 Bromide ($Br^-$) ........ 363
18.6.4 Iodide ($I^-$) ........ 363
18.6.5 Halogen Compounds ........ 363
Bibliography ........ 364
**Chapter 19: Chemical Methods in Environmental Pollution Analysis—Metals ........ 367**
19.1 Speciation of Metals ........ 368
19.2 Alkali Metals ........ 371
19.2.1 Lithium (Li) ........ 371
19.2.2 Sodium (Na) ........ 372
19.2.3 Potassium (K) ........ 372
19.2.4 Rubidium (Rb) and Cesium (Cs) ........ 373
19.3 Copper Subgroup Metals ........ 373
19.3.1 Copper (Cu) ........ 373
19.3.2 Silver (Ag) ........ 374
19.3.3 Gold (Au) ........ 374
19.4 Alkaline-Earth Metals ........ 375
19.4.1 Beryllium (Be) ........ 375
19.4.2 Magnesium (Mg) ........ 376
19.4.3 Calcium (Ca) ........ 377
19.4.4 Strontium (Sr) and Barium (Ba) ........ 377
19.5 Zinc Subgroup Metals ........ 378
19.5.1 Zinc (Zn) ........ 378

19.5.2 Cadmium (Cd) ..... 379
19.5.3 Mercury (Hg) ..... 380
19.6 Boron Subgroup Metals ..... 382
19.6.1 Boron (B) ..... 382
19.6.2 Aluminum (Al) ..... 382
19.7 Gallium (Ga), Indium (In), and Thallium (Th) ..... 383
19.8 Germanium (Ge), Tin (Sn), and Lead (Pb) Elements ..... 383
19.8.1 Germanium (Ge) ..... 384
19.8.2 Tin (Sn) ..... 384
19.8.3 Lead (Pb) ..... 384
19.9 Titanium Subgroup Metals ..... 385
19.10 Nitrogen Subgroup Metalloids ..... 386
19.10.1 Arsenic (As) ..... 386
19.10.2 Antimony (Sb) ..... 388
19.10.3 Bismuth (Bi) ..... 388
19.11 Vanadium Subgroup Metals ..... 388
19.12 Chromium Subgroup Metals ..... 389
19.12.1 Chromium (Cr) ..... 389
19.12.2 Molybdenum (Mo) ..... 390
19.12.3 Tungsten (W) ..... 390
19.13 Manganese Subgroup Metals ..... 390
19.13.1 Manganese (Mn) ..... 390
19.14 Iron Family Metals ..... 391
19.14.1 Iron (Fe) ..... 391
19.14.2 Cobalt (Co) ..... 392
19.14.3 Nickel (Ni) ..... 392
19.15 Platinum Subgroup Metals ..... 393
Bibliography ..... 395

**Chapter 20: Chemical Methods in Environmental Pollution Analysis—Inorganic, Non-metallic ..... 397**

20.1 Silicon (Si) ..... 398
20.1.1 Asbestos ..... 398
20.2 Phosphorus (P) ..... 399
20.3 Selenium (Se) ..... 399
Bibliography ..... 400

**Chapter 21: Chemical Methods in Environmental Pollution Analysis—Organics ..... 401**

21.1 Classification of Organic Pollutants ..... 402

21.2 Conventional Organic Pollutants ..... 403
21.2.1 Oxygen-containing Alkyl Compounds ..... 403
21.2.2 Alkyl Nitrates ..... 405
21.2.3 Oils and Greases ..... 406
21.2.4 Phenolic Compounds ..... 406
21.3 Hydrocarbons ..... 408
21.3.1 Aliphatic Hydrocarbons ..... 408
21.3.2 Mononuclear Aromatic Hydrocarbons ..... 408
21.3.3 Polynuclear Aromatic Hydrocarbons (PAHs) ..... 409
21.4 Synthetic Organic Compounds ..... 410
21.4.1 Halogenated Hydrocarbons ..... 410
21.5 Agrochemicals ..... 413
Bibliography ..... 414

**SECTION VII: PHYSICAL METHODS IN ENVIRONMENTAL POLLUTION ANALYSIS**

**Chapter 22: Physical Methods in Environmental Pollution Analysis—Separation Techniques ..... 421**
22.1 Sedimentation Methods ..... 422
22.2 Barrier Separation Methods ..... 424
22.2.1 Dialysis ..... 424
22.2.2 Ultrafiltration ..... 425
22.2.3 Reverse Osmosis ..... 425
22.2.4 Electrodialysis ..... 425
22.3 Solvent Extraction ..... 425
22.4 Sorption (Electrophoretic) Methods ..... 427
22.4.1 Paper Chromatography ..... 429
22.4.2 Thin Layer Chromatography (TLC) ..... 431
22.4.3 Liquid Chromatography (LC) ..... 432
22.5 Gas Chromatography (GC) ..... 442
22.5.1 Principles of Operation ..... 443
22.5.2 Detector Systems ..... 445
22.5.3 Super Critical Fluid Chromatography (SFC) ..... 453
22.5.4 Hyphenated Techniques ..... 454
22.5.5 Applications ..... 455
22.6 Mass Spectrometry (MS) ..... 455
22.6.1 Ionization Chamber ..... 456
22.6.2 The Mass Analyzer ..... 457

22.7 Electrophoresis ..... 459
22.7.1 Gel Electrophoresis ..... 460
22.7.2 Capillary Zone Electrophoresis (CZE) ..... 461
22.8 Other Methods ..... 461
Bibliography ..... 461
**Chapter 23: Physical Methods in Environmental Pollution Analysis— Electroanalytical Techniques ..... 465**
23.1 Voltammetric Methods ..... 467
23.1.1 Stripping Voltammetry ..... 468
23.1.2 Polarography ..... 470
23.2 Amperometric Methods ..... 471
23.2.1 Coulometry ..... 471
23.3 Conductimetric Methods ..... 472
23.3.1 Gasometry ..... 473
23.3.2 Water Quality Monitoring ..... 473
23.3.3 Bioorganic Analysis ..... 473
23.4 Potentiometric Methods ..... 474
23.4.1 pH Electrode ..... 475
23.4.2 Gas-sensing and Ion-selective Electrodes ..... 475
23.4.3 Enzyme Electrodes ..... 477
23.4.4 Application of Potentiometric Methods ..... 477
23.5 Determination of Oxygen ..... 478
23.5.1 Ion-selective Electrode Method ..... 479
23.5.2 Magnetic Detection ..... 479
23.5.3 Measurement of Dissolved Oxygen (DO) ..... 480
23.6 Electroanalytical Methods in Pollution Monitoring ..... 481
23.6.1 Voltammetry and Polarography ..... 482
Bibliography ..... 483
**Chapter 24: Physical Methods in Environmental Pollution Analysis— Spectroscopic Techniques ..... 485**
24.1 UV-Visible Spectroscopy ..... 486
24.1.1 Spectrophotometry ..... 489
24.1.2 Turbidimetry/Nephelometry ..... 492
24.2 Absorption and Reflectance Spectroscopies ..... 493
24.2.1 Action Spectroscopy ..... 493
24.2.2 Diffuse Reflectance Spectroscopy (DRS) ..... 494

24.2.3 pH-dependent UV absorption Spectroscopy ..... 494
24.2.4 Correlation Spectrometry ..... 495
24.2.5 Differential Optical Absorption Spectrometry (DOAS) ..... 496
24.2.6 Second-derivative Spectroscopy ..... 497
24.2.7 Doppler-limited Spectroscopy ..... 497
24.2.8 Resonance-Ionization Spectroscopy (RIS) ..... 498
24.2.9 Photoacoustic Spectroscopy (PAS) ..... 498
24.2.10 Thermal Lens Spectrometry (TLS) ..... 498
24.3 Fluorescence Spectroscopy ..... 499
24.3.1 Induced Fluorescence ..... 502
24.4 Luminescence-Based Assays ..... 503
24.4.1 Enzyme-mediated Luminescence Assays ..... 504
24.4.2 Chemiluminescence Immunoassays ..... 505
24.5 Infrared (IR) Spectroscopy ..... 506
24.5.1 Applications ..... 507
24.5.2 Non-dispersive (Fourier-transform) Spectrometry ..... 509
24.5.3 Non-dispersive Interferometry ..... 511
24.6 Raman Spectroscopy ..... 512
24.7 Electron Spin Resonance (ESR) Spectroscopy ..... 513
24.8 Atomic Spectroscopy ..... 514
24.8.1 Atomic Absorption Spectrometry (AAS) ..... 514
24.8.2 Atomic Emission Spectrometry (AES) ..... 524
Bibliography ..... 528
**Chapter 25: Physical Methods in Environmental Pollution Analysis— Nuclear Analytical Techniques** ..... **531**
25.1 Radiation Measuring Instruments ..... 532
25.1.1 Gas-filled Radiation Counters ..... 533
25.1.2 Semiconductor Detectors ..... 535
25.1.3 Scintillation Counters ..... 535
25.2 Features of X-ray Spectroscopy ..... 537
25.2.1 Sources of X-rays ..... 540
25.2.2 Filters and Monochromators ..... 541
25.2.3 Detectors ..... 542
25.2.4 Pulse-height Analyzers ..... 543
25.3 X-ray Absorption and Emission (Fluorescence) ..... 543
25.3.1 X-ray Absorption Spectrometry ..... 543
25.3.2 X-ray Emission (Fluorescence) Spectrometry ..... 544

25.3.3 Total Reflection X-ray Fluorescence (TXRF) ........ 544
25.3.4 X-ray Spectrometers ........ 545
25.4 Ion Spectroscopies ........ 547
25.4.1 X-ray Photoelectron Spectroscopy (XPS) ........ 547
25.4.2 Auger Electron Spectroscopy ........ 548
25.5 Ion-Beam Analysis (IBA) Techniques ........ 550
25.5.1 Nuclear Reaction Analysis (NRA) Methods ........ 550
25.5.2 Measurement of Natural Radioactivity ........ 551
25.5.3 Neutron Activation Analysis (NAA) ........ 551
25.6 Choice of Analytical Techniques ........ 554
25.7 Radioimmunoassays ........ 554
25.7.1 Radioimmunoassay (RIA) ........ 555
25.7.2 Immunoradiometric Assay (IRMA) ........ 556
Bibliography ........ 558

**Chapter 26: Physical Methods in Environmental Pollution Analysis— Other Techniques ........ 561**

26.1 X-ray Diffraction Methods ........ 562
26.1.1 Principles of X-ray Diffraction ........ 563
26.1.2 Unit Cell and Space Group ........ 564
26.1.3 Structure Determination ........ 564
26.2 Microscopic Methods ........ 566
26.2.1 Electron Diffraction ........ 567
26.3 Noise Pollution ........ 567
26.3.1 Intensity of Sound ........ 567
26.3.2 Sound Levels ........ 568
26.3.3 Noise Pollution Measurement ........ 568
26.3.4 Effects of Noise Pollution ........ 572
26.3.5 Noise Pollution Control ........ 573
Bibliography ........ 573

**SECTION VIII: ENVIRONMENTAL POLLUTION CONTROL**

**Chapter 27: Air Pollution Control Methods ........ 581**

27.1 Control Strategy ........ 582
27.2 Collection Methods ........ 582
27.2.1 Precipitation Methods ........ 582
27.2.2 Absorption Methods ........ 582
27.2.3 Adsorption Methods ........ 583

27.2.4 Chemical Conversion Methods ... 583
27.2.5 Thermal Oxidation Methods ... 583
27.3 Control of Gaseous Pollutants ... 584
27.3.1 Control of $SO_2$ ... 584
27.3.2 Control of Oxides of Nitrogen ($NO_x$) ... 584
27.3.3 Control of Hydrocarbons and Carbon Monoxide (CO) ... 586
27.4 Control of Industrial Air Pollution ... 587
27.4.1 Power Generation Plants ... 587
27.4.2 Non-conventional Energy Sources ... 588
27.4.3 Metallurgical Industry ... 589
27.4.4 Chemical Industry ... 591
Bibliography ... 592

**Chapter 28: Water Pollution Control Methods ... 595**

28.1 Water and Wastewater ... 596
28.2 Wastewater Treatment Processes ... 597
28.2.1 Primary Wastewater Treatment Processes ... 599
28.2.2 Secondary Wastewater Treatment Processes ... 600
28.2.3 Different Wastewater Treatment Systems ... 601
28.2.4 Tertiary Wastewater Treatment Processes ... 605
28.2.5 Secondary-Tertiary Wastewater Treatment Processes ... 610
28.3 Industrial Wastewater Treatment Processes ... 612
28.3.1 Removal of Heavy Metals ... 612
28.4 Sludge Removal/Disposal ... 613
28.4.1 Sludge Treatment ... 613
28.4.2 Landfill ... 614
28.4.3 Incineration ... 614
Bibliography ... 615

**Chapter 29: Land Pollution Control Methods ... 617**

29.1 Land Pollutants and their Disposal ... 618
29.1.1 Landfill ... 619
29.1.2 Composting ... 620
29.1.3 Incineration ... 620
29.2 Recycling ... 621
29.3 Radioactive Waste Disposal ... 622
29.3.1 Low-level Radioactive Waste Disposal ... 623
29.3.2 High-level Radioactive Waste Disposal ... 623
Bibliography ... 624

**Chapter 30: Environmental Health** ........ **625**
30.1 Environmental Health and Engineering ........ 626
30.1.1 Environmental Ecology ........ 626
30.1.2 Industrial Ecology ........ 627
30.2 Environmental Toxicology Testing ........ 627
30.2.1 Physicochemical Methods ........ 627
30.2.2 Biomonitoring ........ 628
30.3 Environmental Pollution and Health ........ 630
30.3.1 Chemical Pollutants ........ 630
30.3.2 Biological Pollutants ........ 631
30.3.3 Toxins and Poisons ........ 631
30.4 Food Preservation ........ 632
30.4.1 Moisture-Solids Balance ........ 633
30.4.2 High-temperature Treatment ........ 633
30.4.3 Freeze-drying ........ 633
30.4.4 Additives ........ 634
30.4.5 Ionizing Radiation Treatment ........ 635
30.4.6 Role of Genetic Engineering in Food Technology ........ 636
30.5 Biodegradation and Bioremediation ........ 640
30.5.1 Biodegradation ........ 640
30.5.2 Bioremediation ........ 642
Bibliography ........ 644

## APPENDICES

*Appendix I*: System Internationale (SI) Unit Prefixes ........ 648
*Appendix II*: Fundamental and Derived Constants in System Internationale (SI) Units ........ 648
*Appendix III*: Elements and their Atomic Mass (based on atomic mass 12C) ........ 649

*Glossary* ........ **651**

*Index* ........ **661**

# List of Tables


| | | |
|---|---|---|
| 1.1 | Future Energy Sources | 13 |
| 1.2 | Air Quality Standards | 15 |
| 2.1 | Subdivision of Atmospheric Regions | 23 |
| 2.2 | Relative Total Energy Transfer in various Processes | 25 |
| 2.3 | Köppen's Classification of Climate Zones | 31 |
| 2.4 | Relationship between Wind flow and Altitude above the Earth's Surface | 33 |
| 2.5 | Pasquill's Stability Classes of Winds | 34 |
| 2.6 | Beaufort's Classification of Wind Types | 41 |
| 2.7 | Conversion Table for Pressure Measurements | 43 |
| 2.8 | Measurement of Pressure by Manometric Methods | 45 |
| 3.1 | Composition of Dry Air at Sea Level | 48 |
| 3.2 | Air Pollution Episodes | 55 |
| 4.1 | Typical Secondary Pollutants | 73 |
| 4.2 | Classification of Air Pollutants | 74 |
| 4.3 | General Effects of some Common Pollutants | 75 |
| 4.4 | Gaseous Pollutants in the Atmosphere | 76 |
| 5.1 | Relative Reactivities of Hydrocarbons toward HO* Radical | 81 |
| 5.2 | Residence Times of Aerosols in the Atmosphere | 82 |
| 5.3 | Automobile Exhaust-gas Composition | 92 |
| 5.4 | Heating Values of Some Carbonaceous Fuels | 96 |
| 5.5 | Environmental Pollutants and their Effects | 97 |
| 6.1 | Effects of Air Pollutants on Inert Materials | 103 |
| 6.2 | Photosynthesis and Respiration Processes in Plants | 105 |
| 6.3 | Effects of Air Pollutants on Plants | 106 |
| 6.4 | Composition of Aerosol/Particulate Matter | 107 |
| 6.5 | Trace Elements Essential to Human/Animal Metabolism | 113 |
| 6.6 | Adverse Health Effects Associated with some Trace Elements | 113 |

6.7 Preferred Binding Sites of Metal Ions with Nucleic Acid Constituents .................. 114
6.8 HbCO in Blood and Symptoms .................. 120
6.9 Air Pollutants and their Effects on Animals/Humans .................. 120
6.10 Primary and Secondary Standards for some Air Pollutants .................. 121
7.1 Distribution of Earth's Water Supply .................. 128
7.2 Some Physical Properties of Water .................. 136
7.3 Impact of Physical Properties of Water .................. 136
7.4 Vapor Pressure of some Liquids .................. 137
7.5 Major Chemical Constituents in Natural Waters .................. 140
8.1 Maximum permitted Levels (mg/L) of Ions in Drinking Water .................. 147
8.2 Classification of Water Pollutants .................. 147
8.3 Contributions from various Polluting Sources that Affect Water Quality .................. 149
8.4 Physical Characteristics of Water Polluted due to Human Activities .................. 149
8.5 Pollutants in Sewage Discharges .................. 152
8.6 Carcinogenic and Mutagenic Substances .................. 152
8.7 Trace Level Metals Found in Water .................. 153
8.8 Hydrological Effects of Urbanization .................. 154
9.1 Standards for Fresh Surface Waters .................. 162
9.2 Solubility of Oxygen in Water as a Function of Temperature .................. 163
9.3 Effects of Acidic Water on Ecology .................. 164
9.4 Vectors causing Waterborne Diseases .................. 164
10.1 End-Products of Bacterial Decomposition of Organic Matter .................. 168
10.2 Essential Plant Nutrients in Water Bodies .................. 174
11.1 Generic Soil Profile on the Earth's Surface .................. 190
11.2 Classification of Particulate Matter According to Size .................. 190
11.3 Factors Contributing to Land Pollution .................. 195
11.4 Domestic and Municipal Solid Wastes and Patterns of Disposal .................. 195
11.5 Water Solubility and MCL of BTEX Chemicals .................. 197
11.6 Sources and Types of Industrial Waste .................. 199
11.7 Some Industrial Pollutants and their Effects .................. 200
12.1 Industrial Chemicals/Substances and their Effects .................. 204
12.2 Common Pesticides Found in Water Bodies .................. 206
12.3 Persistent Pesticides in the Environment .................. 210
12.4 Species Dependence of Toxicity of Pesticides .................. 211
13.1 Some Primordial Radionuclides and their Half-Lives .................. 214
13.2 Weighting Factors for Ionizing Radiation .................. 219
13.3 Linear Energy Transfer (LET) with respect to X-rays .................. 219
13.4 Radionuclides in a Human Body (70 kg) .................. 220
13.5 Natural Radioactive Isotopes in Common Foods .................. 220

13.6 Some Important Radionuclides in Ocean Waters ..... 220
13.7 Doses Received During Routine X-ray Diagnosis ..... 222
13.8 Radiation Hazards in Nuclear Technology ..... 223
13.9 Some Radioisotopes and their Half-Lives ..... 224
13.10 Important Radioisotopes in the Fallout ..... 226
14.1 Particulate-size Selection in Anderson Sampler ..... 240
14.2 Characteristics of some Airfilter Materials ..... 243
14.3 Efficiency of Cyclones Depending on the Particle Size ..... 245
14.4 Separation Characteristics of some Collectors ..... 247
15.1 Preservation of Water Samples ..... 253
15.2 Characteristics of some Oxidizing Agents ..... 265
15.3 Common Analytical Techniques in Environmental Studies ..... 266
16.1 Qualitative Pollution Monitoring Methods ..... 280
16.2 Indicator Tubes for Detection of some Gases/Vapors ..... 281
16.3 Continuous Monitoring Methods for Air Pollutants ..... 284
16.4 Detection Methods for Gaseous Air Pollutants ..... 294
16.5 Limits of Detection of Gaseous Air Pollutants ..... 295
16.6 Size-Distribution Measuring Techniques for Aerosols ..... 297
16.7 Instrumental Techniques for Measurement of Particulate Matter ..... 300
16.8 Instrumental Techniques for Monitoring Elemental Pollutants ..... 301
16.9 Methods for Determining Toxic Organic (TO) Pollutants in Ambient Air ..... 301
16.10 Radiation Detectors ..... 303
17.1 Important Parameters in Determining the Water Quality ..... 309
17.2 Physical Techniques for Determination of Solutes in Water ..... 315
18.1 Chemical Methods for $SO_2$ Determination ..... 358
18.2 Physicochemical Properties of Halogen Elements ..... 361
18.3 A Summary of Wet-chemical (Colorimetric) Analysis of Gases and Aerosols ..... 364
19.1 Some Metal-chelate Extraction Systems ..... 370
19.2 Use of Masking Agents in the Dithizone Extraction System ..... 370
19.3 Some Physical Properties of Alkali Metals ..... 371
19.4 Some Physical Properties of Copper Subgroup Metals ..... 373
19.5 Some Physical Properties of Alkaline-Earth Metals ..... 375
19.6 General Solubility Rules for Solubility of Salts in Water ..... 376
19.7 Some Physical Properties of Zinc Subgroup Metals ..... 378
19.8 Some Physical Properties of Boron Subgroup Elements ..... 382
19.9 Some Physical Characteristics of Group (IVA) Elements ..... 384
19.10 Some Physical Characteristics of Group V(A) Elements ..... 386
19.11 Uses of Arsenic and its Compounds ..... 387
19.12 Some Physical Properties of Iron Family Metals ..... 391

19.13 Some Physical Properties of the Platinum Group Metals .......... 393
19.14 Some Wet-Chemical Methods for Determination of Metals .......... 394
20.1 Wet-chemical Analysis of Inorganic, Non-Metallic Compounds .......... 400
21.1 Pollutants in Exhaust Gases .......... 403
21.2 Some Commercially Important Halogenated Hydrocarbons .......... 411
21.3 Wet-Chemical Analysis of some Organic Compounds .......... 413
22.1 Classification of Separation (Concentration) Methods .......... 422
22.2 Classification of Chromatographic Techniques .......... 429
22.3 Mobile-Stationary Phase Combination for Paper Chromatography .......... 431
22.4 Filter Papers for Paper Chromatography .......... 431
22.5 Sorbents for Thin Layer Chromatography (TLC) .......... 432
22.6 Sorbents and Eluents use in Solid-phase LC and HPLC .......... 434
22.7 Sorbents for Adsorption Chromatography .......... 436
22.8 Gel Beads Used in Size-Exclusion Chromatography .......... 436
22.9 Commonly Used Ion-Exchanger Matrix Materials .......... 440
22.10 Some Elements that can be Volatilized .......... 443
22.11 Detectors for Monitoring Volatile Organic Compounds (VOCs) .......... 446
22.12 Comparison of HPLC and GC Separation Methods .......... 455
23.1 Gas-Sensing and Ion-Selective Electrodes .......... 476
23.2 Some Pesticides and their Active Group(s) .......... 483
24.1 Spectroscopic Methods based on Interaction of Radiation with Matter .......... 486
24.2 Some Chromophore Moieties and their Absorption Maxima .......... 488
24.3 Detection Limit in Second-Derivative Spectroscopy .......... 497
24.4 Laser-induced Fluorescence Detection .......... 503
24.5 Energy Sources Used for Atomization .......... 516
24.6 Flame Gases used in AAS .......... 517
24.7 Comparison of APDC, DDTC and DZ as Extracting Agents .......... 518
24.8 Comparison of Detection Limits for various AAS Methods .......... 520
24.9 Summary of Atomization Methods in AAS .......... 522
24.10 Wavelength Maxima of Various Elements by AAS .......... 522
24.11 Characteristic Wavelengths and Detection Limits for Metals by AES Method .......... 527
24.12 Sensitivity of Various Methods for Metal Analysis .......... 528
25.1 X-ray Line Spectra of Some Elements .......... 538
25.2 Some Radioisotope Sources for X-ray Spectroscopy .......... 540
25.3 Choice of Detectors for Measuring X-rays .......... 542
25.4 Crystal Analyzers Employed in Wavelength-Dispersion X-ray Spectrometers .......... 545
25.5 Comparison of Some Electron Spectroscopies .......... 550
25.6 Detection Limits of Selected Elements in Typical Environmental Matrices by NAA .......... 553

25.7 Comparison of Some Analytical Techniques for Analysis of Trace Elements ......... 554
25.8 Detection Limits of Some Analytical Methods for Trace Elements ......... 554
25.9 Features of Competitive (RIA) and Non-Competitive (IRMA) Assays ......... 558
26.1 Target-Filter Combination for Monochromatic X-ray Generation ......... 562
26.2 Sound Pressure Levels ......... 568
27.1 Global Wind Power Utilization ......... 589
27.2 Air Pollutants and Control Methods in Metallurgical Operations ......... 590
27.3 Pollutants Due to Inorganic Chemical Processes ......... 590
27.4 Organic Air Pollutants and Control Methods ......... 591
27.5 Ambient Air Quality Standards ......... 592
28.1 Characteristics of Water and Standards ......... 596
28.2 Substances that Impair Water Quality ......... 597
28.3 General Wastewater Treatment Processes ......... 598
28.4 Chemical Coagulants to Precipitate Pollutants in Water ......... 600
28.5 Performance of Wastewater Treatment Processes ......... 606
28.6 Methods for Removal of Organics in Water ......... 608
28.7 Removal of Waste in Industrial Wastewater Treatment ......... 613
28.8 Removal of Trace Metals in Industrial Wastewater Treatment ......... 614
29.1 Land Pollutants and their Control Methods ......... 618
29.2 Solid-Waste Disposal Methods ......... 619
29.3 Methods for Recycling Solid-Wastes ......... 622
29.4 Solid-Waste Management Methods ......... 622
30.1 Substances Associated with Chronic Poisons/Carcinogens ......... 632

# List of Abbreviations

| | |
|---|---|
| AAS | Atomic Absorption Spectrometry (Spectroscopy) |
| AES | Atomic Emission Spectrometry (Spectroscopy) |
| ATMOS | Atmospheric Trace Molecule Spectroscopy |
| BOD | Biochemical Oxygen Demand |
| CARS | Coherent Antistokes Raman Spectroscopy |
| COD | Chemical Oxygen Demand |
| CZE | Capillary Zone Electrophoresis |
| DIAL | Differential Absorption LIDAR |
| DO | Dissolved Oxygen |
| DOAS | Differential Optical Absorption Spectrometry |
| DRS | Diffuse Reflection Spectroscopy |
| ELISA | Enzyme-Linked Immunosorbent Assay |
| EQM | Environmental Quality Management |
| ESR | Electron Spin Resonance |
| FRS | Flame Resonance Spectrometry |
| GC | Gas Chromatography |
| GC-MS | Gas Chromatography-Mass Spectrometry |
| GLC | Gas-Liquid Chromatography |
| GSC | Gas-Solid Chromatography |
| HAP | Hazardous Air Pollutant |
| HPLC | High Performance Liquid Chromatography |
| IBA | Ion-Beam Analysis |
| ICP | Inductively-Coupled Plasma |
| IRMA | Immunoradiometric Assay |
| LC | Liquid Chromatography |
| LED | Light-Emitting Diode |
| LIDAR | Light Detection and Ranging |

| | |
|---|---|
| MS | Mass Spectrometry |
| NAA | Neutron Activation Analysis |
| NDIR | Non-Dispersive Infrared (spectrometry) |
| NRA | Nuclear Reaction Analysis |
| NVOC | Non-Volatile Organic Compound |
| PAGE | Polyacrylamide Gel Electrophoresis |
| PAH | Polyaromatic Hydrocarbon |
| PAN | Peroxyacetyl Nitrate |
| PAS | Photoacoustic Spectroscopy |
| PEG | Polyethyleneglycol |
| PIGE | Particle-Induced Gamma-ray Emission |
| PIXE | Particle-Induced X-ray Emission |
| PMT | Photomultiplier Tube |
| RIA | Radioimmunoassay |
| RIS | Resonance Ionization Spectroscopy |
| SBUV | Solar Backscatter Ultraviolet (spectrometry) |
| SDS-PAGE | Sodiumdodecylsulfate-PAGE |
| SEC | Size-Exclusion Chromatography |
| SEM | Scanning Electron Microscopy |
| SFC | Supercritical Fluid Chromatography |
| SIMS | Secondary Ion Mass Spectrometry |
| STEM | Scanning Transmission Electron Microscope |
| SVOC | Semi-Volatile Organic Compound |
| TLC | Thin-layer Chromatography |
| TLS | Thermal Lens Spectrometry |
| TOC | Total Organic Carbon |
| TOD | Total Oxygen Demand |
| TOMS | Total Ozone Mapping Spectrometer |
| TRRS | Time-Resolved Resonance Raman Spectroscopy |
| TRXF | Total Reflection X-ray Fluorescence |
| VOC | Volatile Organic Compound |
| WQA | Water Quality Index |
| XMA | X-ray Microanalysis |
| XPS | X-ray Photoelectron Spectroscopy |
| XRF | X-ray Fluorescence |

# 1

# Environment: Introduction

The environment literally means the surrounding. The environment is the aggregate of all those things and set of conditions which directly or indirectly influence not only the life of organisms but also the communities at a particular place.

Treshow (1970) defined environment as "environment includes all the factors and forces prevailing internally and externally or around and in the plants and animals." It includes all light, moisture, wind, temperature, soil, organisms, pollutants, insecticides, pesticides, radioisotopes and man. The total environment extends from the microcosm, within every cell to the cosmos of the atmosphere and universe. When the environment includes non-living things like soil, water, wind and temperature, etc., and forces of nature like solar radiation, gravity and molecular energy, it is known as physical environment; on the other hand, when it includes all living things that affect the organisms, it is known as biotic environment. This is how, environment is a complex of many factors which interact not only with the organisms but among themselves. Broadly speaking, the complete environment of the earth can be divided into four subdivisions—Atmosphere, Lithosphere, Hydrosphere and Biosphere.

This Chapter illustrates a brief account of environment, cycles of environment, heat budget of earth, environmental pollution and sources of energy.

Environment includes various organisms and their physical surroundings. Ecology, also called environmental biology, is the study of the relationships between organisms and their environment. Thus, environmental science is a multi-disciplinary subject that encompasses physics, chemistry, biology, geology, and meteorology, hydrology and many other branches of physical and natural sciences.

The socio-economic development of any nation/country/region is closely linked to its industrial progress, with the energy sector being the major driving force. Achieving a sustainable development, without drastically disrupting the environmental balance of nature, is the challenging problem facing the mankind today. Any industrial activity will pose some degree of environmental impact that could lead to environmental degradation and hazards to well-being and health of living organisms. Keeping in mind this major endeavor, all environmental studies incorporate appropriate physical, chemical, biological and mathematical techniques to analyze reactions in the atmosphere, determine specific pollutants, their effects on the environment and their transport. The objective is twofold—(1) understanding the complex dynamic processes that take place in the environmental systems and (2) gathering quality information on the pollutants in the atmosphere, hydrosphere and the lithosphere and understand their effects on the environmental balance in the biosphere. The aim is to take appropriate corrective steps to maintain the quality of the environment.

## 1.1 ENVIRONMENTAL SYSTEMS

The planet Earth, also called the geosphere, consists of air and water vapor, the essential components of the troposphere (the lowest atmospheric layer), water, the essential component of the hydrosphere, and land or soil, the essential component of the lithosphere. The biosphere is a global ecosystem composed of living organisms (biota) and the non-living (abiotic) factors from which they derive energy and nutrients. The physical systems that make up the biosphere are closely interacting systems (Fig. 1.1).

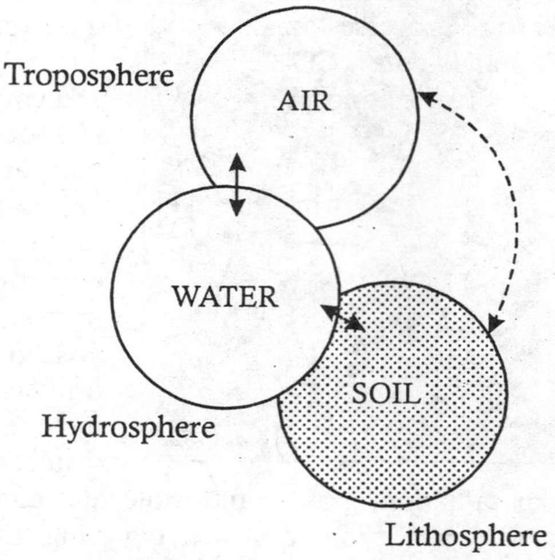


**Fig. 1.1:** Physical Systems that make up the Biosphere

Living communities and nonliving environment are inseparable, interrelated and constantly interact with each other in exchange of matter and energy. Such a combined system is called an ecosystem. An ecosystem must include at least an autotroph (primary producer), a decomposer (consumer), water, a source and sink of energy, and all chemical elements required by the autotroph and the decomposers. Ecosystems may be further subdivided into smaller biotic units called communities.

There exists a tremendous diversity of life in each ecological system (*biodiversity*). Different species are organized at many levels through their interrelationships into complex biological communities.

The continued functioning of the biosphere is dependent not only on the maintenance of the intimate interactions among various species with local communities but also on the crucial interaction of all species and communities around the globe. The richness and diversity of species (biodiversity), evolved over millions of years are irreplaceable. To sustain life on Earth, more than a few animal and plant species that humans consider as essential must be preserved. Even seemingly insignificant species are crucial to the stability of communities and ecosystems.

## 1.2 ENVIRONMENTAL CYCLES

The whole biosphere is a closed loop with interrelated environmental cycles. The major environmental cycles are carbon (C), nitrogen (N) and sulfur (S) cycles.

### 1.2.1 The Carbon Cycle

The carbon cycle is an environmental process that includes photosynthesis, decomposition, and respiration by which carbon as a major component of various compounds cycle between its major reservoirs—atmosphere, hydrosphere, lithosphere and living organisms. It emphasizes the central importance of carbon in the biosphere. Life is evolved on the conversion of carbon dioxide ($CO_2$) into carbon-based organic compounds of living organisms. The slowest part of the carbon cycle involves carbon that resides in sedimentary rocks. The biological cycling of carbon begins with photosynthetic organisms, which assimilate $CO_2$ or carbonates from the surrounding environment. In the terrestrial communities, chlorophyll-containing green plants use sunlight to induce $CO_2$ from the atmosphere to react with $H_2O$ to produce carbohydrates with the release of $O_2$. Plants are, thus, primarily responsible for the generation and continued presence of atmospheric oxygen. When natural decay or high temperature processes oxidize plants and vegetation, $O_2$ in air is consumed and $CO_2$ is regenerated. All animals return $CO_2$ directly to the atmosphere as a byproduct of their respiration. The release of $CO_2$ into the atmosphere or hydrosphere completes the biological part of the carbon cycle (Fig. 1.2).

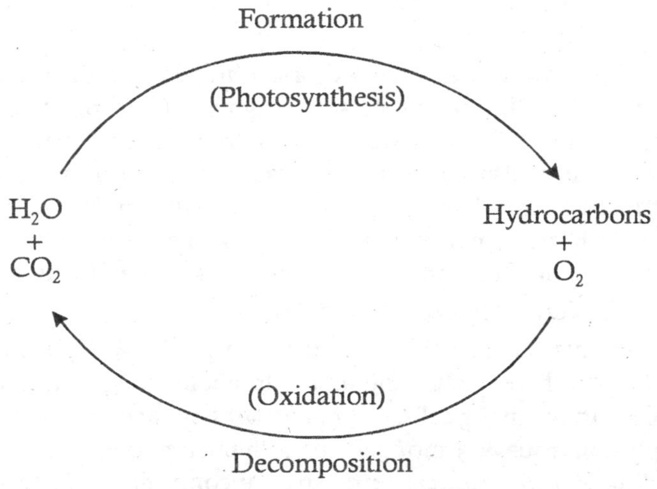


**Fig. 1.2:** The Atmospheric Carbon Cycle

Most of the carbon in terrestrial communities is stored in the tissues of plants. Therefore, the $CO_2$ cycle is disturbed and $CO_2$ in the atmosphere is increased by deforestation. Burning of fossil fuels, and converting limestone into cement and building materials have a similar effect. Most of the $CO_2$ released to the atmosphere comes from industrial burning processes, a major concern to mankind because of the "greenhouse" effect that $CO_2$ creates. The release of carbon monoxide (CO) is due to incomplete burning of carbon-containing compounds. CO has harmful effects on animals and humans.

$$2C + O \rightarrow 2CO \qquad ...(1.1)$$

$$CO_2 + C \rightarrow 2CO \quad \text{(at high temperature)}$$

Increased $CO_2$ in the atmosphere brings about climatic changes through alteration of the Earth's surface temperature. Nearly all the solar radiation reaching the earth is in the visible region of the spectrum. The Earth absorbs part of it and as a *blackbody* it emits radiation back to the atmosphere in the infrared (IR) spectral region (thermal radiation). $CO_2$ absorbs this IR radiation as a one-way filter. The more the $CO_2$ in the atmosphere, the more heat will be retained and hotter the atmosphere becomes. This phenomenon is termed 'greenhouse' effect.

### 1.2.2 The Nitrogen Cycle

Like carbon, nitrogen has its own biogeochemical cycle, circulating through the atmosphere, hydrosphere and lithosphere. The cycle consists of atmospheric nitrogen-forming compounds, which are metabolized by bacteria and plants, and eventually nitrogen is returned to the atmosphere by bacterial decomposition of organic matter. Most nitrogen occurs in the atmosphere as inorganic gas ($N_2$). Most plants cannot use nitrogen in its gaseous form. They can assimilate it only after it is converted to ammonia ($NH_3$) or nitrates ($NO_3^-$) (nitrogen fixation by nitrifying bacteria). Consumers eat plants, and nitrogen re-enters the environment when these organisms die, urinate or produce excrement. Denitrifying bacteria break the nitrogenous compounds into molecular nitrogen ($N_2$), and the cycle continues. Nitrogen monoxide (NO) and nitrogen dioxide ($NO_2$) are also constituents of the atmospheric nitrogen cycle (Fig. 1.3).

$$N_2 + O_2 \rightarrow 2NO \qquad ...(1.2)$$

$$2NO + O_2 \rightarrow 2NO_2$$

Most of anthropogenic nitrogen monoxide (NO) and nitrogen dioxide ($NO_2$) are due to oxidation of atmospheric nitrogen ($N_2$) in burning processes at high temperature, in power generation plants and automobiles. Both NO and $NO_2$ are toxic and are detrimental to materials, vegetation, animals and humans. NO decreases photosynthesis and $NO_2$ causes necrosis of plant tissues, nasal irritation and pulmonary edema in animals. They also take part in atmospheric photochemical reactions. Hydrocarbons reacting with molecular oxygen ($O_2$) and ozone ($O_3$) produce free radicals that are the precursors for photochemical oxidants, photochemical smog and acid rain.

$$2NO_2 + H_2O \rightarrow HNO_2 + HNO_3 \quad \text{(acid rain)} \qquad ...(1.3)$$

Further, organic pollutants, such as chlorofluorocarbons (CFCs) deplete the ozone layer in the stratosphere ("ozone hole"), thereby allowing more ultraviolet (UV) radiation to reach the Earth, causing radiation damage to the living cells. UV radiation is mutagenic (e.g. thymine dimerization). Excessive nutrients (phosphorous and nitrogen-containing compounds) in water bodies affect the quality of water, leading to *eutrophication* (excessive, uncontrolled growth of algae and aquatic life).

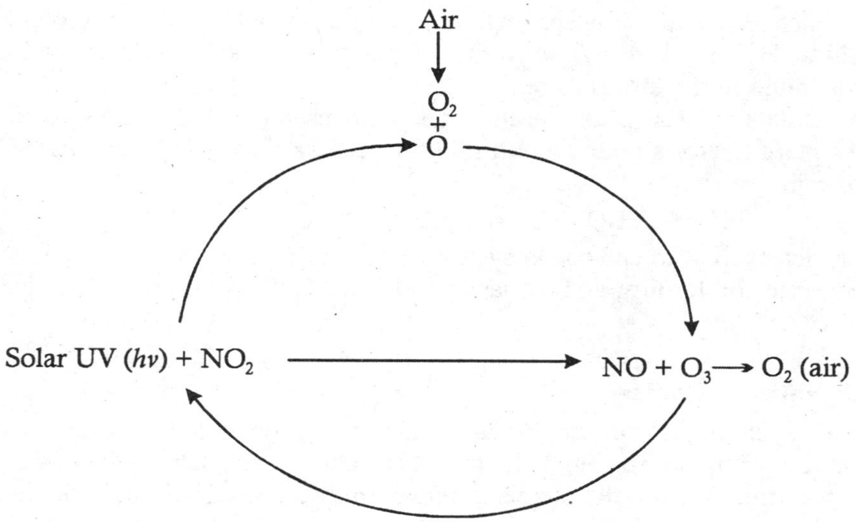


**Fig. 1.3:** The Photolytic Nitrogen Cycle

### 1.2.3 The Sulfur Cycle

Like the carbon and nitrogen cycles, sulfur cycle occurs between the atmosphere, hydrosphere and lithosphere. Unlike the other two elements, nitrogen has major reservoirs in both the atmosphere and the lithosphere. As is the case of the nitrogen cycle, the activities of microorganisms are crucial in the global cycling of sulfur (Fig. 1.4).

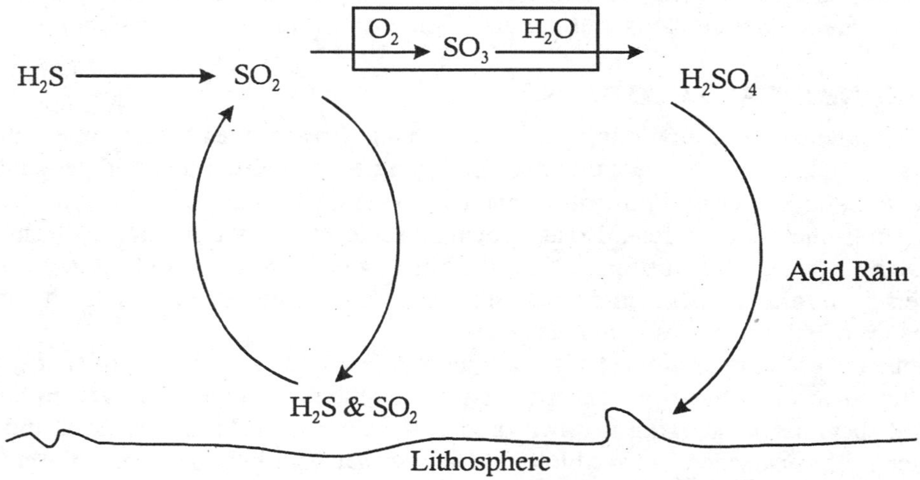


**Fig. 1.4:** The Atmospheric Sulfur Cycle

Since the Industrial Revolution, human activities have contributed significantly to the movement of sulfur from the lithosphere to the atmosphere, primarily due to burning of fossil fuels. Sulfur is now one of the major elements among the atmospheric pollutants. The main pollutant is sulfur dioxide, $SO_2$, which reaches the atmosphere through burning of sulfur-containing ores and materials. Hydrogen sulfide, $H_2S$, released by the rotting of plants and organic matter is another form, in which sulfur is found in the atmosphere.

$SO_2$ is detrimental to objects, plants, animals and humans. It damages building surfaces, corrodes metals, causes plant necrosis and respiratory ailments in animals and humans. $SO_2$ is also the cause for acid rain.

$$2SO_2 + 2H_2O + O_2 \rightarrow 2H_2SO_4 \qquad ...(1.4)$$

Acid rain is detrimental to ecological system by way of mechanical damage to structures and materials, decreased soil fertility, and damage to aquatic life. $SO_2$, $O_3$ and $H_2SO_4$ have synergistic effects.

### 1.2.4 Other Nutrient Cycles

Other nutrients, such as phosphorus, potassium, calcium, enter the environment through the weathering of rocks and soil through the biosphere and hydrosphere and back to lithosphere. Phosphorus and other nonvolatile elements move unidirectionally from land, through aquatic channels into ocean sediments. The addition of phosphorus fertilizers to soils, organophosphor pesticides, and of phosphorus-containing detergents have a great impact on the phosphorus cycle and pollution in many ecosystems.

### 1.2.5 The Hydrologic Cycle

Water itself cycles within the biosphere. However, the hydrologic cycle would continue in some form even in the absence of living organisms. The water available on the planet continually cycles through the atmosphere, oceans, and terrestrial environments, mainly, through evaporation, transpiration (a physical process that allows green plants to expel excess water through their leaves into the atmosphere), and precipitation. This part of the hydrological cycle is driven by solar energy. Atmospheric contaminants and pollutants enter this cycle at various points.

## 1.3 ENVIRONMENTAL POLLUTION

Close links exist between the environmental cycles. Changes in any one of these cycles influence the other cycles as well. Each biosphere unit can contaminate the environment and get contaminated by it. Contaminants dispersed in the atmosphere are eventually transferred back to the ocean or Earth. The atmosphere is, thus, regarded as a potential vehicle for contamination of the hydrosphere and the Earth's surface (soil). Deforestation, the burning of fossil fuels and the increased use of synthetic fertilizers, and industrial and other anthropogenic activities have been the major cause for the disturbance of natural cycles in the biosphere.

Environmental pollution is defined as the addition of any substance or form of energy (e.g. heat, sound) to the environment at a rate faster than at which the environment can accommodate it by absorbing, dispersing or breaking it down, and that would harm humans, flora and fauna or abiotic systems. Physical, chemical and biological factors that disturb the balance of the environment may be termed as pollution and the factors or substances, which cause the deterioration, the pollutants. They may be in the form of solid particles, liquid droplets, gases, or in combination thereof. Environmental degradation is due to contaminants and pollutants arising out of transfer processes (dispersion) of matter and energy. Air pollutants are often grouped in categories for ease in classification; some of the categories are—particulate matter, sulfur compounds, volatile organic chemicals, nitrogen compounds, oxygen compounds, halogen compounds, agrochemicals, radioactive compounds, and odours.

Every pollutant has a source and the receptor is anything that is affected by the pollutant (air, water, soil, vegetation, animal and human life). For example, *eutrophication*, the ecological stagnation of water bodies occurs when organic wastes, carrying nutrients, are discharged into water sources. This would disturb the ecological control mechanisms prevailing in the water bodies and leads to damage to aquatic life. Detection of the disturbed balance between the living organisms and the environment requires efficient ways of measuring the substances affecting the environment.

### 1.3.1 Contaminants and Pollutants

The terms "contaminant" and "pollutant" are sometimes interchangeably used in environmental studies. However, 'contaminants' while being unwanted substances in the atmosphere, do not necessarily cause perceptible adverse effects, while the 'pollutants' do. The presence of contaminants or pollutant substances (microorganisms, chemicals, toxic substances, wastes, or wastewater in a concentration that makes the medium unfit for its next intended use) in the environment interferes with human health and welfare, and/or produces other harmful environmental effects to surfaces of objects, buildings, and various household and agro products.

Most pollutants and contaminants are resources gone to waste or 'resources out of place.' A contaminant is something that causes a deviation from the normal composition of the environment. Contamination can be regarded as the dispersion of a substance in the environment at concentration, which if increased to a higher level, can lead to environmental pollution and have adverse effects. Whether an atmospheric contaminant would become a pollutant depends not only on its concentration level in the atmosphere, but also on the atmospheric conditions. For example, sulfur dioxide is more corrosive and reactive in an aqueous environment (in the humid atmosphere). Similarly, certain substances are more reactive under reducing atmosphere, while others are reactive in an oxidizing atmosphere. That is, an atmospheric contaminant can become a pollutant under opportune conditions even at trace level concentrations. Types of pollutants are:

1. Oxygen-demanding wastes (domestic, municipal and industrial).
2. Infective agents—bacteria, viruses and protozoa.
3. Plant nutrients—nitrates, and phosphates.
4. Organic chemicals—petrochemicals, oils, insecticides, pesticides, herbicides, detergents and carcinogens.
5. Minerals and trace metals.
6. Radioactive minerals and compounds.
7. Thermal pollution.
8. Noise.

#### *1.3.1.1 Environmental Contaminants*

Contaminant is any physical, chemical, biological, or radiological substance or matter that has an adverse effect on air, water, or soil. Atmospheric contaminants are of natural or anthropogenic origin. Natural contamination is due to fermentation and decomposition of organic matter by aerobic and anaerobic micro-organisms, forest fires, airborne pollen spores, fungi, and geological and meteorological processes, such as volcanic activities, lava, hot springs, desert sands and dust bowls. Anthropogenic contaminants are due to human activities in the urban, rural and industrial areas. Urban activities create enormous amounts of domestic waste, sewage, smoke and automobile exhaust. Contaminants from rural activities are chemical fertilizers, fumigators, pesticides, and agricultural and animal wastes. Industrial pollutants are varied. Metallurgical operations let out toxic metals in fine particulate form. Coal and fuel burning produces large quantities of CO, $CO_2$ and $SO_2$. Chemical and petrochemical industries release lot of chemicals, like nitrates, sulfates, phosphates, surfactants and hydrocarbons, and a variety of organic compounds.

#### *1.3.1.2 Environmental Pollutants*

Most of the toxic pollutants found in the atmosphere are from anthropogenic sources. Power plants, commercial and residential heating units and transport vehicles release a variety of gaseous pollutants. Inorganic pollutants are $SO_2$, $Cl_2$, $NH_3$ and CO, ions such as halides, $NO_3^-$, trace metals, such as As, Bi, Co, Cr, Cu, Fe, Hg, Mo, Ni, Pb, Sb, Se, Sn and V. Metal fumes are produced from mining and metallurgical industries (Fig. 1.5). Organic pollutants are hydrocarbons (e.g. polycyclic aromatic hydrocarbons (PAHs), and polychlorinated biphenyls (PCBs), organic solvents, organometallic compounds, pesticides, surfactants and a host of other man-made compounds. Bioaccumulation of trace elements and persistent pesticides has also created serious ecological problems.

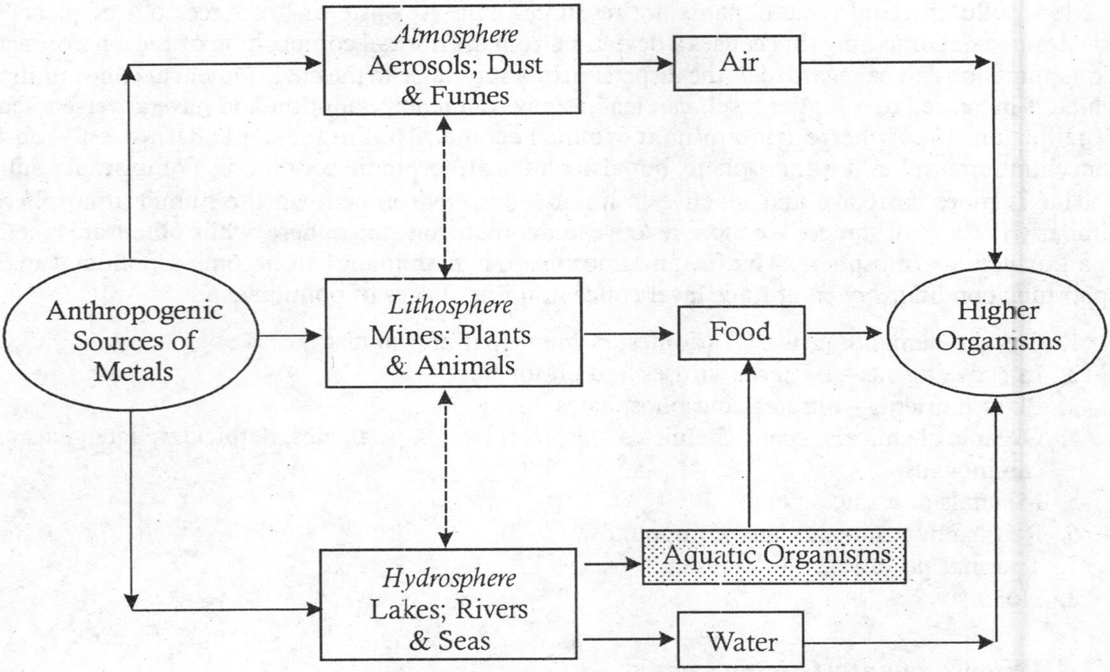


**Fig. 1.5:** Anthropogenic Effects of Heavy Metals Reaching Higher Organisms

Airborne particles can damage crops and plants. They also reflect solar heat back into space. Carbon dioxide ($CO_2$) and methane ($CH_4$) are 'greenhouse gases' that promote global warming. These gases, along with water vapor, are transparent to short-wave radiation from the Sun, but tend to block long-wave heat energy from radiating back toward space from the Earth. Global warming could cause climatic disturbances as well as outbreak of tropical diseases. Halocarbons rising in the atmosphere destroy the ozone layer, which normally absorbs ultraviolet (UV) radiation from the Sun. Thus, a protective blanket against the hazardous effects of UV radiation, to the organisms on the earth, is removed. This would lead to destruction of phytoplankton and zooplankton and a major threat to all ocean ecosystems. Ozone at the ground level, however, causes damage to plants and animals. The formation of ozone at that level is the result of pollutants like $NO_x$ in the air. Discharge of oil-containing brine into rivers and streams can adversely affect aquatic ecosystems. Strip-mining leads to acid mine drainage, destroys the land and its recreational value and wildlife habitats. Power plants are major sources of thermal pollution, leading to natural imbalance of aquatic life in lakes, estuaries, and rivers. Nuclear-fuel power plants also produce such pollution.

## 1.4 THE ENERGY PROBLEM

The driving force behind the industrial activities of all countries is the energy sector. Growing population and developing industries create an ever-increasing demand for energy. As the developing nations proceed from agricultural to industrial economy, demand for energy will be rapidly escalating. At present most of the electricity generation is based on burning of fossil fuel (coal, oil and natural gas). In the later part of the 21st century availability of oil reserves in the world would dwindle and the consumer would face a crunch. Though some countries are endowed with large reserves of coal to last for a few centuries, its availability is not universal and it's mining and transporting would ultimately make energy dearer. Over and above this consideration, there is growing awareness of global warming, already on the horizon, due to atmospheric pollution largely caused by carbon dioxide buildup. There is a concomitant problem of sulfur dioxide pollution from the same source and acid rain. Under these circumstances world attention is gradually drawn to alternate energy sources, which can supply this commodity without adding much to the pollution load. There are intermediate or short-term solutions like reducing the sulfur content of coal (flue gas desulfurization), substituting oil and gas for coal for power generation.

Because our world depends so much on energy, we need to find sources of energy that will last a long time. If the world wants long-lasting development for all of its inhabitants, it must reduce its reliance on fossil energies in favor of energies that are less polluting and consume fewer resources. Only *renewable* and nuclear energies provide sufficient prospects in this area. We should also use energy sources that produce as little pollution as possible. While all energy sources cause some pollution in their creation or their consumption, renewable energy systems generally are less polluting than fossil fuel systems.

### 1.4.1 Alternate Energy Sources

From a long-term perspective several alternate energy sources have been investigated and experimented with. At present, generation of electricity is based, mainly, on the burning of fossil fuels—coal, oil and gas. There are major problems with regard to this strategy. In the later part of the 20th century attention has been focussed on exploiting alternate energy sources for power generation. Renewable energy sources are those that are replenished by natural processes at rates exceeding their rate of use for human purposes, unlike fossil fuels, which are not replenished at a useful rate. Sources considered renewable include solar, biomass, wind, geothermal, hydropower. *Although renewable energies are "free", their recovery is not*! So far no unique solution has been found but at least one, the nuclear option is attractive. The various alternatives for energy generation are briefly presented here.

#### 1.4.1.1 Solar Energy

The Sun is an extremely powerful energy source and it supplies directly (solar radiation) and indirectly (through fossil fuels) all the energy required for sustaining life on the Earth. The enormous potential of solar energy is implied in the fact that each day the Earth receives energy about 200,000 times the total world electrical-generating capacity. However, the negative aspect of this source is its low intensity, as it reaches the earth. Despite this limitation, because of its inexhaustible supply and nonpolluting character, scientists and technologists have expended considerable effort on exploring ways of utilizing it.

For capturing solar energy and converting it into thermal energy, flat-plate collectors and concentrating collectors have been found practical. Flat-plate collectors are useful for heating water or some other fluid to a temperature ranging from 60 to 90° C. The stored hot fluid can be utilized for heating the dwellings. Concentrating or focussing collectors consist of arrays of carefully aligned mirrors. It is possible to focus enough sunlight to heat a target to temperatures up to 2,000° C. These devices, which are normally useful for small experiments, can be extended to operate a boiler of moderate size to generate steam for driving a turbine and produce electricity. In view of the low intensity of solar radiation, both types of collectors must be large in area.

Electricity can be generated directly from solar radiation using *photovoltaic* (PV) cells. PV cells make electricity without moving, or polluting. Major drawbacks are—(1) the electrical voltage generated is small; (2) because PV systems only produce electricity when the Sun is shining, these remote systems need batteries to store the electricity (by connecting large number of cells together, as in modern solar batteries, more than one kilowatt of electric power can be generated in a unit); (3) in spite of its free availability and great efforts, the high cost of collection, conversion and storage of solar energy limits its exploitation.

#### 1.4.1.2 Wind Power

The idea of wind farms, consisting of clusters of wind turbines, has been tested in areas where there is steady prevalent wind (which is not available in all parts of the world). Use of windmills for pumping water and grinding grain is not altogether new. The power obtainable in a windmill varies as the square of the rotor diameter and cube of wind velocity. In 1990, windmills in different parts of USA generated about 2.5 billion kWh of power. In spite of this impressive figure, it formed less than 0.01 per cent of the total electric power generated in USA. It has been estimated that with a wind speed of 14 m/s, it would require 17,000 windmills (with rotor 30 m across), operating at 40 per cent conversion efficiency, to generate 1 million kW electric power. Therefore, it appears that wind turbines alone will not play a major part in making up the power demand of an industrialized nation. The price is still 2 to 4 times greater than that from nuclear or gas energy.

#### 1.4.1.3 Geothermal Energy

We can also get energy directly from the heat in the Earth. Geothermal energy starts with hot, molten rock (called magma) kilometers below the Earth's surface that heats a section of the Earth's crust. The heat rising from the magma warms underground pools of water known as *geothermal reservoirs*. While geothermal energy is a good source of power, we could run out of it by drawing so much energy out of the reservoir that it is not able to replenish itself at the rate we are using it. In addition, water from geothermal reservoirs often contains minerals that are corrosive and polluting.

### 1.4.1.4 Hydropower

The water in rivers and streams can be captured and turned into *hydropower*, also called *hydroelectric power*. The most common form of hydropower uses dams on rivers to create large reservoirs of water. Water released from the reservoirs flows through turbines, causing them to spin. The turbines are connected to generators that produce electricity. Hydropower currently is one of the largest sources of renewable power. Hydropower is also inexpensive, but cause nuisances, environmental disturbances.

### 1.4.1.5 Biomass Energy

The *biomass* is a good means of storing diffuse and intermittent solar energy, but its efficiency is low. Although wood is the largest source of biomass energy, we also use corn, sugarcane wastes, and other farming byproducts. There are four ways to use biomass—

1. *Combustion* (*incineration*) mainly produces heat and electricity, or can be changed to a gas-like fuel such as methane, or changed to a liquid fuel. Liquid fuels, also called *biofuels*, include two forms of alcohol: *ethanol* and *methanol*. Because biomass can be changed directly into a liquid fuel, it could someday supply much of our transportation fuel needs.

(2) *Methanization* is carried out by anaerobic fermentation (decomposition through bacterial action in the absence of air) of very damp substances such as algae, animal excrement or household waste.

(3) *Alcoholic fermentation* is adapted to sacchariferous products, such as beets or sugarcane, or starchy products, including starch such as cereals.

(4) *Thermochemical transformation*, which leads to the gasification of organic and vegetal substances, is particularly well suited to the conversion of lignocellulose products such as wood or straw into energy. The thermochemical system of "*pyrolysis*" and gasification of the biomass includes a combination of pyrolysis and steam reforming process.

### 1.4.1.6 Fuel Cells

Fuel cell is an electrochemical device with no moving parts that converts chemical energy of a fuel directly into electrical energy. The fuel (usually hydrogen gas) and oxidizer (oxygen gas) take part in electrochemical reactions at two separate electrodes (Fig. 1.6). Hydrogen is exposed to the anode, and the anode then draws the electron from the hydrogen leaving a proton ($H^+$). On the opposite side of the cell, the cathode absorbs oxygen from the air. When protons pass through the membrane, it creates a flow of direct current electricity between the terminals. The only byproduct of this process is water, which is formed by electrons, hydrogen ions and oxygen at the cathode. Fuel cells are relatively clean, fuel-efficient and fuel flexible. There is no $NO_x$, CO, HC emissions, because hydrogen is not burned in air.

In the alkali fuel cell (AFC), the electrolyte is a solution of sodium or potassium hydroxide and the electrodes are carbon and nickel. The oxidation product being water, in principle, there is no pollution from a fuel cell. However, the highly corrosive nature of the alkali and the limit on its operating temperature (less than 100° C) restrict the applicability of these cells. Moreover, the water formed in the oxidation of the fuel has to be removed by evaporation from the electrolyte, in a separate step. For operating above 100° C other type of cells have been in limited use. They include the phosphoric acid fuel cell (PAFC) and the molten carbonate fuel cell (MCFC).

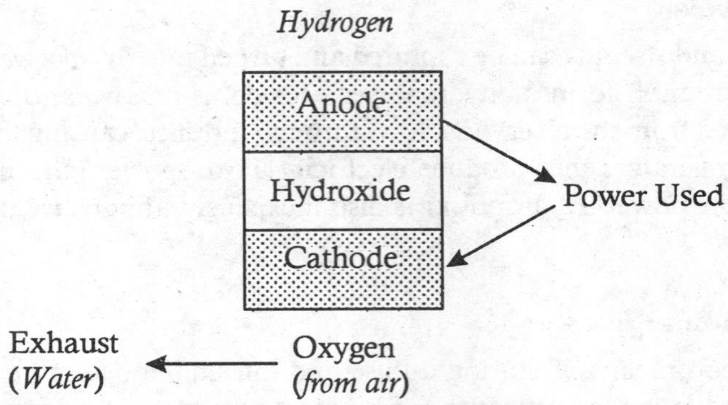


**Fig. 1.6:** Schematic Diagram of Hydrogen Fuel Cell

More promising type fuel cells are being developed. *Proton exchange membrane fuel cell* (PEMFC) is currently the technology being used to power passenger vehicles. These cells operate at relatively low temperatures (about 80° C), have high power density, can vary their output quickly to meet shifts in power demand, and are suited for applications—such as in automobiles, where quick startup is required. They are the primary candidates for light-duty vehicles, for buildings, and potentially for much smaller applications such as replacements for rechargeable batteries.

*Direct methanol fuel cell* (DMFC) is a relatively new technology (operating range 50-100° C) that may also be used in passenger cars and other vehicles. These cells are similar to the PEM fuel cells in that they both use a polymer membrane as the electrolyte. However, in the DMFC, the anode catalyst itself draws the hydrogen from the liquid methanol, eliminating the need for a fuel reformer. This is a relatively low range, making this fuel cell attractive for tiny to mid-sized applications, to power cellular phones and laptops. A major problem, however, is fuel crossing over from the anode to the cathode without producing electricity.

In the *Solid oxide fuel cell* (SOFC) the electrolyte is an oxygen-conducting solid, cubic zirconia stabilized by zirconium oxide. The anode is metallic nickel supported on stabilized zirconia, and the cathode is porous lanthanum manganite. The operating temperature is 700-1,000° C.

With the requirement of special materials and technical problems, fuel cells at present are extremely expensive. They are used only for applications, in which cost is of secondary importance (as in spacecraft). But, they do hold a great promise in the near future for residential power generation, transportation, and portable-electronics (computers, cellular phones). Fuel cells are used at wastewater treatment plants and landfills, proving themselves as a valid technology for reducing emissions and generating power from the methane gas they produce.

### *1.4.1.7 Nuclear Power*

One of the notable developments of the post World War period is the evolution of technology for generating electricity based on nuclear energy. Some experts believe that it is a viable alternative to fossil fuel burning, from the point of view of atmospheric pollution. At the same time there are others who highlight a different set of hazards that nuclear power generation poses to civilization.

The actual nuclear reaction takes place in what is called the 'reactor core'. In most of the present day commercial units uranium is the nuclear fuel. The U-235 isotope, present in natural uranium as a minor component (0.7 per cent, rest being U-238) undergoes nuclear fission by absorbing slow neutrons and releases energy. U-235 is boosted to about 3 per cent before using it as fuel Uranium fuel pellets (usually as oxide) are put into tubes and then placed in the reactor. When neutrons strike the U-235 nuclei they cause a chain fission reaction, while at the same time releasing their own neutrons. Water is used for extracting the heat (cooling) and for slowing down the released neutrons (moderation). Water is converted to steam, which is used to run the turbines for generating electricity. Natural uranium can also be used as fuel if, instead of ordinary water, heavy water ($D_2O$) is used as moderator and primary coolant. The disposition of control rods, placed between the fuel rods, is adjusted to maintain the fission reaction and through that the energy release, at a steady required rate.

The nuclear reaction produces highly radioactive fission products and also some transuranic elements, $^{239}Pu_{94}$, being the most important. Most of the radioactivity remains in the pellets. The reactor core is located inside a thick (20 cm) steel vessel. In a reactor several safety measures are incorporated to prevent release of radioactive material to the environment.

Once the fissionable isotopes are optimally consumed the spent fuel rods are discharged from the reactor and fresh ones are loaded. It is the handling and subsequent storage and processing of these fuel rods, with associated high radioactivity that poses several problems. The permanent storage of long-lived radioactive isotopes in the waste also involves possible risk. It is mainly based on these and on chances of accidental leakage of radioactivity that many voices are raised against nuclear power generation. Another section of the people advocate reasonable risk taking in these matters keeping in view that nuclear energy is clean, and is an inexpensive alternative to other methods of producing electricity.

There exists an urgent need to utilize the existing energy-producing resources with more environment-friendly measures, ("*integrated waste-management*"—source reduction, recycling and better utilization), as well as search for alternate and better methods of energy production and utilization (Table 1.1).

**TABLE 1.1:** FUTURE ENERGY SOURCES

| *Energy Source* | *Process & Limitations* |
|---|---|
| Coal conversion | Manufacturer of gas, HC liquids from coal. It is not completely free from environmental pollution and energy sources are limited. |
| Oil shale | Petroleum-like fuel from oil shale. Poses environmental problems. |
| Geothermal | Underground heat. Technical advances are needed in harnessing. |
| MHD | Electrical generation by passing hot gas plasma through a magnetic field. Materials that can withstand high temperature are not yet available. |
| Solar | Direct conversion of solar energy. Environmentally clean source of energy. The main problem is storage. |
| Wind | Wind energy. Harnessing problem has to be solved. |

Coal can be converted to gaseous, liquid (synfuel) or low-sulfur (flue gas desulfurization), low-ash solid fuels. Oil shale utilization poses the same environmental problems such as land degradation and habitat destruction. Fusion devices are believed to be the best long-term option, because the primary energy source, deuterium, is abundantly present in water (heavy water). Geothermal sources are not available all over the world, and technical advancement is required in harnessing geothermal energy. Magneto-hydrodynamic (MHD) generators convert fuels (natural gas, oil, coal or nuclear heat sources) into a hot, electrified gas, called hot plasma. When passed through a powerful magnetic field, the gas plasma produces an electric current that can be fed into power lines. This eliminates the need for boilers, turbines and generators. The problem is the non-availability of materials that can withstand high temperature (> 2,000 K) needed to produce hot, electrified gas. Sunlight impinging on a photosensitive layer (CdS) is converted to electric energy in solar cells (photovoltaic cells). Photovoltaic cells are self-generating energy sources (no external energy is needed) and hold a great promise because solar energy is the most environmentally clean source of energy. The problem is in storage and integration of large-scale use.

## 1.5 POLLUTION CONTROL METHODS

Sampling and analysis of air (particulate and aerosols), water (ground, marine and wastewater), soil and other pollutants that encompass various sampling matrices are the essential features of environmental analysis and quality control of contaminants and pollutants. In the analysis procedures, the environmental behavior of elements, metals, compounds and radioactive materials, their bioaccumulation and sorption characteristics have to be taken into consideration. Physical, chemical, and biological methods can be employed in qualitative as well as quantitative determination of various environmental pollutants. These range from simple qualitative identification methods, to the use of sophisticated instrumentation for quantitative assessment, or a combination of several methods.

As many environmental pollutants are in trace quantities, in the *environmental quality management (EQM), the quality assurance and quality control* (*QA/QC*) of analytical data must be stringent, maintained and validated in all aspects of measurements, from the collection of samples to reporting the data, to ensure the reliability of data acquisition and measurements. Concepts such as *life cycle assessment* (*LCA*), of both the process and product in a given environmental ambience must be emphasized.

There are standard reference materials (primary and secondary), established to validate analytical methods in order to ensure that the results obtained are consistent with the facts. Standardization, data harmonization, and data validation are the major factors in *QA/QC* analysis. A homogeneous, stable material with certified physical or chemical properties, is called a *reference material*, and is used for calibrating instruments. Blank samples are commonly "reagent blanks", and control samples contain known concentrations of the analyte. "Spiked" samples are prepared by adding known amount of the analyte of interest to blank samples or samples that have already been analyzed in order to provide matrix with a known quantity.

Among quantification standards used in environmental science analysis, a *primary standard* is the one that is used as such for determining an analyte by comparison. A primary standard purity should be > 99.98 per cent. A *secondary standard* is one whose composition is established by comparison with a primary standard. There are environmental quality standards prescribed by regulations to limit the level of pollutants during a given time in a defined area. Among environmental quantification standards for regulation of pollutants, primary standards are designed to protect the human health with an adequate margin for safety, the levels below which certain species are nontoxic to humans (a limit set against adverse health effects). Secondary standards are designed to protect welfare, personal comfort and well-being of humans, animals, and damage to plants, crops, and buildings, etc. (Table 1.2).

**TABLE 1.2:** AIR QUALITY STANDARDS

| *Pollutant* | *Primary Standard* $\mu g/m^3$ *(ppm)* | *Secondary Standard* $\mu g/m^3$ *(ppm)* |
|---|---|---|
| **Gases** | | |
| Sulfur dioxide ($SO_2$) (annual mean) | 80 (0.03) | – |
| Carbon monoxide (CO) (8h conc.) | 10 (9) | 10(9) |
| Nitric oxide (NO) (annual mean) | 100 (0.05) | 100 (0.05) |
| Photochemical oxidants (1h conc.) | 160 (0.08) | 160 (0.08) |
| Non-methane hydrocarbons | 160 (0.35) | 160 (0.25) |
| **Particulate** | | |
| Annual | 75 | 75 |
| Max 24h conc. | 250 | 250 |

## BIBLIOGRAPHY

Barcelo, D. (ed.) (1993), Elsevier: Amsterdam. "*Environmental Analysis: Techniques, Applications and Quality Assurance.*"

Brown, W. (1994), Chelsea House Publishers: New York. "*Alternative Sources of Energy.*"

Carless, J. (1993), Walker: New York. "*Renewable Energy: A Concise Guide to Green Alternatives.*"

Chigier, N. A. *et al.* (1978), Pergamon Press: New York. "*Energy from Fossil Fuels and Geothermal Energy.*"

Cunningham, W. P. (ed.) (1994), Gale: Detroit, IL. "*Environmental Encyclopedia.*"

Herda, D. J. (1991), Watts Pubs: New York. "*Energy Resources: Toward a Renewable Future.*"

Rand, G. M. & Petrocelli, S. R. (eds.) (1985), Hemisphere Pubs: New York. "*Fundamentals of Aquatic Toxicology: Methods and Applications.*"

... (1990), Office of SRMS/Natl. Inst. of Standards & Technology: Gaithersburg, MD. "*Standard Reference Materials for Environmental Research, Analysis and Control.*"

... (1993), McGraw-Hill: New York. "*Environmental Science and Engineering.*"

# SECTION-I

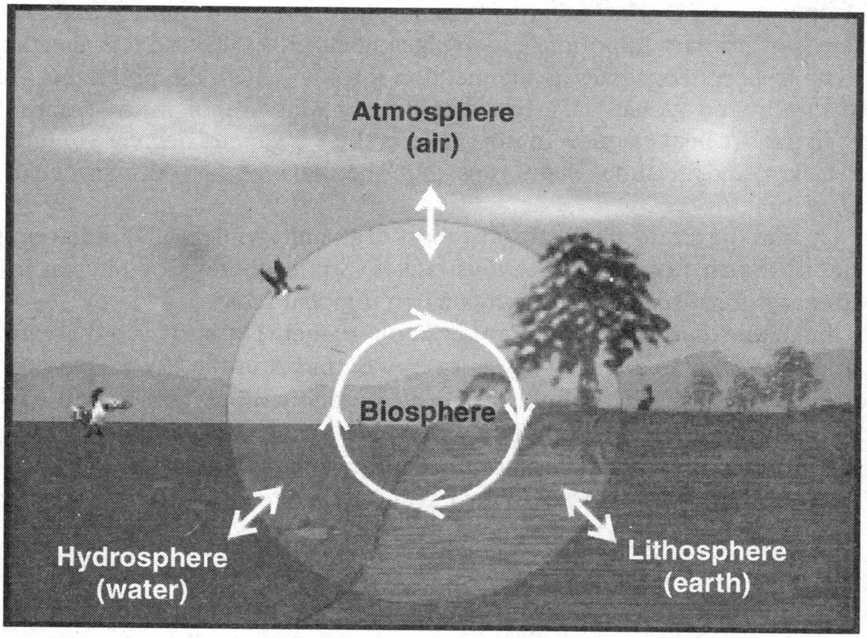


# The Atmosphere

# OVERVIEW

Chapters 2 to 6 deal with the basic features of the atmosphere, namely atmospheric physics, factors responsible for atmospheric pollution, and impact of the pollutants on objects and organisms.

## ATMOSPHERE AND ITS CHARACTERISTICS

Earth's atmosphere provides a protective blanket from the hostile environment of the outer space. It also nurtures life on Earth. Atmospheric pollution has direct impact on vegetation and all living organisms on the Earth.

The density of the atmosphere is unevenly distributed, with 80 per cent of the atmospheric constituents being distributed within 6 kilometers from the Earth's surface. The temperature of the atmosphere varies in a nonlinear manner with altitude, and based on this it is subdivided into five regions—(1) troposphere, (2) stratosphere, (3) mesosphere, (4) thermosphere and (5) exosphere.

Troposphere is of primary importance to living systems on Earth, and it is also the main region of pollution. The troposphere serves as the medium through which the pollutants are transported and dispersed locally and globally. Carbon dioxide and water vapor in the troposphere play an important role in Earth's heat balance. In troposphere, there is general decrease of temperature with altitude (~6.5° C/km) and the atmosphere is unstable. Thermal inversion occurs when the atmosphere resists vertical motion.

The ozone layer in the stratosphere absorbs most of the ultraviolet (UV) radiation from the sun and shields life on Earth from its deleterious effects. Stratosphere is of interest to aeronautics, communications and for dispersion of pollutants on a global scale.

The Sun is the main source of energy transfer to our planet. The surface temperature of the Sun is ~6000 K, and therefore, as a blackbody its energy output is in the UV-visible regions (300-700 nm). On the other hand, the Earth with its mean surface temperature at 260 K radiates, as a blackbody in the infrared (IR) region. This difference in the wavelength regions of solar and terrestrial radiation is of fundamental importance to the energy balance of the Earth-atmosphere system. The heat exchange at the surface of the Earth is not uniform, giving rise to climatic changes.

## AIR POLLUTION

Sources of air pollution are natural as well as anthropogenic and their impact has local, regional and global implications. Mixing and dispersion of atmospheric pollutants occur mainly due to convection and turbulence in the air. Theoretical methods for predicting concentration of non-reacting air pollutants, over a given area from a point source, are based on plume dispersion models (Gaussian type) and available meteorological data. Prediction of air pollution due to reacting constituents needs, in addition, the application of principles of thermodynamics and reaction kinetics. From such studies, disastrous consequences of pollution from newly planned industries can be avoided or at least minimized.

## ATMOSPHERIC POLLUTANTS AND THEIR IMPACT

*Atmospheric pollutants can be of different types*: particulate, aerosol, radioactive, thermal or mechanical (noise). Pollutants can be from point sources (e.g. industrial installations) or from mobile sources (e.g. automobiles and other transport systems). Major atmospheric pollutants are particulate matter (e.g. asbestos, dust, fly ash and soot), and gaseous primary pollutants, such as oxides of sulfur

($SO_x$), oxides of nitrogen ($NO_x$), ammonia ($NH_3$) and carbon monoxide (CO). Secondary pollutants, such as ozone, peroxyacetyl nitrate (PAN) and other oxidants are produced as a result of various photochemical reactions in which the primary pollutants and hydrocarbons take part. They are responsible for photochemical smog.

Atmospheric pollutants have local, regional or global level effects. Fluorides and silicates (e.g. sand, asbestos) are particulate pollutants arising out of mining and construction operations and have impact, mostly, on a local scale. Carbon monoxide (CO) is produced by the incomplete combustion of carbonaceous materials and fossil fuels. The major sources of CO are automobiles and coal-burning operations. The environmental impact of CO pollution is on the regional level. It has greater affinity (~250 times) to bind to hemoglobin (Hb) than to oxygen, and thus it is a powerful asphyxiating gas.

The large-scale transport of $SO_2$ occurs in the troposphere, resulting in the production of acid rain. The environmental impact of $SO_2$ is on the regional and continental levels. Depletion of protective ozone layer in the mesosphere-stratosphere boundary, by scavengers of ozone (e.g. fluorocarbons) would lead to increase in the UV radiation damage to the vegetation and animals on the Earth. Oxides of nitrogen play a central role in the formation of photochemical smog. $CO_2$ by itself is not a pollutant. However, a significant increase in $CO_2$ in the atmosphere would increase the average temperature of the lower troposphere by trapping some IR radiation (heat) from the Earth that would otherwise escape into outer space. This is called the "*greenhouse*" effect. The environmental impact of $CO_2$ and $NO_x$ is on the global scale.

## AIR POLLUTION EFFECTS ON OBJECTS AND ORGANISMS

Effects of air pollutants are wide ranging. They cause damage to materials and buildings, deteriorate soil and affect vegetation and living organisms. Atmospheric haze is caused by the presence of aerosols ($SO_x$, $NO_x$ and hydrocarbons). Aerosols absorb and reflect both solar and terrestrial radiations. They also act as nuclei for cloud formation. The net effect of increased levels of aerosols in the atmosphere is cooling of the Earth's troposphere.

Effects of particulate pollutants depend on their physical as well as chemical nature. Their size and composition are factors influencing their impact, with smaller particulate matter showing greater health hazards. Particulate matter is brought into contact with pulmonary membrane by sedimentation, impaction, interception and diffusion. Sub-micron range particles ($10 \rightarrow 2$ μm; $PM_{10} \rightarrow PM_2$) pose pulmonary health problems.

If the pollutants are chemically reactive or toxic their physiological effects are more serious and lethal. For example, $SO_2$ is a chemical irritant that damages mucous lining of the respiratory tract and causes chronic bronchitis. Powerful oxidants—$O_2$, $O_3$ and $H_2O_2$ are edemagenic agents. They enhance cell damage and premature aging. They generate free radicals, which interfere in many biochemical reactions.

Heavy metals like As, Cd, Hg, and Pb are highly toxic even in traces. Metal chelation affects the structure, and consequently function, of various enzyme complexes. Mercury can bind to both nucleic acid bases and phosphate groups, and thus interfere in the base-pair formation and the genetic code. Arsenic, which has similarities to phosphorus, interferes with enzymatic phosphorylation reactions.

# 2

# The Atmosphere and its Characteristics

The **atmosphere** is the gaseous mantle around the lithosphere and hydrosphere. It constitutes a number of gases such as nitrogen, oxygen, argon, carbon dioxide, helium, neon, xenon, ozone and krypton. These gases have definite percentage in the atmosphere in normal condition. In addition to this atmosphere also includes, water vapor, dust particles, smoke, industrial gases, pollen and several other substances that cause environ-mental degradation. The **lithosphere** is mainly made up of the solid components and consti-tute the rocky substance. **Hydrosphere** includes the liquid components such as lakes, rivers, ponds and oceans. **Biosphere** includes plants, animals, and all the other organisms.

The density of the atmosphere is unevenly distributed, with 80 per cent of the atmospheric constituents being distributed within 6 kilometers from the Earth's surface. The temperature of the atmosphere varies in a nonlinear manner with altitude, and based on this it is subdivided into five regions—(1) troposphere, (2) stratosphere, (3) mesosphere, (4) thermosphere and (5) exosphere.

Troposphere is of primary importance to living systems on Earth, and it is also the main region of pollution. The troposphere serves as the medium through which the pollutants are transported and dispersed locally and globally. Carbon dioxide and water vapor in the troposphere play an important role in Earth's heat balance. The ozone layer in the stratosphere absorbs most of the ultraviolet (UV) radiation from the Sun and shields life on Earth from its deleterious effects. Stratosphere is of interest to aeronautics, communications and for dispersion of pollutants on a global scale. This Chapter presents a detailed account of the atmosphere, its composition and structure.

The atmosphere is the blanket of gases or vapors that surround the Earth, and held together by the force of gravity. Other planets too have atmospheres but that of the Earth, called 'air', is different from them. The composition of the gas mixture in our atmosphere has evolved through millions of years. Several types of probes sent into the atmosphere, which include the rockets and more recently the satellites provide evidence for its extension for several thousand kilometers from the surface of the Earth. At the present stage of evolution, the atmosphere protects life on this globe from the hostile environment of outer space. In fact, it nurtures life on this Earth, providing carbon dioxide to the plants to make food through photosynthesis and oxygen to organisms for respiration and releasing energy from their food. Nitrogen is essential to make living matter; nitrogen fixing bacteria and ammonia manufacturing plants produce chemically bound nitrogen, which is also available to the animals. Atmosphere also transports water from the oceans (due to evaporation and precipitation). Therefore, pollution of such an important system is bound to have direct and indirect effects on living organisms.

## 2.1 COMPOSITION AND STRUCTURE OF THE ATMOSPHERE

The present composition of the Earth's atmosphere, in terms of total molecules, is—diatomic nitrogen (78.08 per cent), diatomic oxygen (20.95 per cent), argon (0.93 per cent), water in various forms (from about 0 to 4 per cent) and carbon dioxide (0.0325 per cent). Inert gases helium, neon, krypton, oxides of nitrogen, compounds of sulphur and minute suspended solid and liquid particles are present in lesser amounts.

The atmosphere extends from the surface of the Earth to heights of thousands of kilometers. Its very presence is the result of the gravitational pull of the Earth exceeding the escape velocity of the gas molecules. The composition and behaviour of the atmosphere are not the same at all heights. While the longer-lived, major constituents of the atmosphere, $N_2$ and $O_2$, are distributed more or less homogeneously around the Earth, the shorter-lived, minor species like carbon monoxide, nitric oxide and ozone vary considerably both in time and space.

It is also an established fact that many of the constituents like $O_2$, $CO_2$, $H_2$ and methane ($CH_4$) are either directly or indirectly under the influence of life on Earth, including the water on it. Some of the linkages are well established. Examples are: oxygen is generated by photosynthesis, which can be represented by the qualitative reaction

$$CO_2 + H_2O + h\nu \text{ (energy)} \rightarrow CH_2O \text{ \{carbohydrate\}} + O_2 \qquad \text{...(2.1)}$$

Aerobic respiration, decay involving bacteria or burning of organic matter releases $CO_2$.

$$CH_2O + O_2 \rightarrow CO_2 + H_2O + h\nu \text{ (energy)} \qquad \text{...(2.2)}$$

Methane, $CH_4$, is generated by anaerobic decay of organic matter. This gas is also generated in the stomachs of ruminants. Molecular hydrogen is a product of oxidation of hydrocarbons by OH radicals. A balance in the composition of atmosphere is maintained by a complex set of reactions generating and consuming the various constituents.

The molecules of the atmosphere experience a force of gravity proportional to their molecular masses. Therefore, heavy gases are bound more closely to the Earth while the lighter ones have a tendency to migrate to higher altitudes. The average molecular mass of the atmosphere, in atomic units, decreases rapidly with increasing altitude. Up to heights of 100 km, however, this value is nearly constant and hence this region is referred to as the 'homosphere'. Above this height the average molecular mass of the gaseous mixture decreases gradually and this region is called the 'heterosphere'. There is a greater abundance of atomic oxygen than nitrogen above about 160 km and the former gradually gives way to helium above 600 km. Hydrogen is the major constituent beyond (above) 1,000 km. About half of the air mass is confined within about 6.5 km height above the sea level and 99.9 per cent within the first 100 km.

Atmospheric pressure, $P_h$, decreases exponentially with altitude and is given by the equation

$$\log P_h = \log P_0 - \left(\frac{Mgh \times 10^5}{2.303RT}\right) \quad \text{...(2.3)}$$

where, $P_0$ = atmospheric pressure at the sea level;
$h$ = height in kilometers;
$M$ = molecular mass of air;
$g$ = acceleration due to gravity;
$R$ = gas constant;
$T$ = absolute temperature.

The Earth's atmosphere is subdivided into five regions on the basis of temperature changes with height—(1) troposphere, (2) stratosphere, (3) mesosphere, (4) thermosphere and (5) exosphere (Table 2.1 and Fig. 2.1). The density and characteristics of the atmosphere decrease sharply with increasing altitude.

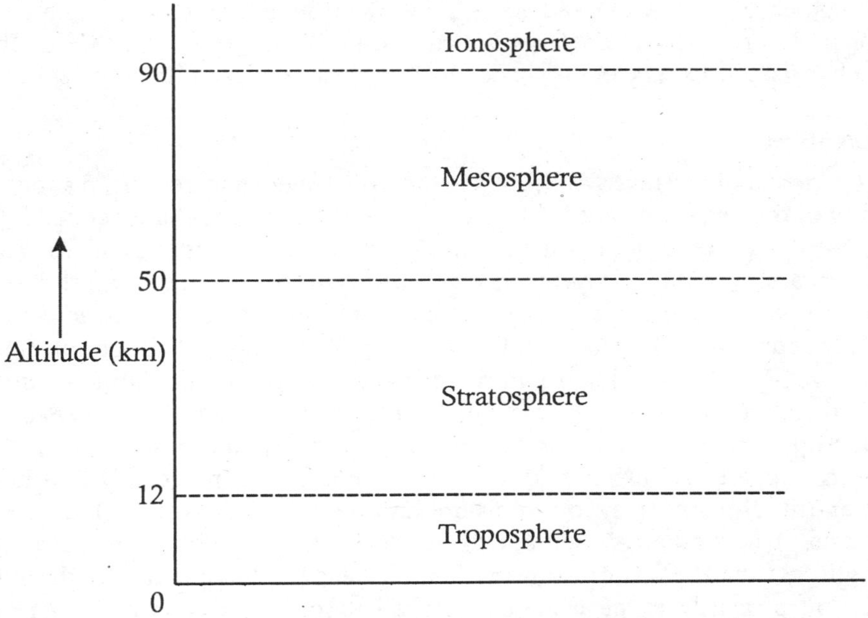


**Fig. 2.1:** Classification of the Atmosphere with Altitude (km)

**TABLE 2.1:** SUBDIVISION OF ATMOSPHERIC REGIONS

| *Region* | *Altitude* (km) | *Temperature* (° C) | *Composition* |
|---|---|---|---|
| Troposphere | 0 to 12 | 15 to −55 | $N_2$; $O_2$; $CO_2$ and $H_2O$ |
| Stratosphere | 12 to 50 | −55 to −2 | $O_3$ |
| Mesosphere | 50 to 90 | −2 to −90 | $O_2^+$; $NO^+$ |
| Thermosphere | 90 – 400 | Very high (−90 → 2,000) | $O_2^+$; $O^+$; $NO^+$; Ionized layers—E, F1 and F2 |
| Exosphere | 4,000 → | -- | Solar wind |

### 2.1.1 Troposphere

The layer of atmosphere closest to the Earth is the 'troposphere,' meaning 'turning ball'. About 80 per cent of the atmospheric mass is contained within it. Its thickness varies between 17.5 km at the equator to about 6.4 km at the poles, averaging about 12 km. This region is of primary importance to the living systems of the earth as practically all the organisms live in it. Most of the clouds and the weather systems are also contained in it. This is also the region most affected by pollution. There is a general decrease of temperature with altitude, –6.5° C per km, resulting in about –55° C at 12 km height. As the Sun's rays heat the surface of the Earth and not the air directly, the air closest to the ground is the warmest. Being lighter this warm air rises up while at the same time denser, cold air tends to sink down. This causes considerable convection in the troposphere.

Water vapor and carbon dioxide in the troposphere play an important role in Earth's heat budget by absorbing and trapping much of the infrared radiation that comes from the Sun. The cold air at the top of the troposphere (tropopause) is responsible for condensation of water to ice particles, preventing water from reaching higher altitudes where it would have dissociated through the action of high-energy ultraviolet radiation. Thus, the tropopause serves as a barrier to prevent decomposition of water, which would have resulted in loss of hydrogen from the Earth. In primordial times, when a barrier of this type did not exist, hydrogen escaped from the atmosphere.

### 2.1.2 Stratosphere

Stratosphere (stretched out or layered) lies above the troposphere and persists to about 50 km. The upper boundary of this region is called "stratopause". Stratosphere is characterized by increase of temperature with altitude, reaching a maximum –2° C. This tendency gives rise to strong thermodynamic stability, with little turbulence and vertical mixing. An air parcel that attempts to rise becomes rapidly colder and denser than the air that it displaces. Motions in the stratosphere are, thus, largely confined to the horizontal, accounting for the layered structure of high altitude stratus clouds. High-speed (~400 km/h), meandering wind currents circulating around the world at altitudes 15 to 25 km, called "jet streams" fill this region. The jet streams can affect the weather patterns in the troposphere and also help in global transport of pollutants.

Air in the stratosphere is thinner than in the troposphere. It is practically devoid of weather phenomenon and it is nearly free from moisture and dust. It is, however, characterized by the presence of ozone, $O_3$ (almost all of the atmosphere), which absorbs most of the harmful ultraviolet rays (wavelength less than 290 nm) from the Sun. Increase in temperature with altitude in the stratosphere is due, primarily, to the absorbed solar radiation. Ozone is formed as a result of short wave (less than 242 nm) solar radiation, which dissociates normal oxygen molecules, $O_2$, into two oxygen atoms. These oxygen atoms then combine with undissociated $O_2$ molecules to form ozone. Once it has been formed, ozone can also be destroyed by solar radiation (less than 300 nm). Ozone concentration is, generally, greatest near 25 km above the surface of the earth, where a balance exists between its formation and destruction by absorption of solar radiation. Stratosphere is of interest to aeronautics for the flight of jet planes and for radio communications.

### 2.1.3 Mesosphere and Thermosphere

In the mesosphere (middle), which is above the stratopause, temperature begins to drop again. At a height of about 85 km from the Earth the temperature is the lowest, –90° C. Inversion takes place and the temperature begins to rise again in the thermosphere, which lies above the mesosphere. The gases continue to thin to an altitude of about 600 km but the temperature rises to about 1,200° C at an altitude about 400 km. The inversion of temperature in this region is, essentially, due to energy extracted from sunlight at wavelengths below 200 nm. Among the important processes taking place in this region are the dissociation of oxygen molecule, ionization of the oxygen atom, oxygen molecule and the nitrogen molecule as shown below:

$$h\nu + O_2 \rightarrow O + O \qquad \text{...(2.4)}$$

(100-200 nm)

$$h\nu + O \rightarrow O^+ + e \qquad \text{...(2.5)}$$

$$h\nu + O_2 \rightarrow O_2^+ + e$$

$$h\nu + N_2 \rightarrow N_2^+ + e \qquad \text{...(2.6)}$$

(less than 100 nm)

Here $h\nu$ represents a photon and e, the electron.

At these high altitudes the dissociation of $O_2$ and $N_2$ molecules is balanced by their reformation in direct or indirect ways. In a situation where the density of the gases decreases with altitude, oxygen atoms diffuse downward while $O_2$ molecules tend to float in the reverse direction. However, the two processes occur at different altitudes. Recombination of oxygen atoms is mainly confined to altitudes below 100 km, as the relatively denser atmosphere there is conducive to that purpose. On the other hand, dissociation of $O_2$ can proceed at any level, being limited only to the supply of oxygen and the radiation.

Due to the photoionization reactions the mesosphere-thermosphere region is rich in electrically charged particles and electrons. Therefore, this region is also called 'ionosphere.' It is particularly significant at altitudes close to 80 km. Ionosphere is the environment for earth-orbiting satellites and ballistic missiles, and is therefore gaining importance in the context of atmospheric pollution.

## 2.2 ENERGETICS OF THE EARTH-ATMOSPHERIC SYSTEM

The Sun is the ultimate source of energy transfer to our planet. Its importance can be seen from the values given in Table 2.2. The energy involved in most of the natural phenomena or man-made incidents (e.g., nuclear explosion) is negligible compared to the solar energy received by the Earth. The later is represented on an arbitrary scale of 1 to 100, taking into account the variations due to time, season and place on the globe.

**TABLE 2.2:** RELATIVE TOTAL ENERGY TRANSFER IN VARIOUS PROCESSES

| *Process* | *Relative Range* |
|---|---|
| Solar energy received by the earth daily | $1–10^2$ |
| Annual consumption of energy | $10^{-2}–1$ |
| Cyclone storm/hurricane | $10^{-4}–10^{-1}$ |
| Nuclear explosion | $10^{-8}–10^{-3}$ |
| Tornado | $10^{-11}–10^{-9}$ |
| Gusty wind | $10^{-17}–10^{-15}$ |

Both the Sun and the Earth emit electromagnetic radiation and in this process behave like ideal 'blackbodies' at characteristic temperatures. The radiant energy emitted from the surface of a body, and the maximum ($\lambda_{max}$) of the spectral density of radiation are given by the Stefan-Boltzmann and the Wien displacement laws.

$$E = \sigma T^4 \qquad \text{Stefan-Boltzmann law} \qquad \ldots(2.7)$$

$$\lambda_{max} \cdot T = \text{constant} \qquad \text{Wien's law}$$

where, $E$ = radiation energy emitted from unit area in 1 sec;
$\sigma$ = Stefan-Boltzmann constant;
$T$ = absolute temperature.

The Sun emits electromagnetic radiation like a blackbody at 6,000 K. The output consists of a wide range of wavelengths, which include *X*-rays, ultraviolet, infrared and visible light. About half this solar radiation is in the visible region, 400-700 nm, with peak intensity at 470 nm. Though the Sun radiates $3.8 \times 10^{23}$ kilowatt energy per second, only a small portion of it reaches the Earth's atmosphere. The radiant energy thus falling perpendicular to the Sun's rays at the Earth's mean distance from the Sun is only $1.34 \times 10^3$ watt/m$^2$. This is called the *solar constant*. About 30 per cent of this energy is not utilized by the Earth or its atmosphere, as it is scattered back to space (approximately, 4 per cent is reflected by ground, 20 per cent reflected by clouds and 6 per cent scattered by air). The radiation thus reflected is called *albedo*. The remaining 70 per cent, which is absorbed by clouds, air, ground and water is sufficient to power the movement of atmospheric wind and the oceanic currents and to sustain all activity in the biosphere. A fraction of short wave radiation is absorbed by the atmospheric gases, particularly in the stratosphere and the thermosphere, resulting in direct warming of the air or in photochemical reactions. About 40-50 per cent of the solar radiation (> 300 nm) reaches the Earth's surface (Figs. 2.2 and 2.3).

Solar radiation directly reaching the Earth depends on cloud cover above. The magnitude of solar radiation reaching the surface of the Earth also depends on latitude, time of the year, time of the day and orientation of land surface with respect to Sun. Part of the scattered and radiated energy from the hot clouds and air also reaches the Earth. Earth, which has a surface temperature of 200-300 K, emits long wavelength (IR) radiation, with a peak at about 12,000 nm appropriate to blackbody radiation at such a temperature. The water vapor and carbon dioxide, present in the atmosphere absorb a large fraction of the long wave radiation. Part of the heat from the Earth goes to evaporate water from the surface. The heat carried by the water vapor is transmitted to the air when the former condenses to form clouds and either rain or snow. Thus, the energy budget of the lower atmosphere is very much controlled by phase changes of water.

Solar radiation reaches the Earth in the UV/visible region while the radiation from the surface of the Earth is concentrated in the infrared (IR) region. *This difference in the wavelength regions of solar and terrestrial radiation is of fundamental importance to the energy balance of the Earth-atmosphere system* (Fig. 2.4).

The atmosphere is essential in maintaining the heat balance of the Earth as it helps to trap the heat of the sun. The Earth is warmed by the sunlight that passes through the atmosphere but the heat that it creates cannot easily escape through the same atmosphere. As a result the atmosphere serves as an important heat stabilizing function and prevents the tremendous temperature extremes on the Earth (some other planets do have, such as Venus and Mars).

The surface of the Earth is the prime heat exchanger (process of transmission of energy from one body to another, without doing work) in the system by conduction, convection and radiation.

Conduction, which is the mode of energy transfer by physical contact, has little effect on the atmospheric heat transfer process. This is because the atmosphere is a poor conductor of heat as the air molecules are relatively far apart from each other. Convection is the heat transfer process by movement (of air bodies). Radiation, which is a direct energy transport process, has also little effect on the atmosphere. This means, convection is the major process by which energy transfer in the atmosphere takes place.

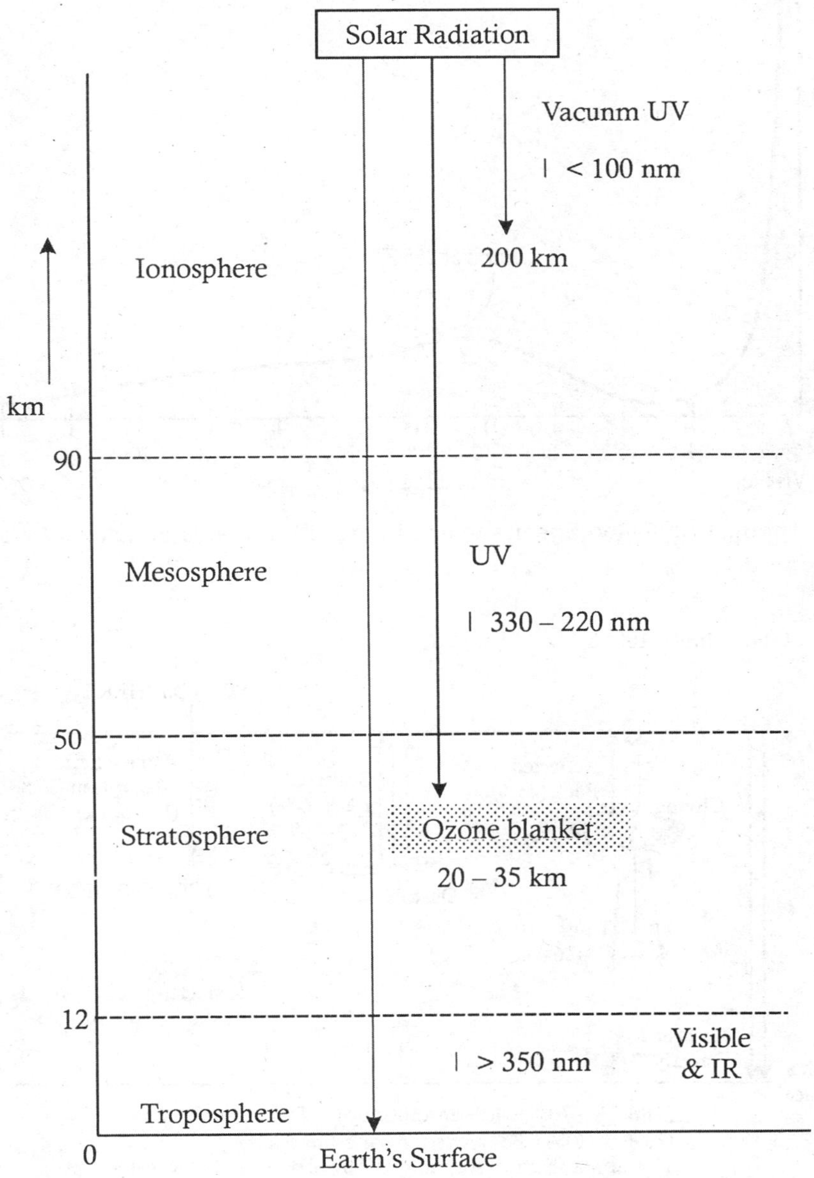


**Fig. 2.2:** Penetration of Solar Radiation through different Regions of the Atmosphere

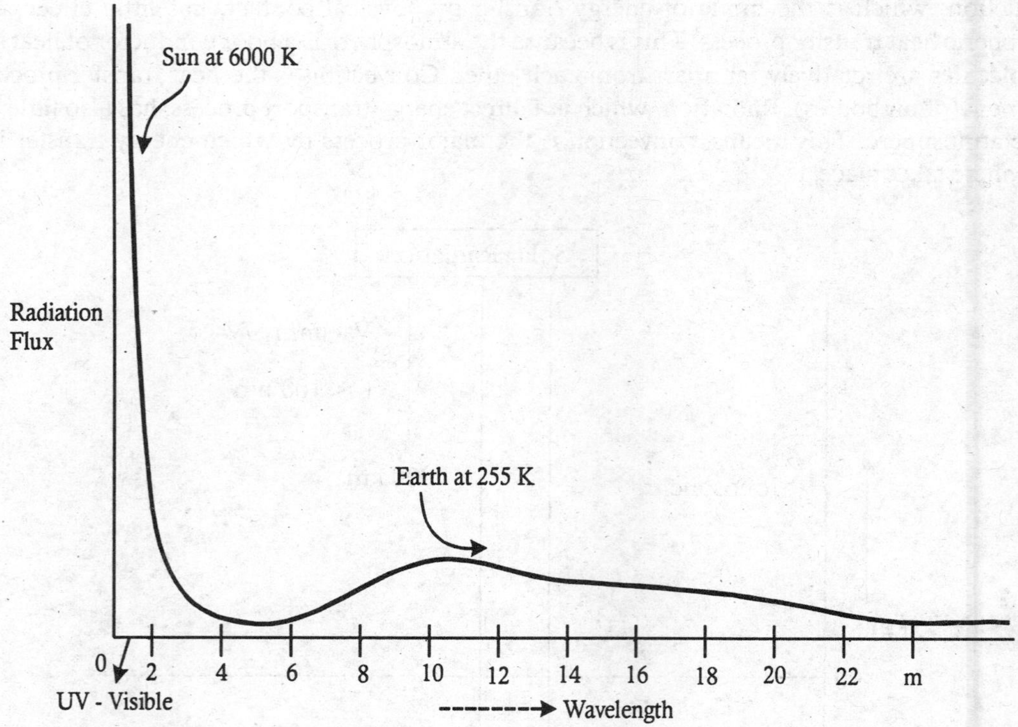


**Fig. 2.3:** Thermal Radiation Spectra to and from the Earth as a Function of Wavelength

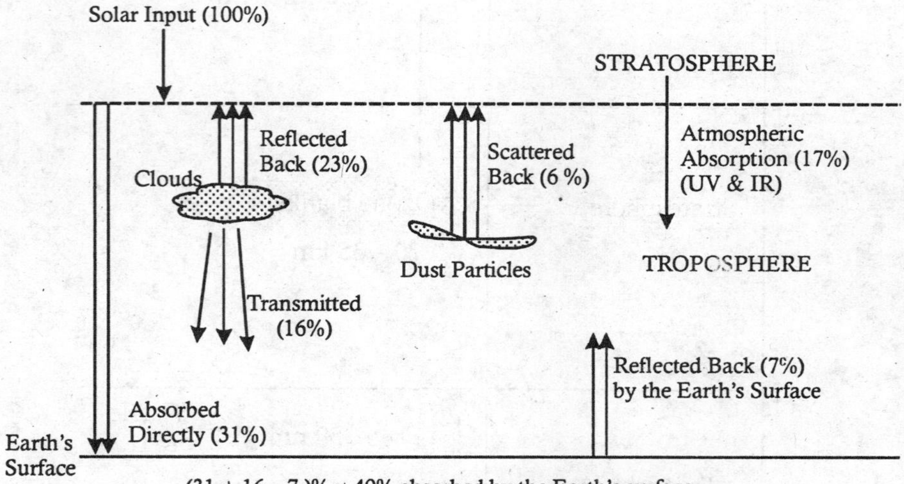


(31 + 16 – 7 )% » 40% absorbed by the Earth's surface
(23 + 6 – 7)% » 36% reflected back into space;
17% atmospheric absorption (UV by ozone and IR by water vapor)

**Fig. 2.4:** Radiation-Energy Balance in the Atmosphere

## 2.3 WEATHER AND CLIMATE

Atmospheric pressure and wind are both significant controlling factors in the Earth's weather and climate. Wind circulation is the result of uneven heating of the atmosphere and the surface of the Earth by energy from the Sun and existence of horizontal and vertical differences (gradients) in pressure. The magnitude of solar radiation reaching the surface of the Earth depends on a combination of latitude, cloud cover, time of the year, time of the day, and orientation of the land surface with respect to the Sun. The air from hot areas expands and rises while air from cooler areas flows to replace the hot air. Atmospheric pressure and wind circulation are closely related. Upper air circulations play a key role in the development and propagation of surface pressure systems. Near the Earth's surface winds, generally, flow around regions of relatively low and high pressure, called cyclones and anticyclones, respectively. The total motion of a parcel of air has two components: (1) motion caused by the force created by horizontal difference in pressure and (2) motion caused as a result of Earth's rotation (Coriolis force). As a result the relative motion of the wind is deflected to the right (clockwise) in the Northern Hemisphere and to the left (anticlockwise) in the Southern Hemisphere.

Winds play a significant role in determining and controlling weather and climate. Among the local wind systems is the so-called sea/land breeze, encountered along coastline along large bodies of water. They are caused by difference in the rates of heating and cooling of the land and water surfaces. Monsoons represent another wind pattern caused by seasonal variations in the mean sea level pressure over continents. They are principally encountered in Southern Asia and parts of Africa. They blow for approximately for six months from the northeast and for six months from the southwest.

### 2.3.1 Weather

Weather is the condition of the wind patterns (horizontal movement of air currents), during a short time (hours), at any geographic location. Description of weather includes information on such atmospheric phenomena like temperature, humidity, precipitation, air pressure, cloud cover and wind pattern and speed. Weather occurs in the troposphere. Weather conditions in an area are affected by local geographical features like mountain terrain, ocean and nearby lakes. A region's weather may change from day to day and profoundly affects the habitat, vegetation and health.

### 2.3.2 Climate

Climate of any region constitutes the total experience of weather and atmospheric behavior over a long period (say, 30 years). However, climate is not just the 'average weather.' It is a complex concept involving data on all concepts of weather and, in addition, their extreme ranges, variability and frequency. Weather in any particular region may change significantly from day to day but its climate remains the same from year to year over several decades.

The fundamental determinant of climate is the geometry of the Earth-atmosphere system (Fig. 2.5). The planet's surface receives, from the Sun, more energy per unit area near the equator than elsewhere. Most parts of the surface of the Earth are not perpendicular to the Sun and the energy they receive depends on the solar elevation angle (90° for the overhead Sun). This angle changes with latitude, the time of the year and the time of day.

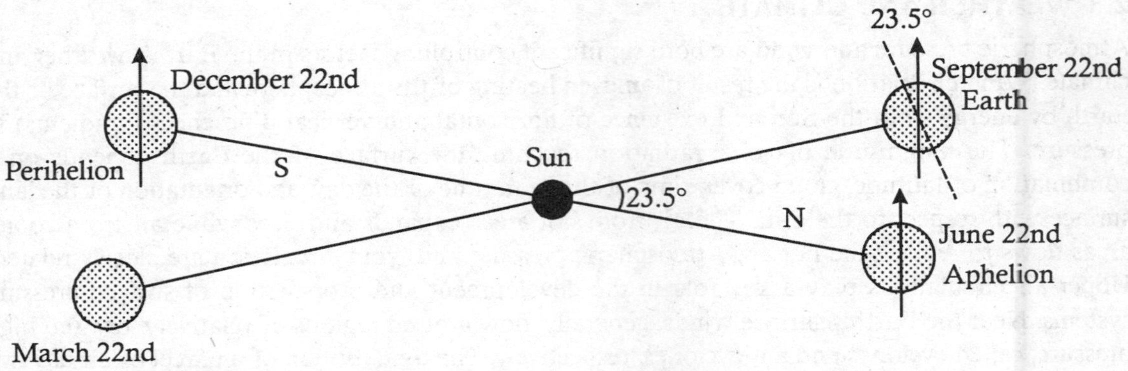


**Fig. 2.5:** Tilt of the Earth's Axis and Orbit around the Sun

The intensity of Sun's radiation flux, $S$, on the surface of the Earth is 9 $J \cdot m^{-2} \cdot min^{-1}$. Radiation flux at altitude $h$ is given by

$$S_h = S \cdot \cos(Z) \qquad \text{...(2.8)}$$

$$\cos(Z) = \sin(\phi) \cdot \sin(\delta) + \cos(\phi) \cdot \cos(\delta) \cdot \cos(\eta) \qquad \text{...(2.9)}$$

where, $Z$ = zenith angle, $\phi$ = latitude, $\delta$ = solar declination, $\eta$ = hour angle (about 15°).

$\phi$ and $\delta$ are positive in the Northern Hemisphere and negative in the Southern Hemisphere. The solar azimuth, $\omega$, is the angle measured between the South and the direction toward the Sun in a horizontal plane and is given by

$$\sin(\omega) = \frac{\cos(\delta) \cdot \cos(\eta)}{\sin(Z)} \qquad \text{...(2.10)}$$

The daily rotation of the Earth around its axis, the eccentricity of the orbit around the Sun and the 23.5° tilt of the Earth's axis from a line normal to its plane of orbit around the Sun (Fig. 2.5) are the basis of diurnal cycles and annual seasons. The inclination, rather than the distance of the Earth from the Sun, controls seasonal temperatures. The four seasons—winter, spring, summer and autumn, represent consistent annual changes in the weather. In the Northern Hemisphere, winter starts with the winter solstice (December 22 or 23), the spring with the vernal equinox (March 20 or 21), the summer with the summer solstice (June 21 or 22) and the autumn with the autumnal equinox (September 22 or 23). In the Southern Hemisphere, summer and winter are reversed as are spring and fall. These characteristics provide the basis for seasonal and diurnal cycles, dependent on sunlight radiation.

Climate zones, based on vegetation, are classified under: (*i*) Tropical (average monthly temperature > 18° C), (*ii*) Temperate and (*iii*) Polar Regions. Modified versions of Köppen's classification sub-divide climate regions based on annual and monthly averages of temperature and rainfall (Table 2.3).

The uneven heating sets up air currents in the atmosphere. Warm air rises and cool air sinks, causing the atmospheric air to circulate. Due to Earth's rotation around its axis, meridional motions of air become zonal like trade winds, temperate and polar winds. Since winds transport heat and moisture they affect an area's temperature, humidity, precipitation and cloudiness.

The circulation of ocean water is an important factor in air temperature distribution. There is a complex two-way interaction between the ocean and the atmosphere. The great ocean currents are set in motion by the drag of winds over vast areas of the sea surface. Ocean currents that have a northward and southward component, such as the warm Gulf Stream in the North Atlantic and the cold Peru Current of South America are responsible for exchange of heat between low and high latitudes. The fast flowing warm Gulf Stream is responsible for the relatively mild climate of Northern Europe. The Labrador Current cools the coast of Canada and New England.

**TABLE 2.3:** KÖPPEN'S CLASSIFICATION OF CLIMATE ZONES

| *Group* | *Features* |
|---|---|
| Af | Tropical wet climate with at least 60 mm of rain in the driest month |
| Aw | Tropical dry climate with at least one month with rain fall <60 mm |
| Am | Monsoon climate; short dry season and heavy seasonal rainfall |
| BS | Semi-arid climate |
| BW | Arid climate |
| Cs | Dry summers and rainfall maximum in winter |

## 2.4 METEOROLOGY

*Meteorology* is the study of atmospheric phenomena including daily or even hourly changes in weather within the troposphere. Meteorology specifies the wind circulation patterns, which result from a combination of air pressure, Coriolis and frictional forces (Fig. 2.6) and temperature. The meteorological elements that have the most direct and significant effects on the distribution of air pollutants in the atmosphere are wind speed, wind direction, solar radiation, atmospheric stability and precipitation.

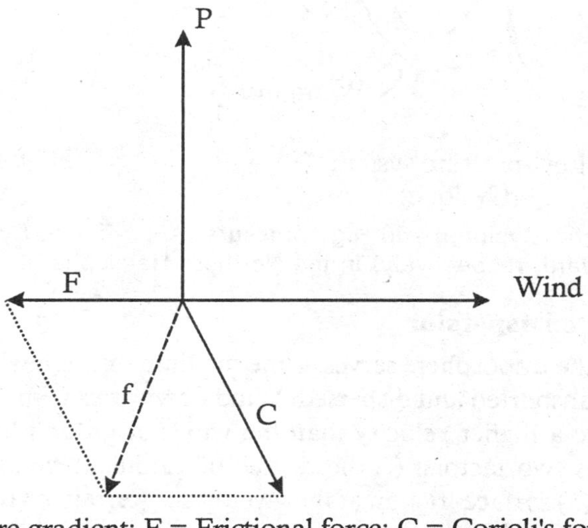


(P = Pressure gradient; F = Frictional force; C = Corioli's force)

**Fig. 2.6:** Resultant Force, f, between Pressure, Coriolis and Frictional Forces

### 2.4.1 Atmospheric Systems

The energy from the Sun is responsible for heating the Earth, but this energy is not received equally over the Earth's surface. This uneven distribution sets up wind systems, which transfer heat from the 'surplus' to the 'deficit' areas and prevent large temperature contrast. The distribution of heat over the Earth's surface is moderated by circulation patterns of air in the atmosphere and ocean currents induced by the prevailing winds. The specific heat (heat required to alter the temperature of a unit mass by 1° C) of water (1.0) is greater than that of land (0.2-0.4). Therefore, the temperature of the water bodies rises and falls more slowly than that of the landmass exposed to the same amount of heat radiation. So, water bodies moderate land temperatures. In addition ocean currents interacting with the prevailing winds transport heat from the equator toward the poles.

Wind is the motion of air relative to the Earth's surface. Upward motion of air into the stratosphere is inhibited by a stable thermodynamic stratification. Therefore, any air transported upward by convection diverges towards the poles in the upper troposphere. Moreover, rotation of the Earth around its axis has an effect on the trajectory of the wind circulation pattern (the *Coriolis* force). Coriolis effect is significant in studies of the dynamics of the atmosphere. In the hydrosphere it affects the rotation of the oceanic currents. On the Earth, the winds tend to follow a path deflecting to the right (clockwise) in the Northern Hemisphere, and to the left (anti-clockwise) in the Southern Hemisphere (Fig. 2.7). The Coriolis deflection is related to the motion of the object, the motion of the Earth, and the latitude. The magnitude of the Coriolis effect, *C*, per unit mass of air is

$$C = (2\omega\sin\varphi) \cdot v \qquad \text{...(2.11)}$$

where, $\omega$ = angular velocity of the Earth;
$\varphi$ = latitude;
$v$ = speed of air motion

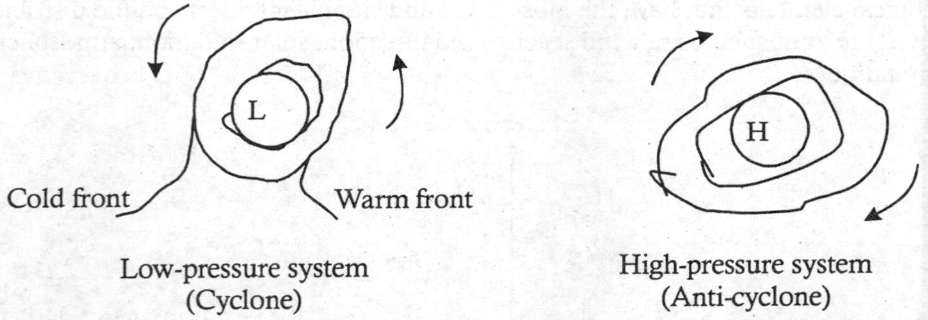


**Fig. 2.7:** Low-pressure (Cyclone) and High-pressure (Anticyclone) Systems and Wind flow (Upward and Downward, respectively) in the Northern Hemisphere

### 2.4.2 Atmospheric Dispersion

The atmosphere in the troposphere serves as the medium through which its constituents, including the pollutants are transported and dispersed. Wind flow varies with altitude (Table 2.4). The winds aloft, generally, have a higher velocity than the winds at ground level. The effect of altitude on wind speed involves two factors: (*i*) the degree of turbulent mixing (characterized by Pasquill stability class), and (*ii*) surface friction at the given place (terrain characteristics). Variation of wind speed, $U_z$, with height (z) is given by

$$\frac{U_z}{U_o} = \left(\frac{z}{z_o}\right)^a \qquad \text{...(2.12)}$$

where $U_0$ = wind speed at the Earth's surface;
$z_0$ = 10 meters;
$\alpha = 0.1 \rightarrow 0.5$

Wind (horizontal air movement) and air currents (vertical air movements) are strongly involved in air pollution phenomena. Wind carries and disperses air pollutants. Therefore, prevailing wind direction and speed are relevant factors in determining the areas affected by the air pollution sources. Dispersion, and thereby dilution, of constituents and contaminants in the atmosphere depends on the mean wind speed and turbulence.

**TABLE 2.4:** RELATIONSHIP BETWEEN WIND FLOW AND ALTITUDE ABOVE THE EARTH'S SURFACE

| *Altitude* | *Atmospheric Wind Flow* |
|---|---|
| Above 1,000 meters | Frictionless, horizontal flow--either geotropic (unaccelerated) or gradient |
| 1,000-50 m | Speed and direction of the wind varies with altitude |
| 50-5 m | Speed of wind varies with height $\sim(z/z_0)^\alpha$ |

Fick's laws of diffusion can be used to obtain an estimate of the dispersion (dilution) of constituents under steady state conditions—

$$\frac{\partial C}{\partial t} = -D\frac{\partial J}{\partial z} \quad \text{(continuity equation)} \qquad ...(2.13)$$

$$J = -D\frac{\partial C}{\partial z} \quad \text{Fick's 1st law of diffusion} \qquad ...(2.14)$$

$$\frac{\partial C}{\partial t} = \frac{\partial J}{\partial z} = D\frac{\partial^2 C}{\partial z^2} \quad \text{Fick's 2nd law} \qquad ...(2.15)$$

For dispersion with chemically reacting constituents,

$$\frac{\partial C}{\partial t} = -\frac{\partial}{\partial z}(J) + \chi \qquad ...(2.16)$$

$$D = \frac{1}{3}U\lambda; \quad \eta = \frac{1}{3}\rho U\lambda \qquad ...(2.17)$$

where, $J$ = Flux (diffusion gradient);
$C$ = concentration;
$z$ = height;
$D$ = diffusion coefficient;
$\chi$ = reaction kinetics parameter;
$U$ = average velocity;
$\lambda$ = mean free path of the particles;
$\eta$ = viscosity coefficient;
$\rho$ = density.

### *2.4.2.1 Atmospheric Turbulence*

Airflow can be laminar or turbulent. In the laminar flow, heat is transferred more efficiently by conduction while in the turbulent flow it is by convection. Turbulence is irregular air movement in which the wind constantly varies in speed and direction. In turbulent atmosphere there are discrete pockets of air known as *eddies*, which move randomly and cause mixing. As the atmospheric air is not a good conductor of heat, diffusion and mixing in the atmosphere are produced primarily by organized motion of eddies (turbulence). At low velocity, laminar flow occurs and the mixing is due to molecular diffusion. Mixing by turbulence is many thousand times rapid than mixing by molecular diffusion. Therefore, it is on turbulence that the mixing of the atmosphere mostly depends.

Turbulence makes low-level winds extremely unstable. Turbulence has a marked diurnal variation, reaching maximum at about midday.

Turbulence is due to a combination of mechanical and thermal factors. Mechanical turbulence occurs when the wind changes with height. Sunlight passing through clear air heats the surface of the Earth, and as the air at the ground level gets heated it rises from the ground, expands and cools (adiabatic cooling). Thus, under normal atmospheric conditions there is gradual temperature decrease with height in the troposphere. Heat convection occurs whenever the temperature decreases rapidly with height and the lapse rate exceeds ~ –1°/100 m. The Richardson number, Ri, characterizes mechanical turbulence and heat convection.

$$Ri = \frac{\text{Turbulence by buoyancy}}{\text{Turbulence by mechanical shear}} \quad ...(2.18)$$

$$\begin{aligned} &Ri = 0 \text{ Mechanical turbulence} \\ &Ri < 0.05 \text{ Heat convection} \\ &Ri > 0.25 \text{ Inversion (no vertical mixture)} \end{aligned} \quad ...(2.19)$$

On clear, calm nights the value of Ri becomes large and positive as radiational cooling results in a temperature increase with height. Turbulence is suppressed by the strong thermal stratification and a nearly laminar flow can result over flat terrain. During windy conditions mechanical turbulence is important.

The amount of turbulence in the ambient air has a major effect upon the rise and dispersion of air pollutant plumes. The amount of wind turbulence can be categorized into Pasquill "stability classes—A, B, C, D, E and F (Table 2.5). Class A is the most turbulent, and class F the most stable (least turbulent). Classes A, B and C represent unstable atmospheric conditions, which indicate various levels of extensive mixing (e.g. lofting, looping). Classes E and F indicate stable air in which stratification strongly dampens mechanical turbulence. These conditions can produce a fanning plume that does not rise much and retains a narrow shape in the vertical dimension for a long distance downwind. Class D stability is neutral, with moderate winds and mixing properties. It produces coning plume.

**TABLE 2.5:** PASQUILL'S STABILITY CLASSES OF WINDS

| *Surface Wind Speed m/s* | *Day Time Incoming Solar Radiation* | | *Night-Time Cloud Cover* |
|---|---|---|---|
| | *Strong* | *Slight* | *>50%/<50%* |
| < 2 | A | B | E/F |
| 2 - 3 | A/B | C | E/F |
| 3 - 5 | B | C | D/E |
| 5 - 6 | C | D | D |
| > 6 | C | D | D |

## 2.5 ATMOSPHERIC STABILITY/INSTABILITY

There are three types of temperature profiles—

(1) A neutral temperature profile occurs on overcast days and everyday at dawn and dusk. In neutral conditions temperature has no influence on friction-driven vertical transport of matter and energy.

(2) During clear summer days the ground is hotter than the air above and temperature decreases with height above of the Earth's surface. Under these conditions the density of air increases

with height. The atmosphere is unstable and vertical transport is enhanced. This is an unstable atmospheric state.

(3) During clear nights, there is greater heat loss from the ground. The air above it is denser and friction-driven. Vertical transport is damped. This represents a stable atmospheric state.

Vertical air motions (convection) affect both weather and mixing process. This is important in monitoring air pollution. The temperature and density of the atmosphere normally decrease with elevation.

Near the Earth's surface, where heat is exchanged between the Earth and the air, the processes are non-adiabatic (Chapter 3, Principles of Thermodynamics). Air gets cooled as it rises and warms up as it descends.

$$-\frac{\Delta T}{\Delta z} = \frac{g}{C_p} \quad \text{...(2.20)}$$

where, $z$ = height;
$g$ = acceleration due to gravity;
$C_P$ = Specific heat of gas at constant pressure;
$T$ = absolute temperature

Since the pressure decreases with height, there is an upward pressure gradient force. The difference between this force and the gravitational force is called the *buoyancy* force. As the dry air rises, it expands due to decrease in pressure and undergoes adiabatic cooling. The rate at which dry air gets cooler with altitude is called *dry adiabatic lapse rate* (or *lapse rate*). It is ~1° C/100 m. Lapse rate is a function of the elevation (Fig. 2.8).

### 2.5.1 Temperature Inversion

Under normal conditions the temperature of the atmosphere decreases with height in the troposphere, as shown in Fig. 2.9(*a*). However, during a thermal inversion there is a region of the troposphere where the temperature increases with altitude (Fig. 2.9(*b*)). A thermal inversion causes a stable atmosphere. Such inversions are due to nocturnal cooling of the ground. As a result the air closer to it is cooler, and the cold air is trapped under the movement of a warm front over a body of cooler air.

Thermal inversion plays an important role in determining cloud forms, precipitation and visibility. An inversion acts as a lid on the vertical movement of air in the layers below. If there are horizontal winds to disperse air pollutants they get trapped overhead. Inversion occurs most often on clear nights, when the ground cools off rapidly by radiation. They are more pronounced in mountain valleys. If the temperature of the surface air drops below its dew point, fog might result. Most of the air pollution disasters have occurred during periods of extended inversions. A subsidence inversion develops when a layer of air descends bodily. High-pressure systems (anticyclones) are generally regions where the air is sinking (subsiding). As the air moves to lower altitudes and higher pressure, it is compressed and its temperature rises. This causes temperature inversion.

### 2.5.2 Lapse Rates and Atmospheric Stability

Ambient and adiabatic lapse rates are a measure of atmospheric stability. Profiles of atmospheric stability are given in Fig. 2.10.

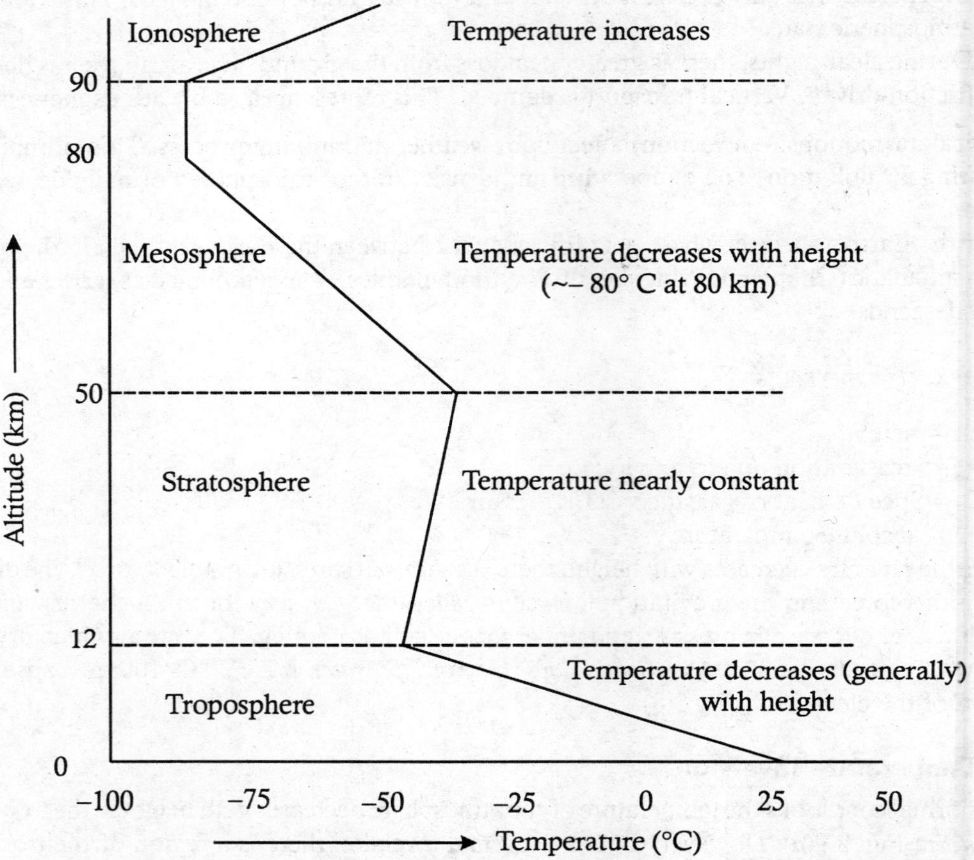


**Fig. 2.8:** Atmospheric Temperature Profile with Altitude

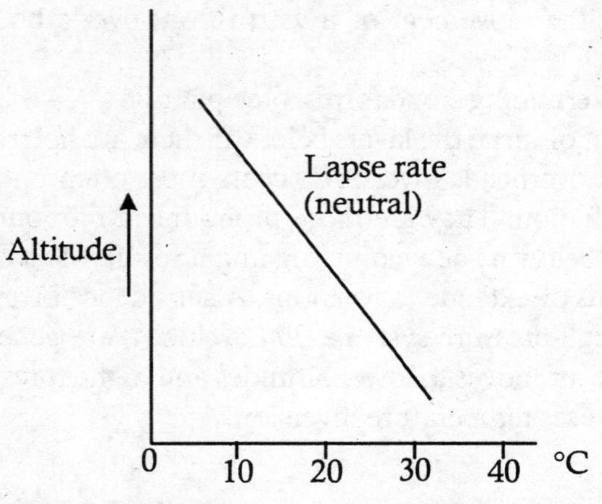


**Fig. 2.9(*a*):** Normal Temperature Profile of the Atmosphere

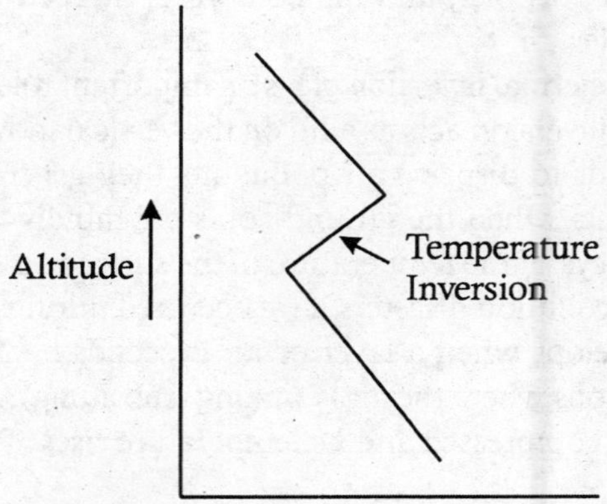


**Fig. 2.9(*b*):** Temperature Profile of the Atmosphere with Temperature Inversion.

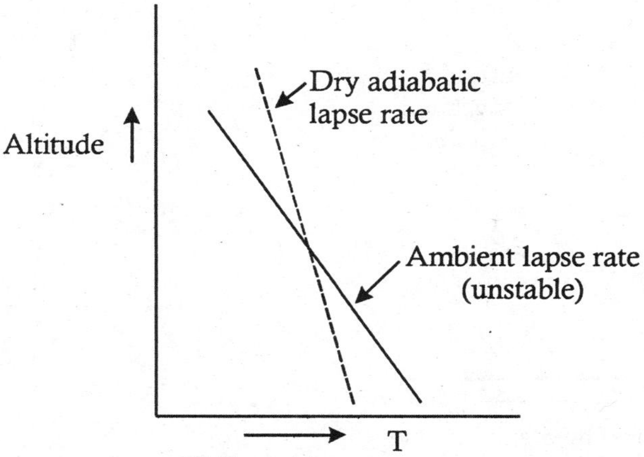


**(i)** Unstable

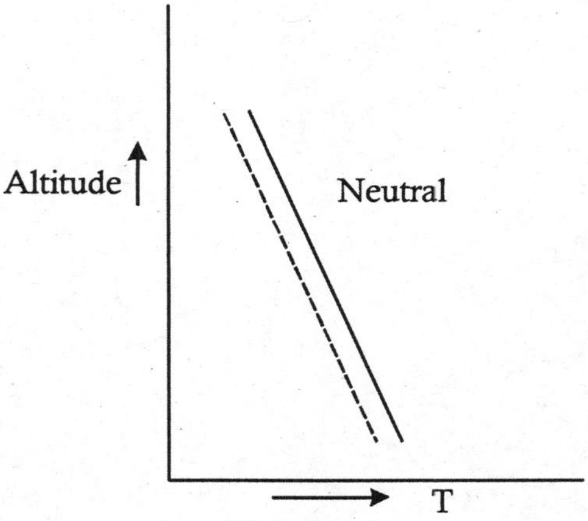


**(ii)** Neutral

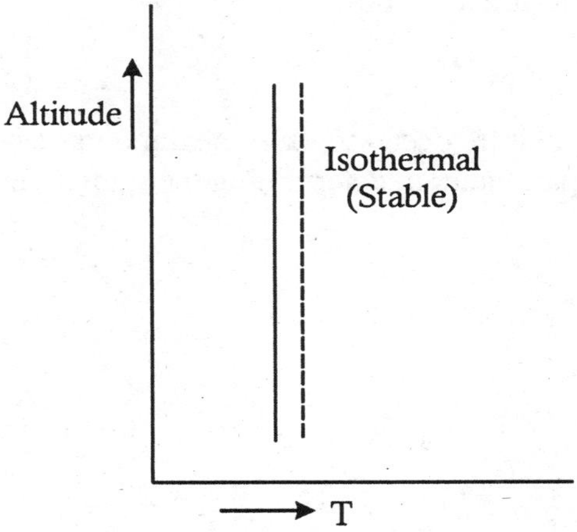


**(iii)** Isothermal (stable)

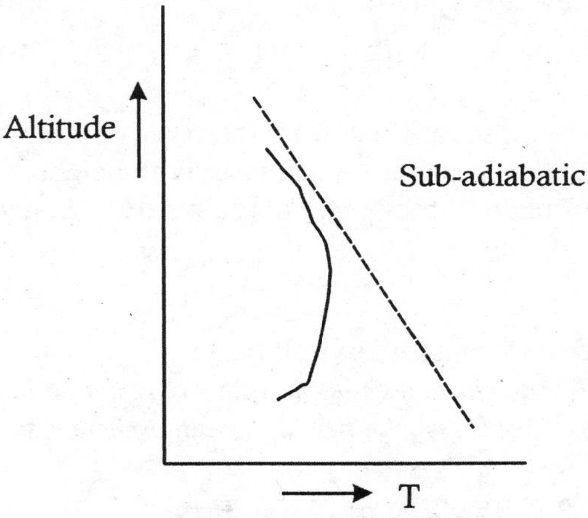


**(iv)** Sub-adiabatic (stable)

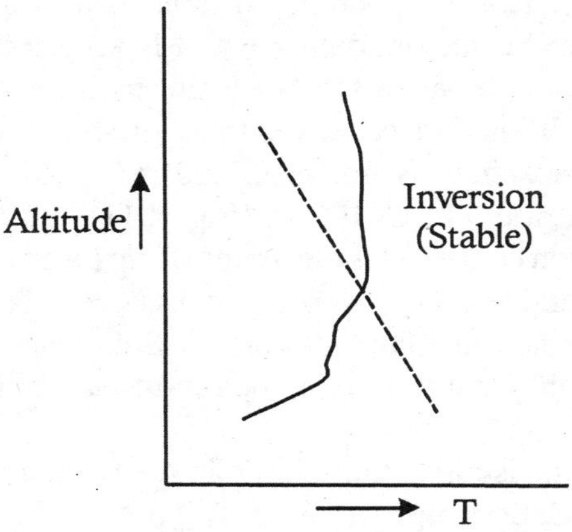


**(v)** Inversion (stable)

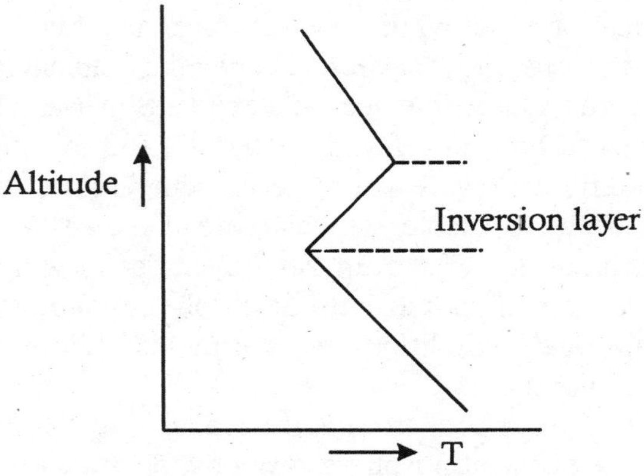


**(vi)** Inversion Layer (stable)

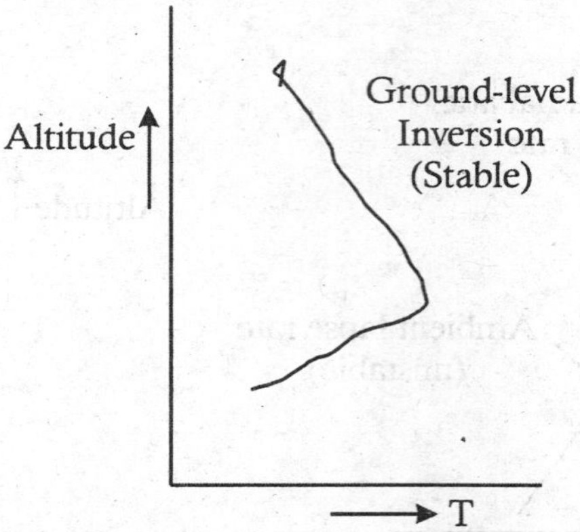


**(vii)** Ground-level Inversion (stable)

**Fig. 2.10:** Profiles of Atmospheric Stability

Any atmospheric system is a stable one if the temperature lapse rate

$$\left(\frac{dT}{dz}\right) > \left(\frac{dT}{dz}\right)_{adia} \quad ...(2.21)$$

The lapse rate is considered positive when the temperature decreases with height and negative when the temperature increases with height. A useful parameter in determining the stability in the atmosphere is the potential temperature, θ, given by

$$\theta = T\left(\frac{1000}{P}\right)^{0.288} \quad ...(2.22)$$

where $P$ = pressure in millibars.

*If θ decreases with height, the atmosphere is unstable.*

*If θ increases with height, the atmosphere is stable.*

### 2.5.2.1 Profiles of Lapse Rates

The relationship between the ambient and dry adiabatic lapse rates essentially determines the stability of the air and speed with which the pollutants disperse. The atmosphere is unstable as long as a parcel of air moving upwards cools at a lower rate than the surrounding air and is accelerated upwards by buoyancy force. When the ambient lapse rate and dry adiabatic lapse rate are the same, the atmosphere has neutral stability. Super-adiabatic conditions prevail when the air temperature drops faster with elevation (>1° C/100 m) and the atmosphere is unstable, and sub-adiabatic conditions prevail when the air temperature drops at a rate <1° C/100 m (Fig. 2.11) and the atmosphere is stable. A special case of sub-adiabatic inversion condition is the temperature inversion, when the air temperature actually increases with altitude and a layer of warm air exists over a layer of cold air. Super-adiabatic atmospheric conditions are unstable and favor dispersion. Sub-adiabatic atmospheric conditions (e.g. thermal inversions) are extremely stable and trap pollutants, inhibiting their dispersion.

The amount of turbulence available to diffuse pollutants is also a function of the temperature profile of the atmosphere, given by the pressure-volume relationship.

$$(PV)^n = \text{constant} \quad ..(2.23)$$

where $n = 1$ for isothermal atmosphere ( $T$ = constant).
$n = 1.4$ for isoentropic atmosphere (constant entropy; a function of height only).

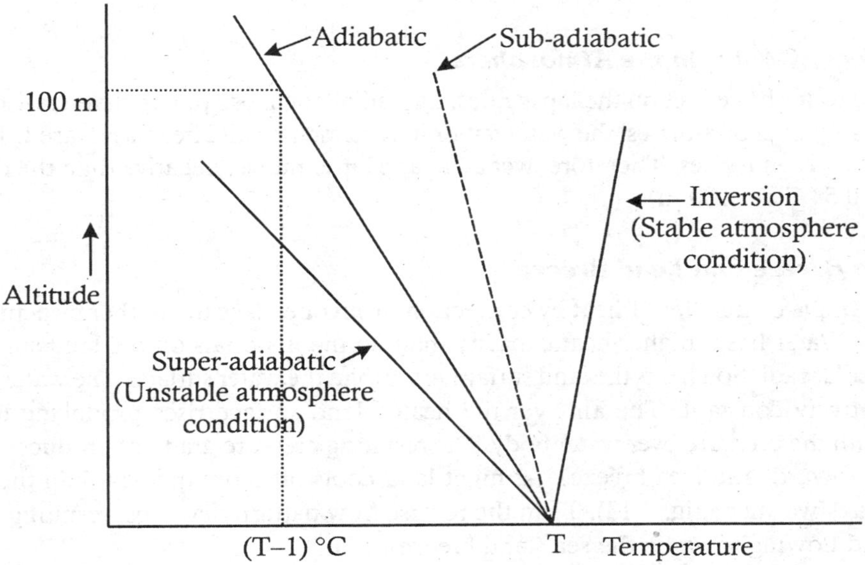


**Fig. 2.11:** Dry Adiabatic Lapse Rate and Ambient Lapse Rates

Profiles of lapse rates are given in Fig. 2.12.

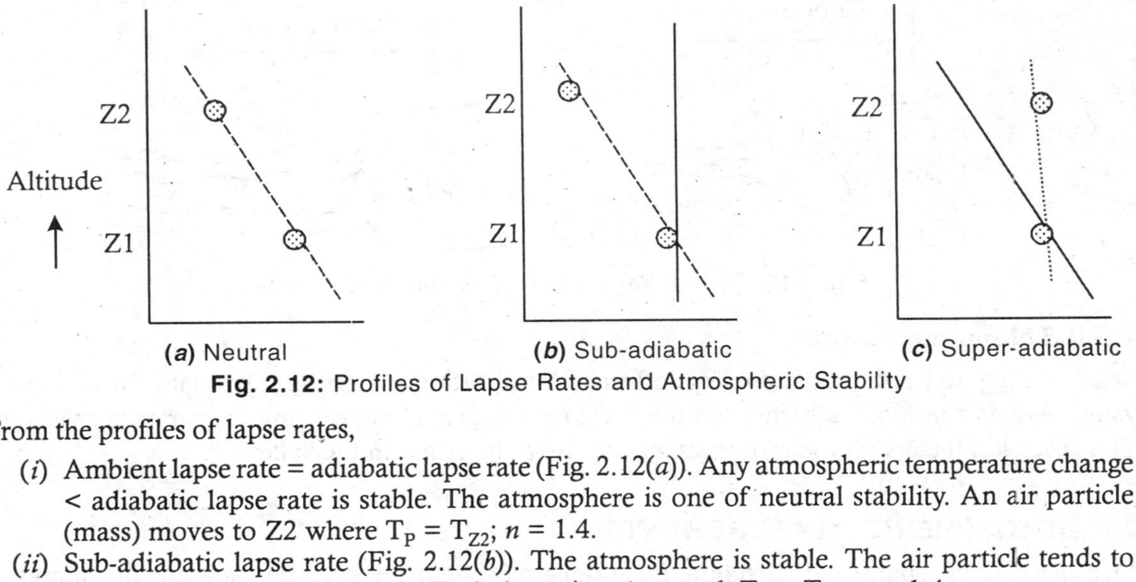


(*a*) Neutral (*b*) Sub-adiabatic (*c*) Super-adiabatic

**Fig. 2.12:** Profiles of Lapse Rates and Atmospheric Stability

From the profiles of lapse rates,

(*i*) Ambient lapse rate = adiabatic lapse rate (Fig. 2.12(*a*)). Any atmospheric temperature change < adiabatic lapse rate is stable. The atmosphere is one of neutral stability. An air particle (mass) moves to Z2 where $T_P = T_{Z2}$; $n = 1.4$.

(*ii*) Sub-adiabatic lapse rate (Fig. 2.12(*b*)). The atmosphere is stable. The air particle tends to come back to its original state. Turbulence is suppressed. $T_P < T_{Z2}$; $n < 1.4$.

(*iii*) Super-adiabatic lapse rate (Fig. 2.12(*c*)). $T_P > T_{Z2}$. Particle moves to Z2; $n > 1.4$. Therefore, density of air mass is less than its surroundings and the air continues to rise. Such an atmosphere is unstable.

### 2.5.3 Topological Influences

For large-scale wind transport, horizontal wind force and Coriolis force are important. Local wind systems, due to topological variations, influence these forces.

#### *2.5.3.1 Effect of Water in the Atmosphere*

Atmospheric water has effect on the lapse rates. Dry adiabatic lapse rate is characteristic of dry air. In general, as a parcel of air rises, the water vapor in it condenses and heat is released. The rising air cools more slowly as it rises. Therefore, wet adiabatic lapse rate is, negative than the dry adiabatic lapse rate (~0.5° C per 100 m).

#### *2.5.3.2 Sea Breeze and Land Breeze*

This is an example of transfer of heat by convection, from one place to another by actual motion of hot material. Water has a higher heat capacity than do the materials on the land surface. During the daytime solar radiation heats the land surface more than the water surface. The water temperature remains relatively constant. The air over the heated land surface rises producing low pressure compared with the pressure over water body. The resulting pressure gradient produces flow of cool air from sea toward land (sea breeze). At night land cools off more quickly than the sea and air above the sea is warmer (Fig. 2.13). Then the reverse flow pattern develops, resulting in cooler air from the land flowing towards the sea (land breeze).

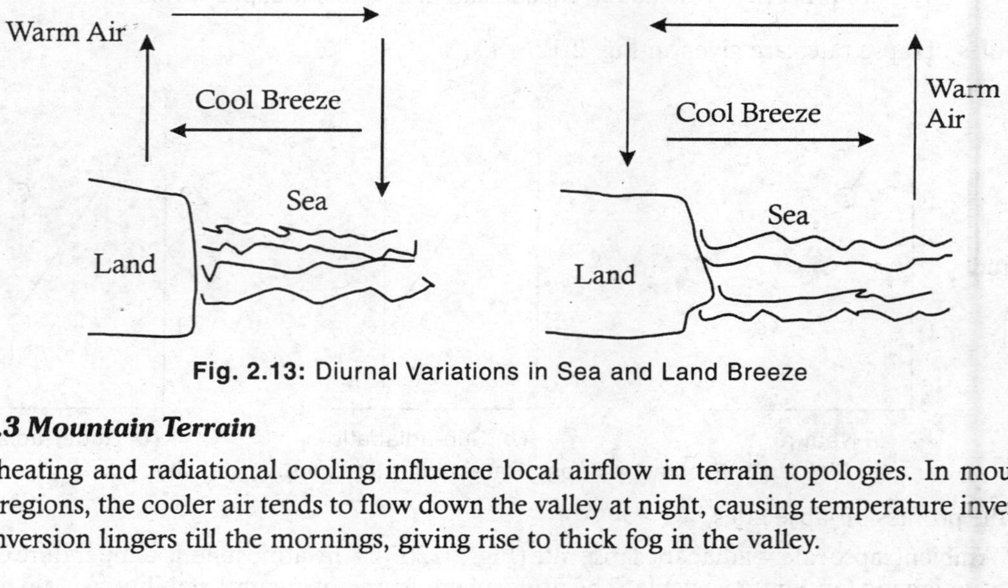


**Fig. 2.13:** Diurnal Variations in Sea and Land Breeze

#### *2.5.3.3 Mountain Terrain*

Solar heating and radiational cooling influence local airflow in terrain topologies. In mountain valley regions, the cooler air tends to flow down the valley at night, causing temperature inversion. This inversion lingers till the mornings, giving rise to thick fog in the valley.

## 2.6 ATMOSPHERIC MEASUREMENTS

Measurement of wind speed, direction, force and atmospheric pressure, temperature and humidity are important parameters for weather forecasting and evaluation and monitoring of atmospheric pollutants. Measurement of air pressure, volume, and temperature forms the basis for analysis of non-reacting gases.

### 2.6.1 Measurement of Wind Speed and Wind Direction

The relationship between the atmospheric pressure and horizontal wind speed is given by

$$f = -\frac{1}{\rho}\frac{\partial P}{\partial Y}; \qquad f_v = -\frac{1}{\rho}\frac{\partial P}{\partial X} \qquad \ldots(2.24)$$

$$f = 2\omega \cdot \sin\varphi \qquad \text{(Coriolis parameter)}$$

where $P$ = pressure;

$\rho$ = air density;

$\omega$ = angular velocity of the Earth's rotation;

$\varphi$ = latitude;

$X$ and $Y$ = east and north directions, respectively

Wind types are classified under twelve categories (from gentle breeze to cyclone), based on the wind speed, according to Beaufort's classification (Table 2.6).

**TABLE 2.6:** BEAUFORT'S CLASSIFICATION OF WIND TYPES

| *Beaufort's Scale* | *Average Wind Speed (km/h)* | *Wind Type* |
|---|---|---|
| 0 | <1 | Calm (still) air |
| 1 | 1-5 | Light wind |
| 2 | 6-11 | Light breeze |
| 3 | 11-19 | Gentle breeze |
| 4 | 20-28 | Moderate breeze |
| 5 | 29-38 | Fresh breeze |
| 6 | 39-49 | Strong breeze |
| 7 | 50-60 | Moderate gale |
| 8 | 61-74 | Fresh gale |
| 9 | 75-89 | Strong gale |
| 10 | 90-104 | Storm (whole gale) |
| 11 | 105-119 | Violent storm |
| 12 | >120 | Violent storm (hurricane) |

Determination of wind speed and wind direction is required for meteorological studies. Near-surface (<10 m from the ground) meteorological instruments are employed to determine wind speed and direction. Wind speed and direction may be determined by mechanical anemometers-(cup- or pitot-tube type), wind-vane and propeller-vane or triaxial anemometer with speed sensors or more accurately with high sensitivity by sonic anemometers.

The vane can measure average wind direction and fluctuations. A cup-anemometer is composed of three or four hollow hemispherical cups that are attached to a central pivot. The speed of the pivot is proportional to the wind speed. Pitot-tube anemometer has two openings, one parallel to and the other normal to the wind flow. A manometer is attached to measure the difference in air pressure between the openings. The wind speed, $U$, is

$$U \approx \sqrt{\frac{2\Delta P}{P}} \qquad \text{...(2.25)}$$

where $P$ = atmospheric pressure;

$\Delta P$ = pressure difference in the manometer.

Mechanical type anemometers have moving parts. They suffer from wear out, weathering and contamination. In addition, wind must provide enough force to overcome friction and accelerate the moving sensing parts. As a consequence these devices have threshold wind speed below which they fail to respond. Also, acceleration and deceleration of these parts are not instantaneous, which results in a delayed response.

Sonic anemometers do not have moving parts, and hence they are free from any of the inertial effects that plague mechanical wind sensors. Sonic anemometers can analyze extremely slow wind speeds accurately and instantaneously (velocity ~20 times/s). They are widely used in the study of atmospheric turbulence and monitoring of air pollution.

Sonic anemometer determines intensity of wind speed and direction by measuring the decrease or increase of velocity of sound between points (transducers) of known separation, due to the effect of moving air mass. The sensors are piezoelectric devices that can serve alternately as transmitter and receiver. This allows a set of transducers to make two measurements, in opposite directions, of the sound propagation velocity. The anemometer determines the propagation of velocity by measuring the time-of-flight (ToF) of a sonic impulse from transmitter to receiver.

Two-dimensional (2-D), and three-dimensional (3-D) sonic anemometers are more sophisticated and employ more elaborate transducer arrangement. A 3-D anemometer measures wind speed and the speed of sound on three non-orthogonal axes. The wind speeds are then transferred into the orthogonal wind components and referred to the anemometer head. The speed of sound, $c$, is the average between the three non-orthogonal sonic axes. The ambient air temperature can be deduced from the sonically determined speed of sound.

Balloon-borne radiosondes, and airborne lidars, aircraft, Doppler sodar (for sound direction and ranging), FM-CW radar and other such instruments can be employed to measure wind speed and direction above the surface. Radiosondes aloft by helium balloons are the primary observation platforms above the boundary layer within the limits of the troposphere (~17 kilometers). They measure atmospheric temperature, dew point and barometric pressure. Since their position is tracked by radar, wind speed and direction are also obtained as function of height. Remote-sensing observations are made with active systems like radar, lidar and acoustic sounders. The radar systems measure back scattering of microwave wavelengths. The $CO_2$ IR lidar (pulsed laser radar) at wavelength 10.6 μm can measure wind structure and turbulence. Atmospheric tracers are generally employed to evaluate the atmospheric dispersion.

Doppler variety instruments provide estimates of wind speed, and direction. Doppler sodar is used for determining wind direction and spotting of storms and tornadoes. If the direction of the wind is towards the instrument the frequency of reflected radio waves, picked by the sodar, increases. If the wind is moving away from the instrument the frequency of the reflected radio waves decreases. The Doppler weather map shows direction of the storm but if there is convergence of winds blowing in opposite directions the position of a tornado is also obtained (see Chapter 16 for more details).

### 2.6.2 Measurement of Pressure

Non-reacting gases at normal pressure and temperature obey gas laws.

$$PM = \rho RT \quad \text{...(2.26)}$$

$$\left(P + \frac{a}{V^2}\right)(V - b) = nRT \quad \text{(Van der Waals eqn.)} \quad \text{...(2.27)}$$

where, $P$ = pressure,
$V$ = volume,
$T$ = absolute temperature,
$\rho$ = density,
$R$ = gas constant,
$a$ and $b$ are empirical constants.

Different methods are used for measuring pressure in different ranges of pressure. Devices for measuring pressure are called manometers. Manometers can be divided in to two categories. The first type (primary type) includes those that measure the pressure as a quantity equal to ratio of force to surface area ($P = F/A$). Manometers of the second type (secondary type) measure a certain quantity associated with pressure rather than the pressure itself.

While Pascal (Pa) is the accepted international (SI) scale for reporting atmospheric pressure, other scales such as atmosphere, bar, mm of Hg, are also very much in use and their mutual conversion is given in Table 2.7.

**TABLE 2.7:** CONVERSION TABLE FOR PRESSURE MEASUREMENTS

| *Unit* | *Atmosphere* | *kPa* | *Bar* | *mm (of Hg)* |
|---|---|---|---|---|
| 1 atm | 1 | 101.325 | 1.01325 | 760 |
| 1 kPa | 0.009869 | 1 | 0.01 | 7.50062 |
| 1 bar | 0.98692 | 100.0 | 1 | 750.0616 |
| 1 mm Hg | 0.001316 | 0.13332 | 0.001333 | 1 |

### *2.6.2.1 Barometers*

Manometers for measuring atmospheric pressure range are called barometers. Various instruments are available to measure atmospheric pressure; the most common ones are the mercury and aneroid barometers. A mercury barometer consists of a glass tube, containing mercury, with a reservoir of mercury at the bottom. The top end of the glass tube is sealed and evacuated. Changes in air pressure cause the mercury in the tube to rise and fall. The pressure, in millimeters or millibars, can be read on the scale fixed beside the tube. The pressure of air on the surface of mercury in the cup holds the mercury in the tube. Before reading, the scale is adjusted so that its zero point is even with the surface of mercury in the reservoir. The sea level pressure is, normally, close to 101.32 kPa (760 μm of Hg).

Atmospheric pressure can also be measured using an aneroid barometer (used in airplanes). Aneroid barometer contains a small metal box from which air has been partially removed. The box has a flexible lid to which a lever is attached. Changes in air pressure move the lid and its lever. The lever is connected to a pointer, which moves on the dial to indicate air pressure in millibars or millimeters. This type of barometer has the advantage of being portable, unlike the mercury barometer. Aneroid barometer is calibrated against a mercury instrument. A type of aneroid barometer, with facility to record air pressure on a paper chart, mounted on a rotating drum, is used to monitor atmospheric conditions at different altitudes. It is called a 'barograph.'

The pressure varies with temperature at the rate of 16.66 Pa per 1° C change. For accurate readings the measured pressure is corrected for temperature and latitude. Taking into account the temperature variation with height, the international barometric formula can be used to calculate pressure as a function of height up to the height of ~11 km (troposphere).

$$P_H = 101.3\ (0.0226 \times H)\ 5.255 \qquad \text{...(2.28)}$$

where $P_H$ = pressure in kPa;
$H$ = height in km

### 2.6.2.2 Low- and High-pressure Measurement

For some purposes, particularly while conducting controlled simulation experiments in the laboratory to study atmospheric phenomena, gas pressures have to be measured in various ranges Different types of instruments, other than the barometers, are available for this purpose.

Pressures, above and below the atmospheric pressure, can be measured using an open manometer (Fig. 2.14). Measured pressure in the system is

$$P = (h + \Delta h)\rho g \qquad \text{...(2.29)}$$

where $h$ = height;
$\Delta h$ = difference in height in the manometer readings;
$\rho$ = density of the liquid; $g$ = acceleration due to gravity.

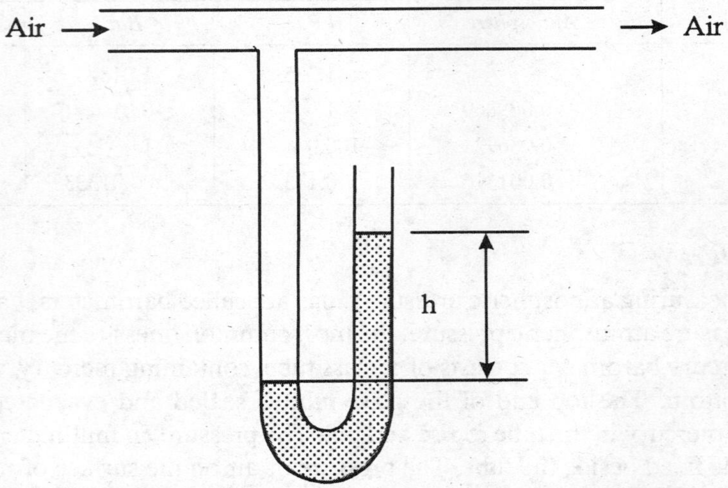


**Fig. 2.14:** Open Manometer Setup for Measuring Over-Atmospheric Pressure

Low pressures are measured with secondary type manometers. *McLeod* manometer (gauge) is based on *Boyle-Mariotte* law. Instruments based on internal friction, thermal conductivity or molecular ionization in gases, at low pressure, are also used.

The quartz filament manometer is based on this principle that the internal friction of a gas is directly proportional to pressure. The instrument consists of a quartz vessel inside which a quartz filament is fixed. A glass tip on to which an iron wire is fused vibrates the filament. The time required for the filament vibration to decrease to one half, due to friction, is proportional to the pressure, which can be calculated from the relationship

$$P\sqrt{M} = a + \frac{b}{t} \quad \ldots(2.30)$$

where, $M$ = molecular mass;
$a$ and $b$ = instrumental constants;
$t$ = time required for decrease of the vibration of the filament to one half value.

Thermal conductivity measurements are also utilized to measure low pressures in the Pirani manometer. The manometer, consisting of a platinum or tungsten filament, heated to ~100° C, as one arm of the Whetstone bridge. The thermal conductivity of the gas leads to heat loss from the filament. This results in a decrease in its resistance that is proportional to gas pressure in the range 1-$10^{-3}$ Pa. Thermal conductivity manometers are also used in vapor pressure measurements. The very high sensitivity to hydrogen and helium gases permits these gases to be used to find leaks in tubes and vessels.

The ionization manometer is a triode. The heated cathode emits electrons, which are accelerated towards the anode and in the process collide with gas molecules leading to ionization (Chapter 25). The rate of ion formation is proportional to the number of gas molecules present i.e. to the gas pressure. This type of manometer is used to measure low pressures in the range $10^{-1}$-$10^{-7}$ Pa (Table 2.8). The radioactive manometer is based on the ionization of gas molecules by α-particles from a radioactive substance (radium).

**TABLE 2.8:** MEASUREMENT OF PRESSURE BY MANOMETRIC METHODS

| *Instrument* | *Pressure Range (Pa)* | *Principle* |
|---|---|---|
| Membrane | $1 \rightarrow 10^{-2}$ | Volume |
| Glass | $10^{-2} \rightarrow 10^{-4}$ | Elasticity |
| McLeod | $10^{-2} \rightarrow 10^{-7}$ | Internal friction |
| Quartz filament | $10^{-4} \rightarrow 10^{-7}$ | Internal friction |
| Pirani | $10^{-3} \rightarrow 10^{-7}$ | Thermal conductivity |
| Thermoelectric | $10^{-1} \rightarrow 10^{-7}$ | Thermal conductivity |
| Ionization | $10^{-4} \rightarrow 10^{-11}$ | Molecular ionization |
| Radioactive | $1^{-1} \rightarrow 10^{-6}$ | Molecular ionization |
| Magnetic discharge | $10^{-6} \rightarrow 10^{-9}$ | Molecular ionization |

### 2.6.3 Measurement of Volume

Graduated burettes and gas displacement vessels such as gas meters, rotameters and manometers can be employed to measure volume of a gas. A rotameter consists of a graduated glass tube with a light metal or plastic float in it. Gas entering the tube lifts (levitates) the float. The vertical displacement of the float is related to the flow-rate and velocity of the gas. Ultrasonic flow meters are used for continuous measurement of mean velocities of flowing gas at 20 to 250° C. They are used to determine gas flow-rates in smokestacks and chimneys. Venturi tubes are differential pressure manometers to determine flow-rates. Thermal conductivity of gases can be used in manometric measure of gases at low pressures. At high pressures this property can be used to measure flow-rates of gases. In this device, the reference element is maintained at a constant temperature.

### 2.6.4 Measurement of Temperature

Temperature is a quantitative measure of the "degree of heating" of a body (see Chapter 16 for more details). The most common are liquid-filled (alcohol, or mercury) glass bulb thermometers (range—200 → 600° C; sensitivity ~0.05°). Resistance thermometers, and thermocouples are widely used for accurate measurement (sensitivity ~0.0001°). Radiant energy emitted by a body can be used to determine its surface temperature (radiometers and pyrometers). A minimum-maximum thermometer is a self-registering device for detection of minimum and maximum temperatures, and is routinely used in atmospheric studies.

## BIBLIOGRAPHY


Barry, R. and Chorley, R. J. (1970), Holt, Reinhart & Wilson: New York. "*Atmosphere, Weather and Climate.*"

Boucher, K. (1976), Wiley: New York. "*Global Climate.*"

Cunningham, W. P. (ed.) (1994), Gale: Detroit, IL. "*Environmental Encyclopaedia.*"

Fritschen, L. J. and Gay, L. W. (1979), Springerverlag: New York. "*Environmental Instrumentation.*"

Hubert, M. K. (1971), Sci. Amer (Sept.). "The energy resources of the earth."

Monteith, J. L. (1973), Edward Arnold: London. "*Principles of Environmental Physics.*"

Morgan, J. M., Morgan, M. D. and Wiersma, J. H. (1973), Little-Brown: Boston, MA. "*An Introduction to Environmental Science.*"

Munn, R. E. (1966), Academic Press: New York. "*Descriptive Micrometeorology.*"

Neuburger, M., Edinger, J. G. and Bonner, W. D. (1973) San Francisco, CA: Freeman, "*Atmospheric Environment.*"

Nriagu, N. O. (ed.) (1992), Wiley & Sons: New York. "*Gaseous Pollutants.*"

Oort, A. H. (1970), Sci Amer (Sept); 223; 54. "The energy cycle of the earth."

Perkins, H. C. (1974), McGraw-Hill: New York. "*Air Pollution.*"

Pettersen, S. (1969), McGraw-Hill: New York. "*Introduction to Meteorology*" (3rd edn.).

Robinson, N. (ed.) (1966), Elsevier, Amsterdam. "*Solar Radiation.*"

Seinfeld, J. H. and Pandis, S. N. (1997), Wiley: New York. "*Atmospheric Chemistry and Physics.*"

Stern, A. C. *et al.* (1984), Academic Press: New York. "*Fundamentals of Air Pollution*" (2nd edn.).

Turk, A. *et al.* (1978), Saunders: Philadelphia, PA. "*Environmental Science,*" (2nd edn.).

Wilson, J. D. (1974), Heath: Lexington, M. A. "*An Environmental Approach to Physical Science.*"

Woodward, F. I. and Sheehy, J. E. (1983), Butterworth: London. "*Principles and Measurements in Environmental Biology.*"

# 3

# Air Pollution

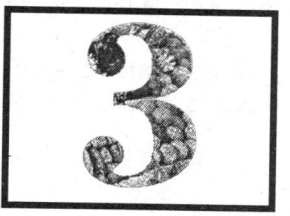

The Earth is surrounded by air that covers the environment upto the height of 1.5 km from the surface of Earth. It contains gases such as oxygen (21%), carbon dioxide (0.3%), nitrogen (78%), argon (0.03%), etc. in a fixed ratio. But, due to some or the other reasons, the ratio of all other gases except oxygen increases in the atmosphere resulting in its pollution. Air pollution is release or occurence of any foreign materials or gases into the atmosphere, which may be harmful to man, animals and vegetation, etc.

The air pollution results mainly from gaseous emissions of industry, thermal power stations, automobiles, domestic combustion, etc.

Sources of air pollution are natural as well as anthropogenic and their impact has local, regional and global implications. Mixing and dispersion of atmospheric pollutants occur mainly due to convection and turbulence in the air. Theoretical methods for predicting concentration of non-reacting air pollutants, over a given area from a point source, are based on plume dispersion models (Gaussian type) and available meteorological data. Prediction of air pollution due to reacting constituents needs, in addition, the application of principles of thermodynamics and reaction kinetics. From such studies, disastrous consequences of pollution from newly planned industries can be avoided or at least minimized.

This Chapter deals with the analysis of air pollutants and the application of physical and chemical principles in determining various parameters relating to air pollutants.

Clean air is a mixture of several gases in different proportions, primarily nitrogen and oxygen. Pollution of air is due to increase in the concentrations of the constituents or due to the presence of extraneous materials as a consequence of natural or man-made activities.

## 3.1 AIR POLLUTION DETERMINANTS

The composition of dry air in the atmosphere contains, in addition to major constituents nitrogen and oxygen, minor and trace level constituents (Table 3.1). Air is considered to be polluted when it contains certain substances in concentrations high enough and for duration long enough to cause harm or undesirable effects. Air pollution is defined as the presence in the atmosphere of air contaminant(s) in such quantities and duration that it (they) may tend to be injurious to life, or property, or it (they) may interfere with the comfortable enjoyment of life or property, health, repose and safety. These contaminants include gases, fumes, vapors, aerosols, and liquids and particulate matter. Troposphere is the region where weather and air pollution arises.

**TABLE 3.1:** COMPOSITION OF DRY AIR AT SEA LEVEL

| *Constituent* | *Concentration ppm and (% by Volume)* |
|---|---|
| **Major** | |
| Nitrogen | 781,000 (78.1%) |
| Oxygen | 209,500 (20.95) |
| **Minor** | |
| Argon | 9300 (0.93) |
| Carbon dioxide ($CO_2$) | 315 (0.032) |
| **Trace Components** | |
| Neon | 20 (0.002) |
| Helium | 5 ($5 \times 10^{-4}$) |
| Methane | 2 ($2 \times 10^{-4}$) |
| Krypton | 1 ($1 \times 10^{-4}$) |
| Hydrogen ($H_2$) | 0.5 ($5 \times 10^{-5}$) |
| Nitrous oxide ($N_2O$) | 0.3 ($5 \times 10^{-5}$) |
| Carbon monoxide (CO) | ~1 ($1 \times 10^{-5}$) |
| Nitrogen dioxide ($NO_2$) | ~1 ($1 \times 10^{-5}$) |

ppm = *parts per million*

There has always been a balance between natural *sources* and *sinks* to air pollution, but human and industrial activities have created pollution problems that overburden the natural removal systems for pollutants. The places from which pollutants emanate are called 'sources,' and destinations to which pollutants reach are called 'sinks' (soil, vegetation, etc.). *Receptors* are inanimate as well as animate bodies that receive and are seriously affected by pollutants.

Sources of air pollution are natural as well as man-made. Natural sources are volcanoes, forest fires and dust storms. Man-made sources are industrial, metallurgical, manufacturing and utility establishments. Most of the anthropogenic atmospheric pollutants originate from combustion processes. The mechanisms whereby pollutants are removed from the atmosphere are called 'scavenging' processes. Oxidation is a prime removal mechanism for inorganic as well as organic gases.

Air pollution problem has local, regional, continental and global ramifications. Local level pollution affects at the local, municipal, town and city levels; regional level pollution affects at the troposphere level and continental pollution at the stratosphere level. Global level pollution affects the entire atmosphere. Sources causing pollution at different levels are different (see Chapter 5 on ecological effects).

1. **Local Level Pollution:** Local agricultural activities, waste, sewage disposal and other utility and service establishments.
2. **Regional Level Pollution:** Power plants and large-scale industries.
3. **Continental Level Pollution:** Acid rain (see Chapter 6).
4. **Global Level Pollution:** Air transport, atmospheric testing of nuclear weapons, depletion of ozone layer, global warming due to "greenhouse" effect (see Chapter 5), reduction of solar flux reaching the earth due to contaminants in the atmosphere.

Air pollution sources are either site-specific, such as industrial power plants, waste disposable units, residential, occupational and recreational places, or diffuse or mobile sources, such as, agriculture, forestry, mining, and vehicular and other transport systems.

Air pollutants from the atmosphere reach the surface of the earth by sedimentation, washout or impaction. If the pollutant is soluble in water, it can enter the food chain directly through or via aquatic organisms. If the pollutant is not soluble in water, it may settle on plant leaves and fruits and thus may enter the food chain.

### 3.1.1 Air Quality Impact Assessment (AQIA)

Concentration of air pollutants in a particular place is primarily dependent on specific sources, such as industrial plants, and non-specific sources, such as residential sites, hospitals and other utility places. In addition, meteorological conditions, land topology and utilization patterns of the resources also play a role. The levels of pollution and lengths of exposure above, which adverse health and welfare effects may occur, are established by air quality criteria. These are determined by air quality impact assessment (AQIA). In air quality impact assessment (AQIA) procedures, mathematical models of plume dispersion, and atmospheric conditions are used to evaluate pollution sources, meteorological and topographical conditions and to arrive at solutions for minimization of air pollution (Fig. 3.1). Air quality standards—the levels of pollutants prescribed by regulations that may not be exceeded during a given time in a defined area, are set from AQIA procedures.

## 3.2 MONITORING OF AIR POLLUTION

Air pollution meteorology specifies what happens to a puff or plume of pollutant from the time it is emitted to the time it is detected at some other place. The motion of the air in the atmosphere causes dilution of the pollutants by dispersion. The atmosphere serves as the medium through which pollutants are transported. While being transported, the pollutants may undergo chemical reactions too.

The meteorological data include: (1) vertical mixing, (2) horizontal wind speed and direction, (3) atmospheric stability, diurnal and local variations, (4) precipitation and (5) rate of dissipation (dispersion) of pollutant. Both monitoring and modeling of air pollution is essential to provide a picture of possible damage to vegetation, organisms and to the environment. Air quality modeling procedures help in predicting concentration of air pollutants in the atmosphere, from the input data on meteorology, terrain and the ambient air quality. Models include transport, turbulence and reaction of pollutant species.

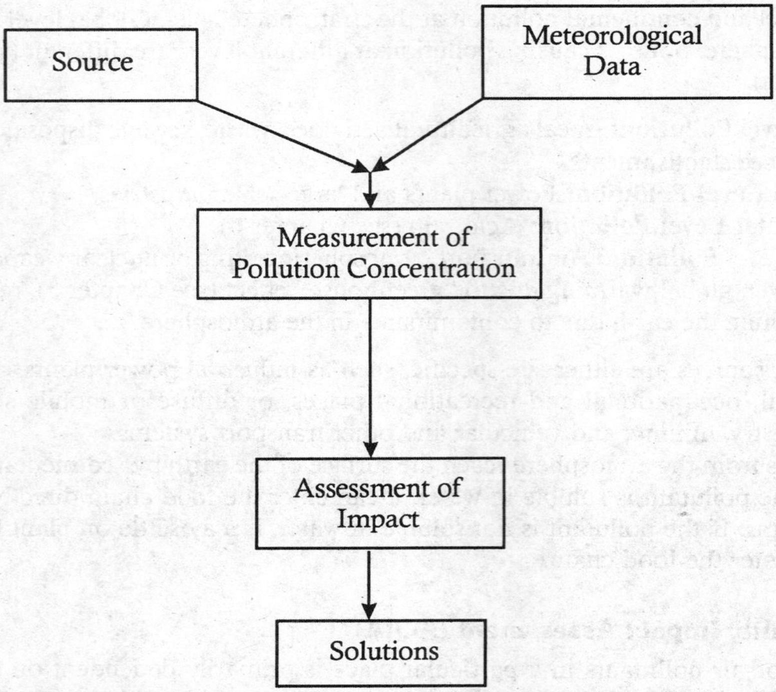


Impact Assessment: (*i*) Systems and processes which influence the transfer of the substances emitted in the atmosphere and duration time; (*ii*) Effects of exposure over time.

**Fig. 3.1:** Flowchart of Air Quality Impact Assessment (AQIA) Procedures

### 3.2.1 Atmospheric Dispersion

Pressure gradients are the primary causes of wind, cloud formation and motion. Other forces are due to Earth's rotation and tilt (Coriolis force) and gravitational force. Pressure changes are related to temperature variations. For an ideal gas, the pressure and volume are linearly related to temperature. For dry air, the relationship between pressure and temperature is given by the *Poisson's* equation.

$$PV = RT \quad \text{(ideal gas equation)} \qquad \ldots(3.1)$$

$$\left(\frac{T}{T_0}\right) = \left(\frac{P}{P_0}\right)^{0.288} \quad \text{Poisson's equation for dry air} \qquad \ldots(3.2)$$

where, $P$ = pressure;
$V$ = volume;
$T$ = temperature (Kelvin).

One of the physical effects of pollutants in the atmosphere is the attenuation of solar radiation. Pollutants do affect atmospheric scattering of light. The molecular scattering of radiation is dependent on the particle size of the components. Molecules of unpolluted air, where particle diameter is less than the wavelength of radiation (clean skies without large-size particulate matter, $d << \lambda$) produce *Rayleigh* scattering of light. However, in a turbid atmosphere, where there are suspended particles comparable in size with the wavelength of the incident radiation ($d \sim \lambda$), there is more forward scattering by particles (*Mie* particles) and Mie theory is more appropriate to explain the scattering. The meteorological visibility range, $L$, is given by

$$L = (4/b) \qquad \ldots(3.3)$$

$b = 1.5 \times 10^{-5}$ for Rayleigh scattering

$b = \pi r^2 Nk$ for *Mie* scattering

where, $N$ = number of particles;
$r$ = radius of the particle;
$k$ = a function of refractive index.

Since pressure decreases with height, there is an upward pressure gradient force. The buoyant force is the difference between the upward force and the gravitational force. The acceleration, $a$, of an air particle under this force in the vertical direction is given by

$$a = g\left(\frac{T - T_e}{T}\right) \qquad \ldots(3.4)$$

where, $T$ = temperature of the ambient air (Kelvin);
$T_e$ = temperature of the surrounding environment;
$g$ = acceleration due to gravity

The density of air is affected by moisture in the air. It is related to thermodynamic parameters.

$$P = \rho RT$$
$$\Delta H = C_P \cdot \Delta T - V \cdot \Delta P \qquad \ldots(3.5)$$
$$C_P = R + C_V$$

where, $\rho$ = Density of air;
$R$ = gas constant;
$\Delta H$ = change in heat quantity (enthalpy);
$V$ = volume;
$C_P$ = specific heat at constant pressure;
$C_V$ = specific heat at constant volume.

Fick's law of diffusion can be invoked to express plume rise (diffusion of hot gases) by incorporating the settling-out term to the Fick's equation.

$$\frac{dC}{dt} = -DU\frac{dC}{dx} + v_z\frac{dC}{dz} \qquad \ldots(3.6)$$

Rate of change Diffusion Settling out

where, $C$ = concentration of the pollutant;
$U$ = horizontal wind speed;
$D$ = diffusion coefficient;
$v_z$ = wind speed in the vertical direction

## 3.3 AIR POLLUTION DISPERSION MODELING

The goal of air quality dispersion modeling is to estimate a pollutant's concentration at a point downwind of one or more emission sources. The dispersion models are useful tools for detecting potential concentration impacts from sources. The plume rises mainly due to two factors: (1) the velocity of exhaust gases, which impart momentum to the plume, and (2) the temperature of the exhaust gas, which gives the plume buoyancy in ambient air. Some meteorological factors that affect plume rise include wind speed, air temperature, atmospheric stability, and turbulence and land topography. The plume models are empirical mathematical models aimed at obtaining a first estimate of plume concentrations of non-reacting atmospheric pollutants. Friction effects and chemical reactions are neglected in the plume models.

The heat transfer by simple diffusion is by conduction, but the process of dispersion is by convection (by turbulence). Therefore, the Fick's laws of diffusion cannot adequately explain the dominant process in atmospheric dispersion, namely dispersion by convection (turbulence). Atmospheric stability is categorized into various classes (Pasquill's "stability classes"). Classes *A*, *B* and *C* are associated with super-adiabatic conditions, class *D* with neutral condition, and classes *E* and *F* with sub-adiabatic (inversion) conditions. Class *G* represents extremely severe temperature inversion. The amount of turbulence in the atmospheric air has a major effect upon the rise and dispersion of air pollutant plumes.

Dispersion by turbulence is the process by which contaminants disperse through the air by convection and a plume spreads over an area, thus reducing the concentration of the pollutants. The plume spreads vertically and horizontally. Dispersion of a contaminant in the atmosphere depends on wind speed, turbulence and atmospheric and terrain conditions. Turbulence mixes and enhances the dispersion of pollutants. Stronger the turbulence, the more the pollutants are dispersed. It decreases the concentrations of the contaminants in the plume and increases their concentrations outside the plume.

The most commonly used models for dispersion of gaseous (non-reacting) air pollutants are based on the Gaussian profile. These models assume an ideal steady state of constant meteorological conditions over long distances, idealized plume geometry, uniform flat terrain. Such ideal conditions rarely occur. More robust statistical methods (e.g. non-parametric statistical methods) have to be applied in addressing such multivariate cases (see Chapter 15). However, such models are useful in predicting the dispersion and dilution of stack gases, and atmospheric pollutants under various atmospheric and terrain conditions.

The concentration of a pollutant, $C_x$, at a distance, *X*, from the source is

$$C_x = \frac{Q}{\pi U \sigma_y \sigma_z} \exp\left(-\frac{y^2}{2\sigma_y^2} - \frac{z^2}{2\sigma_z^2}\right) \quad \text{...(3.7)}$$

where, $Q$ = source strength;
$\sigma$ = standard deviations in $x$, $y$ and $z$ direction;
$U$ = average wind speed

Concentration can be correlated to diffusion coefficients, which gives an estimate of the spread of the pollutants horizontally (along wind direction), laterally and vertically.

$$C = \frac{Q}{4\pi r}\frac{1}{\sqrt{(D_y D_z)}} \exp\left[\left(-\frac{U}{4x}\right)\left(\frac{y^2}{D_y} + \frac{z^2}{D_z}\right)\right] \quad \text{...(3.8)}$$

where, $r = \sqrt{(x^2 + y^2 + z^2)}$ ; $D_y$ and $D_z$ are diffusion coefficients in the lateral and vertical wind directions.

In the case of stable atmosphere (inversion), the stability parameter, $S$, is given by

$$S = \frac{\Delta g}{T}\left(\frac{dT}{dz} + \tau\right) \qquad \text{...(3.9)}$$

where,

$$\tau = \left(\frac{dT}{dz}\right)_{\text{adiabatic}} \qquad \text{...(3.10)}$$

For an adiabatic ambient (neutral) atmosphere, $\frac{dT}{dz} = -\tau$ ; and $\therefore\ S = 0$ ...(3.11)

For stable atmosphere, $\frac{dT}{dz} > \tau$ ; and $\therefore\ S > 0$ ...(3.12)

For unstable atmosphere, $\frac{dT}{dz} < \tau$ ; and $\therefore\ S < 0$ ...(3.13)

### 3.3.1 Temperature Inversion and Atmospheric Stability

Variation in temperature with altitude has a great influence on the dispersion of air pollutants. Under normal conditions, as the warm air rises, it expands adiabatically and cools, due to decrease in atmospheric pressure with height. There is a gradual decrease in atmospheric temperature with height (~ –1° C/100 m). Mixing of air pollutants occurs in this (unstable) system. Temperature inversion is a situation where a layer of warm air prevails above cooler air. This situation occurs (due to convection) whenever the lapse rate exceeds 1° C/100 m. Temperature inversion is a major cause for atmospheric pollutants getting concentrated. Pollutants travel with wind and inversion layers suppress the vertical mixing of pollutants and make them spread horizontally.

Mechanical turbulence and heat convection are characterized by the Richardson number, Ri (Fig. 3.2). Richardson number, Ri, can be estimated from the wind speed, time of the day and year, and the cloudiness. On a clear night, with little wind, the Ri is positive, and with strong winds, the Ri ~ 0.

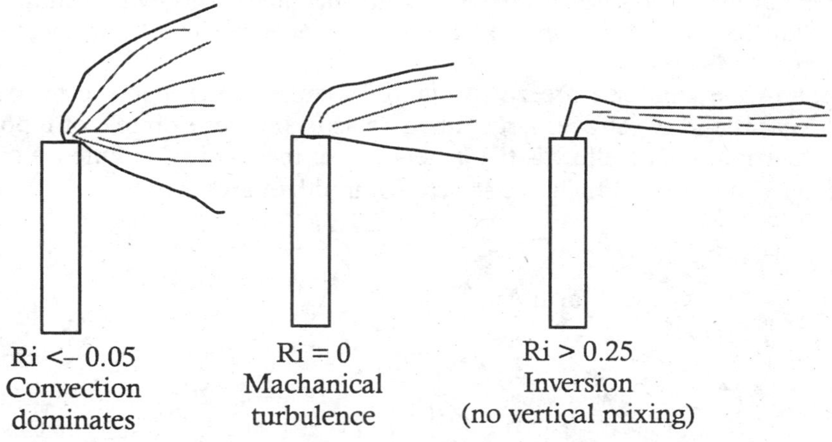


**Fig. 3.2:** Turbulence Characteristics in Plume Dispersion

### 3.3.2 Atmospheric and Topographic Effects

Low and high-pressure systems have different ventilation characteristics. Low-pressure systems generally cover relatively small areas. Low-pressure systems are frequently accompanied by cloudy skies, which may be sources of pollutants. High-pressure systems generally occupy larger areas. Stagnation occurs if the ventilation rate becomes low.

One of the topographic effects is the mountain-valley wind (Fig. 3.3). A town situated in a mountain-valley is shielded from the prevailing wind. At night cooler air flows down the slopes, forcing warmer air to rise. This produces temperature inversion that can persist if the daytime sunshine is not warm enough to cause mixing.

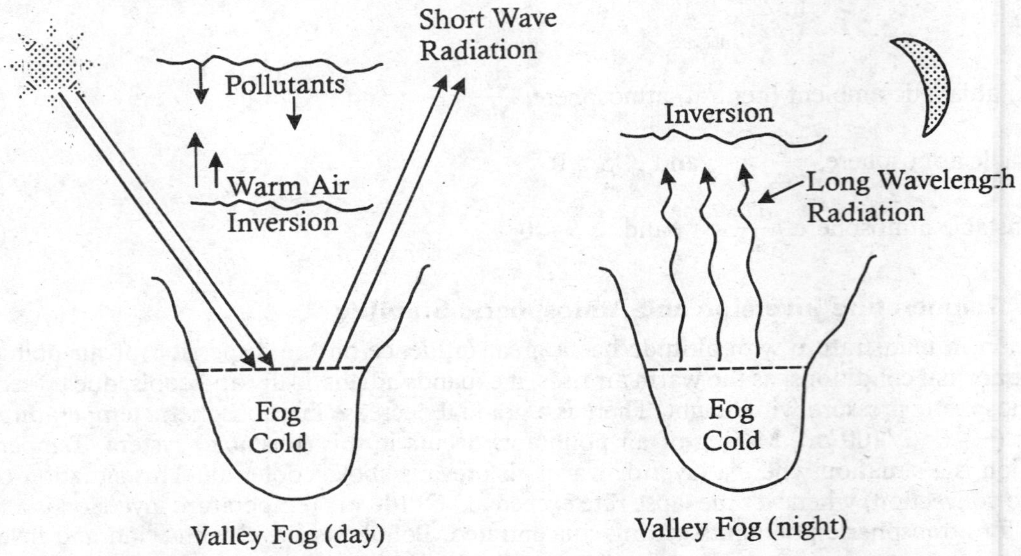


**Fig. 3.3:** Fog Formation in Mountain-Valley Atmospheric System

Serious air pollution episodes are almost always accompanied by fog. Dense fog in a valley reflects shorter wavelength radiation during the day and longer wavelength radiation (heat) from the top of the fog at night. Fogs sit in valleys and stabilize inversions by preventing the Sun from warming the valley floor, thus often prolonging the pollution episodes. Occurrences of fogs tend to be more in urban areas.

Both mountain-valley and sea breeze topographies are important in the meteorology of air pollution. The heat-island effect over large urban areas is another example of topographic effect (Fig. 3.4). Another man-made effect is the generation of mechanical turbulence caused by non-uniform buildings (low-rise and high-rise structures) in urban areas.

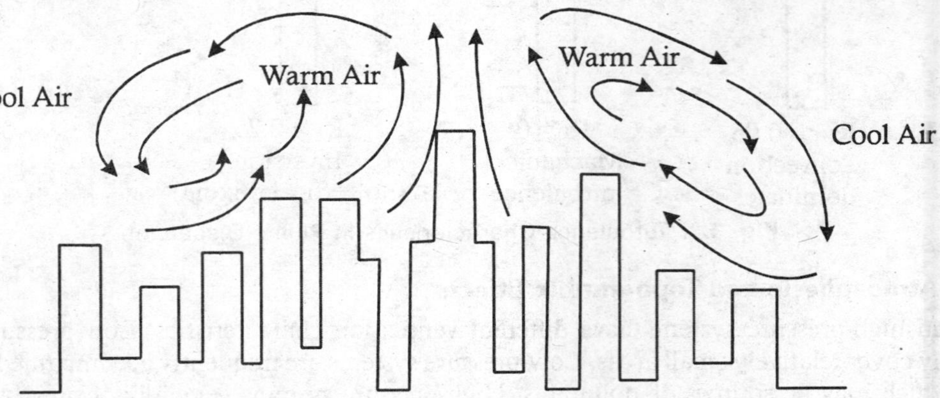


**Fig. 3.4:** Urban Heat-island Effect on Air Circulation

Conditions conducive for poor dispersion are: (1) low wind speed, (2) stable meteorological conditions resulting in limited vertical air movement, (3) large differences between day and night temperatures, (4) trapping of cold air in valleys, resulting in stable conditions, (5) fog, which hinders the Sun from warming the ground surface to break temperature inversion and (6) absence of rain for washing the atmosphere.

### 3.3.3 Diurnal Variation of Air Pollution

Air pollution patterns depend on the time of the day and the geographic region. During the day, the temperature at the ground surface is high both in rural and urban areas. After sunset, the air temperature near the ground surface in rural areas falls rapidly, producing a temperature inversion reaching down to the ground. At night vertical mixing is negligible and the air near the ground is clean, but just after sunrise the polluted air is mixed with clean air, leading to pollution. In urban areas pollution occurs at night. That is, air pollution in the countryside is more in the morning than at night and *vice versa* in the cities (Fig. 3.5).

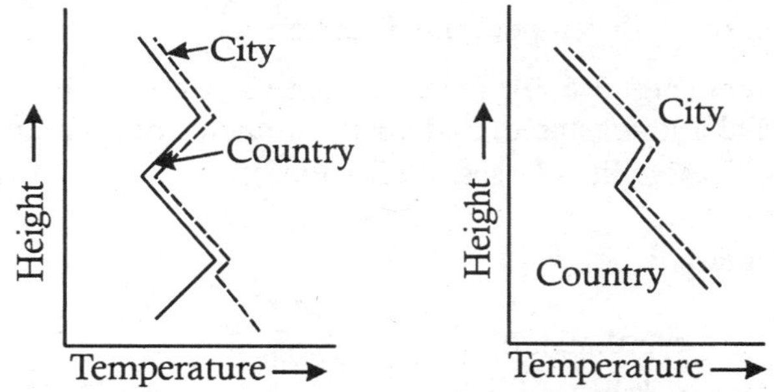


**Fig. 3.5:** Schematic of Vertical Temperature Distribution over City and Country

Cities like Los Angeles in California, USA, suffer from constant air pollution due to temperature inversion. Inversions coupled with weak winds have been responsible for all air pollution disasters. Some of the recorded air pollution episodes, due to temperature inversion and stagnation (due to weak winds) that lasted several days are given in Table 3.2.

**TABLE 3.2:** AIR POLLUTION EPISODES

| *Place* | *Year/Date & Duration* | *Deaths* | *Causes* |
|---|---|---|---|
| London (UK) | 1875 | — | Cattle deaths |
| Meuse valley (Belgium) | 1930 (Decem.) 5 days | ~60 | Respiratory ailments; chest pain cough; lung and eye irritation Cause: $SO_2$ due to coal burning |
| Donora (Penn., USA) | 1948 (25th Oct.) 3 days | 20 | Same as above |
| London (UK) | 1948 | 800 | Cause: Smog and effects due to $SO_2$ pollutants |
| | 1952 | 4000 | |
| | 1956 | 2000 | |
| | 1959 | 300 | |
| | 1962 | 700 | |
| | 1963 | 700 | |

| Place | Year/Date & Duration | Deaths | Causes |
|---|---|---|---|
| Poza Rica (Mexico) | 1950 | 22 | $H_2S$ gas leakage |
| New York City (USA) | 1953 | 160 | Cause: $SO_2$ pollutants |
| | 1963 | 400 | |
| Yokohama City (Japan) | 1956 | — | Report of illness and symptoms similar to Meuse valley episode |
| Denver (USA) | 1975 | — | $H_2S$ gas leakage |
| Seveso (Italy) | 1976 | — | TCCD pesticide |
| Bhopal (India) | 1984 | > 10,000 | Methyl isocyanate (MIC) dispersion |

### 3.3.4 Plume Behavior

The shapes of plumes emitted from smokestacks (Fig. 3.6) are often used to monitor atmospheric stability. Smoke is dispersed by turbulence, in both vertical and horizontal directions. The spread of the plume is directly related to the vertical temperature gradient. Change in the slope of the ambient lapse rate indicates the existence of temperature inversion.

1. **Looping (of Plumes) [Fig. 3.6(*i*)]:** Occurs when vertical temperature gradient, ($dT/dz$), is negative (super-adiabatic; unstable) and air is turbulent (occurs during warm seasons with clear skies) (Pasquill's stability classes: A, B and C).

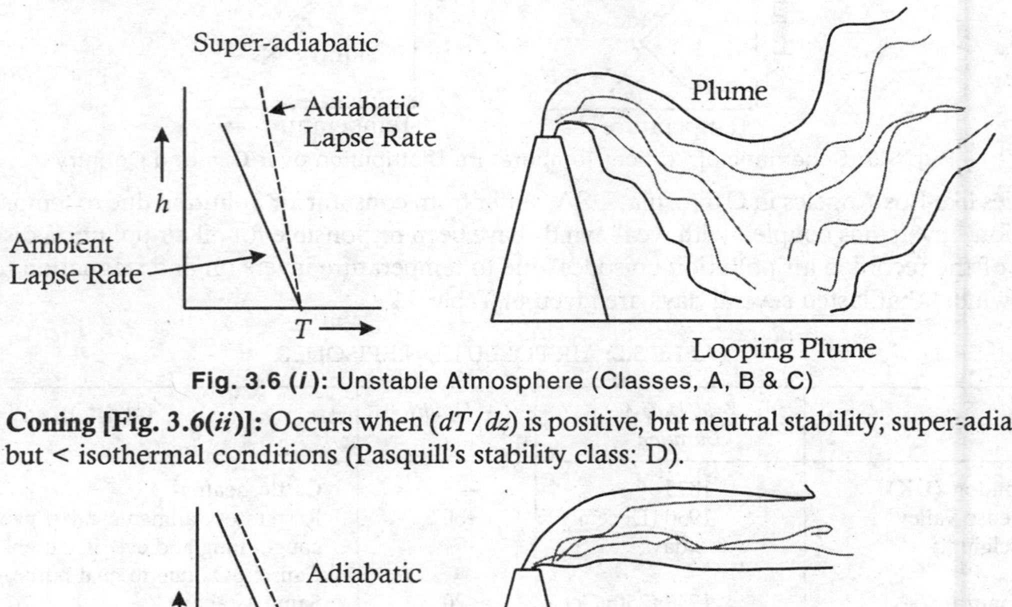


**Fig. 3.6 (*i*):** Unstable Atmosphere (Classes, A, B & C)

2. **Coning [Fig. 3.6(*ii*)]:** Occurs when ($dT/dz$) is positive, but neutral stability; super-adiabatic, but < isothermal conditions (Pasquill's stability class: D).

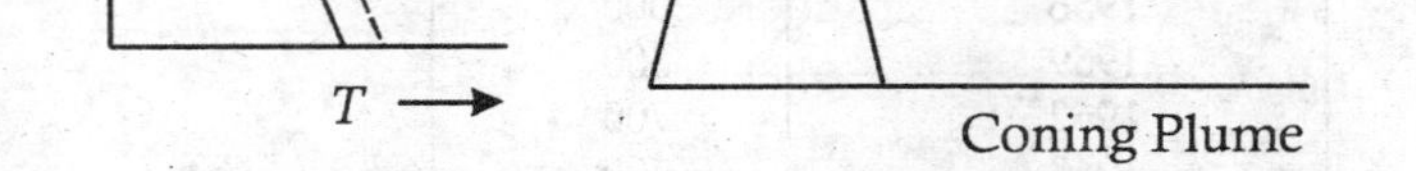


**Fig. 3.6 (*ii*):** Neutral Atmosphere (Class D)

3. **Fanning [Fig. 3.6 (*iii*)]:** Occurs when there is a temperature inversion, and $(dT/dz)$ is positive, above and below the plume. The plume tends to spread out in a single flat layer (Pasquill's stability classes: E and F).

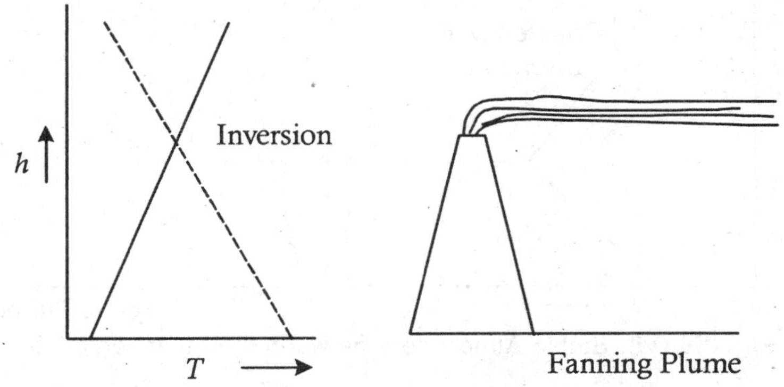


**Fig. 3.6 (*iii*):** Stable Atmosphere (Classes E & F)

4. **Fumigation [3.6(*iv*)]:** Occurs when a temperature inversion occurs over super-adiabatic condition. Pollutants are caught under an inversion and mixed.

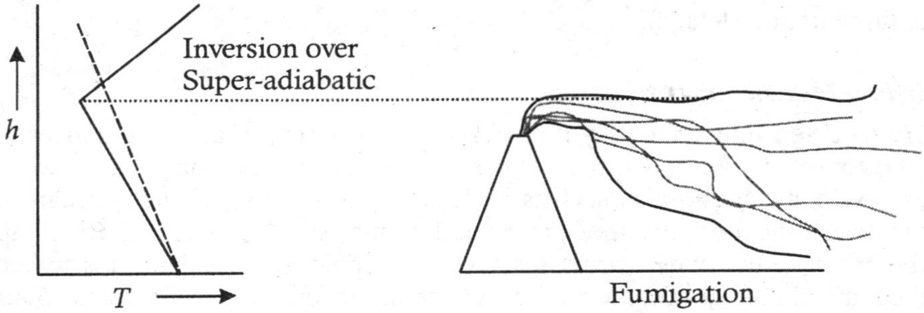


**Fig. 3.6 (*iv*):** Stable Atmosphere above the Stack

5. **Lofting [3.6(*v*)]:** Occurs when a temperature inversion exists only below the plume (occurs only in the evenings) (Pasquill's stability classes: A, B and C).

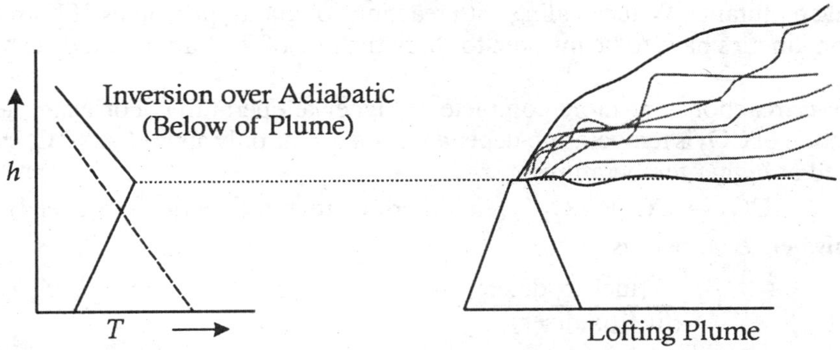


**Fig. 3.6 (*v*):** Stable Atmosphere below the Stack (Classes A, B & C)

6. **Trapping [3.6(*vi*)]:** Occurs if the stack height is between inversions. The plume is caught between two inversions and can diffuse between the inversion layers with a limited vertical spread.

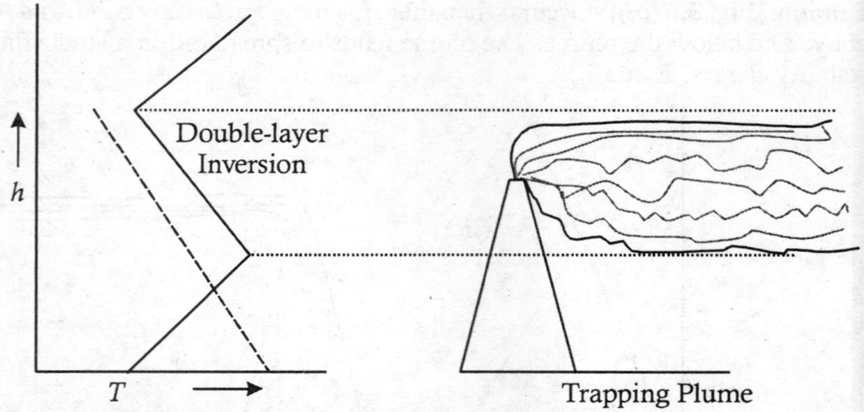


**Fig. 3.6 (*vi*):** Stable Atmosphere between the Inversion Layer

### *3.3.4.1 Tracer Studies*

Tracer studies are very important experimental methods for monitoring atmospheric dispersion. A known quantity of tracer substance (Zn-Cd sulfide; sulfur hexafluoride) is released and is monitored at various locations by gas chromatograph (GC) and other instrumental methods (see Chapters 22-26 for instrumentation details).

### *3.3.4.2 Airflow Measurement*

It is necessary to determine accurately air volumes (flow rates) to evaluate concentration of air pollutants. The total air volume (flow rate) can be determined by using dry or wet test meters, spirometers, gas-velocity instruments, such as gas flow meters, rotometers and thermal anemometers. It is possible to determine the concentration of a pollutant by static samplers. Static sampler systems operate on the principle that when air passes over the sampler surface the gas is sorbed leading to physical and chemical changes on the surface, which are recorded (see Chapters 15 and 16).

## 3.4 AIR POLLUTION DUE TO REACTING CONSTITUENTS

Air pollution modeling (Gaussian and other) procedures address the concentration dispersion of non-reacting pollutants. While dealing with reacting pollutants, principles of both thermodynamics and reaction kinetics have to be invoked to study the rate of formation and distribution of reacting pollutants.

*Endothermic* reactions are more complete at higher temperatures. For example, in combustion the dissociation of $CO_2$ is temperature-dependent and occurs only above 2,000° C, but $NO_x$ pollutants are produced at higher temperatures.

$$CO_2 \rightarrow CO + \tfrac{1}{2}O_2 \quad \text{(endothermic; requires energy to proceed)} \qquad \ldots(3.14)$$

The equivalence ratio, $\phi$, is

$$\phi = \frac{(\text{fuel/oxidizer})}{(\text{fuel/oxidizer})_{\text{Stoichio}}} \qquad \ldots(3.15)$$

$$\phi > 1 \text{ fuel-rich (air-lean).} \qquad \ldots(3.16)$$

$$\phi < 1 \text{ fuel-lean (air-rich).}$$

The effect of $\phi$, the fuel/air ratio is very important in combustion.

CO builds up for $\phi > 1$ (fuel-rich/air-lean). ...(3.17)

NO builds up for $\phi < 1$ (fuel-lean/air-rich).

An air-rich mixture tends to produce minimal CO, but produces NO, and a fuel-rich mixture produces more CO and less NO. Keeping combustion temperature down tends to reduce both CO and NO (by reducing compression ratio in automobiles); but reduced combustion temperature and compression ratio result in reduced efficiency of the engine.

### 3.4.1 Principles of Thermodynamics

Principles of thermodynamics are invoked in understanding and unraveling the mechanisms and the energetics of reacting systems. A many-particle system obeys certain general laws like the law of conservation of energy. The laws are called the laws of thermodynamics. Thermodynamics aims at studying phenomenologically the properties of material bodies characterized by macroscopic parameters, without resorting to the microscopic phenomena under investigation. It deals with energy processes of a reacting system.

Thermodynamics is based upon three universal laws.

**1st law:** Statement of conservation of energy.
**2nd law:** Statement of the direction of the spontaneous processes.
**3rd law:** Relates to the experimental approach to the absolute zero temperature.

#### *3.4.1.1 The 1st Law of Thermodynamics*

The first law thermodynamics is a statement of the law of conservation of energy. The parameters are internal energy $U$, heat $Q$, and the work done $W$. Internal energy (rest energy) of a body is the energy associated with all possible movements of particles constituting a system and their interactions (this does not include the kinetic energy of the motion of the entire center of mass of the system). Heat is a special form of energy, arising out of molecular motion. The quantity of heat, $\delta Q$, is the energy transmitted by heat exchange (a process of transmitting energy from one body to another without dong work). It s a measure of the change in the internal energy of a body in heat exchange.

According to the convention, $\delta Q$ = quantity of heat absorbed by the system, and in such a case $\delta W$ = work done on the system

$$W = -PV \text{ (pressure-volume work)} \quad \text{...(3.18)}$$

(*Note*: in Physics and Engineering work done by the system is taken as positive; and there $W = PV$). For a 'closed' system, where there is no transfer of mass to or from the system, the law of conservation of energy for processes involving heat is

$$\delta U = \delta Q + \delta W \quad \text{The 1st law of thermodynamics} \quad \text{...(3.19)}$$

The 1st law of thermodynamics describes the quantitative relations between the quantities characterizing a system, while the state of the system changes. But the 1st law does not determine the nature of the processes and the direction of the changes. The law does not state which energy state is spontaneous. However, any process occurring in a thermodynamic system must obey the 1st law.

##### *3.4.1.1.1 Enthalpy (H)*

The internal energy, $U$, is at constant volume. *Enthalpy* ($H$), "*energy of the expanded system*", is the heat content at constant pressure, equal to the internal energy plus the expansion work.

$$\delta Q = dU - \delta W \qquad \text{(1st law)}$$

$$(\delta Q)_P = dU + (PdV)_P = d(U + PV)$$

Enthalpy, $H$, is $H = U + PV$ ...(3.20)

If $\Delta H > 0$ The process is *endothermic* (heat is absorbed). ...(3.21)

If $\Delta H < 0$ The process is *exothermic* (heat is released).

### *3.4.1.1.2 Heat Capacity (C)*

When a quantity of heat, $\delta Q$, is supplied to a system, its temperature changes by $dT$. The quantity, $C$, called the heat capacity is

$$C = \frac{\delta Q}{dT} \qquad \text{...(3.22)}$$

The molar heat capacity is called the specific heat.

Heat capacity depends on the condition under which heat is supplied to a body and its temperature changes. The heat capacity at constant volume, $C_V$, is defined as

$$C_V = \left(\frac{\delta Q}{dT}\right)_V = \left(\frac{dU}{dT}\right)_V \qquad \text{...(3.23)}$$

The heat capacity at constant pressure, $C_P$, is defined as

$$C_P = \left(\frac{\delta Q}{dT}\right)_P = \left(\frac{dH}{dT}\right)_P \qquad \text{...(3.24)}$$

$$C_P = C_V + R \quad \textit{Meyer's equation} \qquad \text{...(3.25)}$$

$$\frac{C_P}{C_V} = \gamma \qquad \text{(adiabatic exponent)} \qquad \text{...(3.26)}$$

Processes in which the heat capacity is a constant quantity are called *polytropic*. Isobaric, isochoric, isothermal and adiabatic processes are particular cases of polytropic process.

$$PV^n = \text{constant} \qquad \text{...(3.27)}$$

$$n = \frac{(C - C_P)}{(C - C_V)} \qquad \text{...(3.28)}$$

where, $C$ = constant;

$n$ = polytropic exponent

For $C = C_P$ & $n = 0$; Isobaric process.

For $C = C_V$ & $n = \pm\infty$; Isochoric process. ...(3.29)

For $C = \infty$ & $n = 1$; Isothermal process.

For $C = 0$ & $n = \gamma$; Adiabatic process.

### *3.4.1.2 The 2nd Law of Thermodynamics*

The second law of thermodynamics indicates the direction of a spontaneous process. The 2nd law is defined in terms of the state function, $S$, called *entropy*.

$$dS \geq \frac{\delta Q}{T} \quad \ldots(3.30)$$

$$S \geq k_B \ln Z_i \quad \ldots(3.31)$$

where, $k_B$ = Boltzmann constant;
$Z_i$ = statistical probability of the $i$th state.

The entropy, $S$, is determined by the logarithm of the number of microscopic states through which a given macroscopic state is realized. The larger the number of microscopic states in a system, the more disorder will be in that system. Therefore, the entropy of a system is a measure of disorder in the system. Spontaneous process always results in increase in entropy of the system. In the equilibrium state, the entropy of the system attains its maximum value, since the equilibrium state is the most probable macro-state, accomplished through the largest number of microscopic states.

#### 3.4.1.3 The Free Energy

The 'free energy' of a system is a property of the system at whose expense the system does work. For reversible processes, the 2nd law of thermodynamics assumes

$$TdS = \delta Q = dU + dW \quad \ldots(3.32)$$

$$(\delta W)_{max} = d(U - TS)$$

Helmoltz 'free energy' $A = U - TS$ ...(3.33)

Helmoltz free energy is the work done at constant volume. All work done is not 'useful' work, and the Gibbs 'free energy', $G$, gives the work done at constant pressure.

$$(\delta W)_{useful} = (dW)_{max} - W_{PV} \quad \ldots(3.34)$$

$$G = A - W_{PV}$$

$$G = U - TS + PV = H - TS \text{ (Gibbs free energy)} \quad \ldots(3.35)$$

$$\Delta G = \Delta H - T\Delta S$$

$\Delta G < 0$ The reaction is spontaneous.
$\Delta G > 0$ The reaction is not spontaneous; must be coupled to another reaction so the $\Delta G < 0$.

#### 3.4.1.4 Criteria for Thermodynamic Stability

1. Adiabatically isolated state (no mass transfer; $\delta Q = 0$; $\delta W = 0$).

   $$dS > 0 \quad \ldots(3.36)$$

   The state is stable at the maximum entropy of the system.
2. The state with $V$ = constant and $S$ = constant. $dV = 0$ and $dS = 0$.

   $$TdS > \delta Q; \quad \delta Q < TdS$$

   $$dU + PdV - TdS < 0; \quad dU < 0. \quad \ldots(3.37)$$

   The state with minimum internal energy is stable.

   The state with $P$ = constant and $S$ = constant. $dP = 0$ and $dS = 0$.

   $$d(U + PV) < 0; \quad dH < 0. \quad \ldots(3.38)$$

   The state with minimum enthalpy is stable.
3. The state with $V$ = constant and $T$ = constant. $dV = 0$ and $dT = 0$.

   $$d(U - TS) < 0. \quad dA < 0. \quad \ldots(3.39)$$

   The state with a decrease in Helmoltz free energy is stable.

4. The state with $P$ = constant and $T$ = constant. $dP = 0$ and $dT = 0$.

$$dG - SdT + VdP < 0 \qquad dG < 0. \qquad \text{...(3.40)}$$

The state with minimum Gibbs free energy is stable.

### 3.4.1.5 Heat Engines and Efficiency

Principles of thermodynamics can be applied to determine the working principles and efficiency of internal combustion engines. A cyclic process is one in which the initial and final positions coincide at the same point (an ideal thermodynamical situation in an internal combustion process). The efficiency of such a process, $\eta$, is defined as the ratio of the work, $W$, performed by the engine during one cycle to the quantity of heat, $Q$, supplied to it.

$$\eta = (W/Q) \qquad \text{...(3.41)}$$

The efficiency of an internal combustion engine can be calculated for an ideal situation by applying the principles of the *Carnot* cycle, invoking the laws of thermodynamics. A reversible Carnot cycle (Fig. 3.7) consists of two thermostats, the one with higher temperature, ($T_1$) is called the *source*, and the one with lower temperature ($T_2$) is called the sink. The two isotherms, $1 \rightarrow 2$ and $3 \rightarrow 4$ are at temperatures $T_1$ and $T_2$ respectively. The two adiabats ($\delta Q = 0$) are $2 \rightarrow 3$ and $4 \rightarrow 1$, respectively.

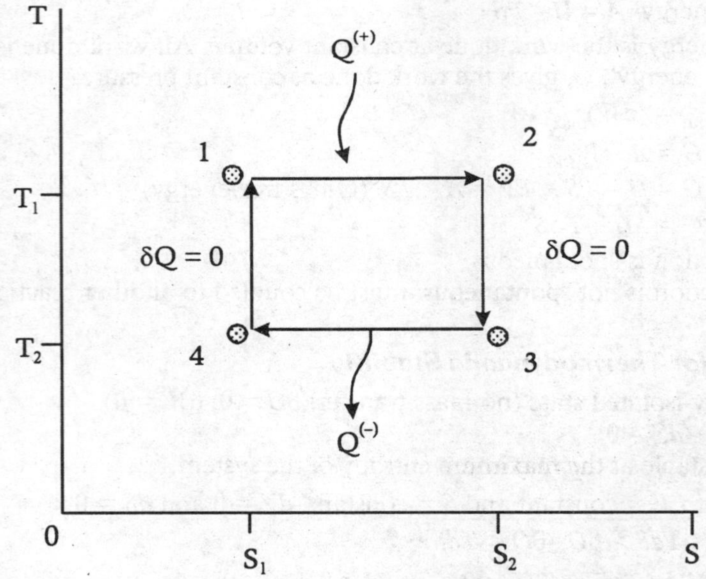


**Fig. 3.7:** Schematic Diagram of the Reversible Carnot Cycle in the *T*, *S* Coordinates

For a reversible Carnot cycle (ideal), the efficiency

$$\eta_{Carnot} = 1 - \left(\frac{T_2}{T_1}\right) \qquad \text{...(3.42)}$$

Thus, the efficiency of an ideal heat engine operating on a reversible Carnot cycle is determined *only by the temperature difference between the source and the sink.* However, the efficiency of a real heat engine is always less than that of an ideal Carnot engine operating in the same temperature interval

($\eta_{real} < \eta_{carnot}$). One of the ways of improving the efficiency of a heat engine is by raising the temperature difference between the source and the sink of the heat engine.

Carnot principles can be extended to determine the efficiency of an idealized four-stroke internal combustion engine (Fig. 3.8).

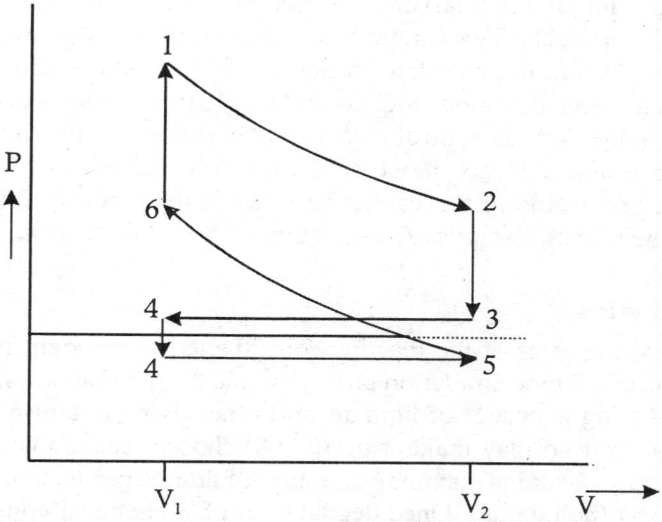


**Fig. 3.8:** Schematic Diagram of an idealized Cycle of a Four-stroke Internal Combusticn Engine

The following processes take place in an idealized cycle of a four-stroke internal combustion engine.

**State 1:** Volume $V_1$ contains gas at high-pressure ($P_1$), after ignition of air-fuel mixture. The working stroke is adiabatic $1 \rightarrow 2$.

**State 2:** Expansion of gas reaches the maximum volume ($V_2$). After the exhaust valve opens, the pressure in the cylinder should drop down to atmospheric pressure and the $2 \rightarrow 3$ is ideally, isobaric.

**State 3:** The exhaust gases remaining in the cylinder are swept out. At the point 4, the exhaust valve closes and intake valve opens. Segment $4 \rightarrow 5$ corresponds to the fuel charging.

**State 4:** The intake valve closes and on the segment $5 \rightarrow 6$ the working mixture is ignited.

**State 5:** The air-fuel mixture is ignited.

In an ideal cycle, the points 3 and 5 coincide, and the segment $3 \rightarrow 4$ coincides with the segment $4' \rightarrow 5$ and no work is done on the segment $3 \rightarrow 4 \rightarrow 4' \rightarrow 5$. In actual case, this is not so. The efficiency of a four-stroke internal combustion engine is

$$\eta = 1 - \left(\frac{V_1}{V_2}\right)^{(\gamma-1)} \qquad \ldots(3.43)$$

$$(\gamma - 1) = \frac{(C_P - C_V)}{C_V} = \frac{R}{C_V} \qquad \ldots(3.44)$$

The ($V_2/V_1$) ratio is called the compression ratio. The larger the compression ratio, higher is the efficiency.

The efficiency of a real internal combustion engine is about 50 per cent of the value calculated from thermodynamical considerations. The reason for this large discrepancy (almost by a factor 2) is due to the fact that the operating conditions of the reactants are quite different in actual situations than those in an idealized cycle, based on the thermodynamical principles. Thermodynamics deals with the reaction states (initial and final) and thermodynamically controlled reactions lead to the equilibrium ratio of the products. This cannot be of much use in understanding the reaction rates of reacting constituents where the products formed are in proportions different from those that prevail at equilibrium between the same products under the same conditions (kinetically controlled reactions). Reactions under kinetic control give a greater amount of the thermodynamically less stable of two possible products. Hence, based on thermodynamic considerations alone, inferences concerning the nature of transition states cannot be drawn of the products. Reaction kinetics has to be invoked to study the rate of formation/dissociation of compounds (from reactants).

### 3.4.2 Reaction Kinetics

Knowledge of the reaction rates of the reacting constituents is important in understanding and controlling the chemical and biodegradation processes, and for the evaluation of the persistence of pollutants and to assessing exposure of humans and other living organisms. Knowledge of the reaction mechanisms involved may make it possible to choose reaction conditions favoring one path over another, thereby obtaining maximum amounts of desired products and minimum amounts of undesirable products (pollutants). Once degradation of a chemical commences, the amount disappearing with time and the shape of the disappearance curve will be a function of the compound in question—its concentration, the organisms responsible, and a variety of environmental factors. The concentration-time relationship characterizes the concentration of the chemical remaining at any time and permits prediction of the levels likely to be present at some future time, and allows assessment of whether the chemical will be eliminated before it is transported to a site at which humans, animals and plants may be exposed.

Kinetic methods are based on relationships quantitatively describing the dynamics of reaction systems and permit determination of reaction mechanisms. This approach requires exact knowledge of the dependence of concentration of reactants as a function of time and of the reaction mechanisms. Reaction kinetics methods deal with these problems. Reaction kinetic formalisms quantitatively describe the dependence of the reaction rate on the concentrations of the components of the reaction system. Reaction rates depend on the nature and concentration of the reactants, temperature, pressure and the presence of catalysts.

The rate of reaction is proportional to the product of the concentrations of those substances that comprise the transition state. If the concentrations of all but one reactant are held constant while the concentrations of that reactant are changed, then variation in rate with the concentration changes will establish how many molecules of that particular reactant are involved in the transition state. This figure is called the order of reaction with respect to the reactant in consideration.

In general, for reactants, $A$ and $B$ forming products $C$ and $D$, in molar stoichiometric ratios, the relationship is given by

$$A + B \rightarrow C + D \qquad \ldots(3.45)$$

The reaction rate, $\nu$, is defined as the change with time in the concentrations of the components in reaction and is given by the relationship

$$v = -\frac{d[A]}{dt} - \frac{d[B]}{dt} + \frac{d[C]}{dt} + \frac{d[D]}{dt} \quad \text{...(3.46)}$$

But, for many simple, irreversible reactions, the reaction rate is

$$v = K[A]^{\alpha}\,[B]^{\beta} \quad \text{...(3.47)}$$

where, $\alpha$ and $\beta$ are reaction order with respect to substances $A$ and $B$; $K$ = rate constant

Reactions are classified according to the order of reaction ($n = \alpha + \beta$, the overall reaction order).

$$\text{0th order } (n = 0, \alpha = 0, \beta = 0): \; -\frac{d[A]}{dt} = K$$

$$\text{1st order } (n = 1, \alpha = 1, \beta = 0): \; -\frac{d[A]}{dt} = K_A[A] \quad \text{...(3.48)}$$

$$\text{2nd order } (n = 2, \alpha = 2 \text{ (or 1)}, \beta = 0 \text{ (or 1)}): \; -\frac{d[A]}{dt} = K[A]^2 \text{ or } K[A][B]$$

$$\text{3rd order } (n = 2, \alpha = 3 \text{ (or 2)}, \beta = 0 \text{ (or 1)}): \; -\frac{d[A]}{dt} = K[A]^3 \text{ or } K[A]^2[B]$$

$$n\text{th order } (nn0, n \neq 1): \; -\frac{d[A]}{dt} = K[A]^n$$

where, $[A]$, $[B]$, $[C]$, etc., are molar concentrations and $K$ = rate constant.

Rate constants vary during the reaction and depend on the initial concentrations of the reactants. For a simple 2nd order reaction, the relationship between activity coefficients is given by

$$\frac{dx}{dt} = K[A][B] = K^0 \frac{\gamma_A \gamma_B}{\gamma_{(AB)^*}}[A][B] \quad \text{...(3.49)}$$

where, $\gamma A$, $\gamma B$, $\gamma(AB)^*$ are the activity coefficients of substances $A$ and $B$ and activated complex $(AB)^*$ respectively, $K^0$ = rate constant of the reaction at infinite dilution,

where ($\gamma_A = \gamma_B = \gamma_{(AB)\#} = 1$). Therefore,

$$K = K^0 \frac{\gamma_A \gamma_B}{\gamma_{(AB)^*}} \quad \text{...(3.50)}$$

The rate of chemical reaction is temperature dependent. It can be increased by an increase in the temperature of the reaction mixture. The dependence of the rate constant on the temperature is given by the *Arrhenius* equation, which is related to entropy and enthalpy factors.

$$K = F \exp\left(-\frac{E^*}{RT}\right) \quad \text{...(3.51)}$$

$$K = \frac{k_B}{h} \exp\left(\frac{\Delta S^*}{R}\right) \cdot \exp\left(-\frac{\Delta H^*}{RT}\right) \quad \text{...(3.52)}$$

where, $F$ = frequency factor, $E^*$ = activation energy, $R$ = gas constant, $T$ = absolute temperature, $k_B$ = Boltzmann constant, $h$ = Planck constant, $\Delta S^*$ and $\Delta H^*$ = entropy and enthalpy of activation, respectively.

#### *3.4.2.1 Zero-order Reactions*

In zero-order processes, the rate is independent of the concentration of the reactants. When the concentration is plotted against time, the concentration decreases at a constant rate in zero-order process (Fig. 3.9). That is, a constant amount is lost per unit time until the entire chemical is gone. Zero-order kinetics or linear biodegradation of organic molecules has been observed frequently (e.g. mineralization of phenols, pesticides).

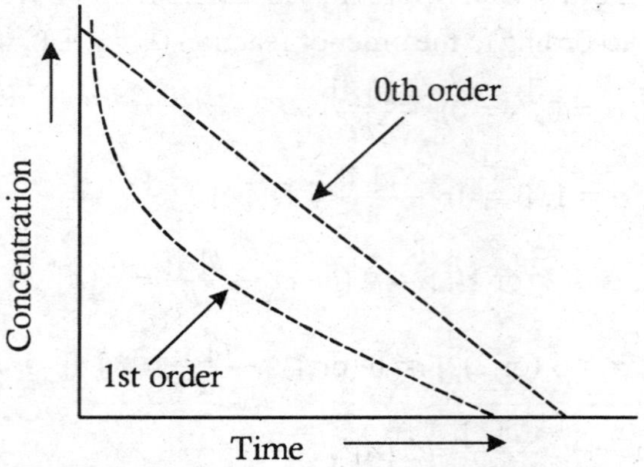


**Fig. 3.9:** Concentration-Time Plot in Reaction Kinetics

### 3.4.2.2 1st Order Reactions

In the 1st order reactions (unimolecular reactions), the rate of reaction is proportional to the concentration of only one of the reactants.

$aA + bB \rightarrow$ Products, where

$$-\frac{d[A]}{dt} = K_A[A]; \qquad \frac{dA}{[A]} = -K_A dt \qquad \ldots(3.53)$$

$$\therefore \quad [A] = [A_0]\exp(-K_A t)$$

A plot of log[$A$] against time $t$, is a straight line and the slope gives the value of $K_A$. The concentration falls quickly initially and then more slowly in 1st order processes. That is, a constant percentage of the chemical disappears per unit time in 1st order reactions.

where, $A_0$ = initial concentration;

$K_A$ = rate constant (not equilibrium constant).

1st order kinetics is sometimes termed half-life kinetics because half of the chemical is degraded in time $t$. For 1st order reactions, the rate constant and half-life, $t_{1/2}$ are inversely related.

$$t_{1/2} = \frac{0.693}{K} \qquad \ldots(3.54)$$

A substrate whose destruction follows 1st order kinetics persists long after the first half-life is over (e.g. metabolism of glucose by bacteria; radioactive decay).

### *3.4.2.3 2nd Order Reactions*

In 2nd order reactions, the rate of reaction is proportional to the 2nd power of concentration.

$aA + bB \rightarrow$ Products,

where, $$-\frac{d[A]}{dt} = K_A[A][B] \qquad ...(3.55)$$

2nd order kinetic reactions are relevant in dealing with the reaction kinetics in catalytic and combustion reactions. Concerted bimolecular elimination reactions are characterized by 2nd order kinetics; they occur readily with powerful nucleophiles.

Catalytic reactions are often systems of two successive reactions. In the first step the catalyst reacts with one of the reactants with formation of an intermediate complex.

$$C + S \underset{K_{-1}}{\overset{K_1}{\Leftrightarrow}} M \qquad ...(3.56)$$

$$M + R \xrightarrow{K_2} P + C$$

where, $C$ = catalyst,
$S$ = substrate,
$M$ = intermediate complex,
$R$ = reactant,
$K_1$, $K_{-1}$ and $K_2$ = rate constants

For combustion reactions, the reaction sequence is

Reactants → Combustion → Intermediary product (M) → Products (P)

Reaction kinetic processes can be addressed in three steps:

1. **Initiation step:** Production of species as radicals gives rise to further reaction.
2. **Propagation step:** Chain reactions resulting in intermediary products.
3. **Termination step:** Removal of radicals and production of final products.

Formation of $H_2O$ can be considered as an example: $2H_2 + O_2 \rightarrow 2H_2O$

$$M + H_2 \rightarrow 2H + M \text{ (initiation step)} \qquad ...(3.57)$$

where $M$ is energy absorber, but chemically not affected.

$$\begin{aligned} H + O_2 &\Leftrightarrow OH + O^* \\ O^* + H_2 &\Leftrightarrow OH + H \qquad \text{(chain-reaction step)} \qquad ...(3.58) \\ OH + H_2 &\Leftrightarrow H_2O + H \end{aligned}$$

where, $H$ is both a reactant and a product, leading to a chain reaction.

$$OH + H + M \Leftrightarrow H_2O + M \text{ (termination step).} \qquad ...(3.59)$$

In the termination step, radicals, OH and H are removed.

Under such conditions, where the intermediate product is a component of the reaction process, the reaction kinetics is

$$A \underset{K-1}{\overset{K_1}{\Leftrightarrow}} M + C \qquad ...(3.60)$$

$$M + B \xrightarrow{K_2} \text{Products}$$

Molar concentration $[M]$ and reaction rate are given by

$$[M] = \frac{K_1[A]}{K_{-1}[C] + K_2[B]} \qquad ...(3.61)$$

$$\text{Rate} = \frac{K_1K_2[A][B]}{K_{-1}[C] + K_2[B]} \quad ...(3.62)$$

Case 1: If $K_{-1}$ $[C] >> K_2$ $[B]$

$$\text{Rate} = \frac{K_1K_2[A][B]}{K_{-1}[C]} \quad ...(3.63)$$

Case 2: If $K_{-1}[C] << K_2$ $[B]$

$$\text{Rate} = K_1[A] \quad \text{1st order reaction} \quad ...(3.64)$$

### ***3.4.2.4 3rd Order Reactions***

In 3rd order the rate of reaction is proportional to the 3rd power of concentration; and products of a reaction can react to produce the original reactants.

$$A + B \underset{K_{-1}}{\overset{K_1}{\Leftrightarrow}} C + D \quad ...(3.65)$$

$$-\frac{d[A]}{dt} = K_1[A][B] - K_{-1}[C][D] \quad ...(3.66)$$

That is, a decrease in $[A]$ depends on both the forward and backward reactions. At equilibrium, the reaction is 2nd order overall and 1st order with respect to each concentration. At equilibrium

$$\frac{d[A]}{dt} = 0$$

$$\therefore \quad K_1[A][B] = K_{-1}[C][D] \quad ...(3.67)$$

The equilibrium constant $K(T)$ is

$$K(T) = \frac{[C][D]}{[A][B]} = \frac{K_1}{K_{-1}} \quad ...(3.68)$$

This relationship is used to determine one of the reaction constants, if the other is known

$$A \xrightarrow{K_A} B$$

$$B \xrightarrow{K_B} C \quad ...(3.69)$$

$$A \rightarrow C$$

where, $B$ is the product of the 1st reaction and is a reactant in the 2nd reaction and the overall reaction is $A \rightarrow C$.

$$\frac{d[A]}{dt} = -K_A[A]$$

$$\frac{d[B]}{dt} = -K_A[A] - K_B[B] \quad ...(3.70)$$

$$\frac{d[C]}{dt} = K_B[B]$$

$$\frac{d[A + B + C]}{dt} = 0; \quad [A + B + C] = \text{constant} \quad ...(3.71)$$

$$\therefore \quad [B] = a_1 \cdot \exp(-K_B t) \text{ (general)} \qquad \ldots(3.72)$$

$$= a_2 \cdot \exp(-K_A t) \text{ (particular)}$$

$$\therefore \quad [B] = a_1 \cdot \exp(-K_B t) + a_2 \cdot \exp(-K_A t) \qquad \ldots(3.73)$$

Assuming at $t = 0$, both $[B] = [C] = 0$, concentration of the end product, $[C]$, can be determined from the total concentration $[A_0]$.

$$[C] = [A_0]\{1 - \exp(-K_A t)\} - \frac{K_A[A_0]}{(K_B - K_A)}\{\exp(-K_A t) - \exp(-K_B t)\} \qquad \ldots(3.74)$$

$$([A_0] = [A + B + C])$$

For rate-limiting steps,

$$[C] \approx [A_0]\{1 - \exp(-K_A t) \quad \text{for } K_B >> K_A \qquad \ldots(3.75)$$

$$\approx [A_0]\{1 - \exp(-K_B t) \quad \text{for } K_A >> K_B$$

These kinetic reaction (successive reaction) mechanisms are applicable to determine reaction rates involving molecules, ions or radicals and the rates of reactions are determined by the rates of decomposition of the intermediates (e.g. redox reactions, enzyme-catalyzed and other catalytic reactions). Applying the reaction kinetics to kinetically limited reactions (e.g. combustion processes), if a reaction is stopped before completion, all the reactant species will be present ($A$, $B$ and $C$) and some of them are pollutants.

Understanding of combustion-generated air pollution requires an understanding of (*i*) stoichiometric combustion processes, (*ii*) the approach towards an equilibrium concentration and (*iii*) the effect of kinetic limitation on the final concentrations.

$NO_x$ starts at the combustion burner.

$$K(T) = \frac{C_{NO}}{\sqrt{[C_{O_2} \cdot C_{N_2}]}} \qquad \ldots(3.76)$$

where, $C$ = concentration

$$C = C_{NO} + C_{O_2} + C_{N_2} = 1 \qquad \ldots(3.77)$$

Substantial amounts of NO occur only at high temperatures. So minimization of peak combustion temperature and excess of air ($O_2$) should be maintained for minimizing air pollution ($NO_x$). Residence time of $N_2$ in the high-temperature zone is another important control parameter. The principal factors affecting $NO_x$ emission are thermodynamic and kinetic effects. The peak-flame temperature, the amount of excess air available and the length of time that the combustion gases are at the peak-flume temperature are the controlling parameters for $NO_x$ production and emission. Control strategy is—(1) minimization of residence time at peak temperature, (2) reduction of the peak temperature, (3) minimization of available $O_2$ for reaction with $N_2$ (see Chapter 27).

(*Note:* Salient details on thermodynamic principles and reaction kinetics have been included to make the chapter self-contained. Readers are advised to refer to relevant books on Chemical thermodynamics and reaction kinetics for more and in-depth details).

## BIBLIOGRAPHY

Andrews, W. (ed.) (1972), Prentice-Hall: Engelwood Cliffs, NJ. "*Environmental Pollution.*"
Cranck, J. (1975), Oxford University Press: Oxford "*The Mathematics of Diffusion,*" (2nd edn.).
Crawford, M. (1976), McGraw-Hill: New York. "*Air Pollution Control Theory.*"

Faith, W. C. and Arthur, A. H. Jr. (1992), Wiley: New York. "*Air Pollution.*"
Gilbert, R. O. (1987), Van Nostrand-Reinhold: New York. "*Statistical Methods for Environmental Pollution Monitoring.*"
Hammes, G. C. (ed.) (2000), Wiley Interscience: New York. "*Thermodynamics and Kinetics for Biological Sciences.*"
Harrison, R. M. and Perry, R. (1984), Chapman & Hall: London. "*Handbook of Air Pollution Analysis,*" (2nd edn.).
Hodges, L. (1973), Holt-Reinhart & Wiley: New Yok. "*Environmental Pollution.*"
Kormonday, E. J. (1977) Engelwood Cliffs, NJ: Prentice-Hall. "*Concepts of Ecology,*" (2nd edn.).
Landsberg, H. E. (1969), Doubleday: New York. "*Weather and Health.*"
Liu, D. H. F. and Liptak, B. G. (eds.) (2000), Lewis Pubs: Boca Raton, FL. "*Air Pollution.*"
Mateev, A. N. (1985), Mir Pubs: Moscow. "*Molecular Physics.*"
Narayanan, P. (2003), New Age Intl. Pubs: New Delhi. "*Essentials of Biophysics*"(2nd Print).
Nriagu, N. O. (ed.) (1992), Wiley & Sons: New York. "*Gaseous Pollutants.*"
Parker, H. W. (1977), Prentice-Hall: Engelwood Cliffs, NJ. "*Air Pollution.*"
Perkins, H. C. (1974), McGraw-Hill: New York. "*Air Pollution.*"
Pfafflin, J. R. and Ziegler, E. N. (1976), Gordon Breach Science: New York. "*Encyclopedia of Environmental Science and Engineering.*"
Preist, J. (1973), Addison-Wesley: New York. "*Problems of our Physical Environment.*"
Reeve, E. N. (1994), Wiley & Sons: New York. "*Environmental Analysis.*"
Scorer, R. (1968), Pergamon Press: Oxford. "*Air Pollution.*"
Seinfeld, J. H. (1975), McGraw-Hill: New York. "*Air Pollution: Physical and Chemical Foundations.*"
Seinfeld, J. H. and Pandis, S. N. (1997), Wiley: New York. "*Atmospheric Chemistry and Physics: From Air Pollution to Climate Change.*"
Stern, A. C., *et al.* (1984), Academic Press: New York. "*Fundamentals of Air Pollution,*" (2nd edn.).
Turiel, I. (1975), Prentice-Hall: Engelwood Cliffs, NJ. "*Physics: The Environment and Man.*"
Turk, A., *et al.* (1978), Saunders: Philadelphia, PA. "*Environmental Science,*" (2nd edn.).
Turner, D. B. (1970), U.S. Health Ser., Pub. 999-AP-26: Washington, DC. "*Workbook of Atmospheric Dispersion Estimates.*"
Wilson, J. D. (1974), Heath: Lexington, MA. "*An Environmental Approach to Physical Science.*"
... (1993), USEPA/600/8-91/038: Washington, DC. "*Selection Criteria for Mathematical Models used in Exposure Assessments: Atmospheric Dispersion Models.*"

# 4

# Atmospheric Pollutants

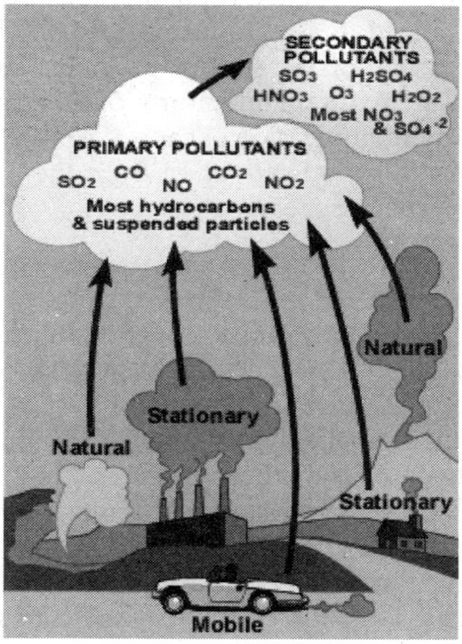


Atmospheric pollutants can be of different types such as particulate, aerosol, radioactive, thermal or mechanical. Pollutants can be from stable sources (e.g. industrial installations) or from dynamic sources (e.g. automobiles and other transport systems). Major atmospheric pollutants are particulate matter (e.g. asbestos, dust, fly ash and soot), and gaseous primary pollutants, such as oxides of sulfur ($SO_x$), oxides of nitrogen ($NO_x$), ammonia ($NH_3$) and carbon monoxide (CO). Secondary pollutants, such as ozone, peroxyacetyl nitrate (PAN) and other oxidants are produced as a result of various photochemical reactions in which the primary pollutants and hydrocarbons take part. They are responsible for photochemical smog.

Thus, a pollutant may include any chemical substance, biotic component or its product, or physical factor that is released intentionally by man into the enironment in such a concentration that may have adverse harmful or unpleasant effects. The various pollutants are deposited matter, such as soot, smoke, tar, dust, grit; gases, such as oxides of nitrogen, sulphur, carbon monoxide, halogens and fluorides, etc., acid droplets; metals, such as murcury, lead, tin, cadmium, chromium; agrochemicals, such as pesticides, herbicides, fungicides, nematicides, bactericides, weedicides, and fertilizers; complex organic substances, viz. benzene, aldehydes, ethylene, ether, acetic acid, benzpyrenes; photochemical oxidants, such as PAN, solid wastes, radioactive waste and noise.

This Chapter illustrates the nature, role and types of various pollutants.

The atmosphere is susceptible to pollution from natural sources as well as from human activities. Among the natural sources of pollution are the volcanoes, which on eruption emit large quantities of sulfur oxides and particulate matter. Equally significant, in this respect, are forest fires. Anaerobic oxidation of plant material, and microbes working in the guts of cattle release methane. Man-made or anthropogenic pollution, in principle, started when Man first used fire. Most air contaminants originate from combustion processes. However, a number of natural factors work to keep the parts of atmosphere in balance. For example, plants use carbon dioxide, generated in the burning of carbonaceous matter, and produce oxygen. Animals, in turn, use up oxygen and produce carbon dioxide through respiration. The gases and particulate matter from volcanic eruptions and forest fires are scattered or washed away by wind and rain. The industrial revolution of the 19th century, which accelerated the use of fossil fuels for domestic and industrial activities, marked a steep rise in both the intensity and frequency of episodes of pollution. In the 20th century air quality problem in the cities, gradually, increased with the advent and popularization of gasoline powered automobiles, trains, airplanes and steam ships. The growth of chemical and metallurgical industry, thermal power generation using coal as the main fuel, burning of garbage and other solid waste and many other industrial operations generate a variety of pollutants. Many of these have known or suspected harmful biological and long-term climatic effects. Pollutants from these sources may not only prove a problem in the immediate vicinity of the sources but can travel long distances and some of them reacting chemically in the atmosphere to produce secondary pollutants such as acid rain and ozone. Starting with mid-twentieth century it is realized, first in industrialized and later in other parts of the world, that environmental pollution is one of the serious problems facing humanity and other life forms. It is also realized that out of the natural and anthropogenic causes of pollution only the latter are subject to mitigation and control. Identification of causes of pollution, its local and global effects and mode of control have attracted serious attention from the world community. Limited success has already been achieved in particular areas.

## 4.1 NATURE OF POLLUTANTS

Air is considered to be polluted when it contains substances in concentration high enough and duration long enough to cause harmful or undesirable effects. The substances might be released by natural or human activity. The adverse effects they cause might lower chemical, physical and biological characteristics of the atmosphere. The resultant damage might affect the health of humans, animals and plants and affect the strength of buildings, roads, dams and other structures. Pollutants might be found in the solid, liquid or gaseous phase. There are six criteria air pollutants—carbon monoxide (CO), sulfur dioxide ($SO_2$), nitrogen dioxide ($NO_2$), ozone ($O_3$), lead (Pb) and fine particulate matter (PM-10, diameter <10 mm). In addition, a large number of hazardous substances are recognized as environmental pollutants at the present time. The sources and the mechanism by which various pollutants are generated are many.

### 4.1.1 Physical Forms of Pollutants

Atmospheric pollutants occur in different physical states—particulate, gaseous, aerosol and radioactive (Fig. 4.1). To this category must be added thermal and noise pollutions.

### 4.1.2 Origin of Air Pollutants

Pollutants can originate from point, non-point (diffuse) and mobile sources. Stationary objects, which release pollutants, are classified as point sources (e.g. factories, smokestacks). Since bulk of the anthropogenic pollutants, both criteria and hazardous types result from industrial activity, the power plants (thermal and nuclear), and oil refineries, metallurgical and chemical industries and refuse incinerating plants fall in this category. Volcanic eruptions release large quantities of particulate matter of complex chemical composition, oxides of sulfur and compounds of fluorine. Therefore, among the major sources of natural pollution volcanoes are also included in this category. Non-point diffusion sources are residential areas, hospitals, utility and waste-disposal units, forestry (forest fires) and agriculture operations. The mobile sources are a special category of non-point emitters. Transport vehicles using coal, diesel or gasoline, contribute a major part of modern day pollutants. Cars and trucks are considered on-road mobile sources. Tractors, lawn-movers, boats, ships, locomotives and airplanes are off-road mobile sources. The major pollutants produced by these sources are hydrocarbons, $SO_x$, $NO_x$, $CO_x$ and smoke.

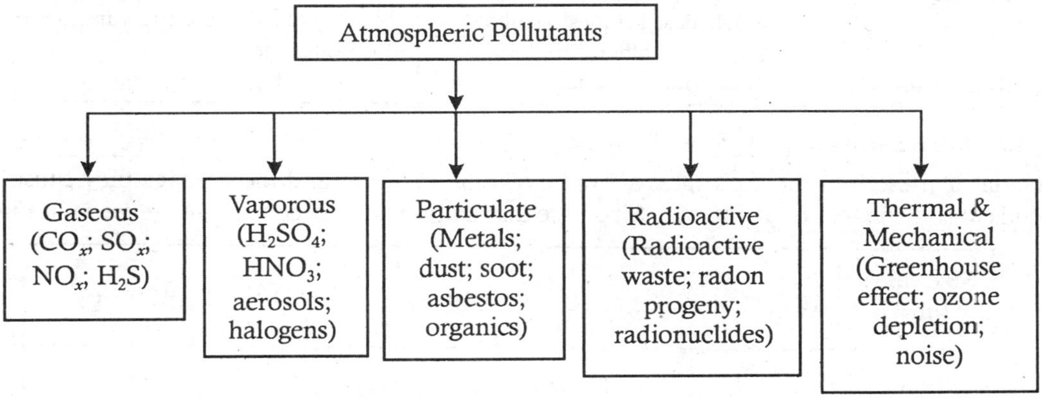


**Fig. 4.1:** Forms of Atmospheric Pollutants

### 4.1.3 Classification of Air Pollutants

Air pollutants can be classified as: (1) primary and (2) secondary types. Primary air pollutants are substances that exist in the same form, such as CO, $CO_2$, $SO_2$, $H_2S$ and metal vapors, as in source emissions and which are released directly into the atmosphere from the source industries or operations. A large number of hazardous chemicals are also released directly. Secondary pollutants are produced in the atmosphere by interaction between various substances present there or as a result of chemical reactions. Ozone, smog and components of acid rain ($H_2SO_4$ and $HNO_3$) are examples of secondary pollutants. The distinction between the two types is important while dealing with their attenuation. Some common secondary pollutants are given in Table 4.1.

**TABLE 4.1:** TYPICAL SECONDARY POLLUTANTS

| *Primary Pollutants* | *Secondary Pollutants* | *Type of Reaction* |
|---|---|---|
| Acid + Alkali | Salt | Acid-Base Reaction |
| $SO_2 + H_2O$ | $H_2SO_4$ | Oxidation |
| $NH_3 + SO_2$ | $(NH_4)_2SO_4$ | Oxidation |
| $NO + O_2$ | $NO_2$ | Photochemical |
| $NO + O_2 + HC^*$ | $O_3$ + Free radicals | Photochemical |

*HC = Hydrocarbons.*

### 4.1.3.1 Organic and Inorganic Pollutants

Atmospheric pollutants can also be classified under three broad categories-- particulate organic, inorganic (Table 4.2).

TABLE 4.2: CLASSIFICATION OF AIR POLLUTANTS

| *Major Class of Pollutants* | *Subclass* | *Examples* |
|---|---|---|
| Inorganic | Gases | $SO_x$; $NO_x$; $CO_x$; $NH_3$; |
| | Acids | HF; HCl; $H_2SO_4$; $HNO_3$ |
| | Mineral compounds | Oxides; chlorides; sulfates; silicates; fluorides; phosphates |
| Organic | Hydrocarbons; acids; aldehydes; ketones; alcohols and many others | Benzene; toluene; hexane; acetone; benzaldyhyde; alcohol and more complex substances |
| Particulate | Dust; smoke; fumes | Mist; fog, etc. |

### 4.1.3.2 Particulates of Different Types

Particulates, present in the atmosphere, can be of different sizes and the damage they cause to life depends to some extent on this factor. Size-wise distribution and nomenclature are given in Fig. 4.2.

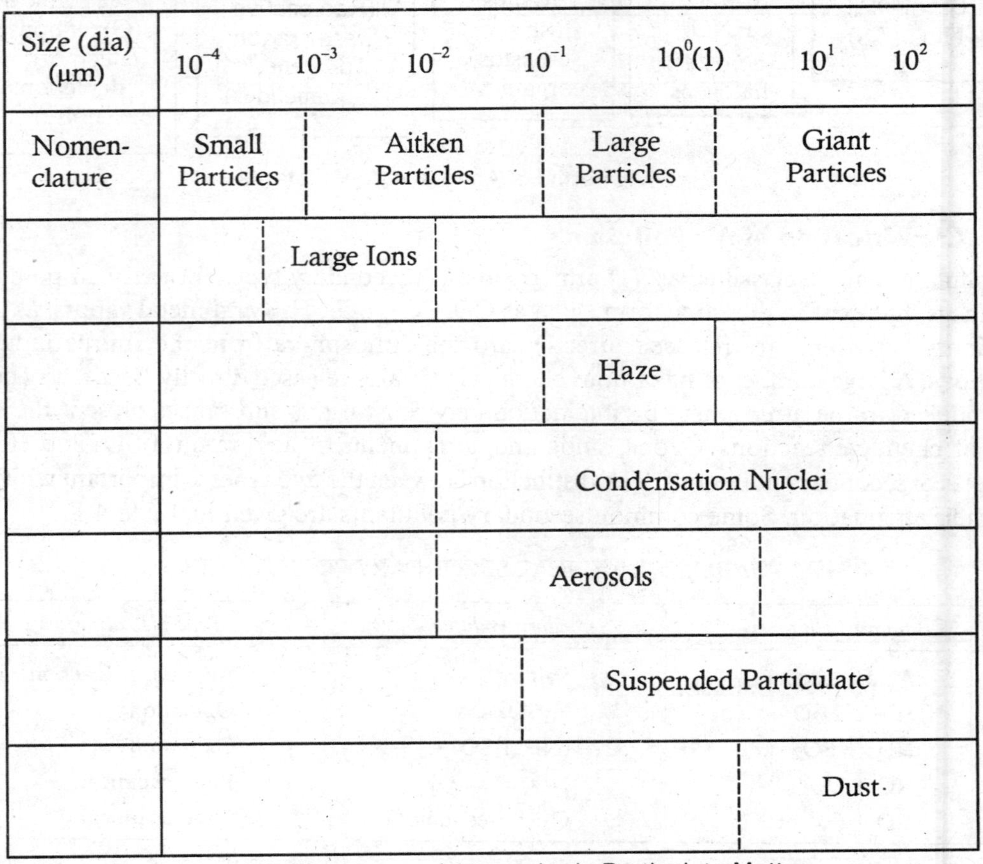


**Fig. 4.2:** Size-range of Atmospheric Particulate Matter

Particulate in the atmosphere can be of different types, differing in their physical composition, as follows:

*Aerosol:* It consists of minute particles, part of which is invariably water.
*Fog:* Aerosol in which the dispersed phase is liquid ($G1 - L2$) is fog.
*Mist:* Dispersed liquid particles are of larger size ($G1 - L2$) than in fog.
*Smoke:* It is the product of incomplete combustion of wood, coal, gasoline or other biomass. It consists mostly of carbon and small liquid and solid particles of less than 1.0 micrometer size ($G1 - S2$).
*Fly ash:* It is emitted in large quantities in coal burning units like power generation installations. The finely dived particles ($G1 - S2$) consist, mainly, of inorganic oxides, silicates and other compounds.
*Fume:* It is made up of metal vapor in 0.03 - 0.3 μm size ($G1 - S2$).
*Dust:* It consists of solid particles of larger than colloidal size (1 mm and above). Often it is soil or rock material (silica and silicates).
*Smog:* It is a combination of smoke and fog ($G1 - L2 + S2$). It occurs when a high concentration of moisture is combined with smoke and oxides of sulfur and nitrogen. Sunlight facilitates its formation. Low ground temperature (<10° C), thermal inversion and absence of wind stabilize it.
*Spray:* It is atomized liquid ($G1 - L2$).

where $G$, $L$ and $S$ refer to gas, liquid and solid phase. 1 and 2 refer to major and dispersed phase, respectively.

Most of the particulate emission is due to fossil fuel combustion (30 per cent), industrial (25 per cent), forest and foliage burning (30 per cent) and vehicular transport (5 per cent).

## 4.2 ROLE OF DIFFERENT POLLUTANTS

The environmental problems created by different pollutants vary in nature and intensity (Table 4.3).

**TABLE 4.3:** GENERAL EFFECTS OF SOME COMMON POLLUTANTS

| *Pollutant(s)* | *Effects* |
|---|---|
| Particulate | Visibility reduction; damage to materials; bronchial illness. |
| Oxides of Sulfur ($SO_x$) | Metal corrosion; material damage; damage to plants and vegetation; damage to upper respiratory tract. |
| Oxides of Nitrogen ($NO_x$) | Damage to plants and vegetation; eye and nose irritation. |
| Carbon monoxide (CO) | Headache; nausea; and death. |
| Hydrocarbons ($HC_S$) | Damage to plants and environment; carcinogens. |
| Oxidants | Damage to materials, plants and vegetation; eyes, throat and lung irritation |

### 4.2.1 Particulates

Particulates in the atmosphere are complex mixtures of organic and inorganic substances. They can be liquids, solids or a mixture of both. They can be of primary or secondary origin. Major sources of primary particles are combustion processes, in particular diesel combustion. A rough distribution is as follows: road transport 25 per cent, non-combustion processes 24 per cent, industrial combustion plants and processes 17 per cent, commercial and residential combustion 16 per cent and public power generation 15 per cent. Secondary particles are formed, typically, when some of the primary pollutants react to form non-volatile products. This is the case when $H_2SO_4$ is formed by the oxidation of $SO_2$. Particulate matter such as asbestos is highly resistant to chemicals. Asbestos can cause serious health problems if it is inhaled as tiny fibers. Inhalation leads to bronchial carcinoma, pleural calcification, asbestosis and tumors.

In the recent past emphasis was laid on particles of size <10 mm (designated PM-10) as the fraction posing significant health risk, because they can be inhaled deep into the lungs and become trapped in the lower respiratory system. However, in more recent years concern has shifted to monitoring smaller particle fraction like PM-2.5 (< 2.5 μm).

### 4.2.2 Gaseous Pollutants

There are a number of gaseous chemicals, which can be present in the atmosphere in different concentrations. A limited number out of them are inorganic materials, such as oxides of sulfur, of nitrogen, and of carbon, and ozone, and the rest are organic type. Organic pollutants include hydrocarbons, hydrohalocarbons; and in addition, there are many 'hazardous' industrial-, and agro-chemicals that are included in an ever-expanding list. Some of the common gaseous pollutants, their source and normal paths for removal are indicated in Table 4.4.

**TABLE 4.4:** GASEOUS POLLUTANTS IN THE ATMOSPHERE

| *Compound* | *Molecular Mass, $M_r$* | *Property* | *Source* | *Removal/Remarks* |
|---|---|---|---|---|
| Ammonia ($NH_3$) | 17.03 | Colorless; pungent | Waste treatment; biogenic processes | Reaction with $SO_2$ to form $(NH_4)_2SO_4$ |
| $SO_2$ | 64.07 | Colorless; irritant | Fossil fuel & Coal combustion | Oxidation to sulfate- $SO_2 \rightarrow H_2SO_4$ |
| $SO_3$ | 80.06 | Colorless; irritant | Oxidation of $SO_2$ | Conversion to $H_2SO_4$ |
| $H_2S$ | 34.08 | Colorless; foul smelling | Chemical processes & sewage | Oxidation to $SO_2$ |
| CO | 28.01 | Colorless & odorless | Incomplete combustion of carbon; auto exhaust | Photochemical reaction with $CH_4$ & OH |
| $CO_2$ | 44.0 | Colorless & odorless | Combustion & biological processes | Photosynthesis |
| NO | 33.01 | Colorless | Combustion & biogenic processes | Oxidation to nitrate |
| $NO_2$ | 46.01 | Red brown | Combustion & biogenic processes | Oxidation to nitrate |
| Ozone ($O_3$) | 48.0 | Faint blue; irritant | Oxidation of $O_2$ & free radical reactions | Reduction to $O_2$ |
| Fluorine ($F_2$) | 38.0 | Colorless; irritant | Burning of fluoride ores | — |
| Hydrogen fluoride (HF) | 20.01 | Greenish yellow; irritant | Burning of fluoride ores | — |
| Hydrogen chloride (HCl) | 36.47 | Colorless | Industrial processes | — |
| Chlorine ($Cl_2$) | 70.9 | Greenish yellow; irritant | Industrial processes; electroplating | — |

Some of the gaseous inorganic pollutants are sulfur dioxide, oxides of nitrogen, oxides of carbon, ozone, halides, chlorofluorocarbons (CFCs) and hazardous trace metals (impact of these pollutants are treated in Chapters 5 and 6). Under this list come radioactive aerosols such as radon and radon progeny.

### 4.2.3 Radioactive Pollutants

Radioactive waste products (produced by nuclear reactors and weapon factories are gaseous as well as particulate. They give rise to various radioactive materials and radon progeny. In addition to radon gas, there are other radioactive materials that can cause pollution. Radioactive waste produced by nuclear reactors, and used-fuel reprocessing plants and nuclear weapon factories pose a threat to atmospheric pollution through release of radioactive gases like iodine, xenon and others. Radioactive materials, such as radioactive iodine and strontium, are concentrated in living tissues and can cause damage even when the general level of environmental contamination is low.

Yet another possible source of radioactive contamination lies in the possible accidents to nuclear power generating and fuel reprocessing installations that are functioning in many parts of the world. The Three Mile Island nuclear power plant accident in USA (1978) and the more serious Chernobyl nuclear power plant explosion in erstwhile USSR (1986) are warnings in this direction.

#### *4.2.3.1 Radon and Radon Progeny*

Radon is a colorless odorless radioactive gas, which chemically belongs to the family of rare gases. It is given off through the decay of uranium in rocks within the Earth. Minerals carrying these radioactive elements are widely distributed in nature in many rocks and soils. Radon is an invisible pollutant that can contaminate any part of the environment. People can be exposed to radon when the gas leaks into basements of homes built over radioactive soil or rock. It is harmful, not because of any direct chemical or physiological effects but due to its radioactivity. Its effects are felt through the series of radioactive disintegration products formed from it. Radon exposure risk is high for those working in uranium mines, milling and processing plants and near ore tailing dumps.

Some of the decay products of radon (radon progeny) are—$^{54}Mn$, $^{60}Co$, $^{85}Kr$, $^{131}I$, $^{133}Xe$, $^{137}Cs$, $^{214}Po$ and $^{216}Po$. The daughter products of radon being solids attach themselves readily on dust particles in the air. When inhaled, the fine dust particles are lodged in the lungs, damaging the sensitive lung tissues. Studies have established a strong link between exposure to radon and its daughter products and lung cancer.

### 4.2.4 Organic Pollutants

Hydrocarbons are the principal constituents of petroleum and natural gas. All of them are combustible. Methane, the simplest of hydrocarbons, is also generated as 'marsh gas' by wet decay of organic matter. There are two main groups of organic compounds of concern in air pollution: (1) volatile organic compounds, VOCs, and (2) polycyclic aromatic hydrocarbons, PAHs. VOCs include pure hydrocarbons like benzene, toluene, 1, 3-butadiene, partially oxidized hydrocarbons and organic compounds containing chlorine, sulfur and nitrogen. They find large-scale application as fuels (e.g., propane and gasoline), solvents, and paint thinners and in the production of plastics (e.g., vinyl chloride). Many VOCs are released in vehicle exhaust gases either as unburned fuels or as combustion products. The aromatic VOC benzene is a minor constituent (about 2 per cent by volume) of petrol. About 70 per cent of atmospheric benzene originates from the combustion of petrol (combustion of olefins). Apart from being emitted in the vehicle exhaust as unburned fuel, benzene is also found among the products of decomposition of higher molecular weight aromatic compounds in the fuel. 1, 3 butadiene is also a known, potent, human carcinogen. Other VOCs are important because of the role they play in the photochemical formation of ozone in the atmosphere.

1, 3-butadiene is also a VOC emitted, as a product of combustion of olefins, from exhaust of vehicles using petrol or diesel fuel. This chemical is widely used in the production of synthetic rubber. Both benzene and 1, 3-butadiene are known potent carcinogens. In addition to these two hydrocarbons there are a number of toxic organic micro-pollutants, comprising a complex range of chemicals. They are present in the atmosphere in small amounts but some of them are highly toxic and carcinogenic.

Hydrocarbons entering the atmosphere as primary pollutants are involved in the production of photochemical oxidants, responsible for photochemical smog. Photochemical oxidants, like *peroxyacetyl nitrates* (PAN, Fig. 4.3) are secondary pollutants, produced as a result of chemical reactions in the atmosphere involving $NO_x$ and hydrocarbons under the influence of sunlight. Halogenated hydrocarbons pose a serious environmental threat because of their persistence in the atmosphere (see Chapters 5 and 21).

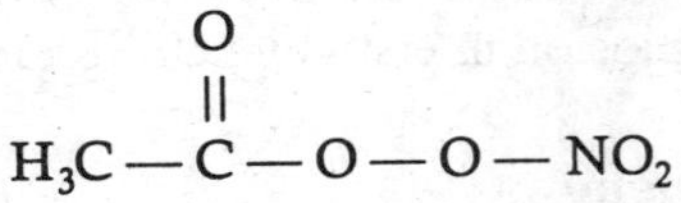

**Fig. 4.3:** Peroxyacetyl Nitrate (PAN)

## BIBLIOGRAPHY


Bockris, J, O'M. (ed.) (1978), Plenum Press: New York. "*Environmental Chemistry.*"

Buffle, J. and Van Leeuwen, H. P. (eds.) (1992), Lewis Pubs: Chelsea, MI. "*Environmental Particles,*" (Vol. 1).

Friedlander, S. K. (1977), Wiley: New York. "*Smoke, Dust and Haze.*"

Goldsmith, J. R. and Landlaw, S. (1968), *Science*, **162**: 1352.

Harrison, R. M. and Perry, R. (1984), Chapman & Hall: London. "*Handbook of Air Pollution Analysis,*" (2nd edn.).

Horne, R. A. (1978), Wiley & Sons: New York. "*The Chemistry of our Environment.*"

Hutzinger, O. (ed.) (1990), Springerverlag: Berlin. "*Handbook of Environmental Chemistry.*"

Kellog, W. W., *et al.* (1972), *Science*, **175**: 587 "The sulfur cycle."

Kouimtzis, T. and Samara, C. (eds.) (1995), Springerverlag: Berlin. "*Airborne Particulate Matter.*"

McGrath, J. J. and Barnes, C. D. (1982), Academic Press: New York. "*Air Pollution: Physiological Effects.*"

Nriagu, N. O. (ed.) (1992), Wiley & Sons: New York. "*Gaseous Pollutants.*"

Perkins, H. C. (1974), McGraw-Hill: New York. "*Air Pollution.*"

Purdom, P. W. (ed.) (1980), Academic Press: New York. "*Environmental Health,*" (2nd edn.).

Rasool, S. I. and Schneider, S. H. (1971), *Science*, **173**: 138. "Atmospheric carbon dioxide and aerosols- Effects of large increase on global climate."

Scorer, R. S. (1968), Pergamon Press: Oxford. "*Air Pollution.*"

Singer, S. F. (ed.) (1970), Springerverlag: New York. "*Global Effects of Environmental Pollution.*"

Stern, A. C. (ed.) (1984), Academic Press: Oxford. "*Air Pollution,*" (Vol. 1).

Stern, A. C., *et al.* (1984), Academic Press: New York. "*Fundamentals of Air Pollution,*" (2nd edn.).

# 5

# Impact of Atmospheric Pollutants

Atmospheric pollutants have local, regional or global level effects. Fluorides and silicates are particulate pollutants arising out of mining and construction operations and have impact, mostly, on a local scale. Carbon monoxide (CO) is produced by the incomplete combustion of carbonaceous materials and fossil fuels. The major sources of CO are auto-mobiles and coal-burning operations. The environmental impact of CO pollution is on the regional level. It has greater affinity to bind to hemoglobin (Hb) than oxygen, and thus it is a powerful asphyxi-ating gas.

The large-scale transport of $SO_2$ occurs in the troposphere, resulting in the production of acid rain. The environmental impact of $SO_2$ is on the regional and continental levels. Depletion of protective ozone layer in the mesosphere-stratosphere boundary, by scavengers of ozone (e.g. fluorocarbons), would lead to increase in the UV radiation damage to the vegetation and animals on the Earth. Oxides of nitrogen play a central role in the formation of photochemical smog. $CO_2$ by itself is not a pollutant. However, a significant increase in $CO_2$ and $N_2O$ in the atmosphere would result in greenhouse effect. The environmental impact of $CO_2$ and $NO_x$ is on the global scale. This Chapter illustrates the impact of various pollutants on the fauna and flora of the globe.

In considering atmospheric pollution and its impact the ionosphere (75 km and above) and the chemosphere (75 to 25 km) regions are important, because of their absorption and scattering of solar radiation. They thereby influence the amount and spectral distribution of solar energy and cosmic rays reaching the stratosphere (50 to 10 km) and troposphere (10 to 0 km). Stratosphere is of interest for global transport of pollution and troposphere is of course of primary importance to all living organisms.

Non-physiological effects of pollutants are: (1) reduced visibility due to fog and smog, (2) damage to materials, structures, organisms and ecosystems (e.g. acid rain; toxic pollutants), and (3) disturbance of global energy balance (e.g. thermal pollution and global warming). A variety of air pollutants have known or suspected harmful effects on human health and the environment (e.g. coal dust, asbestos, $SO_2$ and $NO_2$). These substances are called primary air pollutants. Many of these primary air pollutants produce secondary pollutants such as acid rain, ground-level ozone (reaction of $NO_2$ and VOCs leads to the formation of ground-level ozone, a secondary long-range air pollutant) and peroxyacetyl nitrate (PAN).

When discussing the impact of air pollutants on ecosystems, the geographical scales of their impact are to be taken into account. The geographical spread of different pollutants is different. Atmospheric pollutants have local, regional and global impact. The fluorides and silicates (like asbestos) are pollutants arising from mining, fertilizers, alumina and glass manufacturing. Their impact ("fluoride cycle") is mainly on a local scale.

In the case of sulfur dioxide, long-range transfer of it occurs in the troposphere and lithosphere. The secondary pollutants, sulfur trioxide ($SO_3$) and sulfuric acid ($H_2SO_4$), have their contribution to the "acid rain" (refer to Chapter 6). The impact of the sulfur cycle is, thus, on a regional scale.

Oxides of carbon and nitrogen ("carbon cycle" and "nitrogen cycle") have global impact. In its natural cycle, carbon dioxide ($CO_2$) enters the atmosphere from respiration and decay of organisms and also from atmospheric oxidation of methane. It is removed from the atmosphere by photosynthesis and absorption by the oceans. However, human activity all over the globe (e.g. burning of fossil fuel) is disturbing the balance between the two phenomena. Because of its property of absorbing of infrared radiation, $CO_2$ along with some other constituents of the atmosphere, absorb and retain IR radiation emanating from the Earth. It thus acts as a barrier for heat loss from the Earth and creates a global warming, with all its concomitant ill effects.

After $CO_2$, methane ($CH_4$) is the second most important "greenhouse" gas. On a molecular basis, it is about 25 to 30 times more effective in absorbing IR radiation and thus more effective as a "greenhouse" gas. In addition, $CH_4$ breaks down to form CO, which itself is an important pollutant. Other greenhouse gases are nitrous oxide ($N_2O$) and halocarbons.

Oxides of nitrogen, of which $NO_2$ is the main component, are photochemically active. Under the influence of UV radiation from the Sun $NO_x$ react with hydrocarbons (Table 5.1) in the atmosphere producing secondary pollutants, like hydroxyl radical, $HO^*$, peroxyacetyl nitrate (PAN). These products are responsible for photochemical smog (local or regional effect), and depletion of ozone layer and disturbance of atmospheric energy balance (a global phenomenon).

In summary: Increasing air pollution can modify the climate and atmosphere in many ways—on local, regional and global scales—which may manifest in long-term and long-lasting effects.

## 5.1 ATMOSPHERIC EFFECTS

Pollutants adversely affect the atmosphere. Conversely, the state of the atmosphere has an effect on the pollutants and their products. Clouds normally form when rising, adiabatically cooling air can no longer hold water in the vapor form. Water absorbs IR radiation more strongly than $CO_2$, thus greatly influencing Earth's heat energy balance. Clouds are important absorbers and reflectors of radiant heat in the atmosphere. Clouds reflect light from the sun and have a temperature-lowering effect.

**TABLE 5.1:** RELATIVE REACTIVITIES OF HYDROCARBONS TOWARD HO* RADICAL

| *Reactivity Class* | *Reactivity Range* | *Half-life in the Atmosphere* | *Reactants* |
|---|---|---|---|
| I | < 10 | > 10 days | Methane |
| II | 10 - 100 | 1 - 10 days | Acetylene; ethane |
| III | 100 - 1000 | 2 - 24 hours | Benzene; propane; toluene; ethane |
| IV | 1,000 - 10,000 | 15 - 120 mts | Xylene; propene |
| V | > 10,000 | < 15 mts | Methyl butene |

*Source*: Manahan, S.E., (1988) "*Environmental Chemistry*", 3rd edn, Boston: Willard Grant Press.

Atmospheric humidity makes pollutants, like $SO_2$, more corrosive by the formation of corrosive compounds like $H_2SO_4$. Increase in pollution may often bring about perceptible local changes but may not have discernible effects globally in the short-run. However, small local and regional atmospheric changes may add up to produce cataclysmic events, such as freezing of the oceans or melting of the ice caps. For example, at low altitudes, small changes in atmospheric pressure have little effect on biological species. However, at high altitudes and low atmospheric pressures the situation would be different—reduction of solubility of $O_2$ and $CO_2$ in blood would lead to respiratory problems and reduction of boiling point of water. Besides, pollutants entering the stratosphere region tend to remain for a longer time because of constant temperature distribution with altitude and stable atmospheric conditions, with sparse mixing, leading to long-term adverse effects.

## 5.2 ROLE OF DIFFERENT AIR POLLUTANTS

Air pollutants are particulate, aerosols, gases and gaseous mixtures and radioactive materials. Of these particulate matter, <10 μm (PM-10), carbon monoxide (CO), sulfur dioxide ($SO_2$), nitrogen dioxide ($NO_2$), ground-level (secondary) ozone ($O_3$), and lead (Pb) are considered as criteria air pollutants.

### 5.2.1 Particulates

Particulate matter (0.1-10 μm) is due to mining, metallurgical, industrial, construction, agriculture activities, forest fires, combustion of fossils and fuels, vehicular exhaust (carbon, hydrocarbons and lead compounds, etc.) and metal fumes. Particulates placed in the stratosphere (>10 km) remain there a long time due to poor vertical mixing. They are responsible for secondary pollutants, smog, poor visibility and atmospheric events that linger for years.

Particulate sediments, even if they are not hazardous to health, do soil the surfaces and cause physical and structural deterioration. Cement dust causes eczema of skin; sand dust causes silicosis and asbestos causes asbestosis. Smoke, soot and dust pose respiratory and other health problems. Health hazards depend on the particulate size. Particulate matter of diameter >20 mm is filtered out in the nose and throat and does not enter the lungs. But, sub-micron range particulate matter

(PM-10—PM-2.5 range) poses pulmonary health problems. They are small enough to penetrate deep into the lungs and so potentially pose significant health risks. Particulate matter above the 2.5 µm range (PM-2.5) is removed by ciliary action of epithelial cells. However, particulate matter in the range 0.5-2.0 µm (<PM-2.5) can reach lung lymph nodes and can cause impairment of lung function. Particulates <0.5 µm can cause bronchitis. PM-10 particles may carry surface-adsorbed carcinogenic compounds into the lungs. (Physiological aspects of some of these air pollutants are dealt in Chapter 6).

### 5.2.2 Aerosols

Particulate matter consists of both sediments and 'suspended' particles. Aerosols are "suspended" particles, both liquid and solid. Solid aerosols in the atmosphere are mixtures of dusts, smokes, particles of soot and fumes. Suspended particles reduce the visibility and air quality. The atmospheric visibility can be expressed as

$$L = \frac{1000}{C} \qquad \ldots(5.1)$$

where, $L$ = visibility in km;

$C$ = concentration $\left(\frac{\mu g}{m^3}\right)$

Aerosols absorb and scatter both solar and terrestrial radiations. Thus, they affect the heat-balance of the earth. In addition, they act as nuclei for the condensation of water vapor, and play an important role in cloud formation, by initiating and controlling precipitation. Back scattering of the solar radiation increases the reflectivity of the atmosphere. Aerosols also reduce the terrestrial scattering of infrared (IR) radiation into space and also absorb some solar radiation. The effect of aerosols on solar (visible) radiation is greater than they are on the infrared radiation from the earth to space. Therefore, there is a decrease in the atmospheric temperature with increase in aerosol content in the atmosphere. That is, the net effect of an increase in the particulate matter in the atmosphere is to cool the earth.

Another important aspect of aerosols that is relevant to ecology is their residence times in the atmosphere. Residence time of a substance is the length of time it resides in a particular location before it moves on through a particular process or cycle. Pollutants in the stratosphere and mesosphere remain there over a long period (months and years) (Table 5.2).

**TABLE 5.2:** RESIDENCE TIMES OF AEROSOLS IN THE ATMOSPHERE

| *Stratum* | *Residence Period* |
|---|---|
| Lower troposphere | ~10 days |
| Upper troposphere | ~30 days |
| Lower stratosphere | ~6 months |
| Upper stratosphere | 2-3 years |
| Mesosphere | 5-10 years |

## 5.3 GASEOUS AIR POLLUTANTS

Many of the gaseous air pollutants are small molecular mass inorganic and organic compounds.

There are primary pollutants, such as $SO_x$, $NO_x$; $CO_x$, $H_2S$, HF, and secondary pollutants, such as $H_2SO_4$, hydrocarbons, PAN, lead and arsenic compounds. Of these primary atmospheric pollutants, $SO_x$, $NO_x$, and $CO_x$ (criteria pollutants) are important.

### 5.3.1 Oxides of Sulfur ($SO_x$)

Of the oxides of sulfur, sulfur dioxide is a criteria and primary pollutant whose principal source is the combustion of sulfur containing fossil fuels. Smelting of sulfide-containing ores ($FeS_2$; PbS; HgS), manufacture of elemental sulfur and sulfuric acid, conversion of wood pulp to paper and incineration of refuse are other significant contributors for its release to the atmosphere. It is estimated that 50 per cent annual global emission of this pollutant is from coal burning and 25 to 30 per cent from oil burning. All the remaining sources, including natural ones like volcanoes and forest fires share the remaining 20 to 25 per cent. The contribution from mobile sources is small. Sulfur dioxide in the atmosphere is oxidized to sulfuric acid by hydroxyl radicals (OH) produced by photodecomposition of ozone ($O_3$):

$$\begin{aligned} HO + SO_2 + M &\rightarrow HSO_3 + M \\ HSO_3 + O_2 &\rightarrow SO_3 + HO_2 \\ SO_3 + H_2O &\rightarrow H_2SO_4 \end{aligned} \qquad \ldots(5.1)$$

Other oxidants like $NO_2$, reactive oxygen transients and peroxy radicals, are present in the atmosphere but they are believed to have a minor role in the oxidation of $SO_2$. $SO_2$, being water-soluble, it is also found in cloud droplets. At a pH less than 5, hydrogen peroxide ($H_2O_2$) is the main oxidant for its oxidation in the droplets:

$$SO_2 + H_2O_2 \rightarrow H_2SO_4 \qquad \ldots(5.2)$$

Both wet and dry deposition is responsible for damage to building materials, destruction of vegetation and in the degradation of soil. Sulfuric acid, which is the final oxidized product of $SO_2$, has low vapor pressure and condenses in the atmosphere to form aerosol particles. In turn, these aerosols contribute to visibility reduction. Along with other aerosols, they scatter part of the solar radiation back to space, thus making the climate cool down to some extent.

Major pollution problems with $SO_2$ occur now in cities in which coal is still widely used for domestic heating, in industry and power stations. As there is a tendency to locate power stations away from urban areas, the impact of $SO_2$ pollution can be felt in both rural and urban areas. Sulfate aerosols reach the stratosphere and take part in decreasing ozone levels. Major volcanic eruptions like that of El Chichón (1982) and of Mt. Pinatubo (1991) raised sulfuric acid particulates in the stratosphere by two to five orders of magnitude.

Lichens are very sensitive to $SO_2$ in the air. Lichens are widely used as bioindicators of $SO_2$ pollution (phyto-indicator). If there are no lichens present, the air quality is very poor.

Reduction of $SO_2$ is achieved generally by the use of low-sulfur fuels. A similar alternative is available for liquid fuels. In the case of coal, sulfur may be present partly in a chemically bound state and partly carried as the mineral pyrite, $FeS_2$. While the chemically bound sulfur cannot be removed before burning of coal, pretreatment of coal before feeding it to the furnace or boiler to reduce its sulfur content can eliminate the pyrite part. The cleaning process takes advantage of higher specific gravity of pyrite, compared to the coal and uses standard gravity separation methods.

One of the processes available for reducing $SO_2$ in the off gases from the burners involves the use of fluidized-bed combustion of coal. Magnesium oxide or limestone is added to the coal in the fluidized bed to neutralize $SO_x$, and to precipitate them as solids (refer to Chapter 27).

$$CaCO_3 + SO_2 + \tfrac{1}{2}O_2 \rightarrow CaSO_4 + CO_2$$
$$CaO + SO_2 + \tfrac{1}{2}O_2 \rightarrow CaSO_4 \qquad ...(5.3)$$

### 5.3.2 Oxides of Nitrogen ($NO_x$)

Nitrogen is the primary constituent of the atmospheric air. A major part of atmospheric nitrogen is in the molecular state, $NO_2$. In the biosphere, several types of bacteria convert nitrogen into various compounds. When organic substances decay in the biosphere, a considerable part of the nitrogen they contain transforms into ammonia ($NH_3$) and nitrous oxide ($N_2O$). Under the influence of the nitrifying bacteria in the soil they are oxidized to nitric acid.

$$N_2 + 3H_2 \rightarrow 2NH_3 + 92\ kJ \qquad ...(5.4)$$
$$4NH_3 + 5O_2 \rightarrow 4NO + 6H_2O + 910\ kJ$$
$$4NO_2 + O_2 + 2H_2O \rightarrow 4HNO_3$$

As part of the action of denitrifying bacteria, oxides of nitrogen are transformed to molecular nitrogen, $N_2$: ($NO_3 \rightarrow NO_2 \rightarrow N_2O \rightarrow N_2$). While the denitrifying bacteria liberate nitrogen in the free state, the nitrobacteria assimilate the atmospheric nitrogen. These conversions are summarized in the nitrogen cycle (Fig. 5.1). But, a major source of $NO_x$ pollution is anthropogenic.

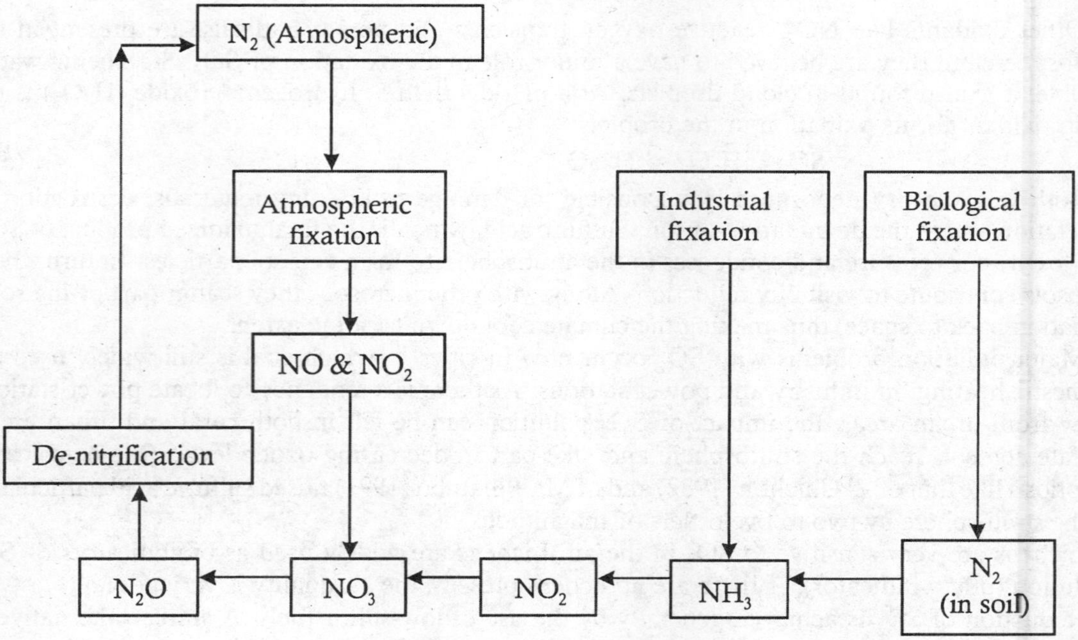


**Fig. 5.1:** The Nitrogen Cycle

Oxides of nitrogen are $N_2O$ (nitrous oxide), NO (nitric oxide) and $NO_2$ (nitrogen dioxide). NO and $NO_2$ are grouped as $NO_x$, a criteria pollutant. While $N_2O$ and NO are primary pollutants, $NO_2$ is a secondary pollutant.

#### *5.3.2.1 Nitrous Oxide ($N_2O$)*

Nitrous oxide is released in the troposphere by bacteria. Being relatively unreactive its concentration

(0.25 per cent) and residence time (4 years), in the unpolluted air of the troposphere are distinctly higher than those of nitric oxide (0.2-2.0 × $10^{-3}$ per cent and 5 days) and of nitrogen dioxide (0.5-4.0 × $10^{-3}$ per cent and several days). $N_2O$ molecules do not photolyze or react in the troposphere and thus, can reach the stratosphere, where they react with electronically excited oxygen atoms $O(^1D)$, to form reactive nitric oxide, NO.

$$N_2O + O(^1D) \rightarrow 2NO \quad ...(5.5)$$

### 5.3.2.2 *Nitric Oxide (NO), and Nitrogen Dioxide ($NO_2$)*

The two oxides of nitrogen, NO and $NO_2$, are usually considered together as $NO_x$. Nitric oxide is formed in the combustion of organic material, including coal and gasoline. Part of the nitrogen needed for the formation of $NO_x$ comes from the nitrogen compounds present in the fuel. Depending on the combustion conditions 18 to 80 per cent of the original nitrogen is oxidized. The rest of the nitrogen for oxidation in and around the flame is derived from the atmosphere. High flame temperatures facilitate formation of reactive oxygen and nitrogen atoms from their molecules and hydroxyl radicals, HO. All three can produce NO through reactions of the type

$$\begin{aligned} N_2 + O^* &\rightarrow NO + N^* - 75 \text{ kcal/mol} \\ O_2 + N^* &\rightarrow NO + O + 31.8 \text{ kcal/mol} \quad ...(5.6) \\ N^* + HO &\rightarrow NO + H + 39.4 \text{ kcal/mol} \end{aligned}$$

On entering the atmosphere NO slowly turns into $NO_2$ through a series of reactions in the photochemical smog. The net result is

$$NO + \tfrac{1}{2}O_2 \rightarrow NO_2 \quad ...(5.7)$$

On the other hand, oxidation of NO by ozone is much faster

$$NO + O_3 \rightarrow NO_2 + O_2 \quad ...(5.8)$$

$NO_2$ is a pungent, irritating gas, which absorbs in the green region of light and therefore displays a reddish-brown color. Of the several oxides of nitrogen $NO_2$ is of most concern as pollutant. It has a variety of environmental and health impacts. It is known to cause pulmonary edema, an accumulation of excessive fluid in the lungs. It can exacerbate asthma and increase susceptibility to infection.

$NO_x$ plays a complex role in the atmosphere. In the troposphere $NO_x$ helps formation of ozone but in the stratosphere leads to the depletion of the same substance. Both are undesirable phenomena. $NO_2$ reacts with hydroxyl radicals in the gas phase to form nitric acid, which easily enters aerosols, finally leading to acid rain

$$HO + NO_2 + (M) \rightarrow HONO_2 + (M) \quad ...(5.9)$$

### 5.3.2.3 *Photochemical Smog*

In the presence of sunlight $NO_2$ reacts with hydrocarbons to produce photochemical pollutants like ozone. Thus, it plays a role in the formation of photochemical smog, a reddish-brown haze that has been experienced by many urban areas before strict pollution control measures were implemented. Though hydrocarbons are not, generally, toxic in low concentration, their participation in smog formation and the irritant properties of their reaction products call for controlling their concentration in the atmosphere. Some of the polynuclear aromatic hydrocarbons (PAHs) are well-recognized carcinogens.

In photochemistry, the initiation step of the reaction of a photon (in the UV-visible region) by an atom, a molecule or a radical, that leads to its subsequent ionization and dissociation. The initial step of photon absorption and dissociation is called the 'primary' photochemical reaction. Subsequent reactions caused by the primary products are called the 'secondary' photochemical reactions. For a species, $A$,

$$A + h\nu \rightarrow A^* \text{ (initiation, "excitation" step)}$$

$$A^* \rightarrow D1 + D2 \text{ (dissociation)} \qquad \text{...(5.10)}$$

where, $A^*$ = species in the excited state; $D1$ & $D2$ are products

It is the combination of light of proper energy (UV-visible), duration, and availability of reactants and stability of the atmosphere that produce photochemical oxidants.

Photochemical smog is a reaction product of $NO_x$ and hydrocarbons in the atmosphere. Neither smoke nor fog is required for photochemical smog formation. Oxides of nitrogen are greatly responsible for photochemical smog formation. The sources are microbial processes, power generation plants, nitric acid factories and automobiles. Microbial processes in soils produce >2/3 of $N_2O$ that enters the atmosphere. $N_2O$ is a good absorber of IR radiation, and thus is a "greenhouse" gas. On a molecular basis, it is ~100 times as effective in this regard as $CO_2$. In the stratosphere it is converted to nitric oxide (NO) and the NO generated from the $N_2O$ interacts with and destroys $O_3$ in the stratosphere allowing more UV radiation to penetrate the atmosphere and reach the earth, leading to skin damage and cancer and detrimental effects on crops and vegetation.

$NO_x$ abstracts energy in the UV radiation range <430 nm from sunlight and dissociated to give NO and reactive oxygen atom ($O^*$). The reactive species initiates further reactions. In a series of steps involving $O_3$ with $NO_x$, HCs and sunlight, photochemical smog is formed.

$$NO_2 + h\nu\ (< 400\text{ nm}) \rightarrow NO + O^*\ \text{(atomic)}$$

$$O^* + O_2 + M \rightarrow O_3 + M \qquad \text{...(5.11)}$$

$$O_3 + NO \rightarrow NO_2 + O_2$$

Net reaction: $$O_3 + NO \rightarrow O_2 + NO_2$$

where, $M$ is an energy absorbing species, generally a hydrocarbon.

Nitric oxide (NO) participates in reactions in the troposphere that result in $O_3$ formation. Smog reduces solar energy input considerably at wavelength ~323 nm, corresponding to the absorption wavelength of $NO_2$ (Fig. 5.2). The dissociation of $NO_2 \rightarrow (NO + O^*)$ requires ~300 kJ (~400 nm).

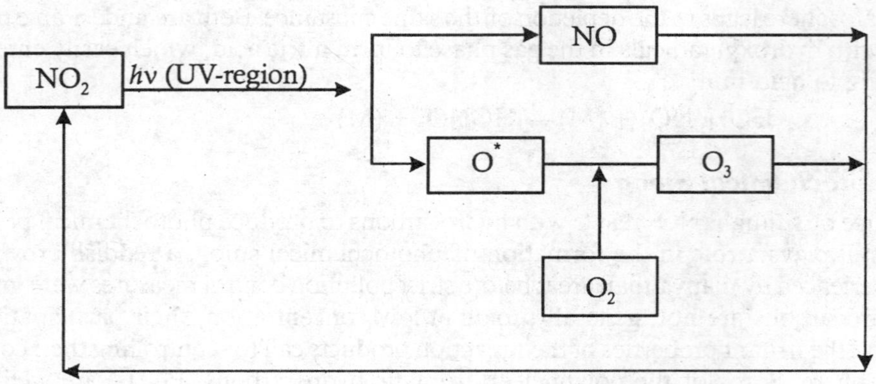


**Fig. 5.2:** Photolytic Cycle of $NO_x$

$NO_x$ contribute to ground-level ozone formation and can have adverse effects on both terrestrial and aquatic ecosystems (e.g. photochemical smog, acid rain and eutrophication of water bodies).

Ozone, which is a powerful oxidant, in turn, reacts with hydrocarbons in the atmosphere to form a variety of organic compounds that are responsible for photochemical smog. The formation of photochemical smog (in the troposphere) is a dynamic process (Fig. 5.3). Weather and geography determine the severity of smog. When temperature inversions occur and winds are calm, smog may stay in a place for days at a time, forming a pollutant air blanket over the place.

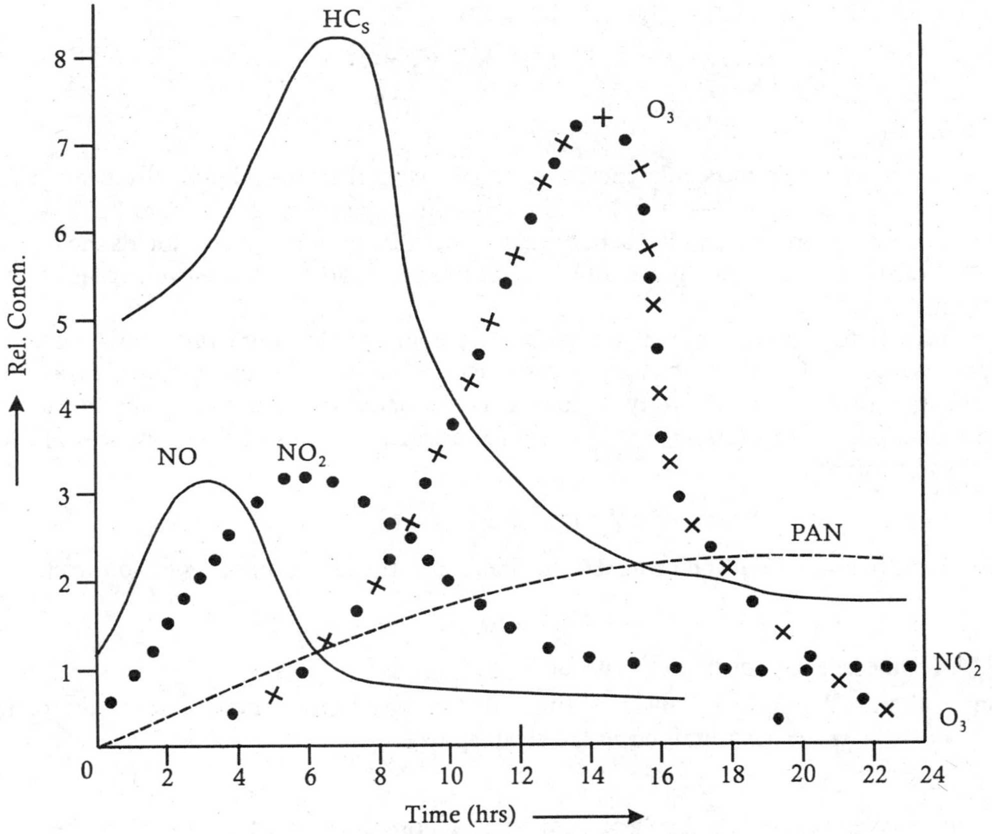


**Fig. 5.3:** Dynamic Process of Photochemical Smog

(After: Peirce, J. J., *et al.* (1998) "Environmental Pollution and Control", 4th edn., New Delhi: Butterworth-Heineman.)

It appears that natural and anthropogenic sources contribute almost equally for the release of $NO_x$ in the atmosphere. Among the natural sources a major part arises from lightning (about 30 trillion grams nitrogen per year), while only a small contribution comes from soils, ammonia oxidation and transport downward from the stratosphere. The anthropogenic contribution of $NO_x$ (estimated as about 34 trillion grams) nitrogen is almost exclusively from burning of fossil fuel and biomass, in power stations, heating plants, industrial processes or mobile vehicles. Urban areas where traffic is the heaviest experience greatest concentrations of NO and $NO_2$, as compared to rural areas. However, $NO_x^-$ triggered ozone formation could be higher in rural areas in contrast to urban locations due to circumstances. Some of these are: (1) $NO_2$ facilitates ozone formation while NO tends to destroy it; (2) due to a high level of NO emitted by the vehicles in the urban areas conditions are not conducive for formation of ozone; (3) by the time the emissions find their way from urban to rural areas NO would have been converted to $NO_2$, which enhances ozone formation in conjunction with the hydrocarbons (also present in vehicular emissions).

Abatement of $NO_x$ pollution involves the use of alternate sources of less-polluting fuels, such as natural gas in automobiles and catalytic conversion (oxidation) of $NO_x$ in exhaust gases (refer to Chapter 27).

$$4NO + 4NH_3 + O_2 \rightarrow 4N_2 + 6H_2O \qquad ...(5.12)$$

$$2NO_2 + 4NH_3 + O_2 \rightarrow 3N_2 + 6H_2O$$

### 5.3.3 Ozone ($O_3$)

Ozone ($O_3$) is a triatomic molecule containing three oxygen atoms (O—O distance = 0.128 nm) with $sp^3$ hybridization (O—O—O = 116.5°). It occurs in significant amounts (>10 ppm) in the lower stratosphere, produced by photochemical reaction. Its presence there shields the living beings on the Earth from the biologically harmful Sun's ultraviolet radiation at wavelengths between 200 and 320 nm.

Ozone in the troposphere is contributed by two sources: (1) downward movement from the stratosphere and (2) direct photochemical production within the troposphere. Ultraviolet (UV) radiation, at <180 nm, is strongly absorbed by molecular oxygen, $O_2$, in the ionosphere (>180 km). Above 50 km (60-80 km), molecular oxygen, $O_2$, absorbs energy at <240 nm, and dissociates to form atomic oxygen, $O^*$.

$$O_2 \xrightarrow{h\nu\ (240\ nm)} O^* + O^* \text{ (above 50 km)} \qquad ...(5.13)$$

Molecular oxygen ($O_2$) in the upper stratosphere absorbs UV radiation (<240 nm) to form $O_3$.

$$O_2 + O + M \xrightarrow{h\nu\ (<240\ nm)} O_3 + M \qquad ...(5.14)$$

where, $M$ is some energy-absorbing molecule

Ozone is thermodynamically unstable and is decomposed photochemically (in the region 220-330 nm) and a balance is maintained in the stratosphere.

$$O_3 \xleftrightarrow{h\nu} O_2 + O^* \qquad ...(5.15)$$

The ozone layer (in the mesosphere-stratosphere boundary) absorbs much of the ultraviolet (UV) rays of the solar radiation (strongly between 300-210 nm). The absorption of UV radiation by $O_3$ (and $O^*$) at the higher regions of the stratosphere provides a protective blanket against the UV rays reaching the earth and causing deleterious effects on the living organisms on the earth.

Depletion of ozone layer by scavengers like chlorofluorocarbons (halons), and engine discharge from supersonic airplanes in the stratosphere would destroy this protective cover, thereby increasing the incidence of UV radiation reaching the troposphere, which is hazardous to living organisms. The relative concentration of $NO_x$ and $O_3$ determine whether the "destruction" or "generation" of ozone takes place in a certain polluted atmosphere. $NO_x$, which is responsible for build up of ozone in the troposphere leads, in fact, to its destruction in the stratosphere, thus interfering with the protection from UV radiation. Mixing ratios of $NO_x$ generally exceed the critical value for ozone production in urban and many rural areas where pollution levels are high.

Ground-level ozone is not emitted directly into the atmosphere, but is a secondary pollutant produced (photochemically) by reaction between $NO_2$, HCs and sunlight (eqn. 5.11). Since $O_3$

itself is photodissociated to form free radicals, it catalyzes its own formation (autocatalysis). Consequently, high levels of $O_3$ are generally observed during hot, still sunny, summertime weather in locations where the air mass has previously collected emissions of HCs and $NO_x$ (e.g. urban areas with vehicular traffic). Ozone destroys materials such as rubber (a qualitative test for $O_3$). It affects bronchial function and is toxic to plants and vegetation.

The relative weakness of the O—O bond (143 kJ/mol) in the ozone molecule, compared to that in the oxygen molecule (498 kJ/mol) is responsible for many oxidation reactions initiated by ozone. Such reactions damage biological tissues. Therefore, the health of living beings, vegetation and natural systems is adversely affected by exposure to high levels of ozone. An additional problem created by ozone lies in its causing the production of many other oxidants, which turn out to be more harmful than ozone, itself. These toxic substances include hydroxyl radical (HO), hydroperoxyl radical ($HO_2$) hydrogen peroxide ($H_2O_2$), peroxy nitric acid ($HO_2NO_2$), organic peroxides (ROOR), peroxy acids, R((O)OOH), and peroxyacetyl nitrates (PANs).

### 5.3.4 Chlorofluorocarbons (CFCs)

Chlorofluorocarbons (CFCs) are family of inert, nontoxic, and easily liquefied chemicals. They do not occur in nature; they are industrially produced chemicals. Industries that produce aluminum expel fluoride dust, and plants that produce plastic foams are a major source of compounds of chlorine, fluorine and carbonaceous organic compounds (chlorofluorocarbons (CFCs)). Because CFCs are not destroyed in the lower atmosphere they drift into the upper atmosphere where their chlorine components destroy ozone, and they pose a serious threat to depletion of $O_3$ and NO. At altitudes above 25 km, fluorocarbons undergo photolysis and produce oxidants, free radicals and other secondary pollutants—

$$\begin{aligned} CF_2Cl_3 &\xleftrightarrow{h\nu} CF_2Cl_2 + Cl^* \\ CF_2Cl_2 &\xleftrightarrow{h\nu} CF_2Cl + Cl^* \qquad \ldots(5.16) \\ Cl^* + O_3 &\rightarrow ClO + O_2 \end{aligned}$$

Species having reactive chlorine and bromine in their molecule take part in such reactions, in addition to the already known $NO_x$ and reactive hydrogen species. The main source gases for the chloro-species are the chlorofluorocarbons, methyl chloroform ($CH_3CCl_3$), carbon tetrachloride ($CCl_4$), methyl chloride ($CH_3Cl$) and hydrochlorofluorocarbons (HCFCs) designed to replace the CFCs.

CFCs, such as Freon, do not readily undergo chemical reactions. After CFCs are released into the atmosphere, they diffuse into the ozonosphere (a region in the upper atmosphere between 10-50 km in altitude), and the Sun's ultraviolet radiation breaks them apart. The chlorine atoms from the chlorocarbons are stripped and are incorporated into inorganic reactive species, collectively called $Cl_y$ (= HCl + $ClONO_2$ + HOCl + ClO + $2Cl_2O_2$ + OClO + BrCl + Cl (radical)). Similarly, the bromine containing hydrocarbons are broken and $Br_y$ species consisting of HBr + $BrONO_2$ + HOBr + BrO + BrCl + Br are formed. From the fluorocarbons fluorine atom is liberated but it is fixed as HF, which does not participate in further chemical reactions because of its stability. Even minor quantities of sulfur hexafluoride ($SF_6$), hexafluoroethane ($C_2F_6$) and carbon tetrafluoride ($CF_4$), having life times of the order of thousands of years in the stratosphere, pose a serious environmental threat. Even a relatively small decrease in the stratospheric ozone layer (*ozone hole*) can result in an increased incidence of skin cancer in humans and genetic damage in many organisms.

All CFCs with an ozone depletion level less than 0.2 are classified as class II substances, and those chemicals with ozone depletion potential 0.2 or higher are classified as class I substances (CFCs, HBFCs, halons, carbon tetrachloride, methyl chloroform and ethyl bromide).

Atomic oxygen, $O^*$, can also react with $NO_x$ and halogen 'free radicals', introduced into the stratosphere. Destruction of the stratospheric ozone layer ('ozone hole') is due to scavengers of atomic oxygen, $O^*$, that results in the lack of recombination to maintain ozone equilibrium (Eqns. 5.11 and 5.14). The scavengers of $O^*$ are NO, Cl, OH 'free radicals' and chlorofluorocarbons (CFCs). CFCs (e.g. $CF_2Cl_2$ = dichlorodifluoromethane—Freon-12) are particularly good scavengers of atomic oxygen, $O^*$, that would result in the depletion of protective ozone layer, which would lead to far-reaching ecological consequences. NO is emitted by supersonic aircraft. Chlorine, Cl, results from the petrochemical breakdown of CFC aerosols,

$$Cl + O_3 \rightarrow ClO + O_2 \qquad \text{...(5.17)}$$

$$ClO + O_2 \rightarrow Cl + O_2$$

### 5.3.5 Oxides of Carbon ($CO_x$)

Oxides of carbon are carbon monoxide (CO) and carbon dioxide ($CO_2$).

#### *5.3.5.1 Carbon Monoxide (CO)*

Carbon monoxide, CO, is colorless and odorless, but asphyxiating gas with physiological effects (see Chapter 6). It is emitted into the atmosphere from natural as well as anthropogenic sources. Among the natural sources are the oxidation of biogenic methane and other hydrocarbons, natural wild fires and oceans. The anthropogenic sources are the combustion of fossil fuels, biomass, anthropogenic hydrocarbons and wood. When the materials mentioned are burned in an atmosphere of restricted oxygen, carbon monoxide is generated along with carbon dioxide, which is the main product.

$$2C + O_2 \rightarrow 2CO \quad \text{(incomplete combustion)}$$

$$C + O_2 \rightarrow CO_2 \quad \text{(complete combustion)} \qquad \text{...(5.18)}$$

In urban areas CO is produced almost entirely (90 per cent) from road traffic emissions and it is the most abundant constituent of them. It has a long residence time (about 3 years) in the atmosphere. Hydroxyl radical is the only species known to react with CO in the troposphere.

$$CO + HO \rightarrow CO_2 + H \qquad \text{...(5.19)}$$

While carbon monoxide (CO) is produced by incomplete combustion of carbonaceous materials, oxides of nitrogen ($NO_x$) are produced in all combustion processes. The most important difference between spark ignition (petrol) and compression ignition (diesel) engines is that the diesel engine operates with excess air. Thus, the combustion in such system is generally more complete and hence emissions of CO and unburned HCs are lower. But, it emits large quantities of particulate matter. In contrast, internal combustion engines, using petrol fuels, produce large amounts of hydrocarbons. In all fuel injection/ignition systems, the effect of equivalence ratio $\phi$ (air/fuel ratio) is important. It has a marked effect on the emission of hydrocarbons (Table 5.3). In processes where fossil fuels are burnt (in automobiles, petrochemical plants), there are several kinetically connected steps and the combustion does not reach an equilibrium state. While production of CO is reduced in fuel-lean (oxygen-rich, $\phi < 1$) and high temperature combination, generation of NO is thermodynamically and kinetically favored. Low-excess air (fuel-rich, $\phi > 1$) firing results in incomplete fuel burnout, resulting in the emission of hydrocarbons, soot and CO. Reduction of ignition temperature also reduces the efficiency of the engine. That is, attempts at reducing one type of pollutant results in enhancing the other type of pollutant (Fig. 5.4).

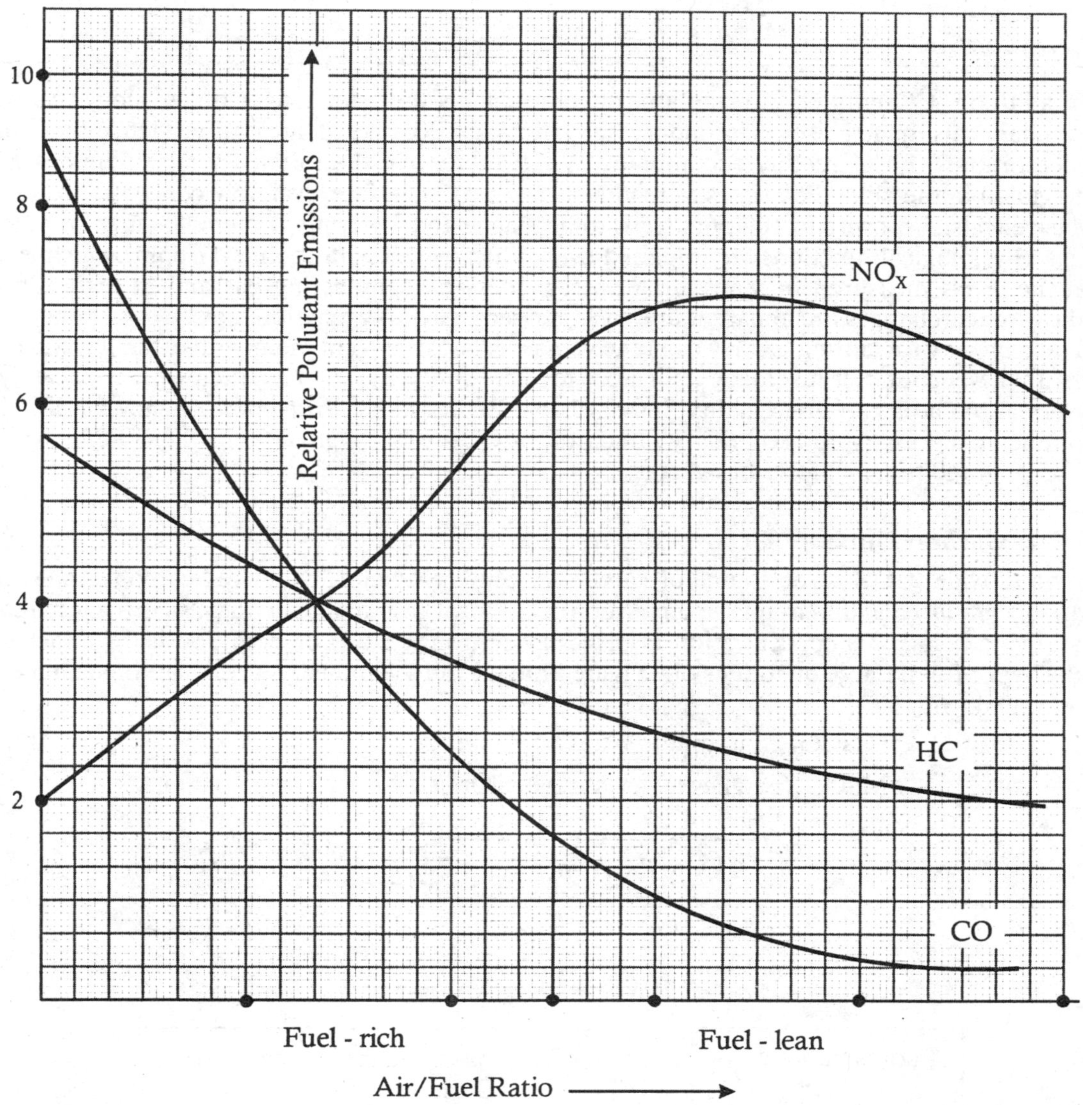


**Fig. 5.4:** Effect of Air/Fuel Ratio on Pollutant Emission from an Internal Combustion Engine

**TABLE 5.3:** AUTOMOBILE EXHAUST-GAS COMPOSITION

| *Mode of Operation* | *$NO_x$ (ppm)* | *CO (ppm)* | *$CO_2$ (ppm)* | *Hydrocarbons (ppm)* |
|---|---|---|---|---|
| Idle | 30 | 500 | 10 | 800 |
| Cruise | 1,500 | 1 | 13 | 300 |
| Acceleration | 3,000 | 5 | 10 | 400 |
| Deceleration | 60 | 4 | 10 | 4,000 |

In addition, CO emitted into the atmosphere remains there for longer times and reacts with other chemicals, giving rise to secondary air pollutants. Because of its long residence time local concentrations of CO readily accumulate. When present in the inhaled air it combines with hemoglobin in the blood 250 times as readily as oxygen. Therefore, even low concentrations have adverse effects. Exposure to 100 ppm leads to headache and lowering of mental acuity. In human beings physiological functions are impaired even when its concentration is 17 ppm, which is less than the levels measured in many urban areas. It is an asphyxiating gas and at higher levels cardiovascular changes are apparent due to decreased tissue oxygenation.

The only way for cutting CO levels in the vehicle exhausts is to fit them with catalytic oxidation units. In many countries it is made obligatory for transport vehicles to be fitted with such catalytic converters. As a part of Air Quality Standards the Environment Protection Agency in USA has recommended primary and secondary standards for CO levels: 9 ppm (10 mg/m$^3$) for 8 hr average and 35 ppm (40 mg/m$^3$) for 1 hr-average.

### *5.3.5.2 Carbon Dioxide ($CO_2$)*

Carbon dioxide ($CO_2$) is the source of carbon for photosynthetically derived carbohydrates. It is invariably produced in the complete combustion of carbonaceous materials and also in the respiration of all living beings. Its concentration in places away from centers of pollution is 0.032 per cent (water vapor free). Large quantities of $CO_2$ are produced in the industry where carbonate minerals are calcified (Fig. 5.5).

$$CaCO_3\text{(limestone)} \rightarrow CaO + CO_2\uparrow \quad ...(5.20)$$

$$MgCO_3\text{(magnesite)} \rightarrow MgO + CO_2\uparrow$$

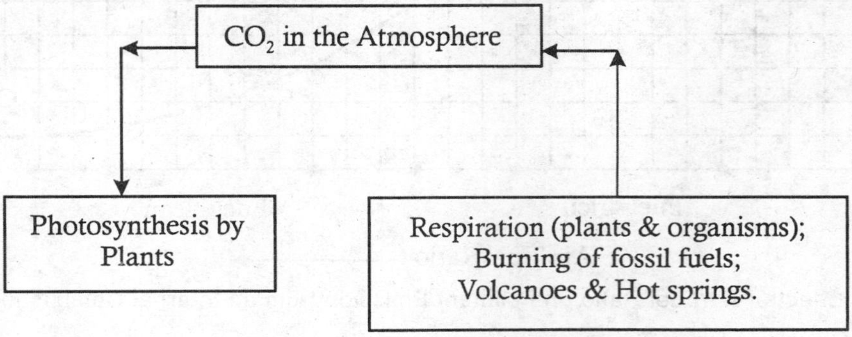

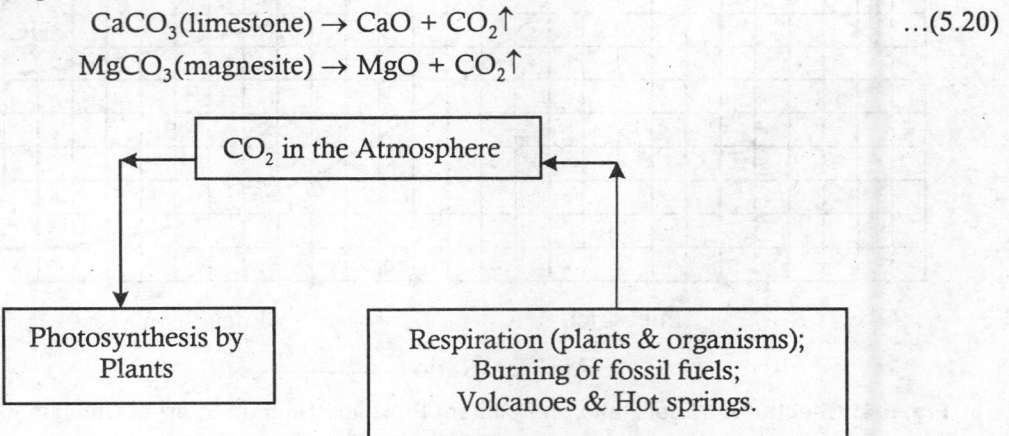

**Fig. 5.5:** Carbon Dioxide Cycle

Carbon dioxide is not a pollutant in the usual sense of the word. In fact, it is essential for plant life by directly taking part in the process of photosynthesis. However, build up of $CO_2$ concentration in the atmosphere can lead to a number of ecological and climatic problems. Carbon dioxide has long-term global effects due to its absorption characteristics. $CO_2$ absorbs infrared radiation at 1.5–15 μm range, a region where the energy that is re-radiated by the earth into space is concentrated. This absorption effectively reduces the heat loss from the earth and causes temperature in the troposphere to increase. This is termed as the '*greenhouse*' effect.

## 5.4 THERMAL POLLUTION

Artificial heating of the environment is termed thermal pollution. Thermal pollution (heat stress) can have both immediate and long-term effects. Thermal pollution has effects on all processes and systems and has pervading consequences, locally, regionally and globally. Natural causes of this type of pollution are volcanic eruptions and man-made causes are urban phenomena and industrialization.

Thermal pollution poses a serious problem that affects all physical, chemical and biological processes and the metabolism of organisms. A rise of 10° C approximately doubles the rate of many chemical reactions and drastically disrupts thermodynamical phase equilibriums. Biochemical reactions are temperature dependent, with exponential increase in the reaction rate with temperature.

$$\text{Reaction rate} = K(T)\ (\text{concentration-dependent term}) \quad \ldots(5.21)$$

$$Log K \propto \frac{1}{T} \quad \ldots(5.22)$$

$$K = A \exp\left(-\frac{E_a}{RT}\right) \quad \ldots(5.23)$$

$$\frac{d(\ln K)}{dt} = \frac{\Delta H}{RT^2} \quad \textit{Clausius-Clapeyron}\ \text{equation} \quad \ldots(5.24)$$

where, $K$ = rate constant;
$A$ = pre-exponential term;
$E_a$ = activation energy;
$R$ = gas constant;
$T$ = absolute temperature

### 5.4.1 Effects on Organisms

Biological reactions are critically dependent on the temperature variations. One of the ways of adjusting to temperature variations is acclimatization. Plant cells can withstand temperatures up to ~ –15° C. Frosts with temperature below –15° C can damage flora irrevocably. Variation in the mean temperatures for different geographical regions, either due to natural or other causes would adversely affect the plant growth and consequently can create irrevocable ecological imbalance.

Cold-blooded organisms are sensitive to temperature changes in their habitats. The metabolism rate increases with temperature. Speeding up metabolic rate, due to increase in temperature, puts a greater demand on respiration rate (oxygen consumption). Increased oxygen consumption at elevated temperatures is compounded by the fact that the dissolved oxygen (DO) is inversely proportional to temperature. The decrease in dissolved oxygen has adverse effects on all organisms, in particular on aquatic organisms, which undergo thermal stress—impairment of reproductive processes and face threat to survival.

Higher temperatures are deleterious to catalytic performance of enzymes and biological processes and growth. Biological molecules (proteins, nucleic acids and biological membranes) get denatured above a critical temperature, resulting in loss of function or leading to abnormal functioning. Fluidity of biological membranes is temperature-dependent. Increase in membrane fluidity results in the loss of membrane compartmentalization and cellular functions.

At the cellular level, harmful damage to the body parts (erythema, diathermy, etc.) and cancer are some of the effects. All organisms suffer under temperature stress. As all biological functions and survival of organisms are temperature-dependent, regional thermal variations influence their functions profoundly.

### 5.4.2 Effects on Flora

The most important chemical process on this globe is the conversion of atmospheric $CO_2$ and water into carbohydrates by chlorophyll-containing vegetation. The energy required for carrying out photosynthesis is derived from the solar radiation. The pigment molecules that absorb sunlight (chlorophylls, carotenes, and phycobilins) have specific wavelength absorptivity. Chlorophyll, Chl *a*, has two absorption bands, at $\lambda = 420$ and 660 nm. In plants, the chromoprotein, phytochrome, is in physiologically active form at $\lambda = 660$ nm, and promotes germination and leaf growth. It is interconverted to an inactive form if plants are exposed to radiation at $\lambda = 730$ nm.

$$\text{Phytochrome (Pr)} \xrightarrow{\lambda=730nm} \text{Phytochrome} \qquad \ldots(5.25)$$

$$\text{(active form)} \xleftarrow{\lambda=630\text{nm}} \text{(inactive form)}$$

The ratio of radiation, $\lambda_{630}/\lambda_{730}$ determines the plant growth response. In fact, the spectral responses of organisms—photomorphogenesis, phototaxis and vision are confined to the 350-750 nm range. The photosynthetic organelles, containing chlorophylls, absorb radiation mainly in the blue and red regions (therefore, leaves look green). Green vegetation absorbs at $\lambda = 440$, 620 and 720 nm.

$$2H_2O + CO_2 \xrightarrow{hv} \{CHO\} + H_2O + H_2O + O_2 \qquad \ldots(5.26)$$

$$H_2D + A \xrightarrow{hv} H_2A + D \qquad \text{(Green equation)}$$

where, $A$ = acceptor;

$D$ = donor; {CHO} = carbohydrate.

Higher plants can also use nitrate as an acceptor.

$$9H_2O + 2NO_3^- \xrightarrow{hv} 2NH_3 + 6H_2O + 9O \qquad \ldots(5.27)$$

The well-being and survival of fauna depend on the well-being of floral ecosystems. Higher temperature drastically disrupts the plant growth response. Deterioration of vegetation and water bodies and subsequent degradation of soil would have drastic adverse effects on vegetation and food production. These would have direct detrimental consequences on animal and human lives.

### 5.4.3 Global Impact of Thermal Pollution

Sun is the primary source of energy for our planet, Earth. The irradiance of the sun, type of soil, water bodies, vegetation, and seasonal and diurnal variations control the temperature of the Earth's surface. Natural causes for thermal variations regionally and globally are volcanic eruptions. Human-related activities, such as coal/oil-based combustion, locomotives and power plants, have now become significant contributors to atmospheric thermal pollution.

Atmospheric characteristics due to pollution—reflectivity, smog formation, mode of water vapor precipitation (cloud-rain formation)—all add to global warming and possibly to other global ecological abnormalities like melting of polar ice-caps, polluted and corrosive air and acid rain. These can make the survival of organisms more and more stressful.

The Earth receives solar radiation at short-wavelength region (in the UV-visible region), and in turn re-radiates toward space about 30 per cent of the energy coming from the Sun. The atmosphere absorbs about another 30 per cent while the remaining (about 40 per cent) radiation finds its way to the Earth's surface. The Earth reflects about 15 per cent of the solar energy toward space at effective 'blackbody' temperature of 255 K, in the 3-80 μm region (infrared region), with a peak at 12 μm (in the far infrared region). *Thus, the Earth's atmosphere is a selective absorber—transparent to all short-wavelength radiations (150-500 nm) and is practically opaque to long-wavelength radiations (7 and 14 μm).* The atmosphere loses heat into space directly through the 'transparent window' between 8-11 μm. Radiation re-radiated by the earth is absorbed by $CO_2$ in the 2.7, 4.3 and 12-18 mm bands) and water vapor in the far infrared region. Major sinks for $CO_2$ are terrestrial and aquatic vegetation during photosynthesis. Increase in $CO_2$ and water vapor in the atmosphere leads to the 'greenhouse' effect, by blocking the dissipation of thermal radiation into space, thus leading to global warming.

The increase in $CO_2$ in the atmosphere does not much affect the solar radiation reaching the earth, because $CO_2$ is essentially transparent to UV-visible radiation regions. However, the radiation emanating from the earth is all in the infrared region, where absorption bands of $CO_2$ lie. Therefore, a significant increase in $CO_2$ tends to increase the temperature of the lower troposphere by trapping some radiation (heat) that would otherwise escape into outer space (Fig. 5.6). This atmospheric phenomenon is called the 'greenhouse' effect, though the label is a misnomer. Glass is also transparent to short wavelength solar radiation. But, unlike $CO_2$, glass absorbs long-wavelength radiation emitted from inside the greenhouse. However, glass also reduces cooling of the plants by convection by blocking the outside air. This is the dominant effect (the effect similar to the condition of people sitting in a car with windows closed).

Maintenance of $CO_2$ balance is very critical in maintaining the heat-balance of the earth. It is calculated that if the natural 'greenhouse' effect were not there the Earth's average temperature would be ~33° C less than what is now. There would not have been liquid water in many places on the Earth. While it appears the greenhouse gases are serving a useful purpose of maintaining Earth's energy balance, a problem is developing in a different direction. Excessive burning of fuels and industrial activities that would release large amount of $CO_2$ into the atmosphere has amplified global implications on the temperature-balance of the earth (Fig. 5.6). The consequences such global warming result in melting of polar ice caps and mountain glaciers, resulting in higher coastal waters, patterns of extreme drought and rain fall in some regions of the world and serious disruption to food production.

Carbon-based energy sources are wood, peat and coal. These not only produce $CO_2$ but also are less Eco-friendly sources. Abatement of $CO_2$ build up in the atmosphere is possible only by drastically cutting down the burning of fossil fuel. This means, in turn, developing and using on a large-scale alternate sources of energy—nuclear, solar, wind, hydrogen and geothermal. However, the implied problems for such a change are many and they include availability of viable options, maturing of technology and economics. Natural gas (methane) is an excellent alternative source of fuel. It creates less pollution, easy to transport through pipelines and has better heating value (= energy production per kilogram fuel combustion), compared to other carbonaceous fuels (Table 5.4).

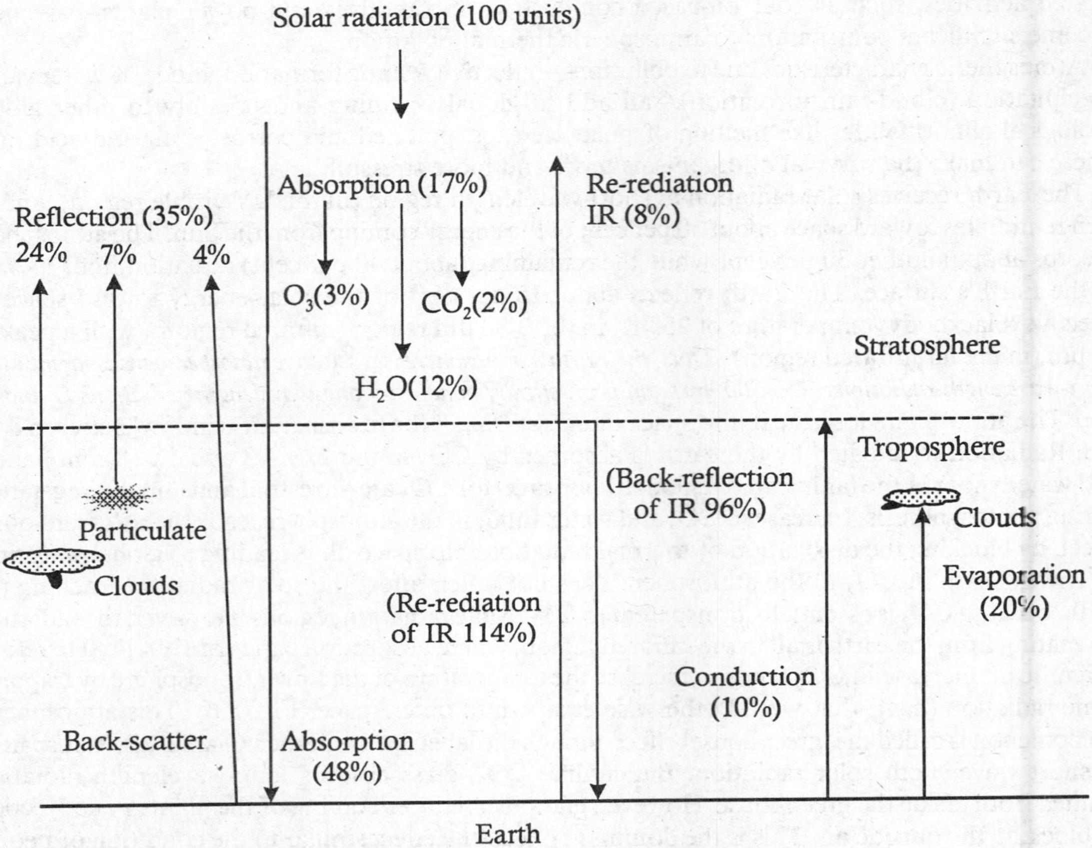


**Fig. 5.6:** Atmosphere-Earth Energy Balance

**TABLE 5.4:** HEATING VALUES OF SOME CARBONACEOUS FUELS

| *Fuel* | *Heating Value (kJ/kg)* | *Fuel* | *Heating Value (kJ/kg)* |
|---|---|---|---|
| Wood | 19,000 | Peat | 23,000 |
| Coal (lignite) | 28,000 | Coal (bituminous; anthracite) | 35,000 |
| Petroleum | 44,000 | Natural gas (methane) | 50,000 |

Water vapor in the atmosphere can produce a greenhouse effect, more so than $CO_2$. Water absorbs infrared radiation <10 μm and acts as the long-wavelength radiation balance of the earth. Clouds reflect sunlight and they also trap infrared radiation. Pollution in these regions alter the reflecting power of the clouds and hence the disruption in the temperature distribution in the atmosphere. Supersonic jet transport has increased the cloudiness in the stratosphere. The long-term effects of supersonic transport, 'smog', formed by interaction of hydrocarbons with ozone and alteration of reflectivity of the clouds are some of the global effects whose consequences are still not thoroughly investigated.

Increase in particulate concentration in the atmosphere tends to reduce the amount of solar radiation reaching the earth, due to absorption and back-scattering of incident radiation (e.g. volcanic eruptions). Such events have the effect of cooling the Earth, an effect opposite to the greenhouse effect.

A list of environmental pollutants and their effects is given in Table 5.5.

**TABLE 5.5:** ENVIRONMENTAL POLLUTANTS AND THEIR EFFECTS

| *Pollutant* | *Source* | *Properties & Uses* | *Effects* | *Abatement* |
|---|---|---|---|---|
| Acids & alkalis | Chemical & Industrial | Acidic & alkaline. Industrial, bleaching and cleaning | Corrosive; damage to skin | Wet-scrubbing |
| Aerosols | Flyash; soot; smoke; dust; metal vapors | Particulate | Atmospheric smog; health hazards | Cyclones; bag hoses; scrubbers |
| Alcohols | Chemical & Industrial & commercial | Industrial, medical and domestic | Neurotoxic; impaired coordination | Chemical treatment & wet-scrubbers |
| Antimony (Sb) | Metallurgy; ($SbS_3$) | Industrial; alloys | Toxic; diarrhea; $SbH_3$ poisonous | Scrubbers; bag collectors |
| Arsenic (As) | Mining; chemical waste | Glass & paint industry; insecticide | Highly toxic; pulmonary edema; diarrhea | Air cleaning devices; bag houses |
| Asbestos | Mining; | Particulate & fibrous silicates; building construction | Asbestosis; lung cancer; silicosis; pleural calcification | Ventilation & wet processes |
| Beryllium (Be) | Mining; Beryl ($Be_3Al_2(SiO_3)_6$; nuclear power | Used in gas turbines & optical devices | Toxicant; berylliosis; acute pnemonitis | Scrubbers; wet cyclones; bag collectors; electrostatic precipitators |
| Barium (Ba) & barium salts | Mining | Silvery-white metal | Toxic; abdominal pain; loose motions; hypokalemia | Solid removal methods |
| Boron (B) | Metallurgy; ($HbO_3$); coal burning | Borone fuel; refractory material | Highly toxic | Elimination of boron additives |
| Cadmimum (Cd) | Mining (Zinc ores) & Metallurgy | Metal used in electroplating, dyeing & glass manufacturing | Highly toxic; kidney damage | Bag filters; cyclones |
| Carbon dioxide ($CO_2$) | Burning of carbonaceous compounds uses | Colorless gas. Industrial & commercial | Toxic | Ventilation; use of alternate energy sources |

| *Pollutant* | *Source* | *Properties & Uses* | *Effects* | *Abatement* |
|---|---|---|---|---|
| Carbon disulfide ($CS_2$) | Industrial | Industrial solvent | Poisonous and flammable | Wet-scrubbers |
| Carbon monoxide (CO) | Furnace and coal gas; automobile exhaust | Colorless, tasteless and odorless | Asphyxiate; headache; vertigo; coma & death | Ventilation; after burners; catalytic oxidation |
| Carbon tetra chloride, $CCl_4$ | Chemical industry | Organic solvent; fire extinguisher | Toxic | Ventilation; chemical cleaning |
| Chlorine (Cl) | Chlor-alkali & bleaching industry | Greenish gas; oxidant | Hazardous; respiratory irritation; pulmonary edema | Water & alkali scrubbers |
| Chromium (Cr) | Mining ($Fe_2Cr_2O_3$) & Metallurgy | Metal used chrome plating; pigment and tanning | Toxic & poisonous as Cr(VI) | Bag filters; scrubbers and electrostatic precipitators |
| Fluorides | Fluoride ores ($CaF_2$) ; phosphate fertilizers; | Industrial | Hazardous & irritant; bone fluorosis; pulmonary edema | Scrubbers; cyclones |
| Fluorocarbons (Freons) | Chemical industry | Industrial and domestic uses. Undergo photolysis | Depletion of $O_3$ – balance | Discontinue manufacture |
| Formaldehyde (Formalin) | Commercial | Organic compound; medical uses | Toxic; irritation to respiratory tract and eyes | General cleaning methods of organic chemicals |
| Hydrocarbons | Petrochemical industry; automobile emissions | Chemical and industrial uses. | Irritants and toxicants; react with $NO_x$ to produce smog | Incineration; activated carbon adsorption |
| Hydrocarbons(polynuclear) | Refuse burning | Industrial and domestic | Carcinogens | Stop open-burning of waste; better waste disposal Management |
| Hydrogen chloride (HCl) | Chemical & Industrial | Colorless & hygroscopic; pungent. Industrial uses | Harmful to organisms | Absorbers & scrubbers |
| Hydrogen Sulfide ($H_2S$) | Sewage plants; petroleum refineries | Foul-smelling gas. Industrial uses | Highly toxic | Oxidation to $SO_2$ & Absorbers & scrubbers |
| Iron (Fe) | Mining & Metallurgy | Metal industry | Toxic; nausea & vomiting; hyperglycemia | Scrubbers; filters; electrostatic precipitators |
| Lead (Pb) | Mining (galena); automobiles; plumbing | Gray metal Storage batteries;. paints | Poisonous; head ache; renal damage; CNS deterioration | Bag filters; scrubbers; electrostatic precipitators |

| *Pollutant* | *Source* | *Properties & Uses* | *Effects* | *Abatement* |
|---|---|---|---|---|
| Manganese (Mn) | Mining ($MnO_3$) & Metallurgy | Hard, brittle metal. Used in steel and alloy industries | Toxicant; CNS deterioration | Scrubbers; filters & precipitators |
| Mercury (Hg) | Mining (cinnabar, HgS) & Metallurgy | Silvery liquid metal. Used in paints, mirrors and industry and laboratory | Highly toxic; CNS disorders; pneumonitis. Organic Hg causes food poisoning | Condensing & water scrubbers |
| Molybdenum (Mo) | Mining ($MoS_2$) & Metallurgy | Industrial uses; alloys | Toxic; diarrhea, anemia | Bag filters; electrostatic precipitators |
| Nickel (Ni) | Mining & Metallurgy | In Nickel-alloy plants; aviation | Toxicant; bronchial cancer; dermatitis | Bag filters; scrubbers; precipitators |
| Nitrogen oxides ($NO_x$) | Power plants (boilers); $HNO_3$ plants; motor vehicles | $NO_2$ is highly corrosive and atmospheric pollutant | Toxic; bronchial edema | Use of natural gas; modification of combustion process; catalytic conversion of NOx in exhaust gases |
| Odorous compounds | Industrial & chemical plants | Unhealthy & malodorous | Unhealthy & unhygienic | Incineration; absorption; chemical & biological control |
| Particulates | Mining; coal & fuel burning; industrial and agricultural processes | Visibility reducers; smog formers | Health hazard pulmonary problems | Gravitational settling; filtration; wet-scrubbers |
| Phosphorus (P)(Inorganic) | Mining; oil-fired boilers | Non-metal. Industrial; fertilizers and commercial | Toxic; gastro-irritation | Scrubbers; cyclones; precipitators |
| Phosphorus (organo) | Chemical industry | Chemicals; insecticides; herbicides | Extremely poisonous | Ventilation & P4-laboratory conditions |
| Photochemical oxidants | Fuel combustion | Secondary air pollutants; smog formers | Health hazard; respiratory pneumonia | More effective combustion processes |
| Radioactive waste | Uranium mines; nuclear installations | Radioactive products for energy and weapons | Ionization and radiation damage | Adsorption and sorption of radon; filters and precipitators |
| Selenium (Se) | Metallurgy; in Cu-refinery | Metal. Industrial; semiconductors | Toxic | Wet-scrubbers; precipitators |
| Sulfur oxides, $SO_X$ | Combustion of S-containing coal; metallurgy & industry | $SO_2$ is oxidant/ reductant; reacts catalytically & photochemically | Irritant; chronic bronchitis; cause for acid rain | Pretreatment of coal to remove sulfur; removal of $SO_2$by lime stone; catalytic conversion to $H_2SO_4$ |
| Vanadium (Va) | Coal-ores; metallurgy | Used as an alloy; automobiles | Toxic; $VaO_5$ leads to conjunctivitis; cough | Use $MgO_2$ in oil-fired burners; cyclones; precipitators |
| Zinc (Zn) | Mining (ZnS) & Metallurgy | Galvanization; alloys; | Toxic | Electrostatic precipitators |

## BIBLIOGRAPHY


Adriano, D. C. and Johnson, A. H. (1989), Springerverlag: New York. "*Acid Precipitation (Vol.2): Biological and Ecological Effects.*"

Allen, S. E. (ed.) (1989), Blackwell Science Pubs: New York. "*Chemical Analysis of Ecological Materials,*" (2nd edn.).

Bockris, J. O'M. (ed.) (1978), Plenum Press: New York. "*Environmental Chemistry.*"

Buffle, J. and Van Leeuwen, H. P. (eds.) (1992), Lewis Pubs: Chelsea, MI. "*Environmental Particles,*" (Vol I).

Butler, J. D. (1979), Academic Press: New York. "*Air Pollution Chemistry.*"

Caddle, R. D. and Allan, E. R. (1970), *Science*, **167**: 243. "Atmospheric chemistry."

Clement, H. E., Yang, P. W. and Koester, C. J. (1999), *Anal Chem.*, **71**: 257. "Environmental Analysis."

Finkel, A. J. and Duel, W. C. (eds.) (1976), Pub. Sciences Group Acton, MI. "*Clinical Implications of Air Pollution Research.*"

Harrison, R. M. and Perry, R. (1984), Chapman & Hall: London. "*Handbook of Air Pollution Analysis,*" (2nd edn.).

Heichen, J. (1976), Academic Press: New York. "*Atmospheric Chemistry.*"

Hutzinger, O. (ed.) (1990), Springerverlag: Berlin. "*Handbook of Environmental Chemistry.*"

Kouimtzis, T. and Samara, C. (eds.) (1995), Springerverlag: Berlin. "*Airborne Particulate Matter.*"

Kebbekus, B. B. and Mitra, S. (eds.) (1998), Blakie Pubs: London. "*Environmental Chemical Analysis.*"

Manahan, S. E. (1993), Lewis Pubs: New York. "*Environmental Chemistry,*" (5th edn.).

Mar Gonzalez, M., Gallego, M. and Valcarcel, M. (2001), *Talanta*, **55**(1): 135. "Determination of arsenic in wheat flour by electrothermal atomic absorption spectrometry using a continuous precipitation-dissolution flow system."

Marr, I. L. and Cresser, M. S. (1983), Intl. Textbook Co: New York. "*Environmental Chemical Analysis.*"

Nriagu, N. O. (ed.) (1992), Wiley & Sons: New York: "*Gaseous Pollutants.*"

Nriagu, N. O. and Simmons, M. S. (1994), Wiley & Sons: New York. "*Environmental Oxidants.*"

O'Neal, P. (1983), Chapman & Hall: New York. "*Environmental Chemistry,*" (2nd edn.).

Parker, H. W. (1977), Prentice & Hall: Engelwood Cliffs, NJ. "*Air Pollution.*"

Peirce, J. J., Weiner, R. F. and Vesiland, P. A. (1998), Butterworth-Heineman: New Delhi. "*Environmental Pollution and Control,*" (4th edn.).

Perkins, H. C. (1974), McGraw-Hill: New York. "*Air Pollution.*"

Perry, R. and Young, R. J. (eds.) (1979), Chapman & Hall: London. "*Handbook of Air Pollution Analysis.*"

Reeve, R. N. (1994), Wiley: New York. "*Environmental Analysis.*"

Robinson, N. (1966), Elsevier: Amsterdam. "*Solar Radiation.*"

Singer, S. F. (ed.) (1970), Springerverlag: New York. "*Global Effects of Environmental Pollution.*"

Turk, A., *et al.* (1978), Saunders: Philadelphia, PA. "*Environmental Science,*" (2nd edn.).

Wark, K. and Warner, C. F. (1976), IEPA: New York. "*Air Pollution: Its Origin and Control.*"

Warneck, P. (1988), Academic Press: New York. "*Chemistry of the Natural Atmosphere.*"

# 6

# A

Effects of air pollutants are wide ranging. They cause damage to materials and buildings, deteriorate soil and affect vegetation and living organisms. Atmospheric haze is caused by the presence of aerosols ($SO_x$, $NO_x$ and hydrocarbons). Aerosols absorb and reflect both solar and terrestrial radiations. They also act as nuclei for cloud formation. The net effect of increased levels of aerosols in the atmosphere is cooling of the Earth's troposphere.

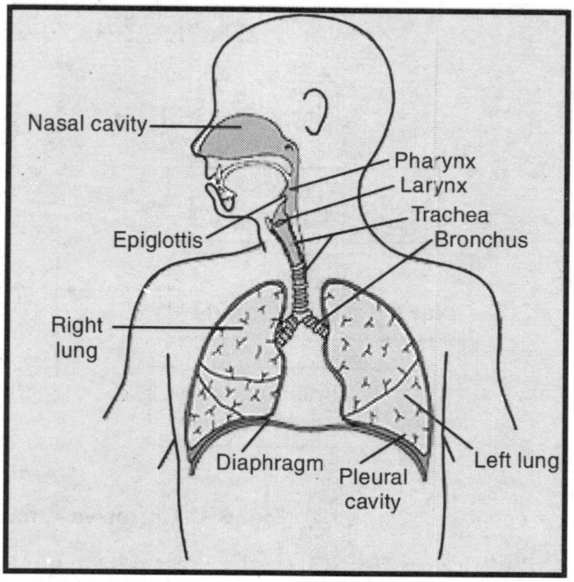


Effects of particulate pollutants depend on their physical as well as chemical nature. Their size and composition are factors influencing their impact, with smaller particulate matter showing greater health hazards. Particulate matter is brought into contact with pulmonary membrane by sedimentation, impaction, interception and diffusion. Sub-micron range particles (10 → 2 μm; $PM_{10} \rightarrow PM_2$) pose pulmonary health problems.

If the pollutants are chemically reactive or toxic, their physiological effects are more serious and lethal, irrespective of the size. For example, $SO_2$ is a chemical irritant that damages mucous lining of the respiratory tract and causes chronic bronchitis. Powerful oxidants—$O_2$, $O_3$ and $H_2O_2$ are edemagenic agents. They enhance cell damage and premature aging. They generate free radicals, which interfere in many biochemical reactions.

Heavy metals like As, Cd, Hg, and Pb are highly toxic even in traces. Metal chelation affects the structure and consequently function, of various enzyme complexes. Mercury can bind to both nucleic acid bases and phosphate groups, and thus interfere in the base-pair formation and the genetic code. Arsenic, which has similarities to phosphorus, interferes with enzymatic phosphorylation reactions.

This chapter deals with the effects of air pollutants on the inanimate and animate objects of this globe.

Effects of air pollutants are wide ranging. They include aesthetic losses, economic losses, safety hazards, and personal discomfort and health effects. They encompass damage to inert materials such as buildings, roads, bridges, monuments and other constructions, to vegetation and ecosystems and to living organisms. Aesthetic effects also include the loss of clarity of the atmosphere as well as the presence of objectionable odorous compounds. Air pollution can also affect biodiversity. Biodiversity impacts occur on local, regional, and global scales. Health effects range from personal discomfort (throat and eye irritation) to acute and chronic health hazards. Ultimately the concerns must be for the *biosphere*, made up of living organisms that inhabit the physical environment. Life forms vary greatly in their response to environmental factors.

## 6.1 EFFECTS OF AIR POLLUTANTS ON INANIMATE MATERIALS

Air pollutants, acids, alkalis and oxidants create a corrosive atmosphere (Fig. 6.1). The effect on metals leads to corrosion and structural damage and change in their electrical properties.

$$H_2SO_4 + Zn \rightarrow ZnSO_4 + H_2 \quad \ldots(6.1)$$

$$2NaOH + 2H_2O + Zn \rightarrow Na_2Zn(OH)_4 + H_2$$

$$\text{Anode:} \quad Zn \rightarrow Zn^{2+} + 2e^- \quad \ldots(6.2)$$

$$\text{Cathode:} \quad 2H^+ + 2e^- \rightarrow H_2$$

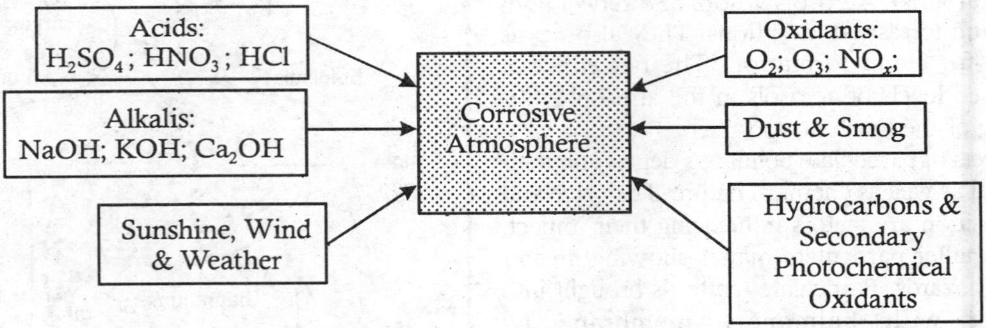


**Fig. 6.1:** Corrosive Atmosphere due to Air Pollutants

Sulfur dioxide, $SO_2$, is the most detrimental pollutant in the corrosion of metals and materials. Under humid conditions, $H_2SO_4$ is formed which is a highly corrosive acid. $SO_2$ attacks rubber (Fig. 6.2), leather, paper, fabrics, and natural and man-made fibers, paints and structures.

Air pollutants abrade, corrode, tarnish, soil, erode, crack, weather and discolor structures and materials. Air pollutants cause not only structural damage to the buildings, bridges and other inert objects, but they also mar the aesthetic quality of the buildings, monuments and life. For example, hydrogen sulfide, $H_2S$, reacts with lead-based paints, causing discoloration.

$$Pb^{2+} + H_2S \rightarrow PbS + 2H^+ \quad \ldots(6.3)$$

The intensity of sensation (the aesthetic quality) is a logarithmic function of stimulus, qualitatively represented by *Weber-Fechner* law.

$$S = K. \log(\text{Stimulus}) \quad \ldots(6.4)$$

where, $S$ = intensity of sensation;

$K$ = a constant.

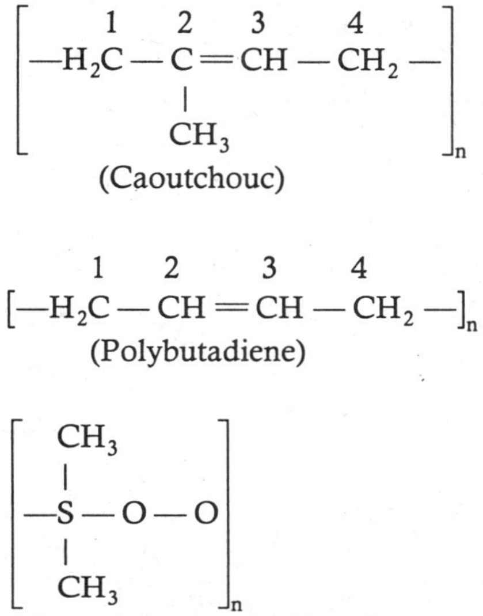


Natural rubber

Synthetic rubber

Silicon rubber

$SO_2$ attacks chemical bonds

**Fig. 6.2:** Chemical Formulae of Natural and Man-made Rubber

The effects of air pollution on inert materials are given in Table 6.1

**TABLE 6.1:** EFFECTS OF AIR POLLUTANTS ON INERT MATERIALS

| *Material* | *Action* | *Effects* |
|---|---|---|
| Metals | Corrosion & damage | Loss of reflectance, material strength and electrical properties |
| Building materials | Leaching; physical and chemical actions | Loss of reflectance, material strength and aesthetic beauty |
| Leather & rubber | Chemical action | Loss of tensile strength; brittleness and cracking |
| Paper | Chemical action | Brittleness & damage |
| Paint | Chemical action | Discoloration; peeling off and cracking |

## 6.2 EFFECTS OF AIR POLLUTANTS ON ECOSYSTEMS

Effects of air pollutants are at all levels—global, regional and local levels.

### 6.2.1 Global Effects

Atmospheric haze is the condition of reduced visibility caused by the presence of fine particles (aerosols) or $SO_x$, $NO_x$ and $CO_x$ in combination with the secondary pollutants in the atmosphere. Hygroscopic particles, called '*cloud condensation nuclei (Aitken particles)*' are 'activated' in the presence of humid air and would shift the atmospheric precipitation processes (Fig. 6.3). If these particles were pollutants, then their effects would be of global concern.

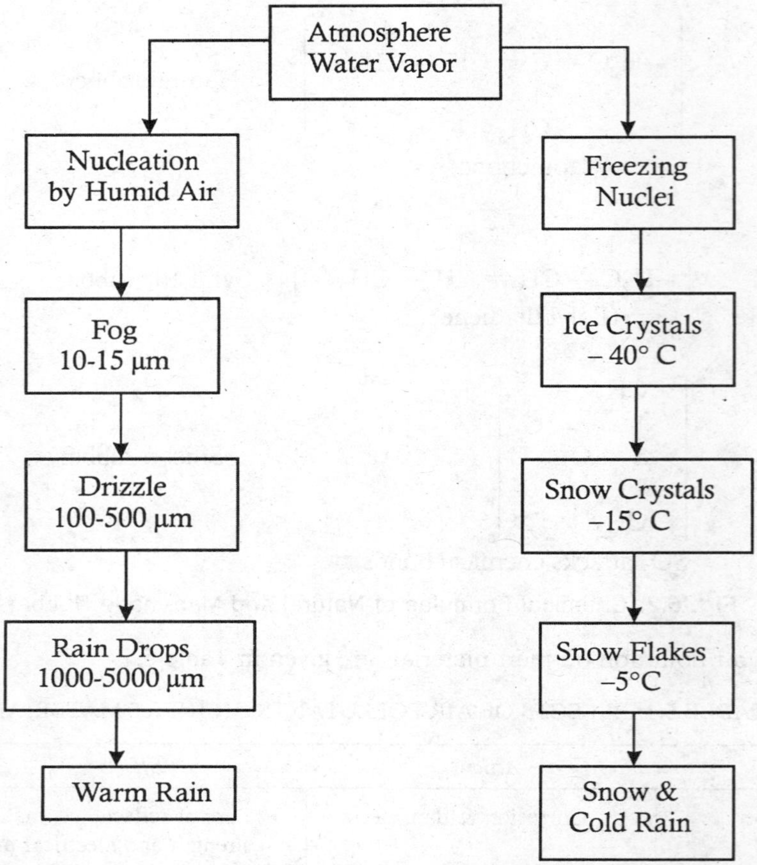


**Fig. 6.3:** Principal Atmospheric Precipitation Processes

### 6.2.2 Effects on Vegetation

Forest- and agro-ecosystems act as sinks as well as sources for atmospheric air pollutants. Forest fires are sources of major air pollutants—$SO_x$, $NO_x$, $CO_x$, $H_2S$ and $NH_3$. These pollutants can lead to change of the forest composition and destruction of soil and vegetation. Degradation of forests and crops, and vegetation, and underlying soils, groundwater and drainage networks not only result in ecological imbalances, but also pose a threat to welfare of the living organisms.

#### *6.2.2.1 Acid Rain*

Acid rain is defined as any rainfall that has an acidity level beyond what is expected in non-polluted rainfall (pH < 5.5). Excessive amounts of acid gases ($SO_x$, and $NO_x$) in the atmosphere are responsible for acid rain. Acid rain is formed when $SO_x$, and $NO_x$ combine with water to form acids $H_2SO_4$ and $HNO_3$. Sunlight increases the rates acid-forming reactions. Through rain, snow, sleet, hail, dew and fog the acidic constituents in the atmosphere reach the Earth. In all the forms acid precipitation has wide consequences.

Effects of acid rain are at regional and global levels, because air pollutants can be carried many hundred kilometers by winds, and acid pollutants emitted in one country may be deposited as acid precipitation in other countries. Acid rain precipitation causes multiple effects on both terrestrial and aquatic ecosystems.

Acid rain can be absorbed both by materials, plants, aquatic organisms and animals, and the effects of acid rain are felt both by aquatic systems and vegetation. Acids may react with sandstone and limestone structures, and form a powdery substance (gypsum) that can be washed away. They also can cause corrosion of structures, roads, and bridges. Metal corrosion is primarily by oxygen and moisture, although sulfur dioxide does accelerate the process.

Acid water—acid rain and acid mine drainage, has distinct effect on soil microbiology and chemistry. It enhances soil degradation and loss of fertility by leaching off nutrients out of the soil. Toxic metals dissolved by acidic water from the soil are absorbed by the roots of the trees and affect the tree growth. Acid rain and leachates lead to forest degradation, necrotic lesions on foliage, and long-term damage to the topsoil. Reduction of rainforest and foliage would lead changes in local climate, and destruction of biological diversity. Aquatic systems on neutral or acidic bedrock are normally very sensitive to acid deposition because they lack basic compounds that buffer the acidification. The effects are: (*i*) carbon sources change from carbonate ($HCO_3$) to $CO_2$, (*ii*) release of toxic metals and (*iii*) disappearance of many algal and aquatic species. Acidification of water bodies reduces or kills the aquatic organisms. At acid pH ($pH < 5.0$), aquatic life (fish) is no longer able to survive and reproduce. Death of aquatic organisms can inhibit nutrient cycling. Acidification of terrestrial soils may make toxic trace metals (Al, Hg) to become more soluble and leach them into water bodies and poisoning them. That is, a decrease in pH is often associated with an increase in metal availability of Al and Hg. Leaching of plant nutrients (Ca, Mg, and K) causes decline in plant growth. Soft-bodied organisms such as leeches, snails and crayfish are early victims. The death of fish at low pH is due to release of toxic metals such as Al. At pH 5 Al is at its most poisonous. Al can damage roots and is implicated in Alzheimer's disease in humans.

The plant leaf structure has three main functions: (*i*) photosynthesis, (*ii*) transpiration and (*iii*) respiration (Table 6.2).

**TABLE 6.2:** PHOTOSYNTHESIS AND RESPIRATION PROCESSES IN PLANTS

| *Photosynthesis* | *Respiration* |
|---|---|
| Only by chlorophyll-containing cells in the presence of sunlight | Occurs at all times in all living cells |
| Oxygen is liberated | Carbon dioxide is liberated |
| Energy is stored | Energy is utilized |
| $H_2O$ & $CO_2$ are raw materials | Sugars & $O_2$ are raw materials |
| Carbohydrates synthesized | Carbohydrates utilized |
| Weight increase | Weight decrease |

(*i*) **Photosynthesis:** Conversion of atmospheric $CO_2$ to carbohydrates, utilizing solar energy.

$$6CO_2 + 6H_2O \xrightarrow{h\nu} C_6H_{12}O_6 + 6O_2 \qquad ...(6.5)$$

(*ii*) **Transpiration:** Movement of water from the root system up to the leaves and its subsequent evaporation to the atmosphere. This process moves nutrients throughout the plant and cools the plant.

**(iii) Respiration:** It is the heat-producing process, resulting from the oxidation of carbohydrates to form $CO_2$ and $H_2O$.

$$C_6H_{12}O_6 + CO_2 \rightarrow 6H_2O + 6CO_2 \qquad \text{...(6.6)}$$

These three functions involve the movement of $O_2$, $CO_2$ and $H_2O$ through the epidermal layers of the leaf.

Gaseous pollutants have direct pathway to the cellular system of the plant leaf structure, and they enter indirectly through the root system. Air pollution interferes with normal metabolism of plants. Leaves provide larger surface area for adsorption, absorption and accumulation of air pollutants. Acid deposition on leaf surfaces reduces the rate of photosynthesis and nutrient stress. Exposure of plants to $SO_x$, $NO_x$ and $O_3$ causes disruption of the thylakoid membranes in chloroplast, loss of chlorophyll, and tissue damage.

The toxic effects of pollutants are directly related to the functioning of the stomata. Stomata openings are related to the physiological activity of the plant and regulate gas-exchange. There is a correlation between the extent of air pollution effects and the degree of opening of the stomata. Factors such as sunlight, humidity and temperature strongly influence the uptake of pollutants by plants. Reduction of crop yield is related to high-level, long-term chronic exposure to air pollutants. The effects of some common air pollutants on plants are given in Table 6.3.

**TABLE 6.3:** EFFECTS OF AIR POLLUTANTS ON PLANTS

| *Pollutant* | *Affected Cells* | *Symptoms* |
|---|---|---|
| Sulfur dioxide ($SO_2$) | Mesophyll cells | Bleached spots; chlorosis; necrosis |
| Hydrogen disulfide ($H_2S$) | Epidermal & mesophyll cells; young leaves | Basal and marginal scorching |
| Sulfuric acid ($H_2SO_4$) | All parts of the plant | Necrotic spots |
| Hydrochloric acid (HCl) | Epidermic & mesophyll cells | Necrotic lesions |
| Hydrofluoric acid (HF) | Epidermic & mesophyll cells | Tip and margin burns; leaf abscission |
| Chlorine ($Cl_2$) | Epidermic and mesophyll cells | Bleaching and leaf abscission |
| Ozone ($O_3$) | Palisade; spongy parenchyma | Flecks & premature aging; necrosis & bleaching |
| Ammonia ($NH_3$) | Complete tissue | 'Cooked green' appearance |
| Nitrogen dioxide ($NO_2$) | Mesophyll cells | White-brown lesions; suppressed growth |
| Peroxyacetyl nitrate (PAN) | Spongy cells | Glazing; silvering |
| Ethylene ($C_2H_4$) | All cells | Sepal withering; flower dropping; abscission |
| Mercury (Hg) | Mesophyll cells | Brown spots; chlorosis; abscission |
| 2, 4-Dichlorophenoxy acetic acid (2-4D) | Epidermis | Swollen stems |

## 6.3 EFFECTS OF AIR POLLUTION ON ANIMAL KINGDOM

Air pollutants in particulate and gaseous state have physiological effects on animal kingdom, depending on their size and state. In animals and humans, the air pollution primarily affects the respiratory, circulatory and olfactory systems, respiratory system being the principal route of air pollution effects. Air pollution on human beings lends to deterioration of the quality of life, sickness and death. Gas poisoning is a possible hazard in most chemical industries.

Effects of pollutants depend on the physical and chemical properties of the particulates. Acids, alkalis and oxidants exert direct chemical action. Inert particulate pollutants induce physiological responses. The particulates of most concern with regard to their effects on human health are particualtes <10 μm (10-PM) in diameter, because they can be inhaled deep into the lungs and become trapped in the lower respiratory system. Certain particulates, such as asbestos fibers, are known carcinogens. Chronic obstructive pulmonary diseases result from chronic irritation of lung tissues. Chronic bronchitis is characterized by excessive mucous secretion in the bronchi. Lung cancer, 'black lung' disease, emphysema, asbestosis are all due to particulate pollutants.

### 6.3.1 Physicochemical Effects

The nasopharyngeal section (throat), tracheobronchial section (conduction airways) and the pulmonary structure (gas-exchange zone in the lung) are the three sections in the respiratory system, taking part in the respiratory processes. The nasal passages lead through the nasopharyngeal structure to the *trachea and bronchi*. The bronchi break up into smaller *bronchioles*, which then terminate in the alveolar sac where gas-exchange ($O_2/CO_2$) process takes place between the air and the blood.

Any particulate matter that collects in the lungs can be dangerous to human lungs (e.g. asbestosis, silicosis, berylliosis, and byssinosis). For particulate pollutants, size as well as composition is a factor affecting toxicity, with smaller size particulate matter (PM-2 size) showing greater toxicity (e.g. Cd and $O_3$ are highly toxic). Aerosols (dust, smoke, mist, spray, fumes) (Table 6.4) deposit in the pulmonary system according to their physical size and density (Aitken nuclei <0.1 μm; small particulate ~0.1-2 μm; large particles >2 μm).

**TABLE 6.4:** COMPOSITION OF AEROSOL/PARTICULATE MATTER

| *Class* | *Example* |
|---|---|
| Organic | Tobacco smoke; soot; oil mist and spray; pollen |
| Inorganic | Acid mists; dust |
| Metallic | Fumes; colloids |
| Fibrous | Asbestos; cotton; wood pulp |
| Radioactive | Radium, uranium and thorium products; radon and progeny |
| Bacterial | Bacteria; viruses, etc. |

The physiological effects of aerosol/particulate pollutants on the nasopharyngeal and pulmonary systems depend on the physiochemical properties of the pollutants (e.g. diffusion, viscosity, and solubility and chemical reactivity). The mobility (flow) and interaction of particulate matter in a medium that depend on the effect of inertial and viscous forces acting on the particles. Reynolds' number $R_e$ gives the relationship between viscous and inertial forces.

$$R_e = \frac{UR}{\eta} \qquad \ldots(6.7)$$

If $R_e < 2{,}000$, viscous forces dominate and the flow is laminar. If $R_e > 2{,}000$, inertial forces dominate, with turbulence and convective mixing of air turbulent eddies.

$$R_W = R\left(\frac{\omega}{\eta}\right)^{1/2} \quad \text{...(6.8)}$$

$R_W < 1$ for steady flow.

where, $U$ = average flow velocity;
$R$ = fluid column radius;
$\eta$ = kinematic (fluid) viscosity;
$\omega$ = breathing frequency;
$R_W$ = airway radius

Particles are brought into contact with the pulmonary membrane by: (1) sedimentation, (2) inertial impaction, (3) interception and (4) diffusion.

#### 6.3.1.1 Sedimentation

All particles of diameter, $d > 0.1$ μm sediment under the influence of gravity.

$$F_g = \frac{\pi d^3}{6}(\rho_p - \rho_a)g \quad \text{...(6.9)}$$

Frictional drag, $F_d$, is given by *Stokes* equation,

$$F_d = 3\pi d\eta v_t \qquad \text{Stokes law} \quad \text{...(6.10)}$$

At equilibrium, the terminal velocity of a particle is

$$V_t = \frac{gd^2}{18\eta}(\rho_p - \rho_a) \quad \text{...(6.11)}$$

where, $F_g$ = gravitational force;
$d$ = particle diameter;
$\rho_p$ = particle density;
$\rho_a$ = density of air;
$g$ = acceleration due to gravity;
$v_t$ = terminal velocity of the particle;
$\eta$ = kinematic viscosity

#### 6.3.1.2 Inertial Impaction

Collision is inertial impaction. Lateral displacement, $L$, of a particle and Stokes number ($S_N$) are given by

$$L = \frac{v_t \sin\theta}{g} \quad \text{...(6.12)}$$

$$S_N = \frac{\rho d^2 v_t}{18\eta R} \quad \text{...(6.13)}$$

#### 6.3.1.3 Interception

Deposition of hygroscopic and charged particles is given by

$$\text{Deposition} = \frac{cq^2}{Ud} \quad \text{...(6.14)}$$

Impaction parameter $\xi = d^2F$ ...(6.15)

where, $c$ = Cunningham slip correction;
$q$ = charge of the particle;
$U$ = average velocity of air;
$d$ = particle diameter;
$F$ = average inspiratory flow rate

$$S_N = \frac{4d^2F}{18\pi\eta(BDS)^3} \quad \text{...(6.16)}$$

Settling deposition efficiency, $\varphi S$, is

$$\varphi s = \frac{v_t z}{Ud} \quad \text{...(6.17)}$$

For a particle of diameter >2 μm (>$PM_2$ size particles), the deposition efficiency, $\varphi$, is

$$\varphi s \cong 0.8 \times S_N \quad \text{...(6.18)}$$

where, $BDS$ = bronchial deposition size;
$\eta$ = viscosity coefficient.

Large, heavy particles tend to be removed through inertial impact in the nasopharyngeal structure. Small particle matter, $d < 0.5$ μm (< PM-2 particulate matter) is deposited in the lower pulmonary tract (Fig. 6.4).

#### 6.3.1.4 Diffusion

Diffusion (<0.25 μm) is the dominant mechanism in the alveolar region. Brownian motion dominates diffusion of finer particles. While particles of diameter, $d < 1$ μm exhibit Brownian motion, Brownian motion predominates for particles of $d < 0.25$ μm. The diffusion coefficient is related to the particle size (size) and the temperature and is given by

$$D = \frac{kT}{3\pi\eta d} \quad \text{...(6.19)}$$

$$\bar{r} = \sqrt{6Dt} = \sqrt{\left(\frac{2kTt}{\pi\eta d}\right)} \quad \text{...(6.20)}$$

where, $D$ = diffusion coefficient;
$k$ = Boltzmann constant;
$T$ = absolute temperature;
$t$ = time;
$d$ = particle diameter;
$\bar{r}$ = mean displacement

### 6.3.2 Physiological Systems

Of all the physiological systems affected by air pollutants, the pulmonary system is the most important.

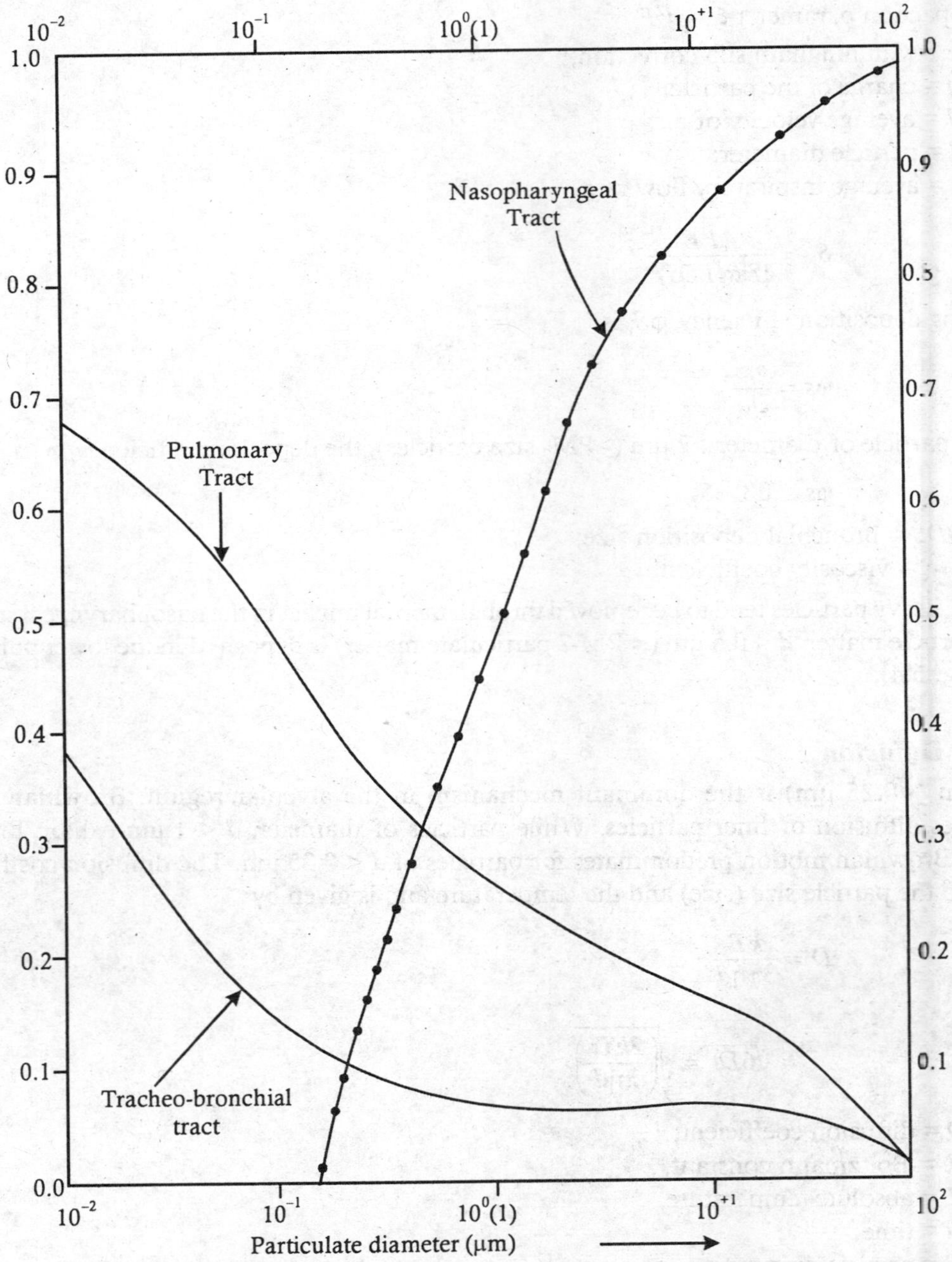


**Fig. 6.4:** Fraction of Particulate deposited in the Respiratory Tract Compartments as a Function of Particle Diameter
(After: Perkins, H. C. (1974) "*Air Pollution*", New York: McGraw-Hill)

### *6.3.2.1 Effect of Pollutants on Breathing*

Airway resistance, $R_{aw}$, is one of the commonly determined parameters in humans for the evaluation of pulmonary processes.

$$R_{aw} = \frac{(P_m - P_a)}{Q} \quad ...(6.21)$$

Specific resistance, $SR_{aw} = R_{aw} \times V_{tg}$ ...(6.22)

where, $P_m$ = pressure in the mouth;
$P_a$ = atmospheric pressure;
$Q$ = flow rate;
$V_{tg}$ = thoracic gas volume.

Forced expiratory volume (FEV) is also measured, and pulmonary resistance, $R_L$, is determined by measuring pressure in the intrapleural space, $P_{pl}$, and the mouth, $P_m$.

$$R_L = \frac{(P_m - P_{pl})}{Q} \quad ...(6.23)$$

### 6.3.3 Physiological Processes

Lung parenchyma consists of four types of cells—endothelial, mesenchymal, epithelial and extrapulmonary. Endothelial cells, lining the capillary network, and mesenchymal cells (fibroblasts, pericytes and interstitial cells) are ~80 per cent of the cell population. Epithelial cells, which line the airspace in the lung, are divided into alveolar type I and type II cells. Type I cells have larger surface area. Type II cells secrete surfactants that coat the lung epithelium. Extrapulmonary cells include blood cells, lymphocytes, mast cells, and alveolar macrophages. These cells are responsible for the defense of the lung.

Cells in the parachyma can respond to changes in their environment. Information about local changes is received via specific receptors on the cell surface (e.g. histamine receptors, corticosteroid receptors) or by phago- or pinocytosis. Alveolar macrophages function as the primary defense against the particulate matter in the lungs. These phagocytes reside as free cells, at the air-tissue interface, a location that exposes them directly to inhaled environmental pollutants, toxins and pollens.

Macrophages are responsible for clearing of particulate matter from the respiratory bronchioles and alveoli and from the ciliated portions of the trachiobronchial tree (air passage). Respiratory airspace sterility is maintained by alveolar macrophages. Mast cells are potent mediators of immediate hypersensitivity (inflammation). They are localized in bronchial submucosa and perivenular connective tissues. The induction of hypersensitivity (inflammation) by mast cells depends on specific antigen-antibody reaction.

Alveolar type II cells play an important role in maintaining normal function and structure of the lung. These secretary cells are essential in the production of surfactants.

The mechanisms of physiological effects of air pollutants on the pulmonary system are: (*i*) the physical irritation of the inhaled particulate (aerosol) on the airway epithelium causes a reflexive narrowing of the airways, effected by smooth muscle-stimulating compounds and histamine-releasing factors, and (*ii*) release of endogenous mediators of inflammation by cells in the lung. Polynuclear leukocytes are the first cells to react to an injurious agent. The alveolar macrophages and mast cells are the primary candidates for such activity. (*iii*) Release of prostaglandins by alveolar macrophages may result in potentiation (cooperativity) of mast cell degranulation and aggregation of platelets.

A large portion of membrane-bound fatty acids in the alveolar macrophage is archidonic acid. The products of archidonic oxygenation (thromboxine, prostaglandins, leukotriches, etc.), released by the alveolar macrophage can mediate inflammation. They (prostaglandin $F_{2\alpha}GA$, prostaglandin

$G_2$, thromboxine $A_2$) cause bronchoconstriction. Neutrophile chemotoxic factors (collagenase, elastase, lysozyme, acid hydroxylases etc.) are released by macrophages by phagocytosis.

Mast cell granules contain four classes of mediators—(*i*) vasoconstrictive and smooth muscle-reactive mediators (e.g. histamines, serotonin), (*ii*) chemotactic mediators, (*iii*) structural proteoglycans (e.g. heparin, chondroitin sulfate, dermatan sulfate) and (*iv*) enzymes (chymase, kallikrein).

Chronic bronchitis is a condition where inflammation of bronchi and an excessive amount of mucus in the bronchi result in recurring cough. In the respiratory system, pulmonary emphysema is closely linked to chronic bronchitis. This is due to loss of elasticity and deterioration of the alveolar walls, resulting in shortness of breath.

### 6.3.4 Toxic Effects of Particulate Matter

Particulates less than 10 μm (PM-10) have to be viewed with concern in relation to their effects on human health because they flow deep into the lungs and get trapped in the lower respiratory system. Dust particles produce a distinctive reaction in the lungs and lead to formation of masses of fibrous tissue and typical nodules of dense fibrosis. Particulate pollutants, even if they are inert, have physiological effects on living systems. If they are chemically reactive or toxic, then the physiological effects are more serious and lethal.

Inhalation of silica and asbestos dust lead to fibriotic lesions, fibriosis, asbestosis or abnormal accumulation of connective tissues (inflammation). Silicosis is a chronic disease of the lungs that is caused by the inhalation of silica dust by the workers exposed to silica dust, particularly while rock drilling. It is a form of pneumoconiosis. The macrophages that engulf the silica particles form nodules and fibriotic masses. Silica-activated macrophages produce a factor that enhances the synthesis of collagenous hydroxyproline by fibroblasts, and increased synthesis of prostaglandins (PGE1 and $PGF_{2\alpha}$). Silica fibrigenicity is due to mechanical, piezoelectric, hydrogen bonding and epitaxis. Coal dust, alone, even if it is low in silica, causes a lung problem known as "black lung" disease, which is a type of pneumoconiosis.

Cotton dust consists of soil, bacteria, fungi, and agricultural chemicals. *Byssinosis*, commonly known as '*brown lung*' disease, is a respiratory disease caused by inhalation of organic dusts (that includes coal dust) generated from cotton, flax and sisal during the production of fiber.

A type of pneumoconiosis is found among workers whose occupation involves asbestos. Asbestos fibers that have been inhaled remain lodged in the lungs for years and eventually cause excessive scarring and fibrosis, resulting in stiffening of lungs that long after exposure ceases. This state results in shortness of breath and inadequate oxygenation of the blood. Secondary heart disease may be induced b the increase cardiac effort needed to expand the stiff lungs. Mesothelioma, a malignant tumor of the pleura is also associated with asbestos inhalation and asbestosis. It is now known that cigarette smoking aggravates the stiffening of lungs due to asbestosis. There is no effective treatment for asbestosis.

### 6.3.5 Metal Toxicity

Metallic trace elements, such as cobalt, copper, iron, manganese and zinc are essential for human and animal growth (Table 6.5). These elements, although present in living tissue in minute concentrations, can have profound effect on health. In the body these essential trace elements are either a part of bone and tissue structures or are complexed with hormones, vitamins and proteins/enzymes. For example, cobalt is a component of vitamin $B_{12}$ and cobalt-containing proteins/enzymes. Chromium is activator of insulin and copper is a part of electron-transport chain enzymes.

Iron is present in heme-proteins (myoglobin, hemoglobin, cytochromes), manganese in mangano proteins (photosynthetic PSII complex) and zinc in insulin and in many enzymes (carboxy peptidase A). But, trace elements, below and above a certain concentration result in metabolic disorders (Table 6.6), and some of then, especially, Al, Cd, Cr, Se, V, may constitute serious health hazards.

**TABLE 6.5:** TRACE ELEMENTS ESSENTIAL TO HUMAN/ANIMAL METABOLISM

| *Element* | *Functions* |
|---|---|
| Aluminum (Al) | Protein transferin in serum |
| Chromium (Cr) | In glucose metabolism enzymes (phosphoglutamase) |
| Cobalt (Co) | Vitamin $B_{12}$; cobalt-containing proteins |
| Copper (Cu) | In electron-transport enzyme complex |
| Iron (Fe) | Myoglobin; hemoglobin |
| Manganese (Mn) | In mangano proteins (PSII complex) |
| Molybdenum (Mo) | In Mo-containing proteins and enzymes |
| Nickel (Ni) | Needed for normal growth |
| Selenium (Se) (metalloid) | Involved in enzyme formation; fat metabolism |
| Silicon (Si) (Non-metal) | For normal growth and bone development |
| Tin (Sn) | For normal growth |
| Vanadium (V) | Needed for normal growth |
| Zinc (Zn) | In insulin, and in many enzymes |

**TABLE 6.6:** ADVERSE HEALTH EFFECTS ASSOCIATED WITH SOME TRACE ELEMENTS

| *Element* | *Deficiency* | *Excess* |
|---|---|---|
| Aluminum (Al) | – | Dialysis encephalopathy; osteomalacia |
| Arsenic (As) | – | Loss of hair; kidney and liver damage |
| Cadmium (Cd) | – | Hypertension; Itai Itai; renal and kidney damage |
| Chromium (Cr) | Impaired glucose tolerance | Skin disease; cancer |
| Copper (Cu) (in serum) | Hypocuprosis—anemia; impaired bone formation; Wilson's disease | Thyrotoxicosis; biliary cirrhosis; malignancy |
| Iron (Fe) | Anemia; renal failure | Liver cirrhosis |
| Lead (Pb) | – | Headache; neuropathy; encephalopathy; memory loss; CNS damage |
| Selenium (Se) | Cardiomyopathy | Lassitude; hair loss |
| Vanadium (V) | Loss of growth of bone | Kidney failure |
| Zinc (Zn) | Growth retardation; skin lesion | Nausea; anemia; diarrhea; dermatitis |

In the chemical sense, all metals are characterized by a comparative readiness to give up their valence electrons and as a consequence metals in free state are reducing agents. They readily form chemical compounds. Metal oxides constitute a major class of inorganic particulate pollutants in the atmosphere. They are formed whenever fuels containing metals are burned. Many trace heavy metals and their vapors are injurious to health. Several of them are tumorogenic. Contemporary concern over the environmental chemistry of trace metals arises from the potential deleterious effects of metal pollution on ecosystems and living beings. Metals like mercury, lead and arsenic are harmful to most forms of life. The threshold limit value (TLV) of metal toxicity is

$$Be < Hg < Cd < Pb < As < V < Ni < Cu < Fe < Zn.$$

While divalent cations such as $Ca^{2+}$ and $Mg^{2+}$ bind preferentially to the phosphate oxygen atoms of the phosphodiester backbone in nucleic acids, other cations of the transition metal group, such as $Mn^{2+}$ bind to both nucleic acid bases and phosphate groups in nucleic acids (Table 6.7). Chronic manganese exposure can damage the brain. Similarly, toxic hazards of cadmium are high. Even small quantities can cause irritation of gastrointestinal tract, nausea, diarrhea, and kidney and lung damage. Mercury preferentially binds to the nucleic acid bases, thus completely disrupting (denaturing) the double helical complex formation in nucleic acids and thus affecting the genetic code. Thus, $Hg^{2+}$ (and $Pb^{2+}$ and other heavy metals) directly interfere, at molecular level, with the structural features and thereby with the functional aspects of protein and nucleic acid complexes.

**TABLE 6.7:** PREFERRED BINDING SITES OF METAL IONS WITH NUCLEIC ACID CONSTITUENTS

| *Metal Ions* | *Nucleic Acid-binding Sites* |
|---|---|
| **Alkali** | Nucleic acid bases + ring nitrogen and keto |
| $Li^+$, $Na^+$, $K^+$, $Cs^+$, $Rb^+$, | Oxygens + sugar hydroxyl groups |
| **Alkaline earth** | |
| $Mg^{2+}$, $Ca^{2+}$, $Sr^{2+}$, $Mg^{2+}$, $Ba^{2+}$ | Sugar hydroxyl + phosphate oxygen groups |
| **Transition metals** | |
| $Mn^{2+}$, $Ni^{2+}$, $Co^{2+}$, $Co^{3+}$, $Zn^{2+}$, $Cd^{2+}$, $Cu^{2+}$, $Ag^{2+}$, $Au^{3+}$, $Pt^{2+}$, | Nucleic acid base + ring nitrogen + phosphate oxygen groups |
| $Hg^{2+}$, $Ru^{3+}$, $Os^{6+}$ | Nucleic acid bases + ring nitrogen and keto oxygen groups |

*Source*: P. Narayanan (2003), New Age Intl. Pubs: New Delhi. "*Essentials of Biophysics*", 2nd print.

All metals are toxic to plants, organisms and animals above the threshold levels. In general, metal toxicity can be explained, arising due to formation metal-chelate coordinate complexes at the active sites of enzymes, thus interfering with biochemical functions. The toxicity of heavy metal ions can be explained at cellular and molecular levels as due to their detrimental interference in enzyme action (Fig. 6.5). The toxicity of Zn-group metals increases sharply in the order of Zn, Cd and Hg. $Pb^{2+}$ binds SH-group or competes for $Zn^{2+}$- or $Ca^{2+}$-binding sites, leading to inhibition of ATPase function. This affects the structure and thereby the function of the cell membrane, which manifests in intracellular and functional abnormalities. Replacement of a particular metal ion, in metalloproteins, by a similar ion can affect the structure and thereby the function of the proteins (e.g. replacement of Zn by Cd in Zn-metallo enzymes).

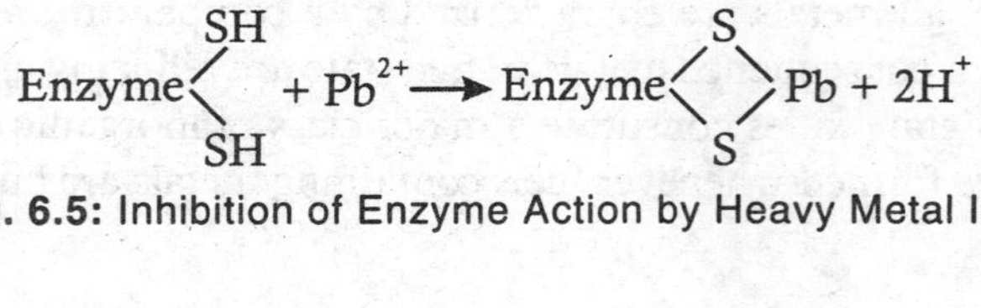


**Fig. 6.5:** Inhibition of Enzyme Action by Heavy Metal Ions

#### 6.3.5.1 Aluminum (Al)

Aluminum accumulation can lead to bone disease and to the dialysis-related disease—*dialysis encephalopathy syndrome (DES).*

#### 6.3.5.2 Cadmium (Cd)

Cadmium and cadmium compounds are biologically nonessential, but highly toxic, leading to kidney damage, renal cancer and *itai-itai* disease. Cadmium from the soil enters animal kingdom through plants. The major sources of cadmium in food items appear to be grains and seafood.

#### 6.3.5.3 Chromium (Cr)

All salts of chromic acid are toxic. Hexavalent (Cr(VI)) compounds are 100 to 1,000 times more toxic than Cr(III) compounds.

#### 6.3.5.4 Copper (Cu)

Copper is an essential trace element, necessary for oxidation and absorption of iron and vitamin C (ascorbic acid) in digestive system. It is a component of oxidative enzymes—hemocyanins (in the blood of lower animals), ascorbic acid oxidase (in plants and animals), tyrosinase, cytochrome oxidase, monoamine oxidase, and ceruloplasmin, a copper-transport protein in plasma. It, thus, serves an important function in the body's oxidative processes. Imbalance in copper uptake can produce disorders, such as Wilson's disease, also known as hepatolenticular degeneration, a progressive degeneration of basal ganglion of the brain. This is caused by a deficiency in the circulation of ceruloplasmin. Liver cirrhosis, stomach and intestinal distress, anemia, impaired bone formation and other malfunctions can also result from deficiency of copper.

#### 6.3.5.5 Zinc (Zn)

Zinc is an essential element and beneficial for human growth. It is found in high concentrations in the red blood cells (carbonic anhydrase). It is also present in insulin and in many enzymes. It plays an important role in the growth process, including cell division and protein synthesis. Zinc is toxic to plants at higher levels. Zinc pollution causes delayed wound healing, diarrhea and dermatitis. The toxicity of the zinc group metals increases sharply with atomic number. Cd and Hg are extremely toxic even at lower levels.

#### 6.3.5.6 Lead (Pb)

Lead is the most widely used non-ferrous metal and has large number of industrial applications. Specific industries in which lead-containing solids, dust and fumes are encountered are the petroleum industry, mining and smelting of sulfide ores, printing, cutlery and manufacture of storage batteries, paints and pigments, ceramics, glass and ammunition.

Lead is probably the most ubiquitous metal poison, and it is considered as criteria pollutant. In the urban atmosphere lead-containing particulate matter from motor exhaust is the chief source of lead. Other possible sources of lead compounds include the agricultural use of insecticides containing lead compounds. Lead is one of the metals, which enters the atmosphere through various pathways.

All soluble compounds of lead are poisonous. Toxicity increases as solubility increases. Solubility of lead increases with acidity of the medium. Therefore, acidified streams and lakes due to mine leaching and or rain would magnify the lead poisoning of aquatic life. Lead pollution in water may occur due to natural causes as well as domestic and industrial discard (plumbing, soldering waste).

Lead is taken up through inhalation of air and ingestion of lead in food, water or dust. It has a tendency to accumulate in the body. Inhaled lead particulate is particularly harmful to children. Lead poisoning causes gastrointestinal and kidney damage. Lead dust is absorbed primarily in the duodenum, bone and soft tissue. Once absorbed, lead is transported in the blood (Fig. 6.6) and binds to erythrocytes. As lead is not easily excreted it interferes with the production of red blood cells and may damage the kidneys, liver and nervous systems. In case of severe poisoning peripheral involvement results in a type of paralysis called "lead palsy". Inhalation lead particulates in the form of fumes and dust are particularly harmful to children, in whom even slightly elevated levels of lead in the blood can be responsible for learning disabilities, seizures and in an extreme case even death.

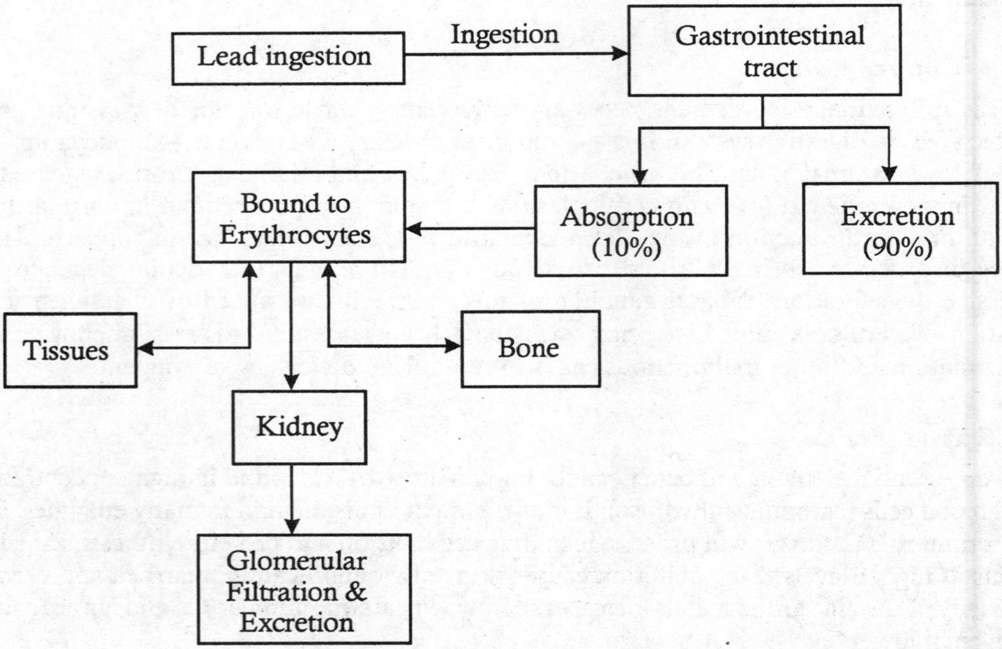


**Fig. 6.6:** Schematic of Lead Transport in Human System

The $Pb^{2+}$ ion inhibits various enzyme systems—acetylcholine esterase (AChE), alkaline phosphatase, ATPase, carbonic anhydrase, cytochrome oxidase. Lead interferes with heme synthesis by inhibiting the enzyme, aminoluvievulinic acid dehydrogenase (ALAD), and in platelet destruction. Anemia by lead poisoning is due to decrease in blood hemoglobin. Nervous system is also affected by exposure to lead. Lead damages peripheral nerves (motor neuron disease). In the central nervous system, behavior dysfunction leads to *encephalopathy*. Low-calcium diet enhances lead retention, thus enhancing lead toxicity. Because of chemical similarities between $Pb^{2+}$ and $Ca^{2+}$, bones serve as a repository for lead accumulated by the body. Both lead and calcium compete for similar binding sites in intestinal mucosal proteins. Lead affects the kidney function (reduced glomerular function and proximal tubular damage). Lead poisoning impairs mitochondrial function. It also suppresses the immune system. If there is no permanent brain damage, a the symptoms of lead poisoning can be eased gradually by removing lead from the body tissues with substances such as calcium salt of ethylenediamine tetraacetic acid (EDTA) and penicillamine.

Metal coordination, acting as bridge between protein-nucleic acid constituents in protein-metal-nucleic acid complexes, is a general structural feature. Formation of proper tertiary and quaternary structures (protein folding) of proteins, protein-nucleic acid and other biological complexes is very sensitive to the local environment, pH and temperature. While $pK_a$ values of hydrated-metal complexes are in higher pH range (>8.5), the pKa of hydrated Pb(II) is in the physiological pH range (pH = 7.4), and water-mediated Pb(II) may take part in the site-specific cleavage of phosphodiester bonds in nucleic acids (Fig. 6.7).

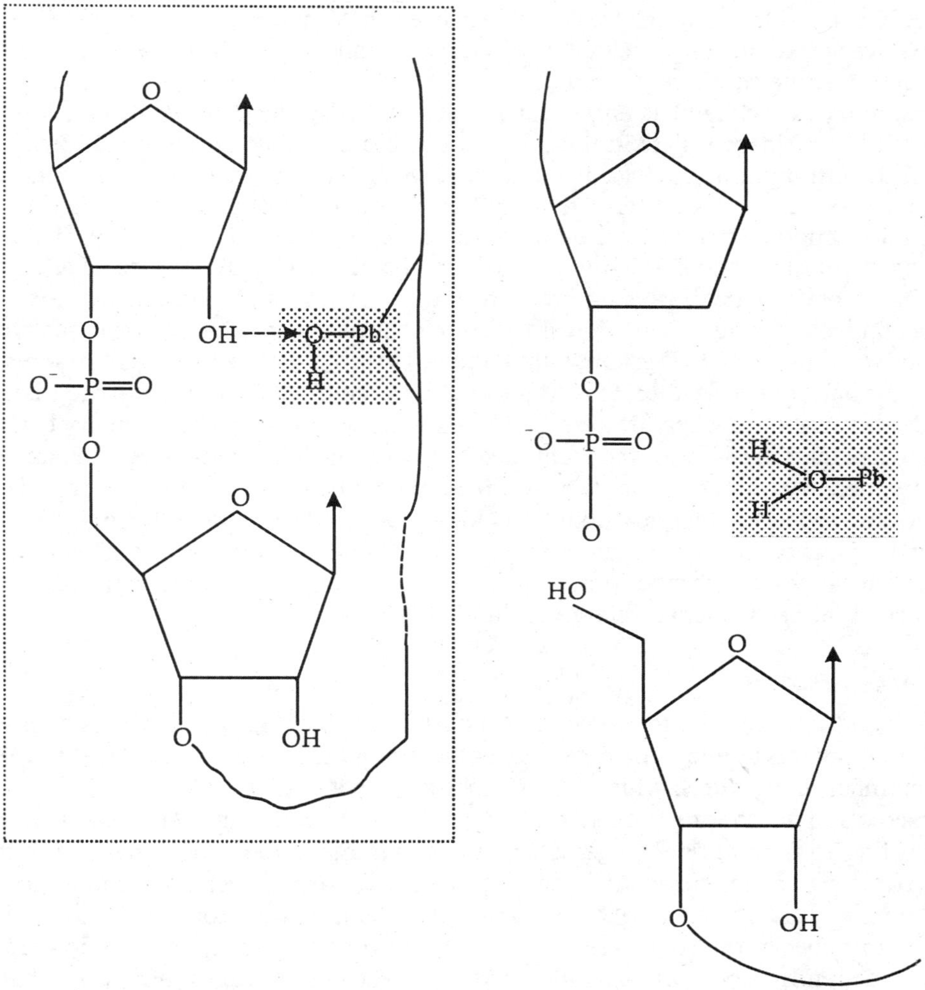


**Fig. 6.7:** Mechanism of Pb(II) Cleavage in Nucleic Acids

*Source*: Brown, R.S. *et al.* (1983), *Nature*, 303; 5430

### *6.3.5.7 Mercury (Hg)*

Mercury is used in many industrial processes and applications. Major sources of mercury pollution are anthropogenic—mining, industrial (chlor-alkali) effluents and pesticides and fungicides. Elemental mercury and mercury compounds are toxic. Because of its high volatility mercury 95 per cent of mercury is in the vapor phase and is widely dispersed in the Earth's atmosphere Chronic mercury poisoning may result from occupational inhalation of mercury vapors, dusts and volatile organic mercurials. Vapors of mercury cause renal damage and injury to the central nervous system. The symptoms of mercury poisoning depend on the type of mercury compound involved and the mode of contact. Symptoms may include intestinal cramps, vomiting, and tremors, paralysis and other neurological disorders and brain damage. Mental and personality changes marked by depression may also be found in some cases. Lesions of the central nervous system characterize poisoning with organic mercurials. This form of mercury poisoning became known as Minamata disease because of a dramatic outbreak that occurred in Minamata, Japan in early 1950s. There was progressive weakening of muscles, loss of vision, impairment of the cerebra functions eventual paralysis, and in some cases, coma and death.

Once mercury is oxidized it enters lakes and ocean by deposition. There, after bacterial methylation, it can bioaccumlate in the biota and be biomagnified along the food chain to levels that limit fish consumption. Ecological and toxicological effects of mercury are strongly dependent on the chemical species present. Mercurous, Hg(I), compounds are less toxic than Hg(II) compounds, and inorganic compounds are less toxic than organic compounds. $HgCH_3$ is readily absorbed by the intestine membranes and is 100 times more toxic than the inorganic species. $HgCH_3$ can readily cross the blood-brain barrier, causing injury to the cerebellum and cortex. Physicochemical and environmental factors influence the transformation of mercury into highly toxic methylated form.

$Hg^{2+}$ binds to sulfhydryl (SH) groups (similar to Pb) in proteins, like hemoglobin, serum albumin, alkaline phosphatase, and lactate dehydrogenase. Organic mercury is more toxic than elemental mercury. Methyl mercury is readily absorbed by membrane lipids. Methyl mercury enters the food chain through planktonic life and is concentrated by fish as it goes through the food chain. Conversion of inorganic mercury to methyl mercury by anaerobic methane-synthesizing bacteria is one of the most distressing aspects of mercury pollution. Alkyl mercurials can affect the brain tissues and can pass through placental barrier and enter fetal tissues. Binding of mercury to cell membranes may cause imbalance in the membrane permeability to $K^+$, affecting the neuronal transmission processes leading to cerebral palsy, mental retardation and chromosome defects.

#### *6.3.5.8 Arsenic (As)*

Arsenic is found in copper, lead, cobalt and nickel ores. Smelting of these ores releases arsenic. Glass, ceramics, paints and insecticides are the other industries contributing arsenic. Arsenicals are used in a number of products, which include insect, rodent and weed killers.

Free arsenic and all its compounds are strong poisons. Arsenic compounds, in general, are skin irritants leading to dermatitis. Inorganic arsenic is extremely toxic, both acute and chronic. Of these, As(III), or arsenite, compounds are the most toxic (e.g. insecticides containing arsenious oxide, calcium or lead arsenate). Most natural waters contain the more toxic inorganic forms of arsenic. Organic species of arsenic are found in aquatic organisms. Organoarsenites are less toxic. The potency depends also on the oxidation state. $H_3AsO_3$ is sixty times more toxic than $H_3AsO_4$. The quaternary arsenium compounds (arsenocholine) are essentially non-toxic.

The potential hazards of arsenic to humans and animals are due to inhalation, or ingestion, usually from drinking water containing high concentration of inorganic arsenic compounds. Exposure may be accidental or an occupational hazard. Inorganic arsenic compounds can be degraded to arsenic or organoarsine ($AsCH_3$) by microorganisms.

Arsenic causes mesothelioma, cancer of pulmonary lining, and damage to many organs; and in some respects its toxicity resembles that of lead and mercury. Arsenic is believed to exert its toxicity by combining with certain enzymes (inhibits SH groups in enzymes). In persons exposed to arsenic, the discernable effects are destruction of red blood cells and damage to kidneys. In the case of chronic exposure, pigmentation and scaling of skin are clearly detected. Nervous manifestations marked by paralysis and confusion may be noted at some stage. Sooner or later the characteristic streaks appear across the nails.

As(III) and As(V) inhibit pyruvate dehydrogenase. As(III) or arsenite compounds are more toxic. Because of the chemical similarities of arsenic to phosphorous, AS(III) interferes with some biochemical reactions involving phosphorylation (citric aid cycle) reactions (no ATP formed). The clinical manifestations of arsenic poisoning are myriad. It is difficult to diagnose early symptoms of arsenicosis because such nonspecific symptoms may also present in many other diseases. Its ingestion may result in internal malignancies, including cancer of kidneys, bladder, liver and other organs.

### 6.3.6 Physiological Effects of Gaseous Pollutants

Physical stimuli (cold, hot, etc.), pharmacological agents and irritant and noxious gases ($SO_2$, $NO_2$, $O_3$ and $Cl_2$, etc.) produce bronchoconstriction. The irritability of the airways is termed airway resistance, which is the reduction in tolerance or increase in the degree of bronchospasm.

Sulfur is present in all forms of fuels—coal, petrochemicals (organic compounds). It is extremely difficult to eliminate organic sulfur. Sulfur dioxide, $SO_2$, is formed from burning of fossil fuels and volcanoes emit many gases including $SO_2$. Exposure to it can cause eye and throat irritation. $SO_2$ acts as an irritant gas in the respiratory tract, damaging the mucus lining of the respiratory tract, and leading to chronic bronchitis. Acting synergically with $O_3$ and $NO_2$ it is responsible for cardiovascular deseases. It is associated with asthma and chronic bronchitis. It also reacts with water to produce $H_2SO_4$, which is the primary cause for acid rain.

Particulate matter is deposited in the pulmonary tract according to the size. Large particles are retained in the nose (hair cilia function as scrubbers), small particles reach into small bronchiole, and still smaller particles are deposited in the alveoli sacs. $SO_2$, because of its high solubility, is absorbed almost totally in the upper airways. $H_2SO_3$ and $H_2SO_4$ may be the actual stimuli for the receptors.

Similarly, nitrogen dioxide, $NO_2$, and photochemical oxidants such as peroxyacetyl nitrate (PAN) can produce eye irritation, pulmonary edema (accumulation of fluid in the lung tissues) and respiratory pneumonia. An influx of edema fluid into the lung could provide an ideal culture medium for microbial growth. Powerful oxidants such as $O_2$, hydrogen peroxide, $H_2O_2$, and $O_3$ are edemagenic agents. They enhance cell damage, and premature aging, etc. they generate free radicals and damage particularly the-SH groups in enzymes.

Carbon monoxide, CO, is a powerful asphyxiating gas, interfering with the normal function of hemoglobin. The function of hemoglobin in blood is to transport molecular oxygen, $O_2$. CO combines with blood hemoglobin (Hb) to form carboxy-hemoglobin (HbCO).

$$HbCO + HbO_2 \rightarrow Hb\ (total) \qquad \text{Haldane principle} \qquad \ldots(6.24)$$

Carbon monoxide has greater affinity to hemoglobin than oxygen ($HbCO \approx 250 \times HbO_2$).

$$\frac{[HbCO]}{[HbO_2]} = M\frac{p(CO)}{p(O_2)} \qquad \text{Haldane equation} \qquad \ldots(6.25)$$

where, $M$ = Haldane constant 250;

$p$ = partial pressure of CO and $O_2$.

Therefore, HbCO is more stable and effectively stifles hemoglobin-oxygen transport ($HbO_2$) system. So, binding of CO to hemoglobin not only reduces the total oxygen-carrying capacity of hemoglobin, but it also shifts the oxyhemoglobin curve to a region where $P_{O2}$ is below that usually found in the tissues. In addition to CO-poisoning, damage to the CNS may also occur (Table 6.8).

**TABLE 6.8:** HbCO IN BLOOD AND SYMPTOMS

| *CO Level (ppm)* | *% Present as HbCO* | *Symptoms* |
|---|---|---|
| 20-50 | 5-10 | Chest pain |
| 50-150 | 10-20 | Headache; visual impairment |
| 150-500 | 20-40 | Nausea; drowsiness; disorientation |
| 500-750 | 40-60 | Stupor; coma |
| >900 | >60 | Death |

Hydrogen cyanide, HCN, is used as a fumigating agent against pests. It interferes with the oxidative phosphorylation process.

Though chemical characteristics of individual pollutants and their effects on vegetation and living organisms have been studied, the synergic (potentiation) effects of mixtures of these air pollutants on plant and animal species are not well understood. Combination of pollutants, $O_3$, $SO_2$, $NO_x$, etc., has synergic (potentiation) effect, that is, a combination of these pollutants will have greater effect than the effect of any one of them separately.

Organisms have evolved defense mechanisms against several individual pollutants. For example—*neutrophils* belong to a group of cells, called *polymorphonuclocites* or *granulocytes*, making up 50 per cent of the white cells of the peripheral blood. They constitute the primary defense against microbial infection in mammals. The main function of the neutrophils at the inflammatory site is the destruction of foreign particles or cells.

The *eosinophils* are also polymorphonuclear leukocytes. They are found in the bone marrow. The macrophages are mononuclear phogocytes. Peroxides are abundantly present in neutrophils, eosinophils and monocytes.

$$2O^* + 2H \rightarrow H_2O_2 \qquad \ldots(6.26)$$

Superoxide dismutase enzyme system scavenges the peroxide molecules. Hydrogen peroxide, $H_2O_2$, is a powerful oxidant (similar to $O_3$) and reacts with lipids that contain unsaturated fatty acids to form lipid peroxides. These lipid peroxides are strong oxidants, detrimental to the cell.

$H_2O_2$ → $H_2O$ ; Reduced glutathione (peroxidase) → Oxidized glutathione ; NADPH → NADH (reductase) ...(6.27)

Some air pollutants and their effects on animals and humans are given in Table 6.9 and air quality standards for some air pollutant are listed in Table 6.10 (however, these defense mechanisms are against individual pollutants, not against the synergic effects of combination of pollutants).

**TABLE 6.9:** AIR POLLUTANTS AND THEIR EFFECTS ON ANIMALS/HUMANS

| *Pollutant* | *Mode of Action* | *Toxic Effects* |
|---|---|---|
| Asbestos | Direct | Asbestosis; mesotheliomas (volcanosis); lung cancer |
| Beryllium compounds | Direct | Berylliosis |

| *Pollutant* | *Mode of Action* | *Toxic Effects* |
|---|---|---|
| Carcinogens (Hydrocarbons) | Direct | Cancer |
| Caustic gases & particulates | Direct & Indirect | Chronic bronchitis; pulmonary edema |
| Cotton dust & fibers | Direct | 'Brown lung' disease; lung cancer |
| Particulate (small) | Direct | Silicosis; pneumonosis |
| Oxides of nitrogen | Direct | Infant death |
| Ozone & other oxidants | Direct | Accelerated aging; cell damage; pulmonary diseases |
| Photochemical oxidants | Direct | Eye and lung irritation and diseases |
| Pollen | Direct | Allergy & asthma |
| Respiratory irritants | Direct | Emphysema |

**TABLE 6.10:** PRIMARY AND SECONDARY STANDARDS FOR SOME AIR POLLUTANTS

| *Pollutant* | *Concentration (Primary)* | *Concentration(Secondary)* |
|---|---|---|
| Carbon monoxide (CO) | 10 ppm (10 μg/m$^3$) (8 hrs) | 10 ppm (10 μg/m$^3$) (8 hrs) |
| Hydrocarbons (HCs) | 0.25 ppm (160 μg/m$^3$) (3 hrs) | 0.25 ppm (160 μg/m$^3$) (3 hrs) |
| Hydrogen sulfide ($H_2S$) | 0.03 ppm | 0.02 ppm |
| Lead (Pb) | 2 μg/m$^3$ (annually) | 1 μg/m$^3$ (annually) |
| Oxides of Nitrogen ($NO_x$) | 0.1 ppm (200 μg/m$^3$) (daily) 0.05 ppm (100 μg/m$^3$) (annual) | 0.2 ppm (200 μg/m$^3$) (daily) 0.05 ppm (100 μg/m$^3$) (annual) |
| Oxides of Sulfur ($SO_x$) | 0.15 ppm (360 μg/m$^3$) (daily) 0.03 ppm (80 μg/m$^3$) (annual) | 0.2 ppm (480 μg/m$^3$) (daily) |
| Ozone ($O_3$) | 0.1 ppm (200 μg/m$^3$) (1 hour) 0.08 ppm (150 μg/m$^3$) (8 hours) | 0.2 ppm (200 μg/m$^3$) (1 hour) 0.08 ppm (150 μg/m$^3$) (8 hours) |
| Particulate | 50 μg/m$^3$ (annually)150 μg/m$^3$ (daily) | 50 μg/m$^3$ (annually)150 μg/m$^3$ (daily) |
| Photochemical oxidants | 0.10 ppm | 0.10 ppm |

Conversion formula concentration from mass to ppm

$$\text{Concentration}\ (\mu g \cdot m^{-3}) = \text{Concentration}\ (\text{ppm}) \times \frac{M_r}{24500} \times 10^{-6}$$

where, $M_r$ = Molecular mass.

## BIBLIOGRAPHY

Adriano, D. C. and Johnson, A. H. (1989), Springerverlag: New York. "*Acid Precipitation* (Vol. 2)*: Biological and Ecological Effects.*"

Agarwal, S. K. (1991), Himanshu Pubs: Udaipur. "*Pollution Ecology.*"

Crawford, M. (1976), McGraw-Hill: New York. "*Air Pollution Control Theory.*"

Edwards, C. E. (1974), CRC Press: Boca Raton, F. L. "*Persisting Pesticides in the Environment.*"

Folk, G.E. Jr. (1969), Lea & Febiger Pubs: Philadelphia, PA. "*Introduction to Environmental Physiology.*"

Harrison, R. M. and Perry, R. (1984), Chapman & Hall: London. "*Handbook of Air Pollution Analysis,*" (2nd edn.).
Horne, R. A. (1978), Wiley & Sons: New York. "*The Chemistry of our Environment.*"
Hughes, D. O. and Latham, J. L. (1968), Butterworth: London. "*Physics for Chemists and Biologists.*"
Hutzinger, O. (ed.). (1990), Springerverlag: Berlin. "*Handbook of Environmental Chemistry.*"
Jaffe, L. S. (1970), Springerverlag: New York. "*Global Effects of Environmental Pollution.*"
Kormondy, E. J. (1977), Prentice-Hall: Engelwood Cliffs, NJ. "*Concepts of Ecology*", (2nd edn.).
Koumitzis, T. and Samara, C. (eds.) (1995), Springerverlag: Berlin. "*Airborne Particulate Matter.*"
Lee, S. D. (1977), Ann Arbor Pubs: Ann Arbor, MI. "*Biochemical Effects of Environmental Pollutants.*"
Manahan, S. E. (1993), Lewis Pubs: New York. "*Environmental Chemistry*", (5th edn.).
Markert, B. (ed.) (1993), VCH Pubs: Weinheim. "*Plants as Biomonitors.*"
McGrath, J. J. and Barnes, C. D. (eds.) (1982), Academic Press: New York. "*Air Pollution: Physiological Effects.*"
Milburn, J. A. (1979), Longmans: London. "*Water Flow in Plants.*"
Monteith, J. L. (1973), Edward Arnold: London. "*Principles of Environmental Physics.*"
Narayanan, P. (2003), New Age Intl. Pubs: New Delhi. "*Essentials of Biophysics,*"(2nd Print).
Newberne, P. M. (1976), Marcel Dekker: New York. "*Trace Substances and Health,*" (vols, 1 & 2).
Nobel, P. S. (1974), Freeman: San Francisco, CA. "*Biophysical Plant Physiology.*"
Peirce, J. J., Weiner, R. F. and Vesiland, P. A. (1998), Butterworth-Heineman: New Delhi. "*Environmental Pollution and Control,*" (4th edn.).
Perkins, H. C. (1974), McGraw-Hill: New York. "*Air Pollution.*"
Saha, J. C. *et al.* (1999), *Crit Rev Envi Sci & Eng.*, **29**(3): 281. "A Review of arsenic poisoning and its effects on human health."
Stern, A. C. *et al.* (1984), Academic Press: New York. "*Fundamentals of Air Pollution,*" (2nd edn.).
Van Wijk, W. R. (ed.) (1963), North Holland Press: Amsterdam. "*Physics of Plant Environment.*"
Waldbott, G. L. (1978), Mosby: St. Luis, MO. "*Health Effects of Environmental Pollutants,*" (2nd edn.).
Willeke, K. and Baron, P. A. (1993), Van Nostrand-Reinhold: New York. "*Aerosol Measurement: Principles, Techniques and Applications.*"
Woodward, F. I. and Sheehy, J. E. (1983), Butterworth: London. "*Principles and Measurements in Environmental Biology.*"

# SECTION-II

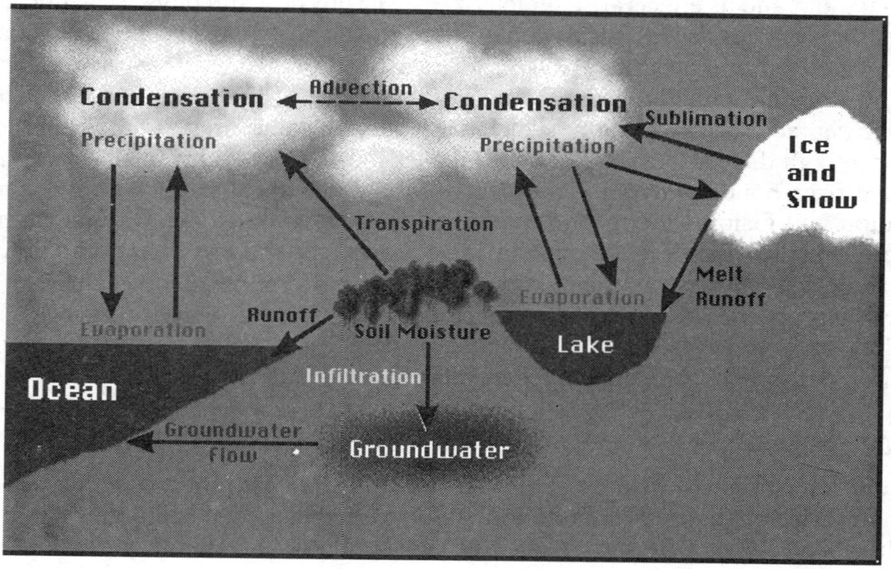


# The Hydrosphere

## OVERVIEW

Chapters 7 to 10 deal with the hydrosphere, its pollutants and their environmental effects. Hydrosphere encompasses about three-fourths of our planet's surface. This includes water sources in lakes, rivers, seas and oceans. In addition, water is found in the atmosphere and to a certain depth under the ground. As the earth's atmosphere is a closed system, the water supply of the earth is continually recycled by rainfall, water percolation and runoffs followed by evaporation. All this constitutes the hydrological cycle. The study of water and water resources are treated under hydrology.

## WATER

The quality of water and its effects are characterized by its physical and physicochemical properties. Water exhibits unique physicochemical properties due to its molecular structure. Water comprise of giant molecular conglomerates of $(H_2O)_n$. Individual $H_2O$ molecules have tetrahedral (spatial) configuration. They are joined together by a spatial network of hydrogen bonding. This network imparts polar characteristics to water, which accounts for some of its unique physicochemical properties. The high dielectric nature of water makes it an almost "universal" solvent. The inter-molecular hydrogen bonding gives unique thermal properties to water, such as high specific heats of vaporization and fusion. The three-dimensional hydrogen bonds give water a clathrate (cage-like) structure, which makes it more open in solid phase (ice) and less dense than liquid phase at that temperature. The physical characteristics of water have profound effect on all living organisms.

The chemical properties of water deal with its salinity and hardness due to dissolved substances, the pH and electrolytic characteristics. Biological attributes of water are due to the presence of organic matter, pathogens, coliforms, trace metals and toxicants.

## WATER POLLUTION

Water pollution is caused by the presence of undesirable and hazardous materials and pathogens beyond certain limits. Much of the pollution is due to anthropogenic activities like discharge of sewage, effluents and wastes from domestic and industrial establishments, particulate matter and metals and their compounds due to mining and metallurgy and fertilizer and pesticide runoffs from agricultural activities.

## ECOLOGICAL EFFECTS

All pollutants, atmospheric and land-based invariably enter water bodies, by direct discharge, precipitation and runoffs. Water bodies, thus, become sinks as well as carriers of pollutants. Water pollution has wide ecological impact, as it is an important raw material in photosynthesis and hydrological processes. Water pollutants impart to it undesirable properties like odor, turbidity and retardation of photosynthesis, deoxygenation and eutrophication and thermal pollution. In addition, thermal stress has profound physical, chemical and biological effects on all organisms. Turbidity blocks sunlight reaching deeper into the water bodies (lakes) and retards photosynthesis in aquatic plant systems. Chemical factors such as acid rain, mining leachates, hazardous chemicals, toxic trace elements deteriorate the water quality.

## BEHAVIOR OF POLLUTANTS IN WATER BODIES

One of the adverse effects of pollutants on a water body is the decrease in dissolved oxygen (DO), which is an essential factor for the survival of aquatic organisms. Decrease in DO is, in fact, a positive indicator of water pollution and stressful survival of aquatic organisms. The primary cause for DO depletion is the presence of organic matter. Biochemical oxygen demand (BOD) and chemical oxygen demand (COD) are quantitative methods for assessing the quality of water and wastewater. COD is the amount of oxygen consumed under specific conditions in the oxidation of organic and oxidizable inorganic compounds.

The biological community in an unpolluted stream or lake is highly diverse. Polluted water bodies diminish or destroy aquatic population and increase algal growth. If the pollutants toxic to aquatic animals, such as pesticides and herbicides, trace elements (As, Cd, Cr, Hg, Pb) and radioactive materials are present, both the type and total number of species will decrease, irrespective of the other ecological characteristics of the water body. The ecology of rivers and streams differs from that of stationary water bodies (lakes and ponds) and, the effects of pollutants in the two are also different. Streams and rivers have inlet as well as outlet for pollutants. Streams also have inherent capacity for re-aeration due to turbulence and mixing. Pollutants are diluted and dispersed due to flow. On the other hand, lakes and ponds have only inflow and no outflow. Hence concentration of pollutants gradually increases in these water bodies, but their effects are localized. Based on the biological activity in them, lakes are of three types, oligotrophic, eutrophic or dystrophic.

Reduced turbidity in lakes and ponds, as compared with streams, encourage growth of certain types of aquatic organisms. Discharge of nutrient compounds into lakes produces uncontrolled growth of plankton in the top layer (epilimnion region) of the lakes. Algae interfere with the aquatic activities. They also act as barrier to the penetration of oxygen into water. There is a temperature-density (depth) relationship that results in the formation of thermal stratification within the stationary water bodies.

Pollution of water bodies results in the disturbance of photosynthesis-respiration cycle (P-R cycle). Net disturbance of P-R balance is indicative of net rate of oxygen production or consumption. There is horizontal separation of P-R function in mobile water bodies and vertical separation of these functions in stationary water bodies. The vertical separation of P-R function in lakes can be induced by nutrients that encourage the algal bloom, depletion of $O_2$. Water flow characteristics in streams, rivers and estuaries can be laminar (placid flow) or turbulent. They determine the dilution and discharge of pollutants.

Environmental risk evaluation is done by statistical methods of analysis, generally by dose-response analysis.

# 7

# Water

Water is an important constituent of biotic community. In nature, it occurs on the land, below its surface, in atmosphere and in the biomass. 97% of the total volume of water available is in the oceans, 2% stored in the form of ice-sheets and less than 1% is available as fresh water.

In the atmosphere most of the water is present in the form of mosture or in vapor form. Water vapor comes from evaporation from the oceans, lakes, rivers, ice-fields and glaciers, transpiration from the plants and animal respiration.

Water plays a significant role in the continuity of life due to its unique qualities. The quality of water and its effects are characterized by its physical and physicochemical properties. Water exhibits unique physicochemical properties due to its molecular structure. It comprises of giant molecular conglomerates of $(H_2O)_n$. Individual $H_2O$ molecules have tetrahedral (spatial) configuration. They are joined together by a spatial network of hydrogen bonding. This network imparts specific polar characteristics to water, which accounts for some of its unique physicochemical properties. The high dielectric nature of water makes it an almost "universal" solvent. The inter-molecular hydrogen bonding gives unique thermal properties to water, such as high specific heats of vaporization and fusion. The three-dimensional hydrogen bonds give water a clathrate (cage-like) structure, which makes it more open in solid phase (ice) and less dense than in liquid phase at that temperature. The physical characteristics of water have profound effect on all living organisms.

The chemical properties of water deal with its salinity and hardness due to dissolved substances, the pH and electrolytic characteristics. Biological attributes of water are due to the presence of organic matter, pathogens, coliforms, trace metals and toxicants.

This chapter illustrates the water cycle in the atmosphere and the physical and chemical properties of water.

Water is an odorless, tasteless, transparent liquid that is colorless in small amounts but when present in large quantity exhibits a bluish tinge. It is undoubtedly the most common but also the most important of chemical compounds. Existing in gaseous, liquid and solid states it covers about 70 per cent of Earth's surface and even finds its way into the atmosphere. It is composed of the two common elements, hydrogen and oxygen; has a simple molecular structure but at the same time, exhibits some extraordinary physical and chemical properties. These properties of water have played a central role in the development and growth of life on Earth. Life is believed to have originated in the waters of the oceans, which have a complicated composition. The occurrence of water as liquid on the surface of the Earth, under normal conditions, makes it invaluable for transport, for recreation and as a habitat for myriad of animals and plants. Again, its anomalous properties like, having lower density in the solid form, as compared to its liquid form, are great boon to aquatic life. Otherwise, its survival would have been repeatedly threatened during freezing winters. In short, water is essential to all living organisms, as a large proportion (over 60 per cent) of their tissues consist of water and they use aqueous solutions like blood and digestive juices as media for carrying out their biological functions. Water is so important to their bodies that they can survive only for a few days without it.

## 7.1 DISTRIBUTION OF WATER AND THE HYDROLOGIC CYCLE

The atmosphere contains water in the form of vapor. Water vapor in the atmosphere absorbs infrared radiation at about 10 μm and contributes towards maintaining the long-wavelength radiation balance of the earth. Water droplets and their size in the clouds influence the distribution of solar radiation by absorption and scattering processes.

Hydrosphere is the discontinuous layer of water at or near the Earth's surface. The liquid and frozen surface water, ground water present in soil and rocks, and the water vapor in the atmosphere are its components. About 1,347 million cubic kilometers of water is distributed between these reservoirs (Table 7.1). But, it is a stark fact that Earth's fresh water reserve is only 2.7 per cent of the total and out of this, 2.2 per cent is locked up in polar ice caps and glaciers. A meager 0.5 per cent is distributed in ground water. The rivers carry only 0.0001 per cent total water reserves.

**TABLE 7.1:** DISTRIBUTION OF EARTH'S WATER SUPPLY

| *Source* | *Volume* ($\times 10^6$ *km*$^3$) | *Per cent of Total* |
|---|---|---|
| Oceans | 1,310.3 | 97.3 |
| Ice | 29.49 | 2.22 |
| Ground water | 6.73 | 0.5 |
| Inland lakes | 0.242 | 0.02 |
| Soil moisture | 0.074 | 0.005 |
| Atmospheric water vapor | 0.014 | 0.001 |
| Rivers | 0.0017 | 0.0001 |

### 7.1.1 The Hydrologic Cycle

The hydrologic cycle (Fig. 7.1) covers the steps by which water is transferred from one reservoir to the other. Many complicated processes are involved in this cycle. Solar radiation causes evaporation of water from the oceans, lakes, rivers, soil, and plants (transpiration). The water vapor rises. As temperature decreases with altitude in the troposphere, clouds form when it becomes sufficiently low for the rising water vapor in the air to become saturated and to form small droplets ($10^{-3}$ to $10^{-2}$ cm diameter). The minute particles of dust, salt (aerosols from the oceans), chemical substances, and soil, which are always freely available in the atmosphere act as nuclei to induce condensation of the water vapor. When the droplets coalesce to a sufficient size (about $10^{-2}$ cm diameter) and weight, the air currents within the clouds can not keep them suspended and they begin to fall down as precipitation—water drops or, if the temperature is below 0° C, as snow flakes or hail. Hailstones measure $10^{-1}$ cm in diameter or much larger.

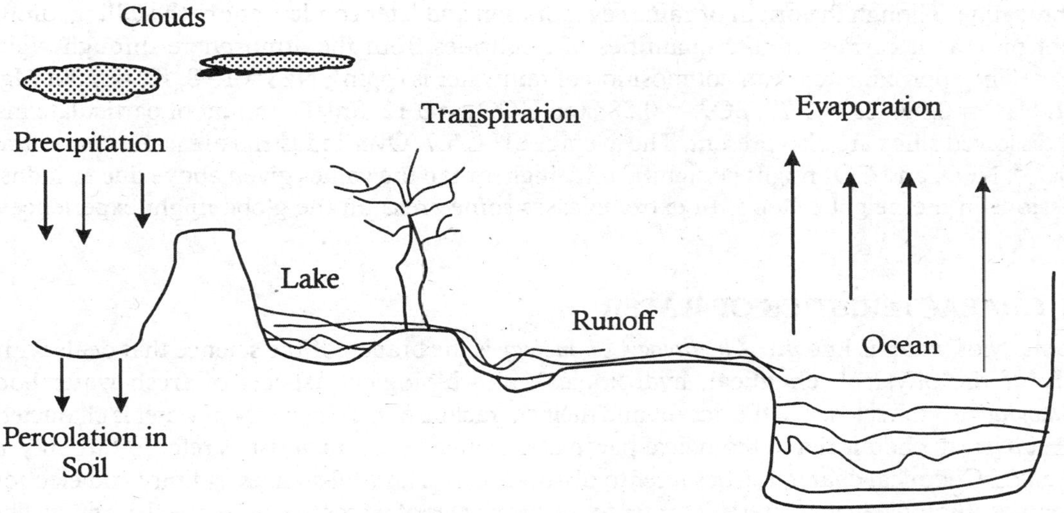


**Fig. 7.1:** The Hydrologic Cycle

Some of the precipitation runs over the Earth's surface and reaches lakes, rivers and marshes, ultimately meeting the oceans. If the surface is covered with dense vegetation, much of the precipitation may be held on different parts of the plants and trees. This process is called 'interception' and it prevents water from reaching the ground. In such a case water may evaporate directly from the plant surfaces into the atmosphere. The precipitation reaching the ground, as rain or melted snow, distributes in different ways: (1) evaporate and reach the atmosphere, (2) infiltrate the soil, (3) detained in catchment areas, including lakes, (4) become over-land flow (a form of run-off). In turn, moisture retained by the soil can be lost by evaporation or being withdrawn by plants (transpiration). The combined process is called 'evapotranspiration'. It can be defined as the sum of water used by vegetation and water lost by evaporation. The energy going into this operation in one day is equivalent to ~10,000 times the yearly energy consumption of all the countries. The main physical factors affecting evaporation are solar radiation, temperature, humidity, and wind speed. From the soil, water can percolate downward into the ground water zone.

### 7.1.2 Other Effects of Water Cycle

The water cycle is driven by solar energy. It is one of the climatic processes that contribute to our weather. Over the course of time long-term weather patterns have created extremes from ice covered continents to vast deserts. While the atmospheric temperature stratifies the major climatic zones of the earth, rainfall (or lack of it), river runoffs and melting of ice also determine the climatic and vegetation zones. Deserts and forests can occur in similar latitudes and temperature zones, depending on these factors. Climatic changes due to lack of seasonal rainfall (monsoons) and changes in the mean temperatures of the oceans (e.g. El Nino southern oscillation) affect the global agricultural production and also bring about ecological and demographic changes. Global warming would have long-term effects like melting of polar icecaps, inundating low-level landmass, and increase of average temperature of the oceans (and thus evaporation).

Fresh water, needed for humans, animals and plants is provided only by rain. Though the oceans contain an enormous amount of water, for non-marine life it is largely useless, as it is highly saline. Though the origin of rain is evaporation and later condensation, the falling rainwater is not pure as it carries minute quantities of impurities from the atmosphere through which it passes. The approximate mean composition of rainwater is (ppm): $Na^+ = 1.98$, $K^+ = 0.30$, $Mg^{2+} = 0.27$, $Ca^{2+} = 0.09$, $Cl^- = 3.79$, $SO_4^{2-} = 0.58$, and $HCO_3^- = 0.12$. Small amount of particulate matter and dissolved silica are also present. The average pH is 5.7. Over industrial areas the concentration of $SO_4^{2-}$, $NO_3^-$ and $CO_2$ might be significantly higher than the values given above due to industrial emissions in the neighborhood. In extreme cases some areas on the globe might experience acid rain.

## 7.2 CHARACTERISTICS OF WATER

The study of water is known as *hydrology. Limnology* is the branch of the science that deals with the study of the physical, chemical, hydrological, and biological aspects of fresh water bodies. *Oceanography* is the science of the oceans and their characteristics. The quality of water is characterized by its physical, chemical and biological parameters. Physical characteristics refer to turbidity, taste and color. Chemical characteristics refer to dissolved gases and substances, salinity and electrolytic properties. Biological characteristics refer to the properties of water that affect health and wellbeing and are determined by the presence of undesirable organisms like algae, bacteria, coliform and pathogens.

### 7.2.1 Properties of Water (General)

Water is a compound with unique physicochemical properties that make it distinct from other molecules of comparable molecular mass. Water, $H_2O$, is not monomolecular, but is a giant conglomerate $(H_2O)_n$. It is an inorganic, polar substance. The water molecule, $H_2O$, has two polar covalent bonds, with two hydrogen atoms bonded to the central, highly electronegative oxygen atom, O, together with two unpaired electron lobes (Fig. 7.2). The molecule is in the $sp^3$ hybridization. The valence angle (H—O—H = 104.5°) is close to the tetrahedral angle 109.5°. Water is characterized by the tetrahedral (spatial) structure of $H_2O$ molecules. The spatial, three-dimensional configuration, and the polar nature of water accounts for its high solvating ability, high surface tension, viscosity, heat of vaporization, and for a number of other properties, all of which can be ascribed to the extensive intermolecular hydrogen bonding network.

### 7.2.2 Hydrogen Bond and its Influence

*Hydrogen bond* is a polar, non-bonded, cohesive interaction. It is primarily an electrostatic attractive force between a positively charged ($\delta^+$) hydrogen atom in one molecule, and an electronegative ($\delta^-$)

atom in a second molecule. A hydrogen bond is formed when a hydrogen atom is between two electronegative atoms (such as O, N, F, Cl, and Br) that are <0.3 nm apart. The atom to which the hydrogen atom is covalently attached is called the *donor*, and the atom(s) that take part in the non-bonded interaction (hydrogen bond formation) is called the *acceptor*(s) (Fig. 7.3). Hydrogen bond is linear and has directionality. Though the energy required for hydrogen bond formation (or breaking) is only ~20 to 40 kJ/mol, in contrast to 400-500 kJ/mol for a covalent bond, it is one of the most important non-bonded interactions. Hydrogen bonds are involved in the structural stability and function of important biological molecules like proteins, nucleic acids, carbohydrates and their complexes (e.g. storage and transmission of genetic code via sequence-specific base-pairing in nucleic acids). They play a role in (driving) many chemical and biochemical reactions: e.g. enzyme-substrate, nucleic acid base-pairing, and protein-nucleic acid interactions.

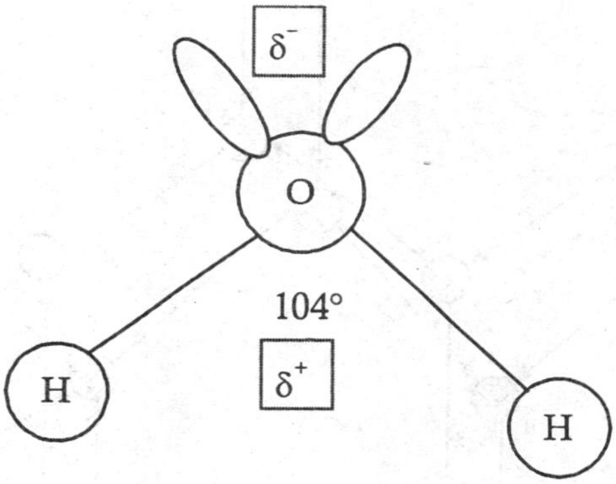


**Fig. 7.2:** Tetrahedral (spatial) Arrangement of Water Molecule, $H_2O$

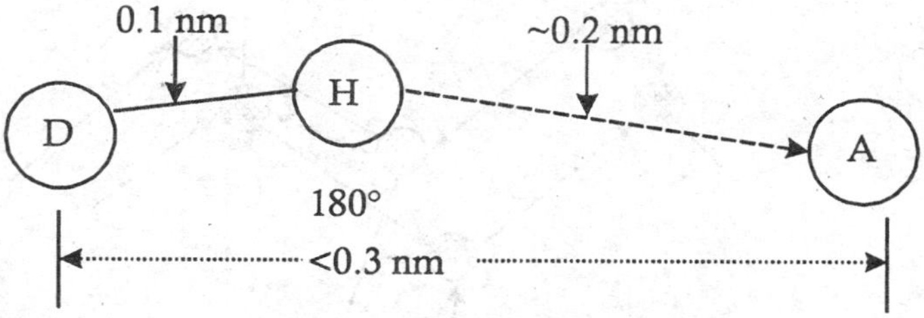


**Fig. 7.3:** Formation of Hydrogen bond between two Electronegative Atoms (D & A)

The properties of water are due to its tetrahedral (spatial) configuration. Each water molecule, $H_2O$, (in liquid and solid state) can form at least (statistically) four hydrogen bonds—two donor bonds, via the two covalently-bound H-atoms, and two acceptor bonds, via the two lone-pair electron lobes, from the donor atoms. As the spatial configuration of water molecule is tetrahedral the hydrogen bond network is also three-dimensional; these give water a clathrate (cage-like) spatial structure (Fig. 7.4).

The hydrogen bond network imparts distinct physicochemical properties to water, which are different from those of other molecules of similar composition (ethyl alcohol, acetone, ether, $H_2S$, $NH_3$, etc). Strong intermolecular forces arising from hydrogen bonding that form the space lattice, characterize liquid water. The unique properties of water are its high surface tension, high dielectric constant (~80 Debye), and distinct thermal properties like high latent heats of fusion and vaporization and vapor pressure, osmotic pressure and other colligative properties.

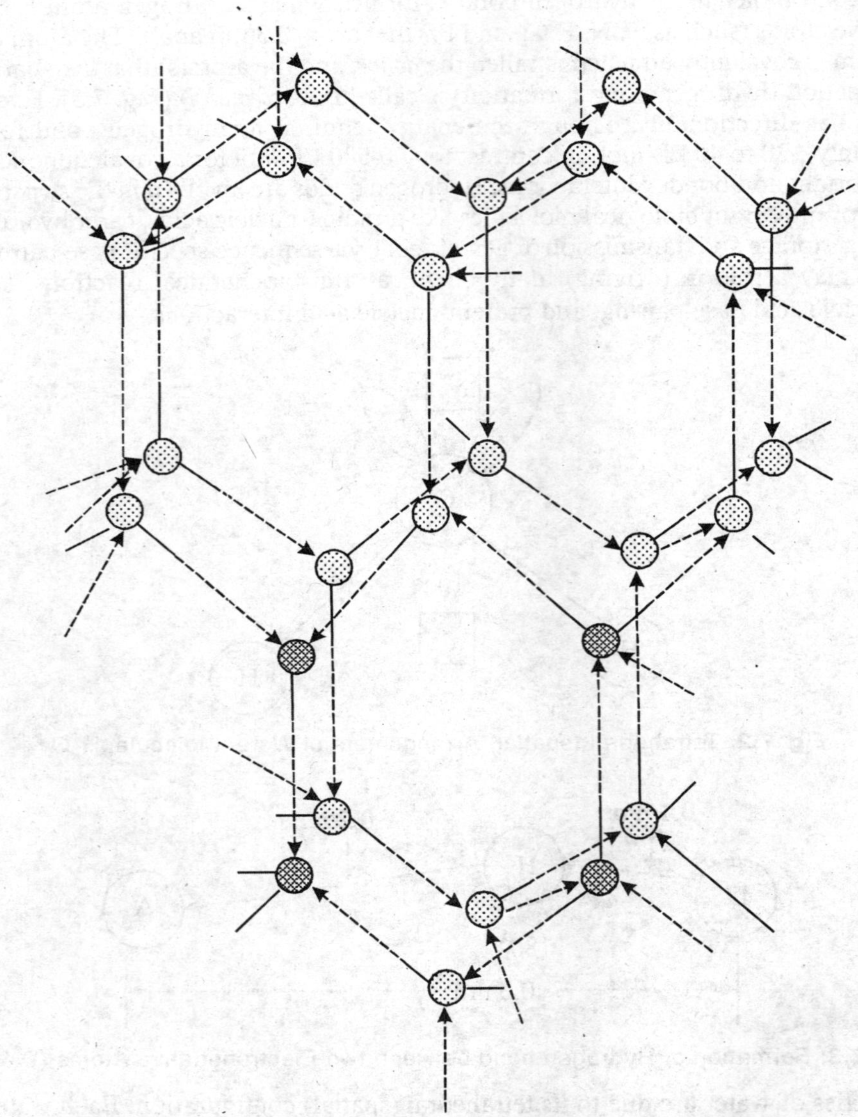

Fig. 7.4: Cage-like (Clathrate) Spatial Structure of Water and Ice

## 7.3 PHYSICAL PROPERTIES OF WATER

Physical properties of water (unique to it) are its anomalous density, surface tension, thermal properties, dipole nature (hydrophilic properties) and viscosity.

### 7.3.1 Density of Water

Most substances contract when they solidify from the liquid state and their densities would be higher in the solid state than in the liquid state. On the contrary, water expands when it freezes.

This is so because the ice has open lattice structure with a low packing efficiency where all $H_2O$ molecules are involved in four straight, tetrahedrally oriented hydrogen bonds. Upon melting, some of these bonds break, while others bend and the structure undergoes a partial collapse. The specific gravity of water (maximum 1.0 at 4° C) is higher than the density of solid water (ice) at 0° C (0.917), because of which ice floats on water. This property of water has important consequences for the survival of aquatic organisms. Ice floats on water and forms an insulating cover between cold air and the water below, so that water bodies even in the arctic areas of the world do not freeze solid. The insulating ice sheet also acts as a thermoregulator so that, aquatic organisms can survive the rigors of winter while living in the water.

Above 4° C, the density of water decreases, reaching 0.998 at 22° C. Density differences in water bodies prevent the wind from circulating bottom water to the surface. Thus, the stratification of zones in water bodies is maintained.

### 7.3.2 Surface Tension and Water Potential

Surface tension is a measure of the strength of the water's surface film. The attraction between the water molecules creates a strong film. It is responsible for the phenomenon called *capillary* action, the rise of liquids in very small diameter tubes. High surface tension of this liquid plays an important role in the transport of water and along with that, the nutrients in plants. In chlorophyll-containing plants, water is used as the source for photosynthetically related reduction of $CO_2$ to carbohydrates.

The *water potential*, $\psi$ is a measure of free energy ($\Delta G$) per mole of the water in the system and is given by

$$\psi = \psi_s + \psi_p + \psi_m \qquad ...(7.1)$$

where, $\psi_s$ = osmotic (solute) potential;
$\psi_p$ = hydrostatic (pressure) potential;
$\psi_m$ = matrix potential.

Organisms must maintain a resistance to water flow between themselves and their external environment, to prevent dehydration of their cells. Reverse is the case in environments of high water potential. The environmental water potential directly modifies the behavior of many organisms. Plant and animal cells have higher water potentials than the terrestrial environment, because of the resistance to water movement in the outer layers of the organisms (skin, cuticle etc.). Lower water potential of the atmosphere exerts suction in the plant. The stomata pores of the leaf are major escape routes for water from the plant to the atmosphere (*transpiration*). Plants strictly control this transpiration to avoid excessive water loss. If the loss of water is greater than the supply, then wilting of leaves and death of the plant occur.

In the plant the leaf cell is at the lowest water potential; therefore, water is forced to move from the soil to the leaf. This potential tends to become more negative during the day through the accumulation of solutes (carbohydrates) and tends to become more positive during the night through utilization of carbohydrates and respiration. In animals, pores of the skin and sweat glands under the skin surface perform this function.

Surface tension also plays a role in the transfer of energy from the wind to water to create waves. Waves are necessary for rapid oxygen diffusion in lakes and seas.

### 7.3.3 Thermal Properties of Water

The melting and boiling points of water are unusually high compared to those of similar compounds. The boiling and freezing characteristics of water are crucial to the life process of organisms, both aquatic-, and land-based. They also have profound effect upon the climate of the land areas near large bodies of water, as well as on wind patterns.

Water has high specific heat of vaporization (4.18 $J/kg^1 . K^1$). The specific heat of a substance defines the relationship between the amount of heat transferred to the change in temperature produced by that exchange. The energy required for evaporating water is much greater than that is required for many common solvents of equivalent molecular mass range (e.g. ether, acetone and alcohol evaporate at lower temperature than water). That is, transfer of heat to and from water produces a far less temperature change than it would in most other substances. At nights and when cold weather sets in, water cools slowly, and in the daytime or when warm weather sets in, water warms slowly. Water, thus, acts like an atmospheric thermostat.

When water is heated, part of the heat is used to break the hydrogen bonds. This explains the high heat capacity of water. These bonds are broken completely only when the water transforms into vapor (steam). Plants and animals utilize this high heat capacity of water for cooling by transpiration and sweating. Vaporization occurs by the escape of rapidly moving water molecules through the surface into the air. Evaporation rate, $\xi$, is

$$\xi = (P_W - P_h) \times (a + bv) \quad \text{...(7.2)}$$

where, $P_W$ = vapor pressure of air in contact with water surface;

$P_h$ = actual air pressure at the same height;

$v$ = wind velocity;

$a$ & $b$ = empirical constants

Evaporation is a physicochemical phenomenon; transpiration is both physicochemical and biochemical.

Latent heat of fusion is the thermal energy required for changing a unit mass of ice to water. *Clausius-Clapeyron* equation relates latent heat, $l_H$, pressure ($P$), volume ($V$) and temperature ($T$).

$$\frac{dP}{dT} = \frac{l_H}{T \cdot \Delta V} \text{ Clausius - Clapeyron equation} \quad \text{...(7.3)}$$

$$\log P = \frac{l_H}{2.303RT}$$

The reverse of evaporation, namely condensation occurs at nights (dew formation), transferring heat to the organisms.

### 7.3.4 The Dipole Nature of Water

Water exhibits polar character. There is asymmetric distribution of charge in the water structure. The electrons forming the O—H bonds are displaced towards the more electronegative oxygen atom; as a result, the hydrogen atoms acquire an effective positive charge. The unshared electron pairs in the $sp^3$ hybrid orbitals set up two negative poles. This asymmetric distribution of charge creates in water a permanent dipole, with a moment = 4.8.

The coulombic (electrostatic) force, $f$, between charges, $q_1$ & $q_2$, separated by a distance, $r$, is

$$f = \frac{q_1 q_2}{\varepsilon \times r^2} \quad \text{...(7.4)}$$

where, $\varepsilon$ = Dielectric constant

The dielectric property of a solvent reduces the electrostatic (*Coulomb*) interaction between the charged ions. The effect of such a solvent on polar solutes (acids, bases and salts) is the formation of relatively independently moving ions in solution. Because of the polar character of the water molecule, it can attach itself to either positive or negative ions, solvating them. In addition, water has a high dielectric constant ($\varepsilon$ = 78; benzene = 2; ethyl alcohol = 24). Electrolytic dissociation takes place more easily in a medium with high dielectric constant, since the interaction between the ions weakens. Due to its high dielectric constant charges in water interact with a force ~80 times less than that in vacuum. The relative ease with which charges can be separated in water makes it a 'universal solvent' to many polar and neutral substances. This property of water makes it an excellent medium by which many mineral nutrients (from soil) and biologically important molecules are transported in living systems. It has unique hydration properties towards biological macromolecules (particularly proteins and nucleic acids) that determine their three dimensional structure, and hence their functions in solution.

### 7.3.5 Viscosity of Water

Water is a relatively non-viscous (Newtonian) liquid and the motion of water (fluid) in non-turbulent state is given by the *Bernoulli's* theorem.

$$P_1 \cdot h_1 \cdot \rho \cdot g + \frac{1}{2}\rho \cdot v_1^2 = P_2 \cdot h_2 \cdot \rho \cdot g + \frac{1}{2}\rho \cdot v_2^2 \quad \text{(\textit{Bernoulli's} equation)} \qquad \ldots(7.5)$$

The motion of viscous fluids can be described by the Navier-Stokes equation. Volume of flow, $V$, due to pressure (in cylindrical conduits) is given the *Poiseuille's* equation.

$$V = \frac{\pi(\Delta P)a^4}{8\eta l} \quad \text{(\textit{Poiseuille's} equation)} \qquad \ldots(7.6)$$

where, $P_1$ & $P_2$ = pressures;
$h_1$ & $h_2$ = heights of fluid columns;
$\rho$ = density of the fluid;
$g$ = acceleration due to gravity;
$v_1$ & $v_2$ = flow velocities;
$\Delta P$ = pressure difference;
$a$ = radius of the cylinder;
$l$ = length;
$\eta$ = viscosity coefficient.

Viscosity changes more drastically than density with temperature. Therefore, coagulation, sedimentation and filtration processes are retarded in cold water. Viscosity decreases as pressure increases. Salient physical properties of water and their impact are listed in Tables 7.2 and 7.3.

## 7.4 PHYSICOCHEMICAL PROPERTIES OF WATER

The physicochemical properties of water are its colligative properties, such as vapor pressure and osmotic pressure (and turbidity, color and other-related properties.)

Colligate properties of solutions are characteristics of solutions under physical equilibria. These are vapor pressure lowering, osmotic pressure, the boiling-point elevation and the freezing-point depression. Colligative properties are caused by the solute, which differentiates a pure solvent and a solution on a macroscopic scale. Colligative properties of a solution depend only on the number of solute particles dissolved in a given amount of solvent and are independent of the nature of the solute. Colligative properties are determined by measurements that compare the solution with the pure solvent.

**TABLE 7.2: SOME PHYSICAL PROPERTIES OF WATER**

| *Property* | *Value* |
|---|---|
| **Physical** | |
| Density (anomalous) | 1 at 4° C (less in solid state) |
| Surface tension (high) | 72.75 $mJ/m^2$ at 20° C |
| Viscosity (low) | 0.89 centi Poise at 25° C |
| **Thermal** | |
| Heat of vaporization (high) | 40.7 kJ/mol |
| Latent heat of steam | 2260 kJ/kg |
| Latent heat of fusion of water | 377 kJ/kg |
| Specific heat capacity (high) | $(C_V + C_P) = 4.18$ kJ/kg · K |
| Thermal expansion (low) | 0.002/° C |
| **Electrical** | |
| Dielectric constant (high) | 78 |
| Dipole moment | 4.8 |

**TABLE 7.3: IMPACT OF PHYSICAL PROPERTIES OF WATER**

| *Property* | *Effects* |
|---|---|
| Transparency to visible radiation | Facilitation of photosynthesis |
| Absorption of IR radiation | Earth's heat energy balance |
| Polar solvent | Solubility of polar substances; transport of nutrients and waste products |
| High dielectric constant | Dissociation of ionic substances; solubility |
| High surface tension | "Capillary" action—important in physiology of the cell. |
| Density anomaly | Ice floats on water; protects organisms from adverse effects of freezing |
| High latent heat of fusion and evaporation | Temperature stability and thermoregulation |

The behavior of ideal solutions is given by Raoult's law, which states that at a given temperature, the vapor pressure of a solvent is proportional to its mole fraction.

$$P \propto \frac{N}{N+n} \quad \text{...(7.7)}$$

$$P = \chi P^0 \quad \text{Raoult's law} \quad \text{...(7.8)}$$

$$\chi = \frac{N}{N+n}$$

$$\Delta G = RT \cdot ln\chi \quad \text{...(7.9)}$$

where, $P$ = pressure;

$N$ = number of moles of solvent;

$n$ = number of moles of solute;

$\chi$ = mole fraction of solvent;

$P^0$ = saturation vapor pressure (at $n = 0$);

$\Delta G$ = free energy

The free energy, $\Delta G$, of the solvent is less in solution than it is in the pure solvent.

### 7.4.1 Vapor Pressure

Vapor pressure is a measure of a substance's propensity to evaporate. It is the force per unit area exerted by vapor in an equilibrium state with surroundings in liquid phase at a given pressure is called the *equilibrium vapor pressure*, or simply the *vapor pressure*. The magnitude of the vapor pressure depends on the liquid and its temperature. Liquids with relatively large cohesive forces, such as the dipole-dipole and hydrogen bonding interactions, have small tendency to escape into the vapor phase and have low vapor pressures. Liquids with small cohesive forces, *non-polar* liquids, like ether and butane have large tendency to escape into the vapor phase and have high vapor pressure (Table 7.4). Substances that have high vapor pressure at room temperature are called *volatile* liquids. Vapor pressure of water (760 mm of Hg = (101.33 kPa) at standard temperature and pressure at sea level) is vital in determining rates of evaporation from water bodies.

**TABLE 7.4:** VAPOR PRESSURES OF SOME LIQUIDS

| *Temperature (° C)* | *Pressure* | | |
|---|---|---|---|
| | *Diethyl Ether* $10^3$ *Pa (torr)* | *Ethyl Alcohol* $10^3$ *Pa (torr)* | *Water* $10^3$ *Pa (torr)* |
| –40 | ~2.27 (17) | – | – |
| –30 | ~5.33 (40) | – | – |
| –20 | ~9.33 (70) | – | – |
| –10 | ~12 (90) | – | – |
| 0 | ~24 (180) | ~1.33 (~10) | 0.61 (4.6) |
| 5 | ~29 (220) | ~2.27 (17) | 0.87 (6.5) |
| 10 | ~35 (260) | ~4 (30) | 1.23 (9.2) |
| 20 | ~49 (370) | ~5.3 (40) | 2.34 (17.5) |
| 30 | ~69 (520) | ~7.3 (55) | 4.24 (31.8) |
| 37 | 101.3 (760) | ~12.27 (92) | 6.5 (48.7) |
| 40 | – | ~20 (150) | 7.38 (55.3) |
| 50 | – | ~32 (240) | 12.33 (92.5) |
| 60 | – | ~47.3 (355) | 19.92 (149.4) |
| 70 | – | ~69.3 (520) | 31.16 (233.7) |
| 78 | – | 101.3 (760) | 44 (330) |
| 80 | – | – | 47.35 (355) |
| 90 | – | – | 70.1 (525.8) |
| 100 | – | – | 101.3 (760) |

Pressure of the water vapor in the atmosphere contributes to the pressure of the atmosphere. The boiling point of a liquid depends on pressure; it is the temperature at which the vapor pressure of the liquid equals the prevailing atmospheric pressure. The *normal boiling point* of a liquid is defined as the temperature at which the vapor pressure of that liquid is equal to 1 atm (101.33 kPa or 760 torr). The boiling points of liquids decrease under lower atmospheric pressures prevalent at higher altitudes. Water boils at 100° C at 101.33 kPa (1 atm = 760 torr) and at 80° C at 47.4 kPa (355 torr); on the other hand water boils at 120° C in an autoclave at 202.6 kPa (2 atm).

Substances that have high vapor pressure at room temperature are called *volatile* liquids. The volatility of component in a solution is referred to as its *fugacity*. Fugacity is partial pressure in ideal solutions. The ratio of the fugacity, *f*, to its standard state is known as activity, *a*.

$$a = \frac{f}{f_s} \qquad \text{...(7.10)}$$

where $f_s$ = fugacity of standard state.

### 7.4.2 OSMOTIC PRESSURE

The passage of solvent molecules in a solution, from a region with little or no dissolved salts, through an osmotic membrane, to a region with high solute concentration is called osmosis. The movement of a solvent (e.g. water) across a semi-permeable membrane depends upon the concentration difference, as well on the size, of the dissolved particles on either side of the membrane. The external pressure, *P*, required just to stop the osmosis process and prevent the net passage of solvent (water) is called the osmotic pressure. Role of osmosis is of great importance in the transport of water in living organisms, and fluid homeostatic mechanisms. In a dilute solution

$$\Pi V = nRT \qquad \text{...(7.11)}$$

$$\Pi / C = CRT \quad (\textit{Van't Hoff} \text{ equation}) \qquad \text{...(7.12)}$$

$$C = n / V \qquad \text{...(7.13)}$$

where, $V$ = volume;
$n$ = number of solute particles;
$C$ = concentration;
$R$ = gas constant;
$T$ = absolute temperature.

Due to osmosis movement of water through the plant membranes occurs. Osmotic pressure forces the fluid and the dissolved nutrients from the roots to the top. Osmotic pressure is enhanced by hydrogen bonds, which allows water to 'wet' the cell walls of tiny channels within the plant and is the basis of capillary action that moves the liquid through them. The hydrogen bonds also provide internal strength to water needed to withstand negative pressures (*turgid pressures*) generated during the water movements.

### 7.4.3 Humidity

Atmospheric water can exist in vapor, liquid or solid (ice) phase. The water content can be expressed as specific, absolute or relative humidity. *Specific humidity* is the number of grams of water vapor per kilogram of moist air. It is constant regardless of the temperature and air pressure. Absolute humidity is the amount of water vapor in a given volume of air at a certain temperature. It is the vapor concentration or density in the air. It varies with air temperature and pressure. Water in the vapor phase can be estimated (measured) according to the formula

$$P_W = \chi \frac{RT}{M} \qquad \text{...(7.14)}$$

$$P_W \approx 0.5\chi \cdot T$$

where, $P_W$ = vapor pressure;
$\chi$ = absolute humidity;
$M$ = molecular mass of $H_2O$;
$T$ = absolute temperature.

In meteorology, the absolute humidity ($\chi$) is expressed in millimeters of mercury column (mm-Hg).

Relative humidity, $R_H$, is the ratio of the vapor pressure, $P$, to the standard (saturated) vapor pressure, $P_S$, at a given temperature

$$R_H = \frac{P}{P_S} \times 100 \qquad \text{...(7.15)}$$

$$R_H = \frac{\text{Amount of water vapor in a volume of air at temperature (T)}}{\text{Maximum of water vapor that same volume of air can hold at (T)}} \times 100$$

Relative humidity is usually determined by hygrometers and *psychrometers*.

Atmospheric humidity constitutes a major factor in climate and weather. Our physical wellbeing is determined partially by humidity, i.e., the presence of water vapor in the air. Air can hold a great quantity of water at high temperatures, but very little at low temperatures. On cold winter days the vapor pressure in the air may be less than 0.6 kPa (4 torr). This is too low for comfort and health. Dry air causes rapid evaporation of moisture from mucous membranes in our breathing passages, causing irritation and increasing the chance of infection. The reverse problem frequently occurs in summers when very moist air prevents necessary elimination of excess heat from our bodies by evaporation.

### 7.4.4 Dew Point

Dew-point temperature of air is the temperature at which the vapor present in a sample of air would just attain saturation (the point of condensation). Below freezing (0° C for water), this index is called the frost point. A substance in vapor state condenses below the *dew point*. The absolute humidity can be determined from the *dew point*. Precipitation of water occurs when air temperature drops below the dew point. The dew point is determined by means of a *hygrometer*.

### 7.4.5 Turbidity and Color

Water turbidity and color are due to impurities in water. Turbidity is due to suspended (colloidal) matter in water. This includes silt, sand and other inorganic and organic particulate matter and pollutants. Turbidity blocks penetration of sunlight and thereby retardation of photosynthesis in aquatic systems. Color is due to presence of metal ions (iron and manganese) and also due to industrial and other organic pollutants discharged into water bodies (see Chapter 17).

## 7.5 CHEMICAL PROPERTIES OF WATER

Chemical properties of water encompass its solubility (solvent action), salinity, hardness, acidity (and alkalinity) and electrolytic properties.

### 7.5.1 Solubility (Solvent Action)

All natural water bodies contain dissolved minerals in more or less quantities. $SiO_2$, $Na^+$, $K^+$, $Ca^{2+}$ and $Mg^{2+}$; chlorides and sulfates make up ~99% of dissolved constituents of water (Table 7.5).

Of the cations found in most freshwater systems, calcium generally has the highest concentration. Primary contributing minerals are gypsum, $CaSO_4 \cdot 2H_2O$; dolomite, $CaMg(CO_3)_2$; and calcite, $CaCO_3$.

The solubility of non-reacting gases (such as $O_2$ and $N_2$) decreases with temperature, according to the *Clausius-Clapeyron* equation.

$$\log(C_2/C_1) = \frac{\Delta H}{2.303R}\left(\frac{1}{T_1} - \frac{1}{T_2}\right) \quad ...(7.16)$$

where, $C_1$ & $C_2$ = gas concentration at temperature $T_1$ and $T_2$;
$\Delta H$ = enthalpy;
$R$ = gas constant.

**TABLE 7.5: MAJOR CHEMICAL CONSTITUENTS IN NATURAL WATERS**

| *Constituent* | *Sources* |
|---|---|
| Silica ($SiO_2$) | Clay; silicates; feldspar |
| Iron (Fe) | Iron minerals; sandstone; igneous rocks; acid mine water |
| Calcium (Ca) | Gypsum; calcite; clay; dolomites |
| Magnesium (Mg) | Magnesium-containing minerals |
| Manganese (Mn) | Soils and sediments |
| Sodium (Na) | Feldspar; clay; sodium chloride |
| Potassium (K) | Feldspar; clay; saltpeter |
| Nitrogen | Minerals; nitrogenous organic matter |
| Phosphorus | Phosphate minerals |

The solubility of $CO_2$ in water depends on the alkalinity of water and presence of bicarbonate, $HCO_3^-$, and carbonate, $CO_3^{2-}$, ion species. Water lacking alkalinity in equilibrium with the atmosphere contains very low level of dissolved $CO_2$. However, formation of $HCO_3^-$ and $CO_3^{2-}$ species greatly increases the solubility of $CO_2$. The $CO_2 - HCO_3^- - CO_3^{2-}$ distribution depends on the pH of the solution (Fig. 7.5).

Bicarbonate is the predominate species in the normal pH range, and with $CO_2$ predominating in the more acidic water.

$$CO_2 + H_2O \leftrightarrow H^+ + HCO_3^- \quad ...(7.17)$$

The solubility of $CO_2$ in water is greatly increased by the presence of a base (alkalinity). The capacity of water to accept protons is called *alkalinity*. Alkalinity is important in water treatment and biomass of natural water. Alkaline water has the capacity for production of biomass through algal activity ("*algal bloom*"). Since alkalinity in most natural waters is due to the presence of $HCO_3^-$, water with high alkalinity has a high concentration of inorganic carbon.

### 7.5.2 Salinity

All natural water bodies contain dissolved salts. The proportion of ions making up total salts (of $Na^+$, $K^+$, $Mg^{2+}$, $Ca^{2+}$, $Cl^-$, $Br^-$ & $SO_4^{2-}$) are constant in seawaters. Therefore, the salinity, $S$, is the amount of inorganic minerals (mostly NaCl and KCl), in seawater. Salinity is expressed in terms of chlorinity, $C_L$, defined as the number of grams of pure silver necessary to precipitate halogens in 32.8523 grams of the seawater.

$$S = 1.8065 \times C_L \quad ...(7.18)$$

$Ca^{2+}$ concentration can be estimated according to

$$Ca^{2+} \approx 0.0213 \times C_L \quad ...(7.19)$$

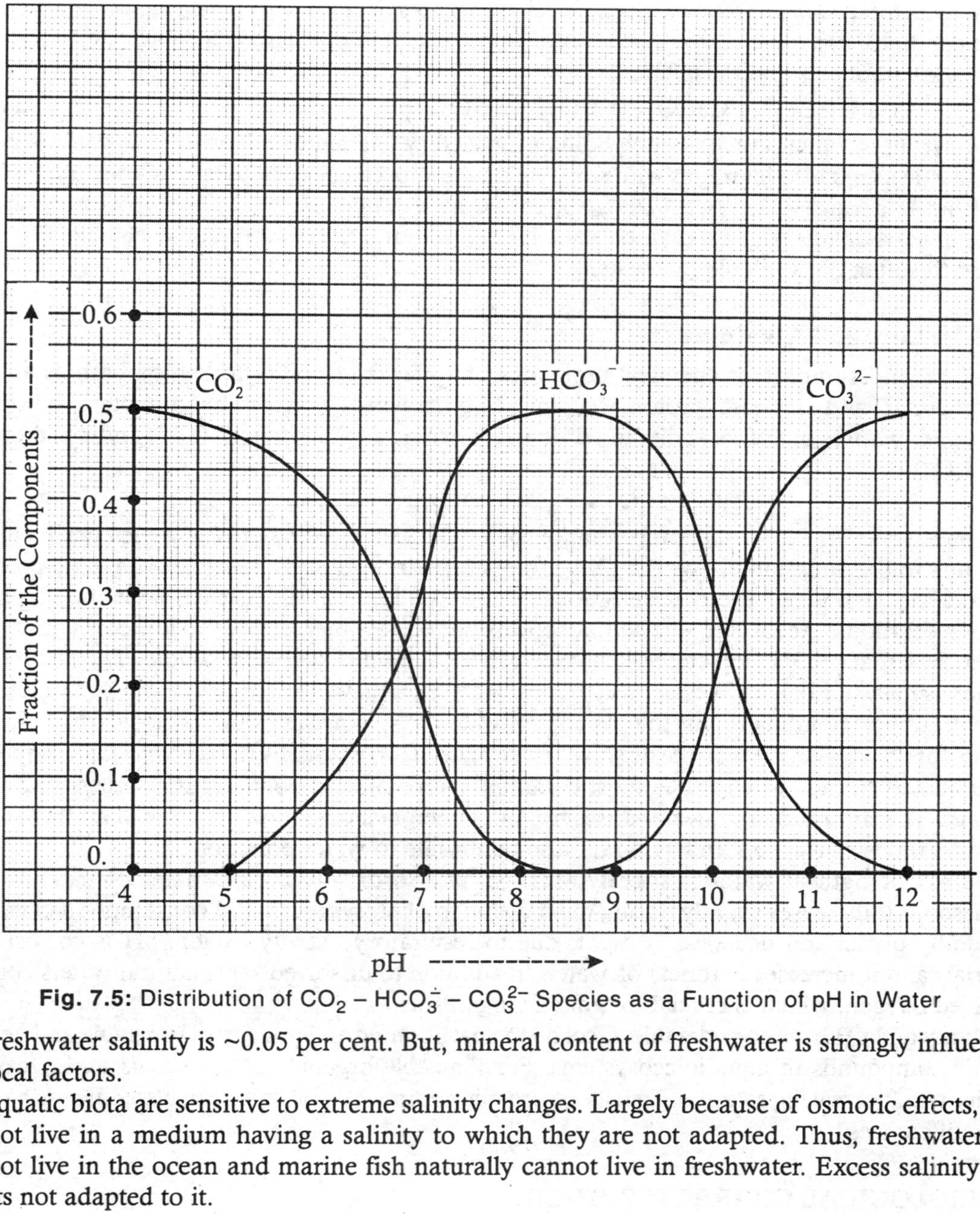


**Fig. 7.5:** Distribution of $CO_2 - HCO_3^- - CO_3^{2-}$ Species as a Function of pH in Water

Freshwater salinity is ~0.05 per cent. But, mineral content of freshwater is strongly influenced by local factors.

Aquatic biota are sensitive to extreme salinity changes. Largely because of osmotic effects, they cannot live in a medium having a salinity to which they are not adapted. Thus, freshwater fish cannot live in the ocean and marine fish naturally cannot live in freshwater. Excess salinity kills plants not adapted to it.

### 7.5.3 Water Hardness

Natural water containing a large amount of dissolved salts of calcium or magnesium (carbonates and non-carbonates of calcium and magnesium) is called *hard water* in contrast to soft water containing only small amount of none of these salts. The overall content of these salts in water is known as its total *hardness*. Non-carbonate salts are sulfates or chlorides of magnesium and calcium. Carbonate hardness is called *temporary hardness*, because when water with carbonate hardness is boiled, the carbonates are precipitated.

$$Ca(HCO_3)_2 \rightarrow CaCO_3\downarrow + CO_2\uparrow + H_2O \quad \ldots(7.20)$$

The hardness remaining after such boiling is called permanent hardness. Hard water is not fit for domestic use. It also interferes with some industrial processes, block pipes (calking) and prevent soap from lathering. Alum is added to soften the hard water. Lime (CaO) is used to react and precipitate carbonates, while soda ash (NaOH) helps removal the sulfates. Ion-exchange resins can be used to soften or even deionize water.

### 7.5.4 Acidity and Alkalinity

One of the most important chemical properties of water is its ability to behave both as an acid (proton donor) and as a base (proton acceptor). In other words, water is an amphoteric substance. This property arises out of its ability to self-ionize, forming a hydroxyl ion ($OH^+$) and a hydronium ion ($H_3O^+$):

$$H_2O + H_2O \rightarrow H_3O^+ + OH^- \quad \ldots(7.21)$$

In pure water at 25° C the concentration of $H_3O^+$ ions is $10^{-7}$ $M$ ($M$ = moles per liter) and according to the equation above, the $OH^-$ ion concentration is also the same. The ionic-product, which is the product of the concentrations of the two ions, is $10^{-14}$. It is maintained so in all aqueous solutions. Accordingly, when an acid is added to water the $H_3O^+$ ion concentration is greater than that of $OH^-$ ion concentration. Such a solution is said to be acidic. The acidity of a solution is measured on a pH scale, which represents the negative logarithm of the hydronium ion concentration. According to this the pH of pure water is 7. Whether water is acidic (pH < 7.0), or alkaline (pH > 7.0), depends on the dissolved electrolytes. $CO_2$ is, usually, the major acidic component of unpolluted surface waters. The most common source of acid pollutant in water is acid mine drainage. The sulfuric acid in water arises from the microbial oxidation of pyrite and other sulfide minerals. Industrial wastes and acid rain also contribute to water pollution.

The pH of natural water is slightly alkaline, due to dissolved carbonates of calcium and magnesium. It increases slightly during the day due to photosynthesis and consumption of carbon dioxide by plants and decrease at night, due to respiratory activity. Acidic pH is corrosive to materials and it increases hardness of water. In addition to dissolved salts, natural waters contain dissolved oxygen, which is crucial for aquatic organisms.

Salinity and pH have considerable effect on the partitioning and toxicity of ionizable and neutral organic compounds in aquatic ecosystems. For ionizable organic acids, there is an increase in toxicity with decreasing salinity. Toxicity of organic compounds in water can also differ as a result of pH differences

## 7.6 BIOLOGICAL CHARACTERISTICS

Biological characteristics of water are due to the presence of biological constituents, like algae and other biological matter that exist even in natural water. All living aquatic organisms release organic matter to the water through excretion of waste products or death. Color and smell and taste are due to minerals and organic matter and aquatic growth and decay of vegetation. Presence of pathogens, toxicants and coliforms in waters pose health hazards to animals and humans.

## BIBLIOGRAPHY

Briggs, G. E. (1967), Blackwell: Oxford. "*Movement of Water in Plants.*"
Hutzinger, O. (ed.) (1990), Springerverlag: Berin. "*Handbook of Environmental Chemistry.*"
Luna, L. B. (1974), Freeman: San Francisco, CA. "*Water: A Primer.*"
Manahan, S. E. (1993), Lewis Pubs: New York. "*Environmental Chemistry,*" (5th edn.).
Meidner, H. and Sheriff, D. W. (1976), Blakie: Glasgow. "*Water and Plants.*"
Milburn, T. A. (1979), Longman: London. "*WaterFlow in Plants.*"
Minear, R. A. and Keith, L. H., (eds.) (1982), Academic Press: New York. "*Water Analysis,*" (Vols. 1 & 2).
Moran, J. M. and Morgan, M. D. (1973), Little-Brown: Boston, MA. "*An Introduction to Environmental Science.*"
Snoeyink, V. L. and Jenkins, D. (1980), Wiley & Sons: New York. "*Water Chemistry.*"
Woodward, F. I. and Seehy, J. E. (1983), Butterworth: London. "*Principles and Measurements in Environmental Biology.*"

# 8

# Water Pollution

Water pollution is the presence of some inorganic, organic, biological, radiological or physical foreign substance in the water that tends to degrade its quality.

Normally, water is never pure in a chemical sense. It contains impurities of various kinds dissolved as well as suspended. These include dissolved gases ($H_2S$, $CO_2$, $NH_3$, $N_2$), dissolved minerals (Ca, Mg, Na, salts), suspended matter (clay, silt, sand) and microbes. These are natural impurities derived from the atmosphere, catchment areas and the soil. They are in very low amounts and normally do not pollute water. All these substances when present in small quantities do not cause any harm and may

even have some positive effects in improving the water quality. However, if their concentration increases substantially, they affect adversely the water quality and make the water unfit for use. Such water is said to be polluted. The polluted water is turbid, unpleasant, bad smelling, unfit for drinking, bath and washing and incompatible in supporting life.

Water pollution is also caused by the presence of undesirable and hazardous materials and pathogens beyond certain limits. Much of the pollution is due to anthropogenic activities like discharge of sewage, effluents and wastes from domestic and industrial establishments, particulate matter and metals and their compounds due to mining and metallurgy and fertilizer and pesticide runoffs from agricultural activities.

This chapter details the various factors that affect the quality of water.

Natural water is never pure in the 'chemical sense'. Natural conditions are partly responsible for contaminating water. The most basic of these is the role of water as a 'universal solvent'. Even as precipitation falls to the Earth water picks up gases, chemicals and particulate matter, which are always present in the atmosphere. After precipitation reaches the Earth water percolates through organic material like the roots and leaf litter, rocks and soil. In this process it reacts with organic matter, living and dead, including, vegetable matter and microscopic organisms to human beings. It picks up minerals from the soil and the rocks it traverses. Therefore, the quality of natural water is controlled by several factors like geology of the area it passes through, with the season of the year and stream discharge.

## 8.1 FRESH WATER

Taking all the factors into account it can be said that 'natural water' is a dilute solution of elements dissolved from the Earth's crust and washed from the atmosphere. The total ionic concentration can vary from as low as 100 mg/L in snow, rain, hail and some mountain streams, to as high as 40,000 mg/L in the saline lakes of internal drainage systems. Freshwater, generally, contains non-toxic and non-hazardous total dissolved solids below 1,000 ppm. The naturally occurring impurities are alkaline and alkaline earth salts, heavy metals iron and manganese, and organic decomposition products of plants and biota. Brackish water contains dissolved solids in the range 1,000-10,000 ppm and salty water (seawater) >10,000 ppm. Of the major ions in water are cations—$Ca^{2+}$ (60 per cent), $Mg^{2+}$ (20 per cent), $Na^+$ (15 per cent) and $K^+$ (5 per cent), and anions—$HCO_3^-$ (55 per cent), $Cl^-$ (25 per cent) and $SO_4^{2-}$ (20 per cent) (Fig. 8.1). The concentration of major ions in natural water is the result of the geochemical balance between source contributions (weathering, leaching, and biological processes) and removal mechanisms.

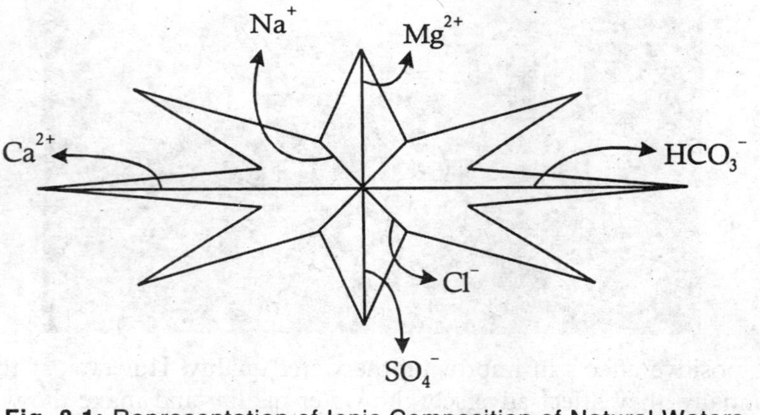


**Fig. 8.1:** Representation of Ionic Composition of Natural Waters

While nature contaminates water to a certain degree, when we talk of 'pollution' we are particularly stressing the degradation of water quality by human activities. In modern times this influence is considerable. In some parts of the world this effect is already felt through the fact that the precipitation inputs to the system are highly polluted. This is the result of presence of acidic gases in the atmosphere, which result from the combustion of fossil fuels in ever increasing quantities. Other recognized causes are the dumping of industrial wastes, discharge of untreated or inadequately treated sewage and the chemicals used extensively in the agricultural activities.

It is difficult to judge whether a particular body of water is polluted or not, in absolute terms. Therefore, in broad terms, water is said to be polluted when it contained enough impurities to make it unfit for a particular use, such as drinking, swimming or fishing. As an example the general specifications for drinking water are given in Table 8.1. For some industrial purposes water of very high purity, with impurities in ultra-trace level, is demanded. It is not only the flowing and stagnant water that is contaminated but even the underground water sources (aquifers) have not altogether escaped the polluting influence of modern industrial, agricultural and community activities.

Some of the aquatic systems are choked with an excess of organic substances and organisms to be poisoned with toxic substances.

**TABLE 8.1:** MAXIMUM PERMITTED LEVELS (mg/L) OF IONS IN DRINKING WATER

| *Contaminant* | *Permitted Level (mg/L)* | *Contaminant* | *Permitted Level (mg/L)* |
|---|---|---|---|
| Chloride | 250 | Sulfate | 250 |
| Sodium | 200 | Magnesium | 50 |
| Nitrate | 50 | Potassium | 10 |
| Nitrite | 3 | Zinc | 3 |
| Copper | 2 | Fluoride (F) | 2 |
| Phosphorus | 2 | Barium | 0.7 |
| Manganese | 0.5 | Boron | 0.3 |
| Iron | 0.3 | Aluminum | 0.2 |
| Molybdenum | 0.07 | Cyanide (CN) | 0.07 |
| Chromium | 0.02 | Nickel | 0.02 |
| Arsenic | 0.01 | Lead | 0.01 |
| Selenium | 0.01 | Antimony | 0.005 |
| Cadmium | 0.003 | Mercury | 0.001 |
| pH | 6.0-9.0 | | |

## 8.2 FACTORS THAT AFFECT WATER QUALITY

Throughout the history, the quality of water has been an important factor in determining the growth of civilizations and welfare of human settlements. Some of the factors that affect the quality of water are suspended and dissolved solids (both inorganic and organic), biological substances, pH, *biological oxygen demand (BOD)*, and *chemical oxygen demand (COD)*. At present, many industrially produced toxic substances pose the greatest hazard to the quality of water.

Water pollutants are varied and are classified under various categories (Table 8.2).

**TABLE 8.2:** CLASSIFICATION OF WATER POLLUTANTS

| *Class of Pollutants* | *Effects* |
|---|---|
| Asbestos | Health |
| Acidity/Alkalinity | Water quality; aquatic life |
| Algal nutrients (P & N) | Eutrophication |
| BOD | Water quality; oxygen levels |

| *Class of Pollutants* | *Effects* |
|---|---|
| Metals | Toxicity |
| Organics & Petrochemicals | Water quality; biological effects |
| Particulate matter | Water quality; aquatic biota |
| Pathogens | Health hazard |
| Pesticides | Toxicity; aquatic biota |
| Sewage | Water quality; oxygen levels; health |
| Trace elements | Aquatic biota; health |

### 8.2.1 Water Quality Index (WQI)

What are the criteria to specify the water quality data, which is influenced by various geographical, natural and anthropological contaminants, an indicator that is understandable and useable by the public? Criteria are based on specific levels of pollutants that would make the water harmful if used for drinking, swimming, farming, fish production, or industrial processes. These criteria are combined to provide a water quality index (WQI). Water quality index (WQI) is based on some very important parameters that can provide a simple indicator of water quality. Though a single number cannot full give the complexity of the water quality and can be over generalized, it does provide a standardized method for comparing the water quality of various water bodies. The WQI translates a wide variety of environmental indicators into a simple system. Easy-to-understand information helps general public to comply with environmental regulations.

The water quality index (WQI) is evaluated taking into account various parameters (DO, fecal coliform, pH, BODs, temperature change, total phosphorous, nitrate, turbidity and total dissolved solids) that influence the water quality. More weightage is given to important factors (e.g. coliform is weighed more heavily, DO is more important than pH etc.).

$$WQI = \frac{(I_i w_i + \ldots I_n w_n)}{(w_i + \ldots w_n)} \qquad \ldots(8.1)$$

where $I_i$ = unit indices;
$w_i$ = weighting of the $i$th parameter;
$n$ = number of parameters.

### 8.2.2 Parameters Influencing Water Quality

Contents of freshwater are strongly influenced by geographic soil conditions, sources of aquifers, climate conditions and anthropogenic sources. Salts from sewage and oil brines, synthetic chemicals, fertilizers and other industrial plants, electroplating chemicals and salts, used for deicing of roads in colder countries, all add to saline contamination. Sulfate contaminants are due to discharge of wastes from paper, timber, textile, tannery establishments and detergents and also from acid rain.

External interference, mostly due to human activities, that affects the quality of water will have a serious effect on the ecology of aquatic life, often by reducing the dissolved oxygen content. Suspended particulate matter and silt prevent sunlight from penetrating the water and limit the access of light to aquatic life within. Some of the factors that affect the water quality are given in

Table 8.3 and physical characteristics of water due to man-made pollutants in Table 8.4. Contributions from various polluting sources are given in Fig. 8.2.

**TABLE 8.3:** CONTRIBUTIONS FROM VARIOUS POLLUTING SOURCES THAT AFFECT WATER QUALITY

| *Factors* | *Quality/Pollution* |
|---|---|
| Meteorological | Runoff water; mud, suspended particles, silt; dissolved gases and minerals |
| Domestic | Sewage waste; bacterial contamination; organic refuse; oils and detergents; sulfates, nitrates and phosphates etc |
| Agricultural | Fertilizer waste; crop residues; organic debris; tannery and animal waste; insecticides, pesticides and herbicides |
| Industrial | Inorganic and organic particulate matter; solid and liquid wastes; trace elements; hydrocarbons and petrochemicals; hazardous chemicals |
| Timber & Paper | Inorganic and organic sediments; minerals; trace elements |

**TABLE 8.4:** PHYSICAL CHARACTERISTICS OF WATER POLLUTED DUE TO HUMAN ACTIVITIES

| *Parameter /Pollutant* | *Source* | *Effects* |
|---|---|---|
| Color | Dissolved inorganic & organic matter | Deterioration of water quality; unaesthetic; hazardous chemicals detrimental to health |
| Inert particulate | Mining & sewage sludge | Turbidity; exclusion of sunlight; detrimental to aquatic life |
| Mining | Mining waste; asbestos; sand and dust; sulfur-bearing particulate | Stream contamination; leaching; silt; and hazardous compounds |
| Nutrients | Domestic, agricultural, mining & industrial | Acid-mine drainage; eutrophication of water bodies |
| Odor | Decay of organic matter; sewage | Unhealthy & unhygienic and health hazard |
| Petroche-micals | Oil drilling & transport | Hydrocarbons detrimental to all organisms; oil spillage in high seas; damage to aquatic ecology |
| Poisons | Industrial & agricultural | Trace elements; radioactive waste; insecticides, pesti-cides and herbicides |
| Taste | Salts & dissolved matter | Deterioration of water quality; unaesthetic |
| Thermal | Discharge of hot effluents from power plants and industrial establi-shments | Affects the physical properties of water; reduction in dissolved gases has drastic effects on the aquatic ecology |

## 8.3 SOURCES OF WATER POLLUTANTS

Water pollutants are from: (*i*) natural and (*ii*) anthropogenic sources, and they may originate from a point source or from a dispersed source. Natural sources are meteorological and geographical like volcanic activity and earthquakes, landslides and streams runoff, dissolved minerals, aquatic growth and decay. Anthropogenic sources are domestic, municipal sewage and other sanitary waste discharge, and agricultural and industrial waste (both liquid and gaseous), mining waste and leachates, and products from other human-related activities. Radioactive substances and heat are also water pollutants.

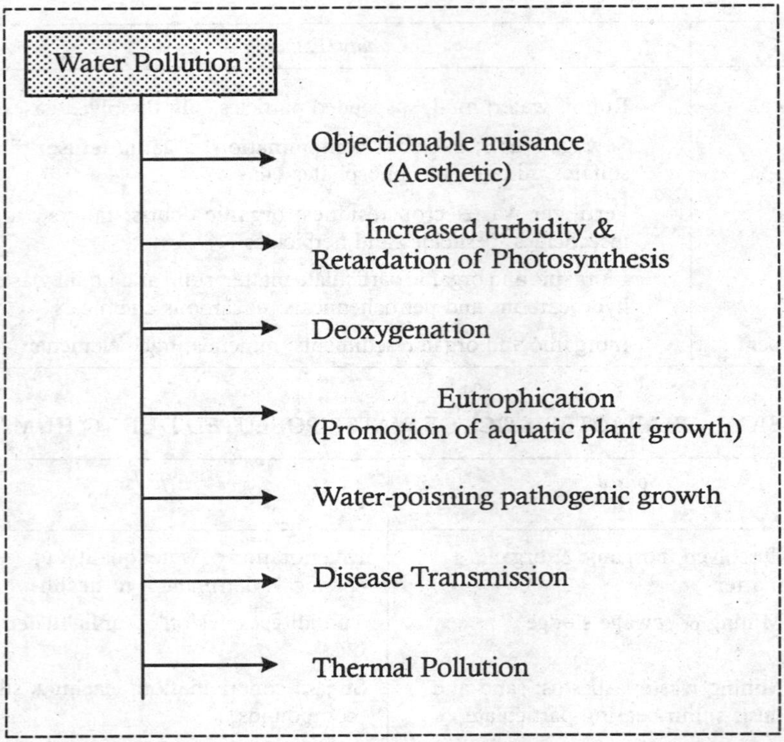


**Fig. 8.2:** Water Pollution Tree

### 8.3.1 Point- and Non-point Sources

Pollutants are from point- as well as from non-point sources. Point sources are domestic, municipal and sewer discharge, power generation plants and industrial waste discharge. Some of them like, breweries, slaughterhouses and sanitary operations, paper mills and wastewater treatment plants contribute major quantities of oxygen-demanding substances. These substances can deplete dissolved oxygen (DO) and create anaerobic conditions in water bodies. Suspended matter also contributes to oxygen depletion in water bodies by blocking penetration of sunlight and interfering with photosynthetic activity. This results in an increase in oxygen demand—biochemical oxygen demand (BOD) and chemical oxygen demand (COD) (see Chapter 10). Nitrogen- and phosphorous-containing compounds (nutrients) can promote accelerated eutrophication of water bodies.

Heat is a universal pollutant, as it drastically alters the ecology of water bodies, by lowering the amount of dissolved oxygen in water; thus, accentuating the oxygen deficiency for aquatic organisms.

Trace metals, hydrocarbons, hazardous chemicals, bacteria and a variety of pathogens are other pollutants that can cause a wide variety of problems in watercourses.

Non-point sources are storm drainage, operations involving agricultural, timber and forest-product operations. Mobile vehicular discharge is also a source of contamination affecting through atmospheric pollution.

### 8.3.2 Domestic Sewage

Domestic and municipal sewage carries used water from houses, offices and other buildings in a city. It is also called sanitary sewage. Most of it (nearly 99.9 per cent) is water. Though the contaminants add up to not more than 0.1 per cent, they contain a wide variety of dissolved and suspended impurities. The nature of these impurities and the large volume of sewage in which they are carried make disposal of domestic wastewater a significant technical problem. Sewage is the primary source of pathogenic organisms, oxygen-demanding waste matter and plant nutrients.

At one time it used to be said that 'the solution to pollution is dilution'. The growth of cities all over the world has given rise to such large volumes of sewage that dilution alone no longer assures the natural processes of stream self-purification. Since early 20th century centralized sewage treatment plants have been set up in all populated areas of the developed world. Instead of discharging sewage directly into a nearby body of water, it is first passed through a series of physical, chemical and biological processes that remove most of the pollutants. However, there are still many countries where such facilities do not exist in all towns and municipalities, where untreated or improperly treated domestic sewage is allowed to flow into water bodies, promoting severe conditions of pollution.

### 8.3.3 Industrial Sewage

Industrial sewage consists of polluted water from industrial and chemical processes. These discharges usually contain specific pollutants, which are related to the nature of products handled in an industry and the processes followed. Industries located along waterways contribute a number of chemical pollutants, some of which are toxic in any concentration. Such pollutants may originate from metallurgical, paper and pulp, cloth and cellulose fibers, and food, beverage and tannery wastes, and detergents, plastics and petrochemicals. Discharge from hospitals and utility sources and power generation plants also comes in this category. In highly industrialized zones, if the industrial sewage were not separately treated before discharge into waterways, serious pollution conditions would develop.

Industries contribute to water pollution through atmospheric pollution also. Hot water from power generating installations, discharged into water streams, cause thermal pollution.

### 8.3.4 Storm Sewage

Storm sewage or storm water is the runoff from precipitation that is collected in a system of pipes or open channels. Such sewage carries organic materials, suspended and dissolves solids and other substances picked up as the water travels over the ground. Sewage discharge from domestic, municipal, food processing and other industrial concerns contain a variety of pollutants detrimental to water quality (Table 8.5).

### 8.3.5 Agricultural Sources

Agricultural wastes, generally, consist of organic products. Fertilizers and other chemicals are spread over agricultural lands. These materials and crop, animal and chemical wastes enter water bodies, mainly in run-off from watershed lands, and cause pollution. The inflow of menures from livestock feed lots also adds to organic pollutants. Many of the pesticides, fungicides, herbicides and other industrial chemicals are highly toxic. They are carcinogenic and mutagenic (Table 8.6).

**TABLE 8.5:** POLLUTANTS IN SEWAGE DISCHARGES

| *Constituents* | *Sources* | *Effects on Water Quality* |
|---|---|---|
| Oxygen-demanding substances | Organic compounds; feces | Oxygen demand (BOD) |
| Household products | Domestic & industrial | Toxic to aquatic life |
| Food, beverages and oils | Domestic | Algal nutrients; harmful to aquatic life |
| Salts | Domestic & industrial | Water salinity |
| Heavy metals | Industrial | Toxicity |
| Particulate matter | All sources | Harmful to aquatic life |
| Pathogens | Human and animal excreta | Biohazards |

**TABLE 8.6:** CARCINOGENIC AND MUTAGENIC SUBSTANCES

| *Carcinogens* | *Mutagens* |
|---|---|
| Benzo(a)pyrene | Trichloroethane |
| Carbon tetrachloride ($CCl_4$) | Methylchloride |
| Chloroform ($CHCl_3$) | Bromoform |
| Vinylchloride | Vinylchloride |
| Methyliodide | Methylbromide |
| Dioxane | Chloroprogane |
| Lindane | Chlorodane |
| DDT, Dieldrin, Aldrin | Acrylonitrile |
| Benzene, Heptachlor | Pyrene |

### 8.3.6 Mining Wastes

Mining, milling, dressing and processing of ores give rise to dust, ore and metal discards (Table 8.7) and large quantities of effluents, which are discharged into streams, ponds and lakes. They not only increase sediments but also release toxic metals into the water sources.

Common trace metals found in sediments and mine effluents are Cd, Cu, Fe, Hg, Mn, Ni, Pb and Zn. Of these, heavy metals Cd, Hg and Pb and metalloids, such as As, are among the most harmful of the elemental pollutants. Most of them have a great affinity for sulfur and attack -SH groups and disulfide bonds in proteins and other biological macromolecules. Cadmium, being chemically similar to zinc, replaces the latter in enzymes and thus affects enzyme action of Zn-containing proteins. Mercury is of great concern as a heavy-metal pollutant. Lead occurs in water in Pb(II) state. It is highly poisonous and causes anemia, central nervous system disorders, kidney and liver dysfunction.

### 8.3.7 Atmospheric Deposition

There is accumulated evidence to establish a close relationship between atmospheric pollution and declining water quality on the globe. Airborne pollutants can be deposited on land or water. This type of deposition can take place, some times, at great distances from its original source. The deposition itself can take several forms: 'wet deposition' occurs when air pollutants fall with rain, snow or fog. 'Dry deposition' takes place as dry particles or gases. These pollutants can fall directly on water or having fallen on land can be washed into a body of water as runoff.

There is evidence showing that atmospheric pollutants can reach even the ground water. Therefore, as pollution falls, part of it might end up in streams, lakes or estuaries and aquifers and affect the water quality there. Presence of secondary porosity and fractures within the rock mass over the aquifers can lead to the movement of ground pollutants through the ground water.

**TABLE 8.7:** TRACE LEVEL METALS FOUND IN WATER

| *Metal* | *Sources* | *Effects* |
|---|---|---|
| Arsenic (As) | Mining; pesticides & wastes | Toxic; carcinogenic |
| Beryllium (Be) | Power & space industries | Toxic; carcinogenic |
| Boron (B) | Coal burning; industrial waste | Toxic |
| Cadmium (Cd) | Mining & Industry | Toxic |
| Chromium (Cr) | Metal platting; industrial discharge | Carcinogenic |
| Copper (Cu) | Mining; domestic and industrial waste | – |
| Iron (Fe) | Mining & industrial | At moderate levels nontoxic, but is a nuisance |
| Lead (Pb) | Mining & industrial | Highly toxic |
| Manganese (Mn) | Mining & industrial | At moderate levels nontoxic, but is a nuisance |
| Mercury (Hg) | Mining & industrial | Highly toxic |
| Molybdenum (Mo) | Industrial waste | – |
| Selenium (Se) | Coal & sulfur burning | Carcinogenic |
| Silver (Ag) | Mining; electroplating | – |
| Zinc (Zn) | Mining & industrial waste | – |

## 8.4 ECOLOGICAL IMPACT

Water pollution has wider ecological impact than just being unsuitable for consumption or posing health hazards. Most of the water withdrawn from resources is used for consumption—household use, and 'water-carriage' of wastes and discards in domestic, sanitary and municipal jurisdiction. Similarly, major consumption of water in industrial plants is for 'water-carriage' of wastes, for removal of byproducts and impurities and as a coolant. Thus, water is the major conduit in direct transmission of toxic agents, trace elements, like As, Cd, Cr, Hg, Pb, and Se and persistent hazardous organic chemicals and infectious agents and vectors for several diseases, such as cholera, gastrointestinal diseases, malaria, schistosomiasis, typhoid fever, filiariasis, encephalitis.

Increase in human settlements, urbanization and population explosion pose a greater demand for water—for domestic use, flush-toilets, washing and bathing, swimming pools, lawns and gardening, recreational activities, automobile and other vehicular uses, construction, sanitation and healthcare centers, and agricultural and industrial operations. But, available water is limited. So, the adverse effects of water pollution on ecological systems at local, regional, continental and global levels would become more and more serious. Some hydrological effects of urbanization are given in Table 8.8.

**TABLE 8.8:** HYDROLOGICAL EFFECTS OF URBANIZATION

| *Operation* | *Effects* |
|---|---|
| Deforestation & agriculture | Decrease in green vegetation; decrease in transpiration; decrease in rainfall; land deterioration; increase in pollution in water bodies. |
| Animal husbandry | Increase in sedimentation and storm runoff; land deterioration; increase in pollutants in water bodies. |
| Mining & construction | Lowering of water table; increase in sediments and pollutants; leaching. |
| Domestic & municipal | Increase in pollutants in water bodies; loss of aquatic life; inferior water quality; sanitary and health hazards. |
| Industrial | Increase in pollutants, toxic and hazardous chemicals; health hazards; increase in temperature of water bodies; loss of aquatic life. |

## BIBLIOGRAPHY


Baier, R. E., Mark, H. B. and Mattson, D. S. (eds.) (1981), Marcel Dekker: New York. "*Water Quality Measurement.*"

Frei, R. W. and Albarge, J. (eds.) (1986), Gordon & Breach: New York. "*Air and Water Analysis: New Techniques and Data.*"

Lamb, J. C. (1985), John Wiley: New York. "*Water Quality and its Control.*"

Minear, R. A. and Keith, L. H. (eds.) (1982), Academic Press: London. "*Water Analysis,*" (Vols. 1 & 2).

Minear, R. A. and Keith, L. H. (eds.) (1994), Academic Press: London. "*Water Analysis: Organic Species.*"

Pfaffin, J. R. and Ziegler, E. W. (eds.) (1976), Gordon & Breach: New York. "*Encyclopedia of Environmental Science and Engineering,*" (Vols 1 & 2).

Purdom, P. W. (ed.) (1980), Academic Press: New York "*Environmental Health,*" (2nd edn.).

Rump, H. H. (ed) (1999), Wiley-VCH: Weinheim. "*Laboratory Manual for the Examination of Water, Wastewater and Soil,*" (3rd edn.).

Wezel, A. (1998), *Environmental Reviews*, **6**(2): 123. "Chemical and biological aspects of ecotoxicological risk assessment of ionizable and neutral organic compounds in fresh and marine waters."

... (1993), WO: Geneva. "*Guidelines for Drinking Water Quality, Vol 1: Recommendations.*"

# 9

# Ecological Effects of Water Pollution

All pollutants, atmospheric and land-based invariably enter water bodies by direct discharge, precipitation and run-offs. Water bodies, thus, become sinks as well as carriers of pollutants. Water pollution has wide ecological impact, as it is an important raw material in photosynthesis and hydrological processes. Water pollutants impart to it undesirable properties like odor, turbidity and retardation of photosynthesis, deoxygenation and eutrophication and thermal pollution. In addition, thermal stress has profound physical, chemical and biological effects on all organisms. Turbidity blocks sunlight reaching deeper into the water bodies (lakes) and retards photosynthesis in aquatic plant systems. Chemical factors such as acid rain, mining leachates, hazardous chemicals, toxic trace elements deteriorate the water bodies.

This chapter illustrates the details of ecological impacts of water pollution in mobile and stationary sources.

All pollutants, atmospheric and land-based invariably enter the water systems. Water bodies become sinks as well as carriers of wastes and pollutants. As water is the most important requirement for vegetation and living organisms, deleterious changes in its quality will have direct short-term and long-term effects on ecosystems.

## 9.1 ECOSYSTEMS

Biosphere is a global ecosystem composed of living organisms (biota) and nonliving (abiotic) factors from which they derive energy and nutrients. The system is characterized by the transformation of solar energy into chemical energy and continuous cycling of matter and accompanying energy, in which water is an important factor. An ecosystem is the smallest unit of biotic system that adequately sustains life. It includes a complex of living organisms, their physical environment, and all their interrelationships in a particular unit of space. A group of ecosystems that are related either by vegetation and/or climate (such as a tropical forest) is called a *biome*. The sum all biomes or biological habitats constitute the biosphere.

All ecosystems exist within the 'ecosphere', which encompasses the totality of the Earth with all its biological components. Energy transfer is from the Sun via each trophic level and this energy (solar) transfer is unidirectional. Nutrient energy flow (chemical energy transfer) in trophic levels is cyclic. The ecological balance of the food chain is called *homeostatis*. Ecosystems are interrelated (Fig. 9.1), and the stability of an ecosystem is proportional to the number of organisms capable of filling various niches (the essence of biodiversity). Pollutants introduced into the ecosystems disturb the ecological balance.

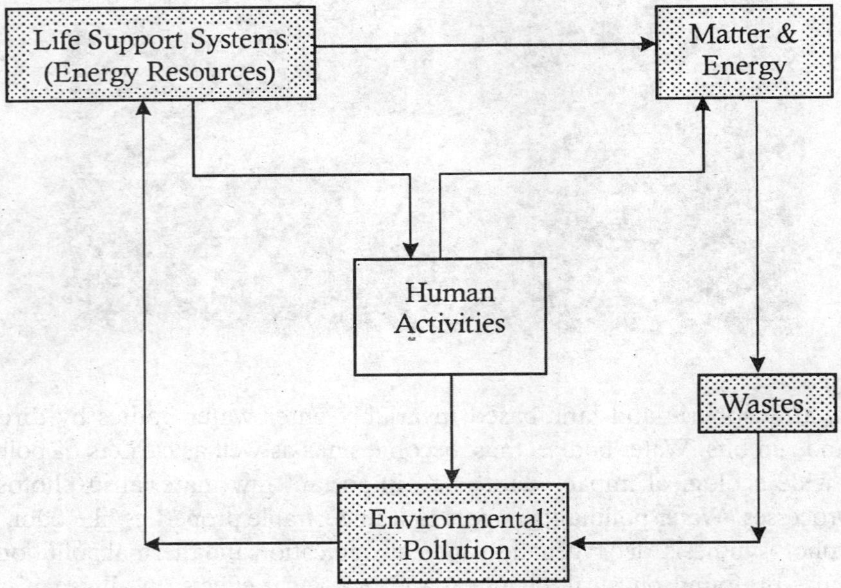


**Fig. 9.1:** The Environmental Interrelationship of Ecosystems

*Ecology* (environmental biology) is the study of the interrelationship between living communities and their nonliving environments. The biotic organisms can be classified into three categories, based on their transfer of energy: (1) primary producers (autotrophs), (2) consumers (heterotrophs) and (3) decomposers (microorganisms).

### 9.1.1 Ecological Cycles (Food Chain)

All life on Earth depends ultimately upon chlorophyll-containing organisms (green plants, cyanobacteria), as well as upon water, for harnessing solar energy by photosynthesis. Organisms capable of photosynthesis are referred to as 'primary' products or 'primary' producers (the plant kingdom). They are *autotrophs*, i.e., producers of their own food from inorganic components. They do not require organic matter, from outside, for their growth. Plants, which are primary producers, take up energy from the sunlight and nutrients from the soil or water and produce high-energy chemical compounds. The rate at which plants photosynthesize depends on the amount of sunlight reaching the leaves, the temperature of the environment, and the availability of water and other nutrients.

*Heterotrophs* are feeding groups that include all non-autotrophs. The consumers (*heterotrophs*) depend on already synthesized organic compounds (chemical energy) as the source of their food. Consumers are graded according to their hierarchy. Primary consumers are herbivores that subsist directly on plants and their products. Secondary and tertiary consumers are carnivores and omnivores. Pests are also consumers. They are living organisms that are hazardous to health or the environment. Some of them compete with human beings for food and survival.

The *decomposers* are decay organisms (special type of consumers), which are *detritivores* (e.g. fungi, bacteria, ants, termites, beetles). They feed directly on detritus (organic garbage found in a natural environment) and in that process convert complex organic molecules into smaller molecules. The sequence is

$$\text{Plants} \rightarrow \text{Herbivores} \rightarrow \text{Carnivores}$$

$$\text{Ingestion (Intake)} \rightarrow \text{Production} + \text{Respiration} + \text{Egestion} \qquad \text{...(9.1)}$$

$$\text{Energy flow (Assimilation)} \rightarrow \text{Production} + \text{Respiration}$$

Such a sequence, namely, the sequence of the processes of consumption and being consumed from one trophic level to another in an ecosystem is called the *food chain*.

Thus, organisms in ecosystems occupy distinct 'niches,' and the overall productivity of the biosphere is limited by the rate at which primary producers convert solar energy into the chemical energy and the subsequent efficiencies at which consumers utilize that stored energy into their biomass. Energy transfer from one trophic level to another higher up is not very efficient. Only a relatively small proportion is transferred up the food train. Water and food supply and ambient conditions of the environment affect these processes. Water quality and aquatic biology are closely intertwined. The growth and proliferation of any type of organism in aquatic systems are dependent on nutrients, chemicals, dissolved gases (e.g. oxygen), pH, sunlight and temperature.

### 9.1.2 Processes of Ecosystems

The constituents of an ecosystem are:

(1) **Simple inorganic compounds:** C, N, $O_2$, $CO_2$, $NO_x$, and $H_2O$.
(2) **Organic compounds:** Peptides, proteins, nucleic acids, carbohydrates and lipids.
(3) **Climate:** Rainfall, temperature, geotopology and hydrology.
(4) **Autotrophs (Producers):** Plants that synthesize food from simple inorganic compounds.
(5) **Phagotrophs: (Consumers):** Organisms that consume (as food) compounds synthesized by autotrophs.
(6) **Saprotrophs:** Organisms (bacteria, fungi, etc.) that breakdown complex molecules into simple compounds (special type of consumers).

The processes involved in an ecosystem are—(1) energy-flow cycles, (2) food chains, (3) biochemical cycles, (4) development and (5) control systems (cybernetics).

Ecosystems develop through a rapid growth stage leading to a steady state. A high production/respiration (P/R) ratio, high yields, low diversity and lack of stability characterize the early growth. In a steady state, there exists high biomass/respiration (B/R) ratio, low yields, high diversity and stability. That is, major energy flow shifts from production to maintenance.

**Growth state:** *Production → Growth → Quantity → Instability*
**Maintenance:** *Protection → Low growth → Quality → Stability*

The limiting factor in human ecology is pollution. Harvest and pollution are stresses that reduce the energy available for self-maintenance. Attempts to force over-productivity from the land (e.g. 'green revolution') may result in short-term gains, but in the long run lead to undesirable consequences. Some of these are: (*i*) need for using heavy doses of fertilizers and insecticides to maintain the crop yield; (*ii*) vulnerability of plants to disease due to monoculture system and (*iii*) greater consumption of fossil fuels to increase production. The latter would increase the carbon dioxide in the atmosphere, a consequence of which is the 'greenhouse' effect and global warming. Ultimately there would be disruption of the ecological balance. Studies indicate that biodiversity is directly correlated with ecological stability. Preservation of biodiversity, rather than overproduction, is a necessity for stability of ecosystems.

## 9.2 PLANT ECOLOGY

Only about 1-2 per cent of the solar energy reaching the Earth is converted into chemical energy. Plants for their metabolic needs use about 20-50 per cent of this chemical energy and the rest is stored in their tissues for further use (net primary productivity).

### 9.2.1 Standing Crop

The quantity of living organisms present at a given time is referred to as the *standing crop*, expressed in terms of energy content. Part of the production removed by man or by other species, is called the *harvest*. The total energy is called *energy flow* and for producers it is called *primary production*.

Water influences almost all the environmental factors. These include: (*i*) amount of radiant energy received from the Sun and scattered back to the atmosphere, (*ii*) temperature of the atmosphere and of land, (*iii*) leaching of minerals and soil, (*iv*) availability of essential elements and nutrients, (*v*) aeration of soil and (*vi*) parasites in the region. No other factor influences the internal and external structures of the plant kingdom, as much as the amount of water present in the air and soil does.

Water is an important raw material in photosynthesis and hydrolytic processes. Water has also an influence on leaf temperature through transpiration and cooling. Water has direct effect on turgor pressure of the plant cells. Closing of the stomata, due to reduced turgor pressure in the guard cells of plants, reduces both transpiration and photosynthesis and can ultimately reduce growth. A reduction in water content of leaves generally results in decrease in the rate of photosynthesis. Nonavailability of sufficient sunlight, due to atmospheric reflectivity where water plays a role, is a major factor in ecological effects on plant kingdom.

The trophic structure of an ecosystem pertains to its nutritional aspects.

1. **Photosynthesis:** It is the biochemical process for the production of organic compounds and oxygen, utilizing inorganic $CO_2$ and $H_2O$ with sunlight as the source of energy (Fig. 9.2). The $O_2$, extracted from $H_2O$ (and not from $CO_2$) is returned to the atmosphere.

$$CO_2 + H_2O \xrightarrow{\text{Sunlight } (h\nu)} \{CH_2O\}_n + O_2 \quad \ldots(9.2)$$

**2. Respiration:** Respiration is the reverse of photosynthesis. It is the biochemical process of utilizing organic compounds and oxygen ($O_2$), and releasing carbon dioxide ($CO_2$), water ($H_2O$) and energy ($h\nu$).

$$\{CH_2O\}_n + O_2 \rightarrow CO_2 + H_2O + h\nu \quad \ldots(9.3)$$

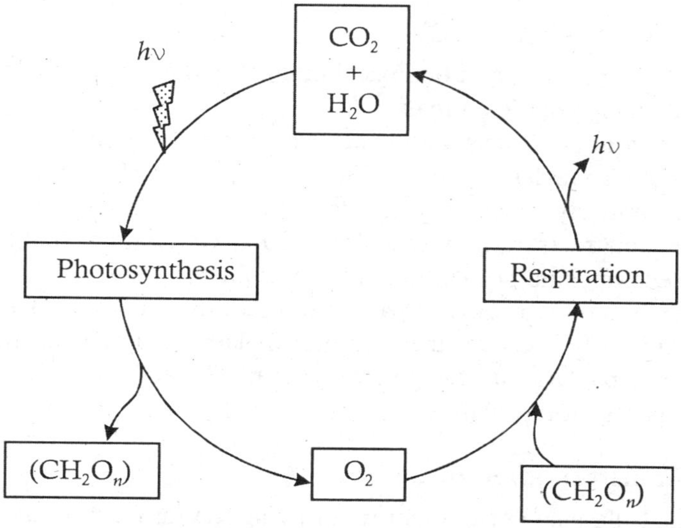


**Fig. 9.2:** The Oxygen Cycle

The ecological principles of pollution can be summarized as the three laws of ecology, namely,

| | |
|---|---|
| (1) Everything is connected to everything else. | 1st law of Ecology. |
| (2) Everything must go somewhere. | 2nd law of Ecology |
| (3) Everything that goes somewhere comes back recycled. | 3rd law of ecology |

Physical and chemical characteristics of water bodies greatly influence their aquatic environment, which in turn affects the type and the number of organisms that develop in the water bodies.

Ecological systems in a clean watercourse usually include many species of organisms, but relatively few of any particular type. That is, this diversity is used as a stable and desirable system.

Growth in flowing streams tends to be largely heterotrophic, because photosynthesis plays a less important role in them than it does in lakes. Also, streams have turbulent conditions, blocking transmission of sunlight through water. In lakes, on the other hand, turbidity in general, is less, allowing sunlight to penetrate. Phytoplanktons synthesize organic compounds through photosynthesis in lakes. That is, lakes are more productive via autotrophic pathway than streams of similar chemical characteristics.

In lakes and in stagnant water bodies, there is gradual accumulation of organisms and their byproducts. This process is called *eutrophication* (refer to Chapter 10). Etrophication can be accelerated drastically through addition of key nutrients and other aquatic plant growth chemicals in wastewater and land run-off.

## 9.3 MICROBIAL ECOLOGY

Bacteria, fungi and algae are living organisms that catalyze vast number of chemical reactions in water and soil. Oxidation-reduction processes of organic matter in water occur through bacterial mediation. Algae are classified as *producers* and fungi and bacteria (with the exception of photosynthetic bacteria) are classified as *reducers*.

**Algae:** They are mainly aquatic, eukaryotic, one-celled or multicellular plants without true stems, roots and leaves, which are typically autotrophic, photosynthetic, and contain chlorophyll. They subsist on inorganic nutrients and produce organic matter from $CO_2$ by photosynthesis.

$$CO_2 + H_2O \xrightarrow{h\nu} \{CH_2O\} + O_2 \uparrow \quad ...(9.4)$$

Algae are the primary producers of biological organic matter (biomass) in water bodies. They are food for fish and small aquatic animals.

**Fungi:** They (molds, mildews, yeasts and mushrooms) are, aerobic, multicellular, and non-photosynthetic (lack chlorophyll) organisms. They breakdown cellulose in wood and other plant materials. Fungi are entirely heterotrophic organisms, deriving carbon and energy from the degradation of organic matter (saprophytes). Along with bacteria, fungi are the principal organisms responsible for the decomposition of carbon in the biosphere.

**Bacteria:** Bacteria are microorganisms and are of three kinds—aerobic, anaerobic and facultative. Aerobic bacteria require $O_2$ as an electron receptor. Anaerobic bacteria function in oxygen-free atmosphere. Facultative bacteria use free $O_2$, when available, and use other oxidants like nitrates when molecular oxygen is not available.

### 9.3.1 Microbial Transformation of Carbon

Carbon is an essential life element, present in all living matter (carbohydrates, proteins, nucleic acids and lipids), and found in all organic compounds. Carbon is fixed as carbohydrates $\{CH_2O\}$ via photosynthesis, and is released to atmosphere via respiration.

$$CO_2 + H_2O \xrightarrow{h\nu} \{CH_2O\} + O_2 \uparrow$$

$$O_2 + \{CH_2O\} \rightarrow CO_2 + H_2O \quad ...(9.5)$$

Anaerobic degradiation of organic matter (fermentation reaction)

$$2\{CH_2O\} \rightarrow CH_2\uparrow + CO_2\uparrow \quad ...(9.6)$$

When $CO_2$ acts as an electron-receptor in the absence of $O_2$, methane is formed.

$$CO_2 + 8H^+ + 8e^- \rightarrow CH_4\uparrow + 2H_2O \quad ...(9.7)$$

This reaction is mediated by methane-forming bacteria. Methane formation is a valuable process responsible for the degrdation of large quantities of organic waste. In the benthal regions of natural waters, methane-formining bacteria degrade organic matter in the absence of oxygen. Methane production is a very effective means for the reduction in oxygen demand.

$$CH_4 + 2O_2 \rightarrow CO_2 + 2H_2O \quad ...(9.8)$$

### 9.3.2 Microbial Transformation of Nitrogen (Nitrogen Fixation)

In *nitrogen fixation* a molecule of $N_2$ is fixed as organic nitrogen. *Nitrification* is the process of converting ammonia to nitrate ($NO_3$). The *denitrification* process is the reduction of nitrates and nitrites to molecular $N_2$. Nitrogen-fixing bacteria mediate these processes. *Azatobacter*, *Clostridium*, and *Rhizobium* bacteria and blue green algae take part in the nitrogen fixation process.

$$3\{CH_2O\} + 2N_2 + 3H_2O + 4H^+ \rightarrow 3CO_2 + 4NH_3 \quad ...(9.9)$$

*Nitrosomonas* and *Nitrobacter* catalyze nitrification. *Nitrosomonas* oxidize $NH_3$ to nitrite ($NO_2^-$) and *Nitrobacter* mediate in the oxidation of nitrite ($NO_2^-$) to nitrate ($NO_3^-$).

$$2NH_3 + 3O_2 \rightarrow 2H^+ + 2NO_2^- + H_2O \qquad ...(9.10)$$

$$2NO_2^- + O_2 \rightarrow 2NO_3^- \qquad ...(9.11)$$

Nitrification is particularly important in nature, bacause nitrogen is absorbed by plants primarily as nitrate. Microbial transformation of chemical fertilizers to nitrate enables maximum assimilation of nitrogen by plants.

Microbial processes also mediate the reduction of nitrogen in chemical compounds to lower oxidation states

$$2NO_3^- + \{CH_2O\} \rightarrow 2NO_2^- + H_2O + CO_2 \qquad ...(9.12)$$

Denitrification is a special case of nitrite ($NO_2^-$) reduction, the reduction product being $N_2$. In nature, it is the mechanism by which fixed nitrogen is returned to the atmosphere.

### 9.3.3 Other Microbial Activities

Acid mine drainage is one of the common problems in the aqatic environment. Acid mine waters arise due to sulfuric acid, produced from the pyrite, $FeS_2$. Micro-organisms are strongly involved in the overall processes. Microbial corrosion can also be brought under this aspect. This activity is responsible for the formation of some aquatic sediments. Some bacterial species produce large quantities of $Fe(OH)_3$ as part of their energy-extracting mediation of the oxidation of Fe(II) to Fe(III). In anaerobic benthal regions, bacteria use sulfate ion as an electron acceptor.

$$SO_4^{2-} \rightarrow H_2S \qquad ...(9.13)$$

$$Fe^{2+} + H_2S \rightarrow FeS + 2H^+$$

Microboes are also involved in the biological reduction of pesticides (e.g. aldrin → dieldrin). Effects of micro-organisms on water bodies are given in Fig. 9.3.

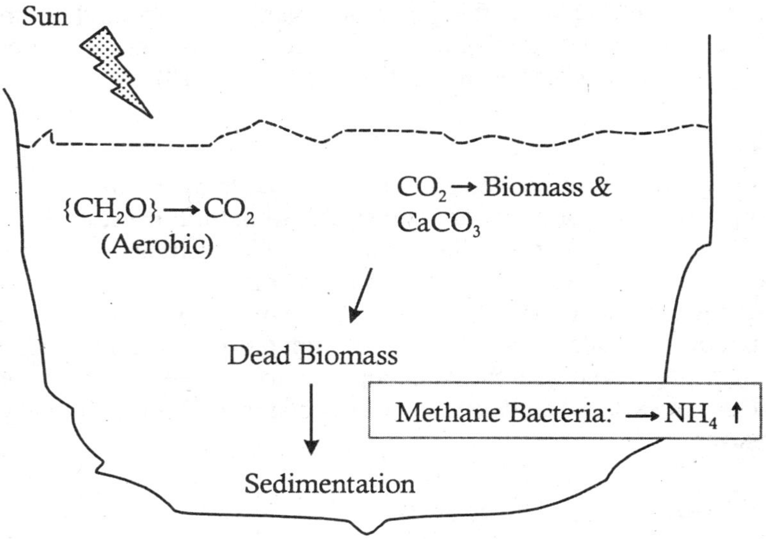


**Fig. 9.3:** Effects of Micro-organisms on Water Bodies

## 9.4 ECOLOGY OF THE WATER BODIES

Efficient usage of water by organisms, animals and humans depends on the quality of water. Factors that affect water bodies are geological factors, dissolved oxygen (DO), coliform (pathogens) and water salinity and alkalinity (pH). Standards for fresh surface waters are given in Table 9.1.

**TABLE 9.1:** STANDARDS FOR FRESH SURFACE WATERS

| *Class* | *Usage* | *DO (mL/L)* | *ColiformNo./l* | *pH* |
|---|---|---|---|---|
| AA & A | Potable | 5.0 | <1,000 | 6.5-8.5 |
| B | Bathing | 5.0 | <25,000 | 6.5-8.5 |
| C | Fishing & Recreation | 5.0 | <25,000 | 6.5-8.5 |
| D | Agriculture & Industry | 3.0 | – | 6.0-9.5 |

### 9.4.1 Geographical Factors

Geographical factors that affect water bodies are vegetation, the salinity, hardness, color, and turbidity. Color in water may result from the presence of metal ions (Fe & Mn), present due to natural geographical terrain or due to pollutants, humus and industrial waste. Color of water is pH-dependent, but intrinsically it is due to effect of these factors as far as ecology is concerned.

### 9.4.2 Physical Factors

The main physical parameters that affect water bodies are dissolved oxygen (DO), turbidity and thermal stress. Color, odor and taste are physical characteristics of drinking water that are more important for aesthetic than for health reasons.

#### *9.4.2.1 Dissolved Oxygen (DO)*

Dissolved oxygen (DO) is fundamental to survival of aquatic organisms. Without DO, water bodies become inhospitable to gill-breathing aquatic organisms. Factors that affect the dissolved oxygen in a water body are: (*i*) geographical terrain, (*ii*) inflow and out-flow conditions, physical, chemical and biological nature of the water bodies (see Chapter 10).

#### *9.4.2.2 Turbidity*

Turbidity in water is caused by suspended particulate and colloidal matter, such as clay, mud, silt, finely divided inorganic and organic matter, either due to natural causes or due to discharge of pollutants into the water bodies. Turbidity blocks the sunlight reaching deeper into the water body, thereby retarding the photosynthesis in aquatic plant systems.

Discharge of sewage into water bodies can produce, in addition to turbidity, depletion of dissolved oxygen (DO) by biological oxidation of matter and stimulation of algal growth (eutrophication). Oil spills interfere with gas exchange at the water-air interface and thus interfere with the respiration process of aquatic organisms. Oil spills destroy beaches and seashores, adversely affect flora and fauna and ruin marine life.

#### *9.4.2.3 Thermal Stress*

Thermal pollution (heat) is a common waste product of many industries. Thermal pollution has profound physical, chemical and biological effects. These effects are short-term as well as long-term, and local as well as regional. While, disruption of atmospheric heat balance due to increase in atmospheric $CO_2$ and water will have global effects (greenhouse effects), thermal pollution of water bodies due to discharge of hot effluents from industrial and other sources will have immediate and long-term effects at local and regional levels.

$$H = M \pm C \pm R - E \qquad \text{...(9.14)}$$

where, $H$ = heat gain/loss;
$M$ = metabolic heat production;
$C$ = heat exchange (by convection and conduction);
$R$ = heat exchange by radiation;
$E$ = heat loss by evaporation.

Solubility of oxygen in water is a function of temperature. Increase in temperature of a water body leads to decrease in the DO content (Table 9.2). This problem is further aggravated by the increase in demand for oxygen by the aquatic organisms, due to increase in their metabolic rates with increase in temperature. Increase in temperature of water bodies will also have direct effect on the reproduction processes of aquatic organisms. Excessive temperature can prevent normal development of eggs.

Commercial fisheries and other food industries reap major harvests from the Great Lakes of the world. Pollution of many of these lakes is gradually reducing the fish catch. The general experience has been the disappearance of more desirable species and the survival of the less desirable ones.

**TABLE 9.2:** SOLUBILITY OF OXYGEN IN WATER AS A FUNCTION OF TEMPERATURE

| *Temperature ofWater °C* | *Saturation Concentrationof $O_2$ in Water (mg/L)* |
|---|---|
| 0 | 14.5 |
| 4 | 13.0 |
| 8 | 12.0 |
| 15 | 10.2 |
| 20 | 8.6 |

### 9.4.3 Chemical Factors

Water is used as a 'courier' to transport chemicals, toxicants, heat and other pollutants discharged from domestic, municipal, agricultural, industrial and rural and urban activities.

Pesticides, insecticides and fungicides and other hazardous chemicals in water, consumed by fish and other aquatic organisms, may become potential health hazard to them. Many polluted lakes have lost their substantial and popular fish population as only the small bodied, rapidly reproducing other organisms, which do not depend on complex food webs, are favored in such waters. The polluted water has indirect impact on species, not necessarily living in water, but depending on the lakes for food and water. Migratory birds depend on water resources for fish. If fish have become scarce in the lakes they are forced to look for other sources of food and this may create a crisis of survival for them. Eating of fish and other shellfish from contaminated water becomes a direct threat to humans, birds and animals. Often bioaccumulative chemicals are retained in an organism's body. If a predator eats such contaminated prey its tissues store those toxic substances. This is called biomagnification.

#### *9.4.3.1 Acid Rain*

Acid rain is a pernicious global problem, created by conversion of chemical pollutants, $SO_x$ and

$NO_x$ to their respective acids. The phenomenon and its effects have been well documented for many parts of Europe and eastern United States. Acid rain destroys large tracts of vegetation, leaches important nutrients such as calcium and magnesium from the soil, thereby degrading its quality. It also draws toxic metals such as aluminum and mercury out of sediments into water bodies, where the toxic metals can harm aquatic life. All aquatic organisms are sensitive to pH changes in water. Acidification of water bodies makes them uninhabitable to aquatic life and pH <3.5 leads to respiratory failure in many organisms (Table 9.3). A water pH below 4.8 leads to extinction of many fish, and practically all mollusks.

**TABLE 9.3:** EFFECTS OF ACIDIC WATER ON ECOLOGY

| *pH* | *Effects* |
|---|---|
| 1.0-3.5 | Damage to plant leaves and vegetation. No aquatic life |
| 3.5-4.5 | Death of fish and most of aquatic life |
| 4.5-5.5 | Salmon and trout fail to breed |
| 5.5-6.0 | Nutrients are leached from the soil |
| 9.4-3.2 | Eutrophication |

### 9.4.3.2 Eutrophication

The process of eutrophication ('abundant food') in a lake or sluggish water starts with a sharp increase in the concentration of phosphorous and nitrogen and other plant nutrients carried into the lakes by land run-off. The latter promote the rapid growth of algae and/or higher aquatic plants, leading finally to 'algal bloom'. Under normal conditions algae generate oxygen through photosynthesis. They also promote fish growth by serving as their food. However, algal bloom crowd out other organisms, overgrow and die, owing to depletion of nutrients in the water. Algal blooms can affect water quality adversely and indicate potentially hazardous changes in local water chemistry. The decaying algae deplete oxygen from water. The water develops a bad taste and color and becomes unfit for human and animal consumption, unless filtered and specially treated. To avoid eutrophication and to maintain healthy waters the cycling of basic nutrients like carbon, nitrogen, sulfate and phosphate is very important.

## 9.4.4 Biological Factors

Water often plays a crucial role as a vehicle for direct transmission, or indirect transmission via food, of microorganisms and pathogens (e.g. coliform organisms) that cause infections and waterborne diseases. The presence coliform organism in water indicates fecal pollution and potentially dangerous bacterial contamination by disease-causing micro-organisms.

Water can also be a habitat for vectors, like mosquitoes, flies, fleas, and ticks, insects and worms, which transmit disease-producing organisms. Some of the vectors responsible for waterborne diseases are given in Table 9.4.

**TABLE 9.4:** VECTORS CAUSING WATERBORNE DISEASES

| *Vectors/Organisms* | *Waterborne Diseases* | *Mode of Transport* |
|---|---|---|
| **Bacterial** | | |
| *Vibrio cholerae* | Cholera | Enteric |
| *Escherichia coli* | Diarrhea; dysentery | Enteric (fecal waste) |
| *Salmonella paratyphi* | Paratyphoid fever | Enteric |

| *Vectors/Organisms* | *Waterborne Diseases* | *Mode of Transport* |
|---|---|---|
| *Salmonella typhi* | Typhoid fever | Enteric |
| *Shigella* | Shigellosis (dysentery) | Via flies & cockroaches |
| *Salmonella* | Salmonellosis | |
| *Yersinia enterocolitica* | Gastroenteritis; botulism | Enteric (contaminated food) |
| *Yersinia pestic* (*Xenopsylla cheopis*) | Plague (bubonic; septicemic; pneumonic) | Rats & fleas |
| **Fungal** | | |
| Compost | Aspergillosis | Contact |
| Ringworm | Dermatophytosis | Bathing |
| **Parasites/Helminthes** | | |
| Blood fluke (Cercarium larve) | Schistosomiasis (Bilharzia) | Water |
| Liver fluke | Clonorchiasis | Fish |
| Guinea worm | Dracontiasis | Water |
| Hookworm | Ancyclostomiasis | Soil |
| Pinworm | Enterobiasis | Feces in water |
| Round worm | Ascariasis | Soil |
| | Trichinosis | Raw meat |
| Tapeworm | Taeniasis | Raw meat |
| Nematodes | Anisakiasis | Food |
| | Filariasis | Mosquito bite |
| | Intestinal capillariasis | Seafood (fish) |
| | Strongyloidiasis | Soil |
| Trematode | Paragonimiasis | Crabs |
| *Plasmodium vivax* | Malaria | Protozoan infection and |
| *Plasmodium falciparum* | Malaria | mosquito bite |
| *Plasmodium malariae* | Malaria | " " |
| *Plasmodium ovale* | Malaria | " " |
| Rats | Leptospirosis | Sewage contact |
| **Protozoa** | | |
| *Entamoeba histolytica* | Amaebiasis; dysentery | Water and soil (enteric) |
| *Giardia lamblia* | Giardiasis (diarrhea); meningoencephalitis; malarial infection | Water and soil |
| **Viral** | | |
| *Aedes aegypti* | Dengue fever; yellow fever | Mosquito bite |
| Enterio virus | Viral gastroenteritis; | Enteric |
| Polio virus | Hepatitis A Poliomyelitis | Enteric |
| **Zoonoses** | | |
| Farm & domestic animals | Anthrax | Soil; tissue; flies |
| | Brucellosis | Blood & milk |
| | Hydrophobia (rabies) | Saliva; animal bite |
| | Leptospirosis (Weil's disease) | Soil; food; water |
| Poultry & Pets | Q-fever | Milk & excreta |
| | Salmonellosis | Contaminated food |

## BIBLIOGRAPHY


Adamson, R. G. (1973), Bellhaven: Ontario. "*Pollution: An Ecological Approach.*"

Adriano, D. C. (1986), Springerverlag: New York. "*Trace Elements in the Terrestrial Environment.*"

Baier, R. E., Mark, H. B. and Mattson, D. S. (eds.) (1981), Marcel Dekker: New York. "*Water Quality Measurement.*"

Barkley, P. W. and Seckler, D. (1972), Harcourt-Brace: New York. "*Economic Growth and Environmental Decay.*"

Clark, J. (1969), *Sci Amer* (March). "Thermal pollution and aquatic life."

Dart, R. K. and Stretton, R. J. (1977), Elsevier: New York. "*Microbial Aspects of Pollution Control.*"

Hanlan, J. J. (1969), Mosby: St Luis, MS. "*Health Effects of Environmental Pollutants.*"

Herber, R. F. M. and Stoeppler, M. (eds.) (1994) Amsterdam: Elsevier. "*Trace Element Analysis in Biological Specimens.*"

Higgins, I. J. and Burns, (1975), Academic Press: New York. "*The Chemistry and Biology of Pollution.*"

Hobbs, P. V. and McCormick, M. P. (eds.) (1988), Deepak: Hampton, VA. "*Aerosols and Climate.*"

Koren, H. (1974), Pergamon Press: New York. "*Environmental Health and Safety.*"

Lamb, J. C. (1985), John Wiley: New York. "*Water Quality and its Control.*"

Markert, B. (ed.) (1993), VCH Pubs: New York. "*Plants as Biomonitors: Indicators for Heavy Metals in the Terrestrial Environment.*"

Meadows, D. H., *et al.* (1972), Universe Books: New York. "*The Limits to Growth.*"

Nriagu, J. D. and Davidson, C. I. (eds.) (1986), John Wiley: New York. "*Toxic Metals in the Atmosphere.*"

Peirce, J. J., Weiner, R. F. and Vesiland, P. A. (1998), Butterworth-Heineman: New Delhi. "*Environmental Pollution and Control.*"

Pfafflin, J. R. and Ziegler, E. N. (eds.) (1976), Gordon & Breach: New York. "*Encyclopedia of Environmental Science and Engineering,*" (Vols. 1 & 2).

Purdom, P. W. (ed.) (1980), Academic Press: New York. "*Environmental Health,*" (2nd edn.).

Turiel, I. (1975), Prentice-Hall: Englewood Cliffs, NJ. "*Physics: The Environment and Man.*"

Waldbott, G. L. (1973), Mosby: St. Luis, MS. "*Health Effects of Environmental Pollutants.*"

Winegar, E. D. and Keith, L. H. (eds.) (1993), Lewis Pubs: Boca Raton, FL. "*Sampling and Analysis of Airborne Pollutants.*"

# 10

# Behaviour of Pollutants in Water Bodies

Pollutants adversely affect the ecology of the water body. Their behaviour depends on the type of water body as different types of water bodies have varied physicochemical dynamism, geographical terrain, pollutants and flora and fauna. Of them, the most important parameter is dissolved oxygen. Decrease in the dissolved oxygen (DO) is one of the adverse effects of pollutants on a water body. The dissolved oxygen is an essential factor for the survival of aquatic organisms. Decrease in it is, in fact, a positive indicator of water pollution and stressful survival of aquatic organisms. The primary cause for DO depletion is the presence of organic matter. Biochemical oxygen demand (BOD) and chemical oxygen demand (COD) are quantitative methods for assessing the quality of water and wastewater.

Polluted water diminishes or destroys aquatic population and increases algal growth. If the pollutants toxic to aquatic animals, such as pesticides and herbicides, trace elements and radioactive materials are present, then both the type and total number of species will decrease, irrespective of the other ecological characteristics of the water body. The ecology of flowing water bodies differs from that of stationary water bodies and, the effects of pollutants in the two systems are also different. The flowing waters have inlet as well as outlet for pollutants and pollutants are diluted and dispersed due to flow. On the other hand, stationary waters have only inflow and no outflow. Hence, concentration of pollutants gradually increases in these water bodies, but their effects are localized.

Reduced turbidity in still water, as compared with flowing water, encourages growth of certain types of aquatic organisms. Discharge of nutrient compounds into still water produces uncontrolled growth of plankton in the top layer of the water body. Algae interfere with the aquatic activities and also act as barrier to the penetration of oxygen into water. There is a temperature-density relationship that results in the formation of thermal stratification within the stationary water bodies.

This chapter deals with the behavior of pollutants in mobile and stationary water bodies.

Behavior of pollutants in a water body depends on the type of water body, physicochemical dynamics of the water body, and the geographical terrain, and the pollutants and organisms present in the water body. The most important parameter in a water body is the amount of dissolved oxygen (oxygen demand). The behavior of pollutants is different in mobile and stationary water bodies.

## 10.1 WATER BODIES

Water bodies comprise: (*i*) mobile systems—run-offs, streams, estuaries and rivers, and (*ii*) stationary systems—ponds and lakes. Geological factors, temperature, hydrological chemistry, soil bacteria, and other environmental factors and anthropogenic activities influence water quality in these bodies. The effects of pollutants on these two types of water bodies and on their ecologies are considerably different. Mobile systems have inflow and outflow of water along with the pollutants. Therefore, such systems dilute the pollutants from the inlet point, but transport them down the stream, ultimately to the seas/oceans. Polluted streams, estuaries and rivers are the major sources of sea and ocean pollution. Ponds and lakes, in general, have only inflow and hence concentration of pollutants gradually increases in these water bodies, but the effects are localized. The most important physical characteristic of lakes is their temperature profile, that is, change of temperature with depth. The thermal stratification plays a major role in the distribution of nutrients and dissolved oxygen (DO) and has an important impact on the ecology of lakes.

Every ecosystem has delicately balanced inputs and outputs. Any disturbance of this balance leads to pollution and as consequence environmental disturbance. That is any unfavorable disturbance of an ecological balance is pollution and the factors causing pollution are pollutants. The biological health of a water body is crucially dependent on its physical and chemical characteristics. A normal healthy stream or lake has balance of plant and aquatic life with species diversity. Pollution disrupts this balance, resulting in the reduction of several species, perhaps desirable ones, and proliferation of other (limited) species. Survival of aquatic organisms (fish, clams, etc) in a water body is an indication of the health of that water body, and complete absence of species is a proof for its degradation.

Conventional water pollutants are suspended particles, sludge, sewage discharge, coliform, nutrients and temperature. One of the adverse effects of pollution of a water body is the decrease in dissolved oxygen (DO), which is essential for the survival of aquatic organisms (minimum level 6 ppm). Decrease in dissolved oxygen is a positive indicator of water pollution, and such a condition also displays the stressful nature of survival of the aquatic organisms in that system. The primary reason for depletion of DO is the proliferation of oxygen-demanding aerobic bacteria. Their action produces $CO_2$ as the main oxidation product. Low levels of dissolved oxygen promote anaerobic fermentation. The effect of anaerobic organisms on organic matter is even more detrimental to aquatic life (Table 10.1).

**TABLE 10.1:** END-PRODUCTS OF BACTERIAL DECOMPOSITION OF ORGANIC MATTER

| *Condition* | *C* | *N* | *S* | *P* |
|---|---|---|---|---|
| Aerobic | $CO_2$ | $HNO_3$ | $H_2SO_4$ | $H_3PO_4$ |
| Anaerobic | $CH_4$ | $NH_3$ | $H_2S$ | $PH_3$ |

## 10.2 OXYGEN DEMAND IN WATER BODIES

The quality of water or wastewater (spent or used water from home, community, farm or industry that contains dissolved or suspended matter) is commonly expressed by an estimate of dissolved oxygen (DO). Two other parameters of interest are the biochemical oxygen demand (BOD), which is a measure of the amount of oxygen consumed in the biological processes that break down organic matter in water, and the chemical oxygen demand (COD), which is a measure of the oxygen required to oxidize all compounds, both organic and inorganic, in water. The greater the BOD (or COD), the greater is the degree of pollution.

### 10.2.1 Biochemical Oxygen Demand (BOD)

The amount of oxygen-demanding contaminants in a water body is a significant parameter that can be estimated via the biochemical oxygen demand (BOD). BOD is the amount of dissolved oxygen (DO) in water consumed by microorganisms in the process of breaking down organic matter (represented by $\{CH_2O\}$) in water. This is usually estimated for a predetermined time, say, for five days. Bio-oxidation of nitrogenous compounds is also included in this.

$$\begin{aligned}\{CH_2O\} + O_2 &\xrightarrow{\text{Microorganisms}} CO_2 + O_2\\ NH_4^+ + 2O_2 &\rightarrow 2H^+ + NO_3^- + H_2O\\ 2SO_3^{2-} + O_2 &\rightarrow 2SO_4^{2-}\end{aligned} \quad \text{...(10.1)}$$

Organisms remove organic matter by chemical and biological oxidation. This process occurs in two stages. The first stage is carbonaceous (C-BOD) and the second stage is nitrogenous (N-BOD) oxidation. C-BOD arises in the case of suspended as well as dissolved matter. N-BOD arises only in the case of dissolved matter, and is not subject to sedimentation.

$$\text{Dissolved oxygen (DO)} + \text{Organic matter} \xrightarrow{\text{Bacteria(Protozoa)}} CO_2 + \text{Biological growth} \quad \text{...(10.2)}$$

$$\text{Dissolved oxygen (DO)} + \text{Ammonia}(NH_3) \xrightarrow{\text{Nitrifying-bacteria}} \text{Nitrates} + \text{Bacterial growth}$$

The BOD test is used as a qualitative measure of pollution, loading on oxygen resources. This is achieved by measuring the oxygen consumed for the oxidation reaction and sewage decomposition. *E. coli* is used as an indicator of pollution loading due to human intestinal waste. Its count in water can be used as an indirect estimate for other pathogens.

Sources of dissolved oxygen (DO) in water are atmospheric re-aeration, photosynthesis and *advection (aeration)*. Aeration promotes biological degradation of organic matter in water. Sinks of dissolved oxygen (DO) are respiration by organisms (fish, bacteria, etc., and algae in benthal deposits in water bodies) and chemical oxidation. The mass-balance equation is

$$\frac{\partial C}{\partial t} = \nabla J \pm \sum s \quad \text{...(10.3)}$$

$$J_r = \underset{\substack{\text{diffusion}\\ \text{flux}}}{D_r \frac{\partial C}{\partial r}} - \underset{\substack{\text{advection}\\ \text{flux}}}{U_r C} \quad \text{...(10.4)}$$

$$\frac{\partial C}{\partial t} = \frac{1}{A}\frac{\partial}{\partial x}\left(D_r A \frac{\partial C}{\partial x}\right) - \frac{1}{A}\frac{\partial}{\partial x}(QC) + \Sigma S \quad \text{...(10.5)}$$

where, $C$ = concentration; $J$ = mass flux;
$S$ = sources and sinks; $D_r$ = diffusion coefficient;
$U_r$ = velocity of the stream; $A$ = area of cross-section;
$Q$ = volume of water flow

The BOD test is standardized. It is run in the dark at 20° C for five days, $(BOD)_5$. The test is an estimation of oxygen used by micro-organisms in the water sample during the five days. Dissolved oxygen (DO) is determined at the beginning and at the end of 5-day period (Fig. 10.1).

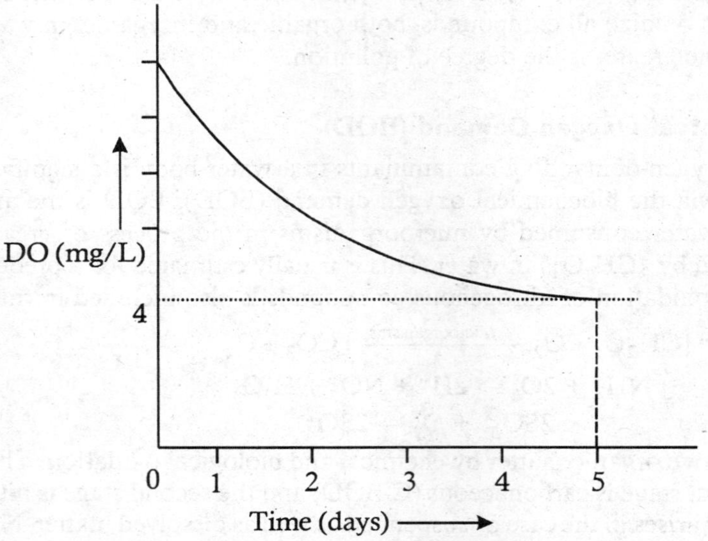


**Fig. 10.1:** Oxygen Uptake Curve in BOD Test

The 5-day $(BOD)_5$ test is (Fig. 10.2).

$$\frac{dy}{dt} = K_1(L - y) \qquad \ldots(10.6)$$

$$y = L(1 - 10^{-K_1 t})$$

$$\frac{y}{L} = 1 - 0.79^1$$

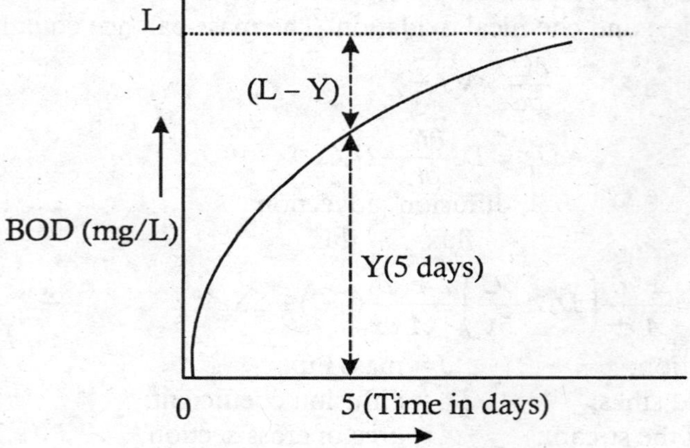


**Fig. 10.2:** 5-Day BOD Test Profile

In the carbonaceous phase, the organisms feed on the more easily degraded compounds. After ten days, nitrogenous phase sets in. Long-term BOD test curve comprises carbonaceous (C-BOD) and nitrogenous (N-BOD) segments (Fig. 10.3).

$$BOD_{ult} = a(BOD)_5 + b(N_K) \quad ...(10.7)$$

where, $N_K$ = Kjeldahl (organic + ammonia) nitrogen;
$a$ & $b$ = empirical constants.

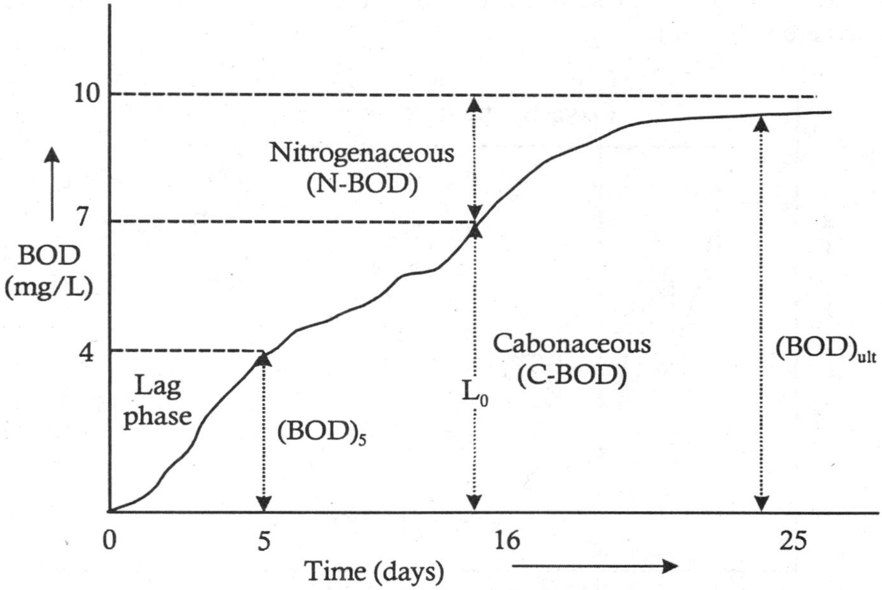


**Fig. 10.3:** Long-term BOD Profiles

Nitrogenous oxygen demand (N-BOD) covers nitrogen fixation of by bacteria.

$$N_2 \rightarrow NH_3 + NO_x \quad ...(10.8)$$

$$2NH_4^+ + 3O_2 \xrightarrow{\text{Nitrosomonas}} 2NO_2^- + 4H^+ + H_2O \quad ...(10.9)$$

$$2NO_2^- + O_2 \xrightarrow{\text{Nitrobacter}} 2NO_3^-$$

### 10.2.2 Chemical Oxygen Demand (COD)

BOD measurement is an indirect method for estimation of non-specific organic pollutants in water bodies. However, this test is time-consuming and is not very accurate. Chemical oxygen demand (COD) is a parameter used to assess industrial wastewater. It is the amount of $O_2$ consumed under specific conditions in the oxidation of organic and oxidizable inorganic compounds. In the COD test, the aerobic bacteria used in BOD tests are replaced with a chemical oxidant. The test involves chemical oxidation of material in water by dichromate ion in 50 per cent $H_2SO_4$.

$$3\{CH_2O\} + 16H^+ + 2Cr_2O_7^{2-} \rightarrow 4Cr^{3+} + 3CO_2 + 11H_2 \quad ...(10.10)$$

After the oxidation reaction, the amount of unreacted dichromate is determined by titration with a standard reducing agent. (Here $\{CH_2O\}$ refers to a carbohydrate).

## 10.3 BEHAVIOR OF POLLUTANTS IN STREAMS

Factors affecting dissolved oxygen (DO) in a natural stream fall into two categories—(*i*) geographical characteristics and (*ii*) biophysical and chemical characteristics—that is, sources and sinks of DO. When raw sewage that contains nutrients (phosphorus and nitrogen compounds) is discharged to a stream, the oxidation of these compounds depletes the dissolved oxygen from the point of inlet (Fig. 10.4). Temperature of the water body influences the solubility of oxygen and rates of both physical and biochemical reactions. When the stream is totally depleted of DO, the stream is treated as an anaerobic system.

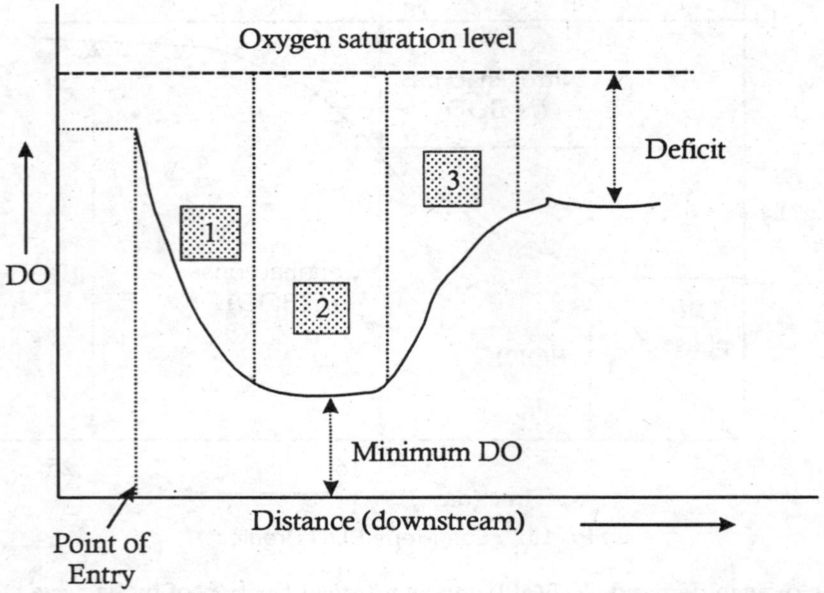


1 = Decomposition zone; 2 = Septic zone; 3 = Recovery zone

**Fig. 10.4:** Profile of Dissolved Oxygen (*DO*) Downstream due to Ingest of Raw Sewage

Relationship between the concentration ($X$) of pollutants and the water-flow can be linear or logarithmic. The concentration of a pollutant, $Y$, is inversely related to flow of water, $Q$ ($Q = 1/X$).

$$Y = a + bX \text{ (linear relationship)} \quad ...(10.11)$$

$$Y = c + aX^b \text{ (Logarithmic relationship)} \quad ...(10.12)$$

The concentration of $O_2$ in a stream as a function of time, $t$, is

$$\therefore \quad \frac{dD}{dt} = (K_1Z - K_2D) \quad ...(10.13)$$

The minimum dissolved oxygen that is absolutely necessary for the survival of aquatic organism is given by the critical deficit, $D_C$,

$$D_C = \frac{K_1L \times 10^{-K_1t_c}}{K_2} \quad ...(10.14)$$

The biological community in unpolluted stream is highly diverse. There are many types of organisms present, but relatively few of each. The diversity of species, $d$, is

$$d = \sum_{i=1}^{s}\left(\frac{n_1}{n_s}\right)\log\left(\frac{n_1}{n_s}\right) \qquad ...(10.15)$$

where, $Z$ = the amount of $O_2$ still required at any time, $t$;
$Y = BOD$;
$K_1$ & $K_2$ = de-oxygenation and re-oxygenation constants;
$L$ = carbonaceous oxygen demand;
$D$ = deficit in dissolved oxygen;
$d$ = diversity index;
$n_i$ & $n_s$ = number of individuals in the $i$th and in all species, respectively

When the rate of oxygen consumption is far greater than oxygen replenishment, the stream may become anaerobic, identifiable by the presence of floating sludge and bubbling of gases. Increased turbidity, particulate matter and low DO all contribute to a decrease in fish life. DO <3 mg/L causes extinction of fish and results in the growth of microorganisms, which produce byproducts that cause foul odors in the stream.

The stream will react differently to inorganic waste (trace metals, etc). *If the waste is toxic to aquatic organisms, both the type and total number of species will decrease (in downstream), irrespective of the availability or lack of dissolved oxygen.*

### 10.3.1 Flow Characteristics in Streams

Water flow characteristics in streams can be analyzed as tubular flow under laminar and turbulent conditions, taking into account hydrodynamic characteristics.

Velocity of water flow, $v$, (stream flow) under non-turbulent (laminar) conditions is given by *Hagen-Poiseulli* equation, and the frictional coefficient, $f$, by *Reynolds* number.

$$v = \frac{\mu s d}{32\eta} \qquad \textit{Hagen-Poiseulli's}\text{ equation} \qquad ...(10.16)$$

$$f = \frac{64}{R_e} \qquad ...(10.17)$$

Under turbulent flow, convection process (eddy current) predominates in flow pattern and the friction coefficient, $f$, is influenced by relative roughness, $d/\varepsilon$

$$\frac{1}{\sqrt{f}} = 2\log\left(\frac{d}{\varepsilon}\right) + 1.14 \qquad ...(10.18)$$

where, $\mu$ = kinematic viscosity;
$\eta$ = viscosity;
$s$ = energy slope = $h/L$;
$d$ = diameter of water column;
$R_e$ = Reynolds number;
$\varepsilon$ = Roughness factor.

## 10.4 BEHAVIOR OF POLLUTANTS IN STATIONARY WATER BODIES

Ponds and lakes are generally stationary water systems, with inflow of water but no outflow. Therefore, the influx of effluents is localized. Lakes and ponds can be *oligotrophic, eutrophic* or *dystrophic*. Oligotrophic lakes are deep, generally clear, deficient in nutrients, and without much biological activity. Eutrophic lakes have more nutrients (Table 10.2), support more life, and are more turbid. During the later stages of eutrophication the water body is choked by abundant plant life due to higher levels of nutritive compounds such as nitrogen and phosphorus. Human activities can accelerate the process. Dystrophic lakes (and ponds) are acidic, shallow bodies of water that contain much humus and/or other organic matter. They are clogged with plant life but few fish.

**TABLE 10.2:** ESSENTIAL PLANT NUTRIENTS IN WATER BODIES

| *Component* | *Source* | *Features* |
|---|---|---|
| Hydrogen | Water | Biomass constituent |
| Oxygen | Water | Biomass constituent |
| Carbon dioxide | Atmosphere; decay | Biomass constituent |
| Nitrogen | Atmosphere; decay | Biomolecules |
| Phosphorus | Minerals; decay; fertilizers | Biomolecules; bones |
| Potassium | Minerals; pollutants | Metabolic |
| Calcium | Minerals; pollutants | Metabolic |
| Magnesium | Minerals; pollutants | Metabolic |

Reduced velocity and reduced turbulence in the lakes and ponds allow much of the particulate matter discharged into them settles. The water, therefore, has a low level of suspended matter and turbidity. This situation in lakes and ponds encourage the growth of certain type of aquatic organisms. For example, greater penetration of sunlight facilitates the photosynthetic activity and growth of algae. The water in the stationary lake gets stratified (thermal stratification) depending on its density with varying temperature along the depth. In summer, the lower density due to warmer temperature of the water surface, and due to lack of mixing (lack of turbulence) causes this thermal stratification. This thermal stratification in lakes plays a major role in the distribution of nutrients and dissolved oxygen and therefore, on the ecology of lakes. Thus, the ecology of stationary water bodies differs from the ecology of rivers and streams (Fig. 10.5).

### 10.4.1 Ecology of Stationary Water Bodies

The blue algae take up carbon, nitrogen and phosphorus compounds from the water and utilizing sunlight produce high-energy compounds. Algae are consumed by zooplanktons, which, in turn, form the food for aquatic animals. Bacteria consume dissolved carbon (organic matter) and produce $CO_2$, which is in turn used by algae. Thus, a food cycle is established (photosynthesis and respiration) within the normal ecology of water bodies.

In the stratified in water body the level (depth) at which photosynthesis is equal to respiration is termed *compensation* level. It roughly corresponds to a depth at which the attenuation of sunlight is ~1 per cent of what is received at the water surface. The region above the compensation level is called the *trophogenic* zone, where photosynthesis is in excess of respiration. The zone below the compensation level is called the *tropholytic* zone, where respiration is in excess of photosynthesis (Fig. 10.6).

The transparency of the water and the depth to which light can penetrate in it are inversely related to its turbidity. Turbidity interferes with photosynthesis, leading to the reduction of the depth of the compensation level (measured from the water surface) and shifting the water-body towards tropholytic environment. This disturbs the ecological balance of the watercourse, with reduction of photoplankton and decrease in dissolved oxygen (DO).

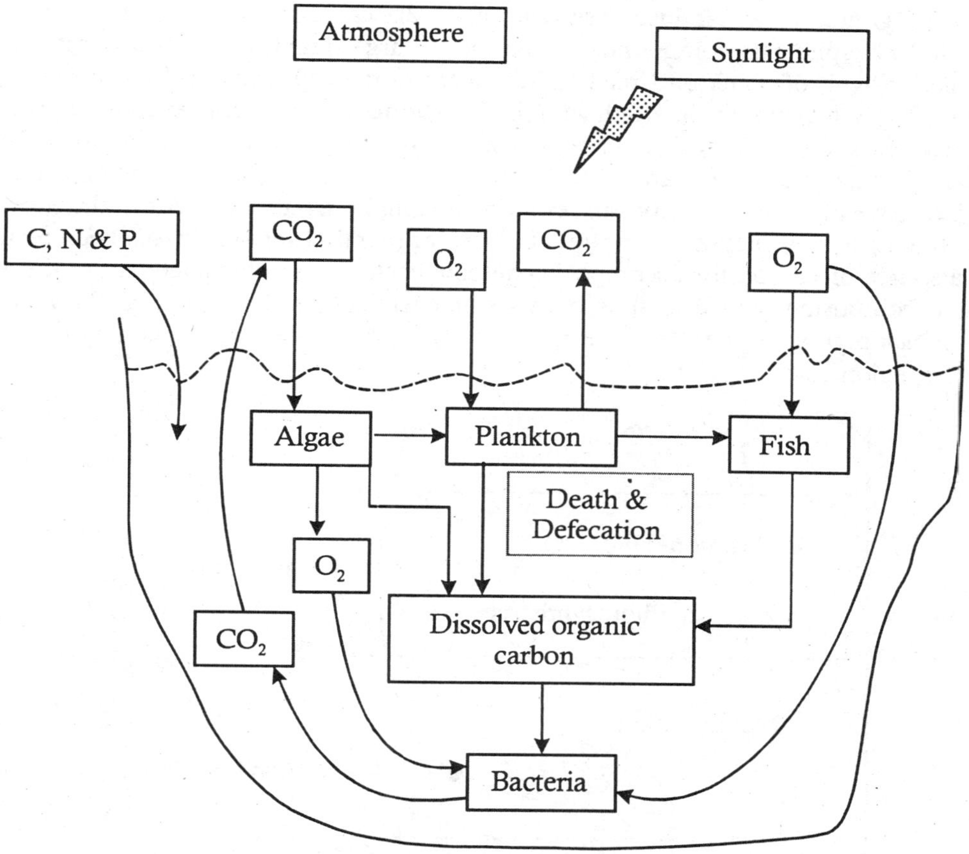


**Fig. 10.5:** Ecology of Stationary Water Bodies

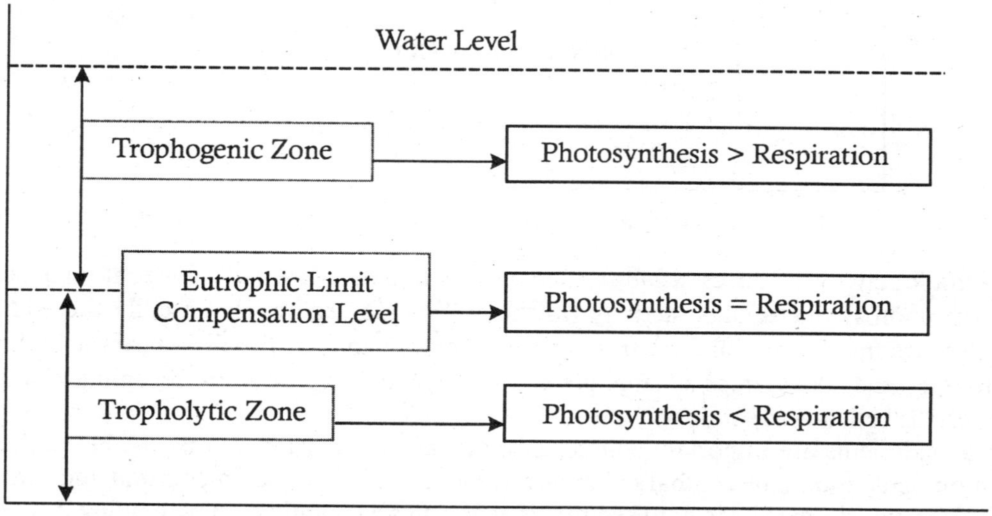


**Fig. 10.6:** Stratification of Stationary Water Bodies

Domestic, agricultural and industrial sewage and sludge contain nutrients in the form of carbon, nitrogen and phosphorous compounds. Dumping of large quantities of such wastes produces uncontrolled growth of plankton (free-floating algae) in the top layer (*epilimnion* region) of the lakes (Fig. 10.7). Algal blooms interfere with the aquatic activities. Dense layer of these organisms would block out sunlight from reaching organisms in deeper parts of the water body. They may also act as a barrier to the penetration of $O_2$ into the water, which may result in depletion of fish species. Excessive algal productivity can result in choking by weeds and odor to the water.

When the algae die, they drop to the lake bottom (benthal region; *hypolimnion* region) and become a source of carbon for decomposing aerobic bacteria. Aerobic bacteria use the available oxygen in decomposing dead algae thus, boosting biochemical oxygen demand in the water. Thus, the *hypolimnion* region can become anaerobic, and subsequently this process may extend to the *metalimnion* region also (Fig. 10.7).

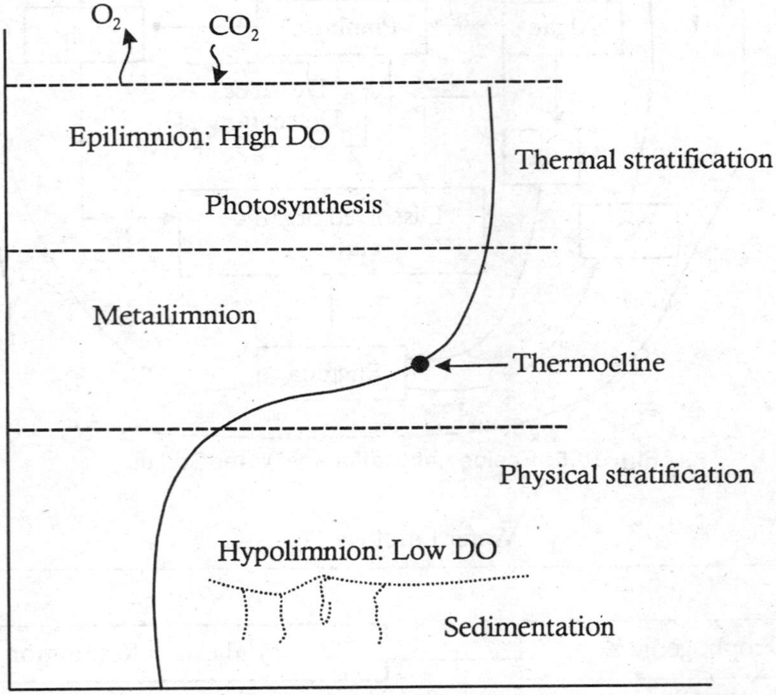


**Fig. 10.7:** Physical and Thermal Stratification of Water Bodies

The aerobic activity produces turbidity, thereby causing a decrease in the penetration of sunlight. This limits photosynthetic algal activity (in the epilimnion region). Eventually the epilimnion region also becomes anaerobic. At this state, all aerobic aquatic life disappears and the algae concentrate on the lake surface as large green mass. The entire process of the aging of stationary water bodies is called *eutrophication*.

Benthal sediments are important sources of inorganic and organic matte in lakes and oceans. The environment around the benthal sediments is anaerobic. Bottom sediments undergo continuous leaching. Phosphorus exchange with sediments aggravates eutrophication by making this essential element available to algae.

The aging of a stationary water body occurs in three phases:

| | |
|---|---|
| The *oligotrophic* phase: | Both variety and number of species grow rapidly during this phase. |
| The *mesotrophic* phase: | Dynamic equilibrium exists among the species during this phase. |
| The *eutrophic* phase: | During this phase, the water body gradually becomes choked with weeds. |

This stratification is temperature-dependent, and there exists a temperature-depth relationship in a water body (Fig. 10.7). The inflection-point of thermal stratification is called the *thermocline*.

## 10.5 ENVIRONMENTAL IMPACT OF WASTEWATER DISCHARGE

Water supports a very complex ecosystem in which intricate physical, chemical and biological processes occur. The hydrological cycle comprises cycles of freezing and thawing, mechanical stress reversals, dynamic equilibrium state that exists between precipitation, percolation, runoff and evaporation.

Precipitation of atmospheric moisture occurs only when temperature of air mass is lowered to or below the dew point (saturation point). Air temperature is lowered due to: (*i*) heat loss by IR radiation, (*ii*) adiabatic cooling as large air masses expand and (*iii*) mixing of hot air (moisture) with cooler air masses. Water loss by evaporation is a physical process. The transpiration process enhances water loss from land. While evaporation is a physical process, transpiration is a physicochemical as well as a biochemical process by which plants and other photosynthetic organisms draw solutions containing essential growth nutrients from the soil.

The ecosystem may be regarded as an '*entropy pump*' that extracts solar energy to produce organized '*open systems*'. The balance between photosynthesis and respiration (P-R cycle) is primarily responsible for regulating the concentrations of $O_2$ and $CO_2$ in the atmosphere (Fig. 10.8).

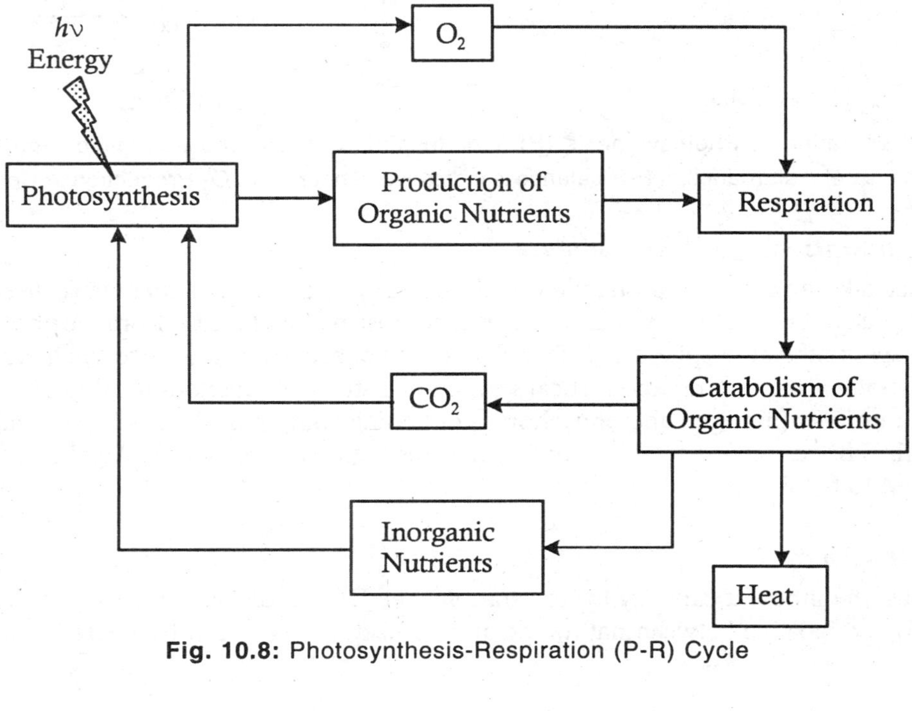


**Fig. 10.8:** Photosynthesis-Respiration (P-R) Cycle

Pollution of surface waters frequently results from a disturbance in the balance between photosynthesis (P) and respiration (R). Adding either an excess of organic waste or an excess of inorganic nutrients (phosphorus and nitrogen fertilizers) also upsets this balance. Photosynthesis << respiration (P << R) and heterotrophic processes tend to dominate and dissolved oxygen (DO) may become exhausted. Case 2: there is progressive accumulation of autotrophic biomass (algae, aquatic plants and weeds). Eventually this biomass decomposes, thereby enhancing rate of respiration (R >> P), which again may reduce DO. Trace metals and other compounds may accumulate in the food chain with deleterious effects on organisms, animals and humans (Plankton → Fish → Man). Dying fish and malodor are indicators of very low oxygen level in a water body. For many natural waters, inorganic fertilizers are often more serious pollutants than biological wastes.

### 10.5.1 Spatial Separation of the *P-R* Cycle

The P-R cycle can also be disturbed by spatial separation of the two processes. There is horizontal separation of P-R functions in rivers and streams and vertical separation in stratified stationary water bodies (Fig. 10.9).

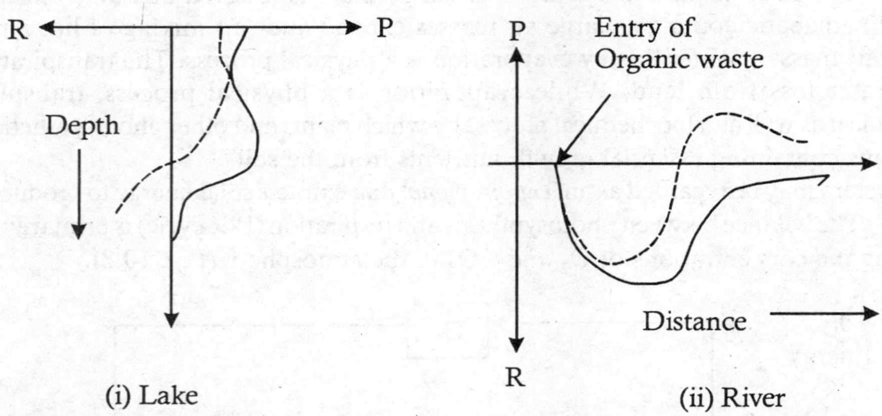


**Fig. 10.9:** Separation of Photosynthesis (P) and Respiration (R) Processes in Polluted Lakes and Rivers (... net disturbance of P-R balance is indicative of net rate of $O_2$ Production or Consumption)

#### *10.5.1.1 In Stationary Water Bodies*

In a stratified lake, an excessive production of algae and $O_2$ in the upper strata may cause anaerobic conditions (lack of DO) in the lower strata. This is because much of the $O_2$ from the photosynthetic process escapes to the atmosphere and, therefore, does not become available to the heterotrophs in the deeper strata of the water body. Vertical separation of the P-R functions in lakes can be induced by organic nutrients (nitrogen and phosphorus-containing compounds), which encourage surface algal growth. The resultant anaerobic condition stifles higher trophic levels. Algal blooms deplete DO and lead to fish-kills.

#### *10.5.1.2 In Rivers*

Streams have the inherent capacity for re-aeration. The DO profile as a function of time or flow (Fig. 10.10) is a balance between natural aeration and de-oxygenation by microorganisms. The

DO in a stream is inversely related to the abundance of microorganisms in water. The latter increases with nutrient supply.

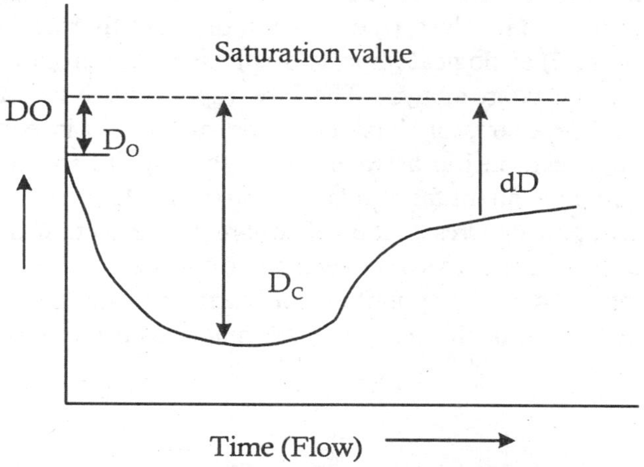


**Fig. 10.10:** Dissolved Oxygen (DO) Profile as a Function of Time (Flow) in a Stream

Longitudinal separation of the P-R process in rivers can be caused by local influx of organic (and inorganic) waste, which stimulates increase of bacterial activity and concomitant $O_2$ consumption. This results in enhanced algal growth downstream, which upon decay exerts an oxygen demand still further along the stream

#### *10.5.1.3 In Estuaries*

Estuaries are regions of interaction between rivers and near-shore ocean waters, where tidal action and river flow mix fresh and salt water. Estuaries have intermediate salinity between streams and seas. Such areas include bays, mouths of rivers, salt marshes, and lagoons. These brackish water ecosystems shelter and feed marine life, birds, and wildlife. Ecological imbalance of estuaries due to wastewater discharge affects the habitat and food supply of fish, clams and shrimp. Pollutants are concentrated by bioaccumulation by a factor of $10^3$ in fish and other aquatic organisms.

## 10.6 ENVIRONMENTAL RISKS

Polluted water bodies diminish and/or destroy aquatic fish population, and make them unusable for recreation, bathing, consumption and other activities. They also pose serious environmental health risks. Industrial detergents, oil and grease make water bodies uninhabitable. In addition, chemicals can interact with aquatic organisms to produce toxicants. As an example, microorganisms in water can transform metallic mercury (Hg) to methyl mercury ($HgCH_3$), which is neurotoxic. Methyl mercury is assimilated by the aquatic life—plankton and fish, and enter into human food chain. If the pollutants are toxic chemicals, such as insecticides, pesticides, and trace elements like As, Cd, Cr, Hg, Pb and Se and radioactive materials, they destroy aquatic life irrespective of the ecological nature of the water body. For example, $NH_3$ is toxic to fish.

### 10.6.1 Environmental Risk Evaluation

The environmental impact—the combined result of technological activities and population change and its effects on the physical environments—is evaluated by environmental risk assessment. It is an analytical process by which the probability, that one or more stressors (physical, chemical or biological entities) will cause adverse effects on human health or ecosystems, is established. General strategy of environmental risk evaluation involves: (*i*) evaluation of stress characteristics, (*ii*) identification of ecosystem(s) at risk and (*iii*) ecological impacts of stressors. Environmental risk assessment is generally carried out by dose-response analysis. The dose-response analysis comprises—(*i*) estimating the potency of a chemical or a stressor (hazard identification), (*ii*) in exposure evaluation, the process of determining the relationship between the dose of a stressor and a specific biological response, and (*iii*) evaluating the quantitative relationship (risk characterization) between dose and toxicological responses. Graphical representation of dose-response relationship (Fig. 10.11) describes the response of biological system to a toxicant over a range of concentrations (risk characterization). Threshold levels, time-weighted average pollutant concentration values, exposure beyond which is likely to adversely affect human health, can be determined from the dose-response relationship graphs.

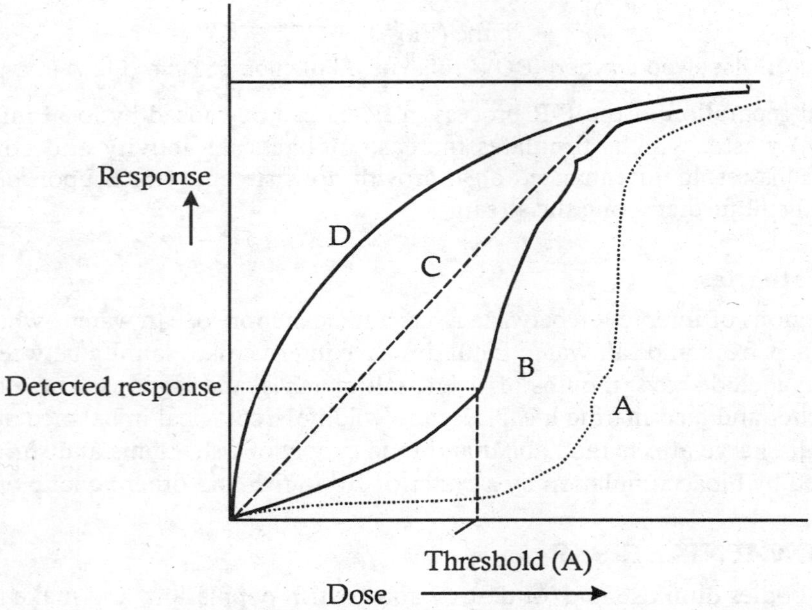


**Fig. 10.11:** Dose-Response Analysis Profiles

1. Curve A: Exhibits a threshold. A threshold is the lowest dose above which there is an observable effect.
2. Curve B: Sigmoid curve; Response is cooperative (allosterism).
3. Curve C: Linear response. The intensity of response is directly proportional to the pollutant dose.
4. Curve D: Supra-linear response. Low doses of pollutants provoke a disproportionately large response (due to *synergism*). Synergism occurs when two or more substances enhance the effects of individual constituents (e.g. smoking and inhalation of coal dust resulting in greater probability of cancer).

Two agents may produce a particular effect—additive, antagonism, protection, synergism, and sensitization (Fig. 10.12). Additive action is sum of both the effects. In antagonism, the combined effect is less than the individual effects. In the case of protection, presence one agent would moderate (minimize) the effect of the other agent. In synergism, the combined effect would be much higher than that expected from an additive action. Sensitization is similar to synergism, with the combined effect much greater than the effect from individual agents.

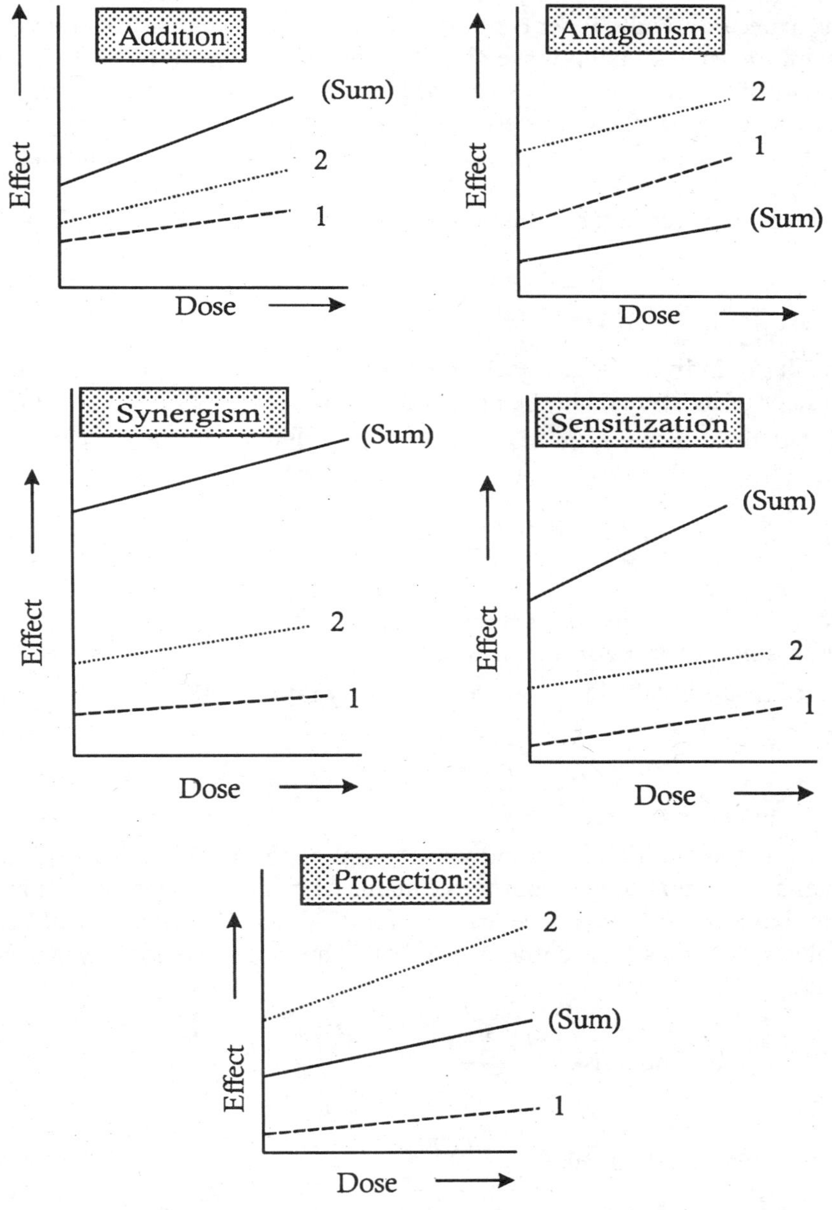


**Fig. 10.12:** Dose-Effect Plots of Multiple Agents

### 10.6.2 Relative Risk

Relative risk evaluation involves determination of administered dose concentrations, and half-life. Physiological half-life of a pollutant in an organism is the time required for the organism to eliminate half of the internal concentration of the pollutant.

$LC_{50}$ (lethal concentration-50) is a calculated (median level) concentration of a pollutant/toxicant through the respiratory route that is expected to kill 50 per cent of the test organisms during a 4 hour exposure (concentration in ppm). $LD_{50}$ (lethal dose-50) is a calculated dose of a pollutant/toxicant that is expected to kill 50 per cent of the experimental animals exposed through a route other than respiration (mg/kg of body weight). The lower the $LC_{50}$ or $LD_{50}$, the more toxic is the substance. Toxicity of a substance can be expressed in terms of hazard ratio, a term used to compare an animal's daily dietary intake of a pesticide to its $LD_{50}$ value. Hazard ratio greater than 1.0 indicates that the animal is likely to consume a dose amount, which would kill 50 per cent of animals of the same species.

The frequency of occurrence (probability) of adverse health effects in a population is given by

$$P = \frac{X}{N} \qquad \text{...(10.19)}$$

where, $P$ = Probability;
$X$ = number of individuals adversely affected;
$N$ = number of individuals in the population.

*Relative risk* is defined as the ratio of the probabilities that an adverse effect will occur in two different populations (e.g. lung cancer in smokers and non-smokers).

$$\frac{P_s}{P_n} = \frac{(X_s / N_s)}{(X_n / N_n)} \qquad \text{...(10.20)}$$

where, $X_s$ & $X_n$ = lung cancer in smokers and non-smokers;
$N_s$ & $N_n$ = smokers and non-smokers.

Relative risk of death is called the standard mortality ratio (SMR).

$$SMR = \frac{D_s}{D_n} = \frac{P_s}{P_n} \qquad \text{...(10.21)}$$

where, $D_s$ & $D_n$ = lung cancer deaths in smokers and non-smokers.

*Unit risk* is defined as the risk to an individual from exposure to a concentration of 1 μg/m³ airborne pollutant or 10 ng/l of waterborne pollutant. *Unit lifetime risk* is defined as the risk to an individual from exposure to these concentrations for 70 years. *Unit occupational lifetime risk* is the exposure (to pollutants of these concentrations) for 2,000 hours per year for 27 years. For waterborne pollutants

$$\text{Unit annual risk} = \frac{(LCF/\text{Y})}{(10^9\ \text{g/l})} \qquad \text{...(10.22)}$$

$$\text{Unit lifetime risk} = \frac{LCF}{(10^9\ \text{g/l})70\ \text{Y}}$$

where, $LCF$ = Latent cancer fatalities

## BIBLIOGRAPHY

Baier, R. E., Mark, H. B. and Mattson, D. S. (eds.) (1981), Marcel Dekker: New York. "*Water Quality Measurement.*"

Frei, R. W. and Albarge, J. (eds.) (1986), Gordon & Breach: New York. "*Air and Water Analysis: New Techniques and Data.*"

Hutzinger, O. (ed.) (1990), Springerverlag: Berlin. "*Handbook of Environmental Chemistry.*"

Jorgensen, S. E. and Johnsen, I. (1981), Elsevier: Amsterdam. "*Principles of Environmental Science and Technology,*" (Vol. 14).

Lamb, J. C. (1985), John Wiley: New York. "*Water Quality and its Control.*"

Manahan, S. E. (1993), Lewis Pubs: New York. "*Environmental Chemistry,*" (5th edn.).

Markert, B. (ed.) (1993), VCH Pub: New York. "*Plants as Biomonitors: Indicators for Heavy Metals in the Terrestrial Environment.*"

McGauhey, P. H. (1968), McGraw-Hill: New York. "*Engineering Management of Water Quality.*"

McKay, G. (ed.) (1996), CRC Press: Boca Raton, FL. "*Use of Adsorbents for the Removal of Pollutants from Wastewaters.*"

Minear, R. A. and Keith, L. H. (eds.) (1994), Academic Press: London. "*Water Analysis: Organic Species.*"

Peirce, J. J., Weiner, R. F. and Vesiland, P. A. (1998), Butterworth-Heineman: New Delhi. "*Environmental Pollution and Control,*" (4th edn.).

Pfafflin, J. R. and Ziegler, E. N (eds.) (1976), Gordon & Breach: New York. "*Encyclopedia of Environmental Science and Pollution,*" (Vol. 1).

Purdom, W. (ed.) (1980), Academic Press: New York. "*Environmental Health,*" (2nd edn.).

Wilson, J. D. (1974), Heath & Co: Lexington, MA. "*An Environmental Approach to Physical Science.*"

Winberg, W. T. *et al.* (1988), EPA/600/4-89/017/USEPA: NC. "*Compendium of Methods for the Determination of Toxic Organic Compounds.*"

... (1996), Water Pollution Control Federation (WPCF): Washington, DC. "*Standard Methods for the Examination of Water and Wastewater,*" (20th edn.).

# SECTION-III

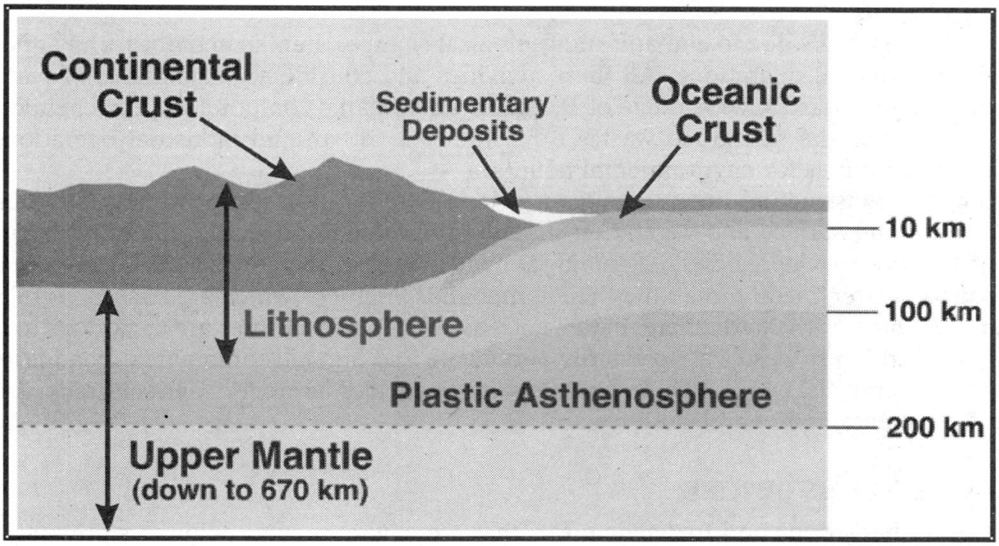


# The Lithosphere

## OVERVIEW

Chapters 11, 12 and 13 give an overview of the soil, and conventional, hazardous and radioactive pollutants and their environmental effects on lithosphere, vegetation and animals and humans.

## SOIL

Earth's surface consists of several layers of soil. Soil pollution deals primarily with topsoil and aquifers. Topsoil consists of clays, sand, minerals and organic matter. Clays are sedimentary minerals, made up of layers of hydrated alumina and silicon dioxide. Organic matter in soil is a heterogeneous mixture of products resulting from the microbial and chemical transformation of organic materials. It is mostly humus.

Soil pollution occurs due to climatic and geological changes, human activities, and agriculture, mining and industrial operations. All these activities add considerable quantities of particulate pollutants, mineral wastes, wide range of inorganic and organic compounds, petrochemicals and toxic substances to the soil. Solid wastes from domestic, urban and industrial operations have provided opportunities for environmental pollution.

Mining and construction operations lead to land deterioration. Strip mining is ecologically destructive to land surface and vegetation. In addition, mine drainage (leachate) is a significant source of pollution of water bodies. All mining and metallurgical operations produce large quantities of particulate matter, metal fumes and their compounds, many of which are hazardous to health. For example, some of the particulate matter, such as borates and silicates are hazardous to health. Agricultural and forest wastes are primarily particulate and organic compounds. In addition, use of industrial chemicals, which include fertilizers and pesticides in modern agricultural operations, poses environmental risks and health hazards.

## HAZARDOUS SUBSTANCES

Practically, all hazardous and toxic chemicals that are environmental pollutants are man-made. These substances include corrosive materials (acids and alkalis), reactants (oxidants and reductants), surfactants, solvents, petrochemicals, explosives, agrochemicals, plastics and a variety of consumer products.

Hazardous and toxic compounds can be classified according to their size and the state of the matter (e.g. flammability, viscosity), and according to their chemical and biological (physiological) effects. Many of the volatile polyaromatic hydrocarbons (PAHs) and polychlorinated biphenyls (PCBs) are known carcinogens and mutagens. Trace heavy elements (As, Cd, Cr, Hg, Pb, Mo, Se) are extremely toxic to human beings. Many organometallic compounds, pesticides, herbicides and fungicides are highly toxic not only to "pests" but to other organisms also. Through water run-off and crops various pollutants enter the food chain.

Discernable physiological characteristics of the pollutants are irritation, asphyxiation and toxicity. Irritant inorganic compounds are $NH_3$, $O_3$, $SO_2$, and $H_2SO_4$. Asphyxiates are either physiologically inert gases ($N_2$, He, $CO_2$, $CH_4$) or chemical asphyxiates (CO, HCN and $H_2S$). Some of them are carcinogenic.

## RADIOACTIVE PRODUCTS

Radioactivity is all-pervasive. Man has been living with it since the creation of the universe. But modern technology has brought several naturally radioactive materials like uranium into the industry and added many hitherto unknown isotopes to man's stock of 'useful' materials. Radioactive pollution is a recent contribution by man to environmental pollution. Radioactive pollutants come under a special category due to their extreme toxicity and long half-lives of many of them. Handling, transport, storage and disposal of these products and their wastes pose short-term and long-term environmental and health risks.

Mining, processing and purification produce radioactive products (e.g. uranium and thorium). They are used for weapons manufacture and in reactors for nuclear energy generation. In addition, a series of radionuclides are produced for nuclear medicine and other applications. The nuclear reactor is the key radioactive waste producer in the nuclear fuel cycle. All radioactive products, their waste products and spent nuclear fuels pose extreme radiation hazard unless properly handled.

## Radiation Hazards

Major modes of exposure to radiation are by body contact (e.g. through diagnostic X-ray radiography, cobalt-needle implantation), injection (e.g. administration of radionuclides) and inhalation (e.g. radon gas). Radium is an $\alpha$-emitter. It is transported via food chain due its chemical similarity to calcium. Radon ($^{222}Rn$) is a gaseous radionuclide and is an $\alpha$-emitter. It is released into the atmosphere during uranium mining operations. $^{222}Rn$, being a dense gas, settles in the lungs on inhalation, and its decay products (its particulate progeny) are absorbed in the lung tissues, leading to lung cancer. Thus, the gas is sobriquetly labeled as the "silent killer".

Risk evaluation is generally carried out by dose-response analyses. But, it is opined that in the case of genetic effects—(*i*) threshold dose is not required, (*ii*) once damage is initiated (at the gene level) it is largely irreversible, and (*iii*) deleterious effects are cumulative.

# 11

# Soil

Soil is the unconsolidated top or superficial layer of Earth's crust lying below any aerial vegetation and undecomposed dead organic remains, and extending down to the limits to which it affects the plants growing about its surface. Beneath the soil, lie the subsoil and unweathered rocks. Soils commonly become stratified into layers or horizons, at different depths. The layers of soil at different depths show different compositions and natures.

Soil pollution deals primarily with topsoil and aquifers. Topsoil consists of clays, sand, minerals and organic matter. Clays are sedimentary minerals, made up of layers of hydrated alumina and silicon dioxide. Organic matter in soil is a heterogeneous mixture of products resulting from the microbial and chemical transformation of organic materials. It is mostly humus.

Soil pollution occurs due to climatic and geological changes, human activities, and agriculture, mining and industrial operations. All these activities add considerable quantities of particulate pollutants, mineral wastes, wide range of inorganic and organic compounds, petrochemicals and toxic substances to the soil. Mining and construction operations lead to land deterioration. Strip mining is ecologically destructive to land surface and vegetation.

This chapter deals with the composition and role of the soil. It also deals about various soil pollutants.

Soil is a thin veneer that covers most of the Earth's land surface. Though its volume and mass are very small, in comparison to the lithosphere, it is of vast importance to man. Soil fulfills a wide range of inter-related functions. It (1) forms a crucial link between the atmosphere, geology, water resources and land use; (2) acts as a reservoir of carbon, which is a key factor in determining concentrations of greenhouse gases; (3) regulates the flow of water from rainfall to water bodies, aquifers, vegetation and the atmosphere; (4) acts as the medium for vegetation, crops and forests and plays a major role in determining the nature and distribution of life; (5) forms the basis for terrestrial ecosystems.

Soil is generally composed of sand, silt and clay particles, organic matter (humus), water and air space. Sand and other mineral parts of the soil result from the weathering of rocks. Soil itself is more complex than an assemblage of clay minerals. It is a vital part of our environment, and an essential resource for life. Its formation takes place over long periods and so it is effectively a non-renewable source. The type and characteristics of the soil in an area make a key contribution to the landscape and agro-ecosystem, and affecting human habitat, wildlife, agriculture and suitability for development.

## 11.1 SOIL COMPONENTS

Soil is a complex assemblage of clay minerals, detrital materials (e.g. quartz), humus and organic compounds, water, trapped gases and living organisms.

The average composition of the Earth's crust, with respect to major components, is (wt. per cent): oxygen 46.6, silicon 27.7, aluminum 8.3, iron 5, calcium 3.63, magnesium 2.09, sodium 2.83, and potassium 2.59. During the weathering process the feldspars and other silicate minerals are broken up by natural elements like water, heat and sunlight, releasing chemical compounds, which include bases, silica and oxides of iron ($Fe_2O_3$) and aluminum ($Al_2O_3$). Percolation of water for a long time through the fine products of weathering results in the leaching of relatively soluble bases, alkaline earths (Ca and Mg) and some silica. Generic disposition of soil over the original rock is indicated in Table 11.1.

**TABLE 11.1:** GENERIC SOIL PROFILE ON THE EARTH'S SURFACE

| *Layers of Soil* |
|---|
| Topsoil ("A" Horizon) |
| Subsoil ("B" Horizon) |
| Weathered parent rock ("C" Horizon) |
| Bedrock |

The particulate matter of soil may be classified according to their physical characteristics (particle size) (Table 11.2) and according to their chemical and biological effects.

**TABLE 11.2:** CLASSIFICATION OF PARTICULATE MATTER ACCORDING TO SIZE

| *Classification* | *Particle Diameter* |
|---|---|
| Clay | 0.25-4 μm |
| Silt | 4-6 μm |
| Sand | 0.05-2 μm |
| Gravel | 2-50 μm |
| Cobbles | 50-250 μm |
| Boulders | >250 μm |

### 11.1.1 Clays

Clays are sedimentary minerals, comprising the smallest particles (<2 μm) in soil, consisting mainly of hydrated silicates of aluminum. The silicon oxide (silica) and aluminum oxide (alumina) combine to form a layered structure consisting of sheets of silica alternating with sheets of alumina. Atoms in each layer are strongly bonded to one another and weakly bound to those in neighboring layers. Therefore, each layer can act as a self-contained structural unit. The clay minerals are of two types: (1) one in which tetrahedral silica and octahedral alumina layers are assembled in a 1 : 1 ratio (-Si.Al.Si.Al.Si.Al-) (1 : 1 clays); (2) the second in which one octahedral alumina layer is sandwiched between two tetrahedral silica layers in a 2 : 1 ratio (-Si.Al.Si.Si.Al.Si.Si.Al-) (2 : 1 clays). As the leaching proceeds the 1 : 1 type changes slowly to the 2 : 1 type.

### 11.1.2 Humus Substances

In addition to clays, soil consists of a complex of organic compounds, living organisms, water, and included gases that give its characteristic properties and its value as the abode of life. Soil is in a continual state of change and evolution. Organic matter in soil is a heterogeneous mixture of products resulting from the microbial and chemical transformation of organic materials. The organic portion of the soil remaining after prolonged microbial decomposition, called '*humus*', imparts to it most of the life sustaining attributes. It accumulates in the soil, essentially, as a result of biological processes. It is derived from microbial decomposition of plant and animal substances. Carbon (60 per cent) and nitrogen (6 per cent) are the major elements in the humus while phosphorous and sulfur are in minor amounts. The roots that die in the Earth and the vegetable and organic matter that fall on the surface of the soil decay and the products are leached down by water. As part of the residual matter gets mixed with the soil, the microorganisms convert them into the complex organic substance, humus.

Humus substances are amorphous, polymorphic and brown-colored. They are classified according to solubility into humic acids (soluble in alkalis but insoluble in acids), humin (insoluble in alkalis) and fulvic acids (soluble in acids and alkalis). Fulvic and humic acids can be extracted with base, but not humins. Humic acids can be precipitated with acid, but not fulvic acids. The average molecular mass of humic acids is 10,000 to 50,000 Dalton and fulvic acids 500 to 700 Dalton. There are two types of humus: (1) 'mild humus', which is dark in color, rich in high molecular weight humic acids and saturated with bases like calcium; (2) 'raw humus', which is red in color, less basic and rich in fulvic acids.

Humic substances can cause decrease in the rate of flocculation. High-molecular mass molecules are generally retained on clay surfaces by non-bonded interactions (e.g. hydrogen bonding). For low-molecular mass organic compounds, which are important as environmental pollutants, ion exchange is the process of interaction. Cationic organic molecules may be sorbed (see Chapter 22) to the cation exchange sites of clay minerals and humic surfaces. Anionic substances may be retained by clay minerals and complex organic matter of soil is often the chief sorbent for monoionic organic compounds. A measure of the concentration of the humic substances can be estimated (the quotient of absorbance at 468 and 644 nm ($Q_{4/6}$)) by spectroscopic methods (see Chapter 24).

## 11.2 ROLE OF THE SOIL

In the scheme of life on the Earth soil plays a vital part and it can be said that 'it is an essential resource for life'. The varied roles that soil plays are:

1. Environmental interaction
   - It is a critical link between the atmosphere, geology, water resources and land use.
   - It receives precipitation of various types from the atmosphere.
   - It is a reservoir of carbon. It emits and, at the same time, removes some atmospheric gases.
   - In the water cycle on the globe it has a role in regulating the flow of this precious material from rainfall to watercourses, aquifers, vegetation and the atmosphere.
2. Source of food and materials
   - For the human society as well as for the animal world soil is the medium for growth of food and energy crops (timber). It is the basis for livestock production.
   - It is the source of minerals like peat.
   - It is a natural reservoir for huge amounts of water.
   - It is also a natural seed bank.
3. Providing habitat
   - It is the support for terrestrial ecosystems, providing water and nutrients for the entire plant-kingdom. It is a habitat for myriad living beings from microorganisms to bigger animals.
4. A platform for growth
   - It is the platform on which civilization thrives, through construction of dwellings and roads for transport.

Soil, at any given time, supports more than one function. For example, an agricultural soil will provide food, fiber and fuel. At the same time, it interacts with other environmental media, supports biodiversity, and protects cultural heritage (archeological remains). Therefore, the sustainable use of such a versatile resource, like soil, implies that we minimize damage to it so that it can continue to be used and can support the widest range of functions.

## 11.3 SOIL POLLUTANTS

Soil pollution involves accumulation on land of substances in dispersed solid or liquid form that are injurious to life and ecosystems. Both inorganic (P- and N-containing fertilizers) and organic matter (fertilizers, pesticides) influence its physical, chemical and biological properties. This has been highlighted in the case of agro-chemicals that are widely used in modern agricultural practice and persistent pesticides (DDT) and insecticides. Though they are spread for the purpose of exterminating pests, they accumulate to the extent that they can do damage to many other forms of life. The problem of pesticide accumulation arises because the microorganisms in soil are not capable of degrading many of them at rates sufficiently high to prevent soil and water pollution.

Soil can be polluted not only by materials directly applied to it, deliberately or by land spreading of wastes, but even by toxic materials present in the atmosphere, brought down by precipitation. In turn, pollutants from land are carried by water flowing over it or percolating through it. The pollutants that affect soil and water include sediments, plant nutrients, inorganic salts and minerals, organic wastes, infectious agents and industrial and agricultural chemicals.

Contaminants affect soil quality, reducing its fertility and crop yields, and damaging ecosystems dependent on soil. Acidification of soil due to pollutant gases like $SO_2$ and $NO_x$ is another phenomenon becoming more common in recent times. Acidification of mineral soils, and mining leachates can release metals like aluminum, which damage plants, for example, by affecting fine root systems. Soil acidification leads to acidification of water, causing damage to aquatic systems. In addition to direct health and environmental problems land contamination can cause economic and financial damage.

Soil erosion/pollution occurs due to natural causes, such as volcanic eruptions, earthquakes, landslides, dust storms, etc., as well as due to human habitation, construction, and agriculture and overgrazing, mining, industrial and transportation activities. Though soil erosion cannot be classified as pollution, it is a type of damage that has several repercussions. The process can be enhanced by human activities such as inappropriate cultivation and crop management, overgrazing, deforestation and construction activity. Soil receives huge quantities of waste products each year—sulfates and nitrates, etc.

Erosion has two aspects. It removes soil from where it is; perhaps, best required and deposits it at sites, like watercourses, where it becomes pollution. Loss of agricultural soil by erosion can affect farm profitability. Erosion may increase as a result of climate change. Increased temperature and reduced moisture content are likely to lead to loss of carbon from the soil making them erode faster. Climate and geological features add to soil erosion and pollution (Fig. 11.1).

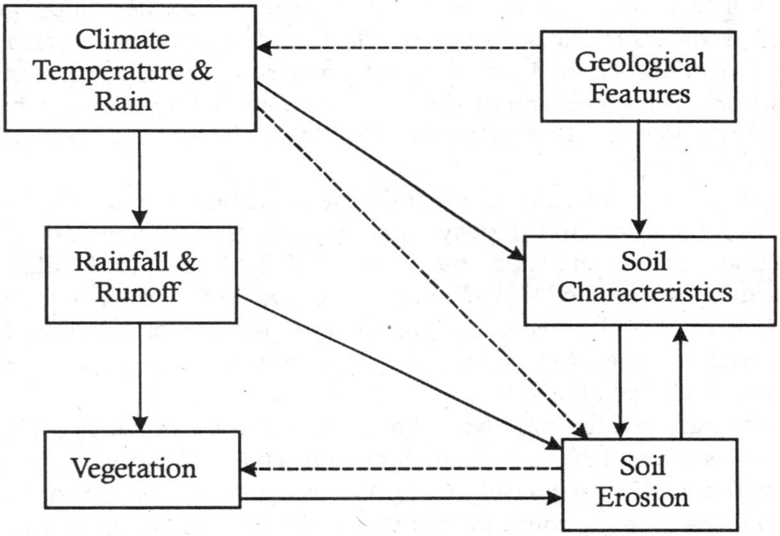


**Fig. 11.1:** Climate and Geological Features Influencing Soil Erosion

Sediments, such as clay, silt, sand, minerals and other particles and debris (leaves, grass and vegetation) due to stream run-offs and soil erosion are transported via water systems. Sediment is a resource out of place whose dual effect is to deplete the land from which came and impair the quality of the water it enters. Suspended sediment impairs the dissolved-oxygen balance in water. Bacteria are mainly found in sediments. They are dependent on the content of organic substances and the particle size of the sediment. The smaller the particle size, the higher the bacterial counts.

Mining, road and building construction operations and power-generation installations add considerable amount of particulate matter and radioactive waste, causing the degradation of soil and vegetation. Agricultural activities add dust, mud, husk, dairy waste and other particulate matter, in addition to increased plant nutrients and toxic chemicals (fertilizers, insecticides, fungicides, herbicides). Nutrients of plants become resources out of place when they appear in groundwater, and surface water. In fact, they become serious pollutants. Nitrates and nitrites in groundwater, which can poison human beings and livestock, result from both agricultural and industrial operations. Industrial operations contribute enormously to soil degradation by discharging inorganic particulate matter, mineral and metal wastes and a wide range of organic and petrochemicals, oils and detergents.

The uptake and behavior of pollutants depend on the types of clays as well. In 1 : 1 type clays, the layers are tightly held together and lattices are non-expanding and swell very little when moistened. Such clays have lower capacities for adsorption and pollutants are held only on the outer surfaces. On the other hand the 2 : 1 type clays, the lattice structure is more open and swells easily when moistened, and hence adsorption of organic substances takes place on both external and internal surfaces. Pollutant molecules that are held on 2 : 1 clays might not be available for biodegradation.

## 11.4 SOLID WASTE POLLUTION

The solid waste that we generate represents one of the major environmental challenges. The industrialization of modern societies resulted in vast increase in the amount of refuse generated per person. In late 1970s it was estimated that the refuse produced per person per year in USA was about 25,000 kg. Part of it is municipal waste, which is commonly known as trash or garbage. This consists of every day items like food scraps, packing material, grass clippings, furniture, clothing, bottles and plastics of various types and appliances, big and small. Some, among the items discarded by household are hazardous too. Common items like paints, cleaners, oils, batteries, pesticides and chemicals come in this category.

There are four main ways of disposing of solid waste of the type mentioned: (1) ocean dumping, which creates water pollution and destroys marine habitats; (2) incineration, which leads to atmospheric pollution, if not conducted under controlled conditions; (3) landfill, which has been by far the most widely used method till its limitations are realized in recent years and (4) recycling, which is the least offensive of the methods with respect to spreading of pollution. However, limited types of materials can be subject to this operation. Moreover, prior sorting is necessary for picking up items suitable for such operations.

Landfill has been the most common way of disposing of non-hazardous solid wastes from the society. All industrial societies have resorted to this mode in the past. Since most of these operations were carried out without adequate precautions against leaching or the escape of landfill gases, their safety is being reviewed in many countries and remedial measures are underway.

Finding of fresh areas for landfill involves destruction of marshland and swamps that have high biological value. In addition, the disposal itself has to be carried out under safe conditions prescribed by local authorities.

*Solid waste is classified as:*

**Garbage:** Decomposable biological domestic waste (food-, vegetable-, animal- and beverage-related wastes).

**Rubbish:** Waste excluding decomposable part of garbage (e.g. paper, beverage and plastic cans etc.)

**Solid Waste:** Solid wastes are municipal solid wastes, other than hazardous and radioactive materials. These include sewage sludge, ash, dirt, rubble, farm wastes, wood and paper products and industrial products—metals, ceramics, paints and pigments, plastic, rubber, leather, glass, etc.

**Refuse:** All solid wastes—garbage, rubbish and trash.

**Litter:** Street and highway refuse.

Solid waste is another environmental pollution that has been greatly enhanced by human-centered activities. Contaminated land sites are among the commonest environmental problems. Aqueous transport is a major route by which pollutants are spread beyond the area initially affected.

Airborne transport can be by gases, vapors and particulate matter. A summary of factors contributing to solid waste is given in Table 11.3.

**TABLE 11.3:** FACTORS CONTRIBUTING TO LAND POLLUTION

| *Contributing Sources* | *Waste Constituents* |
|---|---|
| Domestic discharge (Sewage) | Organic matter (garbage); raw wastes; hazardous chemicals; pathogens and fecal contamination. |
| Municipal/Urban | Discharge of sewage, rubbish, garbage and sludge. |
| Industrial | Inorganic and organic particulate matter; oils, grease, detergents and hazardous chemicals; metals, and salts. |
| Agricultural | Silt and solid particulate matter; fertilizers; insecticides, herbicides and fungicides; farm and crop debris and organic waste; dairy waste |
| Mining | Dust, sand and rock debris; asbestos and silicate particulates; toxic metals and radioactive waste; leachates and land deterioration. |

### 11.4.1 Domestic and Municipal Wastes

Solid pollutants can be sources of air and water pollution. Open dumps are a cause of land pollution, and solid wastes discharged into water can be hazardous to aquatic life directly and to other organisms directly or indirectly. Some of the solid wastes, sources and patterns of disposal are given in Table 11.4.

**TABLE 11.4:** DOMESTIC AND MUNICIPAL SOLID WASTES AND PATTERNS OF DISPOSAL

| *Source* | *Waste* | *Composition* | *Disposal* |
|---|---|---|---|
| Domestic | Garbage; rubbish; ashes; litter | Food and beverages; paper; wood; rags; plastic; furniture; appliances and bric-a-brac | Landfill & incineration |
| Commercial | Garbage & litter | Street refuse; fly ash; sewage discharge residues | Landfill & incineration |
| Construction | Housing and road construction materials | Asphalt materials; lumber; pipes; cement; plastic; rubber; asbestos, etc. | Landfill & incineration |
| Hazardous waste | Chemicals & pathogens | Chemicals; explosives; toxicants; pathological wastes | Proper waste disposal procedures |
| Radioactive waste | Radioactive compounds | Radioactive particulate and chemicals | Special waste disposal regime |

### 11.4.2 Mining and Soil Pollution

All the metallic and nonmetallic minerals required by the industry are recovered by different mining techniques. Most mineral resources are covered by soil that has to be removed before extraction can proceed. This is particularly the case with opencast mining. This means a loss of precious soil unless it is carefully removed, handled, stored and reused. Mine tailings and metal pollution of soils is a major problem globally. Heaps of mining wastes that accumulate outside the mines carry traces to moderate quantities of metals, which get leached by rain and snow. At times this is hastened by biological activity in the dumps, especially the sulfur oxidizing bacteria. The outflowing solutions, carrying the metals, some of which are hazardous, sink into the soil around spreading the pollutants far and wide and even reaching the water sources. Winds blowing over the tailings spread the metal carrying dust over vast areas, contaminating agricultural soil and watercourses.

Though many metals are essential trace elements for plants and animals, at higher concentration they are detrimental to both aquatic and terrestrial ecosystems.

### 11.4.3 Agriculture and Soil Pollution

Soil is the main platform for agricultural activity. Unlike air or water, soil is very diverse. A country of moderate size can have several types of soil and accordingly, the crops that can be raised on them also vary. Agriculture accounts for a large proportion of land use and therefore has a major impact on soil. Along with environmental factors, human activity such as change of land use and management influences soil behavior. Once soils are modified by activities like drainage, plowing and the application of artificial fertilizers, they are less able to support their characteristic biological communities. Good agricultural practice is needed for protection of soil.

Nutrients associated with the plants are lost from the system when crops are removed from the soil. It is necessary to replace these nutrients since they are needed for further plant growth. However, there is a risk that over-application or poor timing of nutrient additions may compromise the ability of the soils to retain these nutrients and make them available to the plants. This results in pollution of water and air. Inappropriate application of organic manure can pollute the soil through: (1) the nutrients within the organic material, (2) the contaminants like metals and organic chemicals and (3) the microbes present in the material. Physical damage to the soil may be caused by way of compaction or water logging. Offensive odor and visual changes can be the other effects. Inorganic fertilizers like phosphates may also contain hazardous trace metals like cadmium, which comes from the rock-phosphate, the raw material for such fertilizers. Pesticides, herbicides, fungicides and other chemicals are environmental contaminants when spread over the globe by wind and water. They, along with some metals can suppress microbial activity, which is essential for healthy functioning of the soil. The contaminants introduced into the soil may be transferred to the food chain with implications for human and animal health, and may lead to other forms of pollution as water pollution and odor nuisance.

Agricultural wastes are primarily organic and particulate compounds. They comprise crop, animal, livestock and dairy product wastes. Crop wastes are corn stalks, husk, bagasse, fruit and vegetable scraps. Most crop and animal wastes are either plowed back into the soil or composted. Forest wastes are primarily organic and particulate compounds, and are predominantly due to timber felling and logging operations and forest fires.

## 11.5 CHEMICAL POLLUTION

A variety of industrial chemicals can pollute the soil, of which an important group is the BTEX chemicals (B = benzene, T = toluene, E = ethylbenzene, X = xylenes). Common sources of BTEX contamination are the spills and leakage of gasoline, diesel fuel and lubricating and heating oils from storage tanks built underground. Their polarity and appreciable water solubility make these chemicals particularly prone to enter the ground water systems, causing serious pollution problems. For the same reason, they are easily washed from the atmosphere onto the ground by rain. BTEXs form about 15 per cent of standard gasoline blends.

Out of the petroleum products BTEX group attracts particular attention, in respect of pollution, because they are volatile, have moderate solubility in water and are easily absorbed by the organic components of the soil. They are toxic to living beings (Table 11.5). They can enter the body by ingestion of contaminated crops, inhalation of vapor from the soil, intake of contaminated drinking water and skin exposure while bathing. They are all toxic but benzene is a recognized carcinogen. When inhaled for long periods in low concentration they can cause brain damage.

**TABLE 11.5:** WATER SOLUBILITY AND MCL OF BTEX CHEMICALS

| *Chemical* | *Solubility (mg/L)* | *MCL (mg/L)* |
|---|---|---|
| Benzene | 1,700 | 0.005 |
| Toluene | 515 | 1 |
| Ethylbenzene | 152 | 0.7 |
| o-Xylene | 175 | 10 |
| m-Xylene | – | 10 |
| p-Xylene | 198 | 10 |

*MCL* = Maximum Contamination Level in drinking water (prescribed).

## 11.5.1 Chemical Reactions in the Soil

Chemical reactions in the soil may influence the mobility and toxicity of many pollutants that are introduced into the soil and ground water systems. The reactions mentioned below may take place simultaneously.

### *11.5.1.1 Oxidation-Reduction*

Oxidation and reduction (redox) reactions occur simultaneously in the soil-water system. A redox reaction can be considered a pair of coupled half reactions, namely a half reaction of oxidation and a half reaction of reduction.

Metal ions can exist in more than one oxidation state in the soils. Depending on the redox status (Eh value) of the soil one or the other state will be the dominant one. Soils under reduced conditions have high Eh value, and oxidized soils have a low Eh value. Water logged soils are under reduced conditions while the reverse is true of well-aerated soils. The behavior of elements in the soil depends on the Eh and pH of the environment. Eh-pH diagrams can provide information on potential fixation of elements in the soil.

Redox reactions control the mobility and toxicity of inorganic and organic pollutants in soils. Under reduced conditions Mn and Fe oxides are reduced to $Mn^{2+}$ and $Fe^{2+}$. $SO_4^{2-}$ is reduced to sulfide, in the absence of more favorable electron acceptors, and then heavy metals are precipitated as sulfides, making them immobile. Chromium in the soil can exist as Cr(III) or Cr(VI). The former is stable and innocuous while the latter is mobile and toxic. Under oxidizing conditions chromium is present mostly as chromate ($Cr^{6+}$) but in the presence of $Fe^{2+}$ or FeS or organic matter it is stabilized in 3+ state and easily immobilized. A similar situation prevails in the case of plutonium, which can be present in 3+ and 6+ oxidation states. The difference, however, is in plutonium being toxic in both states of oxidation. Acidification of mineral soils can release forms of aluminum, which may damage plants, and soil acidification can lead to the acidification of surface waters, with serious impacts on aquatic ecosystems.

Organic contaminants can also be degraded by redox reactions. Oxides of iron and manganese can oxidize aromatic pollutants. The effectiveness depends on the nature of the functional group in the aromatic ring. The greater their electron donating power the easier they are oxidized.

#### *11.5.1.2 Precipitation*

The metal ions in soils can be immobilized by precipitation in three different ways. They are: (1) precipitation due to super-saturation in the soil solution, (2) surface precipitation and (3) co-precipitation.

A salt of the type $A^+B^-$ in solution phase in the soil may be considered as a simple example of the first type. Its solubility product $K_{sp} = [A^+][B^-]$, where [ ] is molar concentration of the ion in solution. If the product of molar concentrations of $A^+$ and $B^-$ in solution phase exceeds the $K_{sp}$ value the salt will be precipitated. Since carbonate, sulfate and hydroxide are present in more or less concentration in solutions percolating through the soils there is a chance of the following metals getting precipitated in one form or the other: Al, Fe, Si, Mn, Ca and Mg.

Surface precipitation is the three dimensional growth of a solid on the surface of a soil particle. This is a more common mode of precipitation in the soil system. Trace metals like Fe, Cr, As and others will precipitate on a soil particle and become immobilized and less toxic.

Co-precipitation is the incorporation of trace elements into a mineral structure during solid solution formation. A trace element can enter the precipitate phase even when its ionic product has not exceeded the solubility product in a given solution. However, only elements that have similar ionic radii as the elements composing the mineral can be accommodated in the growing solid phase. Examples of this type are the presence of $Cr^{3+}$, $Mn^{3+}$ and $V^{3+}$ in oxides of iron and aluminum and $Mn^{2+}$, $Cd^{2+}$ and $Fe^{2+}$ in calcite ($CaCO_3$) mineral.

#### *11.5.1.3 Chelation*

Chelates are a type of compounds formed by complexing metal ions with organic molecules (Fig. 11.2). Soil surface bears, generally, a negative charge. Metal ions, being positively charged, are held by strong adsorption forces to such surfaces. However, the chelates of metals have net negative charges and hence are not retained by the soil surface. Therefore, chelating agents transfer the metals to solution phase. The common reagents present in the soil environment are glycine, citrate, tartrate and gluconate. As a consequence of this type of reaction, soils with a high amount of organic matter tend to have minimal amounts of metal ions within them because of the chelating agents in the organic matter.

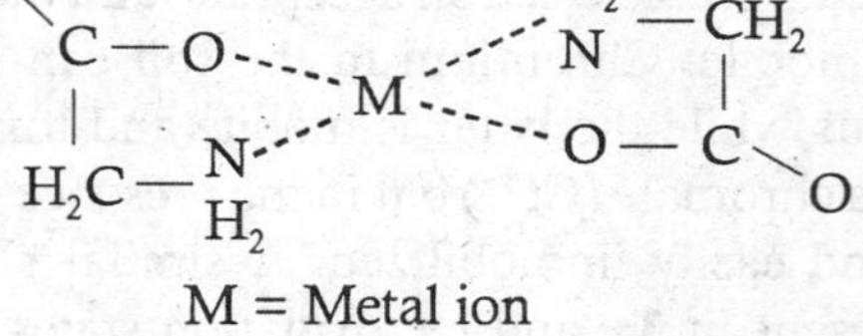


**Fig. 11.2:** Example of Metal Chelation Complex

Increased dependence on chemical fertilizers, to increase crop production, brings about pH imbalance in soil, stimulation of weed growth in fields and nutritional imbalance. Pesticides are not only harmful to 'pests' but also to other organisms. Their concentration and magnification in the biological systems may lead to untoward ecological imbalance.

## 11.6 INDUSTRIAL WASTES

All industrial wastes affect in some way the quality of soil and water bodies. Sources of pollutants are—building and road construction, lumber and paper mills, textile products, food and beverage processing, metallurgy and appliances, transport, chemicals and drugs and petroleum products (Table 11.6). Common industrial pollutants are—inorganic salts, acids, alkalis, detergents, suspended particulate, organic compounds, hazardous and toxic chemicals, trace metals and metal fumes, radioactive materials, heated water and microorganisms. A list of industrial pollutants and their effects is given in Table 11.7.

**TABLE 11.6:** SOURCES AND TYPES OF INDUSTRIAL WASTE

| *Source* | *Processes* | *Wastes* |
|---|---|---|
| Building & road construction | Home buildings; community centers; roads; highways; bridges | Dirt; silt; sand; silicates; cement; tar; glass; rubber; wood-, metal-, and plastic-based materials & sanitation wastes |
| Lumber & paper mills | Saw mills; paper mills; wood and paper products | Wood chips; sawdust; dirt; minerals; paper waste; paints; ink and solvents and organic and inorganic wastes |
| Textile mills | Cloth; leather; dyeing & processing | Cotton fiber; wool and leather scrap; synthetic yarns; plastic wastes; dyes and chemicals |
| Food & beverage | Food processing; packaging; storage & transport | Foods; fats; vegetable wastes; beverage products; package materials, boxes, tins, bottles and cans |
| Metallurgy | Mining, processing & dressing; smelting; forging and alloying | Ore dust; metal scrap; slag; coke; ceramics; acids; alkalis; solvents and lubricants etc. |
| Appliances | Home & industrial appliances and equipment and machines | Metal scrap; carbon products; glass; ceramics; wood; plastics; rubber; resins; paints and chemicals |
| Chemicals & drugs | Manufacturing; processing; packing and transporting | Inorganic and organic compounds; hazardous trace metals and chemicals; metal foils; packing materials; glass; plastic; paints and solvents |
| Petroleum products | Drilling, processing, transport and utilization | Petrochemicals; synthetic materials; resins; plastics; tars; lubricants; fuel; & hazardous chemicals |
| Vehicular transport | Motor vehicles; trains; ships; aircraft & accessory industries | Metal scrap; wood; leather; rubber; plastics; grease; solvents; petroleum products; additives |

**TABLE 11.7:** SOME INDUSTRIAL POLLUTANTS AND THEIR EFFECTS

| *Pollutant* | *Source* | *Effects* |
|---|---|---|
| Inorganic compounds | Mining; ore dressing; chemical plants & oil refineries | Water and land pollution; vegetation and health hazards |
| Acids/Alkalis | Chemical industry; industrial solvents | Corrosive and irritative; destructive to soil, water and organisms |
| Organic compounds | Food & beverage and agro industries; tanneries; paper mills | Increase in BOD; unhygienic; proliferation of pathogens; health hazards |
| Particulate matter | Mining & metallurgy; timber and paper & agro and beverage industries | Degradation of water bodies; health hazards |
| Detergents | Soap & detergent industry | Interferes with biota; degradation of soil and water bodies |
| Toxic chemicals | Metallurgy & metal plating; steel mills; pesticides and man-made agro-chemicals | Interferes with biota; degradation of soil and water; health hazards |
| Microorganisms | Food & beverage industries; tanneries; livestock industry | Harmful and health hazards |
| Radioactive materials | Mining & processing and use of radioactive materials | Harmful and health hazards |

## BIBLIOGRAPHY

Agarwal, S. K. (1991), Himanshu: Udaipur. "*Pollution and Ecology.*"

Horowitz, A. J. (1991), Lewis Pubs: Chealsea, MI. "*A Primer on Trace Metal Sediment Chemistry.*"

Monteith, J. L. (ed.) (1977) Academic Press: New York. "*Vegetation and the Atmosphere.*"

Pafflin, J. R. and Ziegler, E. W. (1976), Gordon & Breach: New York. "*Encyclopedia of Environmental Science & Engineering,*" (Vol: 2).

Peirce, J. J., Weiner, R. F. and Vesiland, P. A. (1984), Butterworth-Heineman: New Delhi. "*Environmental Pollution and Control.*"

Purdom, P. W. (ed.) (1980), Academic Press: New York. "*Environmental Health.*"

Rump, H. H. (1999), Wiley-VCH: Weinheim. "*Laboratory Manual for the Examination of Water, Wastewater and Soil,*" (3rd edn.).

Smith, K. (1991), Marcel Dekker: New York. "*Soil Analysis: Modern Instrumentation Techniques.*"

Sposito, G. (1989), Oxford University Press: New York. "*The Chemistry of Soils.*"

Yaron, B., Calvet, R. and Prost, R. (1996), Springerverlag: Berlin. "*Soil Pollution, Process and Dynamics.*"

# 12

# Hazardous Pollutants

Practically, all hazardous and toxic chemicals that are environmental pollutants are man-made. These substances include corrosive materials (acids and alkalis), reactants (oxidants and reductants), surfactants, solvents, petro-chemicals, explosives, agro-chemicals, plastics and a variety of consumer products.

Hazardous and toxic compounds can be classified according to their size and the state of the matter (e.g. flammability, viscosity), and according to their chemical and biological (physiological) effects. Many of the volatile polyaromatic hydrocarbons (PAHs) and polychlorinated biphenyls (PCBs) are known carcinogens and mutagens. Trace heavy elements (As, Cd, Cr, Hg, Pb, Mo, Se) are extremely toxic to human beings. Many organometallic compounds, pesticides, herbicides and fungicides are highly toxic not only to "pests" but to other organisms also. Through water run-off and crops various pollutants enter the food chain.

Discernable physiological characteristics of the pollutants are irritation, asphyxiation and toxicity. Irritant inorganic compounds are $NH_3$, $O_3$, $SO_2$, and $H_2SO_4$. Asphyxiates are either physiologically inert gases ($N_2$, He, $CO_2$, $CH_4$) or chemical asphyxiates (CO, HCN and $H_2S$). Some of them are carcinogenic.

This chapter deals briefly with some hazardous chemicals such as agrochemicals and industrial substances.

Hazardous and toxic substances are those, which are injurious to health and life in an environment, which is otherwise conducive for normal functions of organisms. Hazardous wastes are classified on the basis of their physical, chemical and biological properties. General criteria of hazardous and toxic pollutants are flammability (explosives and volatile petrochemicals), corrosivity (acids and alkalis), reactivity (powerful oxidants and reductants) and toxicity (toxic chemicals, trace metals and metallo-compounds). Hazardous and toxic substances are, generally, man-made and radioactive materials. There are about twenty-eight priority target compounds and 234 hazardous air pollutants.

## 12.1 TOXIC SUBSTANCES

Toxic compounds are classified according to their physical characteristics (e.g. particle size), physicochemical reactions and biological effects. Many of the volatile organic compounds (VOCs) that are aromatic and chlorinated and have boiling point <200° C are characterized as hazardous, due to their carcinogenic and mutagenic effects. They are persistent in the environment owing to their low volatility and chemical stability. Hazardous atmospheric pollutants (HAPs), hazardous to health, have been classified into various categories—oxygen-containing compounds, nitrogen-containing compounds, aromatic hydrocarbons, halogenated hydrocarbons, pesticides (insecticides, fungicides and herbicides), and inorganic trace elements. Primary routes of entry and action of toxic materials are: (*i*) inhalation, (*ii*) ingestion and (*iii*) contact.

### *12.1.1 Physical Classification*

Physical classification is based on the physical characteristics of the pollutants—generally on its size and the state of the matter. Suspended particulate (liquids and solids) in a gaseous medium (aerosols) (Fig. 12.1) include chemical dusts, smokes, mists (atomization of liquids), sprays and fumes. Fumes are vaporized metals or oxides.

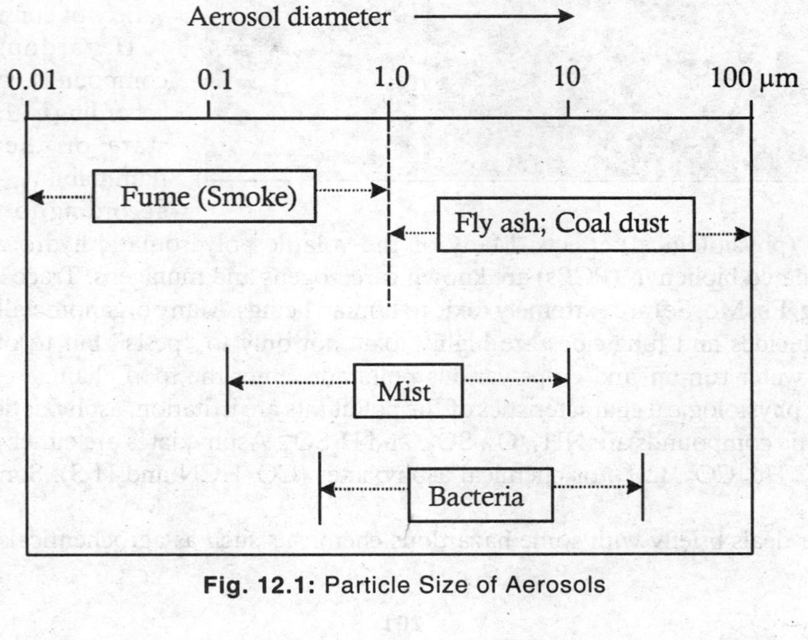


**Fig. 12.1:** Particle Size of Aerosols

### 12.1.2 Physiological Classification

Physiological classification of toxic compounds can be dealt under irritants, asphyxiates and poisons.

Irritants are corrosive to eyes, mucous membranes, and skin. Action of the corrosive compounds (irritants) depends on the solubility in aqueous medium. Sparingly soluble toxicants affect lungs and eyes. Highly soluble compounds, such as ammonia ($NH_3$) and mineral acids (HCl, $H_2SO_4$) affect mainly the upper respiratory and pulmonary tracts. Particulate matter, such as silica, asbestos and coal dust cause lung damage.

The main action of asphyxiates is to deprive oxygen to tissues. Simple asphyxiates—$CO_2$, $H_2$, $N_2$, $CH_3$, He—are physiologically inert gases. Chemical asphyxiates—CO, HCN, $H_2S$—render body tissues incapable of utilizing oxygen supply.

Systemic poisons are toxic chemicals, which include plant, animal and bacterial poisons. They cause damage to various systems of the body. Some of the hazardous and toxic chemical poisons are, heavy metals such as As, Bi, Cd, Cr, Hg, Pb, Mo, Sb, Se and Sn, and organic solvents such as benzene, toluene, carbon disulfide ($CS_2$), carbon tetrachloride ($CCl_4$), acetonitrile, and vinyl chloride and organophosphorous compounds. Plant poisons are alkaloids, aconite, mushroom poisons, aflatoxin and other mycotoxins. Animal poisons are bee, toad, scorpion and snake venoms. Bacterial poisons are botulism, colic and other pathogens.

## 12.2 HAZARDOUS CHEMICALS

A variety of chemicals are manufactured for industrial as well as domestic purposes for ever-increasing needs of the human race. These industrial chemicals (acids, alkalis, solvents, petrochemicals, fertilizers and pesticides etc.) are used for industrial purposes, food preservation and storage, timber, paper, leather, glass, ceramics, plastics and other household furniture industries, sanitation products (detergents and cleaning agents), chemical fertilizers and pesticides and other petrochemical products. Many of these substances are harmful to health and well-being of life and environment. Many of the industrial chemicals and petrochemicals (hydrocarbons) seriously endanger the environment. Though chlorofluorocarbons (CFCs; and halons) are inert and nontoxic to humans, they are potential atmospheric ozone depletion chemicals. Some of the chemicals hazardous to terrestrial organisms and aquatic life are produced by chemical reactions or microbial action. As examples—production of peroxyacetyl nitrates (PAN) due to interaction of atmospheric nitrogen with hydrocarbons in the atmosphere. Mercury poison due to methyl mercury ($Hg_2CH_3$) produced by aquatic planktons and consumed by fish, thus entering into the human food cycle. Minute concentrations of hydrocarbons endanger the quality of the atmosphere and can compromise the quality of water or foods. Polyaromatic hydrocarbons are carcinogenic and mutagenic. Interaction of these chemicals with the constituents of the environment may produce more toxic substances. Some of the industrial chemicals/substances and their effects are given in Table 12.1.

Some highly hazardous substances/chemicals are: Acrylonitrile; arsenic (As); asbestos; benzene; beryllium (Be); cadmium (Cd); chlorinated solvents; chlorofluorocarbons; chromates; coke emissions; diethylstilbesterol; dibromochloropropane; ethylene dibromide; ethylene dioxide; lead (Pb); mercury (Hg); nitrosamines; ozone ($O_3$); polyaromatic hdrocarbons (PAHs); polychlorinated biphenyls (PCBs); radiation; sulfur dioxide ($SO_2$); vinyl chloride. Largest known group of carcinogenic substances are—benzo[a]anthracene, dibenzo[a,h]anthracene, benzo[b]fluoranthene, benzo[k]fluoranthene, benzo[a]pyrene, chrysene, etc.

**TABLE 12.1:** INDUSTRIAL CHEMICALS/SUBSTANCES AND THEIR EFFECTS

| *Chemical/Substance* | *Effects* |
|---|---|
| Acetylene ($C_2H_2$) | Firehazard |
| Acrylamide | On skin, eyes and nervous system |
| Acetonitrile | Lung and bowel cancer |
| Alkanes ($C_5$-$C_8$) | On skin and nervous system |
| Ammonia ($NH_3$) | Pulmonary irritation and nervous system |
| Arsenic (As) | Poisonous; dermatitis; lung cancer; lymphatic tissue damage |
| Asbestos | Asbestosis; volcanosis; pleural cavity and lung cancer |
| Azides | Poisonous and explosive |
| Benzene ($C_6H_6$) | Leukemia (hematopoietic tissue) |
| Beryllium (Be) | Toxic; bone and lung cancer |
| Boron trifluoride | Respiratory system |
| Butadiene | Leukemia (hematopoietic tissue) |
| Cadmium (Cd) | Poisonous; lung and kidney damage; *Itai-itai* disease |
| Carbon dioxide ($CO_2$) | Respiratory effects; asphyxiation |
| Carbon disulfide ($CS_2$) | Heart and nervous system; poisonous and flammable |
| Carbon monoxide (CO) | Chemical asphyxiation |
| Carbon tetrachloride ($CCl_4$) | Liver cancer |
| Chloroform ($CHCl_3$) | Liver, kidney and central nervous system (CNS) |
| Chloroprene | Cancer |
| Chromic acid | Nasal ulceration |
| Chromium (Cr) | Toxic; skin and lung cancer |
| Coal tar and products | Lung and skin cancer; black lung disease |
| Cotton dust | Pulmonary (black lung) disease |
| Cyanides | Chemical asphyxiation; respiratory and blood systems |
| DDT | Liver damage |
| Dioxane | Liver and kidney cancer |
| Ethylene dibromide | Carcinogenic and mutagenic |
| Ethylene dichloride | CNS; heart and lung |
| Fluorides | Kidney and bone cancer |
| Fluorocarbons | Lung damage |
| Formaldehyde (Formalin) | Skin and lung cancer |
| Heptachlor | Liver cancer |
| Hydrazine ($N_2H_4$) | Poisonous |
| Hydrogen fluoride (HF) | Skin, eye and airway irritation |
| Hydrogen sulfide ($H_2S$) | Irritation; CNS and respiratory toxicant |
| Kepone | Liver cancer; CNS effects |
| Lead (Pb) | Kidney, blood and nervous system damage |
| Malathion | Affects nervous system |

| *Chemical / Substance* | *Effects* |
|---|---|
| Mercuric chloride ($HgCl_2$) | Corrosive and poisonous |
| Mercury (Hg) | CNS and spinal cord; neurological '*Minimata*' disease |
| Methelene chloride | Affects CNS |
| Methyl alcohol ($CH_3OH$) | Blindness; metabolic acidosis |
| Nickel (Ni) | Lung, nasal and skin cancer; liver, kidney and nervous system |
| Nicotine | Lung cancer and nervous system effects |
| Nitric acid ($HNO_3$) | Nasal and lung; dermatitis |
| Nitrogen oxides ($NO_x$) | Airway effects |
| Organophosphates | Cholinesterase crisis |
| Ozone ($O_3$) | Oxidation and mutation |
| Parathion | Affects nervous system |
| Pesticides | Poisonous; liver and nervous systems |
| Petroleum solvents | Skin, lung and nervous system |
| Phenol | Skin, eye, liver, kidney and CNS |
| Phosgene | Airway effects |
| Phosphorus (P) | Poisonous |
| Polychlorinated biphenyls (PCBs) & PAHs | Skin, liver and pancreas cancer. Carcinogenic |
| Silica ($SiO_2$) | Lung disease (Silicosis) |
| Sodium hydroxide (NaOH) | Skin and airway irritation; dermatitis |
| Sulfur dioxide ($SO_2$) | Respiratory and skin irritation bronchitis and emphysema |
| Sulfuric acid ($H_2SO_4$) | Pulmonary, skin and eyes; dermatitis |
| Tetrachloroethane | Heart, liver, respiratory and nervous systems |
| Tetraethyl lead ($Pb(C_2H_5)_4$ | Poisonous liquid (used in petrol) |
| Toluene ($C_6H_6CH_3$) | Affects central nervous system (CNS) |
| Trichloroethelene | Affects CNS and carcinogenic |
| Tungsten (W) | Lung and skin damage |
| UV radiation | Eye and skin damage |
| Vanadium (V) | Eye, skin and lung |
| Vinylchloride | Liver cancer |
| Xylene | Airway irritation and CNS effects |
| Zinc oxide (ZnO) | Metal fume fever |

## 12.3 AGROCHEMICALS

There are a large number of agrochemicals available in the market—chemical fertilizers (phosphates and nitrates), pesticides, insecticides, fungicides, and defoliants. Many of the organo-pesticides (that include insecticides, herbicides, and fungicides) are: halogenated-hydrocarbons (aldrin, dieldrin, DDT, chlordane, heptachlor, endrin, and lindane); organophosphorus (malathion, parathion, diazinon); organosulfur; carbamates (carbaryl, aldicarb, cabofuran); and phenol-based (2, 4-dichlorophenol (2, 4-D); 2, 4, 5-trichlorophenoxy acetic acid (2, 4, 5-T)) compounds.

### 12.3.1 Mode of Action of Pesticides

There is a glut of industrially produced pesticides in the market. Pesticides are used to kill pests and they are neurotoxic. Many of the agrochemicals are not only toxic to the organisms against which they are supposed to act on, but also to other organisms and ecological systems, directly or indirectly. Chemical fertilizers affect the soil pH-balance and stimulate weed growth. Insecticides, herbicides and fungicides are agrochemical compounds that are highly toxic and bring about drastic ecological imbalance in plant kingdom. Most of them come under banned chemicals. They cause paralysis, loss of balance of motor functions. Many polynuclear organic hydrocarbons are carcinogenic. Pesticides cause heavy loss to pollinators, like honeybees and butterflies. They interfere with avian ecology—growth, egg production, shell-thickness, hatching ability and fertility.

Most direct avenues of pesticide contamination, which are harmful to animals and humans, are through ingestion of residues on food products and consumption of seafood and meat, already contaminated. Some common pesticides and their effects are given in Table 12.2.

**TABLE 12.2:** COMMON PESTICIDES FOUND IN WATER BODIES

| *Pesticide* | *Tolerance(mg/L)* | *Effects* |
|---|---|---|
| Aldrin | 0.003 | Chloro-insecticide; persistent and toxic. Organisms convert it to dieldrin |
| Carbaryl (Sevin) | 0.03 | Carbamate insecticide; dangerous to honeybees |
| Chlordane++ | 0.01 | Chloro-insecticide against termites. Persistent and carcinogenic |
| DDT++ | 0.01 | Chloro-pesticide. Neurotoxic and carcinogenic |
| Demeton++ | 0.1 | Organophosphorus insecticide |
| Diazinon++ | – | Organophosphorus pesticide; highly toxic to mammals |
| Dieldrin++ | 0.003 | Chloro-insecticide. Carcinogenic |
| Diquat | – | Herbicide |
| Endosulfan++ | 0.003 | Pesticide; toxic to mammals |
| Endrin++ | 0.004 | Chloro-pesticide; toxic |
| Heptachlor++ | 0.001 | Chloro-pesticide; highly toxic; carcinogenic |
| Lindane++ | 0.001 | Chloro-insecticide; toxic |
| Malathion++ | 0.1 | Organophophorus insecticide |
| Methoxychlor++ | 0.03 | DDT substitute |
| Paraquat++ | – | Herbicide |

++ Banned pesticides

Much of the environmental harm from pesticides results from the use of broad-spectrum pesticides (organochloro-, organophosphorus-carbamates, etc.). Major pesticides are creosote, DDT, aldrin, dieldrin, chlordane, heptachlor, carbaryl (1-napthyl-N-methylcarbamate), chlorophenols, lindane, malathion, parathion. Volatilization is a major factor in the dispersion of pesticides in the atmosphere. Agri-spraying operations by aircraft produce widespread air, water and soil pollution. The water environment provides the ultimate sink for the pesticide residues. Pesticides in soil are due to agriculture operations. Pesticides in food are due to successive accumulation of lipid-soluble, but water-insoluble, toxic substances by fish and other aquatic organisms, land animals, birds and consumption of them by humans. Such an accumulation increases the concentration of DDT and dieldrin in fish by $10^6$-fold. In addition, pesticides in cereals, vegetables and fruits are a major cause of health hazards.

### 12.3.1.1 *Organochlorines*

DDT (dichlorodiphenyltrichloroethane) (Fig. 12.2) is the oldest of the organochlorine insecticides. DDT interferes with the action of Na/K-ion channels, and prevents the normal transmission of nerve impulses. It inhibits the absorption of $Ca^{2+}$ by bones and it produces enzymes, which decompose sex hormone steroids, thus creating hormonal imbalance.

DDT has several drawbacks and environmental impact. (1) Persistence: It is one of the most persistent pesticides (half-life ~5-10 years). DDT is not readily absorbed through the skin when it is in a powder form. Its main entry points are respiratory and digestive tracks. As it is fat-soluble, large amounts accumulate in the brain, liver, and kidneys. (2) Bioaccumulation: Effects of DDT are amplified due to fat-solubility and its accumulation in the living tissues. The chemical accumulates in fish, and, when birds eat such fish, the chemical accumulates in their fat tissue, resulting in fragile eggs. The thinning of eggshells and breakage of eggs are attributed to the effects of DDT upon the enzyme *carbonic anhydrase*.

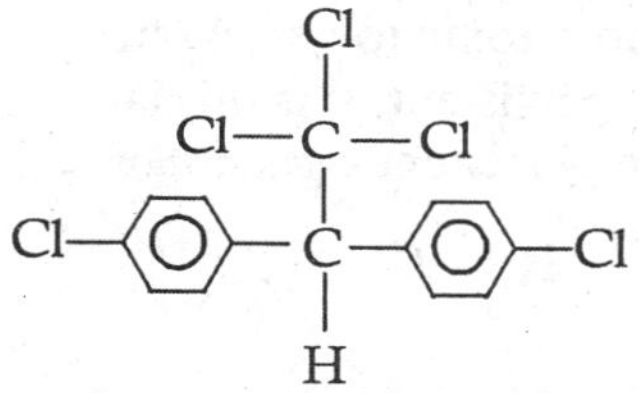

(dichlorodiphenyltrichloroethane)
(DDT)

**Fig. 12.2:** Structural Formula of DDT

Some other organochlorine pesticides are aldrin, chlordane, dieldrin, endrin, methoxychlor (Fig. 12.3). Aldrin, dieldrin and other chloro-pesticides are metabolized in a manner similar to DDT. Aldrin is oxidized to dieldrin in the metabolic process and the latter is even more toxic to fish. That is, in some cases pollutants are converted metabolically in the environment to more toxic chemical species. Kepone exhibits acute, delayed, and cumulative toxicity to birds, mammals and humans. Chlordane is a more potent pesticide than DDT. Like DDT, chlordane is not readily decomposed and persists for long periods of time in the soil, foodstuffs, etc. Unlike DDT, it can enter the body through the skin, respiratory and digestive tracks. Its effect is cumulative. Derivative chlordane is heptachlor, which is fat-soluble. Endrin is the most toxic of all the chlorinated hydrocarbons. It is about fifteen times as toxic as DDT to mammals, thirty times as poisonous to aquatic life, and about three hundred times as poisonous to birds.

$Cl—CH_2CH_2—S—CH_2CH_2—Cl$ Mustard gas

Bis(chloroethyl)sulfate

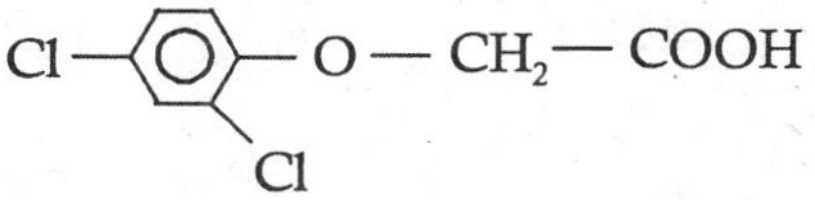

Chlorophenoxy herbicide (2,4-D)

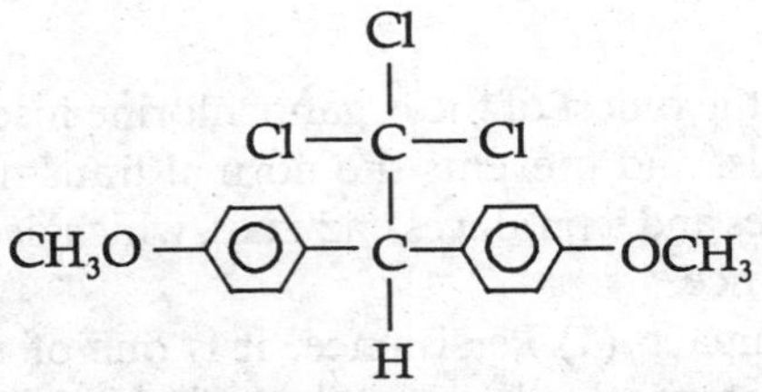

Methoxychlor (DDT substitute)

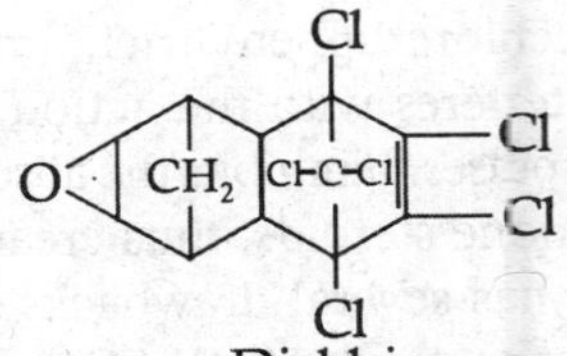

Dieldrin

**Fig. 12.3:** Structural Formulas of some Halogenated Hydrocarbon Pesticides

#### *12.3.1.2 Organophosphates*

Organophosphates (OPs) are the currently used generic term that includes all insecticides containing phosphorus. All organophosphates are derived from one of the phosphorus acids, and as a class are generally the most toxic of all pesticides to vertebrates (Fig. 12.4). The OPs have two distinctive features: they are generally much more toxic to vertebrates than other classes of insecticides, and most are chemically unstable or non-persistent. It is this latter characteristic that brought them into agricultural use as substitutes for the persistent *organochlorines*.

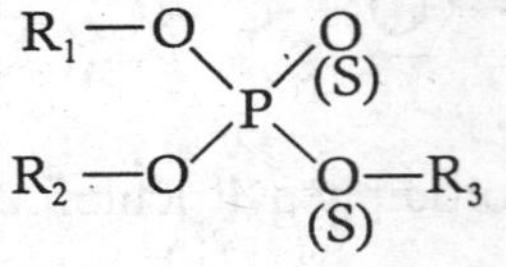

(S = Sulfur; $R_1$, $R_2$ and $R_3$ = Alkyl or aryl groups)

**Fig. 12.4:** General Formula of Organophosphorus Pesticides

Some of the organophosphorus pesticides are diazinon, malathion, manazon, metasyston and parathion (Fig. 12.5). Because of the similarity of OP chemical structures to the "nerve gases," their modes of action are also similar. The nerve gases (insecticides) destroy the acetylcholine esterase (AChE) enzyme activity, namely the hydrolysis of acetylcholine, thereby interfering with the nerve impulse transmission processes. For this reason parathion (diethyl p-nitrophenylthiophosphate), the organic phosphate insecticide is among the most poisonous chemicals. It decomposes in about two weeks. Malathion (and demeton) is less toxic to the higher animals, because of its enzymatic degradation in the liver. Malathion is widely used to control insects from pets and livestock. Fly ash shows a significant capacity for adsorption of organophosphorus compounds and it can be used for removal of pesticides from aqueous solution.

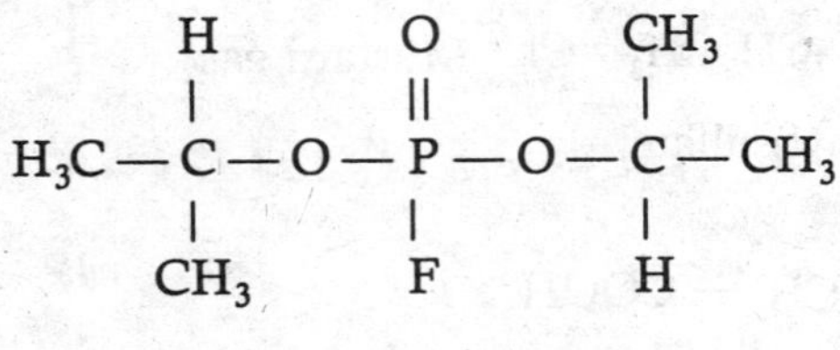

DFP (Nerve gas)
(Diisopropyl fluorophosphate)

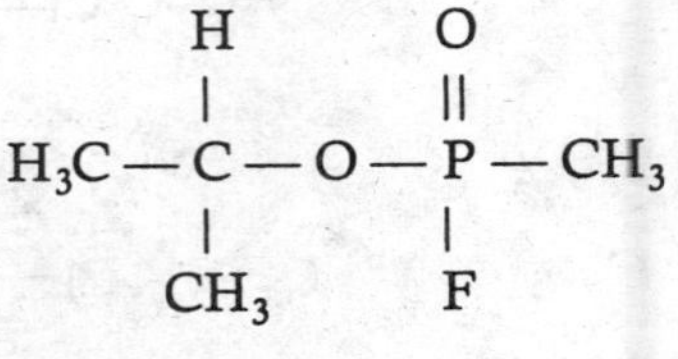

Sarin (Nerve gas)

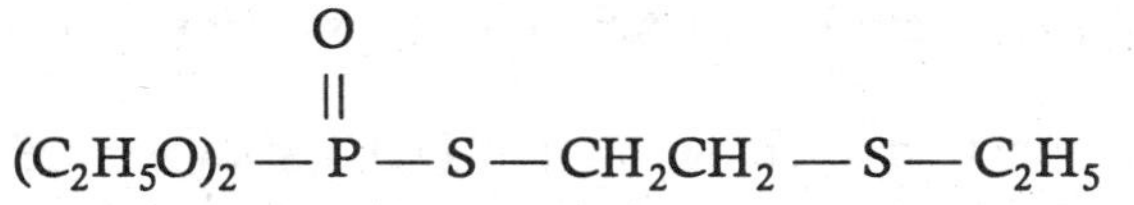

$$\text{(C}_2\text{H}_5\text{O)}_2 - \overset{\overset{\displaystyle O}{\|}}{P} - S - CH_2CH_2 - S - C_2H_5$$

Demeton (pesticide)

$$O_2N-\!\!\bigcirc\!\!-O-\overset{\overset{\displaystyle O-CH_2CH_3}{|}}{\underset{\underset{\displaystyle O-CH_2CH_3}{|}}{P}}=S$$

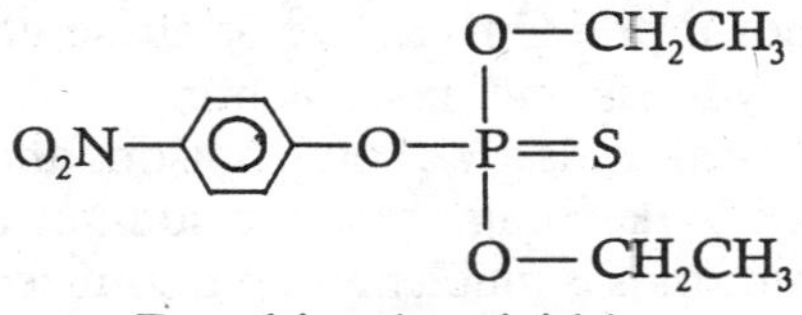

Parathion (pesticide)

**Fig. 12.5:** Structural Formulas of some Organophosphorus Pesticides

### 12.3.1.3 Carbamates

The carbamate insecticides (e.g. carbaryl (sevin)) (Fig. 12.6) are derivatives of carbamic acid (as the OPs are derivatives of phosphoric acid). And like the OPs, their mode of action is that of inhibiting the vital enzyme *cholinesterase* (ChE).

### 12.3.1.4 Phenolic Pesticides

Dinitrophenol (Fig. 12.7) and its derivatives have broad range of insecticidal, herbicidal and fungicidal toxicities.

Herbicides, such as Na-arsinite, Na-chlorate, paraquat (1, 1'-dimethyl-4-4'-bipyridynium ion), simazin, mecocrop, dichlocrop, dalapon, monuron, alachlor, 2, 4-D (2, 4-Dichlorophenoxy acetic acid), and 2, 4, 5-T (2, 4, 5-Trichlorophenoxy acetic acid; *Agent Orange*) are used to kill 'unwanted' herbs and foliants. These chemicals interfere with photosynthesis causing plant death. Defoliation of vegetation and forests lead to increased soil erosion and leaching of nutrients. The ecological implications are long ranging and far-reaching. The pathological effects on plants of herbicides such as Agent Orange, and malathion (O, O-dimethyl S-(1, 2-dicarbethoxyethyl) phosphorodithioate)) in chemical warfare by the USA forces in Vietnam had drastically affected the ecology in that country. Forests have been defoliated, agricultural crops were completely destroyed and fauna has been affected. Spraying of mangroves with 2, 4, -D and 2, 4, 5-T has completely destroyed the vegetation and fauna.

Fungicides are used to kill fungal pathogens of plants. Some of these are alkylmercury, benomyl, captan, carboxin, karathene, maneb, misterm, rovral, thiram, zineb and ziram. All fungicides are toxic.

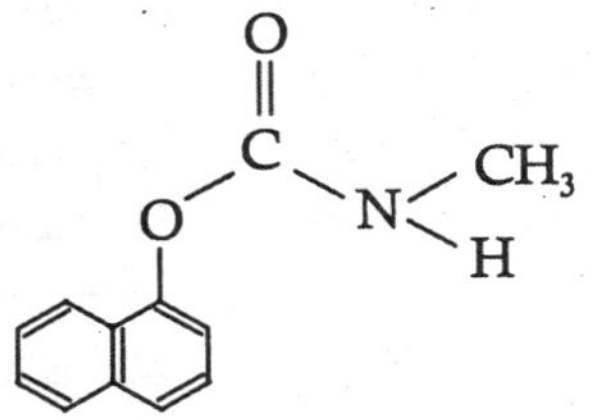

Carbaryl
**Fig. 12.6:** Structural Formula of
2,4- Carbaryl (Sevin)

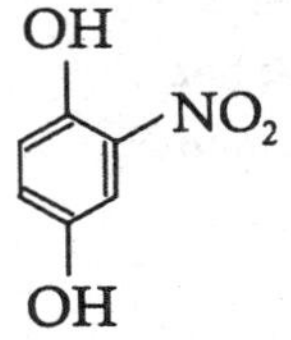

2, 4-Dinitrophenol
**Fig. 12.7:** Structural Formula of
Dinitrophenol

### 12.3.1.5 Botanicals

There has been a perceptive trend towards search for natural pesticidal compounds. These botanical compounds are organic molecules (or mixtures) extracted from plants. These botanicals are toxic to insects but relatively harmless to other organisms. They decompose readily, so bioaccumulation is not a problem. Some of these compounds are pyrethroids, rotenoids, nicotinoids, pyrroles, benzylureas and azadirachtin.

Pyrethroids are axonic poisons and the mode of action resembles that of DDT. They apparently work by keeping open the sodium channels in neuronal membranes. The stimulating effect of pyrethroids is much more pronounced than that of DDT. The nicotinoids act on the central nervous system of insects, causing irreversible blockage of postsynaptic nicotinergic acetylcholine receptors.

Rotenoids are inhibitors of respiratory enzymes, acting between $NAD^+$ (a coenzyme involved in oxidation and reduction in metabolic pathways) and coenzyme Q (a respiratory enzyme responsible for carrying electrons in some electron transport chains), resulting in failure of the respiratory functions. Pyrroles are inhibitors of oxidative phosphorylation, thus preventing the synthesis of energy molecule, adenosine triphosphate (ATP).

Benzylureas act as growth regulators. They act on the larval stages of most insects by inhibiting or blocking the synthesis of chitin, a vital and almost indestructible part of the insect exoskeleton.

Azadirachtin has shown some rather sensational insecticidal, fungicidal and bactericidal properties, including insect growth regulating qualities. Azadirachtin disrupts molting by inhibiting biosynthesis or metabolism of ecdysone, the juvenile molting hormone.

All pesticides that are dangerous for the pests are also dangerous to other organisms. Organophosphorus pesticides kill the pests by attacking the nervous system and causing the inactivation of the synaptic enzyme, acetylcholinesterase (AChE). Organophosphorus esters, carbamates and ureas are metabolized by enzymes in the human liver and are converted to predominantly water-soluble products and eliminated in the urine. But, organochlorine compounds, lacking metabolizing enzymes, are more persistent in the environment. Pesticides that persist in the environment for longer periods are, naturally, more dangerous. For example, pesticides like parathion and diazinon are more dangerous (Table 12.3). Many pesticides come under banned substances. Some of the banned pesticides are alkylmercury fungicides, kepone, depone, 2, 4, 5-T, heptachlor, aldrin, dieldrin, DDT, endrin, chlordane, silvex and paraquat.

**TABLE 12.3:** PERSISTENT PESTICIDES IN THE ENVIRONMENT

| *Pesticide* | *Duration for 90% Loss* | *Pesticide* | *Duration for 90% Loss* |
|---|---|---|---|
| Paraquat | ~8 hours | Amitrole | ~1 week |
| Malathion | ~2 weeks | Captan | ~3 weeks |
| Phorate | ~3 weeks | Mecocrop | ~4 weeks |
| Dalaphon | ~3 months | Prophan | ~3 months |
| Diazinon | 3-6 months | Parathion | 3-6 months |
| Carbaryl | ~6 months | Carbofuran | ~6 months |
| 2, 4, 5-T | ~6 months | Aldrin | 2-3 years |
| Monuron | ~3 years | Chlordane | 3-5 years |
| Dicamba | ~4 years | Picloram | ~5 years |
| Heptachlor | 3-6 years | Paraquat | ~6 years |
| Dieldrin | 4-8 years | Mirex | ~10 years |
| Chlordane | ~15 years | Endrin | ~15 years |
| DDT | 10-20 years | Dieldrin | ~20 years |
| Simazine | ~20 years | Chlorobenzenes | ~30 years |

While all pesticides are toxic, some of them are more deadly to some species than to other (Table 12.4). Carbofuran, diazinon, endrin, parathion and zectran are acutely poisonous to birds. Aldrin, dieldrin, eldrin, heptachlor, chlorodane and DDT are extremely toxic to fish.

**TABLE 12.4:** SPECIES DEPENDENCE OF TOXICITY OF PESTICIDES

| *Pesticide* | *Rats (mg/kg) (body weight)* | *Fish (mg/kg) (body weight)* | *Birds (mg/kg) (body weight)* |
|---|---|---|---|
| Aldrin/Dieldrin | 50 | 0.006 | 500 |
| Carbaryl | 400 | 2 | 2000 |
| Carbofuran | 5 | 0.25 | 0.5 |
| Chlordane | 400 | 0.02 | 1000 |
| DDT | 100 | 0.01 | 2000 |
| Diazinon | 100 | 0.01 | 3 |
| Dicamba | 3000 | 35 | – |
| Endrin | 10 | 0.003 | 5 |
| Heptachlor | 100 | 0.01 | 2000 |
| Lindane | 100 | 0.01 | 1000 |
| Malathion | 1000 | 0.15 | 1500 |
| Parathion | 10 | 3 | 10 |
| Zectran | 15 | 8 | 3 |

*Note*: readers are advised to refer to relevant books to obtain further information on the latest additions to the list of agrochemicals and pesticides.

## 12.4 PHARMACEUTICALS AND PERSONAL CARE PRODUCTS (PPCPs)

The (over) use of antibiotics and health care products in modern societies and the discharge of anthropogenic chemicals like, bio-active metabolites, drugs, genotoxic, chemotherapeutic, and personal care and hygiene products into terrestrial and aquatic environments pose a great pollution concern. Large quantities of a wide spectrum of PPCPs and their metabolites can enter the environment following discharge to sewage treatment plants. No municipal sewage treatment plants are engineered for PPCP removal. The PCCPs may cause hormonal disruption in aquatic organisms and overuse of antibiotics may cause accelerated resistance among naturally occurring pathogens. There are several factors that make quantitative analysis of PPCPs difficult. Very low (trace) concentration and more polar nature of PPCPs necessitate the use of non-standard instrumentation and make the analysis more difficult.

## BIBLIOGRAPHY

Agarwal, S. K. (1991), Himanshu: Udaipur. "*Pollution Ecology.*"

Baier, R. E., Mark, H. B. and Mattson, D. S. (eds.) (1981), Marcel Dekker: New York. "*Water Quality Measurement.*"

Harrison, R. M. and Perry, R. (1984), Chapman & Hall: London. "*Handbook of Air Pollution Analysis,*" (2nd edn.).

Hutzinger, O. (ed.) (1990), Springerverlag: Berlin. "*Handbook of Environmental Chemistry.*"

Jorgsen, S. E. and Johnsen, I. (1981), Elsevier: Amsterdam. "*Principles of Environmental Science and Technology,*" (Vol. 14).

Lee, S. D. (1977) Ann Arbor Pub: Ann Arbor, MI. "*Biochemical Effects of Environmental Pollutants.*"

Minear, R. A. and Keith, L. H. (eds.) (1994), Academic Press: New York. "*Water Analysis: Organic Species.*"

Nriagu, J. O. and Simmons, M. S. (1994), Wiley & Sons: New York. "*Environmental Oxidants.*"
Pfafflin, J. R. and Ziegler, E. N. (1976), Gordon & Breach: New York. "*Encyclopedia of Environmental Science and Engineering.*"
Purdon, P. W. (ed.) (1980), Academic Press: New York. "*Environmental Health,*" (2nd edn.).
Waldport, G. L. (1978), Mosey: St. Luis, MO. "*Health Effects of Environmental Pollutants,*" (2nd edn.).
Weinberg, W. T. *et al.* (1988), EPA/600/4-89/017/: USEPA, N. C. "*Compendium of Methods for the Determination of Toxic Organic Compounds.*"

13

# Radioactive Pollution

Of all the hazardous pollutants, radioactive substances are the most toxic. Radioactivity is all-pervasive. Man has been living with it since the creation of the Universe. But modern technology has brought several naturally radioactive materials like uranium into the industry and also added many hitherto unknown isotopes to man's stock of 'useful' materials. Radioactive pollution is a recent contribution by man to environmental pollution. Radioactive pollutants come under a special category due to their extreme toxicity and long half-lives of many of them. Handling, transport, storage and disposal of these products and their wastes pose short-term and long-term environmental and health risks.

Mining, processing and purification produce radioactive products (e.g. uranium and thorium). They are used for weapons manufacture and in reactors for nuclear energy generation. In addition, a series of radionuclides are produced for nuclear medicine and other applications. The nuclear reactor is the key radioactive waste producer in the nuclear fuel cycle. All radioactive products, their waste products and spent nuclear fuels pose extreme radiation hazard unless properly handled.

This chapter is devoted to the various sources of radioactivity, the measurement of radiation and their biological effects.

Radioactivity is the process of emitting radiation spontaneously. Natural radiation is all round us in the environment. Low-level radiation is found in the air that we breathe, the water that we drink, the food that we eat, and the rocks, soil and the plants around us. The materials with which we construct our buildings emit radiation and expose us all the while. Ultimately, our own bodies contain radioactive materials. There are some unstable atoms in all the substances in the world and they emit energy as radiation when they decay to more stable atoms. This phenomenon is called 'radioactivity' and the atoms that emit radiation are called 'radioactive atoms' or 'radionuclides' or 'radioisotopes'. Many radionuclides were formed long before our solar system came into existence. Some are being continually created in the Earth's atmosphere by the action of cosmic rays and reach the geosphere. Cosmic rays permeate all space and are composed of high energy positively charged particles as well as high-energy photons. By interaction of these particles with atoms in the upper atmosphere many radioactive atoms are generated. While the atmosphere attenuates a large part of cosmic rays, a small part reaches the Earth.

Utilization of naturally occurring primordial radionuclides, like uranium, and production of artificial radionuclides have expanded the list of radionuclides, increased their concentration several fold and brought them closer to human and animal world. The questions, how much natural radioactivity we are exposed to, are we safe in spite of this exposure, and how much additional risk is involved in anthropogenic activity dealing with and applying radioactive materials, need to be addressed in environmental studies.

## 13.1 SOURCES OF RADIOACTIVITY

Primordial radionuclides are left over from when the world and the universe were created. They are typically long lived, with half-lives in the order of hundreds of millions of years. A list of some of these nuclides is given in Table 13.1. The more important of them are—uranium (235 and 238), thorium (232), radium (226), radon (222) and potassium (40). Cosmic radiation, the source of which is primarily outside the solar system, permeates the universe. It consists of high-energy heavy particles and photons and muons. Among the particles ~85 per cent are protons and ~14 per cent α-particles. They interact with nuclei in the upper atmosphere and give rise to a series of radioactive nuclides, which can have long lives but a majority have shorter half-lives than the primordial nuclides. The important cosmogenic radionuclides are $^{3}H$ ($t_{1/2}$,= 12.3 yr.), $^{7}Be$ ($t_{1/2}$,= 53.3 days) and $^{14}C$ ($t_{1/2}$,= 5,730 yr.). Primordial and cosmogenic radionuclides make a major contribution to the background terrestrial radiation. Some of them produce secondary radionuclides by radioactive decay; and some of them like radon, being gases find their way into the atmosphere and biosphere.

**TABLE 13.1:** SOME PRIMORDIAL RADIONUCLIDES AND THEIR HALF-LIVES

| *Radionuclide* | *Half-life (Years)* | *Radionuclide* | *Half-life (Years)* |
|---|---|---|---|
| $^{142}Ce$ | $>5 \times 10^{16}$ | $^{174}Hf$ | $2 \times 10^{15}$ |
| $^{115}In$ | $6 \times 10^{14}$ | $^{148}Sm$ | $>2 \times 10^{14}$ |
| $^{180}Ta$ | $1 \times 10^{12}$ | $^{187}Os$ | $4 \times 10^{12}$ |
| $^{190}Pt$ | $7 \times 10^{11}$ | $^{138}Ce$ | $1 \times 10^{11}$ |
| $^{138}La$ | $>1 \times 10^{11}$ | $^{87}Rb$ | $\sim 4.8 \times 10^{10}$ |
| $^{232}Th$ | $1.4 \times 10^{10}$ | $^{238}U$ | $4.8 \times 10^{9}$ |
| $^{40}K$ | $1.3 \times 10^{9}$ | $^{235}U$ | $7 \times 10^{8}$ |
| $^{226}Ra$ | $1.6 \times 10^{3}$ | | |

Humans have used radioactivity for nearly 100 years. During this period they have added many radionuclides to the already existing list. The amounts are small compared to the naturally occurring stock, but they are used in a very much more concentrated form and hence pose special problems in their handling, storing and disposal. Radioactive materials find many applications in industry and research. In diagnosing diseases and in the treatment of cancer several of these nuclides find special place. Perhaps the elements those are best known and used in large quantity are uranium, plutonium and thorium. They are the raw materials for nuclear power generation and for weapons. In addition, a variety of radioisotopes are used in food preservation and radiopharmaceuticals. The latter find place in medical diagnostics and radiation therapy.

### 13.1.1 Radionuclides

Radionuclides are of natural ($^{3}H$ and $^{14}C$ in the atmosphere) as well as of anthropogenic origin. Nuclei of elements of higher atomic number than lead (Z = 82), are inherently unstable. They achieve a more stable state through natural nuclear disintegration, by emitting high-energy particles and radiations—α-particles, β-particles (electrons and positrons) and γ-rays and internal conversions (protons, to neutrons and *vice versa*). The process is called radioactivity, and the unstable nuclei are called radionuclides.

| | | |
|---|---|---|
| α-Particles: | Helium nuclei ($^{4}_{2}He^{2+}$) = 4 amu ($6.642 \times 10^{-7}$ kg) | (Particles) |
| β-Rays: | Electrons and positrons ($9.13 \times 10^{-27}$ kg) | (Particles) |
| γ-Rays: | High-energy photons (X-rays) | (Radiation) |

α-Particles are massive (two protons + two neutrons) and loose their energy after travelling only a few centimeters in air. A sheet of paper or the outer layer of skin can block them. They are harmful only when taken into the body through eating, drinking or breathing. Too much mass leads to a nucleus to emit α-particles. Uranium, thorium, radium and plutonium emit these particles.

$$^{238}_{92}U \rightarrow ^{234}_{90}Th + ^{4}_{2}He \qquad (\alpha\text{-Particle}) \qquad ...(13.1)$$

$$^{226}_{88}Ra \rightarrow ^{222}_{86}Rn + ^{4}_{2}He$$

β-Rays consist of electrons (and positrons). They travel ~3.8 m in air and can penetrate skin. Shielding is needed to prevent their impact. They can be stopped by glass, wood, plastic or metal. Too many neutrons in a nucleus lead it to emit β-rays, converting neutrons into protons (internal conversion). Many radionuclides formed by nuclear fission and those present in low-level radioactive waste emit these rays. Among them are $^{90}Sr$, $^{35}S$, $^{40}K$, $^{3}H$ and $^{14}C$.

$$^{14}_{6}C \rightarrow ^{14}_{7}N + ^{0}_{-1}\beta \qquad (\text{electron})$$

$$^{40}_{19}K \rightarrow ^{40}_{20}Ca + ^{0}_{-1}\beta \qquad ...(13.2)$$

$$^{89}_{38}Sr \rightarrow ^{89}_{39}Y + ^{0}_{-1}\beta$$

γ-Rays consist of high-energy photons. They can penetrate matter. Human body can be damaged by this radiation. Lead or concrete shielding has to be used for preventing them from harming. Isotopes of cobalt, zinc, cesium and manganese present in low-level wastes emit this radiation.

#### *13.1.1.1 Natural Radioactivity Series*

A radioactive element may achieve stability in a single decay, or it may decay through a series of states before it reaches a truly stable state. There are three distinct chains of primordial (natural) radioactive elements: (1) the uranium series ($^{238}U \rightarrow {}^{206}Pb$), (2) the actinium series ($^{235}U \rightarrow {}^{207}Pb$), and (3) the thorium series ($^{232}Th \rightarrow {}^{208}Pb$) (Fig. 13.1). $^{235}U$, the parent isotope of the actinium series, is the principal nuclide utilized in the fission process.

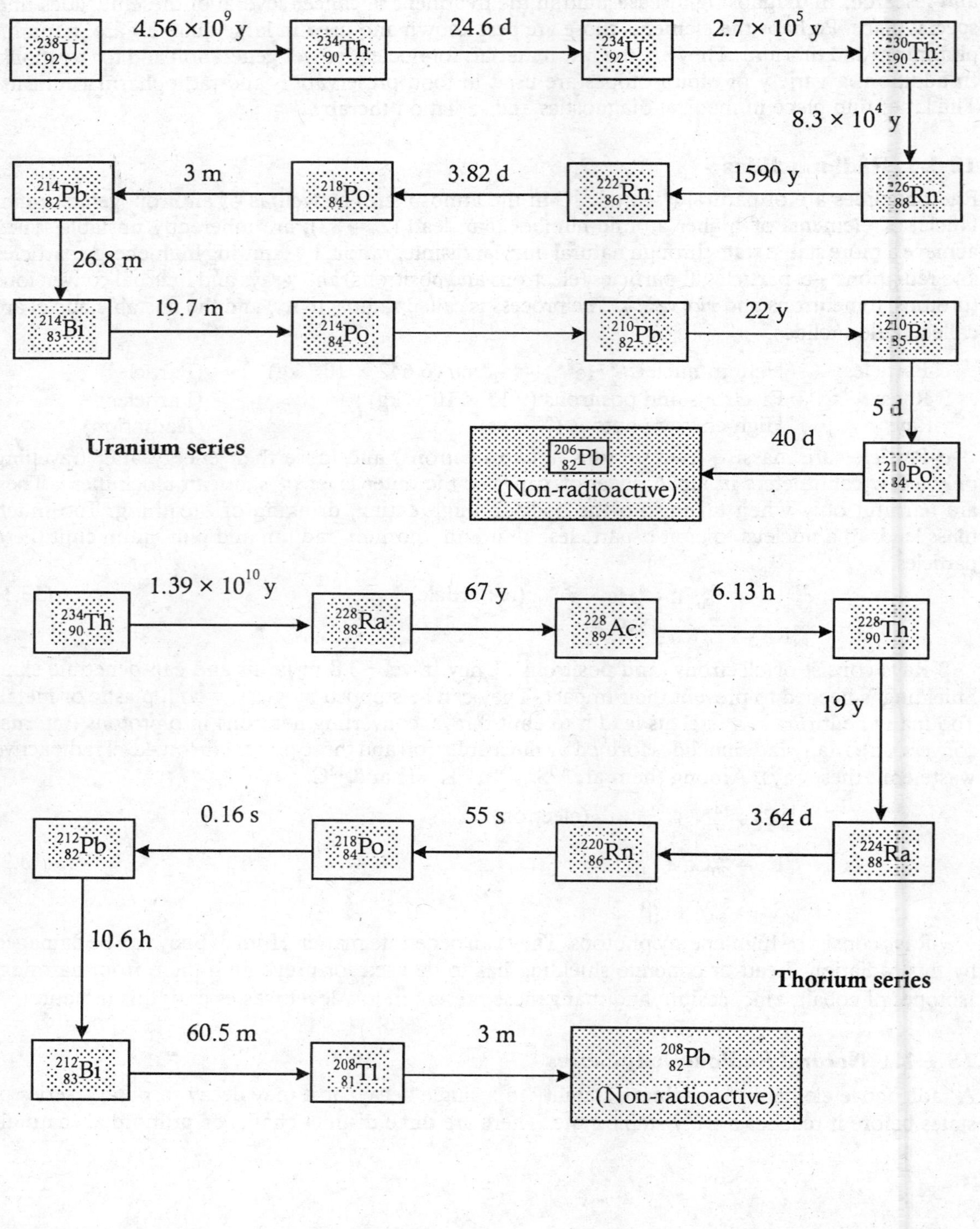

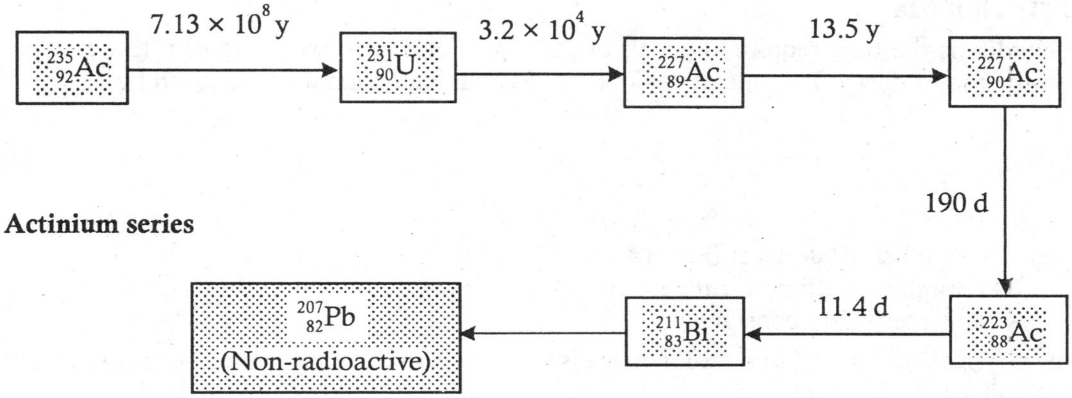


**Fig. 13.1:** The Uranium, Thorium and Actinium Series

### *13.1.1.2 Artificially-Produced Radionuclides*

Radionuclides can be produced artificially by nuclear fission, in a cyclotron with accelerated charged particles or by bombardment of a target with neutrons. Some means of producing artificial radionuclides are according to

Target → Products

$${}^{Y}_{Z}A + n \rightarrow {}^{Y+1}_{Z}A + \gamma \qquad \text{(neutron bombardment)} \qquad \ldots(13.3)$$

$$\underset{\text{(neutron)}}{{}^{1}_{0}n} \rightarrow \underset{\text{(proton)}}{{}^{1}_{1}p} + \underset{\text{(electron)}}{{}^{0}_{-1}\beta} \qquad \text{(internal conversion)} \qquad \ldots(13.4)$$

Nuclear fission:

$${}^{235}_{92}\text{U} + {}^{1}_{0}\text{n} \rightarrow {}^{236}_{92}\text{U} \rightarrow {}^{140}_{54}\text{Xe} + {}^{94}_{38}\text{Sr} + 2\,{}^{1}_{0}\text{n} + 200 \text{ MeV} \qquad \ldots(13.5)$$

## 13.2 MEASUREMENT OF RADIATION

The term radiation refers to a wide range of things. The radiation that causes damage and is of particular concern is the 'ionizing radiation'. This type of radiation creates ions when it interacts with matter, and can affect living tissues seriously. Electromagnetic radiation like X-rays and $\gamma$-rays as well as cosmic rays can cause such damage. Ionizing particle radiation includes $\alpha$-particles, $\beta$-rays and neutrons (via protons). The main forms of radioactivity associated with nuclear power and explosives are $\alpha$-particles, $\beta$-rays and $\gamma$-rays (see Chapter 25 for details of measurement). Radiation has many benefits for humans but too much of any radiation is harmful. This is true of solar radiation too.

Radioactivity is a physical, not a biological, phenomenon. A measure of radioactivity of a radionuclide is its disintegration rate (number of disintegrations per second). The (SI) unit of radioactivity is Becquerel (Bq).

$$1 \text{ Bq} = 1 \text{ disintegration/second} \qquad \ldots(13.6)$$

$$\text{Curie} = 3.7 \times 10^{10} \text{ Bq}$$

### 13.2.1 Half-life

The half-life is the time required for half of the atoms of a radioactive element to undergo self-transmutation or decay. The rate of radioactive decay is exponential and is given by

$$In\left(\frac{N}{N_0}\right) = -\lambda.t \qquad ...(13.7)$$

$$\therefore \qquad N = N_0 \exp(-\lambda \cdot t)$$

where, $N_0$ = number of atoms at time $t = 0$;
$N$ = number of atoms at time $t = t$;
$\lambda$ = disintegration constant.

The physical half-life, $\tau_p$, of a radionuclide is the time period during which the radioactive decay falls to half its initial value.

$$\tau_p = t_{1/2} = \frac{\ln 2}{\lambda} = \frac{0.693}{\lambda} \qquad ...(13.8)$$

The biological half-life, $\tau_b$, is the time taken by an organism to discard half of an administered radionuclide on biological discharge basis. The effective half-life, $\tau_e$, incorporates both the physical half-life and biological half-life and is given by

$$\tau_e = \frac{(\tau_p \times \tau_b)}{(\tau_p + \tau_b)} \qquad ...(13.9)$$

Effective half-life is always less than physical or biological half-life.

### 13.2.2 Absorbed Dose, *D*

Radiation dose is a measure of the administered or absorbed radioactive material. Absorbed dose, *D*, is the energy absorbed by unit mass of a material (example, body tissue).

$$D = (\text{Energy imparted/mass}) \qquad ...(13.10)$$

SI unit of absorbed radiation dose, *D*, is *Gray (Gy)*. It is the quantity of ionizing radiation that results in absorption of one joule of energy per kilogram of absorbing material.

$$\text{Gy} = \text{J} \cdot \text{kg}^{-1} \qquad ...(13.11)$$

$$1 \text{ Gy} = 100 \text{ rad}$$

While the absorbed dose is the energy deposited in the mass, the alternate quantity, *Kerma* (kinetic energy released in a material) is the kinetic energy released from the mass.

### 13.2.3 Dose Equivalent, *H*

High-energy particles/radiation, absorbed by a material, are responsible for radiation damage and health hazards. However, the biological damage caused by different ionizing particles and radiation is not the same on all organs and tissues. The effective dose is the product of the quality factor, *Q* (relative biological effectiveness in different tissues) (Table 13.2), and the absorbed dose, *Gy*. Dose equivalent (*H*) is the unit of biologically effective dose. The SI unit of dose equivalent, *H*, is *Sievert* (Sv).

$$H = D \cdot Q \qquad ...(13.12)$$

$$Sv = Gy \cdot Q$$

Taking into account the tissues weighting factors also, the effective dose, $E$, is

$$E = \Sigma w_T \cdot H_T \cdot Sv \qquad \ldots(13.13)$$

where, $H_T$ = equivalent dose in a tissue;
$w_T$ = tissue weighting factor; $\Sigma w_T = 1$

Thus, the total dose received due to intake is

$$E = \int E(t) \cdot \exp(-\lambda_{eff})\, dt \qquad \ldots(13.14)$$

**TABLE 13.2:** WEIGHTING FACTORS FOR IONIZING RADIATION

| *Ionizing Radiation* | *Weighting Factor (Q)* |
|---|---|
| β-rays, γ-rays and *X*-rays | 1-2 |
| Protons, fast neutrons | 10 |
| α-particles | 20 |

### 13.2.4 Linear Energy Transfer (LET)

X-rays (and γ-rays) are high-energy photons and they can penetrate deep into the tissues/organs. They, being uncharged, do not directly cause damage to the tissues. However, as they are ionizing radiations, they produce secondary electrons (δ-electrons) of sufficient energy in their path. These secondary electrons, being charged particles, interact with matter, thus creating a cascade process of ionization.

The biological effects of ionizing radiation are due to interaction of primary electrons (β-rays), secondary electrons (δ-electrons), α-particles, protons and neutrons with biological tissues/organs. The radiation damage to biological matter does not depend on the penetration power of the radiation, but on its ionizing power. For example, the penetration power of α-particles is less than that of β-rays and X-rays. α-particles can be stopped by a thin sheet of tissue. However, being heavy nuclei, they possess very high ionizing potential and once embedded in a tissue can cause lot of damage within a short distance. The Linear Energy Transfer (LET) is defined as the energy deposited per unit length of track ($dE/dx$) with respect to X-rays (normalized to 1) (Table 13.3). It gives an estimate of the effective ionization (damage) potential to the biological specimen.

**TABLE 13.3:** LINEAR ENERGY TRANSFER (LET) WITH RESPECT TO X-RAYS

| *Radiation* | *Dose Equivalent (H)* | *LET* |
|---|---|---|
| γ-rays | 0.7 | 0.3 |
| β-rays | 1.0 | 0.4 |
| *X*-rays | 1.0 | 3.0 |
| Protons | 2 | 15 |
| Neutrons | ~10 | 20 |
| α-particles | 10 | ~100 |

## 13.3 RADIATION EXPOSURE

Man is living in a sea of radioactivity but of a low intensity. There is no escape from exposure as human body too carries a suite of radionuclides. Traces of primordial and cosmogenic nuclides are part of all living beings. They are irradiating the body cells from within. The important radionuclides present in the human body are listed in Table 13.4. Also food we consume (Table 13.5) and environment we live in (Table 13.6) contain radioactive materials.

**TABLE 13.4:** RADIONUCLIDES IN A HUMAN BODY (70 KG)

| *Nuclide* | *Total Mass Founding the Body (μg)* | *Total Activity of Nuclide (Bq)* | *Daily Intake of Nuclide (μg)* |
|---|---|---|---|
| Carbon-14 | 95 | 15,000 | 1.8 |
| Uranium | 90 | 1.1 | 1.9 |
| Thorium | 30 | 0.11 | 3 |
| Potassium-40 | 17 | 4,400 | 0.4 |
| Radium | $30 \times 10^{-6}$ | 1.1 | $2.3 \times 10^{-6}$ |
| Polonium | $0.2 \times 10^{-6}$ | 37 | --- |
| Tritium | $0.06 \times 10^{-6}$ | 23 | $0.003 \times 10^{-5}$ |

**TABLE 13.5:** NATURAL RADIOACTIVE ISOTOPES IN COMMON FOODS

| *Food* | *K-40 (Bq/kg)* | *Ra-226 (Bq/kg)* |
|---|---|---|
| Red meat | 111 | 0.019 |
| Carrot | 126 | 0.022 → 0.08 |
| Potatoes | 126 | 0.037 → 0.093 |
| Banana | 130 | 0.04 |
| Lima Beans | 170 | 0.075 → 0.185 |

**TABLE 13.6:** SOME IMPORTANT RADIONUCLIDES IN OCEAN WATERS

| *Nuclide* | *Activity (Bq/L)* |
|---|---|
| Tritium (H-3) | 0.001 |
| Carbon-14 (C-14) | 0.005 |
| Potassium 40 (K-40) | 10 |
| Rubidium-87 (Rb-87) | 1 |
| Uranium | 0.04 |

### 13.3.1 Natural Radioactive Hazards

Radioactivity in nature comes from two sources, terrestrial and cosmic. Terrestrial radioisotopes that came into existence with the creation of the planet are now distributed in rock, soil, water and air. They find their way in our bodies through food, water and air.

From the point of view of potential ionizing radiation exposure, naturally occurring radioactive substances radium (Ra-226) and its daughter products are of special concern. Radium is an α-emitter and decays to radon ($^{222}Rn$). Radium is chemically similar to calcium, and is absorbed from the soil by plants and is transported through food chain to animals and humans. In the body, 70-90 per cent of radium is contained in bone.

The largest contributor to our daily exposure of radiation is radon ($^{222}Rn$, $^{220}Rn$) with its progeny, which are from the decay series of uranium and thorium. Radon is a colorless, naturally occurring, gaseous radionuclide. It is normally trapped in the rocks but in mining and other such operations, in which the Earth's crust is disturbed, it escapes into the air. From the soil and water too it is continuously escaping. From building materials and through cracks in the building foundations it enters the buildings and work places. Since the uranium distribution in the Earth's crust varies from place to place, so does the prevalence of radon. Radon is an α-emitter.

$$^{226}_{88}Ra \rightarrow ^{222}_{86}Rn + \alpha \quad \ldots(13.15)$$

Radon, being an inert gas, by itself poses very little biological threat. Though it has very short half-life (~3.8 days), the radiation damage from radon gas is due to its decay products (Fig. 13.2) and α-rays. On inhalation of radon, the decay products ($Po^{218}$; $Po^{214}$; $Bi^{214}$; $Pb^{214}$) are absorbed in the trachea-bronchial tissues, leading to damage to lung tissues and bronchial cancer.

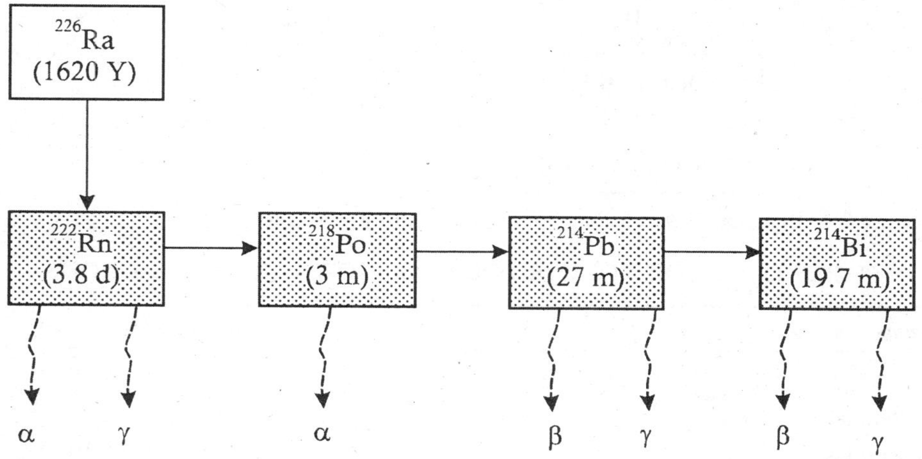


**Fig. 13.2:** Radon and its Progeny

Other sources of radiation are the cosmic rays. The intensity of cosmic radiation varies with latitude and increases with altitude. The radioactive isotopes of carbon ($^{14}C$) and hydrogen ($^{3}H$), which are among the many nuclei created by the interaction of cosmic rays with matter in the atmosphere. Radioactive isotope ($^{14}C$) is produced continuously in the upper atmosphere from $^{14}N$ as a result of the fast neutron flux from cosmic ray interactions and is oxidized $^{14}CO_2$ (Fig. 13.3). $^{14}C$ and $^{3}H$ readily exchange with stable isotopes of carbon and hydrogen, respectively. They occur in various compounds on the globe and become part of hydrosphere, plant and animal kingdom, and therefore, their containment is difficult.

### 13.3.2 Man-made Radioactive Hazards

In addition to natural background radiation, people are exposed to radiation from man-made sources, the largest and oldest of which is the application of X-ray in medical diagnosis. Nearly 90 per cent of human radiation exposure is due to X-rays for diagnostic radiography. In the developed world this diagnostic tool, in various forms, is so frequently used that the average annual dose per capita is not insignificant compared to the natural background radiation (Table 13.7).

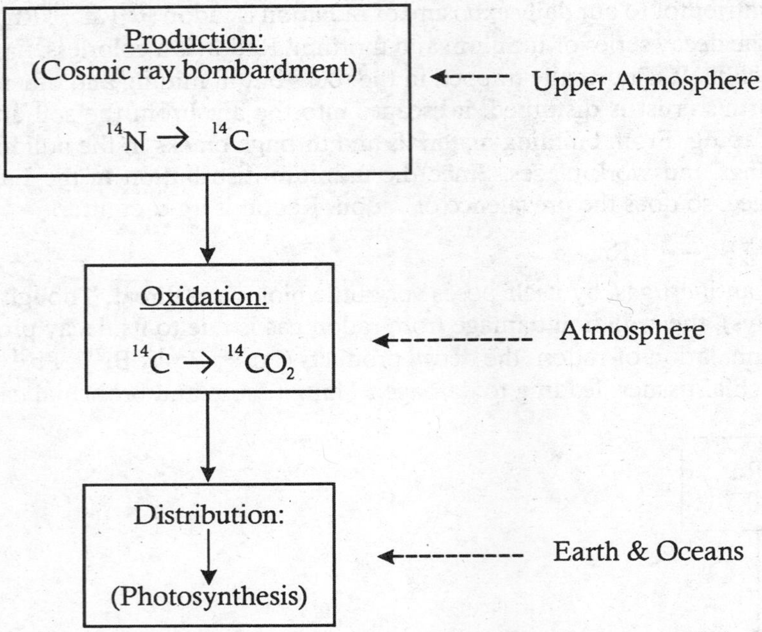


**Fig. 13.3:** Production and Distribution of Radioisotope $^{14}C$

**TABLE 13.7: DOSES RECEIVED DURING ROUTINE X-RAY DIAGNOSIS**

| *Examination* | *Dose per Exposure (mGy)* |
|---|---|
| Chest | 0.4-10 |
| Abdominal | 10 |
| Extremities | 2.5-10 |
| Fluoroscopy | 100-200 per minute |
| CAT Scan | 50-100 per examination |

Less conspicuous but not negligible is the contribution from radioactive minerals in crushed rock, building materials, phosphate fertilizers (uranium in phosphate rock). Consumer items, which add to radiation, are the smoke screens, radiation emitting components of TV sets and various other products.

### *13.3.2.1 Nuclear Technology*

The growth of nuclear technology in the second half of the 20th century has opened up a host of operations involving low to very high levels of radioactivity. The initial stage of this field is the mining of uranium ores. The steps that follow are chemical processing of the ores, purification of uranium, fabrication of nuclear fuel, operation of reactors for generating electric power, discharge and reprocessing of the fuel carrying a high dose of fission products. Each step has its own radiation problems. In addition to electric power the reactor operation also gives rise to plutonium and other transuranic elements, which are truly man-made.

A problem, which is particular to uranium mining and which is not so severe with other ore mining operations is the release of radon gas from the mined rock that is crushed and ground before chemical processing. After the uranium is recovered at the milling stage most of the ore body, with virtually all the radioactive elements (radium and associated nuclides) ends up in the tailings dam. Hence, radiation levels and radon emissions from tailings are potentially significant. For this reason, the tailings are often covered with water while they are accumulating.

Uranium and plutonium are radionuclides that are of public concern, as they are used in both nuclear fuel cycle and nuclear weapons production. Uranium contamination in food products comes from the soils. The separation and handling of plutonium is an activity beset with special problems due to its extreme toxicity. While uranium is chemical toxin, plutonium is a radiological toxin. It is genotoxic, damaging chromosomes, causing cell death and cancer. Both Pu-238 and Pu-239 are emitters of α-particles. Pu-241 is β-emitter with a short half-life (14 years), decaying to Am-241, which has a longer half-life (430 years) is an α-emitter. Some of the hazards involved in this field are summarized in Table 13.8.

**TABLE 13.8:** RADIATION HAZARDS IN NUCLEAR TECHNOLOGY

| *Operation* | *Radioactive Isotopes* | *Hazards* |
|---|---|---|
| Mining & milling of uranium ores | $^{238}U$; $^{232}Th$; $^{228}Th$; $^{228}Ra$; $^{226}Ra$; $^{222}Rn$, $^{220}Rn$ and their short-lived progeny | Inhalation of radioactive dust and gaseous $^{222}Rn$, $^{220}Rn$ and their progeny released in the atmosphere |
| Nuclear fuel fabrication | $^{238}U$; $^{234}Th$; $^{239}Pu$; $^{230}Th$ | Inhalation of radioactive dust |
| Reactor operations | $^{3}H$ & $^{41}Ar$ | Inhalation of $^{3}H$ & $^{41}Ar$. |
| Nuclear fuel fabrication and reprocessing | Fission products—$^{137}Cs$; $^{131}I$; $^{90}Sr$; $^{85}Kr$ and Pu, etc. | Inhalation and external exposure |
| Reactor-based isotope production | Radiopharmaceuticals | Radioactive waste products; inhalation and ingestion |
| Nuclear waste management | Fission products; $^{137}Cs$, $^{90}Sr$ and other fission materials | Potential ecological hazard & public health problem |
| Nuclear reactor & weapons management | Fission materials | Reactor meltdown and accidents are potential ecological disasters (e.g. Chernobyl) |

The nuclear reactor is the key radioactive waste generator in the nuclear fuel cycle as it produces a variety of radioactive fission products.

$$^{235}_{92}U + ^{1}_{0}n \rightarrow ^{140}_{54}Xe + ^{94}_{38}Sr + 2\,^{1}_{0}n + 200 \text{ MeV} \quad \ldots(13.16)$$

$$^{235}_{92}U + ^{1}_{0}n \rightarrow ^{139}_{57}La + ^{95}_{42}Mo + 205 \text{ MeV}$$

It is a device designed to carry out a controlled nuclear-fission chain reaction so that the energy released can be harnessed to generate electricity. The fuel is generally enriched uranium (~3 per cent $^{235}U$). The type and mode of using the coolant designates the reactor. Light-water reactors ($LWR_S$) use ordinary water as the moderator and coolant and heavy-water reactors ($HWR_S$) use heavy water for this purpose. In a *PWR* high pressure and high temperature water removes heat from the core of the reactor and then passes through a steam generator. Here, steam is raised which runs the turbine.

A breeder reactor produces more fissionable material than it consumes to generate energy. Using enriched uranium or plutonium as reactor fuel, the neutron economy in the reactor can be such that more fissionable $^{239}Pu$ is produced from the non-fissionable $^{238}U$ than the equivalent original fuel consumed. Excess neutrons from the fission of $^{239}Pu$ can also be used in breeder reactors to convert non-fissionable $^{232}Th$ into fissionable $^{233}U$.

$$^{238}_{92}U + ^{1}_{0}n \rightarrow ^{239}_{92}U \text{ (24 mts.)} \rightarrow ^{239}_{93}Np + ^{0}_{-1}e \qquad \text{...(13.17)}$$

$$^{239}_{93}Np \rightarrow ^{239}_{94}Pu \text{ (2.33 days)} + ^{0}_{-1}e$$

$$^{232}_{90}Th + ^{1}_{0}n \rightarrow ^{232}_{90}Th \xrightarrow{\beta^-} ^{233}_{91}U \xrightarrow{\beta^-} ^{233}_{92}U \qquad \text{...(13.18)}$$

Though most of the radioactivity produced in nuclear reactor-associated operations is safely contained, a small percentage escapes through stack gases and liquid effluents. Moreover, extreme toxicity and long half-life of radioactive wastes containing plutonium present a problem in handling, transport and storage. Their permanent disposal is a much-debated issue and some accidental leakages are inevitable.

The production of reactor-based radioisotopes for medical and other applications is another area that gives rise to minor leakage and generates some waste to be disposed. Radioisotopes released into the atmosphere from nuclear weapon tests contain C-14 (70 per cent), Cs-137 (13 per cent), Sr-90 (3 per cent), Zr-95 (~3 per cent), Mn-54 (2 per cent), I-131 (2 per cent) and H-3 (1 per cent). Of the radioactive pollutants released in the atmosphere, those with longer half-life will persist over longer periods and pose health hazards. Some of the radioisotopes and their half-lives are listed in Table 13.9. In all operating plants maximum permissible levels of radiation are prescribed.

**TABLE 13.9:** SOME RADIOISOTOPES AND THEIR HALF-LIVES

| *Isotope* | *Half-life* | *Isotope* | *Half-life* |
|---|---|---|---|
| $^{26}Al$ | $7.4 \times 10^5$ yrs | $^{140}Ba$ | 12.8 days |
| $^{7}Be$ | $1.45 \times 10^{-1}$ yrs | $^{14}C$ | $5.73 \times 10^3$ yrs |
| $^{144}Ce$ | 290 days | $^{36}Cl$ | $3.1 \times 10^5$ yrs |
| $^{58}Co$ | 71 days | $^{60}Co$ | 5.25 yrs |
| $^{144}Ce$ | 290 days | $^{134}Cs$ | 2.3 yrs |
| $^{136}Cs$ | 13 days | $^{137}Cs$ | 28 yrs |
| $^{3}H$ | 12.3 yrs | $^{129}I$ | $17 \times 10^6$ yrs |
| $^{131}I$ | 8 days | $^{40}K$ | $1.26 \times 10^9$ yrs |
| $^{140}La$ | 40.5 days | $^{54}Mn$ | 300 days |
| $^{22}Na$ | 2.6 yrs | $^{32}P$ | $3.92 \times 10^{-2}$ yrs |
| $^{239}Pu$ | $2.46 \times 10^4$ yrs | $^{226}Ra$ | $16 \times 10^3$ yrs |
| $^{287}Rb$ | $4.8 \times 10^{10}$ yrs | $^{222}Rn$ | 3.8 days |
| $^{37}S$ | $2.38 \times 10^{-1}$ yrs | $^{90}Sr$ | 27.7 yrs |
| $^{123}Te$ | $1.2 \times 10^{13}$ yrs | $^{132}Te$ | 78 hrs |
| $^{232}Th$ | $1.4 \times 10^{10}$ yrs | $^{134}Th$ | 24 days |
| $^{235}U$ | $7.04 \times 10^8$ yrs | $^{2238}U$ | $4.47 \times 109$ yrs |
| $^{50}V$ | $6.0 \times 10^{15}$ yrs | $^{91}Y$ | 57.5 days |

The question of 'more than normal' background radiation levels around nuclear power plants and the activity released through gas stacks have attracted the attention of many environmentalists. The activity from stack emissions cannot be altogether eliminated by substituting coal as fuel for power generation, instead of uranium. Many sources of coal contain natural radioactivity due to the presence of uranium and daughter products, which find their way out through stack gases along with some toxic chemicals. An average coal burning power plant is a greater source of air borne radioactive material than a nuclear power plant, if we compare them on the basis of unit power they generate.

### 13.3.3 Biological Effects of Radiation

Radiation effects have been studied on many human groups exposed to radiation by accident, warfare (e.g. nuclear bombing of Hiroshima and Nagasaki), and for medical purposes. From the point of view of cell damage α-radiation is most dangerous, followed by β and last by γ. But from the point of view of approach to the target, materials emitting γ-rays are the most dangerous.

Ionization of water is the most important reaction that takes place when a radioactive particle enters a living cell. Hydroxyl radical (OH) and hydrogen peroxide ($H_2O_2$) are among the many products formed. The latter is highly toxic to biological molecules in the cell. Massive radiation exposure (1-2 Sv) results in cell death. However, $H_2O_2$ can damage the cell machinery, enzymes and the like, or damage the DNA itself. If the damage level is beyond the normal repairing ability of the internal system incorrect replication of DNA takes place.

If the genes on the DNA, responsible for turning off cell division, are damaged, the cell and its off spring replicate in an uncontrollable way and such a situation is 'cancer'. The frequency of a given gene mutation increases in proportion to the dose of irradiation in the low-intermediate dose range. Chromosomal damage by irradiation can break both strands of the DNA molecule, leading to rupture of chromosome fiber. This can interfere with the normal segregation of duplicate sets of chromosomes to daughter cells at the time of cell division.

Different organs and tissues in the body respond differently to irradiation dose. The injury is apparent soonest in those organs like skin, lining of the gastrointestinal tract and bone marrow, which proliferate rapidly. Cancer induction is the most important somatic effect of radiation at dose levels <1 Gy. Studies on severely exposed persons showed dose-dependent increase in the incidence of certain types of cancer. The induced cancers have not appeared until years after exposure and at that stage it becomes, at times, difficult to trace the culprit causing the disease. For doses less than 0.015 Sv incidence of cancer has not significantly increased. There is some indication that less than 3 per cent of all cancers in the general population are the result of natural or background radiation.

Radiation in any dose can be dangerous because it has the inherent ability to damage living cells. However, the biological effects of exposure to radioactivity can be classified under two types: (*i*) deterministic and (*ii*) stochastic (random). Deterministic effects arise out of exposure to relatively high radiation doses and they include eye lens cataract, organ damage and skin burn. Stochastic effects occur under low to moderate doses of radiation. When the damage is minor the internal mechanisms may be able to carry out the cell repair.

Attacked cells can be somatic (they do not survive the life span of the individual) or germinal (they transmit genetic information). At low-level doses, somatic and genetic effects are stochastic (probabilistic) processes, due to mutagenic effect of ionizing radiation. Increasing the radiation dose increases the probability of the effect. A major impact upon health and fitness is from heterozygous mutants. Somatic effects are due to acute and heavy doses received over relatively short periods of time. Some of the somatic effects are radiation sickness, hair loss, nausea and systemic breakdown. Genetic (chromic) effects result from long-term exposures, due to mutation rate of the genes. For genetic effects: (*i*) there is no threshold dose for deleterious effects, (*ii*) initiation step may be effected by radiation or a carcinogen, (*iii*) once damage is initiated, the propagation is largely irreversible and (*iv*) the effects are cumulative. General biological effects of radiation are given in Fig. 13.4.

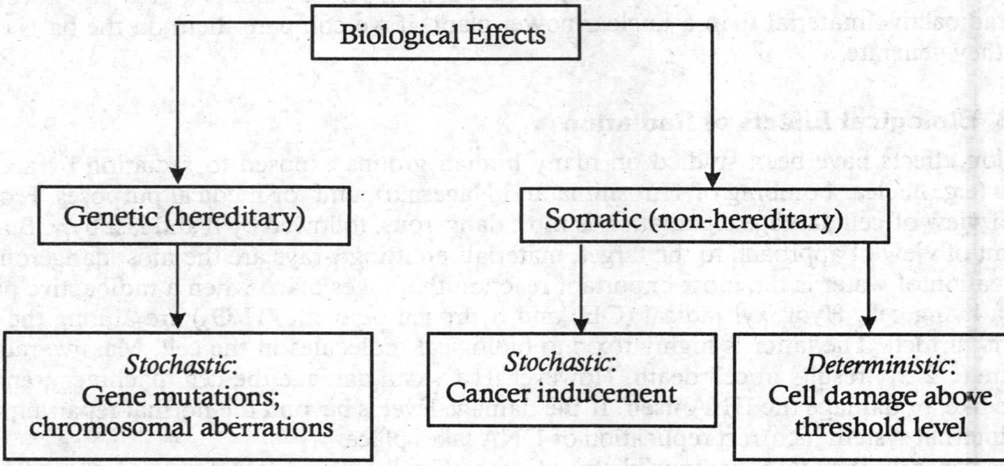


**Fig. 13.4:** Schematic Representation of Biological Effects of Radiation

### 13.3.4 Unusual Exposure Situations

In spite of the general situation that the background radiation all over the world is far from being dangerous, being approximately 350 mrem, there are situations in which radiation is beyond these values. Such situations need timely review and if possible, remedied. Some of these are radiation fallout from nuclear explosions, tailings dam bursts and reactor accidents.

The primary source of radioactivity in a nuclear explosion is the fission reaction. While the short-lived isotopes are responsible for the intense activity released soon after the explosion, the long-lived ones release energy over long periods of time, creating radiation that is much less intense but more persistent. The isotopes that are of special concern are I (131), Sr (89 and 90) and Cs (137) due to their relatively high abundance and special biological affinity. Carbon (14) and tritium are also products of nuclear fallout and are important for possible genetic injury. They are generated in the atmosphere by the action of neutrons from fission-fusion reaction. Tritium is an important product of the fusion reaction. Some information about them is provided in Table 13.10.

**TABLE 13.10:** IMPORTANT RADIOISOTOPES IN THE FALLOUT

| *Isotope* | *Radiation Emitted* | *Half-Life* | *Yield Bq/kt* | *Effects* |
|---|---|---|---|---|
| I(131) | β- and γ | 8.05 d | $6.1 \times 10^{15}$ | Most dangerous in the weeks immediately after explosion (due to low $t_{1/2}$). Readily absorbed by the body and concentrated in the thyroid |
| Sr(89) | β- and γ | 52 d | $1.4 \times 10^{15}$ | Absorbed fairly well and stored in bone |
| Sr(90) | β- | 28.1 yr | $7 \times 10^{12}$ | Sr(90) hazardous for centuries |
| Cs(137) | β- and γ | 30 yr | $7.4 \times 10^{12}$ | Long term γ-emitter; hazardous for centuries |

After recovery of uranium from the ore the mill tailings are normally dumped as sludge in special ponds and piles. In the extraction process the long-lived radionuclides like, Th (232) and Ra (226) are not removed and hence, they go with the tailings. A part of uranium (about 10 per cent) of the original present in the ore also remains with them. In addition, depending on the nature of the ore, trace metals like arsenic, molybdenum and manganese remain in undissolved or partly dissolved condition. The common hazards expected from tailing dumps are: (*i*) radon diffusion, (*ii*) dust carried by air blowing over the dumps and (*iii*) seepage of solutions carrying traces of radioactive and non-radioactive metals, contaminating surface and ground water. In spite of the understanding that dam construction should meet the criterion of integrity for hundreds of years, failure of dams has occurred.

One of the power reactors (Unit-2) in Three Mile Island, USA, met with an accident in March 1979. This is considered the most serious accident in the history of American nuclear power reactor industry. The worst accident to any nuclear reactor, occurred to one of the four 1,000 Mwe units in Chernobyl, Soviet Union, in April 1986. Here again, the man-made errors dominated the list of causes. Millions of acres of forest and farmland were contaminated. In the long run several hundred radiation-based illnesses are expected.

## BIBLIOGRAPHY

Koch-Steindl, H. and Pröhl, G. (2001), *Radiat. & Environ. Biophys.*, **40**(2): 93. Considerations on the behavior of long-lived radionuclides in the soil: A review.

Krishnamoorthy, T. M. and Nair, R. N. (1999), *IANCAS Bulletin*, **15**(3): 25. "Environmental radiation exposure: Natural and man-made."

Mettler, J. R., F. A. and Upton, A. C. (1995), Saunders: Philadelphia, PA. "*Medical Effects of Ionizing Radiation,*" (2nd edn).

Narayanan, P. (2001), Bhalani Pubs: Mumbai. "*Clinical Biophysics: Principles and Techniques.*"

Narayanan, P. (2003), New Age Intl. Pubs: New Delhi. "*Essentials of Biophysics,*" (2nd Print).

Peirce, J. J., Weiner, R. F. and Vesiland, P. A. (1998), Butterworth-Heineman: New Delhi. "*Environmental Pollution and Control.*"

Togyessy, J. and Klehr, E. H. (1987), Wiley & Sons: New York. "*Nuclear Environmental Chemical Analysis.*"

United Nations Scientific Committee on the Effects of Atomic Radiation (UNSCEAR) (1993), United Nations: Vienna. "*Sources and Effects of Ionizing Radiation.*"

... (1990), Office of SRMS/Natl. Inst. of Standards and Technology: Gaithersburg, MD. "*Standard Materials for Environmental Research, Analysis and Control.*"

# SECTION-IV

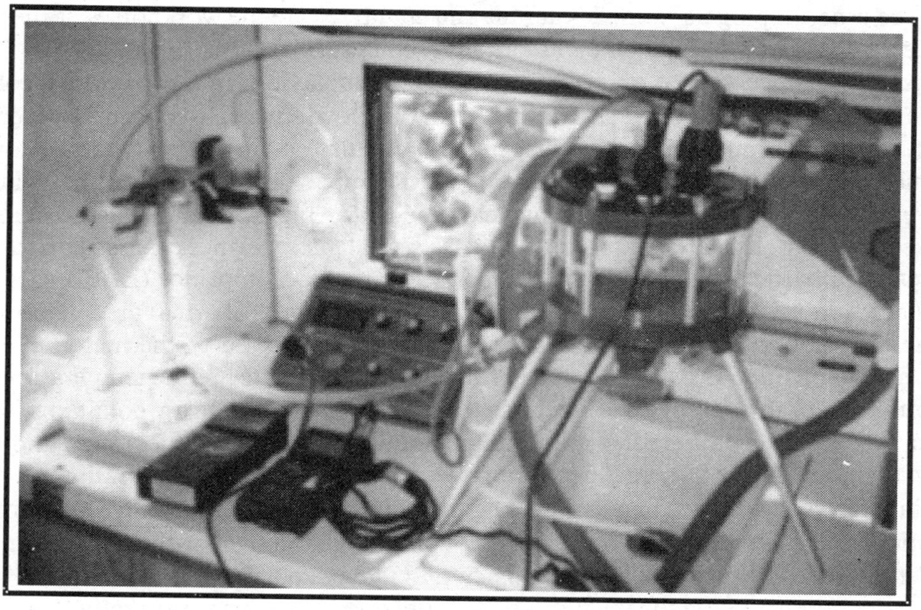

# Collection and Sampling

## OVERVIEW

Chapters 14 and 15 deal with the procedures for collection of samples, handling, preparation, and sample analysis.

## SAMPLE COLLECTION METHODS

Collection methods of pollutants depend on the physical and chemical properties of the pollutants. General methods of collection include impingement (filtration), sorption, thermal and electrostatic precipitation and freeze-out. Impactors, filters, cyclones and wet-scrubbers or combination of these devices can be used. Impingers, wetted-columns and cold-traps are the workhorses of the collection devices. Electrostatic and thermal precipitators are used in special cases.

There are several collectors of air pollutants based on impaction. In these devices, collection is based upon particles not following streamline paths. High-volume (Hi-vol) air sampler operates like a vacuum cleaner. It measures total suspended particulate matter. Low-volume (Low-vol) air sampler is used before analysis of metals (fumes). Inertial impactors are most efficient for collection of large size particles (>10 μm). Cascade impactors, such as Anderson multi-stage samplers can be used to collect different size particulate matter. Quartz crystal microbalance cascade impactor (QCM) is a very sensitive particle size separator. Electrical low-pressure impactor (ELPI) uses electrical mobility of aerosols to obtain particle size distribution (2-0.01 μm). Thermal precipitators separate particle size based on the principle that an aerosol particle subjected to thermal gradient moves from warmer to cooler regions, with a velocity (therefore mass) related to the temperature gradient.

Filtration is one of the widely used methods for collecting particulate air pollutants. Collection efficiency depends on the particle size. Filtration methods depend on sedimentation and diffusion. Gravitational chambers are simple devices to collect large size particles. Filtration of pollutants through fabric filters (dust bags) involves passing the polluted air through a filter medium made of porous fabric. Filter bags can be so arranged that continuous removal of the collected material is possible. Fabric filtration provides high efficiency (99%) dust collection.

Centrifugal separators (cyclones) provide an effective and economic method of screening particulate pollutants. Polluted air is passed through a spiraling air. The centrifugal force pushes the heavy particles against the walls of the cyclone chamber and the particulate pollutants slide downward, while purified air exits at the top outlet.

Sorption is the method of choice for collection of gases. Activated carbon, alumina and silica gel are commonly used adsorbents. Soluble gases ($SO_2$, $H_2S$, $NH_3$, HCl and Cl) are collected by absorption by wet-collectors (Venturi scrubber). In these devices, soluble gases are transferred to liquid phase and the solution is flushed out. Freeze-out is the method of choice for collection of volatile hydrocarbons.

## SAMPLE PREPARATION METHODS

While collecting and handling samples, ambient conditions have to be maintained and there should be minimal external interference. Collected samples should be transported in proper containers to avoid any contamination and preserved and stored till analysis as per the standard procedures.

Sampling is testing of a minute quantity of a substance to extrapolate the results to the whole situation. Sample preparation methods include separation, concentration and pre-concentration prior to analysis. As environmental pollutants are in trace quantities, pre-concentration is invariably necessary. Pre-concentration of pollutants in liquids is necessary to enrich the trace level pollutants and minimize matrix effects. Pre-concentration with complete separation of the matrix proves to be useful in the analysis of toxic, radioactive and explosive materials.

Evaporation (distillation), sublimation (volatilization), precipitation, co-precipitation, solvent extraction, filtration, dialysis, centrifugation, sorption and electrolytic methods are the generally employed methods to achieve separation, concentration and pre-concentration of environmental pollutants.

Evaporation (single-step) and rectification (multi-step) are routinely employed methods for separation and concentration of volatile compounds, petrochemicals and isotopes. Sublimation is volatilization where solid substances are evaporated on heating directly into gaseous phase.

Concentration by precipitation is based on the solubility differences of solutes in the solvent mixtures. Co-precipitation is a method that ensures a high degree of concentrations of precipitants. Co-precipitation by chelate formation is a well-established technique for pre-concentration of trace elements. Solvent extraction (e.g. Soxhlet method) is probably the most extensively employed method for separation and pre-concentration. The method is universal, simple, rapid and efficient.

Filtration is a simple and effective method for concentration of particulate matter. Ultrafiltration is commonly employed for separation and concentration of biological molecules.

Sorption methods—absorption, adsorption and chemisorption—are widely used pre-concentration methods (e.g. chromatographic methods).

Electrolytic methods of pre-concentration are electrodeposition, electrodialysis, electroosmosis and electrophoresis. Electrodeposition is a part of stripping voltammetry. Electrophoresis is a separation method, extensively used in separation and characterization of biological species.

Mineralization (oxidation of organic species) is the process to obtain concentrates of impurities when substances of organic and biological species are analyzed. Dry-mineralization (ashing and pyrolysis) and wet-mineralization (oxidative decomposition with mineral acids) are the methods employed.

Practically, all concentration methods can be combined with spectrophotometry (UV-visible, IR and Raman), and fluorimetry (fluorescence and phosphorescence). Sorption methods with suitable detectors are generally employed in chromatography, mass spectrometry (MS) and hybrid (GC-MS) methods.

## Soil Pollutants

Sampling of soil pollutants often involves two-stage pre-concentration—oxidation of organic compounds by dry- or wet-mineralization followed by concentration by another suitable method (e.g. derivatization). Derivatization method is a convenient method of pre-concentration, used to separate trace elements from their matrices and then pre-concentrate them by hydride generation (e.g. As, Bi, Sb, Se and Sn), ethylation and phenylation.

## Liquid Pollutants

Mass flow and molecular diffusion (dialysis) methods may be employed to extract pollutants in liquid phase. For volatile compounds, polarity and volatility dictate the sampling techniques. Trap-and-purge methods are most commonly used for volatile compounds, and solvent extraction methods for less-volatile compounds. Cryogenic pre-concentration method is efficient, but is not selective.

### Gaseous Pollutants

Particulate matter in gaseous pollutants, after collection, is dissolved in acids and filtration is the standard method employed. Reactive gases must be converted into stable derivatives during the collection step. Diffusion denuders are used for sampling inorganic gases, such as $O_3$, $H_2O_2$, $NH_3$, $NO_x$, and $SO_x$. The selection of the sampling method for organic contaminants relies on the vapor pressure, polarity and reactivity. Adsorption, sorption-trap and solvent extraction methods are generally employed for pre-concentration. Volatile organic compounds (VOCs) are pre-concentrated on sorbents such as activated carbon, alumina, silica gel and molecular sieves. Semi-volatile organic compounds (SVOCs) are pre-concentrated on solid sorbents (polyurethane foam) and by solvent extraction.

# 14

# Sample Collection

Collection methods of pollutants depend on the physical and chemical properties of the pollutants. General methods of collection include impingement (filtration), sorption, thermal and electrostatic precipitation and freeze-out. Impactors, filters, cyclones and wet-scrubbers or combination of these devices can be used. Impingers, wetted-columns and cold-traps are the workhorses of the collection devices. Electrostatic and thermal precipitators are used in special cases.

There are several collectors of air pollutants based on impaction. In these devices, collection is based upon particles not following streamline paths. Inertial impactors are most efficient for collection of large size particles (>10 μm). Cascade impactors, such as Anderson multi-stage samplers, can be used to collect different size particulate matter. Qua-rtz crystal microbalance (QCM) cascade impactor is a very sensitive particle size separator. Electrical low-pressure impactor (ELPI) uses electrical mobility of aerosols to obtain particle size distribution (2-0.01 μm). Thermal precipitators separate particle size, based on the principle that an aerosol particle subjected to thermal gradient moves from warmer to cooler regions, with a velocity (therefore mass) related to the temperature gradient.

Filtration is one of the widely used methods for collecting particulate air pollutants. It depends on sedimentation and diffusion. Gravitational chambers are simple devices to collect large size particles. Centrifugal separators (cyclones) provide an effective and economic method of screening particulate air pollutants.

Sorption is used for the collection of gases. Activated carbon, alumina and silica gel are commonly used adsorbents. Soluble gases ($SO_2$, $H_2S$, $NH_3$, HCl and Cl) are collected by absorption by wet-collectors (Venturi scrubber). In these devices, soluble gases are transferred to liquid phase and the solution is flushed out. Freeze-out is used for the collection of volatile hydrocarbons.

This chapter deals with some of the sample collection methods.

Collection of samples and their preservation are the essential steps prior to sampling and qualitative and quantitative analysis of pollutants. Ultimately, the quality of analysis and reliability depend on the proper methods of sample collection, transportation and storage. Various collection methods are available and some are better suited for laboratory testing and evaluation, while some are well suited and employed on industrial scale for collection/disposal of pollutants. Collection methods that are applicable for air pollutants are employed with suitable modifications for collection of water and soil pollutants as well.

## 14.1 COLLECTION METHODS

For the analysis and control of air pollutants, the stepwise procedures involved are: (1) collection of air pollutants, (2) sampling and (3) determination of their properties by various physical, chemical or biological methods. Air pollutants consist of gases, aerosols and particulate matter of various sizes (Fig. 14.1).

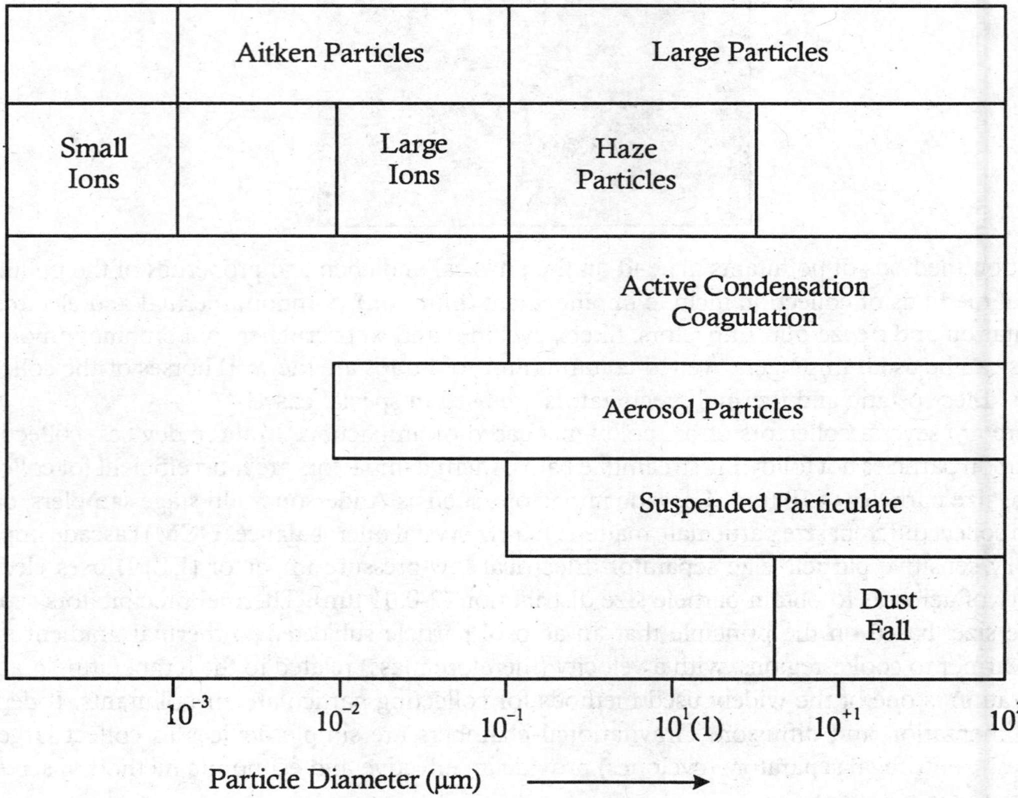


**Fig. 14.1:** Classification of Air Pollutants according to their Size

Collection procedures depend on the physical and the chemical characteristics of the pollutants. Essentially all sample collection methods are based on condensation. The process of particles coming into contact and adhering together is called *coagulation*. Coagulation can be due to gravitation (settling down upon charge neutralization) as well as due to thermal effects (*Brownian motion*).

$$\frac{dn}{dt} = -kn^2 \text{ (Thermal coagulation)} \qquad \text{...(14.1)}$$

$$n = \frac{n_0}{(1 + kn_0t)} \qquad \text{...(14.2)}$$

where, $n$ = number of particles;
$n_0$ = particles at time,
$t = 0$;
$k$ = constant.

The purpose of environmental sampling and analysis is to obtain data on pollution levels at a particular site, at a specific point of time, to see if any remedial action is called for. In this process, the collection of valid samples is the vital first step. For gases 'whole air' samples or samples, which selectively adsorb the gases of interest are collected by passing them through a chamber in which a force acts on the particulates and removes them from the gas stream (by gravitational, electrostatic, thermal, centrifugal or inertial). For analysis of trace elements by chemical and physical techniques, atmospheric aerosols are collected by means of impactors, filters, cyclones, scrubbers, electrostatic precipitators or a combination of these devices. Preference for a particular device depends on the nature of the pollutant (shape and size, density), the level of its undesirability (corrosivity), collecting efficiency, cost effectiveness of removal and other considerations.

General collection methods for gases, aerosols and particulate matter include sorption—adsorption on a solid surface, absorption in a liquid and condensation to a liquid—diffusion, freeze-out, impingement, and mechanical filtration, thermal, chemical and electrostatic precipitation methods.

Sorption is the method of choice for the collection of gases. Activated carbon, alumina and silica gel are the commonly used adsorbents—activated carbon for hydrocarbons, alumina for $SO_2$ and silica gel for polar gases. Absorption of pollutant gases is accomplished by using a selective liquid in a wet-scrubber, packed tower, or bubbler tower. Pollutant gases, commonly controlled by absorption, include $SO_2$, $H_2S$, HCl, Cl, $NH_3$, and oxides of nitrogen and low-boiling hydrocarbons. Water is the most popular solvent used in absorption devices. Water may be acidic or alkaline to enhance the removal of a specific gas pollutant. In cases where water is a poor scrubbing solvent, other solvents are used (e.g. $SO_2$ is readily soluble in alkaline solutions, such as ammonia).

There are two broad categories of aerosol sampling methods—integral and size-selective. In the integral collection method, the goal is to collect a single sample in which the sizes and types of collected particles accurately represent those in the air. In the size-selective methods, sampling is designed to provide information on particle size as well as particle amount. Among the size-selective samplers, high-volume (hi-vol) cascade impactors (e.g. Anderson impactors) and the diffusion battery produce sufficient data to generate complete particle spectra for the chemical species of interest.

Impingers, wetted-columns, cold-traps, bottles and bags are the workhorses of the collection devices. Electrostatic and thermal precipitators are used for special sampling of particulate matter. Filtration is the most widely used method for sampling airborne particles when the total ambient concentration of elements is to be determined. In order to ensure proper results, the target gaseous constituents should not react irreversibly on the surface of the filter used to collect aerosol particles.

Particulate collection depends on the particle size of the pollutants, physical and chemical characteristics, such as concentration, volume, temperature, and collection efficiency.

The sedimentation (terminal) velocity, $v$, of a particle is

$$v = \frac{2(\rho_{part} - \rho_{med}) r g \overline{V}}{D_c A_p \, Re \, \eta} \qquad ...(14.3)$$

The *Reynolds* number, $Re = \frac{v d \rho_\infty}{\eta}$ ...(14.4)

For a spherical particle, $D_c = \frac{24}{Re}$, and the sedimentation velocity is

$$v = \frac{(\rho_{part} - \rho_{med}) r^2 g}{9\eta} \qquad ...(14.5)$$

where, $\rho$ = density;
$r$ = radius of the particle;
$g$ = acceleration due to gravity;
$\overline{V}$ = particle volume;
$D_C$ = drag coefficient;
$A_p$ = particle frontal area;
$\eta$ = viscosity of the medium (air)

For non-spherical particles, the stokes-radius, $r_S$ is the radius of a sphere with the same terminal velocity and density of the non-spherical particle (equivalent aerodynamical radius).

*Particle of size* <1 μm *exhibits Brownian motion*

*Particle of size* <0.25 μm *Brownian motion dominates.*

Collection devices can be classified broadly under laboratory scale and industrial scale operations. Inertial impactors, freeze-out collectors, ultrasonic agglometers can be classified under laboratory scale collection devices, while gravitational chambers, centrifugal separators, wet scrubbers come under industrial collection devices. Filters, electrostatic precipitators are used both in laboratory and industrial scale.

## 14.2 INERTIAL IMPACTORS

Impactors are widely used in ambient air sampling. The mechanism by which particles are collected (mass-size distribution) is based upon particles (of all sizes) not following streamline paths. Cyclones, bafflers, rotating impellers operate on the principle that the aerosol material in the carrying gas stream has greater inertia than the gas. Since the drag forces on the particle are proportional to $d^2$, while the inertial forces are proportional to $d^3$, as the particle diameter increases, the inertial (removal) forces become relatively greater. Inertial impactors are most efficient for large size particles (>10 μm). Impactors collect particles from a relatively high-velocity air stream directed at a surface. The impactor consists of a nozzle through which the aerosol to be size-separated is passed. The output stream is directed against a flat impaction plate (glass, metal, plastic) that deflects the flow to form an abrupt 90-degree deflection in the streamlines. Particles with sufficient inertia (large-size particles) are unable to follow the streamline and impact on the flat plate, while smaller-size particles follow the streamline and escape out of the impactor (nozzle). Hole size in each stage progressively decreases so that the same air volume passing through each stage impinges at an increased velocity. The result is that coarse particulates are deposited on the first stage and successively finer particles are removed at each subsequent stage. Impaction surfaces are coated with an adhesive layer to avoid bouncing off previously deposited particles. Acidic and neutral particles are collected by placing a diffusion denuder upstream of the impactor to remove ammonia.

The collection efficiency is

$$S = \frac{\rho . d_p^2 v C_c}{9 \eta d_j} \qquad \text{...(14.6)}$$

The impaction efficiency, $\zeta$, is

$$\zeta = \frac{N + v \rho d_p^2}{19 \eta d_j} \qquad \text{...(14.7)}$$

where, $S$ = Stokes number;
$\rho$ = particle density;
$v$ = mean flow velocity;
$C_c$ = Milliken-Cunningham correction factor;
$\eta$ = viscosity of the fluid (air);
$d_p$, $d_j$ = particle and nozzle diameters.

To avoid bounce-off of particles, the impactor plates are coated with water, oil, grease or paraffin.

A large variety of air collectors are commercially available. They all have: (*i*) an inlet funnel for capturing airborne particulate matter, (*ii*) a filter and accessories for particulate collection, (*iii*) a gas meter for the determination of gas volumes, (*iv*) reagent bottles to collect soluble gases (e.g. $SO_2$, $CO_2$), and (*v*) a suction pump. The major factors for selection of air collectors are portability, durability and maintenance.

The dichotomous air sampler is capable of separating the particles by means of virtual impaction into two fractions, the fine fraction <2.5 μm, and the coarse fraction >2.5 μm (<10 μm). The particles are collected on Teflon membrane filters, which are ideal for gravimetric analysis of the fine, coarse and inhalable fractions as well as for X-ray fluorescence (XRF) analysis.

Cascade impingers can be employed to collect particulate of different sizes (Fig. 14.2). Cascade impactor directs the air stream against collection plates through progressively narrower orifices and enables stepwise collection of smaller size particulate matter.

Some of the cascade impactors are High- and Low-volume air samplers, Anderson multi-stage (8-stage & 14-stage) sampler, low-pressure impactor, quartz-crystal microbalance, electrical impactor and α-impactor (for radon progeny).

### 14.2.1 High-volume (Hi-vol) Air Sampler

Filtration is one of the most common techniques for sampling particulate matter. Materials collected on filters may be analyzed by chemical and physical methods. Hi-vol air sampler is a vacuum cleaner, which draws air through a filter. It consists of a filter holder, air suction assembly and a flow meter (Fig. 14.3).

Hi-vol air sampler operates like a vacuum cleaner by pumping air at high rate through a filter. The filter holder is connected to a suction assembly and airflow is measured by a flow meter. Analysis is by gravimetric method. The filter is weighed before and after the sampling period and the mass of the particulate collected is the difference between two weights.

The hi-vol sampler measures total suspended particulate matter. As the filter collects dirt during its operation, less air passes through it during later periods than in the beginning. Therefore, the airflow must be measured periodically. Fine particulate matter (<10 μm) is measured with a series of stacked filters of different pore diameter.

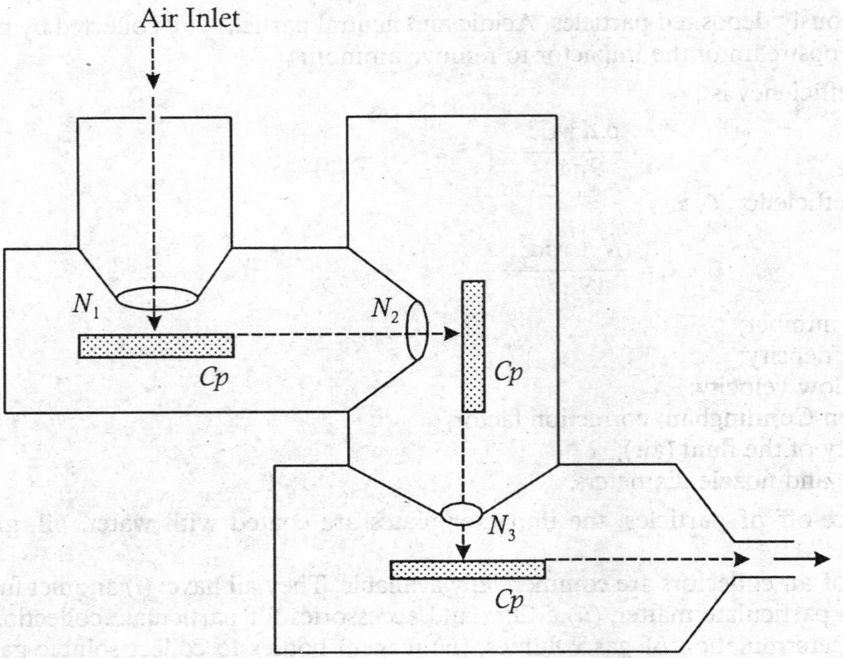


**Fig. 14.2:** Schematic of a Cascade Impactor (*Cp* = Collection plates; $N_1$, $N_2$, $N_3$... are Nozzles of decreasing orifices)

### 14.2.2 Low-volume Air Sampler

Low-volume air sampler (Fig. 14.4) is used for collection and partial analysis (particle size sampling) of metals. The sampler consists of a filter holder, flow-rate measuring device and a suction pump

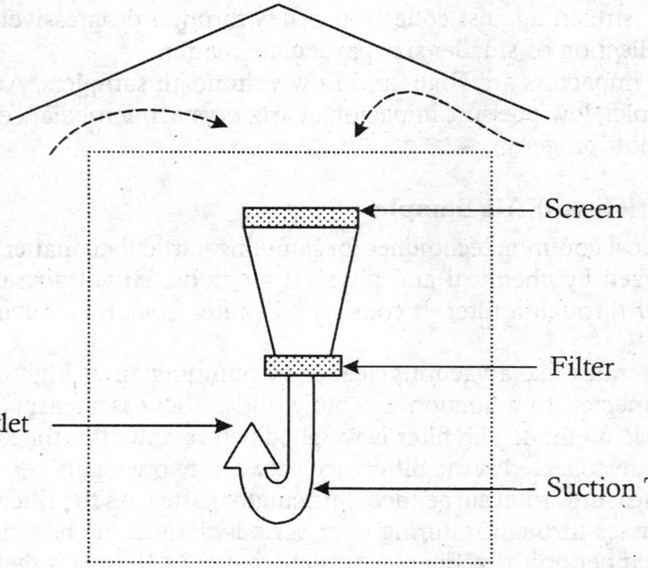


**Fig. 14.3:** (*a*) Components of Hi-vol Air Sampler

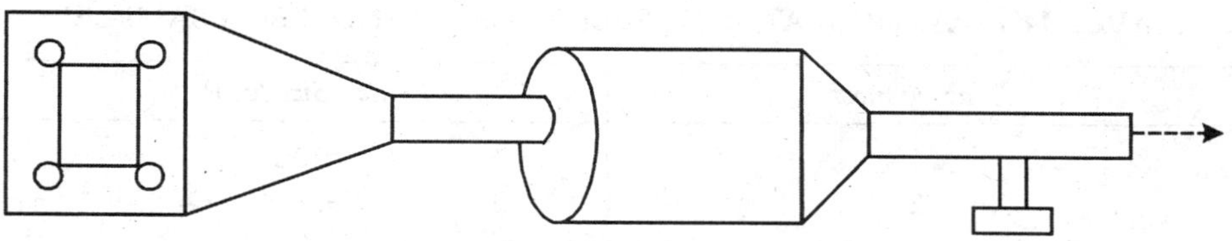

**Fig. 14.3:** (*b*) Hi-vol Air Sampler

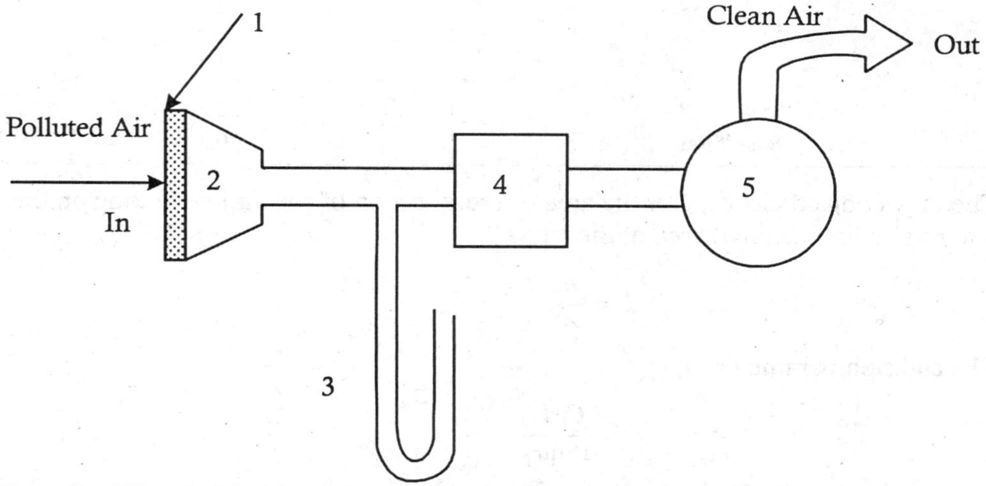


1 = Filter; 2 = Filter holder; 3 = Manometer; 4 = Flow-rate meter; 5 = Vacuum pump

**Fig. 14.4:** Schematic of Low-volume Air Sampler

Glass fiber, nitrocellulose and other membrane filters (of 1-3 μm pore sizes) are used. The concentration of the suspended particulate, *C*, is

$$C = \frac{(W_e - W_s) \times 10^3}{V} \quad \text{...(14.8)}$$

where, $W$ = weight;
$V$ = flow rate

The efficiency of glass fiber filters is 99 per cent for particulate of $d$ ~0.8 μm. Glass filters are not useful for the analysis of inorganic pollutants as metals (silicates) are leached from the glass filter. Polystyrene filters are very good for elemental analysis, but they are not useful for organic samples. So, filter selection depends on the pollutants analyzed.

*Nephelometry* is an indirect method for determining particle size (see Chapters 17 and 24). Intensity of scattered light by fine particles is measured and correlated to the concentration of particles.

### 14.2.3 Anderson Multi-stage Sampler

Anderson multi-stage (8-stage or 14-stage) impactor consists of aluminum sieve trays (8 or 14), stacked one above the other and a base filter. At the base is a collecting disk made of glass or stainless steel. The pore-size in each tray is progressively less to collect different particulate sizes (Table 14.1).

**TABLE 14.1:** PARTICULATE-SIZE SELECTION IN ANDERSON SAMPLER

| *Tray Number* | *Particle Size* (μm) |
|---|---|
| 1 | >10 |
| 2 | ~10 |
| 3 | 7 |
| 4 | 5 |
| 5 | 3 |
| 6 | 2 |
| 7 | 1 |
| 8 | 0.7 |
| Base filter | <0.5 |

The impacting efficiency, $\zeta$, is the area of cross-section of the air jet working on the particulate ($A_p$) to total cross-sectional area of the jet ($A_j$).

$$\zeta = \frac{A_p}{A_j} \quad \text{...(14.9)}$$

The collision parameter, $\psi$, is

$$\psi = \frac{Cvd_p^2}{18\eta d_j} \quad \text{...(14.10)}$$

$$d_p = \sqrt{\frac{18\psi\eta d_p}{Cvp}} \quad \text{...(14.11)}$$

where, $C$ = coefficient;
$d_p$ = particle diameter;
$d_j$ = air jet diameter;
$v$ = jet velocity;
$\eta$ = viscosity of air.
$\psi$ = 0.28 for 95 per cent efficiency. Anderson 14-stage impactor can separate particles of 30-0.08 mm range

### 14.2.4 Quartz Crystal Microbalance Cascade Impactor (QCM)

The quartz crystal microbalance (QCM) impactor can separate aerosol samples to ten sizes, in the range 25-0.05 mm. The deposited particles on the stages are monitored using *piezoelectric* quartz crystal transducers, for their mass analysis (see Chapter 16).

### 14.2.5 Radon Progeny Detectors

The α-impactor is suited for measuring radon progeny deposited on the different stages (plates) of a low-pressure impactor. Each stage has a detector to monitor α-decay isotopes ($^{218}$Po, $^{214}$Po) during collection and sampling. The α-particles from radon and radon progeny interact with the ZnS(Ag) particles in the detector to produce scintillations, which are amplified, and converted to electric signals by a phototube. It is not a size-separation device, but a radioisotope-detecting device (see also Chapter 16).

Micro-orifice uniform deposit impactor (MOUDI) is used for determining the particle size distribution of the decay products of radon and/or thoron gas (50-500 nm). The MOUDI consists of two basic assemblies—a cascade impactor (an air inlet, impactor stages and a backup filter) and a detection assembly. Graded screen array (GSA) diffusion battery is used for measuring the particle size of radon progeny <20 μm. Continuous radon monitors are available for measuring varying concentrations of radon over a long period of time.

## 14.3 ELECTROSTATIC PRECIPITATORS

Electrostatic precipitators are used both in small scale (laboratory) and large-scale (industrial) operations. Electrostatic precipitation uses the forces of an electric field on electrically charged particles to separate solid or liquid aerosols from a gas stream. The method combines electrical distribution with the size of the aerosol. Electrical mobility is utilized to obtain size distribution (0.01-2.0 μm). It is commonly used for removing fine particulate from air streams. The principle of electrostatic precipitation involves charging of the dust with ions and then collecting the ionized particles on a surface. For cleaning and disposal, the particles are then removed from the collection surface. The device consists of a cylindrical chamber with a conducting wire along the central axis of the cylinder (Fig. 14.5). A dc voltage (5-50 kV) is applied, with the central wire kept negative (or positive) with respect to the outer tube. The very high electric field gradient near the wire electrode produces a corona discharge around the wire, and air passing through the tube is ionized. Under the influence of the voltage gradient the particulate ions of opposite charge are attracted towards the cylinder walls (collector electrodes). In a negative corona, positive ions are attracted toward the negative wire electrode, and electrons are attracted toward the positive electrode. Gas molecules beyond the corona region are ionized by electron impact. Particulates deposited on the plates are removed periodically by rapping. Collection efficiency, $\zeta$, is

$$\zeta = 1 - \exp(-AV/Q) \quad \ldots(14.12)$$

where, $A$ = area of cross-section;
$V$ = voltage;
$Q$ = flow rate

Collection efficiency of the device is quite high and wide range of particle sizes can be collected (0.01-2.0 μm). The device can handle large volumes of gases and has tolerance to high temperature gases (flue gases). Electrostatic precipitators with negative coronas are commonly used in industrial applications—coal-burning electric generating plants, smelters and incinerators. Positive coronas are used for cleaning air in inhabited spaces, as negative coronas are accompanied by ozone generation.

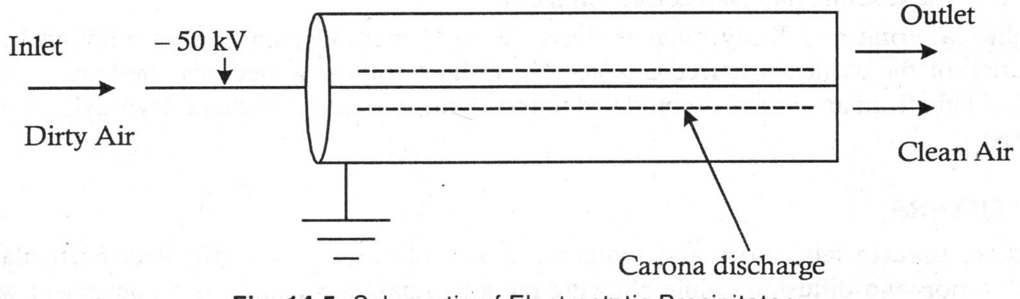


**Fig. 14.5:** Schematic of Electrostatic Precipitator

## 14.4 THERMAL PRECIPITATORS

Thermal precipitation is based on the thermophoretic effect. Particle separation is based on the principle that a particle subjected to thermal gradient moves from warmer to cooler region, with a velocity related to the temperature gradient. Polluted air is passed through two glass plates, between which is maintained a heated wire, establishing a thermal gradient. Particles collect on the cooler glass plate. Thermal precipitators are very effective for the collection of particulate matter over extremely wide range of sizes (0.005-5 μm).

### 14.4.1 Freeze-out Collection

Freeze-out collection is primarily used in collecting hydrocarbon pollutants for laboratory testing and evaluation. All pollutants may be removed from air by freezing or liquefying them in collectors maintained at a low temperature. An effective method for collecting hydrocarbon vapors for analysis is to freeze them by low temperature condensation, using a freeze-out trap. These are U-tubes packed with glass wool and immersed in a Dewar flask containing a refrigerant.

## 14.5 GRAVITATIONAL SETTLING CHAMBERS

Gravitational settling chambers (dustfall bags) are simple devices to collect large-size particulate matter, $d > 50$ μm. They are commonly used in industrial operations and serve as preliminary screening devices for more efficient control devices. Collection efficiency, $S$, is

$$S = 1 - \exp\left[\frac{v_t}{v_g}\frac{L}{H}\right] \quad \text{...(14.13)}$$

$$v_t = \frac{d^2 \rho C_c g}{18\eta} \quad \text{...(14.14)}$$

where, $v_t$ = the terminal velocity;
$v_g$ = the uniform gas velocity;
$L$ = chamber length;
$H$ = chamber height;
d = diameter of the particle;
ρ = particle density;
η = viscosity of the medium;
$g$ = acceleration due to gravity;
$C_c$ = the Cunningham slip correction factor

Light scattering and X-ray microanalysis (XRMA) methods can also be employed in the estimation of the particle size (see Chapter 25). In light scattering methods, the intensity of the scattered light from an aerosol is correlated to the diameter of the particulate (by Rayleigh or Mie analysis).

## 14.6 FILTERS

Filters are used in laboratory and industrial scale collections. Filters collect particulate by sedimentation and diffusion, while allowing the air to pass through. It is a convenient way to remove particulate matter. Removal of particulate from a gas stream by fabric filtration (dust bags) involves passing the polluted air through a filter medium made of porous fabric. Filtration medium is called *baghouse* (in industry, 'Baghouse' refers to the room where a set of bags are located). Collection mechanisms are by: (*i*) interception, (*ii*) impaction and (*iii*) diffusion. For larger particles, sedimentation and impaction procedures are efficient, while diffusion effects predominate for collection of particles in the submicron range. Fabric filtration provides high efficiency (99 per cent) dust collection of submicron size particulate. Fabric filters (baghouses) consist of large number of filter bags, arranged so that continuous removal of the collected material is possible. This device operates like the bag of an electric vacuum cleaner, passing the air and smaller particles while entrapping the larger ones. Materials collected by filters may be analyzed chemically, microscopically and by particle sizing methods.

Selection of filter materials is based on several factors: (*i*) good collection efficiency for submicron particles (99 per cent for particulate of diameter >0.3 μm), (*ii*) high particle and mass loading capacity, (*iii*) high mechanical strength, (*iv*) low flow-resistance, (*v*) low hygroscopicity, (*vi*) temperature stability, (*vii*) absence of interfering impurities, (*viii*) low cost, and (*ix*) availability in a variety of sizes and in large quantities. Filter bags are generally made out of cotton fabrics. In addition, cellulose filters (cellulose acetate & cellulose nitrate), synthetic fibers (Nylon, Rayon, Dacron, polyester, Teflon), sintered glass and other materials are also used as filter materials. Cloth fabrics are suited for temperatures <80° C, Nylon and Rayon fabrics <90° C, Teflon <200° C and glass <300° C. Teflon and glass are fabrics of choice at 200-300° C. A widely used fibrous mat filter type is the high-volume filter, referred to as hi-vol filter. High efficiency particulate attenuation (HEPA) filter is a micropore glass frit filter (combination of cellulose and mineral fibers), used to control toxic and radioactive particulates. Organic particulates are usually sampled by using fibreglass filters. Teflon-coated filter exhibits best properties on account of its low surface activity. While cellulose filters and glass filters contain impurities, silver membrane filters are 99.99 per cent pure. Characteristics of some filter materials are given Table 14.2.

**TABLE 14.2:** CHARACTERISTICS OF SOME AIRFILTER MATERIALS

| *Filter Type* | *Temp Range °C* | *Resistance to* | | |
|---|---|---|---|---|
| | | *Acids (organic)* | *Acids (inorganic)* | *Alkalis* |
| Cotton | 80-105 | G | P | G |
| Nylon | 90-120 | F | P | G |
| Polypropylene | 90-120 | E | E | E |
| Rayon | 90-120 | E | E | E |
| Dacron | 135-160 | G | G | G |
| Polyester | 135-160 | G | G | G |
| Teflon | 230-280 | E | E | E |
| Glass | 280-320 | E | E | P |

E = Excellent; G = Good; F = Fair; P = Poor.

*Absorption* through a solvent, by bubbling the polluted gas through a liquid, is a very common method for the collection of gaseous pollutants (e.g. HF and HCl). Generally, alkaline gases are retained in acidic solutions and acidic gases are collected in alkaline solutions. *Adsorption* is particularly useful for the collection of samples to be analyzed by gas chromatography (GC).

## 14.7 CENTRIFUGAL SEPARATORS

Separation (collection) by centrifugal separators (e.g. cyclones) is an effective and economical way of screening particles.

### 14.7.1 Cyclones

The process in cyclones involves passing of an aerosol through a spiraling air mass (Fig. 14.6). The separator operates with two vortexes. Dirty air is fed off-center into the cyclone in a tangential direction at the outer wall of a cylindrical device. An air vortex is formed by airflow tangentially to the vertical axis of the vortex and air flows in a helical path. Swirling air creates centrifugal force and the heavy particles, because of their greater inertia, move outward and are forced against the wall of the cyclone. Friction between particulate and cyclone walls slows down the particulates and they slide down into the collector at the bottom of the cyclone, while clean air swirls upward and exits at the top outlet. Cyclones are best at removing relatively coarse particulate (>10 μm). Cyclone collectors are generally used at the first stage (precleaners) in industrial dust collection.

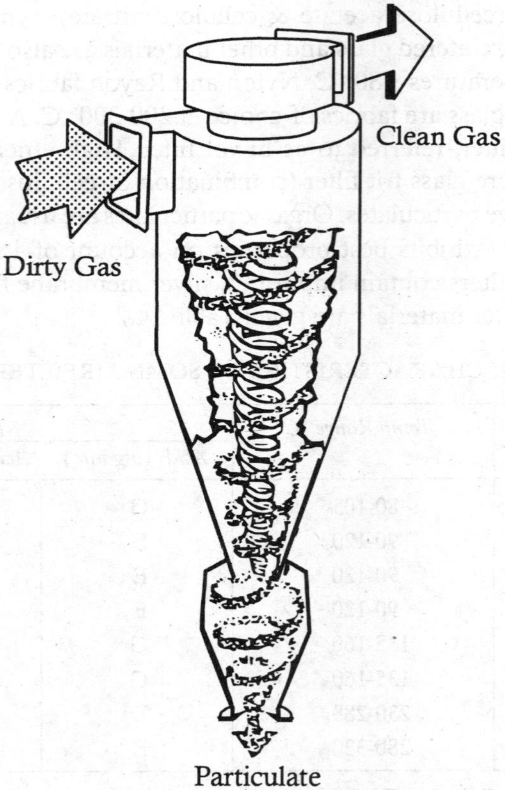


**Fig. 14.6:** Operation of a Cyclone

The sedimentation velocity, $v_s$, is

$$v_s = \frac{d^2 g\rho}{18\eta} \quad \text{...(14.15)}$$

The radial velocity, $$v_r = \frac{d^2 v_t^2 \rho}{18r\eta}$$ ...(14.16)

The sedimentation efficiency, $S$, is

$$S = \frac{v_r}{v_s} = \frac{2v_t^2}{dg}$$ ...(14.17)

The particle diameter at which the collection efficiency is 50 per cent is given by

$$d_{50}^{0} = 1.16\sqrt{\frac{\eta w^2 hD}{\rho v_i V}}$$ ...(14.18)

where, $v_t$ = tangential velocity of air;
$d$ = particle diameter;
$g$ = acceleration due to gravity;
$\rho$ = particle density;
$\eta$ = viscosity of air;
$w$ = cyclone inlet width;
$h$ = cyclone inlet height;
$D$ = cyclone barrel diameter;
$v_i$ = air velocity at inlet;
$V$ = cyclone volume.

Efficiency of cyclones as a function of particle size is given Tablc 14.3.

**TABLE 14.3:** EFFICIENCY OF CYCLONES DEPENDING ON THE PARTICLE SIZE

| *Particle Size* (μm) | *Conventional Cyclone* | *High-Efficiency Cyclone* |
|---|---|---|
| 5 | – | Low (70%) |
| 5-20 | Low (80%) | Medium (80-95%) |
| 40 | High (5%) | High (99%) |

### 14.7.2 Goetz Particle Spectrometer

This instrument is simple and versatile. It consists of a helical channel fixed on a rotating cone. When polluted air is passed through the helical channel, the suspended particles are moved outward by centrifugal force and collected on a metal foil wrapped around the cone. There is gradation of separation of particulate depending on the particle size. The particles can be observed microscopically and their numbers counted. The particle size can be calibrated and estimated from their horizontal distribution on the foil.

## 14.8 SCRUBBERS

Scrubbers are wet collectors (Fig. 14.7). Spray towers, cyclone scrubbers, Venturi scrubbers and packed-bed scrubbers are some of the wet collectors. The underlying principle is to remove the particulate or gas by absorbing the material into liquid droplets (packed scrubbers) and flush them out. The primary collection mechanisms for removal of particulate matter are inertial impaction, interception and diffusion. Selection of a scrubber system depends on the particle size and mass of the particulate matter.

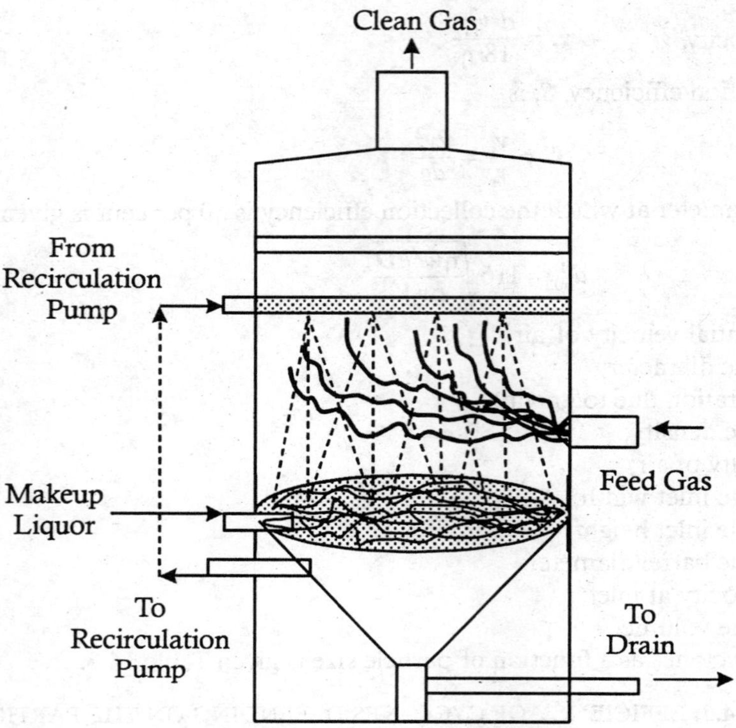


**Fig. 14.7:** Schematic of a Wet Scrubber

Spray towers (Fig. 14.8) and cyclone scrubbers consist of a system with downward flow of water spray injected into the tower and an upward flow of polluted air. Spray tower can effectively remove larger particulate (>10 μm). In the cyclone scrubber, the gas is tangentially injected around, just as in the dry scrubber. Water is sprayed from the top.

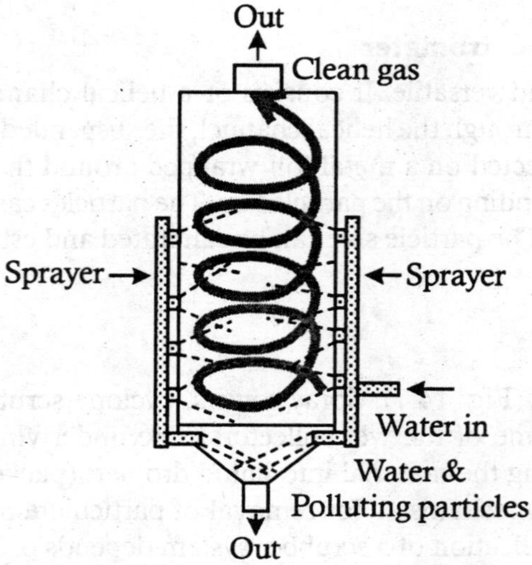


**Fig. 14.8:** Schematic of a Spray Tower

Venturi scrubber is the most widely used scrubber for mist elimination (>0.5 μm). Inertial impact is the primary collecting mechanism. In the Venturi scrubber (Fig. 14.9), water under high pressure is injected upstream of the Venturi throat through which particulate-laden air is passed at high speed. Gas stream forms droplets that collect dirt.

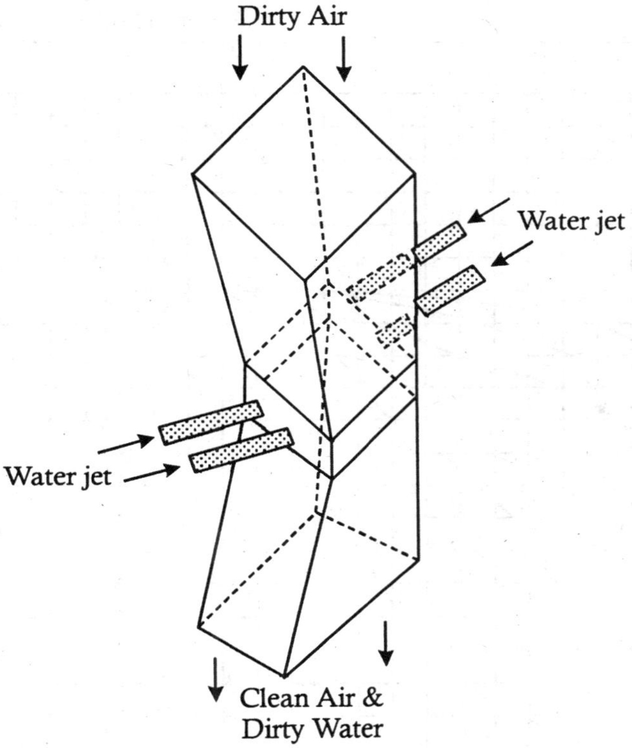


**Fig. 14.9:** Schematic of Venturi Scrubber

Separation characteristics of some collectors are given in Table 14.4, and removal efficiency of some collectors is given in Fig. 14.10.

**TABLE 14.4:** SEPARATION CHARACTERISTICS OF SOME COLLECTORS

| *Collector Type* | *Efficiency* |
|---|---|
| Settling chamber | >50 mm (80%) |
| Cyclones | 10-1 mm (50% on 20 mm) |
| High-efficiency multi-tube cyclones | 80%-90% on 10 mm |
| Impeller cyclones | 98% on 10 mm |
| Cascade cyclones | 2-0.5 mm |
| Impingement separator | 90% on 10 mm |
| Centrifugal sampler | 5-0.1 mm |
| Spray tower | 10-1 mm (90%) |
| Anderson cascade impactor | 10-0.5 mm (8-stage) 30-0.08 mm (14-stage) |
| Quartz crystal microbalance | 25-0.05 mm |
| Electrical low-pressure impactor | 2-0.03 mm |

| Collector Type | Efficiency |
|---|---|
| Electrostatic precipitator | 5-0.05 (99%) |
| α-Impactor (for radon progeny) | 15-0.005 mm |
| Thermal precipitator | 0.5-0.005 mm |
| Venturi scrubber | 5-0.5 mm (98%) |
| Fabric filter | 2-0.3 mm (99.8%) |

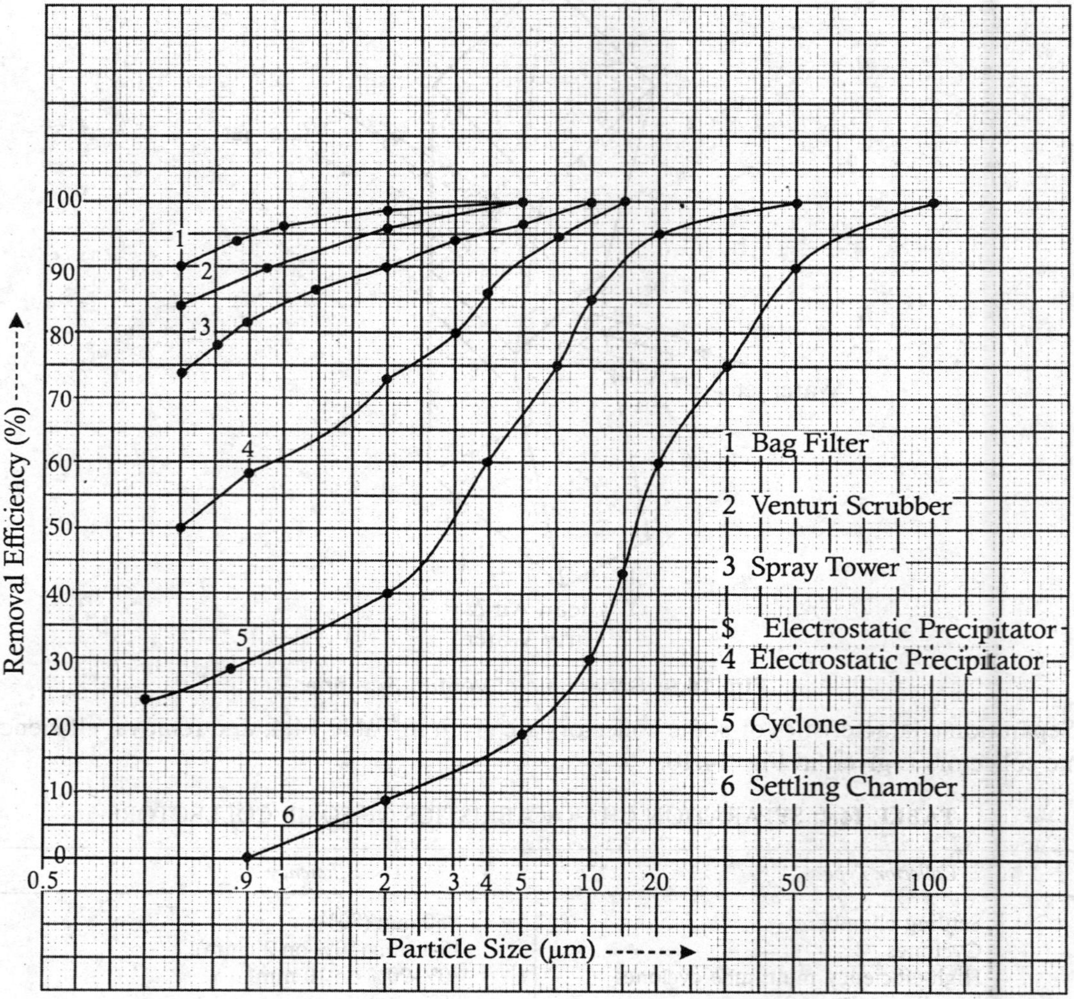


**Fig. 14.10:** Removal Efficiency of some Collectors

(After Peirce, J. J. *et al.* (1998), Butterworth-Heineman: New Delhi. "*Environmental Pollution and Control*").

## BIBLIOGRAPHY

Adriano, D. C. (1986), Springerverlag: New York. "*Trace Elements in the Terrestrial Environment.*"

Crawford, M. (1976), McGraw-Hill: New York. "*Air Pollution Control Theory.*"

Harrison, R. M. and Perry, R. (1984), Chapman & Hall: London. "*Handbook of Air Pollution Analysis,*" (2nd edn.).

Hinds, W. C. (1982), New York: John Wiley. "*Aerosol Technology.*"

Hutter, R. (1995). *Health Phys.*, **68**: 835. "A method for determination of soil gas $^{220}Rn$ (Thoron) concentrations."

Jorgensen, S. E. and Johnsen, I. (1981), Elsevier: Amsterdam. "*Principles of Environmental Science and Technology,*" (Vol. 14).

Keith, L. H. (ed.) (1991), Lewis Pubs: Chelsea, MI. "*Environmental Sampling and Analysis: A Practical Guide.*"

Lioy, P. J. and Lioy, M. J. Y. (eds.) (1989), ACGIH: Cincinnati, OH. "*Air Sampling Instruments for Evaluation of Atmospheric Contaminants,*" (7th edn.).

Lodge, J. P. (1991), Lewis Pubs: Chelsea, MI. "*Methods for Air Sampling and Analysis.*"

Markert, B. (ed.) (1994), VCH Pubs: New York. "*Environmental Sampling for Trace Analysis.*"

Marple, V. A., Rubow, K. L. and Beham, S. M. (1991), *Aerosol Sci. & Technol.*, **14**: 434. "A micro-orifice uniform deposit impactor (MOUDI): description, calibration and use."

Parker, H. W. (1977), Prentice-Hall: Engelwood Cliffs, NJ: "*Air Pollution.*"

Peirce, J. J., Weiner, R. F. and Vesiland, P. A. (1998), Butterworth-Heineman: New Delhi. "*Environmental Pollution and Control,*" (4th edn.).

Perkins, H. C. (1974), McGraw-Hill: New York. "*Air Pollution.*"

Stern, A. C. (1976), Academic Press: New York. "*Air Pollution,*" (Vol, 3).

Stern, A. C., *et al.* (1984), Academic Press: New Yotk. "*Fundamentals of Air Pollution*" (2nd edn.).

Thomas, J.W. (1972), *Health Phys.*, **23**: 783 "Measurement of radon daughters in air."

Warner, P. O. (1976), Wiley-Inter Science: New York. "*Analysis of Air Pollutants.*"

Winegar, E. D. and Keith, L. H. (eds.) (1993), Lewis Pubs: Boca Raton, F. L. "*Sampling and Analysis of Airborne Pollutants.*"

... (1997), Environmental Measurement Lab, USDOE: New York. "*HASL-300,*" (28th edn.).

# 15

# Sampling Analysis Methods

While collecting and handling the samples, ambient conditions have to be maintained and there should be minimal external interference. Collected samples should be transported in proper containers to avoid any contamination and preserved and stored till analysis as per the standard procedures.

Evaporation (distillation), sublimation (volatilization), precipitation, co-precipitation, solvent extraction, filtration, dialysis, centrifugation, sorption and electrolytic methods are the generally employed procedures to achieve separa-tion, concentration and pre-concentration of environmental pollutants. Evaporation (single-step) and rectification (multi-step) are routinely employed methods for separation and concentration of volatile compounds, petrochemicals and isotopes. Sublimation is volatilization where solid substances are evaporated on heating directly into gaseous phase.

Sampling of soil pollutants often involves two-stage pre-concentration—oxidation of organic compounds by dry- or wet-mineralization followed by concentration by another suitable method. Derivatization method is a convenient method of pre-concentration, used to separate trace elements from their matrices and then pre-concentrate them by hydride generation, ethylation and phenylation.

Mass flow and molecular diffusion methods may be employed to extract pollutants in liquid phase. Polarity and volatility influence the sampling techniques in volatile compounds. Trap-and-purge methods are most commonly used for volatile compounds, and solvent extraction methods for less-volatile compounds. Cryogenic pre-concentration method is efficient, but is not selective.

Particulate matter in gaseous pollutants, after collection, is dissolved in acids and filtration is the standard method employed. Reactive gases must be converted into stable derivatives during the collection step. Diffusion denuders are used for sampling inorganic gases, such as $O_3$, $H_2O_2$, $NH_3$, $NO_x$, and $SO_x$. The selection of the sampling method for organic contaminants relies on the vapor pressure, polarity and reactivity. Adsorption, sorption-trap and solvent extraction methods are generally employed for pre-concentration.

This chapter presents some pollution sampling methods.

Application of various instrumental and analytical methods for data collection, together with the *quality assurance and quality control (QA/QC)* procedures form an integral part of the sampling and analysis of the environmental sample matrices.

Sampling is a pre-preparation procedure prior to analysis. Therefore, necessary care should be taken all aspects of involved protocols, from sample collection, transportation, storage and sample preparation. General steps in sample preparation procedures include mineralization, pre-concentration and separation by various physicochemical methods, such as precipitation, filtration, sublimation, solvent extraction, sorption and electro-analytical techniques. Standard procedures of air sampling of air pollutants are: (1) removal of filterable materials from the gaseous pollutants, (2) collection of inorganic and organic samples separately and (3) determination of filtrate and gaseous pollutants by suitable analytical methods. The same general procedures, with suitable modifications, are applicable to sampling of water and soil pollutants. Mathematical models for quality control and quality assurance also come under sampling methods.

## 15.1 SAMPLING PROCESS

Sampling is the process that covers sample collection, sampling design and sub-sampling procedures. Sub-sampling is the process of extracting from a large bulk of material a small fraction that is representative of the whole yet of convenient size for analysis. The properties of the sample, as ascertained by suitable methods can be extrapolated to the bulk material. It also includes concentration and sample preparation to suit the analytical method. The purpose of environmental sampling and analysis is to obtain experimental methods that describe a particular pollution site, at a specific point of time from which an evaluation can be made as a basis for possible action. While collecting a sample, ambient conditions such as particulate concentration, temperature, humidity and other parameters should be maintained to obtain reliable information after analysis. Separate samples should be collected for physical, chemical and biological analyses. As air pollution depends on geological, meteorological and other local conditions (spatial variation), which may vary with time (temporal variation), sampling should be done over an extended period, taking into account wind speed, wind direction, temperature inversion, lapse rates etc., to measure dispersion of pollutants from sources to recipients. Utmost care should be taken to ensure that: (*i*) the samples are minimally altered, (*ii*) they do not interact with implements and other constituent materials do not contaminate containers during collection and preservation.

In the case of water samples determination of temperature, pH, and dissolved gases should always be performed at the site. Because of extreme dilution, the material to be analyzed must be concentrated prior to analysis. Carbon adsorption method is commonly employed for this purpose. The method is simple, fast and efficient. Freeze concentration, and solvent extraction (liquid-liquid) are useful methods for isolating organophylic compounds from water (e.g. isolation of chlorinated hydrocarbon pesticides). Ion exchange is another method to separate polar pollutants.

Very few environmental samples are analyzed immediately after collection. They are transported to a place of analysis and also stored till they are analyzed. Sampling procedure of contaminants includes both sample collection and preservation. There are a variety of methods available for the preservation, out of which chemical preservation, temperature control and selection and separation and storage in suitable containers are the most common. Water samples have to be preserved by adding preservatives prior to analysis (Table 15.1). Acidifying samples to $pH < 2$ is one of the frequently used procedures for stabilizing metals in solution (by avoiding their precipitation as oxides or hydroxides). NaOH stabilizes water samples containing $CN^-$ or sulfides; and ammonia in water is stabilized by $H_2SO_4$. Sodium thiosulfate is used for preserving organic compounds. Samples should be refrigerated (at 4 or –20° C) immediately after collection. Polyethylene bottles are especially suitable for sampling waters continuing Na or silica. Water containing trace metals is best collected in Teflon bottles, and biota samples in glass bottles.

**TABLE 15.1:** PRESERVATION OF WATER SAMPLES

| *Preservative* | *Purpose* | *Use* |
|---|---|---|
| Nitric Acid, $HNO_3$ | To keep metals in solution; to destroy bacteria | Metal-containing samples |
| Sulfuric acid, $H_2SO_4$ | To destroy bacteria | Biodegradable samples |
| Sodium hydroxide, NaOH | Formation of salts | Volatile organic acids |
| Sodium thiosulfate | To preserve organic compounds | Organic compounds |
| Mercuric chloride, $HgCl_2$ | To destroy bacteria | Biodegradable samples |
| Freezing | Inhibition of bacterial growth; retention of volatile materials | Micro-organisms; volatile compounds |

Air pollutants are of various kinds and are classified under several categories. Sampling methods differ, depending on the physical and chemical nature and the nuisance and toxic and hazardous nature of the pollutants. For example, radioactive pollutants, volatile and explosive materials and toxic chemicals must be handled according to well-defined procedures.

**1st Category:** General atmospheric pollutants for which ambient standards have been set (see Chapter 16)—CO, $SO_2$, $NO_2$, $O_3$.
**2nd Category:** Hazardous to human health for which air quality standards are not applicable—Asbestos, Be and Hg.
**3rd Category:** Regulated installations—Coal, cotton and paper industries; dust, particulate, $NO_x$, $SO_x$ and $H_2SO_4$.
**4th Category:** Emissions by mobile sources, hydrocarbons, $CO_x$ and $NO_x$.
**5th Category:** Trace elements—As, B, Ba, Cd, Cl, Cr, Cu, F, HCl, $H_2S$, Li, Mn, Ni, P, Se, Ti, V, Zn; and PAHs and PCBs, pesticides and radionuclides.

Air sampling methods should take into consideration, the size of the sample, duration of sampling and various other parameters to avoid bias. Any sample should contain a mixture of aliquots (random sampling) to minimize bias. A general scheme of sampling steps for air pollution is given in Fig. 15.1.

## 15.2 SAMPLING OF GASEOUS POLLUTANTS

For sampling and analysis of air pollutants, general protocol is: (1) removal filterable materials, (2) separation of inorganic and organic gaseous pollutants and analysis of filtrate and gases by suitable analytical techniques.

### 15.2.1 Sampling of Particulate Matter

Filtration is the usually employed method for collecting particulate air pollutants and sampling them. Deposit of analytes on a filter pad of known porosity allows the examination of particulate matter microscopically. Membrane filters, soluble in non-aqueous liquids can be dissolved. After evaporation of the solvent the residual particulates can be analyzed for their physical and chemical characteristics. While glass filters contain impurities, silver filters are 99.9 per cent pure.

After collecting samples (filtrate), the analysis of atmospheric particulate samples by most analytical methods requires processing—prior dissolution of the samples. The dissolution procedure involves solubilizing the sample by acids HCl, $HNO_3$, $H_2SO_4$ or perchloric acid. Procedures involve fusing the aerosol sample in an alkali or sodium carbonate, followed by acid dissolution or resort to high- or low-temperature ashing and subsequent acid dissolution.

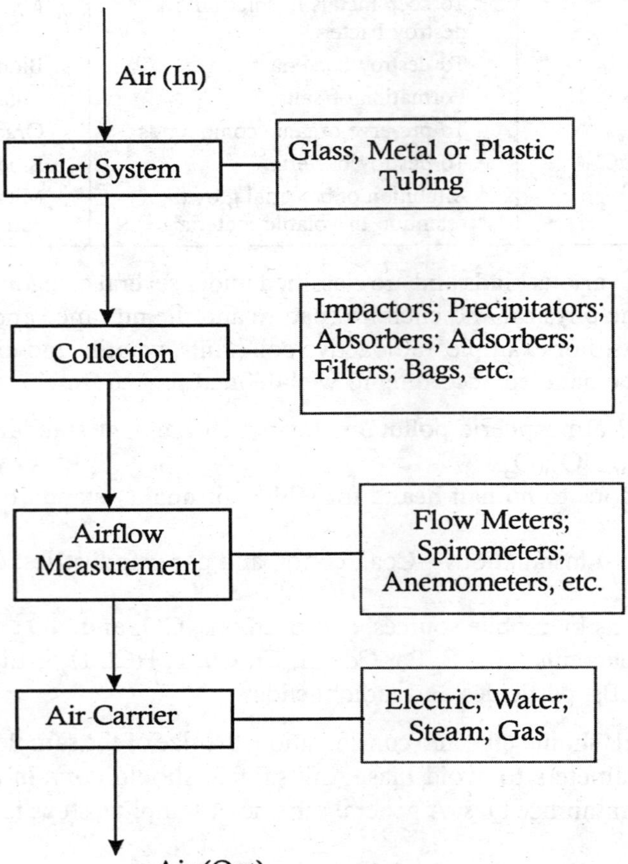


**Fig. 15.1:** Flowchart of Ambient Air-sampling System

### 15.2.2 Air Flow Measurement

Accurate measurement of air volumes of sampled analytes is essential to determine pollution concentrations. Total volume of gas (air) sampled may be determined by 'dry' or 'wet' test meters, gas-velocity instruments (gas flow meters, rotameters), spirometers and thermal anemometers. In certain type of air flow meters (Fig. 15.2) entry of air from the lower-end of the nozzle lifts the float and the float comes to rest at a position where the pressure difference between the two ends of the float is zero.

The volume of airflow, $Q$, is

$$Q = CA\sqrt{\frac{2gV_t}{A_f}\left(\frac{W_f}{W_0}-1\right)} \qquad \text{...(15.1)}$$

where, $C$ = flow coefficient;
$A$ = surface area;
$V_t$ = volume of the float;
$A_f$ = effective area of the float;
$W_f$ = float weight;
$W_0$ = fluid weight;
$g$ = acceleration due to gravity.

The float position is calibrated to readout the airflow directly from the graduated scale.

Velocity measurements in dusty gases are made with type *S Pitot* tube and draft gauge. The velocity, $v$, of gas is given by

$$v = C\sqrt{2gh_L\left(\frac{\rho_L}{\rho_g}\right)} \qquad \text{...(15.2)}$$

where, $C \sim 1$; $h_L$ = height of the liquid column;
$\rho_L$ = density of the liquid in the manometer;
$\rho_g$ = density of the gas.

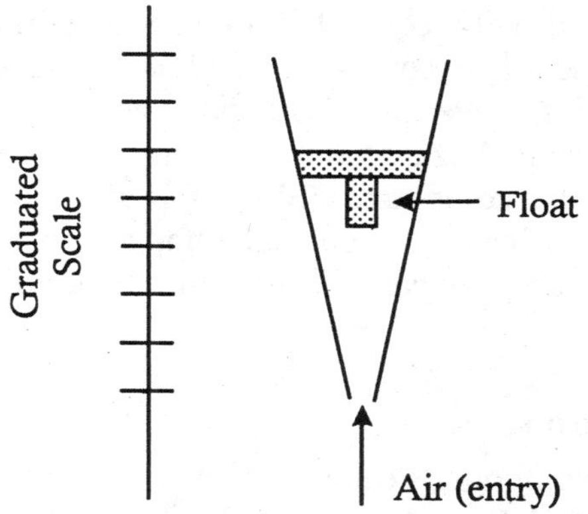


**Fig. 15.2:** Principle of Air-flow Meter

### 15.2.3 Sampling of Inorganic Gases

Pre-concentration (sampling) of gaseous atmospheric pollutants is necessary as these occur in trace levels. Diffusion denuders are used for sampling inorganic gases, $O_3$, $H_2O_2$, $NH_3$, NO, $NO_2$, $HNO_2$, HCl, $SO_2$, etc. A diffusion denuder consists of one or more tubes through which the air sample is drawn. The inner surfaces of the tubes are coated with a material that retains the target gas by sorption. After the gaseous fraction has been collected on the wall coating, gases are eluted or thermally desorbed.

In permeation denuders, also known as "diffusion scrubbers", the gas collection elements are gas-permeable membranes that isolate the sampled gas stream from a liquid absorbent solution (hydrophobic microporous tubing with annular jackets). The air sample is forced through the annulus between the tube and the jacket, and a stream of liquid absorbent is continuously pumped through the microporous tube. Gaseous species diffuse successively through the membrane and into the sorbent solution, where they are finally tapped. Diffusion scrubbers are used for the determination of $H_2O_2$, $NH_3$, and $HNO_3$.

Filters or tubes that are impregnated with a suitable chemical reagent can produce a characteristic color when contacted with the pollutant. This can be the basis for qualitative or semi-quantitative estimation of the pollutant. Samplers containing a fluorescent reagent coating have been frequently used for determination of $O_3$, $NH_3$, $NO_2$, and $N_2O$. Ozone can be detected by its chemiluminescence on a silica gel surface treated with rhodamine-B.

### 15.2.4 Sampling of Gas-phase Organic Pollutants

The selection of a sampling method for organic aerosol contaminants relies on vapor pressure, polarity and reactivity. Reactive gases must be converted into stable derivatives during the collection step. The most important volatile compounds ($>10^{-2}$ kPa) in the atmosphere are aromatic hydrocarbons, halocarbons, aldehydes and ketones. Semi-volatile organic contaminants ($10^{-2}$-$10^{-8}$ kPa) are oligo-nuclear aromatic hydrocarbons and their nitro-, and chloro-derivatives, polychlorinated biphenyls (PCBs) and pesticides.

(*i*) Adsorption, (*ii*) sorbent trap, and (*iii*) solvent extraction methods are used to accomplish pre-concentration. Most commonly used sampling method for volatile organic compounds (VOCs) is pre-concentration of the samples on sorbents. Inorganic sorbents like silica gel, alumina and molecular sieves are used to collect some organic acids, phenols, amines and other polar organic compounds. A polar solvent is usually required for desorption of the adsorbates. Organic polymer sorbents (e.g. Tenax-TA) are used for pre-concentration of non-polar organic contaminants. Activated carbon exhibits higher adsorbing power than do organic polymer sorbents and it allows efficient pre-concentration of important and hazardous volatile organic compounds such as vinyl chloride. Sampling of semi-volatile organic compounds (SVOCs) is usually accomplished by solid sorbents (e.g. polyurethane foam). The universally accepted method for isolating SVOCs from their matrices is solvent extraction. Denuder tubes are probably the most powerful tools for separating gases from particles.

#### *15.2.4.1 Sorbent-Trap Method*

Pre-concentration of organic compounds can be achieved by sorbent-trap method. A hollow tube is packed with a sorbent material that selectively retains the desired species as the fluid sample is passed through. The organic substances can be eluted later by solvent extraction or thermal desorption (if they are volatile (Fig. 15.3)). Most commonly used sorbents are polyurethane foam, non-polar copolymers of styrene-divinylbenzene and activated carbon. Sorbent traps are more effective than filters since they remove volatiles from gaseous samples much more readily.

## 15.3 SAMPLING OF SOIL POLLUTANTS

Soil pollutants are minerals, nitrates, nitrites, sulfates, phosphates, humic substances, and anthropogenic pollutants. The analysis of soils, plants and organic substances (humic substances) and other carbon-, and sulfur-containing materials often involves two-stage pre-concentration—oxidation of organic components in the samples by wet or dry ashing followed by concentration by another suitable method.

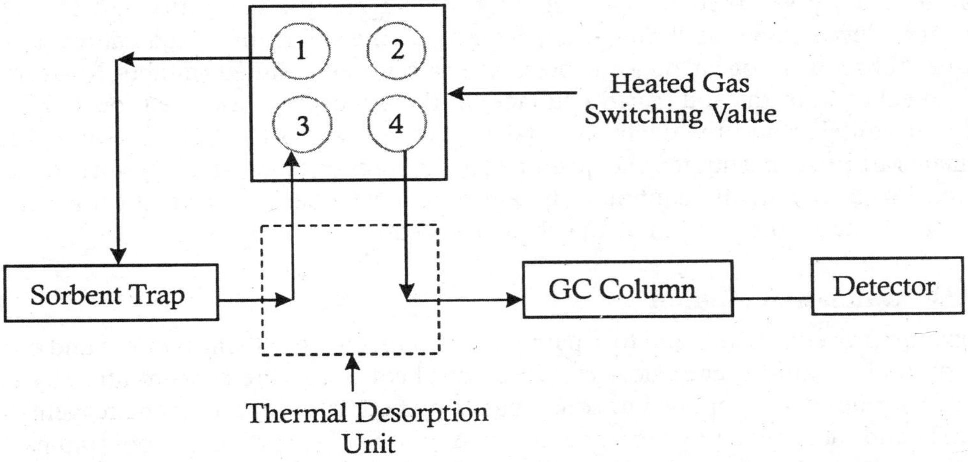


**Fig. 15.3:** Thermal Desorption Setup for Organic Compounds Coupled to a GC Unit

The soil samples are collected from the site, dried, ashed and size reduced (<250 μm) (Fig. 15.4) and analyzed by wet-chemical and physical techniques. Samples should be stored at <4° C and drying temperature should not exceed 40° C, as it has a marked effect on the nitrogen and phosphorus measurements. In addition, As, Hg and Se and some organometallic compounds are easily lost upon air-drying. In such cases, freeze-drying is preferable.

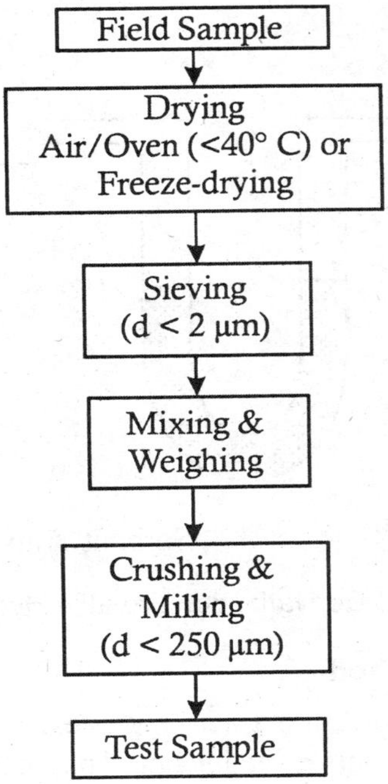


**Fig. 15.4:** Soil Sample Pretreatment Protocol

Soils are usually sampled for (*i*) assessing their agricultural quality, (*ii*) for evaluating contamination levels in polluted sites and (*iii*) for preparing soil maps. Electrical resistivity of soil is used to establish hydrological data and detect the presence of polluted aquifers. Measurements of contaminant flux from the soil (earth's surface) to the atmosphere are used for monitoring and assessing air emission and the environmental impact of waste sites. Samples are collected and contaminants are pre-concentrated and quantified by: (*i*) sorption (sorbent trap), where contaminants are collected by a sorbent and continuously purged in a chamber and the sorbent is subsequently analyzed by (*ii*) derivatization, and (*iii*) solvent extraction.

### 15.3.1 Derivatization Method

Derivatization procedures are used to separate trace elements from their matrices and concentrate analyte species by hydride generation, ethylation or phenylation. Pre-concentration by formation of volatile hydrides can be applied to some metals present at trace levels. The reagent is sodium borohydride and the metals pre-concentrated are As, Bi, Sb, Se, and Sn. A cool trap packed with glass beads or GC packing material (Fig. 15.5) can be used to trap the hydride. The volatile hydrides are carried by a gas like helium and held in a cold trap packed with glass beads or a suitable GC packing material (Fig. 15.5). The hydrides are determined by atomic absorption or emission spectrometry (AAS and AES) with electron capture (ECD), flame ionization or inductively coupled plasma (ICP) detectors (refer to Chapter 24 for details).

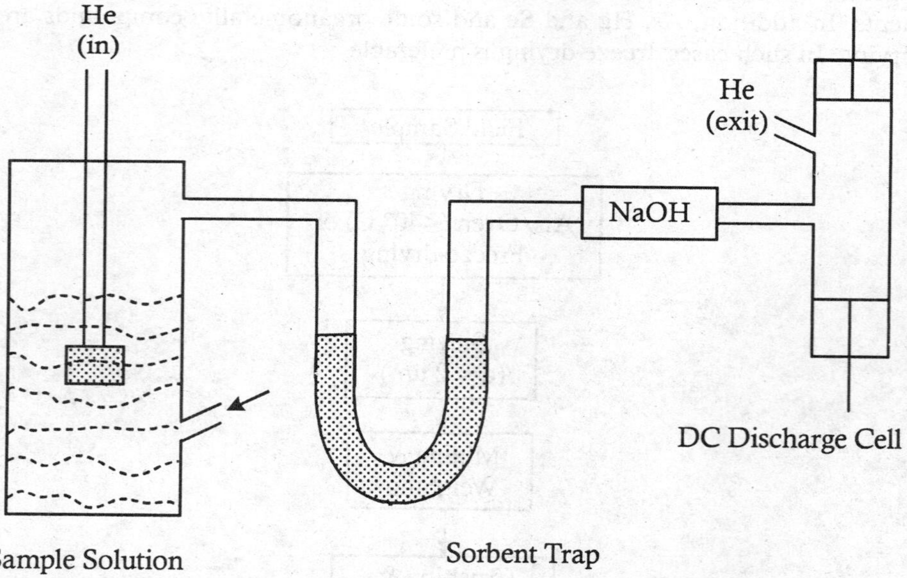


**Fig. 15.5:** Schematic of Derivatization (Volatile Hydride Formation) Setup

### 15.3.2 Soxhlet Extraction Method

Soxhlet extraction is a traditional solvent extraction process. In this method, the sample is placed in a cavity that is gradually filled with the extracting liquid phase by condensation from the distillation matrix. When the liquid reaches a preset level, it is completely evacuated to the distillation flask, with the extracted analytes by means of a siphon system. Other solvent extraction methods, in particular supercritical fluid extraction (SFE) method (see Chapter 22), are better suited for sampling atmospheric contaminants in trace levels. SFE avoids the evaporation step required by the Soxhlet method.

## 15.4 SAMPLING OF LIQUID-PHASE POLLUTANTS

Pre-concentration is necessary to enrich analytes present in trace levels and to minimize matrix effects in their analysis. Solutes in liquid phase are transported by molecular diffusion and mass flow, made use of in equilibrium dialysis samplers (*peepers*). Solvent removal and extraction methods are used in liquid-phase sampling methods.

### 15.4.1 Equilibrium Dialysis Method

In equilibrium dialysis, semi-permeable membranes (peepers) are used to separate a receiver solution from the sediment (sample) and associated water. Dissolved constituents diffuse across the membrane and, at equilibrium, the composition of water in the chamber is the same as that of the adjacent water. Peepers are multi-chambered equilibrium dialysis samplers in which each chamber is filled with de-oxygenated, de-ionized water and covered with semi-permeable membranes. Peepers have been used to sample metals, nutrients and gases.

### 15.4.2 Solvent Removal Methods

Solvent removal (evaporation, volatilization) method is the simplest method for pre-concentration of species in solution. Metals (Ag, Al, Cd, Co, Cr, Cu, Fe, Mn, Ni, Pb, Sb, V) can be pre-concentrated by evaporation in a graphite furnace and determined by AES-ICP method (Chapter 24). Samples in nitric acid are continuously introduced into the nebulizer chamber by a peristaltic pump. A quartz halogen lamp in the nebulizer chamber creates a spray (nebulization) that is lead into ICP torch (Fig. 15.6). Solvent extraction methods (liquid and solid phases) and other separation methods are dealt in Chapter 22.

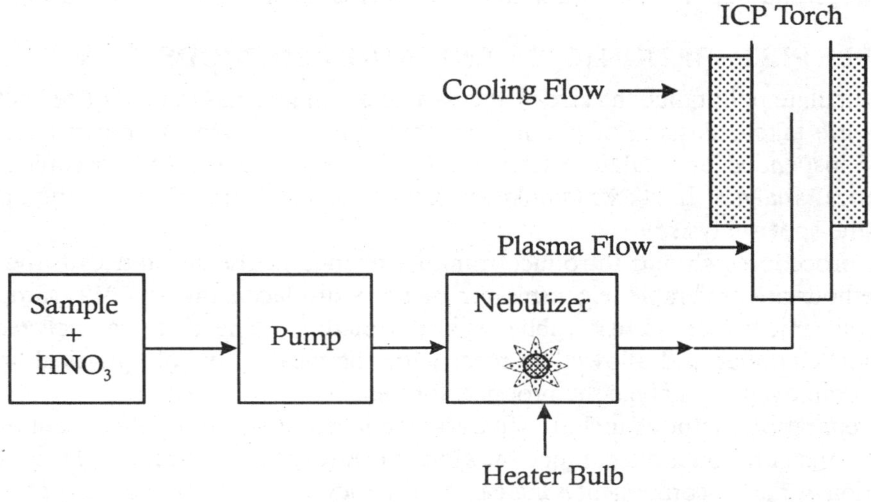


**Fig. 15.6:** Sampling and Detection of Metals at Trace Levels (by ICP-AES)

For organic pollutants, polarity and volatility dictate the sampling technique. Trap-and-purge methods are the most commonly used for volatile compounds. Cryogenic pre-concentration of volatile trace contaminants is efficient, but it is non-selective. Solvent extraction is the most frequently used choice for less-volatile compounds. For example, trihalomethanes (THMs) in water can be determined by using purge valve-trap combination, followed by thermal desorption (Fig. 15.7). An inert gas (He) is dispersed through the sample solution and passed through a sorbent trap, which retains organic compounds. The apparatus can be heated to volatilize organic constituents, which can be determined by GC.

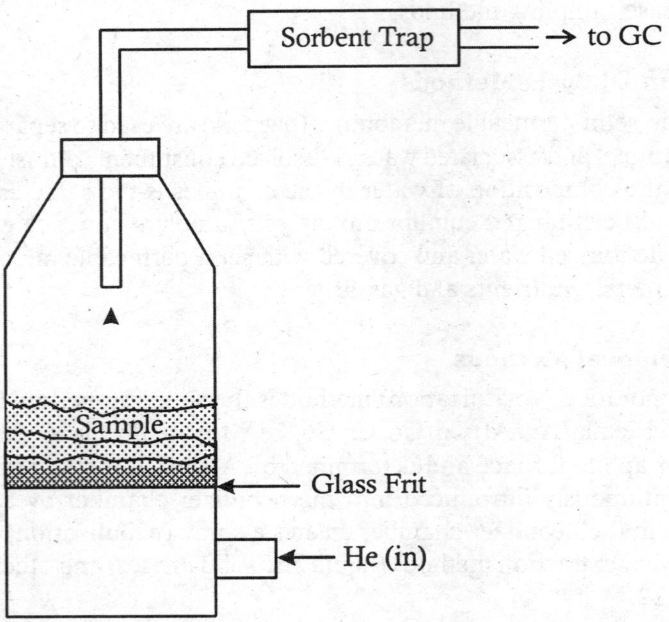


**Fig. 15.7:** Purge-and-trap Method for Extracting Volatile Organic Compounds (VOCs) from Liquids

## 15.5 SAMPLE PREPARATION (CONCENTRATION) METHODS

With the particulate pollutants, the size of the particulate matter has to be reduced before mixing, and aliquots are taken. Passing of polluted air through a filter pad of known porosity permits collection of suspended particulate matter for physical examination (microscopically), and for further chemical analysis. In stacks (smoke emissions), samples are drawn through a hole in the stack for on-the-spot analysis.

Sampling procedures should introduce minimal change in the substances to be tested. Dry sampling methods are preferable (e.g. aspirator, mercury displacement etc.). Wet sampling is done by liquid displacement method (e.g. rubber aspirator method). The air to be analyzed is pumped into a reaction chamber and allowed to react with chemicals. Suitable physical and chemical methods are employed to analyze the products formed.

Sample preparation methods include separation, concentration and pre-concentration prior to analysis. Pre-concentration is a technique by which the ratio of concentration of trace components to concentration of macro-component is increased. Trace levels of some of the atmospheric pollutants can endanger the atmosphere; spoil the quality of water and food products. Pre-concentration of trace levels of pollutants prior to analysis is necessary for trace analysis. Gravimetric methods are seldom used. Evaporation (distillation), sublimation (volatilization), precipitation, co-precipitation, filtration, dialysis, centrifugation, flotation, extraction, sorption, and electrochemical techniques are some of the methods employed to achieve separation and pre-concentration of analytes (Chapters 22 and 23). Pre-concentration minimizes or even eliminates matrix effects and reduces sampling error. Pre-concentration with complete separation of the matrix proves useful in the analysis of toxic, radioactive and explosive materials. Pre-concentration is widely used in spectrophotometric and fluorimetric methods (Chapter 24).

$$\text{Pre-concentration} \rightarrow \text{Separation} \rightarrow \text{Identification} \rightarrow \text{Determination} \qquad \ldots(15.3)$$

Practically all concentration methods can be combined with photometry (UV-visible, IR and Raman) and fluorimetry (fluorescence and phosphorescence) (Chapter 24). Gas chromatography (GC) is the most often-applied chromatography method for the separation of volatile compounds. Column chromatographic methods are used for separation and determination of non-volatile compounds. Hybrid methods are separation and determination, combined together in one analytical cycle (e.g. GC-MS, HPLC-ICP, LC-ECD etc. Chapter 22) and they are more versatile quantitative techniques. Commonly used concentration methods in the descending order of usage are:

Extraction = Sorption > Electrodeposition = Volatilization > Precipitation > Evaporation > Flotation.

### 15.5.1 Evaporation

Evaporation (simple distillation) is a one-step separation and concentration method. Rectification is a multi-step method employed in the purification and concentration of petrochemicals and isotopes. Distillation is well suited for separation and pre-concentration of liquid mixtures with broad boiling range. Steam distillation is a commonly employed method. Azotropic distillation is for liquid mixtures with a constant boiling point (vapor and liquid phases will have the same composition at the azotropic point). Vacuum distillation is employed to avoid thermal decomposition and in this method oxidative reactions are considerably reduced. Distillation in high vacuum (~0.1 Pa) is called *molecular distillation*. The relative volatility, $\alpha_T$, of a binary mixture in molecular distillation is given by the equation

$$\alpha_T = \frac{P_1^0}{P_2^0}\sqrt{\frac{M_2}{M_1}} \qquad \ldots(15.4)$$

where, $P^0$ = vapor pressure of the pure component;
$M$ = molecular mass of the component

Concentration can be achieved by forming volatile compounds such as halides and hydrides—$AsH_3$, $BiH_3$, $GeH_4$, $PbH_4$, $SbH_3$, $SeH_2$, $SnH_4$ and $TeH_2$ and $SiCl_4$ (Chapter 19). Another form of evaporation is vacuum drying (lyophilization). This method is good for analysis of highly volatile trace elements in water. Automatic analysis of many elements (Ag, Al, Cd, Co, Cr, Cu, Fe, Mn, Ni, Pb, Sb and V) in water is possible by evaporating water directly in a graphite crucible and atomizing the dry residue for determination by atomic absorption spectrometry (AAS) (Chapter 24). For neutron activation analysis (NAA) water sample is first evaporated and then the dry residue is irradiated with neutrons. The activated trace elements are separated and analyzed by radiochemical methods (Chapter 25).

### 15.5.2 Sublimation

Sublimation is volatilization where solid substances are evaporated, on heating, directly into the gaseous phase (Fig. 15.8). A measure of the sublimation ability of a substance is its 'sublimation point', defined as the temperature at which the vapor pressure of the solid substance is equal to the outer pressure. Fractional sublimation is carried out by *sublimography*. Sublimation serves as a separation and pre-concentration procedure.

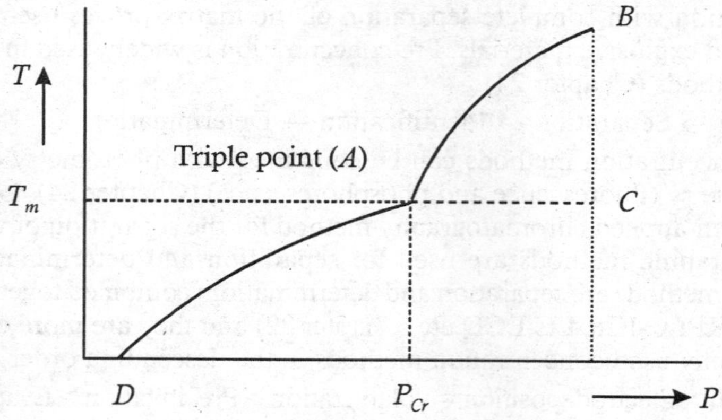


**Fig. 15.8:** Pressure-Temperature Phase Diagram

At the triple-point (*A*), gas, liquid and solid phases are in equilibrium. Curve *AB* shows the dependence of the boiling point on pressure. Curve *AD* is the sublimation region. Line *AC* is independent of pressure at the melting point (at $T_m$). Above the critical pressure ($P > P_{Cr}$) the substance melts on heating at temperature, $T_m$, and boils if $T > T_m$. If pressure is below the critical pressure ($P < P_{Cr}$), the substance passes directly into the gaseous phase (sublimation).

### 15.5.3 Precipitation and Co-precipitation

Concentration by precipitation is based on the solubility difference of the components in aqueous or aqueous-organic solution. Selective precipitation of matrix, by leaving behind the trace elements to be concentrated, in solution is a technique borrowed from gravimetric analysis.

Co-precipitation ensures a high degree of concentration of the trace constituent. The chemistry of co-precipitation depends on the location of the elements in the Periodic Table. When components have opposite donor-acceptor properties, a chemical compound is formed. This is one of the main causes of co-precipitation. Hydroxides, sulfides and phosphates are inorganic co-precipitants. In determining As, Cd, Cu, Fe, Mn, Ni, Pb, Sb and Zn in water, the trace elements are co-precipitated at pH 10-12.

For organic co-precipitation, the precipitant is introduced as solution in an organic solvent (ethanol or acetone). When it mixes with water the reagent (diethyldithiocarbamate (DDTC), ammonium pyrrolidine dithiocarbamate (APDC), dithizone etc) gets precipitated, carrying with it the complexes of the trace metals (chelation method; see Chapter 19). The precipitate can be dissolved in an organic solvent or incinerated to recover the trace metal.

### 15.5.4 Filtration

Filtration is a process of motion of a liquid or a gas through a porous medium, accompanied by the isolation of suspended solid particles. This method is used for concentration of solid particles from air, aerosols and colloidal solutions. Paper, graphite, porous glass, synthetic materials have been used as filtering materials. Ultrafiltration is commonly employed for separating high- and low-molecular mass compounds in a liquid phase. Reverse osmosis is a dialysis method, employed for filtration of solutions through semipermeable membranes, which allow solvent molecules to pass through them.

### 15.5.5 Flotation

Flotation is a process by which substances can be brought to the surface of an aqueous phase. The substance to be floated must be hydrophobic. Special reagents (surfactants) are used to isolate hydrophilic substances and make them hydrophobic. Adding sodium dodecylsulfonate, sodium salts of fatty acids (cationic surfactants) or organic solvents that do not mix with water are some of the flotation methods. Flotation and extraction methods are generally employed, with suitable quenching reagents for determination of various elements to enable differential precipitation.

### 15.5.6 Extraction

Solvent extraction is probably the most extensively employed method for separation and pre-concentration (Chapter 22). The method is universal, simple, rapid and efficient. Chelates are among the widespread classes of compounds used for pre-concentration of trace elements. Chelates are often colored and this permits spectrophotometric determination. Organic sulfides are extensively used for selective extraction of Au, Ag, Hg, Pd and other metals, which are normally precipitated as sulfides. They are readily available, practically insoluble in aqueous media, combine well with different organic solvents, and are characterized by high metal uptake.

Newly developed solid-phase extraction (SPE) methods are capable of isolating (~60 %) water-soluble organic compounds from aerosol samples. These methods facilitate the physical and chemical characterization of water-soluble organic compounds without the interference of inorganic constituents.

### 15.5.7 Sorption

Sorption methods—absorption, adsorption and chemisorption methods (column chromatography and ion exchange) are widely used for separation and pre-concentration of analytes (Chapter 22). Absorption methods rely on wet scrubbers to make the anlaytes soluble in suitable reagents (e.g. $SO_2$ in aqueous $Na_2CO_3$; $H_2S$ in $Cd(OH)_2$; $CO_2$ in KOH; $NO_2$ in $H_2O_2$; NO in $CH_3COOH$ + sulfonic acid) and later determined by titration or spectrophotometry. Adsorption is a very versatile method, specially suited for sampling organic vapors, which have critical temperatures above the laboratory temperature. Adsorption method forms the basis of dosimetry for monitoring pollutants in the atmosphere. Aerosol components can be sampled by adsorption on the surface of suitable porous granular adsorbents (active carbon, silica gel and organic polymers, such as Porapak, Tenax). Rapid heating to temperature 100-200° C, can desorb adsorbed components. They can also be released by washing the adsorbent with a suitable solvent or by evacuation. The desorbed components can an analyzed by wet-chemical (Chapters 16, 18 and 19), physicochemical (thermochemical, electrochemical and other methods, see Chapter 23), and physical methods (chromatography, mass spectrometry (Chapter 22), and spectroscopic (Chapter 24) and other methods (Chapters 25 and 26). Inorganic sorbents (oxides and hydroxides and salts of metals), a wide choice of synthetic ion exchangers, cellulose and activated carbon are used as ion exchangers. Activated carbon is well suited for pre-concentration of organic complexes. Oxidized activated carbon occupies a special place. It is a selective multi-functional cation exchanger.

### 15.5.8 Electrochemical Methods

Electrochemical methods of pre-concentration are electodeposition, electrodialysis, electroosmosis and electrophoresis (Chapter 23). Electrodeposition is the most commonly used method of electrochemical concentration. Electrodeposition is a part of stripping voltametry. Electrolysis at controlled potential (potentiostatic conditions of polarization) is the most commonly used variant of electrodeposition. Electrodialysis is a method for separation of impurities from electrolyte solutions by means of an electric field. Electroosmosis is transferring of solvent by electric field (instead of by hydrostatic pressure). Electrophorsis is based on different mobilities of charged species by the action of an external electric field. This method is basically a separation method, extensively employed in separation of biological species.

### 15.5.9 Mineralization (Ashing)

Mineralization (oxidation of organic compounds) is the process to obtain concentrates of impurities when substances of organic nature are analyzed—carbon and sulfur-containing compounds, plants, animal tissues, plastics, rubbers and petrochemicals. Dry-mineralization (ashing, thermal decomposition or pyrolysis and plasma destruction) and wet-mineralization (with mineral acids) are the methods employed.

#### *15.5.9.1 High-temperature Ashing*

Dry ashing is generally carried out at moderate temperatures. First the sample (pollutant) is air dried (at 20-30° C) and then it is placed in a crucible and the organic components of the sample are volatilized by ashing it at high temperature (~500° C). A small quantity of sulfuric acid is added to the sample before heating to suppress the loss due to volatilization during ashing process. Care should be taken as high-temperature ashing leads to loss of volatile species (As, Cd, Hg, Pb, S, Sb, and Zn). Use of glassy carbon ashing is preferred in such cases. It allows higher temperature ashing and volatile elements like mercury will not seep through, and volatile species like As, Hg, Sb, Se and Sn can be preconcentrated, which are of interest in toxicological and environmental pollution analysis and control. Plants of terrestrial origin may be mineralized using dry ashing procedure without As and Se losses. But, for the analysis of aquatic plants, a separate wet digestion with a mixture of nitric, perchloric and hydrofluoric acids is necessary.

#### *15.5.9.2 Low-temperature Ashing*

Low-temperature ashing is employed for extrusion (forcing a semi-solid through and orifice) of organic compounds by oxygen and high-frequency discharge (Fig. 15.9).

When high-frequency discharge is conducted through molecular oxygen, $O_2$, it is excited to form atomic $O^*$, which oxidizes the sample and accelerates decomposition. Therefore, ashing can be achieved at a relatively low temperature, ~150° C.

The sample is placed in an evacuated oxidation chamber that is subjected to high radio-frequency discharge. $O_2$ is introduced at low pressure (1 mm Hg). The composition in the oxidation chamber, due to high-frequency discharge is

$$O_2 \rightarrow 2O^* \qquad \ldots(15.5)$$

$$2O^* + O_2^* \rightarrow O_2 + h\nu$$

Low-temperature ashing is a better method, because the recovery rates of many volatile metals and inorganic substances that burn above 150° C are 100 per cent.

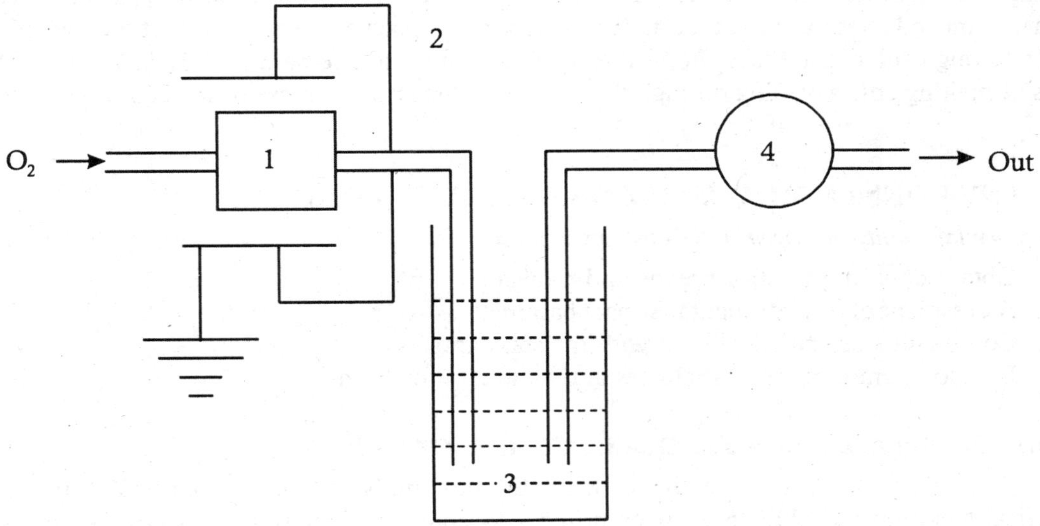


1 = Oxidation Chamber; 2 = RF Generator; 3 = Cold Trap; 4 = Vacuum Pump

**Fig. 15.9:** Low-temperature Ashing Setup

#### *15.5.9.3 Wet Oxidative Decomposition (Wet Mineralization)*

*Wet oxidative decomposition* (wet ashing) is a thermal decomposition with strong oxidizing agents—$H_2SO_4$, $H_2O_2$, $O_3$ (Table 15.2). In most cases wet ashing is preferred and is carried out in a closed vessel.

TABLE 15.2: CHARACTERISTICS OF SOME OXIDIZING AGENTS

| *Oxidant* | *Characteristics* |
|---|---|
| $H_2SO_4$ | Dehydration and charring |
| $HNO_3$ | Decomposition of leaves, cellulose and proteinous matter |
| $HClO_4$ | Powerful oxidant |
| $H_2O_2$ | Powerful oxidant |
| $O_3$ | Powerful oxidant |
| $H_2SO_4$ + Chromic acid | Powerful oxidant |

Al, Cr, Mo, Si, Ti, which are sparingly soluble in acids are made acid-soluble by fusion agents. Al, Cu, Mg and Si are components of glass filters. Therefore, glass should not be used for the analysis of these elements.

### 15.5.10 Zone Melting

Zone melting is used primarily to obtain substances of high purity (e.g. germanium and selenium for semiconductor use). It can also be used for purification and pre-concentration of crystalline contaminants and pollutants (e.g. hydrocarbons—anthracene, pyrene and chrysene). This method is being applied at trace level environmental pollution analysis. In this procedure, a small heating furnace is moved over the solid mixtures to be separated, placed in a tube, so that the melted zone passes through the entire mass. Impurities from the substance remain mainly in the melt. With series of melting zones passing through the substance, the impurities are pushed continually to one end.

## 15.6 ENVIRONMENTAL QUALITY MANAGEMENT (EQM)

*Environmental quality management* (EQM) procedures cover

1. Obtain data on pollution trends and transport.
2. Assessment of environmental impact parameters.
3. Compliance of analytical data with the standards.
4. Development of control strategies and validation methods.

### 15.6.1 Quality Assurance and Quality Control (QA/QC)

Quality assurance (QA) and quality control (QC) of analytical data, obtained from various analytical techniques (Table 15.3), is of prime importance in environmental sample analysis in view of the compositional characteristics, and variability in their concentration and concentration. Quality control (QC), which is included within QA, involves all those planned and systemic actions necessary to provide adequate confidence that a facility, structure, system, or component will perform satisfactorily and safely in service. It comprises all those actions necessary to control and verify the features and characteristics of a material, process, product, or service to specified requirements, from the collection of samples to the reporting data.

**TABLE 15.3:** COMMON ANALYTICAL TECHNIQUES IN ENVIRONMENTAL STUDIES

| *Method(s)* | *Technique(s)* | *Analysis* |
|---|---|---|
| Spectroscopic | Atomic; Molecular; Auger electron; & Ion beam analysis | Identification and elemental analysis |
| Electro-analytical | Voltammetry; Polarometry; Ampe-rometry; & Coulometry | Organo-metallic analysis |
| X-ray methods | Energy- and Wave-dispersive; Total reflection; & Particle-induced X-ray emission | Identification and elemental analysis |
| | X-ray diffraction | Identification and spatial structure |
| Activation analysis | Neutron activation; Particle activa-tion; & Prompt gamma activation; | Identification and elemental analysis |
| Mass spectrometry (MS) | GC-MS; LC-MS | Separation, identification and analysis |
| Nuclear spectrometry | Alpha ray; Beta ray; & Gamma ray | Radioactivity analysis |

GC = Gas chromatography; LC = Liquid chromatography.

The reliability of analytical measurements—reproducibility and accuracy, are necessary for effective environmental control. Basic aspects in environmental analysis are: (1) planning, (2) primary objectives, (3) sampling (4) measurements, (5) precision and accuracy (6) quality control, and (6) validation (Fig. 15.10).

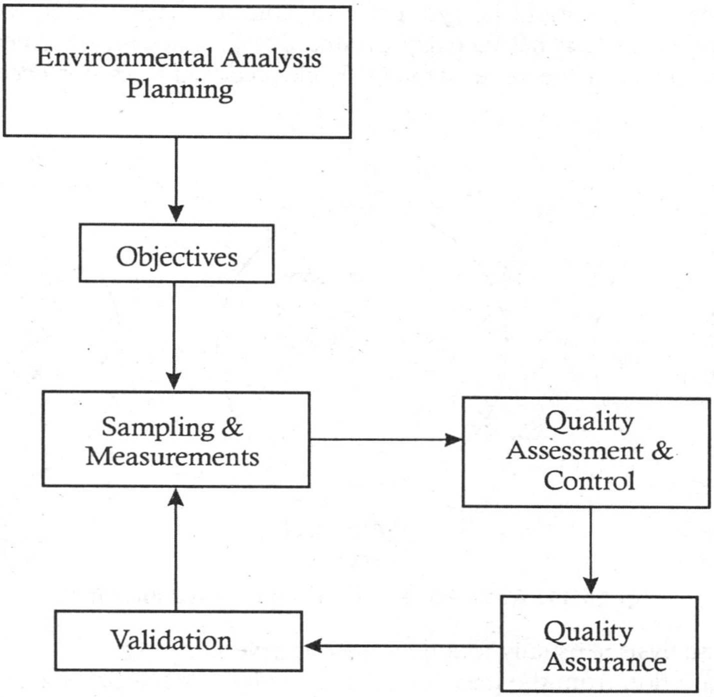


**Fig. 15.10:** A Protocol for Environmental Analysis Planning and Validation

In quality control (QC) methods, the *blank samples* (reagent samples) are analyzed to give a measurement of any contamination of the sample that is occurring during the course of the collection, preparation and analysis. The matrix of the blank samples should be the same as that of the samples being analyzed. Control samples contain known concentrations of the analyte. If possible, they should be of the same matrix as the routine samples. *Replicate samples* are obtained by repeating the collection of sample. The *reference material* is a material or substance whose properties are well established and can be used to calibrate the measurement method or instrument. *Spiked samples* are prepared by adding a known amount of the constituent of interest to blank samples to provide samples with known concentrations. It is useful for determining precision and accuracy. A *standard* is a substance or material, the properties of which are known with sufficient accuracy to permit its use to evaluate the same property of another. Standard is useful for calibration and accuracy determinations. A *primary standard* is a substance whose purity and the stability are well established and the value of it can be accepted without question. While specifying tolerance limits for pollutants primary standards are set to protect against adverse health effects and secondary standards are set to protect against welfare effects (damage to buildings, crops and vegetation).

The sampling model and accuracy of analytical data are the two most important parameters as regards reliability of the results. In quantitative analysis, the precision and accuracy are of the greatest importance. Accuracy is the degree of consistency between the mean of the measured data and the true value (Fig. 15.11). The accuracy of a measurement is a measure of the deviation of the measurement from the actual "correct" value, subject to only random errors (errors that vary in a non-reproducible way around the limiting mean). Random errors can be treated statistically, but (generally) not systematic errors (errors that are reproducible and tend to bias a result in one direction). The distribution should be symmetrical (normal distribution) around the "correct" value. Precision is the consistency between data produced by repeated measurements (reproducibility of replicate measurements of the same sample). High precision does not necessarily mean high accuracy.

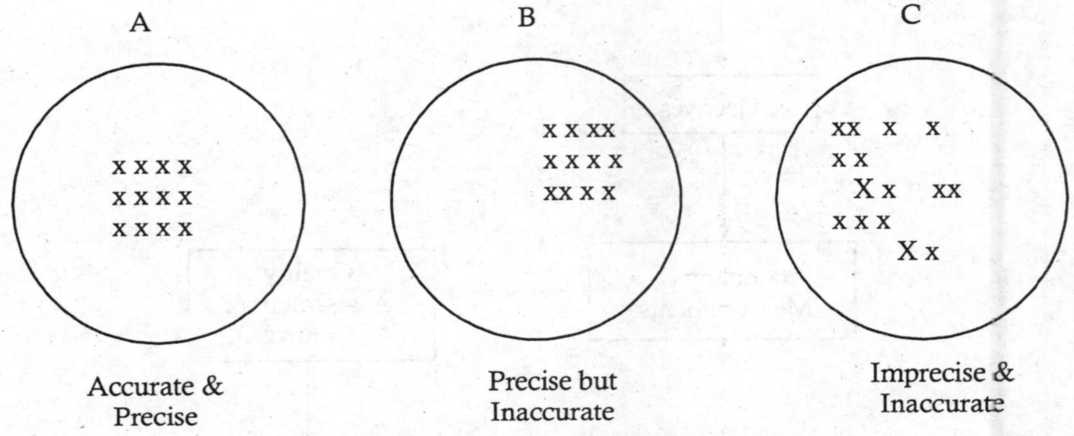


**Fig. 15.11:** Accuracy and Precision in Measurements

Precision of an analysis is usually found by making several determinations, $a_n$, and finding out the mean $\bar{a}$. The variation from the mean is determined by calculation of standard deviation, δ.

$$\bar{a} = \frac{1}{n}\sum_{i=l}^{i=n} a_n \quad \text{...(15.6)}$$

$$\delta = \sqrt{\frac{\sum_{i=l}^{i=n}(a_1 - \bar{a})^2}{n-1}} \quad \text{...(15.7)}$$

$$\delta_m = \frac{\delta}{\sqrt{n}} \quad \text{...(15.8)}$$

The precision can also be quoted in terms of the percentage of variation, $F$.

$$F = \frac{100 \times \delta}{\bar{a}} \quad \text{...(15.9)}$$

Validation is the process by which a sample measurement method or a piece of data is deemed to be useful for a specific purpose. Validation is necessary to confirm that the required specificity, precision and accuracy can be realized. Quality assurance is a part of environmental planning and execution that incorporates quality control and quality assessment.

### 15.6.2 Statistical Methods in Environmental Analysis

Mathematical, statistical and logic-based methods are applied to experimental data collection to derive as much information from the data as possible. They provide key insights into sources, migration pathways and the effects of pollutants on the ecosystems and human health. Invocation of statistical methods in environmental analysis is necessary for various reasons.

1. Most natural and anthropogenic environmental processes involve multidimensional changes in compounds belonging to different environmental compartments. Analyses for a large numbers of species in a large number of samples produce massive sets of highly complex data that make it extremely difficult to understand the nature of the system concerned.
2. Multi-elemental analyses have facilitated experimental data on large number of elements in a single sample. Number of substances (pollutants) determined at trace level also increased due to the application of modern experimental techniques.
3. Environmental data are intrinsically variable in nature. Consequently, a single analysis can reveal little about the pollutant distribution in a given sample. In addition to the spatial and temporal variability in the analytes, one must consider the variability resulting from experimental error.
4. Data processed are not of normal distribution (normal distribution is rarely the case with environmental data).

Regression, correlation, cluster, and discriminant analyses, and factorial methods are some of the statistical methods employed in environmental pollution analysis.

#### *15.6.2.1 Regression Methods*

Regression and correlation analysis methods allow one to establish a quantitative mathematical relationship between one specific variable and others. There are different regression methods of variable complexity. In a multivariate regression method (one dependent and several independent variables), the dependence of a variable, $y$, on several influencing variables is modeled: $y = f(x)$, where $x$ is real, independent, and uncorrelated. If the variables are correlated, the method generally applied is principal component regression. Method of simultaneous equations (correlated independent variables) can be applied, where correlation among independent variables is explicitly accounted for. This method is suited for semi-dependent variables. Partial least squares method (several dependent and several independent variables) allows one to model the dependence of a set of independent variables.

Some of the environmental applications of regression analysis methods are: (*i*) the detection of contributing pollutants in a complex mixtures of sources (e.g. predicting the concentrations of metals in water courses), (*ii*) the determination of ecological, spatial and temporal distribution and influences of pollutants, (*iii*) the assessment of biological responses to perturbations and influences.

#### *15.6.2.2 Correlation Methods*

The purpose of correlation analysis is to compare one or more functions and compute their relationship with respect to a spatial or temporal lag ($\tau$). Correlation between $x(t)$ and $y(t)$, which are dependent on lag, $\tau$, is considered. In environmental matrices the concentrations of many parameters change at the same time. Therefore, local variations in environmental pollutants must be analyzed by using a multivariate correlation method. Multivariate autocorrelation of $x(t)$ contains all possible relationships between the (different) variables within $x(t)$ and also those of $x(t)$ that are dependent on lag ($\tau$).

Multivariate auto- and cross-correlation analyses are frequently used in environmental analysis to describe time and distance series. Multivariate cross-correlation analysis methods can be applied for calculation of transport rate of metals and other elements in rivers. Generally tracers such as salts, colored compounds and radioactive isotopes are used to monitor pollution distribution in water flow systems (streams, rivers, and seacoasts). In addition to high cost these experimental techniques, the tracers themselves may become pollutants. Correlation analysis can reduce the analytical costs through optimized sampling plans by monitoring the concentration of any compound in a stream, present without the need to a tracer.

### 15.6.2.3 Cluster Analysis

Cluster analysis provides a useful means for detecting groups of singular objects in the multi-dimensional space. Hierarchical, non-hierarchical and "fuzzy" cluster algorithms are available for cluster analyses. In hierarchical and non-hierarchical analyses, each object is assumed to belong to a single cluster. Fuzzy cluster analysis allows objects to parts of more than one cluster at a time. Hierarchical cluster analysis (HCA) in the agglomerate mode is the most widely used clustering technique for analysis of environmental data. Such analyses are suitable for determining the similarities between sample sites and seasonal variations in air pollutants, anthropogenic activities on water bodies, group sampling along rivers according to their pollution patterns, pollution stratification in water bodies, determination of polluted soil layers. HCA analysis of the distribution of aerosol patterns enables one to identify particle sources, emission patterns and particle dynamics and particle agglomeration and distribution of elements according to particle size.

### 15.6.2.4 Discriminant Analysis

Discriminant analysis is employed to distinguish object classes. Discriminant functions are combinations of the pattern variables with weight coefficients.

$$df = \sum_{j=1}^{m} e_j . x_j \qquad \text{...(15.10)}$$

where, $x_j$ = variable;

$e_j$ = weight of the variable;

$m$ = number of variables.

Discriminant analysis may be used to derive information about the impact of the particulate emissions from the size and number of sedimented airborne particulate matter. It can also be employed to obtain objective information about the size and soil horizons based on their metal distribution and their impact.

### 15.6.2.5 Factorial Methods

Factorial methods attempt to reproduce maximally the matrix correlations. The principle of principal component analysis (PCA) is the transformation of the original features into uncorrelated new variables. Factorial methods have been used to identify source contributions and the spatial extent of their influence over the region of interest and air pollution control measures (in urban and rural regions). Factorial analysis has found wide application in the identification of atmospheric pollution sources, measurement of atmospheric concentrations of CO, $SO_2$, $NO_x$, $O_3$, and PAN, and their contributions in terms of particulate size. Aquatic systems undergo series of processes that alter their hydrochemistry and environmental quality. Such processes, which can be due to anthropogenic contributions (e.g. eutrophication in lakes), can be analyzed by factorial methods.

## BIBLIOGRAPHY

Alfussi, Z. B. and Wai, C. M. (eds.) (1992), CRC Press: Boca Raton, FL. "*Preconcentration Techniques for Trace Elements.*"

Barcelo, D. (ed.) (1993), Elsevier: Amsterdam "*Quality Assurance in Environmental Analysis.*"

Barth, D. S., *et al.* (1989), USEPA Report, EPA/600/8-89-046: Washington, DC. "*Soil Sampling Quality Assurance User's Guide.*"

Einax, J. W. (ed.) (1995), Springerverlag: Berlin. "*The Handbook of Environmental Chemistry.*"

Frei, R. W. and Albarge, J. (eds.) (1986), Gordon & Breach: New York, "*Air and Water Analysis: New Techniques and Data.*"

Gilbert, R. O. (1987), Van Nostrand-Reinhold: New York. "*Statistical Methods for Environmental Pollution Monitoring.*"

Gonzalez, M. M., Gallego, M. and Valcarcel, M. (2001), *Talanta*, **55**(1): 135. "Determination of arsenic in wheat flour by electrothermal atomic absorption spectrometry using a continuous precipitation-dissociation flow system."

Hoenig, M. (2001), *Talanta*, **54**(6): 1021. "Preparation steps in environmental trace element analysis—facts and traps."

Jorgensen, S. E. and Johnsen, I. (1981), Elsevier: Amsterdam. "*Principles of Environmental Science and Technology,*" (Vol., 14).

Keith, L. H. (ed.) (1988), Amer. Chem. Soc: Washington, DC. "*Principles of Environmental Sampling.*"

Keith, L. H. (ed.) (1991), Lewis Pubs: Chelsea, MI. "*Environmental Sampling and Analysis: A Practical Guide.*"

Lodge, J. P. (1991), Lewis Pubs: Chelsea, MI. "*Methods for Air Sampling and Analysis.*"

Markert, B. (ed.) (1994), VCH Pubs: New York. "*Environmental Sampling for Trace Analysis.*"

Peirce, J. J., Weiner, R. F. and Vesiland, P. A. (1998), Butterworth-Heineman: New Delhi. "*Environmental Pollution and Control.*"

Stern, A. C., *et al.* (1984), Academic Press: New York. "*Fundamentals of Air Pollution,*" (2nd edn.).

Thain, W. (1980), Pergamon Press: Oxford. "*Monitoring Toxic Gases in the Atmosphere for Hygiene and Pollution Control.*"

Van Loon, J. C. (1982), CRC Press: Boca Raton, FL. "*Chemical Analysis of Inorganic Constituents of Water.*"

Varga, B., *et al.* (2001), *Talanta*, **55**(3): 561. "Isolation of water-soluble organic atmospheric aerosols."

Vassileva, E., *et al.* (2001), *Talanta*, **54**(1): 187. "Revisitation of mineralization modes for arsenic and selenium determinations in environmental samples."

Warner, P. O. (1976), Wiley Inter Science: New York. "*Analysis of Air Pollutants.*"

Winegar, E. D. and Keith, L. H. (eds.) (1993), Lewis Pubs: Boca Raton, FL. "*Sampling and Analysis of Airborne Pollutants.*"

Zendelovska, D., *et al.* (2001), *Talanta*, **54**(1): 139. "Electrothermal atomic absorption spectrometric determination of cobalt, copper, lead and nickel in aragonite following flotation and extraction."

Zolotov, Yu. A. and Kuzmin, N. M. (eds.) (1990), Elsevier: Amsterdam. "*Preconcentration of Trace Elements*", "*Comprehensive Analytical Chemistry*", XXV, Svehla, G. (ed.).

# SECTION-V

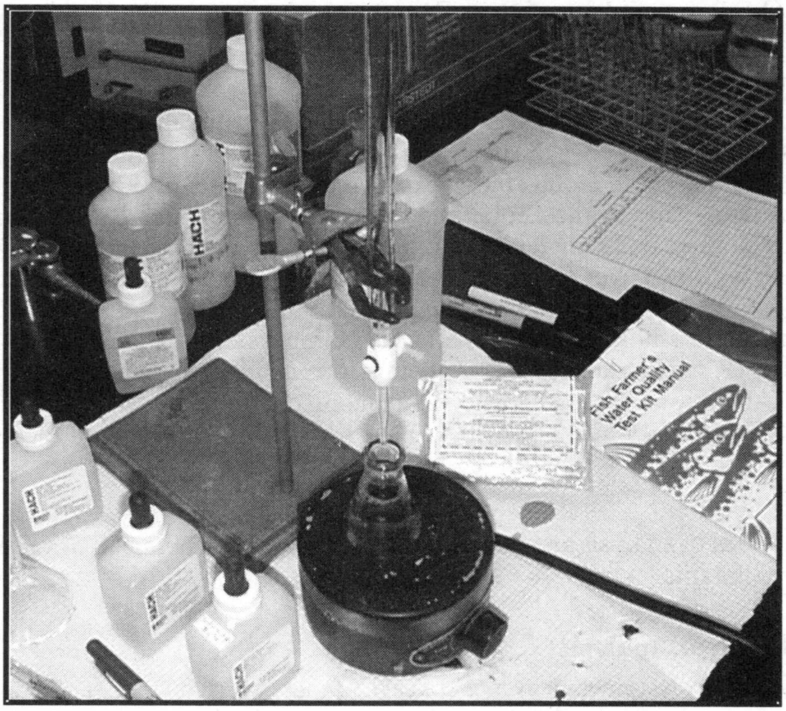


# Quality Analysis

## OVERVIEW

Chapters 16 and 17 give an overview of qualitative and quantitative methods for analysis of environmental pollutants. These include chemical and physical methods. The major aims of monitoring pollutants are: (1) identification and quantification of pollutants, (2) confirming that pollution level in the environment or work place is within statutory limits and (3) previewing the effects of allowing any new emission source in a certain region.

## AIR QUALITY ASSESSMENT

Air pollutants are classified as primary pollutants (e.g. $NO_x$, $SO_x$, $NH_3$, and $H_2S$) and secondary pollutants (e.g. $O_3$, photo-oxidants such as PAN, aldehydes, and free radicals).

### Qualitative Analysis

Qualitative methods of analysis rely on simple tests and help in detecting the presence of certain persistent pollutants. The presence of such pollutants can also be inferred from some physical and chemical changes taking place in the objects exposed to environment. Physical methods are based on changes in structure, appearance, color and other physical characteristics of materials.

It is possible to estimate pollutant level semi-quantitatively by indicator methods (static samplers). When polluted air is passed over an indicator surface it causes color changes. The intensity of such a change is proportional to the amount of the pollutant present.

Air pollution due to smoke from smokestacks, chimneys and exhausts can be monitored by visual comparison (Ringelmann test) or by photometry (nephelometry). Ringelmann test is based on the opacity of the smoke. By referring to Ringelmann chart numbers the per cent opacity can be assessed: number 0 = 0%, 1 = 20%, 2 = 40%, 3 = 60%, 4 = 80% and 5 = 100% opacity.

### Quantitative Analysis

Quantitative methods of analysis are carried out, based upon the chemical, physical and biological behavior of the pollutants.

### Chemical Methods of Analysis

Chemical methods of analysis include wet-chemical methods, which quantify pollutant concentrations by gravimetric, volumetric, titrimetric, electrochemical and spectrophotometric and other techniques.

In gravimetric techniques, the mass of the analyte (pollutant) is determined directly or indirectly. Volumetric methods aim at determining the volume of the analyte (gas). *Orsat* apparatus is used for analysis of gas mixtures. The method is based on step-wise removal of the individual gases from the total sample volume.

Electroanalytical methods (voltammetry, amperometry, polarography, potentiometry and coulometry) involve the use of electrically conductive probes to measure a chosen electrical parameter of the solution. The parameter so measured is related to the quantity of analyte in solution. Some of these techniques like polarography and stripping voltammetry can be employed for the analysis of trace elements.

Spectrophotometry is a common method of quantification among wet-chemical methods of analysis because of its inherent sensitivity. The sample is treated with suitable reagents to form a colored complex, and the characteristic color (wavelength) and its intensity are measured by a spectrophotometer. Turbidimeters and nephelometers are used to determine turbidity (opaqueness).

### Physical Methods of Analysis

Physical methods of analysis aim at determining parameters characteristic of the pollutant, without altering its chemical composition. Some of these parameters are thermal (radiation, temperature, conductivity), electrical and optical (refractive index). Elemental analysis by nuclear methods and mass, size, number and surface composition are included in this category.

Thermal methods employ passive radiation sensors (radiometers, interferometers, laser heterodyne amplifiers) and active sensors. These instruments are used to analyze the intensity and spectral distribution of radiation in the stratosphere and also the nature of atmospheric pollutants.

Temperature can be determined by contact thermometers (mercury, electrical resistance, thermistors, thermocouples and thermopiles) and non-contact thermometers (IR and microwave sensors). Non-contact thermometers are well suited for aerial survey. Combustion can be correlated with change in electrical resistance of a conductor (platinum wire) to determine flammable substances. Thermal conductivity detectors (catharometers) are employed to determine gaseous substances. They are employed in safety control (remote control) devices and as universal detectors (flame ionization detectors, FIDs) in gas chromatography.

Electrical conductivity methods are used to determine gaseous pollutants such as $SO_2$, $NO_2$, $CO_2$ and $H_2S$. The conductivity change of a solution before and after reacting with the pollutant is a measure of the quantity of pollutant gas. The method is highly sensitive and can be automated. It is used to determine $CO_2$ content of the flame in combustion processes.

Refractive index is a physical property that can be used to monitor the pollutants. For gases, interferometric method is used to determine the refractive indices of the test and reference gases. The method is very sensitive, fast and accurate, and is used in the monitoring flue gases, and explosive and combustible gases such as methane and $CO_2$ (firedamp interferometers).

Elemental analysis can be carried by nuclear analytical methods. These methods—X-ray fluorescence (XRF), ion-beam analysis (IBA), particle-induced X-ray/$\gamma$-ray emission (PIXE/PIGE)—are non-destructive, multi-element analytical techniques widely used in the sophisticated analysis of atmospheric aerosols.

Sophisticated techniques for mass concentration determination of aerosols are nephelometer (Tyndallometer), oscillation microbalance sensor, piezoelectric microbalance impactor, and $\beta$-attenuation mass sensor.

Particle size can be determined by microscopic observation ($d > 0.5$ μm), scattering-based light attenuation, and relaxation time-based and diffusion-based particle sizing methods. Scattering-based light attenuation method of particle sizing is based on Rayleigh/Mie theories on molecular scattering of electromagnetic radiation. Relaxation-based particle sizing is the basis on which cascade impactors function. Mobility-based particle sizing depends on particle size and shape. Diffusion batteries, sedigraph analyzer, electrical aerosol analyzer (EAA), Coulter counter and differential mobility analyzers (DMA) are some of the particle sizing instruments.

Number concentration of aerosol particles is determined by condensation nucleus counter (CNC), such as Pollack counter and photometers, such as optical particle counter (OPC).

Determination of surface composition of aerosol particles is of great importance because the surface components and the characteristics of aerosols determine their physiological effects. Secondary ion mass spectrometry (SI-MS), diffuse reflectance, atomic emission, Auger and photoelectron spectroscopies and microscopies are used in the analyses of surface characteristics of aerosols.

Vehicles, aircraft and remote sensing methods can be employed to carry out mobile monitoring of air pollutants. Differential absorption LIDAR (DIAL), and infrared DIAL (IR-DIAL) are used for remote sensing of gases in the atmosphere.

## WATER QUALITY ANALYSIS

Water quality can be determined from physical, chemical, biological and aesthetic points of view. Physical parameters are color, odor, turbidity, particulate matter and conductivity. Chemical parameters that affect the quality of water are pH, dissolved oxygen (DO), biochemical oxygen demand (BOD), chemical oxygen demand (COD), and dissolved substances, metals and redox chemical reactants. Industrial operations are the main sources of metallurgical and chemical pollutants. Biological parameters that affect the water quality are the presence of algae, viruses, coliforms, pathogens and other vectors that are responsible for health hazards and diseases.

### Physical Methods of Analysis

Organoleptic properties of water are related to odor, taste and color. Odor in water is a sign of decaying organic matter. Contaminants that contribute to odor are generally volatile organic compounds (aldehydes, ketones, and phenols). Methods of determination of odor and taste are qualitative and subjective.

Transparency of water is measured by the depth of penetration of light. This parameter depends on the coloring substances, turbidity due to suspended particles, and metal ions and their complexes. Turbidity is an optical phenomenon. It is a measure of attenuation of a beam of light as it passes through a non-absorbing medium. Turbidity in water is due to suspended particulate matter—clay, silt, coagulants, organic matter and photoplanktons. Nephelometers and tubidimeters are used to determine turbidity. Generally, turbidimeters are used to determine dense suspensions.

Humidity is the content of water vapor in air and its measurement is generally relevant in atmospheric studies. However, it is necessary in the determination of water content in soil samples and field studies. Relative humidity can be determined by hygrometers, wet-and-dry bulb thermometers (psychrometers) or by electrical conductimeters. The most sensitive device for determining water vapor concentration is IR-hygrometer or gas analyzer.

The dew point is the temperature below which moisture in a gas condenses. Instruments to measure the dew point (dew cells) are based on water condensation and the Peltier effect.

Pure water is a poor conductor of electricity. Presence of solutes increases the conductivity of water. Electrical conductimetric methods can be used to determine the moisture content, relative humidity and salinity.

### Chemical Methods of Analysis

Chemical parameters for determination of water quality are dissolved solutes, salinity, hardness, pH, DO, BOD, and COD.

Water salinity is mainly due to dissolved salts of sodium and potassium (NaCl and KCl). It can be determined by hygrometric, and titrimetric methods, in addition to optical (AAS), conductometric (salinometer) and ion-selective electrode (ISE)) methods. Water hardness is due to carbonates and sulfates of calcium and magnesium, and can be determined by titrimetric methods.

Acidity or alkalinity (pH) of water can be determined by potentiometric method (pH meter). Ion-selective electrodes (ISEs) can be used to determine ions and gases dissolved in water. Most of the inorganic pollutants in water (nitrates, sulfates, chlorides, hydrated metal complexes) can be determined by wet-chemical methods, monitored by spectrometry or by biological methods. Many of these methods can be automated for continuous monitoring. The automated methods are generally based on wet-chemical methods combined with physical methods of detection and quantification.

Ion-chromatography (IC) is a versatile technique that can be automated for environmental pollution analysis.

The determination of organic pollutants in water can be done by monitoring dissolved oxygen (DO) by BOD and COD methods. The popular method for continuous monitoring of DO is by membrane electrodes (Clark cell). Respiratory methods can also be used to determine $O_2$ uptake of organisms. Wet-chemical methods in combination with physical techniques for quantification are standard methods for analyses of volatile, semi-volatile and non-volatile organic, and humic organic compounds in water. Activated carbon is the most common adsorbent for organic species.

## Biological Methods of Analysis

Biological methods for assessing water quality rely on the estimation of biomass, coliform bodies, micro-organisms and nutrients. Estimation of biomass includes determination of total carbon, nitrogen, phosphorus, lipids, carbohydrates and benthal sediments. Sediments are usually the most appropriate candidates for long-term monitoring of contaminants in aquatic systems.

Algal proliferation is a useful indicator of hydrospheric and atmospheric pollution. Measurement of chlorophyll content is a suitable parameter to evaluate algal biomass. Chlorophyll determination is carried out spectrophotometrically (action spectroscopy). Determination of photosynthesis is carried out by measuring the absorption of infrared radiation (IR) of $CO_2$ in a non-dispersive IR (NDIR) analyzer.

Determination of adenosine triphosphate (ATP) and nucleic acid (DNA) is also used to estimate living biomass. The estimation is based on luminescence of enzymatic reaction of luciferase with ATP as substrate. The intensity of luminescence is proportional to the concentration of ATP.

Water pollution from domestic, food, beverage, sewage, dairy and agricultural wastes contain pathogens (bacteria, protozoans and viruses and disease carrying vectors). Determination of pathogens by instrumental methods is not practical. Instead, estimation of non-pathogenic (coliform) bacteria has been taken as a measure of the presence of waterborne pathogens. Estimation of coliform bacteria by most probable number (MPN) analysis depends on the detection of gas evolved when coliform bacteria ferment lactose. Living bacteria can also be detected and monitored by chemiluminescence ($\lambda$ = 550 nm). Fluorescence antibody technique is a serological method for detecting many pathogens, such as *Salmonella* and *Vibrio cholerae*.

# 16

# General Techniques in Air Quality Monitoring

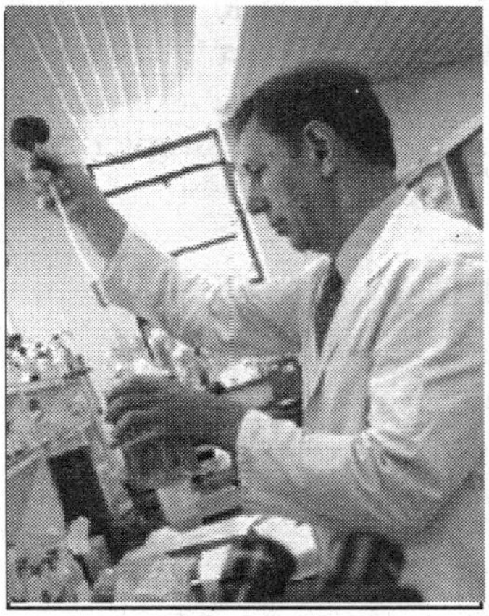

The air pollutants are classified as primary pollutants (e.g. $NO_x$, $SO_x$, $NH_3$, and $H_2S$) and secondary pollutants (e.g. $O_3$, photo-oxidants such as PAN, aldehydes, and free radicals).

Qualitative methods of analysis rely on simple tests and help in detecting the presence of certain persistent pollutants. The presence of such pollutants can also be inferred from some physical and chemical changes taking place in the objects exposed to environment. Physical methods are based on changes in structure, appearance, color and other physical characteristics of materials.

It is possible to estimate pollutant level semi-quantitatively by indicator methods (static samplers). When polluted air is passed over an indicator surface it causes color changes. The intensity of such a change is proportional to the amount of the pollutant present. Air pollution due to smoke from smokestacks, chimneys and exhausts can be monitored by visual comparison.

Chemical methods of quantitative analysis include wet-chemical methods, which quantify pollutant concentrations by gravimetric, volumetric, titrimetric, electrochemical and spectrophotometric and other techniques.

Physical methods of quantitative analysis aim at determining parameters characteristic of the pollutant, without altering its chemical composition. Some of these parameters are thermal (radiation, temperature, conductivity), electrical and optical (refractive index). Elemental analysis by nuclear methods and mass, size, number and surface composition are included in this category.

Sophisticated techniques for mass concentration determination of aerosols are nephelometer (Tyndallometer), oscillation microbalance sensor, piezoelectric microbalance impactor, and β-attenuation mass sensor. Diffusion batteries, sedigraph analyzer, electrical aerosol analyzer (EAA), Coulter counter and differential mobility analyzers (DMA) are some of the particle sizing instruments.

Vehicles, aircraft and remote sensing methods can be employed to carry out mobile monitoring of air pollutants. Differential absorption LIDAR (DIAL), and infrared DIAL (IR-DIAL) are used for remote sensing of gases in the atmosphere.

This chapter deals with the general techniques in air quality monitoring.

Earth's atmosphere can be regarded as a massive photochemical reactor where solar UV radiation catalyses production of free radicals and molecules, which give rise to pollutants. In addition, pollutants are added from Earth-based sources.

Atmospheric pollutants can be classified according to their vapor pressure as—(*i*) very volatile, (*ii*) semi-volatile and (*iii*) non-volatile or particulate. Most commonly occurring gaseous pollutants are: the primary pollutants such as $NO_x$, $SO_x$, $CO_x$, HCl, $NH_3$, $H_2S$ and HF, and secondary pollutants that are produced by chemical reaction in the atmosphere (e.g. photooxidants such as $O_3$, peroxyacetyl nitrate, PAN). Particulate pollutants are metal fumes and aerosols.

The pollutants of public concern are turbidity, inorganic (that include toxic metals) and organic chemicals, microorganisms, and radioactive substances, and thermal pollution mainly due to power generation and metallurgical and industrial operations. The major aims of environmental monitoring are: (*i*) identification and quantification of pollutants in ambient air to determine if they exceed their permissible levels, (*ii*) to make sure that the corrective measures taken have yielded the desired results, that is, have brought down the level of pollution and (*iii*) to detect unsuspected sources of pollution. Measurement of pollutants is carried out by chemical, physical, biophysical and biological techniques. The most desirable method for monitoring environmental pollution is continuous and automatic monitoring (e.g. by non-dispersive infrared analyzer for measuring CO and $CO_2$). However, since instrumentation for this purpose is costly, on many occasions simpler and conventional methods are used for monitoring pollution.

## 16.1 QUALITATIVE METHODS FOR POLLUTION MONITORING

Qualitative methods of monitoring air pollutants depend on observation as well as determination of physical changes as well as chemical changes, spatially and temporally. Physical changes such as tensile strength and structural deterioration owing to the interaction of atmospheric pollutants, chemical changes such as change of color, presence or absence of certain chemical groups are taken as indicators of the presence of certain types of substances and their qualitative detection. In addition, qualitative elemental analysis can provide valuable information for quantitative analysis by confirming the presence of the element to be determined and the order for magnitude of the amount present. Simple techniques are available for rapid and semi-quantitative estimation of some pollutants.

### 16.1.1 Indicator Methods

It is possible to determine pollutant levels by indicator methods (static samplers). Static sampler systems operate on the principle that when air is passed over the sampler surface, the pollutants are sorbed, leading to physical (structural) and chemical changes on the materials. Some general qualitative tests for the presence of pollutants are given in Table 16.1.

**TABLE 16.1:** QUALITATIVE POLLUTION MONITORING METHODS

| *Pollutant* | *Method* |
|---|---|
| Fluorides | Lime paper; glass etching |
| Sulfur and oxides of sulfur | Lead peroxide candles |
| Hydrogen sulfide, $H_2S$ | Silver tarnishing; discoloration |
| Hydrogen peroxide, $H_2O_2$ | Lead acetate paper |
| $CO_x$, $SO_x$ etc. | Colorimetric indicators |
| Particulate matter | Sticky papers; grease plates; dustfall jars |

Diffusion tubes are passive sampling devices in which a pollutant of interest is allowed to diffuse along a tube onto an absorber. The absorbed gas is determined by a color change. Diffusion tubes are used for area screening and in gas monitors. Diffusion tubes for volatile organic compounds (VOCs) carry polymer adsorbents such as Tenax and Porapak. For rapid semi-quantitative estimation of gaseous pollutants, the support material in the tube is coated with special reagents that react with pollutants giving characteristic colors. The intensity of color is proportional to the amount of the substance. Similarly, pencil detectors are used to detect poisonous gases such as HCN and phosgene. These pencils can be used to make marks on paper, wood, and walls. In the presence of a poisonous gas, the mark assumes a characteristic color. Reagent-impregnated strips (Fig. 16.1, & Table 16.2) are also in use for semi-quantitative analysis. These methods are simple, cheap and fast and are commonly used in quick identification of pollutants in the air.

**TABLE 16.2:** INDICATOR TUBES FOR DETECTION OF SOME GASES/VAPORS

| *Gas/Vapor* | *Reagent* | *Detection Limit (ppm)* |
|---|---|---|
| Acetone | Dinitrophenylhydrazine | 100 |
| Ethanol | $K_2Cr_2O_7 + H_2SO_4$ | 100 |
| Ammonia | $HgNO_3$ | 50 |
| Aniline | Furfurol | 5 |
| Arsine | Hg-$AuCl_3$ | 0.1 |
| $Cl_2$, $Br_2$ | o-toludine | 1 |
| CO | $I_2O_5$ + oleum | 10 |
| $CS_2$ | Cupric thiocarbamate | 30 |
| Formaldehyde | Xylene-$H_2SO_4$ | 5 |
| $H_2S$ | Lead(II) salt | 50 |
| HCl | $AgNO_3$ + dithizone | 5 |
| HCN | $HgCl_2$ (<pH 7) | 10 |
| Hydrocarbons | Via $Cl_2$ | 20 |
| Mercaptans | Cupric salt | 10 |
| $NO_2$ | Diphenylbenzidine | 1 |
| Ozone | Decolored indigo | 0.1 |
| Phenol | Indophenol | 1 |
| $SO_2$ | KI + starch | 1 |
| Styrene | $H_2SO_4$ | 50 |

### 16.1.2 Visibility Monitoring

Suspended particulate matter in the air affects its clarity. Visibility in the air can be monitored by Ringelmann test and instrumentally by photometers, nephelometers and light scattering devices (treated in Chapter 17) and data from both methods can be combined to obtain a better picture.

#### *16.1.2.1 Ringlemann Smoke Charts*

A smoke plume can be characterized according to its opacity (optical absorbancy). *Ringlemann* charts, which are a series of shaded bars (Fig 16.2), can be used to measure the opacity of air pollution emissions, ranging from light gray through black; used to set and enforce emissions standards.

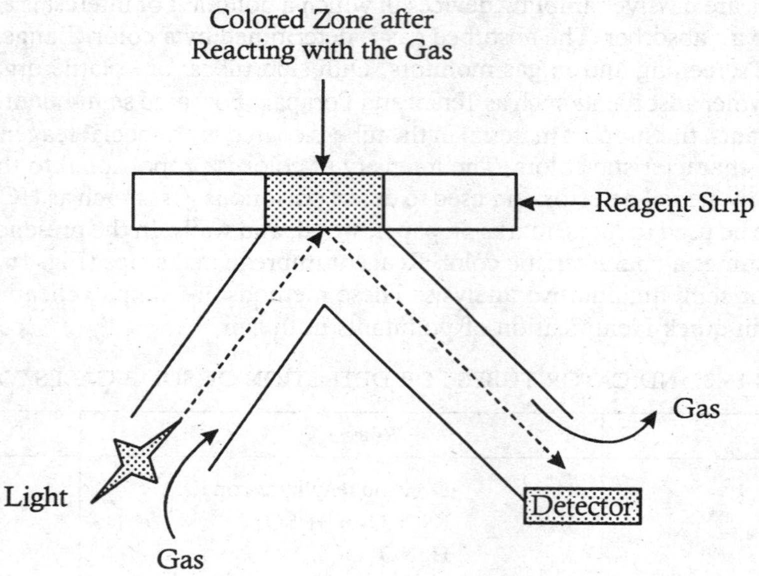


**Fig. 16.1:** Schematic of Colorimetric Analyzer with Reagent Strips

grayness of the smoke. The chart consists of four grids of black lines on a white background, having fractional black areas 20, 40 60 and 80 per cent (numbers 1 to 4). On the Ringlemann scale 0 = completely transparent (no visible smoke), 5 = completely opaque, and 1 = 20 per cent, 2 = 40 per cent, 3 = 60 per cent and 4 = 80 per cent opacity.

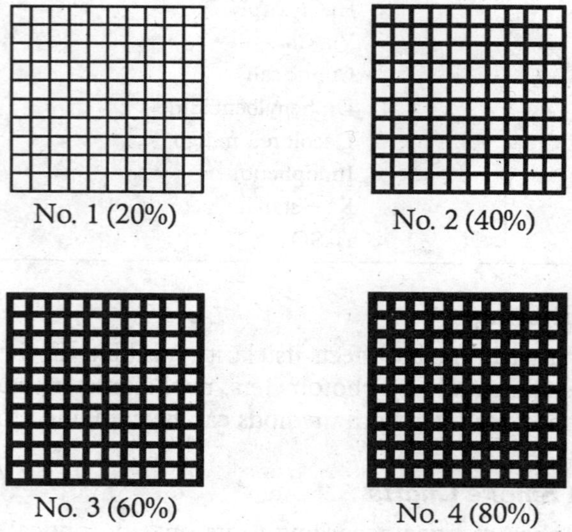


**Fig. 16.2:** Ringlemann's Smoke Charts

### 16.1.3 Odor

Odor is a sensation produced by stimulation of the chemoreceptors in the olfactory epithelium of the nose. Odor is produced due to the presence of volatile substances (aromatic compounds). For each odor, there exists a threshold concentration, below which it is not detected. The odor quotient, $Z$ (Zwaarder number) is

$$Z = \frac{C_o}{C_t} \qquad \text{...(16.1)}$$

General physiological method of analysis is by stepwise addition of odor-producing substance to purified air until the odor could be 'sniffed'. The relationship between odor intensity, $I$, and the concentration of the substance, $C$, is

$$I = K \log C \qquad \text{...(16.2)}$$

where $C_o$ = odor concentration of a sample;
$C_t$ = odor concentration at threshold;
$K$ = a constant.

Concentration in the range of 100 ppm for acetone, 1 ppm for acetic acid and ~0.02 ppm for pyridine can be detected by odor tests. Odor estimation is subjective. Human olfactory system is less sensitive than that of some animals.

Quantification of odorous substances can be determined by analytical methods—adsorption on activated carbon, and subsequent detection by flame combustion.

## 16.2 QUANTITATIVE METHODS FOR POLLUTION MONITORING

Quantitative analysis of pollutants in air is carried out by chemical, physical and other methods. An ideal analytical technique for measuring environmental pollutants must offer: (*i*) very low detection limits, (*ii*) a wide linear dynamic range, (*iii*) multi-element analysis, (*iv*) data free from matrix interference, and (*v*) low cost.

Chemical methods of analysis include wet-chemical methods, gravimetric, volumetric, colorimetric, electrochemical (e.g. voltammetry, conductimetry, potentiometry, and polarography) techniques (Chapters 19-23). Physical techniques comprise various spectroscopic (molecular and atomic absorption, emission and fluorescence), microscopic, electrical, thermal, chromatographic and mass spectrometry methods (Chapters 22-24). Determination of radionuclides is carried out with or without radiochemical separations (Chapter 25). Continuous monitoring for rapid multi-component analysis is carried out by various instrumental techniques (Table 16.3).

### 16.2.1 Chemical Methods

Chemical methods include standard wet-chemical analyses for the determination of elements and compounds. Evaporation, precipitation, extraction, sorption and electrodeposition are generally the methods employed for pre-concentration of trace level pollutants. All gases that react chemically with solutions or dissolve in them can be absorbed and determined by gravimetric, volumetric, manometric and titrimetric methods.

#### *16.2.1.1 Gravimetric Methods*

In this procedure the pollutant is converted into a suitable derivative and its weight is determined. The method is simple and straightforward. However, it needs a large quantity of sample. For example, this method is useful for the direct determination of $CO_2$ in air after precipitation as carbonate, and for indirect determination of carbon monoxide after converting it into $CO_2$. $H_2S$ can be determined by converting it to insoluble PbS.

**TABLE 16.3: CONTINUOUS MONITORING METHODS FOR AIR POLLUTANTS**

| *Pollutant* | *Monitor Type* | *Sensitivity/Specificity* |
|---|---|---|
| Carbon monoxide (CO) | NDIR | 1-10 ppm; specific |
| Carbon dioxide ($CO_2$) | NDIR | 1-10 ppm; specific |
| Hydrogen chloride (HCl) | NDIR | 1-10 ppm; specific |
| Oxides of nitrogen ($NO_x$) | Chemiluminescence | 5 $\mu g/m^3$; specific |
| Oxygen ($O_2$) | Paramagnetic; Electrocatalytic | Specific |
| Ozone ($O_3$) | Chemiluminescence | 5 $\mu g/m^3$; specific |
| Sulfur dioxide ($SO_2$) | NDIR; AAS; NDUV; Fluorescence | 1-10 ppm |
| Sulfur trioxide ($SO_3$) | Colorimetry | 10 $\mu g/m^3$ |
| Hydrocarbons (HCs) | NDIR; FID | Specific |
| Organics | GC-FID; GC-ECD; GC-PID;IR absorption spectroscopy; UV absorption spectroscopy | |

NDIR = Non-dispersive IR spectrometry; NDUV = Non-dispersive UV spectrometry;
AAS = Atomic absorption spectrometry; GC = Gas chromatography;
FID = Flame ionization detector; ECD = Electron capture detector.

### *16.2.1.2 Volumetric Methods*

Volumetric analysis is aimed at determining the volume of a gas absorbed in a solution. At constant pressure and temperature, the partial volume of each component is proportional to its mole fraction in the mixture.

$$\frac{V_i}{V} = \frac{n_i}{\sum_j n_j} \qquad \text{...(16.3)}$$

The method is based on the determination of volume changes before absorption of the gas in a suitable reagent, and after. The volume fraction, φ, is

$$\varphi = \left(\frac{V_a - V_b}{V_a}\right) \times 100 \qquad \text{...(16.4)}$$

where, $V_i$ = volume of the component gas;
$V$ = total volume of the mixture;
$n_i$ = number of molecules of the component gas;
$V_a$ & $V_b$ = gas volume prior to and after absorption.

Gas volume can be measured by burettes, pipettes and also by gas meters and rotameters (rate of flow). The apparatus (*Orsat* instrument) for complete analysis of gas mixture consists of a calibrated burette and a system of absorption pipettes and a combustion unit, based on step-wise removal of the individual gases from the total sample volume. Volumetric analysis can be automated (Fig. 16.3). It is very important to maintain the exact sequence of analysis order in the analysis of gas mixtures. The sequence in order of analysis is:

$CO_2 \rightarrow$ unsaturated HCs $\rightarrow O_2 \rightarrow$ CO and then removal of $CH_4$, $C_2H_6$ etc. by combustion.

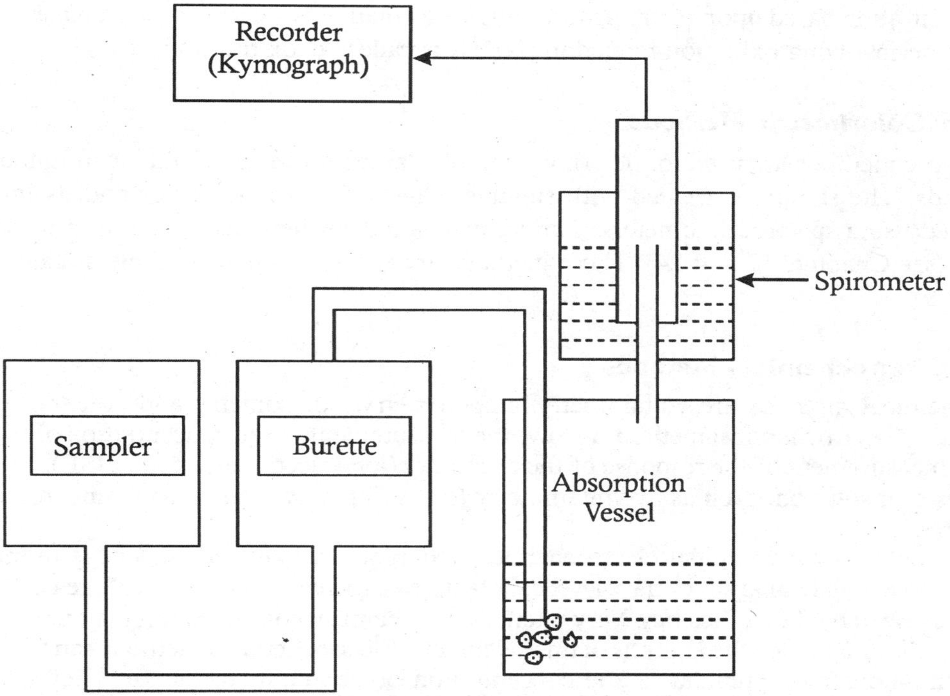


**Fig. 16.13:** Block Diagram of Single-component Absorption Volumetric Analyzer

### *16.2.1.3 Manometric Analysis*

For any given gas, the partial pressure of each component is proportional to its mole fraction in the mixture. Manometric analysis is based on the relationship

$$\frac{P_i}{P} = \frac{n_i}{\sum_j n_j} \quad \text{...(16.5)}$$

Manometric analyses are determination in which the individual components of the mixture are separated by sequential absorption and the quantitative composition is determined from pressure changes before and after absorption. Measurements are carried at constant volume and temperature. The absorbed portion of gas (φ in percent) at constant temperature and volume is

$$\varphi = \frac{\Delta P}{P_0} \times 100 \quad \text{...(16.6)}$$

where, $\Delta P$ = pressure decrease after absorption,
$P_0$ = initial pressure.

### *16.2.1.4 Titrimetric Methods*

Tritrimetric methods make use of reactions in solution (e.g. acid-base reactions) between standard reagents and the analyte. The solution of the reagent of known composition is the *titrant, and the solution of the analyte, the titrand*. The point at which all of the analyte has reacted is known as the end point. Titration based upon electrochemical generation of the reagent is known as *coulometric* titration. Titration based upon metal-ligand complex formation is known as *complexometric* titration and titration involving oxidation-reduction process is called redox titration.

#### *16.2.1.5 Colorimetric Methods*

Colorimetric methods are used for determination of a large number of inorganic ions and organic compounds. The sample is treated with suitable reagents to produce color and its intensity is determined using spectrophotometers. Turbidimeters and nephelometers are used to determine turbidity (see Chapters 17 and 24). These methods are easy and simple and are suitable for field analysis too.

### 16.2.2 Electrochemical Methods

Electrochemical methods are voltammetry, amperometry, polorometry and potentiometry (see Chapter 23). Electrochemical methods involve the measurement of the concentration of the analyte by direct measurement of the response of the electrodes (ion-selective electrodes, ISE), or electrical properties of a solution, such as by voltammetry (e.g. stripping voltammetry), amperometry, and coulometry.

Voltammetry is a dynamical electrochemical technique, where the current is monitored at controlled potential. In analysis of trace level pollutants, electrochemical methods of pre-concentration are characterized by their efficiency. Very often, electrochemical concentration is an integral part of stripping voltammetry for the analysis of trace elements. Electrochemical methods can be employed in the continuous measurement of the concentration of pollutants in the air. They are used as personnel monitors. Heavy metals (Cd, Cu, Pb, Se, Zn) in biological systems can be determined by differential pulse stripping voltammetry (anodic (DPASV) and cathodic (DPCSV)), with hanging mercury drop electrode (HMDE).

In polarographic methods, a voltage applied to the cell is regulated by means of a potentiometer and the current passing through the cell is read by a deflection galvanometer. The method is applicable for those constituents that are reducible at the mercury cathode (see Chapter 23 for details). This method may be employed for the analysis of trace elements—antimony (Sb), arsenic (As), cadmium (Cd), chromium (Cr), cobalt (Co), copper (Cu), lead (Pb), manganese (Mg), nickel (Ni), tin (Sn), vanadium (V) and zinc (Zn). The method is also used for continuous determination of oxygen in gases and water, and in the study of physiological processes (Clark cell sensors).

In ion chromatography (Chapter 22), the conduction of a solution can be monitored, as it is eluted from an exchange column. When an electric potential is applied to electrodes immersed in an electrolytic solution, a current flow is generated. The accumulated products build up a galvanic cell. When the electrolytic e.m.f. equals the applied e.m.f. no current will flow. The potential of the indicator electrode is measured with respect to a reference electrode. Potentiometry is the best method because of its specification (pH, redox).

### 16.2.3 Physical Methods

Physical methods of analysis are aimed at determining parameters specific for the pollutant without altering its chemical composition. Replacement of wet-chemical methods by physical sensors has enabled the elucidation of rapid temporal changes in the concentration of air pollutants. Standard physical techniques are—optical methods (microscopy, spectroscopy, refractometry, particle counting and photometry), chromatographic, and mass spectrometric methods, and nuclear analytical techniques (Chapter 22-25). Some of the physical parameters determined are absorption, emission, fluorescence, and thermal conductivity (Chapter 24). Laser optoacoustic spectroscopy gas analyzers now provide a single system that can automatically analyze a mixture of gases with high sensitivity. HPLC and GC-MS techniques can be employed to carry out routine pollutant measurements in ambient air (Chapter 22). Sound meters are used to determine noise pollution. Determination of radioactive materials forms a separate category involving radiophysics and radiochemistry (Chapter 25).

### 16.2.4 Thermal Methods

Methods based on measurements of temperature, conductivity and combustion are used in quantitative determination of the composition of gases. Temperature of the medium (air, water or soil) can be determined by thermometry. Liquid-filled thermometers, resistance (ohmic and semiconductor type) thermometers, thermocouples, thermopiles and other devices are available. The most common thermometers are liquid-filled thermometers (range –200 → 600° C). Mercury thermometers are accurate to 0.05 °.

#### *16.2.4.1 Ohmic Resistance Thermometers*

These devices are based on the change in electrical resistance with temperature. Both conductor (ohmic resistance) and semiconductor devices are used. The conductor thermometers are based on the ohmic resistance of a conductor (Pt, Cu, Ni and Ag). The resistance to passage of a current through a conducting wire is proportional to the temperature. The empirical relationship is

$$R_T = R_0(1 + a\Delta T + b\Delta T^2) \qquad \ldots(16.7)$$

where, $R_T$ & $R_0$ = resistance at temperature, $T$ and 0 (° C);
$a$ & $b$ = constants

The Pt-resistance thermometer (range –200 → 1,110° C) is most widely used and the resistance of the conducting wire is usually determined with a whetstone bridge (precision ~0.0001°).

#### *16.2.4.2 Thermistors*

The resistance of a semiconductor decreases rapidly with temperature, with a large negative coefficient with temperature (~ –5°/° C).

$$R = R_0 \exp\left(\frac{1}{T} - \frac{1}{T_0}\right) \qquad \ldots(16.8)$$

where, $R_0$ & $T_0$ = resistance & temperature at 298 K

Thermistors are made from mixtures of metallic oxides of Co, Fe, Mn, and Ni. Si and Ge-diodes are also in use. Semiconductor thermometers (thermistors) have high sensitivity (rate of temperature variation is an order of magnitude higher than for ohmic conductors) and fast response to temperature changes (precision ~0.0001°). Consequently they are well suited as temperature sensors. The resistance of a thermistor can be readout by a whetstone bridge, similar to conducting wire thermometer

$$V = a + T(b + c \cdot \ln I) \qquad \ldots(16.9)$$

where, $V$ = voltage;
$I$ = current;
$a$, $b$, and $c$ = empirical constants;
$T$ = ° C

#### 16.2.4.3 Thermocouples

Junctions of two dissimilar metals form a *thermocouple*. Thermocouples are differential temperature sensors, with at least two sensing junctions. One junction of the bimetal device is kept at constant temperature (reference temperature). Temperature measurements with the help of thermocouples are reduced to measuring the potential difference. The electromotive force (e.m.f), $E$, at the test junction is a function of the temperature difference.

$$E = a + bT + cT^2 \qquad \text{...(16.10)}$$

Noble-metal thermocouple (*Pt/Rh*) is the most reliable, and commonly used (0 → 1,600° C) one. Other bimetal thermocouples are Cu-constantan (–200 → 600° C), Fe-constantan (–200 → 900° C), Cr-Ni (0 → 1,200° C) and Cr-alumel (–200 → 1,200° C).

While, thermocouple is a thermogenerative device, *thermopile* is a photogenerative device, and where the current generated is proportional to the incident (absorbed) quanta of radiation. Thermopiles measure the temperature differential between two surfaces.

#### 16.2.4.4 Thermal Conductivity Measurement

At atmospheric pressure, the thermal conductivity of a gas depends on the temperature, which is the basis for the measurement of thermal conductivity of gaseous samples. Thermal conductivity of gases can be used to evaluate the concentration of a substance. A thermal conductivity meter consists of a cylindrical cell (kept in a thermostat) and a metal filament stretched along its axis. The filament is surrounded by the test gas and is heated by a constant electric current, and thermal conductivity measured (see Chapter 22 for details). Thermal conductivity detectors (also called *catharometers*) are employed to identify substances in gaseous state, in safety control devices (remote control). Thermal conductivity detector (TCD) is used as a universal detector in chromatographic methods because of its universal nature (all substances give a signal), non-destructive nature and broad linear dynamic range.

#### 16.2.4.5 Combustion Methods

Combustion methods are used primarily for analysis of combustible gases. These instruments are used to monitor flammable gases, and as safety devices in fire hazard areas. All these instruments are based on total combustion on a heated catalyst (platinum wire). The heat released increases the temperature of the wire and thus also its resistance. The change in resistance is measured by connecting the circuit in one branch of a Whetstone bridge, and the other branch consists of a filament in reference gas chamber.

Combustion methods convert non-absorbable gases into readily absorbable $CO_2$. By introducing combustion, *Orsat* method can be extended to some components for which it is not otherwise useful. Fractional combustion procedure is used for determination of hydrogen, methane and carbon monoxide. The sample is heated first to 150° C to determine $H_2$ and CO (CO → $CO_2$), and then the temperature is raised to 450° C to convert $CH_4$ to $CO_2$.

## 16.3 OTHER METHODS

Electrical conductivity, interferometric, photoacoustic and nuclear analytical methods are some other methods used in air pollution monitoring.

### 16.3.1 Electrical Conductivity

Electrical conductivity is the reciprocal of electric resistance.

$$C \approx 0.01\ K \qquad \ldots(16.11)$$

where, $C$ = sum of ion in $m.mol.L^{-1}$;
$K$ = conductivity.

Electrical conductivity can be used to determine atmospheric pollutant gases such as $SO_2$, $NO_2$, $H_2S$, and $CO_2$. The method is based on chemical reactions involved in absorption of these gases. The solution conductivity before and after chemical reactions is measured by a differential method. For example, $CO_2$ can be determined by absorbing the gas in aqueous $BaCO_3$; $H_2S$ by reaction with bromine and $SO_2$ by $H_2O_2$ reaction ($H_2SO_4$ produced increases the conductivity of the solution). The method is sensitive and no special electrodes are needed. It can be automated. It can be used to determine $CO_2$ content of the flame in the case of combustion of hydrocarbons and to indicate the appearance of other combustion products (intermediates) (Chapter 24).

Electrothermal-AAS (ET-AAS) methods are highly sensitive techniques for the estimation of trace elements in blood, urine and other body fluids and tissues (ET-AAS for As, Cr and Mn; AAS for Fe and Mg) (see Chapter 24).

### 16.3.2 Radiation Monitoring Methods

The optical properties of air offer scope for the application of instrumental methods for the evaluation of particulate matter for air quality standards. Vehicles or aircraft and remote sensing satellites can be used to carry out mobile monitoring of air pollutants. The open-path, optical remote sensing techniques measure the interaction of radiation with pollutants in real-time. Sonic anemometers, radiometers, UV-videocons, interferometers, and laser heterodyne amplifiers are passive sensors, while laser absorption spectrometers, and *LIDARs* (*Light Detection and Ranging*) are active sensors. Active radiation sensors have built-in radiation source. They are used to study atmospheric phenomena in 5-10 km altitudes.

#### *16.3.2.1 Radiometers*

*Radiometer* is a generic name for any device that measures radiation by absorbing the radiation and transmitting it as thermal energy (*radiative transducer*). The principle of these sensors is based on the *Stefan-Boltzmann* law. The flux of the radiation

$$S = \varepsilon \sigma T^4 \qquad \ldots(16.12)$$

where, $S$ = total energy from the surface (flux);
$\varepsilon$ = emissivity (~1);
$\sigma$ = Stefan-Boltzmann constant;
$T$ = absolute temperature.

Infrared (IR) and microwave sensors are non-contact thermometers and are well suited for aerial surveys and medical uses. They are used for remote non-contact measurements of the surface temperature of water, land and foliage. Infrared thermometer is operated at 8-13 μm bandpass, to exclude solar radiation and absorption by water and $CO_2$. Photoconductive cells (CdS & PbS) and photovoltaic cells (Se cell) are the most commonly used in IR-thermometry and IR-gas analyzers. IR-thermometer is the most satisfactory method for determining surface temperature (e.g. surface temperature of skin, face and other thermography studies). Radiometers may be used to measure thermal radiation, with a suitable radiation sensor plate (Fig. 16.4).

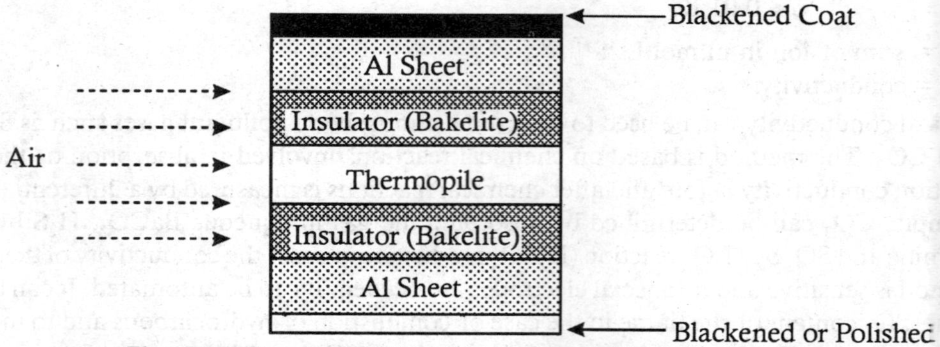


**Fig. 16.4:** Functional Elements of a Radiation Sensor Plate

The energy input at the interface of the sensor plate, which acts as a heat exchanger, at midday with clear conditions, is

$$S_i - S_o + I_i - I_o \pm G \pm L_V \cdot E \pm H = 0 \qquad \text{...(16.13)}$$

where, $S$ = flux of solar (short wave) radiation; $I$ = flux of IR radiation; $G$ = flux of heat into the soil; $L_V$ = latent heat of vaporization; $E$ = rate of evaporation; $i$ = input; $H$ = flux of heat by convection into the air stream; $o$ = output.

Solar meters may be used to measure the flux, $S$.

In the sensor plate, if the top is blackened and the bottom is polished, the instrument is used as an 'all-wave radiometer' ($S_i + I_i$); if both the top and bottom surfaces are blackened (blackbody), then the instrument is a net radiometer (IR analyzer) ($S_i - S_O + I_i - I_O$).

### *16.3.2.2 Interferometric Method*

Refractive index is a physical property; it is the ratio of the velocity of light through vacuum to its velocity through the medium ($n = C_{vac}/C_{med}$). Refractometers are used to identify and determine the concentration of substances. However, refractometers cannot be used to determine the refractive indices of gases as they lack sensitivity (refractive index of gases is ~1.0). Interferometry is a physical (optical) technique used to determine the difference in the refractive indices of the test and reference substances using the shift in the interference spectrum (interferogram). The method is very sensitive (0.0001 μg). Interferometric measurements are very fast and accurate, and they are used in the analysis of flue gases, explosive and combustible gases (*firedamp interferometers*), and impurities in gas mixtures and atmospheric pollutants.

In an interferometer setup (Fig. 16.5), one light beam passes through the reference cell and the other through the test cell. If the reference and test cells contain substances with different refractive indices, then a phase shift occurs with a magnitude proportional to the difference in refractive indices of the media.

$$L = \frac{100N\lambda}{c(n_2 - n_1)} \qquad \text{...(16.14)}$$

where, $L$ = path length of the light beam in the medium;
$N$ = number of interference lines;
$\lambda$ = wavelength;
$c$ = concentration of the test substance;
$n$ = refractive index

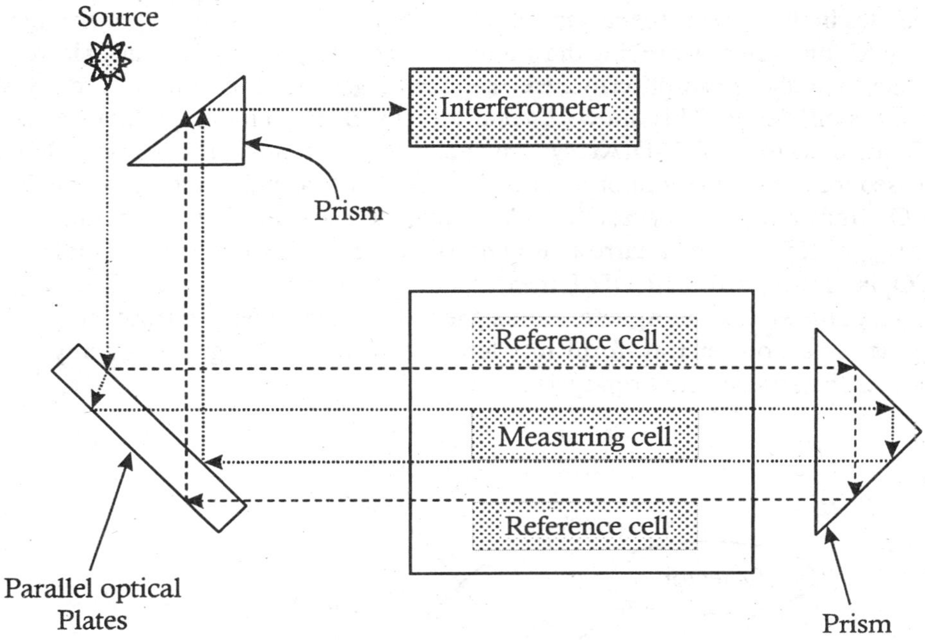


**Fig. 16.5:** Measurement of Refractive Index by Interferometry

### *16.3.2.3 Long-path Monitorng*

In these techniques, a pulse of light from a high-power laser source is sent into the atmosphere. The light pulse is scattered by aerosols (by Mie scattering) and particles (by Rayleigh scattering) and the portion of the light that is reflected back is detected. The position of the absorbing species can be calculated from the time elapsed between the pulse emission and the return signal. Diode lasers have been used for measuring atmospheric gaseous constituents such as CO, $CO_2$, $C_2H_4$, NO and water vapor by long-path monitoring (Fig. 16.6). In order to minimize the effects of atmospheric turbulence on system sensitivity, a derivative spectroscopic technique (refer to Chapter 24 for details) is employed in which lasers are frequency modulated. The long-path diode laser system permits unattended monitoring around the clock. It can also be used for monitoring combustion processes. The chemical composition of the upper atmosphere can be analyzed by atmospheric trace molecule spectrometer (ATMOS), which is a Fourier transform interferometer. These radiation sensors provide real-time atmospheric monitoring.

The most common active radiation sensors are laser absorption spectrometers, LIDARs (single-wavelength, differential Raman or fluorescence type). In LIDAR method, back-scattered pulses are collimated by a lens system and detected by a photometer. Intensity data of in the 3-14 μm spectral ranges are measured by radiometer, grating spectrometer, and interferometer. In *correlation spectroscopy* method, light quanta, collected from the source by a telescope are dispersed by a double-grating spectrometer and recorded electronically. The spectral characteristics of the source are compared with those of known pollutants to detect their presence and concentrations in the source.

Differential LIDAR (DIAL) provides range-related pollutant measurements and is therefore a true *in situ* remote sensing technique, providing the chemical composition of the atmosphere. Differential absorption lidar (DIAL) is used for remote sensing of gases in the atmosphere. DIAL usually operates in the UV/visible region, as atmospheric $CO_2$ and water vapor strongly absorb IR radiation. DIAL measures by probing the atmosphere at two wavelengths (one at lower absorption and the other at higher absorption) where the sample gas and the other efficiently absorb one wavelength less efficiently. This allows three-dimensional mapping of pollutant concentrations. $SO_2$ and $O_3$ are detected by UV-DIAL system operating at 300 nm. LIDARs (UV-DOAS) systems have been used for near-field monitoring of emissions of criteria pollutants such as NO, $NO_2$, $CO_2$, $SO_2$, and $O_3$. Infrared gas analyzer (IRGA) is used to measure $CO_2$ in the atmosphere. $CO_2$ absorbs at $\lambda_{max}$ = 425 μm. So, a narrow IR bandpass filter is used to admit this radiation. Tunable infrared $CO_2$ lasers are used in IR-DIAL measurements to measure $SO_2$, CO, HCl, $CH_4$, $CO_2$, $H_2O$, $NO_x$, $NH_3$ and $H_2S$ in the atmosphere. Atmospheric trace molecule spectroscope (ATMOS) is an IR spectrometer (a Fourier transform interferometer) that is designed to study the chemical composition of the atmosphere from space.

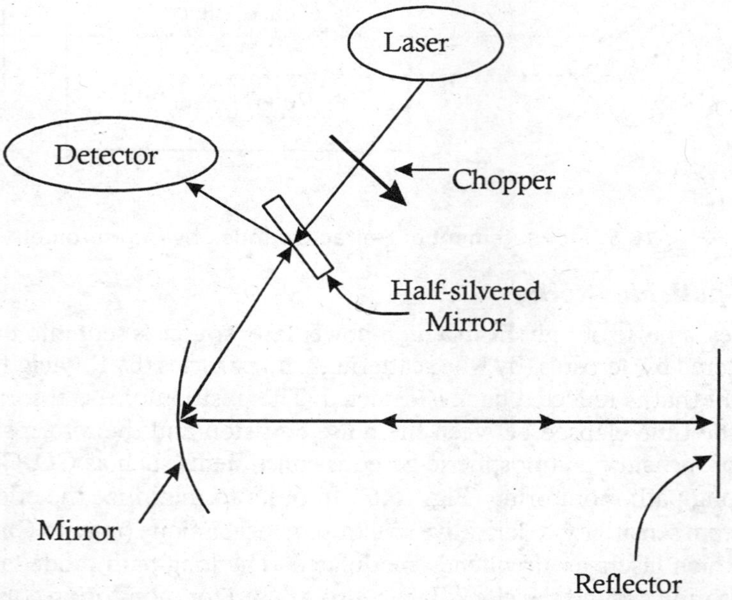


**Fig. 16.6:** Schematic of Long-path Optical Monitoring System

Simultaneous measurement of ozone and water vapor by Raman differential absorption Lidar is based on the simultaneous measurement of the Raman back-scattering of UV radiation by $O_2$, $N_2$ and $H_2O$, using a single pump beam at 266 nm. The ozone concentration is retrieved from the differential absorption of the $O_2$, and $N_2$.

### *16.3.2.4 Remote Heterodyne Monitoring*

The concentration of pollutant gases in the atmosphere as a function of altitude can be monitored by heterodyne monitoring systems. The measurement is based on the effects of pressure on broadening of spectral absorption lines. It could be achieved in a (*i*) passive mode or (*ii*) by reflection mode (Fig. 16.7). In passive solar heterodyne radiometry, the sun serves as the source of radiation and heterodyne receivers record the amount of transmitted radiation on the earth's surface. In the reflection mode, a laser beam that is reflected by a satellite is monitored on the earth's surface by an IR detector. Infrared differential absorption Lidar (IRDIAL) techniques allow remote sensing of gases (air pollutants) in the atmosphere from distances up to several kilometers. Differential optical absorption spectroscopy (DOAS) is capable of *in situ* measurements with high sensitivity.

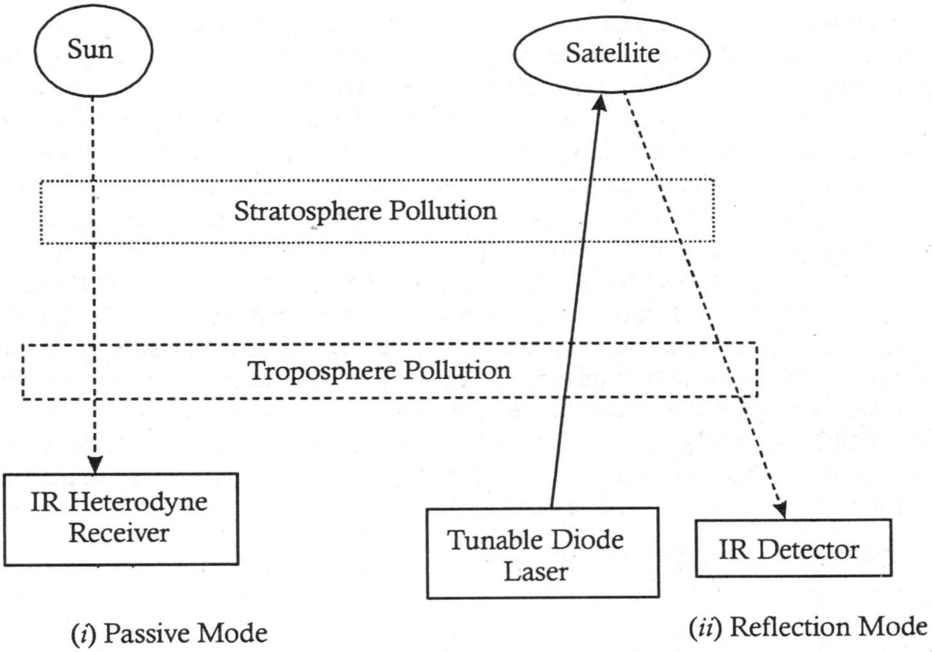


**Fig. 16.7:** Remote Heterodyne Monitoring System: (*i*) Passive Mode, (*ii*) Reflection Mode

### 16.3.4 Nuclear Analytical Methods

Nuclear analytical techniques include X-ray fluorescence (XRF), ion beam analysis (IBA), neutron activation analysis (NAA), particle-induced X-ray/γ-ray emission (PIXE/PIGE), for the elemental analysis. These are non-destructive, multi-element analytical techniques for particulate matter and are widely used in the analysis of atmospheric aerosols. XRF is for the analysis of elements above atomic number, $Z > 10$. Neutron activation analysis (NAA) is used to determine trace elements like As, Be, Br, Cd, Hg, V, and other toxic elements at μg – ng level. Although about 30-40 elements can be determined by NAA as compared to about 15-20 by XRF and PIXE, these two techniques provide speed and the advantage of low cost and small sample requirements (see Chapter 25 for details).

## 16.4 ANALYSIS OF AIR POLLUTANTS

Methods used for the determination of pollutants in air are equally valid for their determination in other media—water and soil, with suitable changes in procedure for sample preparation and concentration. Analyses include determination of size, mass and number, and surface characteristics of aerosols and particulates, determination of air toxics that include hazardous air pollutants (HAPs) and volatile organic compounds (VOCs) and radioactive pollutants.

### 16.4.1 Analysis of Particulate Pollutants

Airborne particulate matter "aerosols" are behind most air pollution problems. They affect global climate directly by scattering solar radiation and indirectly by altering the *albedo* (the ratio between the solar radiation falling upon a surface and the amount reflected) and occurrence of clouds. Aerosols contribute to reduced visibility and acid deposition over wide areas on the planet. Analysis of gaseous air pollutants (aerosols) includes the three important parameters—determination of the (*i*) particle size, (*ii*) concentration (number range) and surface characteristics of particulates, and (*iii*) chemical composition and biological effects of the particles.

Measurement of airborne particles with aerodynamic diameter up to 10 μm (PM-10) or 2.5 μm (PM-2.5) has become increasingly important as the health effects of these fine particles are widely recognized. Composition of gaseous air pollutants (aerosols) is determined by several chemical and physical techniques (Table 16.4). Composition of single particles can be determined by electron probe X-ray microanalysis (EPXMA), microparticle-induced X-ray emission (μ-PIXE), electron microscopy—scanning electron microscopy with energy-dispersive (SEM-EDX) or wavelength-dispersive (SM-WDX) X-ray detection, laser microprobe mass spectrometry (LMMS), FTIR, micro-Raman and electron-energy loss spectrometry (EELS). EELS has high sensitivity for elements of low atomic number (Z); information on electronic state and chemical bonding can be obtained. In LMMS, a high-power pulse laser beam is used to vaporize and ionize a small sample area. The time-of-flight (TOF) mass analyzer separates ions with different mass/charge (m/ze) ratios according to flight times. This technique can detect every element. Ion beams collimated to nanometer size can be very useful for individual particle analysis (see Chapter 25 for details).

**TABLE 16.4:** DETECTION METHODS FOR GASEOUS AIR POLLUTANTS

| *Pollutant* | *Instrumentation* |
|---|---|
| Ammonia ($NH_3$) | Colorimetry; Conductimetry; Potentiometry; Spectrophotometry; Wet-chemistry |
| Carbon oxides ($CO_x$) | Electrochemical; NDIR spectroscopy; Correlation spectroscopy; GC; UV |
| Hydrocarbons (HCs) | Amperometry; Chromatography; Colorimetry; GC-MS; LMMS; Flame ionization; IR; UV absorption; Wet-chemistry |
| Hydrogen chloride (HCl) | Electrical conductivity; Wet-chemistry |
| Hydrogen sulfide ($H_2S$) | Chromatography; Colorimetry; Conductimetry; Electrochemical; NDIR spectroscopy; Wet-chemistry |
| Nitrogen oxides ($NO_x$) | Amperometry; Chemiluminescence; Colorimetry; Potentiometry; IR; UV absorption spectroscopy |
| Oxidants ($O_2$ and $O_3$) | Amperometry; Chemiluminescence; Colorimetry; UV absorption spectroscopy |
| Polycyclic aromatic hydrocarbons (PAHs) | GC-MS; HPLC; LMMS |
| Sulfur oxides ($SO_x$) | Amperometry; Colorimetry; Conductimetry; Flame ionization; Potentiometry; GC; IR; UV (absorption and emission) spectroscopies |

IR = Infrared; NDIR = Non-dispersive Infrared; GC = Gas Chromatography; MS = Mass Spectrometry LMMS= Laser Microprobe Mass Spectrometry; UV = Ultraviolet; HPLC = High Performance Liquid Chromatography.

Automobile exhaust pollutants are monitored by gas chromatography (GC). Inorganic gas chromatography (IGC) has been developed as a method of analysis of volatile compounds that are obtained by chemical transformation or simple distillation of volatile compounds (metal chlorides)

and detected by AAS or AES. Volatile organic pollutants (benzene, ethane, ethylene, methane, toluene, xylene, etc.) can be determined with flame ionization detector (FID) in combination with GC. Semi-volatile organocholorides (e.g. DDT), and non-volatile organic compounds (e.g. polycyclic aromatic hydrocarbons, PAHs) can be determined by reversed phase HPLC, with an UV detector.

Flame ionization detector (FID) is ideal for the determination of hydrocarbons (toxicants; pesticides), because it detects compounds with C-H bonds. The air sample with hydrocarbon contaminants is passed through an adsorption tube that contains a liquid phase held on a solid support. The adsorbed hydrocarbons are eluted with a carrier gas at elevated temperature and passed through a separating column. The separated components are detected by the FID.

Limits of detection depend on the nature of the pollutants and the analysis methods (Table 16.5).

**TABLE 16.5:** LIMITS OF DETECTION OF GASEOUS AIR POLLUTANTS

| *Analysis Method* | $CO_x$ *(ppm)* | $NO_x$ *(ppm)* | $SO_x$ *(ppm)* | $O_3$ *(ppm)* | HCs *(ppm)* |
|---|---|---|---|---|---|
| Electrochemical | 10 | 0.05 | 0.05 | – | – |
| FID | – | – | – | – | 0.02 |
| GC-MS | – | 10 | 0.005 | 0.1 | 0.02 |
| IR | 0.1 | 0.02 | 0.02 | 0.1 | – |
| Luminescence | 100 | 0.005 | 0.005 | 0.002 | – |
| UV | – | 0.05 | 0.003 | 0.15 | – |
| Wet-chemistry | – | 0.01 | 0.01 | 0.02 | – |

### *16.4.1.1 Particle Mass Concentration Determination*

Manual measurement of mass concentration of an aerosol usually includes, collecting the aerosol on a filter paper, followed by gravimetry. More sophisticated techniques use *Tyndallometer* (turbidity tests), *oscillation microbalance sensor, piezoelectric microbalance impactor and β-attenuation sensor.*

*Nephelometers* (Tyndallometers) have been widely used for aerosol mass measurements (see Chapters 17 and 24). Many particles of interest fall in the Mie scattering range. Detection efficiency is maximum for particles of size 0.5-2 μm. Nephelometric data are often used as indicators for the respiratory dust concentration. These instruments are simple, inexpensive and provide real-time indication of aerosol concentration.

*Oscillation microbalance sensor* (Fig. 16.8) is an inertial microbalance. It uses the resonant frequency of a tapered oscillating tube, on the end of which is a particle collection element, to determine the mass of the collected particles. Deposition of particles on the collection element alters the tube mass, and thus its resonant frequency. The change in oscillation frequency is directly proportional to the added mass on the tube.

*Piezoelectric microbalance impactor* (Fig. 16.9) consists of a piezoelectric crystal on to which aerosol particles are deposited by electrostatic precipitation or impaction. The collected mass is measured through the change in the resonant frequency of the crystal.

In a *β-attenuation mass sensor* (Fig. 16.10), aerosol particles are collected on a filter and the particles are irradiated by a β-particles. The mass collected on the filter is proportional to the attenuation difference in β-radiation as it passes through the filter. A β-source ($^{14}C$) and a β-ray detector (fast-response silicon semiconductor) are used.

#### *16.4.1.2 Particle Size Determination*

Particle size strongly affects the physical properties of the aerosols. Particle size (diameter >0.5 μm) can be determined by microscopic observations, which provide geometric size. Light scattering, relaxation time-based techniques such as impactors, particle sizers, and centrifugation devices, and mobility-based detectors can also be used. For particles of <0.1 μm, spatial counters or relaxation time-based techniques are suitable. For sub-micron aerosols mobility-based techniques are employed (Table 16.6).

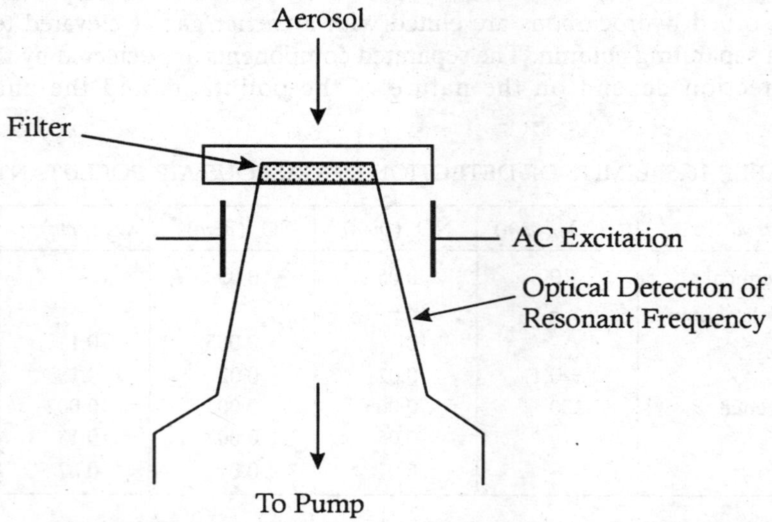


**Fig. 16.8:** Schematic of Oscillation Microbalance Sensor

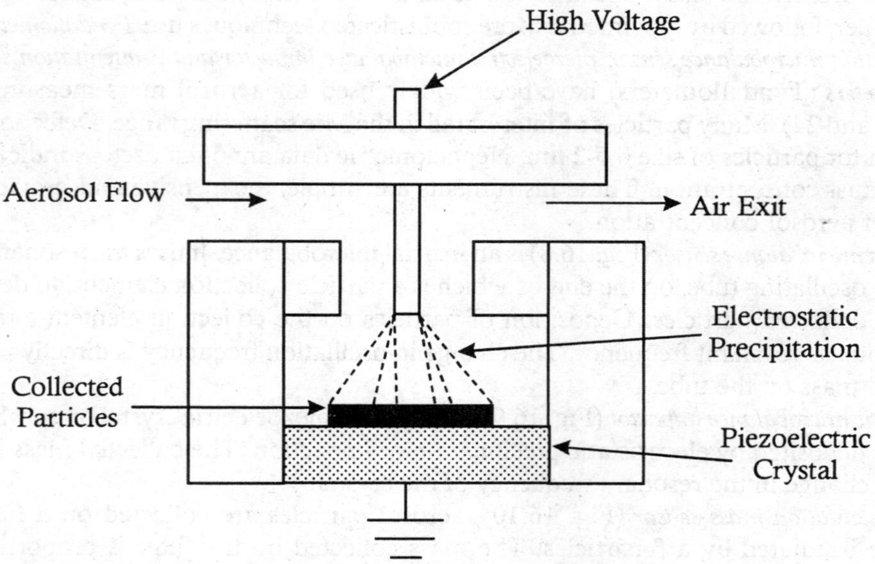


**Fig. 16.9:** Schematic of Piezoelectric Microbalance Impactor

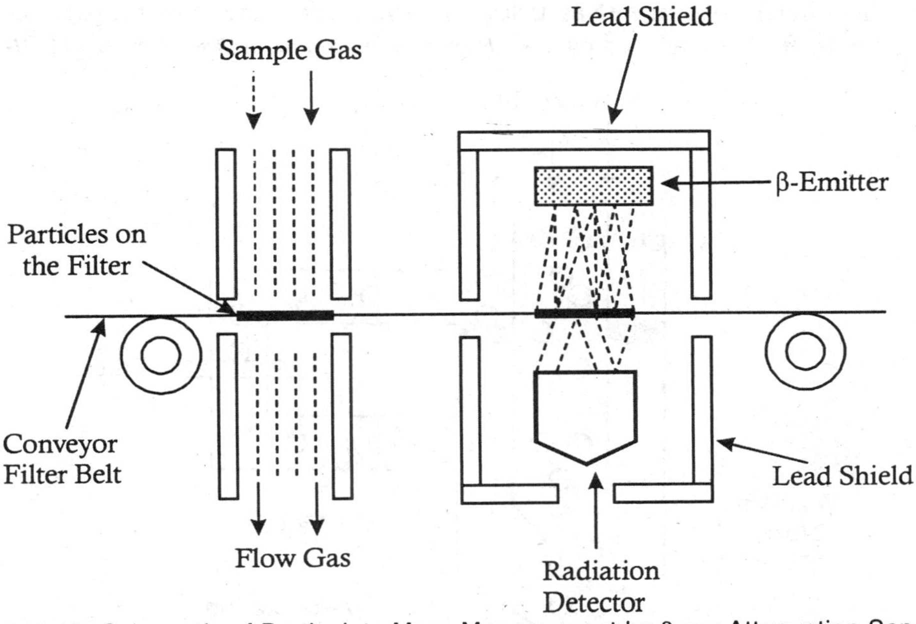


**Fig. 16.10:** Schematic of Particulate Mass Measurement by β-ray Attenuation Sensor

**TABLE 16.6:** SIZE-DISTRIBUTION MEASURING TECHNIQUES FOR AEROSOLS

| *Physical Property* | *Technique* | *Size Range* (μm) |
|---|---|---|
| Light scattering | Mie or Rayleigh scattering | 0.05-15 |
| Relaxation time-based | Impactors; Particle sizers | 0.5-20 |
| Mobility-based | Diffusion; Differential mobility | 0.005-0.5 |

*Scattering-based particle sizing*: Determination of particle size (diameter >0.1 μm) by light scattering is based upon the Rayleigh or Mie's scattering theory of electromagnetic radiation. In an optical particle counter (OPC), an air stream through an illuminated (and viewed) volume in the instrument carries individual particles. The intensity of scattered light and its angular dependence are recorded by a photodetector and the particle size is determined form the data by applying scattering or diffraction principles.

Photometric determination of particle size (diameter > 0.1 μm) is by light-blocking method. Air containing aerosol particles flows through the view volume at a constant velocity and blocks the light beam (He-Ne laser, λ = 632.8 nm) focussed on a photovoltaic cell. The light attenuation is proportional to the particle size.

Aerodynamical particle sizing can be carried out by *aerodynamic particle sizer (APS)* (Fig. 16.11) (the *dual laser* method). In APS, particles are accelerated through a nozzle for time Δt. As a particle passes through the first laser beam ($Ar^+$ laser beam at 488 nm), the scattered radiation provides a stop pulse, and the time between two pulses is recorded. If the ratio of the relaxation time ($\Delta t/\tau$) < 1, then the particle velocity at the nozzle exit is determined by measuring the time-of-flight (TOF) of particles to travel between a pair of laser beams. The velocity is proportional to the square of the aerodynamical diameter of the particle. The aerodynamic particle size can be determined from the

TOF, upon calibrating the system with particles of known particle size. Once the particle is detected and sized, it can be further analyzed by *laser desorption ionization mass spectrometry* (*LDI-MS*).

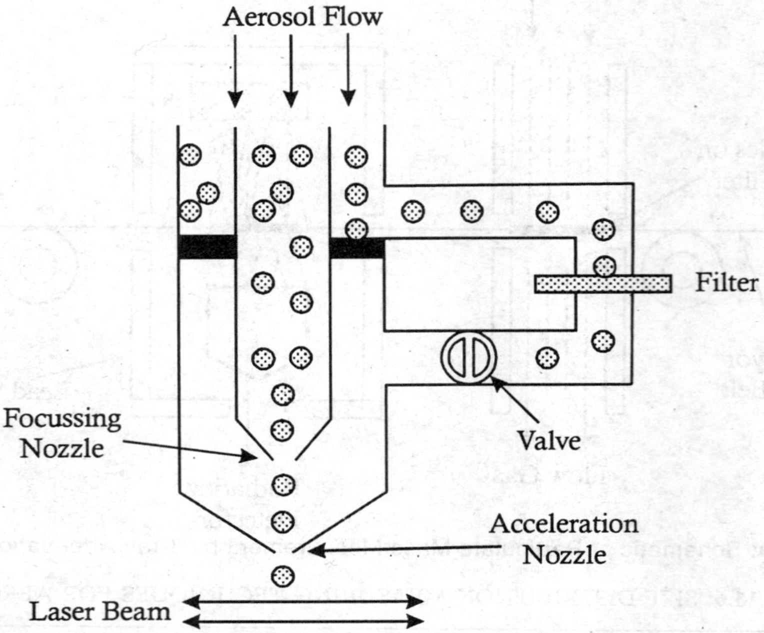


**Fig. 16.11:** Schematic of an Aerodynamic Particle Sizer (APS)

*Relaxation time-based Sizing*: Relaxation time-based methods determine the aerodynamic diameters of aerosol particles. The *aerodynamic* diameter is the equivalent diameter for all particles of identical aerodynamic behavior. For spherical particles of density 1 g/cm$^3$, the aerodynamic diameter is equal to the geometric diameter. Cascade impactors can be used for particle sizing. Impactor devices, where particle size can be determined in situ, consist of a series of orifices, arranged to give gas jets of increasing velocity, proportional to decreasing nozzle diameter, impinging on collector plates. Successive stages collect smaller and smaller size particles.

*Mobility-based Particle Sizing*: Mobility-based measurements depend on particle size and shape. *Diffusion batteries* (Fig. 16.12) use the diffusional deposition of small particles for size distribution measurement. The aerosol sample is passed through a set of capillary tubes in series (diffusion battery) and a condensation nucleus counter is employed to determine the upstream and downstream aerosol concentration and size distribution.

In a *sedigraph analyzer* a sample, dispersed in a liquid, is introduced into a sample cell and allowed to settle. Mass concentration is continuously monitored by attenuation of an X-ray beam passing through the cell. The particle size is a function of X-ray beam location and time. The separation range in this method is 0.1-100 μm.

Electrical mobility-based sizing is employed in the *electrical aerosol analyzer (EAA)*, the *differential mobility analyzer (DMA)* and *Coulter counter*. In the electrical aerosol analyzer (EAA), the aerosol to be measured is first electrically charged in a chamber by unipolar positive ions. The charged aerosol is then lead into the mobility analyzer, which operates on a low-pass filter that precipitates the high-mobility particles while allowing the low-mobility particles to pass through. By altering the voltage and measuring the current carried by the charged particles a voltage-current profile can be recorded from which particle size distribution can be determined.

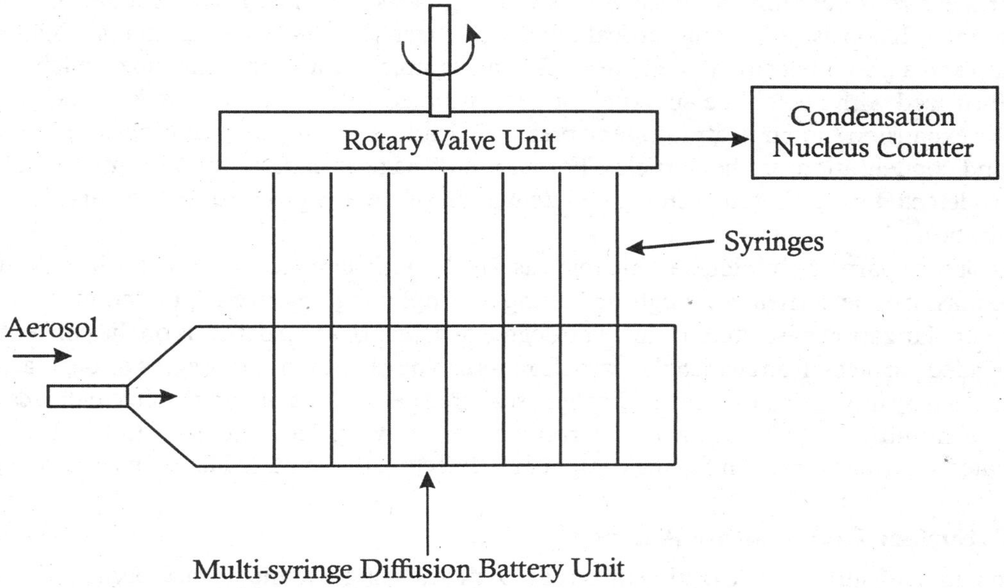


**Fig. 16.12:** Schematic of Multi-syringe Auto-diffusion Battery Setup for Size-distribution Analysis

The *differential mobility analyzer (DMA)* allows high-resolution particle sizing (20-500 nm), by use of a bipolar charge instead of a unipolar one. It operates on the principle that when an aerosol particle passes through a bipolar ionic environment; it undergoes neutralization and attains a steady-state charge distribution. The fraction of particles carrying different units of charge is a function of particle diameter. The instrument consists of a radioactive neutralizer, a cylindrical separation unit into which the aerosol is pumped, a high-voltage power supply and a *condensation nucleus counter (CNC)* for counting the separated particles.

The *Coulter counter* is an electrical sensing zone instrument. In this method, the test sample is dispersed in an electrolyte solution, which is then passed through a small orifice. A constant current or potential is applied between electrodes on either side of the orifice. Passage of a particle through the orifice momentarily reduces current to the extent that it is a function of particle size (volume).

By *on-line single-particle analysis*, the chemical composition and the size of individual particles can be determined *in situ* on a continuous, real-time basis. Advantages are minimal contamination or decomposition and reduction in loss of volatile compounds. *Laser desorption ionization (LDI)* coupled with time-of-flight mass spectrometry (TOF-MS) (Chapter 22) enables an entire mass spectrum to be obtained for each particle.

### *16.4.1.3 Number Concentration Determination*

Number concentration of particulate matter is carried out with condensation nucleus counters (CNC) and photometers (optical particle counters). CNC is based on the property of nucleation of particles for measuring the total number of concentration of particles, in the size range, $d > 0.005$ μm. The operation of an optical particle counter is similar to that of a nephelometer.

The CNCs operate by saturating a carrier air stream with a vapor and leading to super-saturation, either by adiabatic expansion or by contact-cooling, to cause vapor condensation on the particles. The particle size is determined optically by scattering measurements. *Pollack* counter is one such CNC counter. It consists of a long vertical tube, with water-saturated ceramic lining, a light source at the top and a photodetector at the bottom. The air sample is drawn into the tube, which is sealed and pressurized with nuclei free air. After air gets saturated with water, a valve is opened to allow adiabatic expansion (down to atmospheric pressure), leading to a super-saturation condition of the vapor and condensation on the particles. Formation of haze attenuates the transmission of light, which is detected by a photodetector. The attenuation of light is proportional to particle number concentration.

In an optical particle counter, particle mass as well as number concentration can be determined. Particle mass can be correlated to light scattering by employing polarized light and measuring the ratio of depolarization of scattered light. The degree of depolarization depends on the concentration of suspended particles, from which the particle mass can be estimated. The degree of depolarization is determined by measuring the intensity of light scattered perpendicular and parallel to the scattering axis. The number concentration is determined by allowing individual particles through an illuminated viewing volume in the instrument and detecting the scattered light on a photodetector.

### 16.4.2 Surface Composition Analysis

Determination of surface composition of aerosol particles is of great importance, because the surface components and characteristics of aerosols predominantly determine their physiological effects. Secondary ion mass spectrometry (SIMS), diffuse reflectance, atomic emission (AES), electron (Auger, X-ray photoelectron (XPS)) spectroscopies (Chapter 24), and microscopy (visible and scanning electron, (SEM)) (Chapter 26) are used in the surface characterization of aerosol particles. The SIMS techniques give ppm or ppb level details and ion-sputtering AES provides the highest spatial resolution for depth analysis. Spectroscopic methods provide structural and chemical composition, and surface interactions of pollutants. Instrumental methods for measurement of particulate matter are given in Table 16.7 and instrumental techniques for monitoring elemental pollutants are given in Table 16.8.

**TABLE 16.7:** INSTRUMENTAL TECHNIQUES FOR MEASUREMENT OF PARTICULATE MATTER

| *Particulate Parameter* | *Measurement Methods* |
|---|---|
| Mass | β-ray attenuation; Capacitance; Gravimetry; Nephelometry; Piezoelectric microbalance; Thermal precipitation |
| Size | Condensation nuclei counters (CNC); Electrostatic precipitation; Impactors; Light transmission; Microscopy; Sedimentation |
| Number | Optical particle counters; Condensation nucleus Counters (CNC) |
| Surface | SIMS; Electron spectroscopy; Microscopy |
| Velocity | Doppler-shift; Pitot tubes |
| Opacity | Light scattering; Light transmission |

SIMS = Secondary ion mass spectrometry.

**TABLE 16.8:** INSTRUMENTAL TECHNIQUES FOR MONITORING ELEMENTAL POLLUTANTS

| *Element* | *Instrumental Techniques* |
|---|---|
| Aluminum (Al) | AAS; AES; ICP-MS; XRF; NAA |
| Antimony (Sb) | AAS; AES; ICP-MS; Colorimetry; XRF; NAA |
| Arsenic (As) | AAS; AES; ICP-MS; XRF; NAA |
| Boron (B) | AAS |
| Beryllium (Be) | AAS; AES; ICP-MS; Colorimetry |
| Cadmium (Cd) | AAS; AES; ICP-MS; Colorimetry; XRF; NAA |
| Chromium (Cr) | AAS; AES; ICP-MS; XRF; NAA |
| Copper (Cu) | AAS; AES; XRF; NAA |
| Iron (Fe) | AAS; AES; ICP-MS; XRF; NAA |
| Lead (Pb) | AAS; AES; ICP-MS; Colorimetry; XRF; NAA |
| Mercury (Hg) | AAS; AES; ICP-MS; Colorimetry; NAA; XRF |
| Manganese (Mn) | AAS; AES; Colorimetry; XRF; NAA |
| Nickel (Ni) | AAS; AES; XRF; NAA |
| Phosphorus (P) | AAS |
| Selenium (Se) | AAS; AES; ICP-MS; Colorimetry; NAA |
| Tin (Sn) | AAS; AES; ICP-MS; XRF; NAA |
| Vanadium (V) | AAS; AES;ICP-MS; Colorimetry; XRF; NAA |
| Zinc (Zn) | AAS; AES; ICP-MS; XRF; NAA |

*AAS* = Atomic absorption spectrometry; *AES* = Atomic emission spectrometry;
*ICP-MS* = Inductively-coupled plasma-Mass spectrometry; *XRF* = X-ray fluorescence.
*NAA* = Nuclear activation analysis;

### 16.4.3 Analysis of Hazardous Air Pollutants (HAPs)

The United States Environmental Protection Agency (USEPA) has suggested compendium methods for quantitative methods of analysis of hazardous air pollutants (HAPs) in ambient air (Tabie 16.9 and Fig. 16.13). For VOCs that boil below 100 °C, the volatile organic sampling train (VOST) method can be used. VOST procedure involves drawing of the gas sample through multiple adsorbent tubes, and pollutants adsorbed in these tubes are desorbed and analyzed by suitable physical techniques (e.g. GC-MS method) (refer to Chapters 21 and 22 for details).

**TABLE 16.9:** METHODS FOR DETERMINING TOXIC ORGANIC (TO) POLLUTANTS IN AMBIENT AIR

| *Method No.* | *Type of Compounds Determined* | *Sample Collection (sorbent)* | *Analytical Technique(s)* | *Comments* |
|---|---|---|---|---|
| TO-1 | VOCs (80 to 200° C) | Tenax solid sorbent | GC-MS | |
| TO-2 | VOCs (–15 to 120° C) | Molecular sieve (carbon) | GC-MS | Similar to TO-1; more volatile organics (such as vinyl chloride) can be determined. |
| TO-3 | VOCs (–10 to 200° C) | Cryotrap | GC-FID (ECD) | |
| TO-4 | Pesticides/PCBs (Semi-volatile) | Polyurethane; solid-phase extraction | GC-MD | |

| *Method No.* | *Type of Compounds Determined* | *Sample Collection (sorbent)* | *Analytical Technique(s)* | *Comments* |
|---|---|---|---|---|
| TO-5 | Specific to aldehydes /ketones | Impinger | HPLC | Detection at 370 nm |
| TO-6 | Specific to phosgene | Impinger | HPLC | Detection at 254 nm |
| TO-7 | Specific to anilines | Adsorbent | GC-MS | |
| TO-8 | Specific to phenols/ cresols | Impinger | HPLC | Detection at 274 nm |
| TO-9 | Dioxins(Semi-volatile) | Polyurethane; solvent extraction | HRGC-HRMS | |
| TO-10 | Pesticides/PCBs (Semi-volatile) | Polyurethane | GC-MD | Similar to TO-4 |
| TO-11 | Specific to aldehydes /ketones | Adsorbent | HPLC | Similar to TO-5; $O_3$ monitoring |
| TO-12 | Non-methane organic compounds (NMOCs) | Canister or on-line | FID | Similar to TO-3; coupled to PDFID |
| TO-13 | PAHs (Semi-volatile) | Polyurethane | GC-MS | |
| TO-14 | VOCs (Non-polar) (–158 to 170° C) | Specially-treated canister | GC-MS & GC-MD | |
| TO-15 | VOCs (Polar & Non-polar) (–50 to 170° C) | Specially-treated canister | GC-MS | |
| TO-16 | VOCs (Polar & Non-polar) (80 to 200° C) | Open-path monitoring | FTIR | |
| TO-17 | VOCs (Polar & Non-polar) (–158 to 200° C) | Single and multi-bed adsorbents | GC-MS, GC-FID etc. | |

GC-MS: Gas chromatography-Mass spectrometry.
GC-FID: Gas chromatography-Flame ionization detector.
GC-MD: Gas chromatography-Multi-dectector.
HRGC-HRMS: High resolution Gas chromatography-High resolution Mass spectrometry.
HPLC: High performance liquid chromatography.
FTIR: Fourier transform infrared spectrometry.
*Source*: EPA's "*Compendium Methods for the Determination of Toxic Organic Compounds in Ambient Air*", (1999), 2nd edn., EPA/625/R-96/010b.

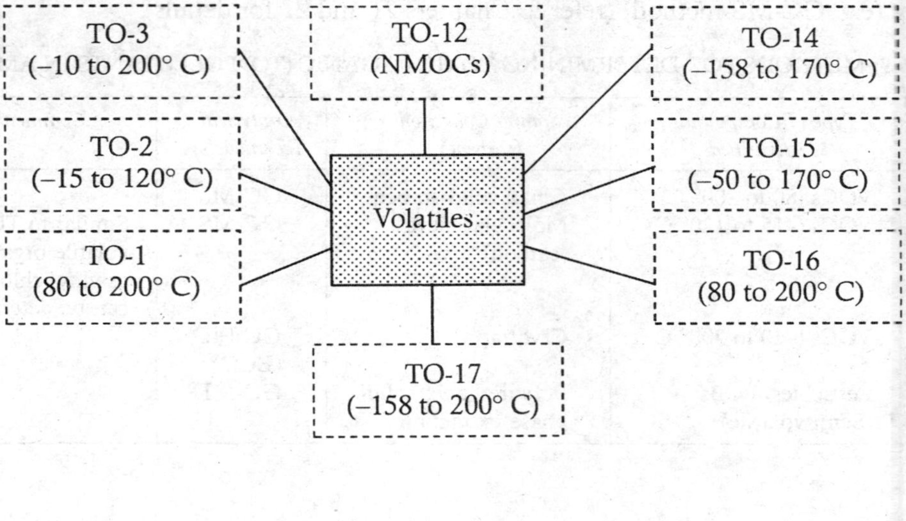

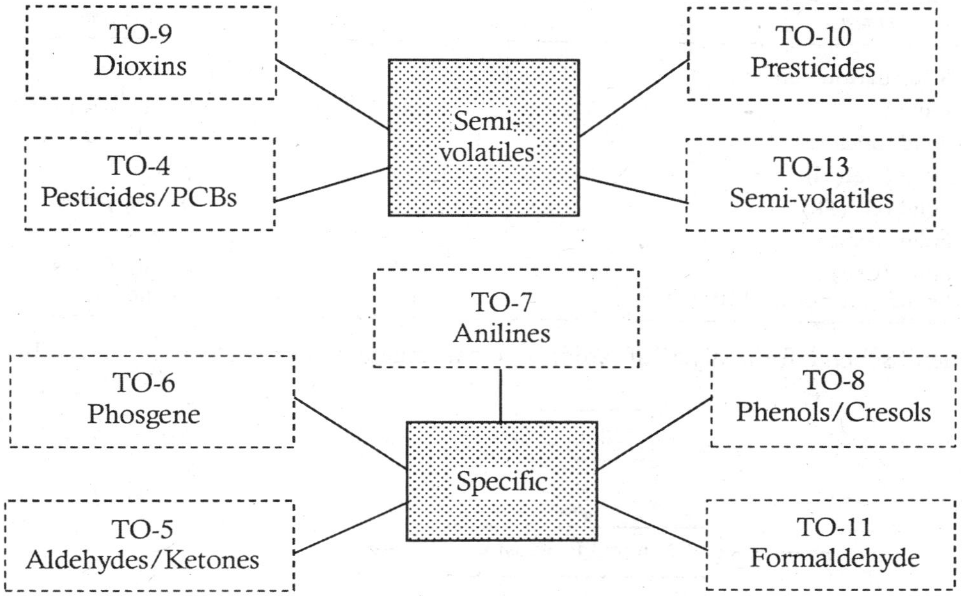


(NMOCs = Non-methyl organic compounds)

**Fig. 16.13:** Compendium Methods for Determination of Toxic Organics (TOs) in Air

(*Source*: EPA's "*Compendium Methods for the Determination of Toxic Organic Compounds in Ambient Air*", EPA/625/R-96/010b (1999))

### 16.4.4 Analysis of Radioactive Pollutants

Radioactive pollution is due to natural as well as anthropogenic causes. However, anthropogenic causes that are becoming of increasing concern to humanity are due to nuclear weapons, power plants, nuclear waste disposal and occupational exposure to radioactive materials and high-energy radiations. Nuclear power reactors and plants discharge $^{3}H$, $^{90}Sr$, $^{131}I$, and $^{137}Cs$. Coal-based power plants too release $^{14}C$, $^{60}Mn$, $^{85}Kr$, $^{131}I$, and $^{133}Xe$. Radioactivity is measured by various types of radiation detectors—photographic films, various particle and scintillation counters and indirectly by thermoluminescent detectors (TLD) (used to measure the dose received in personal dose monitoring) (Table 16.10). Particle detectors/counters are used to determine radioactivity by measuring the number of particles emitted by radioactive material in a given time (e.g. intensity of darkening of the photo film) that can be excited by ionizing radiation) (refer to Chapter 25).

**TABLE 16.10:** RADIATION DETECTORS

| *Detector* | *Detection* |
|---|---|
| Ionization chamber | α- and β-rays |
| Proportional counter | α- and β-rays |
| Geiger-Müller counter | α- , β- and γ-rays |
| Photographic film | General |

| *Detector* | *Detection* |
|---|---|
| **Scintillation counter** | |
| Liquid | α- , β- and γ-rays |
| Solid (organic) | α- , β- and γ-rays |
| Solid (NaI(Tl)) | γ-rays |
| Solid (CsI(Tl)) | α- and γ-rays |
| Solid (ZnS) | α-rays |
| Solid (CsF) | α- and β-rays |
| Semiconductor (Si(Li)) | β- and γ-rays |

A general scheme for analysis of radioactive particulate matter is given in Fig. 16.14.

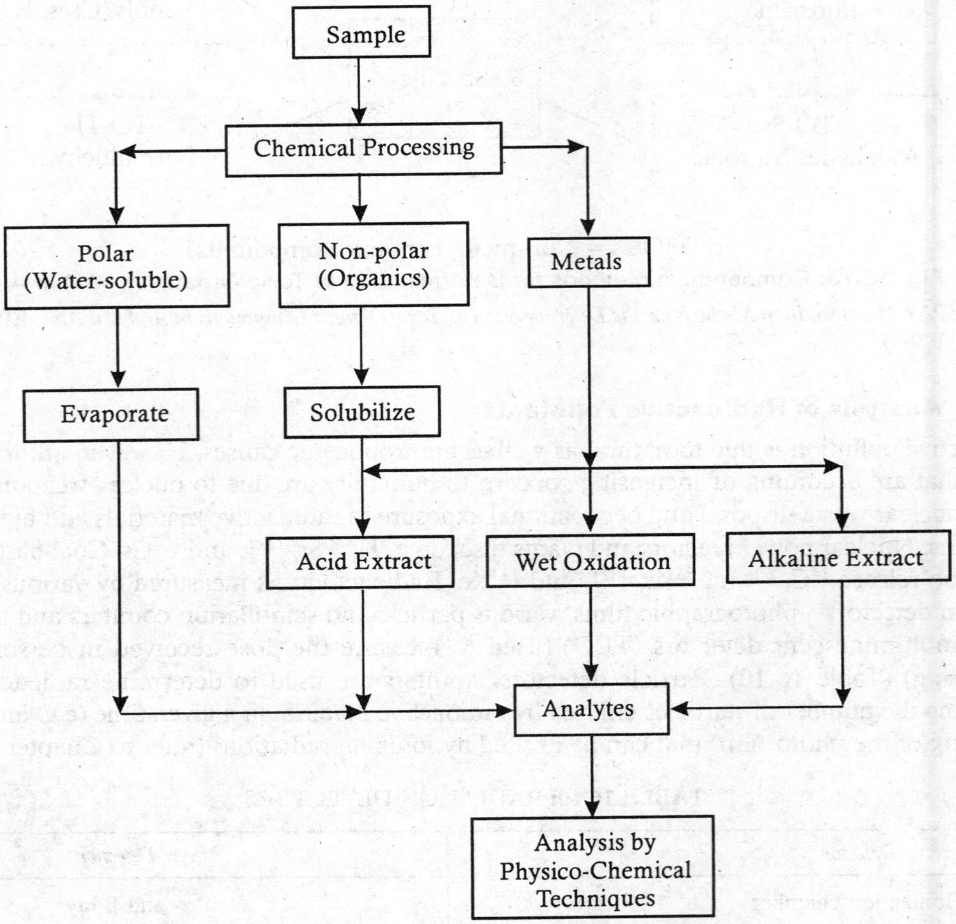


**Fig. 16.14:** Flowchart of Radioactive Particulate Analysis

## BIBLIOGRAPHY


Albarges, J. B., (ed.). (1980), Pergamon Press: Oxford. "*Analytical Techniques in Environmental Chemistry.*"

Allen, T. (1977), Chapman & Hall: London. "*Particle Size Measurement.*"

Buffle, J. and Van Leeuwen, H. P. (eds.) (1992), Lewis Pubs: Chelsea, MI. "*Environmental Particles,*" (Vol. 1).

Clement, H. E., Yang, P. W. and Koester, C. J. (1999), *Anal Chem*, **71**: 257. "*Environmental Analysis.*"

Gondal, M. A. and Mastromarino, J. (2000), *Talanta*, **53**(1): 147. "Lidar system for remote environmental studies."

Fergusson, (1990), Pergamon Press: Oxford. "*The Heavy Elements—Chemistry, Environmental Impact and Health Effects.*"

Frei, R. W. and Albarge, J., (eds.). (1986), Gordon & Breach: New York. "*Air and Water Analysis: New Techniques and Data.*"

Harrison, R. M. and Perry, R. (1984), Chapman & Hall: London. "*Handbook of Air Pollution Analysis,*", ( 2nd edn.).

Hinds, W. C. (1982), Wiley & Sons: New York. "*Aerosol Technology.*"

Jorgensen, S. E. and Johnsen, I. (1981), Elsevier: Amsterdam. "*Principles of Environmental Science and Technology,*" (Vol. 14).

Kebbekus, B. B. and Mitra, S., (eds.). (1998), Blakie: London. "*Environmental Chemical Analysis.*"

Keith, L. H., (ed.). (1991), Lewis Pubs: Chelsea, MI. "*Environmental Sampling and Analysis: A Practical Guide.*"

Kouimtzis, T. and Samara, C. (eds.) (1995), Springerverlag: Berlin. "*The Handbook of Environmental Chemistry,*" (Vol. 4D).

Liu, D. H. F . and Liptak, B. G. (eds.) (2000), Lewis Pubs: Boca Raton, FL. "*Air Pollution.*"

Lodge, J. P., J.r. (1991), Lewis Pubs: Chelsea, MI. "*Methods for Air Sampling and Analysis.*"

Nriagu, N. O., (ed.). (1992), Wiley & Sons: New York. "*Gaseous Pollutants.*"

Peirce, J. J., Weiner, R. F. and Vesiland, P. A. (1998), Butterworth-Heineman: New Delhi. "*Environmental Pollution and Control.*"

Seinfeld, J. H. (1986), Wiley & Sons: New York. "*Atmospheric Chemistry and Physics of Air Pollution.*"

Sigrist, M. W., (ed.). (1994), Wiley: Chichester, UK. "*Air Monitoring by Spectroscopic Techniques.*"

Smolkova-Keulemansova, E. and Fetl, L. (1991), Elsevier: Amsterdam. "*Analysis of Substances in the Gaseous Phase: Comprehensive Analytical Chemistry,*" Vol. XXVIII, Svehla, G. (ed.).

Thain, W. (1980), Pergamon Press: Oxford, UK "*Monitoring Toxic Gases in the Atmosphere for Hygiene and Pollution Control.*"

Willeke, K. and Baron, P. A., (eds.). (1993), Van Nostrand-Reinhold: New York. "*Aerosol Measurement—Principles, Techniques and Applications.*"

Winegar, E. D. and Keith, L. H., (eds.). (1993), Lewis Pubs: Boca Raton, FL. "*Sampling and Analysis of Airborne Pollutants.*"

Woodward, F. and Sheehy, J. E. (1983), Butterworth: London. "*Principles and Measurements in Environmental Biology.*"

...(1999), USEPA/625/R-96/010b: Cincinnati, OH. "*Compendium Methods for the Determination of Toxic Organic Compounds in Ambient Air,*" (2nd edn.).

# 17

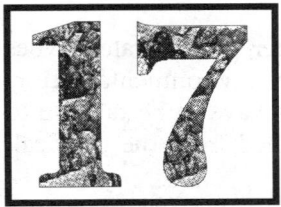

# General Techniques in Water Quality Monitoring

Water quality can be determined from physical, chemical, biological and aesthetic points of view. Physical parameters are color, odor, turbidity, particulate matter and conductivity. Chemical parameters are pH, dissolved oxygen (DO), biochemical oxygen demand (BOD), chemical oxygen demand (COD), and dissolved substances, metals and redox chemical reactants. Biological parameters that affect the water quality are the presence of algae, viruses, coliforms, pathogens and other vectors that are responsible for health hazards and diseases.

Organoleptic properties of water are related to odor, taste and color. Odor in water is a sign of decaying organic matter. Turbidity in water is due to suspended particulate matter—clay, silt, coagulants, organic matter and photoplanktons. Nephelometers and tubidimeters are used to determine turbidity.

Humidity is the content of water vapor in air and its measurement is generally relevant in atmospheric studies.. Relative humidity can be determined by hygrometers, wet-and-dry bulb thermometers (psychrometers) or by electrical conductimeters.

Water salinity is mainly due to dissolved salts of sodium and potassium (NaCl and KCl). It can be determined by hygrometric, and titrimetric methods, in addition to optical (AAS), conductimetric (salinometer) and ion-selective electrode (ISE)) methods.

The determination of organic pollutants in water can be done by monitoring dissolved oxygen (DO) by BOD and COD methods. The popular method for continuous monitoring of DO is by membrane electrodes (Clark cell).

Biological methods for assessing water quality rely on the estimation of biomass, coliform bodies, microorganisms and nutrients. Estimation of biomass includes determination of total carbon, nitrogen, phosphorus, lipids, carbohydrates and benthal sediments. Sediments are usually the most appropriate candidates for long-term monitoring of contaminants in aquatic systems.

This chapter deals with the general methods for water quality monitoring.

Quality of water depends on the type of source (e.g. rainwater, surface water, groundwater), types of terrains and habitats, dissolved substances, chemical, biological and other environmental factors that transported into it. All these varied but inter-dependent parameters have to be taken into account in water quality monitoring methods. Water quality is established from the physical, chemical and biological parameters.

## 17.1 FACTORS AFFECTING WATER QUALITY

Various sources and processes pollute water bodies. Natural water contains high concentrations of $Na^+$ and $K^+$; surface waters generally contain low concentration of metals, trace elements from precipitation and leaching. Groundwater can be contaminated by pollutants and from surrounding landmass. The growth of the primary producers (bacteria, algae and water plants) is affected mainly by nutrient substances (N- and P-containing compounds).

Industrial establishments are primary sources of water pollutants. Automobiles add to the pollution load. Agricultural operations are source of hazardous and toxic chemicals. Power generation establishments, nuclear weapons production, and nuclear power generation operations are sources of thermal and radioactive pollutants. Once the pollutants reach the aqueous environment, water acts as a conduit for transmitting them to other places. Some of the pollutants may be incorporated into the food cycle via micro-organisms, vegetation, marine life and animals.

Large concentrations of metals in water may cause minor irritations (e.g. color due to the presence of Fe and Cr), or may cause major disruptions to ecosystems (e.g. Hg & Pb in water). Trace elements in water, such as As, Cd, Hg and Pb can be deleterious to plants. The trace element contamination of plants and vegetation has a major impact on both the environmental cycling of nutrients and the quality of foodstuffs. Bio-amplification by plankton, and biotransformation by bacteria in the water-sediment interface can strongly influence elemental toxicity through out the remaining food chain. For example, As, Hg, Pb, and Sn undergo biomethylation in the water-sediment interface, resulting in the production of more toxic species.

Chlorinated hydrocarbons, such as DDT (organochlor pesticides), are extremely resistant to biological degradation and have acute or chronic health effects on birds, animals and humans. As they are so persistent and are fat-soluble, they accumulate in membrane tissues (*bioaccumulation*) and have amplified effects on organisms.

Water quality can be judged from physical, chemical, biological and aesthetic points of view. Physical parameters for water quality are color, turbidity, particulate matter, temperature, conductivity and odor. Chemical parameters are pH, dissolved oxygen (DO), BOD, COD, dissolved ions, and chemicals. Biological parameters are bacteria, algae, viruses, coliform and other biological pathogens. Of these, the important parameters in evaluating water quality index are turbidity, dissolved oxygen, pH, BOD, and ammonia (organic matter) (Table 17.1).

## 17.2 PHYSICAL PARAMETERS IN WATER QUALITY ANALYSIS

In water bodies the water currents, eddies and waves influence the distribution of constituents. Therefore, flow pattern should be taken into consideration in assessing the distribution of pollutants in water and its quality, especially in streams and estuaries.

### 17.2.1 Water Flow Patterns

The speed of water flow can be determined by dipping an L-shaped open-ended glass tube

(Fig. 17.1). The shorter arm is in the water, horizontal to the water flow. The velocity, $v$, of water flow is given by

$$v = 0.977\sqrt{2hg} \quad \text{...(17.1)}$$

where, $h$ = height of the water column in the longer arm;
$g$ = acceleration of due to gravity.

**TABLE 17.1:** IMPORTANT PARAMETERS IN DETERMINING THE WATER QUALITY

| *Quality* | *Turbidity (mg/L)* | *DO (mg/L)* | *pH* | *$BOD_5$ (mg/L)* | *$NH_3$ (mg/L)* | *Remarks* |
|---|---|---|---|---|---|---|
| Very good | <20 | >8 | 6.5-8.5 | <1 | <0.1 | Class I: Saturated with $O_2$; low in nutrients |
| Good | 20-40 | >6 | 6.0-9.0 | 2-6 | <0.3 | Class II: Low nutrients. |
| Contaminated | 40-100 | <4 | 5.0-9.0 | 7-13 | 0.3-1.0 | Class III: $O_2$-depleting pollutants |
| Highly Polluted | 100-275 | <3 | 4.0-10.0 | >15 | 1.0-2.7 | Class IV: Putrefying; only bacteria (no higher organisms) |

*Source*: Rump, H. H. (1999), Wiley-VCH: Weinheim. "*Laboratory Manual for the Examination of Water, Wastewater and Soil.*"

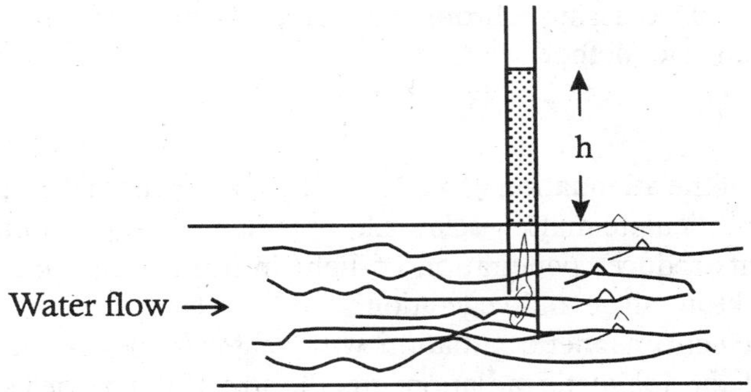


**Fig. 17.1:** Water Flow Measurement

The speed as well as the direction of flow in a stream can be determined by 'float' method. It is determined by the time, $t$, taken by a magnetic float (a compass on a cork) to traverse the distance, $d$, along the path of water flow.

$$v \approx \frac{d}{1.2t} \quad \text{...(17.2)}$$

This method is simple and gives adequate information on the speed and the direction of water flow, in non-eddy, gentle streams. Flow meters provide more accurate data on velocity as well as on rate of water flow.

### 17.2.2 Organoleptic Properties

Organoleptic properties of water are related to odor, taste and color. Transparency of water is related to coloring materials and turbidity due to suspended and colloidal matter. Certain inorganic as well as organic compounds change the organoleptic properties of water or even make it unfit for human or industrial uses.

#### 17.2.2.1 Odor and Taste

Odor and taste determinations are qualitative and subjective. In addition to chemical and biological effects of foul smelling and coloring constituents, they make the water aesthetically unacceptable. Odor in water is a general sign of pollution by decaying organic matter. Compounds that contribute to odor are generally volatile organic compounds, while chemicals that contribute to taste and odor are ketones, phenols, aldehydes and some other organic and inorganic compounds. It is possible to quantify concentrations of analytes responsible for odor and taste by techniques such as GC. But, quantitative data alone do not indicate the perceptive nature of odor and taste. It is still a common practice to employ a panel of individuals who have developed a degree of perception to taste and smell (similar to wine tasters and gourmet connoisseurs) to estimates these characteristics.

#### 17.2.2.2 Color (Hue)

Transparency (visibility) of water is a measure of depth of penetration of light. This parameter depends on the presence of coloring matter and turbidity due to suspended matter. Color of water (hue) can be due to organic or inorganic contaminants. It can also be pH-dependent. Color of water, free from suspended matter, can be estimated semi-quantitatively by comparing samples with standard solutions of potassium chromate of different dilutions. It can be more accurately determined by a colorimetric method.

#### 17.2.2.3 Turbidity

Turbidity is a measure of the attenuation of a beam of light as it passes through a medium. In water it is due to suspended particulate matter—clay, silt, organic matter, photoplankton. High turbidity in aquatic environments reduces penetration of light into the water and consequently reduces photosynthesis by plankton, algae and vegetation.

The interaction of electromagnetic radiation with matter depends on the ratio between the wavelength, $\lambda$, of the electromagnetic radiation, and the size of the particles. If the diameter of the particle, $d > \lambda$, then the interaction follows the laws of geometric optics (reflection and refraction—Tyndall scattering). If the diameter of the particle $d < \lambda/2$, diffusion of light occurs due to diffraction.

Very small particulate matter ($10^{-4}$ μm) scatters blue light more effectively than light of longer wavelength (Rayleigh scattering). Also, if the particle size is small, the intensity of scattered light is more in the forward direction (at $\theta \approx 0°$) than in vertical directions ($\theta \approx 90°$; $270°$) and the scattered light is completely polarized. As the particle size increases, light scattered from different parts of the same particle causes interference. Due to interference effects the forward scattering of light (low-angle scattering) tends to be more in-phase and the backward scattering (high-angle scattering) tends to be out-of-phase and destructive. Therefore, intensity of the scattered light would be greater at low angles (Mie theory). Thus, turbidity measurements can be useful not only to estimate the 'cloudiness' of water samples, but also to determine the size of the particulate pollutants.

Turbidity of water due to homogeneous colloidal particulate ($d \approx \lambda$) can be determined in either of two ways. If the apparatus is designed so that the detector is aligned with the cell and the radiative source ($\pm 10°$), the detector responds to the attenuated intensity of the incident radiation that is caused by scattering in the cell. Measurements of the decreased intensity are turbidimetric measurements; the technique is called turbidimetry. The measurements are completely analogous to absorption measurements. The only difference is in the phenomenon that causes the decreased radiative intensity. As with absorption measurements, the decreased intensity is related to the concentration of the scattering species in the cell at a constant wavelength. If the intensity of the scattered radiation is measured, rather than the decrease in intensity of the incident radiation, the method is known as nephelometry. The apparatus used for nephelometric measurements differs from that used for turbidimetric measurements in the placement of the detector. In nephelometry the detector is not aligned along the radiation source and the cell; normally it is placed perpendicular to the path of the incident radiation, and scattered light is measured at right angle to the incident light, $\approx 90°$ (Fig. 17.2). If the intensity of the scattered radiation is measured, quantitative analysis is performed by preparing a working curve of intensity as a function of concentration of a series of standard solutions (i.e., solutions containing known concentrations of the component being analyzed). The intensity of the scattered radiation in the analyte is measured and compared to the working curve. The concentration of the analyte corresponds to the concentration on the curve that has intensity identical to that of the analyte. Generally, turbidimetry is reserved for measurement of denser suspensions. Sedimentation, coagulation and filtration are the methods generally employed to remove turbidity of water.

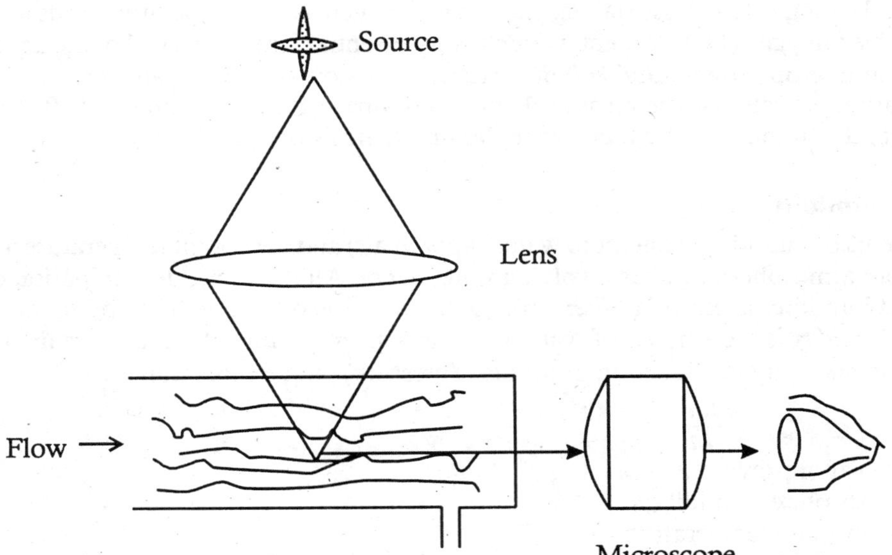


**Fig. 17.2:** Tyndall Effect-based Flow Microscope

In field studies, the transparency of a placid water body can be estimated by lowering a graduated rod with a circular disk fixed to it, with four quadrants of the disk alternately painted black and white. Average depth between readings where the disk is just visible and just not visible gives an estimate of the transparency of the water body. Underwater photometers give more accurate values of attenuation of light at various depths. They also perform under non-placid (turbulent) conditions of water bodies.

Spectro-optical techniques are also employed for determination of water quality by means of remotely-sensed and locally-acquired optical data, and correlating the inorganic turbidity (suspended solid concentration) and organic turbidity (Chlorophyll-a concentration). Suspended inorganic and organic concentrations of a water body can be determined by simultaneous measurements of the transmission and volume reflectance of the water column at any wavelength.

Suspended minerals act essentially as reflectors and non-absorbers of incident radiation, while chlorophyll-a (Chl-a) acts essentially as an absorber and non-reflector of the incident radiation. Digital data are acquired from the terrain by means of a passive multispectral scanner system. The optical transmission, $T$, of a given water column of length, $l$ is given by

$$T = \exp\{(a + b)l\} \qquad \text{...(17.3)}$$

where, $a$ = absorption coefficient;
$b$ = scattering coefficient;
$(a + b)$ = total attenuation coefficient

The absorption coefficient, backscatter coefficient and vertical attenuation coefficient may be either directly and indirectly measured, or inferred from regression curves. That is, *in situ* water quality assessment of water masses is feasible on the basis of their optical properties alone.

### 17.2.3 Temperature

Temperature is an important parameter controlling many of the physical (viscosity, surface tension), chemical and biological processes taking place in water. Increase of temperature leads to: (*i*) decrease in the dissolved oxygen (DO) content, which is very essential for the survival of the aquatic system, (*ii*) reduction in crop growth, and (*iii*) deleterious effects on aquatic organisms.

Temperature of water can be measured with an ordinary glass thermometer (±0.2° C), or more accurately and continuously by thermocouples and thermistors (±0.05° C).

#### *17.2.3.1 Humidity*

Measurements of humidity (content of water vapor in air) and dew point temperature are generally treated under atmospheric studies involving water vapor. All the same, determination of moisture content and humidity is essential in environmental analysis of soil, fodder and other samples.

*Absolute humidity* is the amount of water vapor in a given volume of air at a certain temperature. Atmospheric humidity is a function of altitude (pressure), and temperature.

$$P_w = \chi \frac{RT}{M} \qquad \text{...(17.4)}$$

where, $P_W$ = Vapor pressure;
$\chi$ = absolute humidity;
$T$ = absolute temperature;
$M$ = molecular mass of water

*Relative humidity*, $R_H$, is the percentage ratio of water vapor present at a given temperature to the standard vapor pressure at the same temperature.

$$R_H(\%) = \frac{\text{Amount of water vapor present at temperature, } T}{\text{Saturated water vapor at the same temperature}} \times 100 \qquad \text{...(17.5)}$$

Measurement of relative humidity is carried out by hygrometers or wet- and dry-bulb thermometer (*psychrometer*) or by electrical conductometry method.

#### *17.2.3.2 Hygrometers*

Hygrometric measurements are based upon absorption, condensation or other thermodynamic parameters. Absorption hygrometers are hair-cell and dew-cell types. In the hair-cell hygrometer, the elongation of hair, which changes with humidity, is correlated.

The most sensitive device for measuring water vapor concentration is the infrared (IR) hygrometer or gas analyzer. Water vapor is introduced into the sample tube that contains appropriate band filters (1.1, 1.3, 1.87, 2.7 and 6.3 μm) at the end of the tube. An IR detector (PbS photo cell; Golay cell) records IR absorption as the water vapor passes through the sample tube.

### 17.2.3.3 Psychrometer

A thermodynamic device for measuring water vapor concentration is the psychrometer, which measures the equilibrium temperature of wet and dry surfaces located in the same air stream. Psychrometer consists of two glass-mercury thermometers (dry- and wet-bulb thermometer setup). The bulb of one of the thermometers (wet-bulb) is covered with a fiber (muslin wick), wetted and dipped in a reservoir of distilled water (Fig. 17.3). Evaporation of moisture on the wick cools the wet-bulb thermometer to a temperature (wet-bulb temperature) below that of air temperature (dry-bulb temperature).

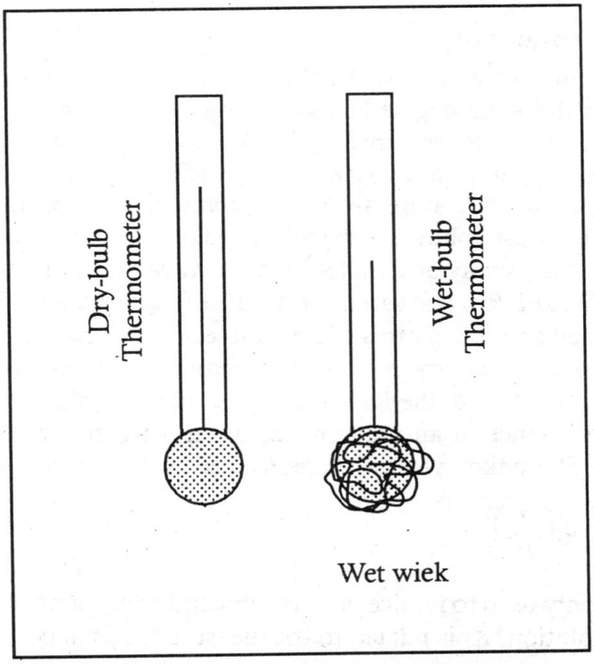


**Fig. 17.3:** Dry- and Wet-bulb Thermometer (Psychrometer)

At equilibrium, the relative humidity (%) can be calculated from

$$P = P_S - \gamma(T_a - T_w) \quad \text{...(17.6)}$$

$$\gamma = \frac{C_P P}{\lambda\varepsilon} \quad \text{...(17.7)}$$

where, $P$ = vapor pressure; $P_S$ = saturated vapor pressure; $\gamma$ = psychrometric constant; $T_a$ & $T_w$ = temperatures of atmospheric (saturated) and water; $C_P$ = specific heat of water at constant pressure; $\lambda$ = latent heat of vaporization of water; $\varepsilon$ = ratio of density of water vapor and dry air.

Generally, the relative humidity is calculated from the wet- and dry-bulb meter readings, employing psychrometric charts. If the dry- and wet-bulb thermometer temperature readings are the same, the relative humidity is 100 per cent.

#### 17.2.3.4 Dew Point Temperature

The temperature below which moisture condenses is called the dew point. The dew point temperature is very useful and the absolute humidity ($\chi$) can be determined from the dew point temperature. The dew point instruments (*dew cells*) are based on water condensation and employ *Peltier* effect. LiCl dew cell can determine the dew point in the range from –30° C to 50° C. The sensor consists of a wick of glass wool wound around a cylinder embedded with winding of silver wire (heating element). The wick is coated with saturated LiCl. When the wick is wet, the passage of current flow is high through the silver-heating element because of low electrical resistance between the windings of the heating element, and water evaporates from the wick. The drying of the wick continues until LiCl absorbs water vapor from the atmosphere. At equilibrium, the heating element temperature is determined with a thermocouple.

### 17.2.4 Electrical Conductivity

Pure water is a poor conductor of electricity. Presence of salts increases conductivity. The value depends on the concentration and degree of dissociation of electrolytes. The migration velocity of the ions in the electric field and temperature affect conductivity. However, the results provide general information on electrolytes, but do not give any information on the nature of the ions.

Specific conductivity of water can be used as a parameter for continuous monitoring of total electrolytes in water. For most solutions, specific conductance consists of the measurement of electrical resistance. This is usually accomplished by a wheatstone bridge. In wheatstone circuit, two known resistances ($R_1$ and $R_2$), one variable resistance ($R_3$) and the analyte resistance (unknown resistance $R_4$) are arranged as parallel arms of a circuit, each with two resistors in series ($R_1$ and $R_3$ in parallel with $R_2$ and $R_4$). A battery is connected across the parallel arms and a galvanometer (detector) from the $R_1R_3$ junction to the R2R4 junction. The variable resistance ($R_3$) is adjusted so that it is equal to the resistance of an unknown solution between two electrodes (galvanometer shows zero deflection). The unknown resistance, $R_4$, can be found from the relation

$$R_4 = R_3\left(\frac{R_2}{R_1}\right) \qquad (17.8)$$

AC current is commonly used to reduce polarization and concentration effects at the electrodes. The conductivity of a solution, $K$, is related to the measured resistance

$$K = \frac{C}{R} \qquad ...(17.9)$$

where, $C$ = cell constant.

Conductimetric methods can be used to determine moisture content and relative humidity (see Chapter 23). In this method, the sensor consists of two porous gold electrodes with a polymer foil in between them as the frequency-controlling element. With changing humidity, the capacitance, and so the frequency of the oscillator changes. The frequency is then mixed with the frequency of a reference oscillator and the difference is fed into a third oscillator that is measured. As the capacitance of the sensor increases non-linearly with relative humidity, the frequency of the final oscillator is decreased gradually with increasing frequency change of capacitance to make the relative humidity-capacitance relationship linear.

Some of the physical techniques employed in the determination of major solutes in water are given in Table 17.2.

**TABLE 17.2:** PHYSICAL TECHNIQUES FOR DETERMINATION OF SOLUTES IN WATER

| *Solute* | *Physical Techniques* |
|---|---|
| Arsenic | EC; AAS; ICP-MS |
| Boron | Colorimetry; ICP |
| Cyanide | Colorimetry; Ion-chromatography (IC) |
| Halides | Colorimetry; Gravimetry; Ion-chromatography (IC); NAA; Titrimetry; XRF |
| $Na^+$, $K^+$, $Ca^{2+}$ and $Mg^{2+}$ | AAS; AES; Colorimetry; Ion-chromatography (IC); ICP; ISE; XRF |
| Sulfate | Colorimetry; Gravimetry; Ion-chromatography (IC); Turbidimetry |
| Total dissolved solids | Hydrometry; Refractometry |
| Total salinity | Conductometry (salinometry); Hydrometry |

EC = Electron capture; AAS = Atomic absorption spectrometry; AES = Atomic emission spectrometry; ICP = Inductively coupled plasma; MS = Mass spectrometry; ISE = Ion-selective electrode; NAA = Neutron activation analysis; XRF = X-ray fluorescence.

## 17.3 CHEMICAL PARAMETERS IN WATER QUALITY ANALYSIS

Chemical parameters for determination of water quality are solutes, salinity and hardness, pH, dissolved oxygen (DO), biochemical oxygen demand (BOD) and chemical oxygen demand (COD). Evaporation, precipitation, extraction, sorption and electrochemical methods are employed for pre-concentration procedures.

### 17.3.1 Dissolved and Suspended Matter in Water

The amount of suspended matter (undissolved, filterable substances) in water is obtained as the difference between the filter weight prior to and after filtration (dried at 105° C; and ashed at 600° C). Wet-chemical methods, and physical methods are used to determine dissolved and suspended matter, cations and anions in water. Most of the wet-chemical reactions are based on measurement of color, emission, and absorption, electrical and any other physical characteristics that could be quantified by spectroscopic, radiation counting and other physical techniques.

Suspended solids refer to the residue that is retained (on a filter paper) after filtration of a water sample and drying at 105° C. Fixed suspended matter is the *non-volatile* solid that remains after heating at 550-600° C (Fig. 17.4).

$$\text{Total suspended solids (after 105° C)} = W_S/V \qquad \text{...(17.10)}$$

$$\text{Volatile suspended particles (after 600° C)} = (W_S - W_f)/V$$

$$\text{Total dissolved solids} = (\text{Total solids} - \text{Fixed suspended solids})$$

where, $W_S$ = mass of total suspended particulate (volatile + non-volatile);
$W_f$ = mass of fixed suspended particulate (non-volatile);
$V$ = volume of the sample in liters.

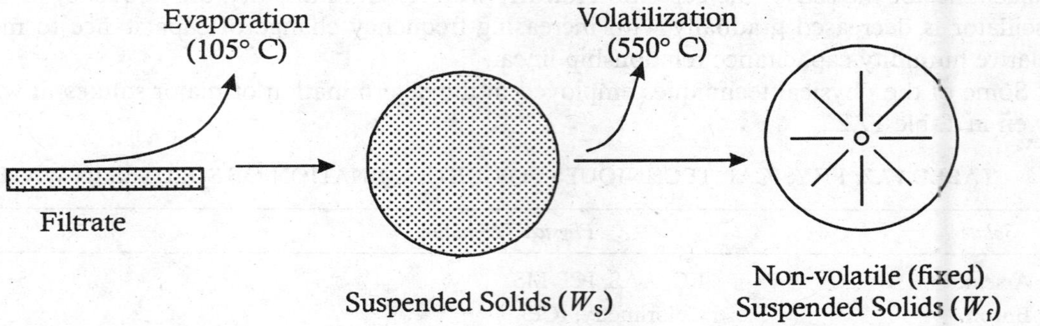


**Fig. 17.4:** Separation of Suspended Particulate Matter in Water Samples

### 17.3.2 Water Salinity and Hardness

Salinity of water is mainly due to dissolved chlorides (NaCl; KCl). It can be determined by conductometry (see Chapter 23), using a *salinometer*. Salinometer consists of a cell containing the sample, an *AC-transformer Bridge*, temperature compensation device and null indicator. The cell has usually platinum electrodes or inductive type electrodes. Inductive salinometers are preferred, as organic films do not foul them. For determining the salinity of brine waters, it is advantageous to use aprotic organic solvents such as acetone that are miscible in water.

Other methods of determining the salinity and other solutes in water are—(*i*) hydrometric method that is based on direct determination of specific gravity of the water sample, (*ii*) argentometric method that involves titration of halides with $Ag^+$ ion in the presence of $K_2Cr_2O_7$ as indicator, (*iii*) optical methods that rely on the refractive index of the solution, (*iv*) atomic absorption and (*v*) electrochemical methods, employing ion-selective electrodes (ISEs).

Water hardness is due to dissolved inorganic salts (carbonates and sulfates) of calcium and magnesium. Water hardness may be determined by titration with EDTA, buffered to pH = 10 (to keep it in a non-protonated form), and by capillary zone electrophoresis (CZE) methods (see Chapter 19).

### 17.3.3 pH and Ion-sensitive Electrodes

Ion-selective electrodes preferentially respond to a single chemical species. The potential between the indicator electrode and the reference electrode varies as the concentration or activity of that particular species varies. Unlike the inert indicator electrodes, ion-selective electrodes do not respond to all species in the solution. An internal reference electrode dips into a reference solution containing the assayed species and constant concentrations of the species to which the internal electrode responds. The internal reference electrode and reference solution are separated from the analyte solution by a membrane that is chosen to respond to the analyte. A second external reference electrode is also dipped into the analyte solution. The selectivity of the ion-selective electrodes results from the selective interaction between the membrane and the analyte. The electrodes are categorized according to the nature of the membrane. The most common types of ion-selective electrodes are the glass, liquid-ion-exchanger, solid-state, neutral-carrier, field-effect transistor, gas-sensing, and biomembrane electrodes. The glass membranes in glass electrodes are most often used for pH measurements, where the hydrogen ion is the measured species. Liquid-ion-exchanger electrodes utilize a liquid ion exchanger that is held in place in an inert, porous hydrophobic membrane. Solid-state ion-selective electrodes use a solid, sparingly soluble, ionically conducting substance, either alone or suspended in an organic polymeric material, as the membrane.

Acidity or alkalinity of water greatly influences the solubility of substances, ionic species present and various chemical reactions that take place. Acidity and alkalinity of water may be determined by titration with 0.02N NaOH and with 0.02N $H_2SO_4$, respectively. Generally it is determined by measuring the pH. Thus, pH is an important parameter in the evaluation of water quality—acidity, alkalinity, acid-base reactions, buffering capacity and dissolved $CO_2$. For example, at acidic pH water contains predominantly gaseous $CO_2$ species; at neutral pH bicarbonate species ($HCO_3^-$) and at alkaline pH carbonate ($CO_3^{2-}$) species predominate in water. pH of a solution is a measure of $[H^+]$ ion concentration.

$$H_2O \leftrightarrow H^+ + OH^- \qquad \text{...(17.11)}$$

$$\therefore \quad K_W = \frac{[H^+ +][OH^-]}{[HA]} \qquad \text{...(17.12)}$$

$$-\log[H^+] = -\log K_W + \log[OH^-]$$

$$pH = pK_W - pOH \qquad \text{...(17.13)}$$

$K_W$ is dependent on temperature, pressure, and ionic strength of the solution. In general, for any polar solution,

$$HA + \leftrightarrow H^+ + A^- \qquad \text{...(17.14)}$$

$$\therefore \quad K_a = \frac{[H^+][A^-]}{[HA]} \qquad \text{...(17.15)}$$

$$pH = pK_a + \log\frac{[A^-]}{[HA]} \qquad \textit{Henderson-Hasselbalch} \text{ equation} \qquad \text{...(17.16)}$$

where, $[A^-]$ = conjugate base;
$[HA]$ = conjugate acid

The buffering capacity of a solution, ξ, is defined as

$$\xi = \frac{d[B]}{d(pH)} = -\frac{d[A]}{d(pH)} \qquad \text{...(17.17)}$$

$d$[A] and $d$[B] are the moles per liter of strong acid or base needed to cause a change in pH by $d$(pH). The slope of the acid-base titration curve gives the buffer capacity (intensity).

The pH of water depends on chemical and biological contributions—$CO_2$ from air, carbonate salts in water and decaying of biomass, $SO_2$ and $H_2SO_4$ or other acids from any source. Acid precipitation (acid rain) lowers the pH of water bodies. Changes in pH and salinity have important effects on the toxicity of ionizable and neutral organic compounds in aquatic ecosystems.

*Note*: The measurement of pH is an important parameter in acid-base biochemical reactions. In the blood plasma, the bicarbonate $[H_2CO_3^-]$ is directly related to the partial pressure of $CO_2$ ($P_{CO_2}$).

$$pH \propto \frac{[H_2CO_3^-]}{P_{CO_2}} \qquad \text{...(17.18)}$$

A high $P_{CO_2} > 45$ mm Hg (blood pH < 7.35) leads to respiratory acidosis, resulting in hypercapnia (alveloar hypoventilation); a low $P_{CO_2} < 35$ mm Hg (blood pH > 7.35) leads to and respiratory alkalosis, resulting in hypocapnia (alveloar hyperventilation).

### *17.3.3.1 pH Meter*

Determination of acidity or alkalinity of a solution can be carried out by titration. The pH of a solution can be determined accurately by a $[H^+]$ ion-sensitive electrode and a pH meter (Fig. 17.5).

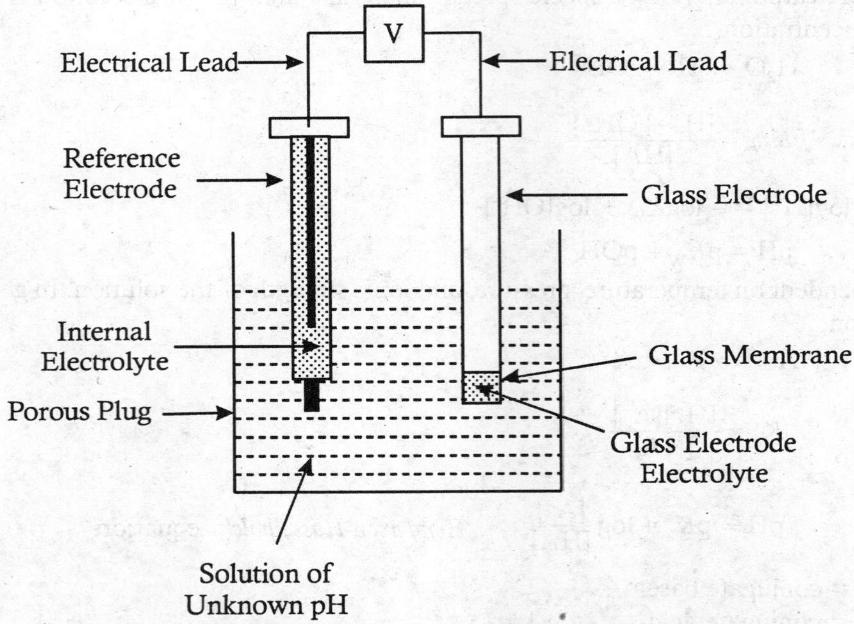


**Fig. 17.5:** Schematic of a pH Electrode Setup

Glass membrane electrode is the most convenient and versatile one for measuring pH. It consists of an internal electrode submerged in a buffer solution in contact with a glass membrane, permeable to $H^+$ ions, separating the electrode from the analyte solution. The inner electrode (reference electrode) is an Ag/AgCl or calomel ($Hg_2Cl_2$) electrode in HCl or buffered potassium chloride (KCl) solution. Calomel electrode is the most commonly used reference electrode. It consists of a sensing electrode made of paste of Hg and $Hg_2Cl_2$ in electrical contact with saturated KCl solution. The entire system is completely sealed in glass. In such a system the voltage of the glass membrane electrode is a logarithmic function of the difference in $H^+$ ion activity of the test and reference solutions.

The pH of natural waters is 4 to 9. It is generally alkaline owing to the presence of carbonates and bicarbonates of alkali and alkaline earth metals.

### *17.3.3.2 Other p-Ion Electrodes*

Ion-selective electrodes are used due their speed, sensitivity and reliability. Other p-ion sensing electrodes are specific to certain ion species. Specific ion electrodes are of three types—modified glass electrodes, membrane electrodes and solid-state electrodes. Ion-exchange properties of silicate glass surfaces can be modified to make them responsive to ions other than $H^+$. The selectivity of the modified glass electrode depends on the composition of the glass membrane.

The solid-state electrode has a single crystal membrane or a membrane consisting of a powder embedded in an inert silicone rubber matrix. The liquid membrane is made up of some water immiscible organic phase containing an ion exchanger and/or and ion carrier (ionophore). The analyte ion forms a more-specific ion pair with the exchanger or complexes with the ionophore. For example, the polyvinyl chloride (PVC) is the membrane material for the nitrate ion. As the nitrate ion is a strongly hydrophilic anion, the exchanger used is a strongly hydrophobic cation, such as tetraalkyl ammonium salt. Ionophore valinomycin has high selectivity for $K^+$. Other ions, such as $F^-$, $Na^+$ etc., can be detected with suitable ion-sensitive membranes.

Gas-sensing electrodes are designed to monitor dissolved gases. For example, an ammonia-selective electrode can be constructed by using an internal glass pH electrode and an ammonium chloride solution between the membranes. The ammonia from the sample diffuses into the ammonium chloride solution between the membranes and partially dissociates in the aqueous solution to form ammonium ions and hydroxide ions. The internal pH electrode responds to the altered pH of the solution caused by the formation of hydroxide ions. *Clark's* cell is $O_2$-specific and *Severinghaus's* electrode is $CO_2$-specific. In oxygen-sensitive membrane electrode setup (Clark's cell), which is typical of a gas sensing electrode system, specific to oxygen, the membrane electrode setup consists of two metal electrodes in an electrolyte solution, and an oxygen-permeable membrane (Fig. 17.6). Normally, the electrodes are imbedded in plastic with one electrode at the center, surrounded by the second electrode, which is mounted as a sleeve. The sensing-portion of the couple consists of a disc-shaped electrode (Pt), surrounded by a ring-shaped anode. Oxygen-permeable membrane (polyethylene, Teflon, Mylar) covers the tip. The current flowing through the cell is proportional to the amount of $O_2$ passing through the membrane.

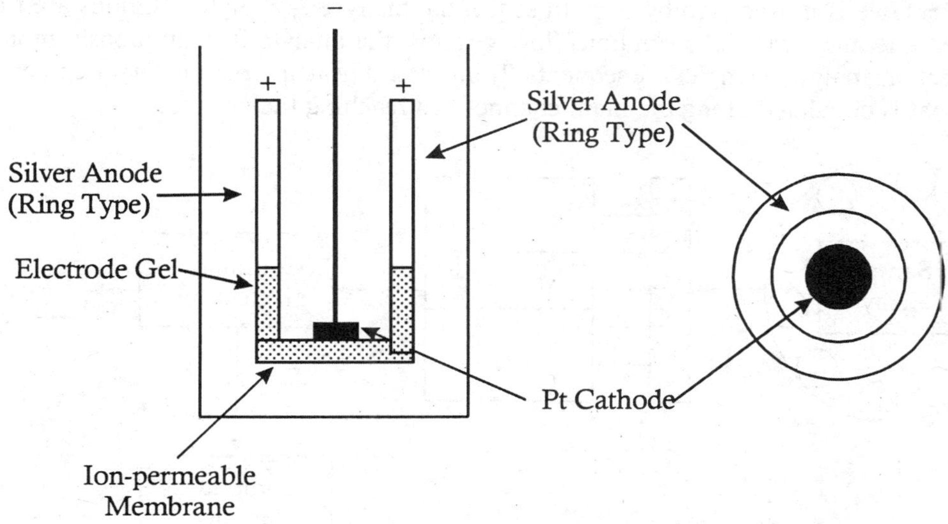


**Fig. 17.6:** Schematic of a p-Ion Electrode Setup

### 17.3.4 Determination of Inorganic Pollutants in Water

Major inorganic pollutants in water are nitrate, sulfate, chloride and hydrated metal oxides, dissolved salts and gases, and suspended matter. Wet-chemical methods and physical techniques as well as biological methods are employed to carry out the identification and determination of pollutants (Chapters 18 to 26). Direct potentiometry with ion-selective electrodes (ISEs) can be used for measurement of free aqueous ions in solution. For determining trace metals (Ag, Al, Cd, Co, Cr, Cu, Fe, Mn, Ni, Pb, Sb and V), water is evaporated in a graphite furnace followed by atomization and detection by AAS and AES (see Chapter 24). Iron introduces matrix errors in AES determination of some metals. Removal of iron by precipitation with ammonium benzoate method would enable the determination of Mo, Pu, U and W by AES in water. Electrodeposition of trace elements on solid graphite electrodes proved useful for analysis by AAS and AES.

Stripping voltammetric method (see Chapter 23) is employed to determine many metals (Bi, Cd, Cu, In, Li, Pb, Sb, Sn, and Zn) in water. The method can be automated to monitor continuously trace metals in natural and sewage waters by using flow injection analysis (FIA) system. Arsenic can be determined by hydride generation and determination by AAS. Organoarsenic compounds (pesticides and herbicides) are first converted to arsenate by on-line photo-oxidation and then to arsine by reduction with sodium borohydride. The hydride is detected by electron capture (EC), AAS, ICP-MS and molecular fluorescence methods.

#### *17.3.4.1 Automatic Monitoring of Inorganic Pollutants*

Automatic analyses are preferred to monitor many inorganic pollutants in water. Automatic analyses rely on the same physical and chemical principles as the manual methods, but they are subject to smaller variability and systematic errors. Also, the automated processes exclude sample collection, transport and storage. Autoanalyzers are irreplaceable in environmental studies and monitoring.

An autoanalyzer comprises the following units—sampler, peristaltic pump, manifold, continuous detector and data processing unit (Fig. 17.7). The sampler is a turntable tray (or a conveyor belt) that contains sample tubes. The sampler is programmed to aspirate a single sample tube at a time, as the turntable is moved step by step. In sequential analyzers, analyte solutions are treated and measured one at a time. In continuous-flow systems, the analyte is continuously monitored. In flow-injection analysis, samples are sequentially injected at peak intervals in an unsegmented carrier stream that is circulated along the main channel and reaching the detector.

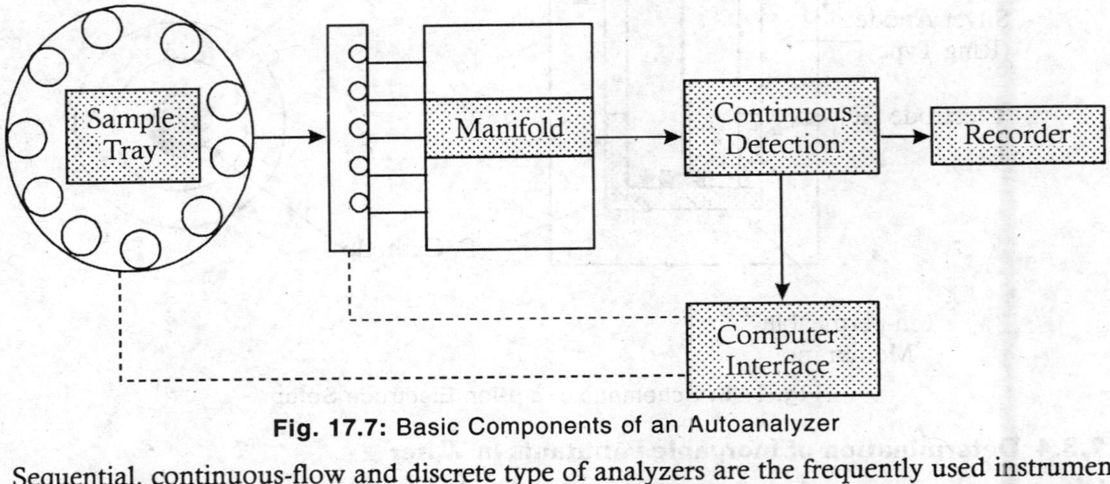


**Fig. 17.7:** Basic Components of an Autoanalyzer

Sequential, continuous-flow and discrete type of analyzers are the frequently used instruments for monitoring water quality. In the continuous-flow method, the analyte concentration is measured without the need to stop the flowing liquid or gas. In discrete sample method, discrete samples (aliquots) are successively passed through the analyzer tube and reagents are added at strategic points, and the signals are recorded in the flow cell uninterruptedly.

Autoanalyzer-based monitoring of dissolved oxygen (DO) is accomplished with a gas separation unit and an $O_2$-detection unit ($O_2$-sensitive galvanic cell, a paramagnetic $O_2$-detector or thermal conductivity detector). $Cl^-$ monitoring is based on wet-chemical method. $Cl^-$ releases $SCN^-$ from mercury thiocyanate that reacts with Fe(III) to give a red complex, whose color is monitored spectrometrically. The determination of $PO_4$ is based on the Molybdenum Blue method. Single detector is used in the analysis of nitrite and nitrate. A cadmium column is used to reduce $NO_3^-$ to $NO_2^-$, which is then determined by a wet-chemical method (Griess-Saltzman method). Metals can be quantified in autoanalyzers by wet-chemical methods. For example, Fe(III) can be determined by bathophenanthroline spectrophotometric method ($\lambda$ = 533 nm). However, physical techniques such as inductively coupled plasma (ICP) and stripping voltametric detectors (SVD) are among the more frequently used techniques for the determination of multiple metals in water.

Ion chromatography (sorption method; see Chapter 22) is another automated method in environmental pollution analysis (Ba, Se, Sr, P etc). The system has a sampler, a measuring device, FIA lines and channels for ion chromatography and detectors (spectrophotometry, conductometry, AAS, AES). FIA-AES-ICP method is used for determination of phosphorous in water.

### 17.3.5 Determination of Organic Pollutants in Water

Polyaromatic hydrocarbons (PAHs), which are potential carcinogens, together with nitrosamines and micotoxins pose health hazards from contaminated water resources. The main sources of organic water pollutants are domestic, agricultural and industrial activities. Sewage is characterized by levels of sediments, colloids, volatile substances and biochemical oxygen demand (BOD). Dissolved oxygen (DO), BOD and chemical oxygen demand (COD) and total oxygen demand (TOD) are important parameters in the assessment of water quality. These are indirect and non-specific methods of monitoring water pollution due to aerobic decomposition of microorganisms. Wet-chemical methods and physical techniques are employed to determine volatile, semi-volatile and non-volatile and humus organic compounds in water. Activated carbon adsorption is a common pre-concentration method for organic species prior to their determination.

#### *17.3.5.1 Dissolved Oxygen (DO)*

Low levels of DO in water bodies lead to the death of fish and other oxygen-dependent organisms. It results in the proliferation of anaerobic microbial population, whose end products are methane, $H_2S$ and offensive sludge. Increase of water temperature drastically decreases DO, leading to deleterious consequences.

A commonly used method for continuos monitoring of DO is based on membrane electrodes (voltammetric and galvanometric). Clark electrode with oxygen-permeable membrane is generally used in monitoring DO. Another electrochemical method for DO determination is the device, consisting of thallium electrode together with a reference electrode. The dissolved oxygen (DO) reacts with the thallium electrode forming Tl-ions. The electrical potential is a measure of the DO concentration.

#### *17.3.5.2 Biochemical Oxygen Demand (BOD)*

The biochemical oxygen demand (BOD) is perhaps the most important measurement for evaluating the character of wastewater, predicting the impacts on receiving water bodies. BOD also gives an estimate of the amount of waste discharged into the water, thus allowing monitoring of the overloading of the sewage treatment plant.

BOD defines the respiratory needs of a biological community over a specific period (mg of $O_2$/L). BOD is taken up in two distinct phases. In the first phase, organic compounds are decomposed and in the second phase, known as the nitrification phase, ammonium ion is oxidized to nitrite and then to nitrate by the action of the bacteria species, *nitrosomonas and nitrobacter*, respectively.

Aerobic microorganisms utilize organic substances as source of energy and consume $O_2$.

$$C_6H_{12}O_6 + 6O_2 \rightarrow 6H_2O + 6O_2 + \text{energy} \quad ...(17.19)$$

A plot of BOD verses time for waters containing organic wastes (Fig. 17.8) is expressed by

$$Y = L(1 - 10^{-Kt}) \quad ...(17.20)$$

Oxygen-depletion curve is related to the standard growth curve

$$N_t = N_0e^{Kt} \quad ...(17.21)$$

where, $Y$ = BOD at time, $t$;
$L$ = Ultimate oxygen demand;
$K$ = kill rate constant;
$N_t$ & $N_0$ = number of organisms at time, $t$ and at $t$ = 0.

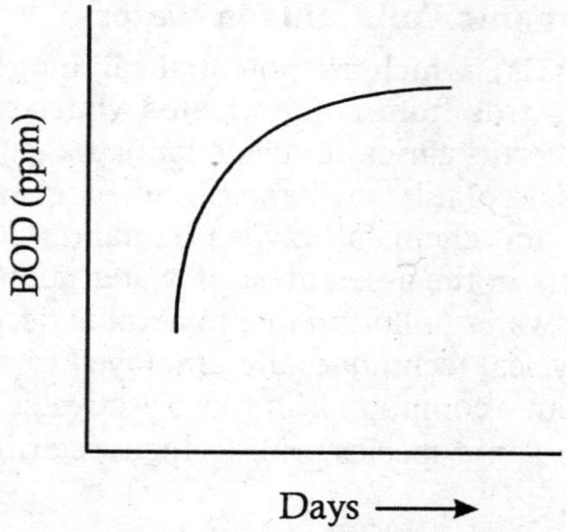


**Fig. 17.8:** BOD Profile

BOD determination depends on temperature. The temperature effect on the utilization of biodegradable substances by organisms is expressed as the $Q_{10}$ value, which is the ratio of oxygen consumption for a temperature T to that at $(T - 10)°$ C. $Q_{10}$ can be predicted from the equation

$$K = A\exp(-E/RT) \quad ...(17.22)$$

where, $K$ = rate constant;
$A$ = constant

The BOD test involves incubating a given amount of water sample over a period of time at 20 °C in the dark and measuring the amount of oxygen consumed by the microorganisms at different intervals. It is an empirical test and valid for studying the natural decomposition of organic matter. In the five-day BOD procedure $(BOD)_5$, samples of wastewater are taken in BOD bottles; seed micro-organisms are added if necessary and the bottles are filled with aerated water containing organic and inorganic nutrients. While the wastewater provides organic matter, the aerated water provides the dissolved oxygen for aerobic decomposition. Dissolved oxygen is measured at the beginning and at the end of the incubation period by Winkler titration (iodometric) method or by oxygen-membrane probes (Clark cell). Reduction in dissolved oxygen is taken as the measure of BOD of the sample.

$$\text{BOD(mg/L)} = \frac{(D_1 - D_2)}{P} \qquad \ldots(17.23)$$

where, $D_1$ & $D_2$ = dissolved oxygen of the sample at the beginning and at the end of the test;
$P$ = sample dilution.

Continuous DO measurement is possible by membrane electrodes. Equivalent BOD (E-BOD) can be estimated according to

$$\text{E-BOD} = 1.3\ \text{Protein} + 0.89\ \text{Fat} + 6.69\ \text{Carbohydrates} \qquad \ldots(17.24)$$

If present in sufficient number, nitrifying bacteria exert a secondary oxygen demand by the oxidation of ammonia (Fig. 17.9). BOD is more sensitive indicator of nitrogenous organic pollutants, which include dairy and food-processing wastes.

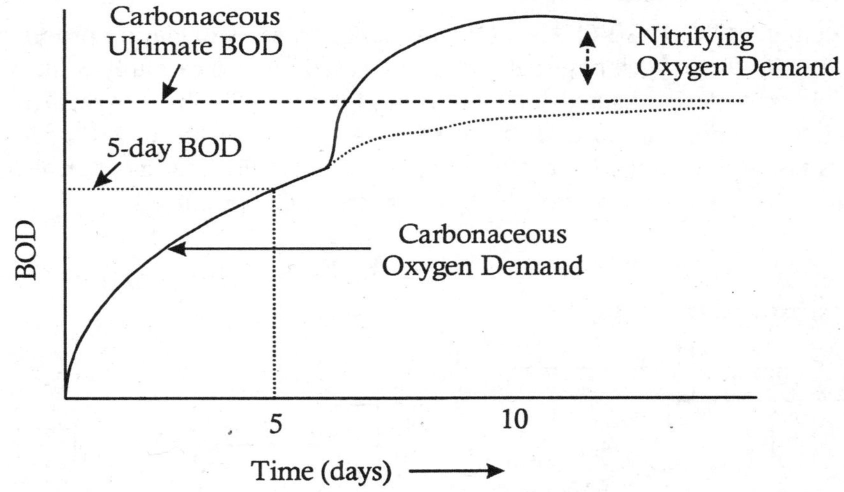


**Fig. 17.9:** BOD of Carbonaceous and Nitrifying Reaction Processes

Replacement of oxygen with a synthetic electron acceptor (e.g. ferricyanide-mediated) in microbial catabolism is a possible rapid method for the determination of BOD. Micro-organisms, *pseudomonas putida, bacillus licheniformis* and *trichosporon cutaneum*, known for their broad range organic substrate utilization, can be used for utilizing microbially produced ferricyanide ion as an alternative electron acceptor, in place of oxygen. Using glucose-glutamic acid (GGA) as a standard, and monitoring of the amount of microbially produced ferricyanide indicates that the catabolic degradation efficiencies approaching those of the conventional 5-day BOD assay could be achieved in one hour.

### *17.3.5.3 Chemical Oxygen Demand (COD)*

BOD determination is time-consuming. A faster method of estimation of oxygen demand (in 2 hours), which can be automated, is by chemical method, known as chemical oxygen demand

(COD). COD is defined as the quantity of oxygen consumed in the oxidation of organic and inorganic matter in water under defined conditions. COD relies on chemical oxidation of all carbon compounds rather than bacterial oxidation.

$$\text{Organic matter} + \text{Oxidant} \rightarrow CO_2 + H_2O \quad \ldots(17.25)$$

A boiling mixture of chromic acid and sulfuric acid is used (for 2 hours at 148° C) to oxidize the organic matter. After the reaction, the solution is diluted and the excess of $K_2Cr_2O_7$ is titrated with ferrous ammonium sulfate in the presence of a color indicator (orthophenanthroline ferrous complex) that can be measured by a photometer.

The COD test is not a direct substitute for the BOD test; however it can be related empirically to the BOD test. Generally, COD is preferred to BOD in process control applications, because results are more reproducible and are available in just a few hours rather than five days. In many industrial samples, COD testing may be the only feasible course because of the presence of bacterial inhibitors or chemicals, which would interfere with a BOD determination.

### *17.3.5.4 Total Oxygen Demand (TOD)*

Total oxygen demand (TOD) is BOD + COD and other types of oxidation processes. In a TOD analyzer (Fig. 17.10), the oxygen required for burning a sample in a carefully controlled air stream is measured. The sample is injected into a burning tube (~900° C) lined with a catalyst. The analyzer calculates the decrease in $O_2$ in the carrier gas ($N_2$) by means of $ZrO_2$ detectors or Pt-Pt cells. The $O_2$ is measured prior to and after injecting the sample. $CO_2$ gas can also be used as a carrier and a non-dispersive IR detector (NDIR) can detect CO produced.

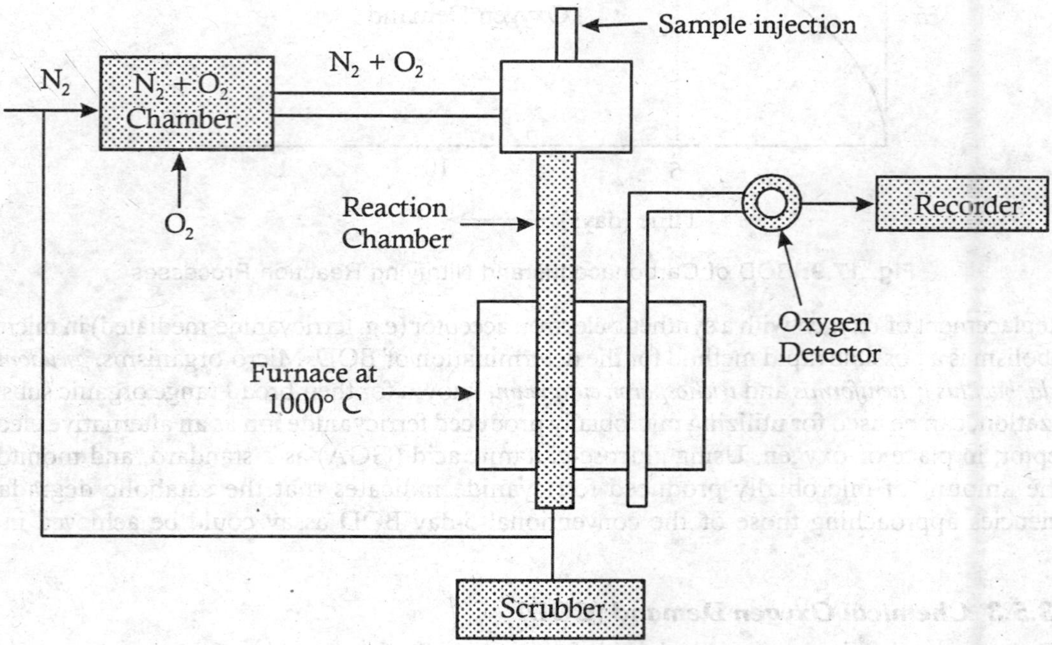


**Fig. 17.10:** Schematic Diagram of Total Oxygen Demand (TOD) Analyzer

#### *17.3.5.5 Measurement of Oxygen Demand by Respirometric Methods*

Respirometric methods enable direct measurement of $O_2$-uptake of organisms. These methods provide continuous measurement of oxygen demand and are useful for assessing biodegradation of specific chemicals, wastewater quality etc. Monometric, volumetric, electrolytic and direct-input respirometers are commercially available.

**Manometric:** Change in pressure but maintains constant volume.
**Volumetric:** Change in $O_2$-uptake at constant volume.
**Electrolytic:** Constant oxygen pressure during electrolysis.

**Direct-input Respirometers:** These respirometers are used in pulmonary oxygen uptake in humans and animals. In this method, the subject is asked to make a maximum inspiration and then exhale as rapidly as possible. Pulmonary resistance, RL, is determined by measuring the pressure difference between intrapleural space ($P_{Pl}$) and the mouth ($P_M$)

$$R_L = \frac{(P_M - P_{Pl})}{V} \qquad \text{...(17.26)}$$

## 17.4 BIOLOGICAL INDICATORS OF WATER QUALITY

Pollutants affect organisms, and thus organisms are indicative of the nature and degree of pollution. Some specialized plants absorb specific contaminants from the soil their roots and foliage. Such plants can be used as ecological indicators of the environment that, when measured, quantifies magnitude of stress, habitat characteristics, degree of exposure to a stressor, or ecological response to exposure. In addition to biological surveillance, such as histological and morphological changes of aquatic organisms, bioassays may be used to determine both acute and chronic impact of pollutants on water bodies. Biochemical estimation of biomass, coliform bodies and microorganisms in a water body are other factors in the estimation of water quality, and nutrients (pollutants) that are present in the water and the chemical and ecological state of the water body. Estimation of biomass includes determination of total carbon, nitrogen, oxygen, phosphorus, lipids, carbohydrates and bottom sediments. Bottom (benthal) sediments reflect the quality of an aqueous ecosystem and its contamination history. Sediments are usually the most appropriate candidates for long-term monitoring of contaminants in aqueous systems. Plants floating on water surfaces can be useful for studying hydrospheric and atmospheric pollution. Lichens and mosses can provide an indication of pollution arising out of wet and dry deposition. Chlorophyll content (photosynthesis) is a parameter to evaluate algal biomass. Determination of ATP, DNA and some other crucial proteins (e.g. rubisco) is also used to estimate living biomass.

### 17.4.1 Photosynthesis Measurement

Measurement of photosynthesis, for the estimation of algal matter and vegetation, should incorporate the measurements of soil water, plant water potential and sap flow.

#### *17.4.1.1 Measurement of Soil Water*

Measurements of soil water can be carried out by gravimetric method, wherein a known quantity of soil is weighed, with water content and after drying. Resistance-block device measures the electrical resistance of a matrix porous block (gypsum, fiberglass or nylon) at equilibrium with the surrounding soil. Nucleonic density gauges can be used, based on $\gamma$-ray attenuation for soil density and soil compactness testing. Neutron scattering method can be used to determine moisture content. This method correlates the amount of soil water of the source with the rate of conversion of fast neutrons from a nuclear reactor source ($^{241}$Am-Be) to slow neutrons as a result of their collision with hydrogen atoms of the soil water. The density of slow neutrons can be related to the soil water by calibration with gravimetric analysis (see Chapter 25).

Water flow rate in soil can be measured by a *permeameter* (Fig. 17.11) by measuring the coefficient of permeability of a soil sample applying *Darcy* equation

$$\text{Flow rate, } Q = Av \quad \text{...(17.27)}$$

$$\text{Energy loss} = \frac{\Delta v}{\Delta L} \quad \text{...(17.28)}$$

$$\therefore \quad Q = KA\frac{\Delta h}{\Delta L} \quad \text{Darcy equation} \quad \text{...(17.29)}$$

$$Q = \frac{\pi K(\Delta h)^2}{\text{In}(\Delta r)} \quad \text{for a cylinder} \quad \text{...(17.30)}$$

where, $A$ = area of the soil sample;
$K$ = coefficient of permeability;
$\Delta h$ = difference in water levels;
$\Delta L$ = length of the soil sample.

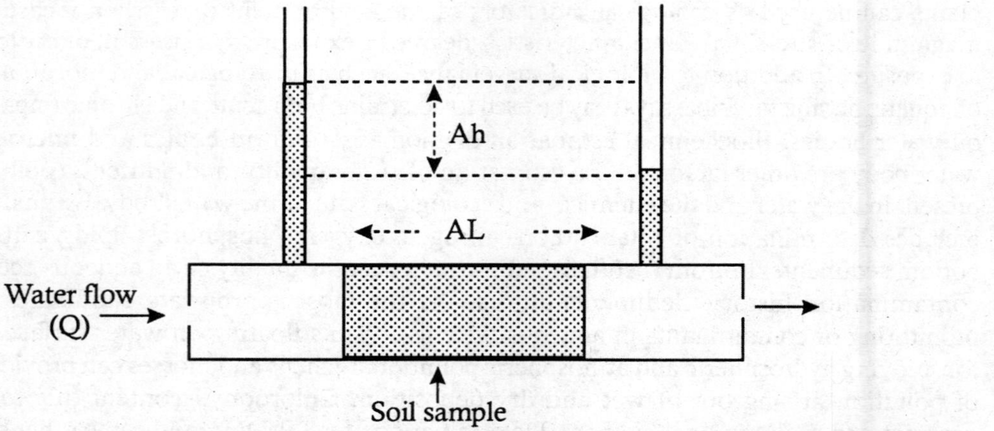


**Fig. 17.11:** Schematic of a Permeameter Setup

Moisture content, $M$(%), water-holding capacity ($H$) of soil samples can be determined gravimetrically.

$$M\% = \frac{(m_1 - m_2)}{m_1} \quad \text{...(17.31)}$$

$$H = \frac{(m_1 - m_2) - (m_2 - m_3)}{(m_2 - m_3)}$$

where, $m_1$ = mass of wet soil + container;
$m_2$ = mass of dry soil + container;
$m_3$ = mass of the container.

Moisture and salt content in soils, fodder, and agriculture produces can be determined by measuring electrical conductivity by a capacitive moisture meter. For determining salt content in soils, two identical soil samples are selected and barium salt is added to one sample (forming a precipitate) and magnesium standard to the other sample. The difference of the electrical conductivity of the two samples is measured.

### *17.4.1.2 Measurement of Plant Water Potential*

The plant water potential is determined by the pressure-bomb method. The water potential of a plant is negative and is therefore under suction. If a stem is cut, the xylem sap will rapidly retreat from the cut end. Applying pressure at the intact end of the tissue, with the cut-end at the atmospheric pressure to balance the xylem water potential, $\psi_W$, can be correlated.

Xylem water potential, $\psi_W$,can be determined by sap flow velocity in xylem by thermoelectric measurements (Fig. 17.12 a & b).

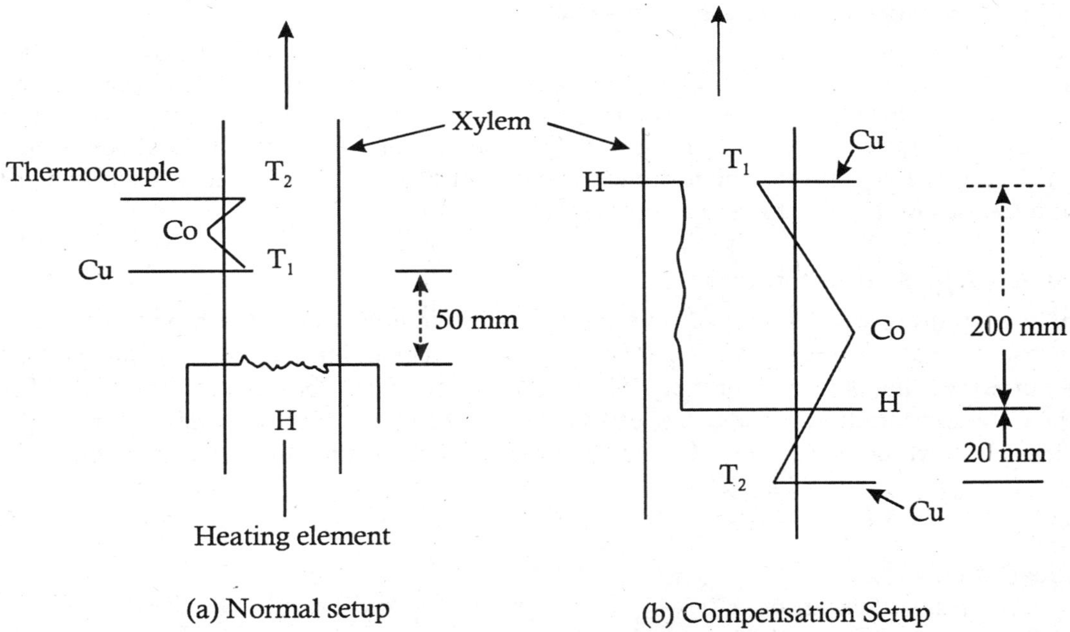


**Fig. 17.12:** Measurement of Sap Flow Velocity

In the normal setup, the heating element, *H*, is heated electrically and the arrival of heat wave at a site is recorded by a thermocouple, T. Reversal of galvanometric deflection is proof of passage of heated fluid (Fig. 17.12 a). In the compensation method (Fig. 17.12 b), a thermocouple is placed asymmetrically above and below the heating element (200/100 mm) and $T_2$ is heated first. Faster the flow, faster is the galvanometric deflection; and the galvanometric deflection is reversed so that $T_1$ becomes warmer than $T_2$. Volume of fluid flow can be obtained from *Hagen-Poiseuille* equation.

$$V = \frac{\pi}{8\eta}\frac{\Delta P}{l}tr^4 \qquad \ldots(17.32)$$

where, $\Delta P/l$ = pressure gradient; $\eta$ = viscosity;
$t$ = time; $r$ = radius.

#### 17.4.1.3 Chlorophyll Determination

Chlorophyll estimation is carried out spectrophotometrically. To determine *chlorophyll-a (Chl-a)* concentration, the pigment is first extracted from the plankton concentrate. The *pheophytin-a (Pheo-a)* present in the sample can be estimated by converting *Chl-a* into *Pheo*-a with acid.

Chlorophyll-a (*Chl-a*) is estimated by absorbance spectroscopy at wavelengths, $\lambda = 630, 645, 663$ and 750 nm and absorbance at 750 nm is subtracted to correct for turbidity. Degradation products of *Chl-a* can be distinguished by their spectral absorption profile. When *Chl-a* is converted to *Pheo-a* by acidification, the absorbance peak is reduced by 60% and shifts from 663 to 665 nm. The ratio of the absorbance peaks, $A_{663}/A_{665} = 1.7$, is a good indicator of the metabolic activity of the phytoplankton.

Chlorophyll-a might also be determined fluorimetrically with excitation at $\lambda = 430$ nm and emission at $\lambda = 663$ nm. The fluorimeter is precalibrated with *Chl-a* solution of known concentration.

#### 17.4.1.4 Measurement of Photosynthesis

Photosynthesis is determined by measuring the absorption of infrared radiation by carbon dioxide, in a non-dispersive IR (NDIR) analyzer. The analyzer is filled with $CO_2$ and a chopper allows IR radiation to pass down the sample tube and reference tube alternately. The displacement of membrane diaphragm in detector, monitored electronically, is a measure of the amount of IR radiation absorbed by the $CO_2$ in the sample tube and reference tube. The instrument is calibrated by using known $CO_2$ concentrations in the detector and sample tubes (see Chapter 24 for details).

#### 17.4.1.5 Leaf Photosynthesis

Leaf photosynthesis is determined by enclosing a leaf in a chamber, and a portion of air with known concentration of $CO_2$ is passed through this chamber and through the analysis tube. At the same time the other potion of air is passed through the reference tube directly. From the difference in the $CO_2$ concentration in the two streams the extent of photosynthesis can be estimated. Thus, the leaf photosynthesis, *LP*, (net $CO_2$ exchange per unit leaf area per unit time) is estimated.

$$LP = \frac{\Delta CO_2 \cdot Q}{A} \quad \text{...(17.33)}$$

where, $Q$ = flow rate;
$A$ = leaf surface area.

Quantum sensors generally measure the relationship between leaf photosynthesis and irradiance. A quantum sensor consists of a photoelectric cell with optical filters to block violet and IR radiation and allow radiation of wavelength between 400-700 nm. The response of the sensor as a function of $\lambda$ (400-700 nm) is monitored (action spectra). The relationship between photosynthesis and irradiance depends on various factors. Some of these are—(*i*) the rate of dark respiration (the rate of $CO_2$ released from the leaf in the dark), (*ii*) quantum efficiency ($CO_2$ uptake per unit increase in the irradiance = $\Delta CO_2/\Delta$irradiance), (*iii*) saturation irradiance (above which net photosynthesis is zero) and (*iv*) heat production.

$CO_2$ and $H_2O$ are the end products of aerobic decomposition. These are used in plant photosynthesis. Anaerobic decomposition is performed by different set of microorganisms for which $O_2$ toxic.

$$C_XH_YN_Z \rightarrow CO_2 + CH_4\uparrow + \text{products (anaerobic)} \quad \text{...(17.34)}$$

#### 17.4.1.6 *ATP Determination*

In many studies related to water quality, it is desirable to measure total biomass. One biomass indicator that has the most potential is ATP. ATP occurs only in living cells and it is a fairly labile intermediate. So, determination of ATP is a reliable estimation of living biomass. The assay is based on bioluminescence with enzymatic reaction of *luciferase* with ATP. The emitted light, measured spectrophotometrically or by scintillation counting method, is proportional to the ATP concentration.

$$E + ATP + LH_2 \leftrightarrow E\text{-}LH_2\text{-}AMP + Pi \qquad \ldots(17.35)$$

$$E\text{-}LH_2\text{-}AMP + O_2 \leftrightarrow E + AMP + CO_2 + T + h\nu$$

$$2E + ATP\text{-}AMP + O_2 \leftrightarrow 2E\text{-}LH_2 + Pi\text{-}CO_2 + T + h\nu$$

$$ATP + O_2 + LH_2 \leftrightarrow AMP + Pi + T + h\nu \ (\sim 550 \text{ nm})$$

where, $E$ = luciferase;
$LH_2$ = reduced luciferin;
$AMP$ = adenosine monophosphate;
$T$ = thiazoliname

### 17.4.2 Bacterial Estimation of Pathogens

Microorganisms are vital components of all ecosystems whose diversity and abundance are greatly influenced by a wide range of factors. For example, presence of "sewage fungus" community in water bodies is indicative of high levels of organic pollutants. Water pollution due to domestic, food, dairy, agricultural and industrial wastes contain pathogens (disease-causing microorganisms). Determination of individual pathogens (bacteria, protozoans, helminthes, and viruses) by instrumental methods is not practical. Instead, non-pathogenic coliform (fecal) bacteria (microorganisms that are found in the intestinal tracts of humans and animals) have been used as indicators of potential presence of pathogens and quality of water for usage. For example, *Escherichia coli* (*E. coli*) species are a subset of the coliform bacterial groups that is a part of normal intestinal flora of humans and animals. Therefore, presence of E.coli is a direct indicator of fecal contamination of water and presence of pathogens. Similarly, Enterococci are enteric bacteria used to indicate fecal contamination and the possible presence of pathogens in water. These coliform bacteria are aerobic, facultative and non-spore forming, and the number of coliform group of bacteria present in water provide an estimate of the presence of waterborne pathogens. A high coliform count (coliform index) usually indicates sewage pollution. Microbiological tests involve the determination of total count of these coliform bacteria.

The most probable number (MPN) and membrane filter (MF) methods are commonly employed for enumerating coliform bacteria in water. The MPN methods of estimation of coliform bacteria (estimation of concentration of target organisms) include culture and enzyme-substrate techniques. Culture methods rely on the detection of gas evolved when coliform bacteria (*E. coli*) ferment lactose or the presence of turbidity (enterococci), colony formation or color to detect the target organism. The multiple-tube fermentation methodology is useful for detecting low concentrations of organisms, toxic compounds. Enzyme-substrate test uses chromogenic or fluorogenic substrates that react with specific enzymes to produce color changes or fluorescence to detect target organism.

Membrane filter (MF) test is a direct plating method of estimating bacteria in water based on the development of colonies on the surface of the filter. These filters are made of a mixture of cellulose esters of pore size ~0.45 μm (bacteria is of the size ~0.5-5 μm). After separation, the filter is incubated on a nutrient medium to allow growth of desired coliform bacteria and the resulting colonies are counted. A second substrate medium is used in two-step MF procedures to confirm and/or differentiate the target organisms. Plates are incubated and target colonies are counted or detected (golden-green metallic sheen). Another coliform test for pathogens depends on the detection of hydrogen released during the metabolism of lactose (35-45° C). Radiolabeling of fermentable sugars is a sensitive assay method. The presence of bacteria can be estimated by chemiluminescence also.

$$\text{Luminol} + H_2O_2 \xrightarrow{\text{Porphyrins}} \text{Luminol}^* + h\nu \quad \ldots(17.36)$$

Living bacteria can be detected by monitoring emitted light ($\lambda$ = 550 nm) during a chemiluminescence reaction (Eqn. 17.36).

Fluorescence antibody technique is an important serological method of detecting *Salmonella* (by a fluorescence microscope), *Vibrio cholerae* and other bacteria. Specific reagents are used for isolation of Shigella strains that show red colonies.

The bacterial tests for the estimation of pathogens present in water can be summarized under:

1. **Electrochemical:** Measurement of $O_2$ with growing cultures (Electron transfer process).
2. **Radiometric:** Radiolabeling of fermentable sugars (Radiochemical method).
3. **GC-MS:** Instrumental physical technique.
4. **Serological:** Fluorescence antibody technique & endotoxin assays (Immunoassays).
5. **Luminescence:** Chemiluminescent immunoenzymatic assays— luminol-bacterial iron porphyrin assays; firefly luciferase-ATP assays.

## BIBLIOGRAPHY


Albarges, J. B. (ed.) (1980), Pergamon Press: Oxford. "*Analytical Techniques in Environmental Chemistry.*"

Baier, R. E., Mark, H. B. and Mattson, D. S. (eds.) (1981), Marcel Dekker: New York. "*Water Quality Measurement.*"

Bates, R. G. (1973), John Wiley: New York. "*Determination of pH: Theory and Practice,*" (2nd edn.).

Bohren, C. F. and Huffman, D. R. (1983), John Wiley: New York. "*Absorption and Scattering of Light by Small Particles.*"

Botsford, J. L. (1999), *Environ. Toxicol.*, **14**(2): 285. "A simple method for determining the toxicity of chemicals using a bacterial indicator organism."

Catterall, K., *et al.* (2001), *Talanta*, **55**(6): 1187. "The use of microorganisms with broad range substrate utilization for the ferricyanide-mediated rapid determination of biochemical oxygen demand."

Fifield, F. W. and Haines, P. J. (2000), Blackwell Science: Oxford, UK. "*Environmental Analytical Chemistry.*"

Frei, R. W. and Albarge, J. (eds.) (1986), Gordon & Breach: New York. "*Air and Water Analysis: New Techniques and Data.*"

Grace, J. and Ford, E. D. (eds.) (1981), Blackwell: Oxford, UK. "*Plants and their Atmosphere.*"

Guilbault, G. G. (1976), Marcel Dekker: New York. "*Handbook of Enzymic Methods of Analysis.*"

Lamb, J. C. (1985), John Wiley: New York. "*Water Quality and its Control.*"

Lintz, J., Jr. and Simonett, D. S. (1976), Addison-Wesley: Reading, M A. "*Remote Sensing of the Environment.*"

Luttge, U. and Higginbotham, N. (1979), Springerverlag: New York. "*Transport in Plants.*"

Markert, B. (ed.) (1993), VCH Pubs: New York. "*Plants as Biomonitors: Indicators for Heavy Metals in the Terrestrial Environment.*"

McKay, G. (ed.) (1996), CRC Press: Boca Raton, F L. "*Use of Adsorbents for the Removal of Pollutants from Wastewaters.*"

Milburn, J. A. (1979), Longmans: London. "*Water Flow in Plants.*"

Minear, R. A. and Keith, L. H. (eds.) (1982), Academic Press: New York. "*Water Analysis,*" (Vols. 1 & 2).

Minear, R. A. and Keith, L. H. (eds.) (1994), Academic Press: New York. "*Water Analysis: Organic Species.*"

Nobel, P. S. (1974), Freeman: San Francisco, C A. "*Biophysical Plant Physiology.*"

Pfallin, J. R. and Ziegler, E. N. (1976), Gordon & Breach: New York. "*Encyclopedia of Environmental Science and Engineering,*", (Vol. 2).

Quinby-Hunt, M. S., McLaughlin, R. D. and Quintanilha, A. T. (1986), Wiley & Sons: New York. "*Instrumentation for Environmental Monitoring*," (Vol. **1** & **2**; 2nd edn.).
Rump, H. H. (1999), Wiley-VCH: Weinheim. "*Laboratory Manual for the Examination of Water, Wastewater and Soil*," (3rd edn.).
Skoog, D. A., Holley, J. F. and Nieman, T. A. (1998), Harcourt-Bruce: Philadelphia, PA. "*Principles of Instrumental Analysis*," (5th edn.).
Van Loon, J. C. (1982), CRC Press: Boca Raton, FL. "*Chemical Analysis of Inorganic Constituents of Water*."
Wezel, A. (2000), *Environmental Reviews*, **6**(2): 123. "Chemical and biological aspects of ecotoxicological risk assessment of ionizable and neutral organic compounds in fresh and marine waters."
Woodward, F. I. and Sheehy, J. E. (1983), Butterworth: London. "*Principles and Measurements in Environmental Biology*."
Zimmermann, M. H. and Milburn, J. A. (eds.) (1975), Springerverlag: Berlin. "*Encyclopedia of Plant Physiology*."
...(1986), Wiley Interscience: New York. "*Instrumental Methods for Environmental Monitoring, Vol. 2, Waters*", (21st edn.).
...(1990), Office of SRMS/Natl. Inst of Standards and Technologies: Gaithersburg, MD. "*Standard Reference Materials for Environmental Research, Analysis and Control*."
...(1998), American Public Health Assn: Washington, DC. "*Standard Methods for the Examination of Water and Wastewater*," (20th edn.).
...(2000), EPA 600-R-00-013: Cincinnati, OH. "*Membrane Filter Method for the Simultaneous Detection of Total Coliforms and Escherichia coli in Drinking Water*."

# SECTION-VI

# Chemical Methods in Environmental Pollution Analysis

## OVERVIEW

Environmental pollutants are present in air, water and soil. They can be of inorganic or organic type and be present in as particulate matter, gaseous residues, aerosols, metal fumes and their compounds, and host of other man-made chemicals. To add to the difficulty of their monitoring some are highly reactive with short life-times and some quite stable and they remain and accumulate in the environment. Their monitoring might be local or global interest. With all these variations the analysis of environmental pollutants is a highly complicated task. Chapters 18 to 21 outline the chemical methods for determining environmental pollutants. They include gases and aerosols (Chapter 18), metals (Chapter 19), inorganic and non-metallic constituents (Chapter 20) and organic compounds (Chapter 21).

Irrespective of where they are present (air, water or soil) most of the recognized pollutants exist in very low concentration. This necessitates their pre-concentration before subjecting them to analysis proper. In the case of air and water samples the particulate matter is, usually, separated first by suitable means and pollutants associated with the residue are analyzed separately. The choice of the pre-concentration method depends on the matrix in which pollutants are present and their nature and concentration. The detection devices are many—UV-visible spectrometry, fluorimetry, mass-spectrometry (MS), AAS, AES, flame ionization, etc.

## ANALYSIS OF GASES AND AEROSOLS

In Chapter 18, chemical methods for analyzing pollutants in the form of aerosols and gases are outlined. The groups of pollutants considered are oxidants and radicals (include OH radical and ozone) carbon compounds (carbon monoxide and carbon dioxide), nitrogen compounds (oxides of nitrogen, ammonia, cyanides), sulfur compounds (sulfides, sulfur dioxide) and halogens and halides.

The distribution, accumulation, bioavailability, biodegradability and toxicity of elements and compounds depend not only on their concentration but also on their physicochemical associations with the surrounding systems. Generally, all chemical methods of analysis rely on suitable reagent reactions, followed by quantification by gravimetric, volumetric or any other physicochemical or physical techniques. In most cases, the analyte (pollutant) is converted by wet-chemical methods and its detection and concentration are carried out by spectrophotometric methods. Characteristic color (wavelength) and intensity of the colored complex gives a quantitative estimate of the concentration of the pollutant present.

Air borne particulate matter, after collecting on a filter paper, can be analyzed semi-quantitatively by the ring-oven technique. The color and the intensity of the complex are quantitatively determined by colorimetry. This method is simple and it is useful for field surveys and spot-checks. However, it is not suitable for continuous monitoring of air for pollutants.

Automated instrumental setup for monitoring air pollution by wet-chemical methods requires a series of interconnected modules that automate step-wise wet-chemical reactions. Detection and quantification are generally by titrimetric or colorimetric (spectrophotometric) methods. In the case of pollutants of global consequence, like $NO_x$ and $CO_x$, monitoring of the atmosphere is done more or less continuously using instruments located on several platforms like airplanes and satellites, using suitable spectroscopic techniques.

## DETERMINATION OF METALS

In Chapter 19, methods for the determination of metals are summarized. In the case of metals and metalloids, determination of their concentrations in various oxidation states and forms (speciation) is necessary, because the intensity of toxicity of many metals and their compounds (e.g. As, Cd, Cr, Hg, Pb, Se and V) depends on the oxidation state and the chemical forms. Except for the preliminary treatment and pre-concentration procedures, the methods for determination do not fundamentally change whether the original sample is air or water or soil. In the case of air samples it is a common practice to analyze air particulate for metals. Since wet-chemical methods still find some place in the determination of individual pollutants they are also mentioned in this chapter for most of the metals. Spectrophotometric methods, based on the color formation between the metals and chelating agents, are among the traditional procedures used for trace metal determination. In modern times these molecular spectroscopic methods have been largely supplanted by AAS and ICP emission spectroscopy in view of their enhanced sensitivity and precision and multi-element capability and hence fastness. The later, coupled with mass-spectrometric detectors has proved to be a highly sensitive and precise method of analysis.

There are instances of individual metals being determined by exploiting their special properties. Thus, traces of mercury in air or water can be determined taking advantage of the characteristic UV absorption of its vapor. The elements arsenic, antimony, bismuth, selenium, tellurium, tin, germanium and lead form volatile covalent hydrides. Taking advantage of this property these elements can be separated from many interfering metals and determined with high sensitivity by AAS, ICP-AES or ICP-MS technique.

## INORGANIC, NON-METALLIC COMPOUNDS

In Chapter 20, procedures for determination of some selected inorganic, non-metallic constituents are given. They are silica, asbestos, phosphorus and selenium. Solid inorganic particulate and coagulated substances can be determined by gravimetric and volumetric as well as more sophisticated physical techniques. Some of the physical techniques of determination are ion-chromatography (IC), electrical conductometry, optical and electron microscopy, X-ray fluorescence (XRF) spectroscopy and neutron activation analysis (NAA) and X-ray diffraction and other methods. IC method is suitable for rapid, sequential determination of many common ions, such as $F^-$, $Cl^-$, $Br^-$ nitrites, nitrates, sulfates and phosphates.

## ORGANIC POLLUTANTS

In Chapter 21, procedures for analysis of organic pollutants are outlined. Hundreds of organic compounds, most of them of anthrapogenic origin, are recognized to be pollutants. They are present in air, water and even soil. From the later they reach even groundwater. Their number and variety are staggering.

Organic pollutants can be classified under—(1) conventional pollutants (aldehydes, ketones, alcohols) (2) aromatic hydrocarbons, (3) polynuclear aromatic hydrocarbons (PAHs), and (4) synthetic organic compounds that include organopesticides, fungicides and herbicides. Some of them like the clorofluoro-hydrocarbons are involved in ozone depletion in the stratosphere. Their manufacture and use is being regulated under various International Protocols. Their determination is generally carried out by one or the other variation of GC (like capillary GC) and detection is also by various means, the important one being electron capture. For components, which are nonvolatile HPLC is a frequently employed technique. To tackle the variety of compounds different stationary phase materials are in use. Mass spectrometry is a very frequently used detection system in all these separations.

# 18

# Chemical Methods in Environmental Pollution Analysis—Gases and Aerosols

In the analysis of gases and aerosols, the groups of pollutants considered are oxidants and radicals (include OH radical and ozone), carbon compounds (carbon monoxide and carbon dioxide), nitrogen compounds (oxides of nitrogen, ammonia, cyanides), sulfur compounds (sulfides, sulfur dioxide) and halogens and halides.

The distribution, accumu-lation, bioavai-lability, biodegrad-ability and toxicity of elements and compounds depend not only on their concentration but also on their physicochemical associations with the surrounding systems. Generally, all chemical methods of analysis rely on suitable reagent reactions, followed by quantification by gravimetric, volumetric or any other physicochemical or physical techniques. In most cases, the analyte is converted by wet-chemical methods and its detection and concentration are carried out by spectrophotometric methods. Characteristic color and intensity of the colored complex gives a quantitative estimate of the concentration of the pollutant present.

Air borne particulate matter, after collecting on a filter paper, can be analyzed semi-quantitatively by the ring-oven technique. The color and the intensity of the complex are quantitatively determined by colorimetry. This method is simple and it is useful for field surveys and spot-checks, but is not suitable for continuous monitoring of air for pollutants.

Automated instrumental setup for monitoring air pollution by wet-chemical methods requires a series of interconnected modules that automate step-wise wet-chemical reactions. Detection and quantification are generally by titrimetric or colorimetric (spectrophotometric) methods. In the case of pollutants of global consequence, like $NO_x$ and $CO_x$, monitoring of the atmosphere is done more or less continuously using instruments located on several platforms like airplanes and satellites, using suitable spectroscopic techniques.

This chapter outlines the chemical methods for analyzing pollutants in the form of aerosols and gases.

Environmental pollutants include particulate matter, gaseous residues ($SO_x$, $NO_x$, $CO_x$, $H_2S$, $NH_3$), aerosols, metal fumes and their oxides, hydrocarbons, organics, pesticides, toxins and biological wastes. Chemical methods of analysis of these pollutants take advantage of the variation in their chemical properties. These substances can be estimated by gravimetric, volumetric, titrimetric, colorimetric or any other suitable physicochemical or physical techniques. In many cases the analyte is converted to a characteristic compound that can be determined by measuring its absorption/emission characteristics by spectroscopic methods.

Gaseous pollutants are often analyzed after the particulate matter is trapped by filtration or by passing the sample through water or a liquid. Chemical estimations of gaseous pollutants can be semi-quantitative or quantitative. Pre-concentration of some of the gaseous constituents is achieved by adsorption or absorption in suitable reagent solutions before the procedures suitable for the determination are undertaken. Spectrographic techniques, which include absorption or emission of radiation, play a crucial role in the final analysis.

## 18.1 GENERAL ANALYSIS PROTOCOLS FOR GASES AND AEROSOLS

General protocols for chemical analyses include filed survey tests, laboratory methods and continuous monitoring methods. Various semi-quantitative and quantitative methods are available tailor-made to determine pollutants, based on their physiochemical properties.

### 18.1.1 Determination by the Ring-oven Technique

The ring-oven method is a semi-quantitative method, well suited for determining traces of aerosols and airborne particles collected on a filter paper. The determination is performed either visually by comparing the color developed on the sample filter paper against the standards prepared in a similar manner, or colorimetrically. It is often possible to perform four or even more separate determinations on one sample filter paper by cutting as many sectors of the ring as there are determinations to be performed (Fig. 18.1). This method is simple and is useful for field surveys and spot-checks, but not useful for routine monitoring.

The steps involved in the procedure are given below:

**Step 1:** A filter paper containing the analyte (pollutant) is placed over an annular heating block. Temperature of the block is maintained depending on the solvent. An annular ring-zone is formed on the filter paper.

**Step 2:** The analyte is solubilized and washed to the ring zone and when the solvent evaporates, the analyte complex is found in a sharply defined ring.

**Step 3:** Suitable reagents are sprayed on the ring zone to develop colored complexes and analyzed comparatively with standards or colorimetrically.

### 18.1.2 Continuous Flow Chemical Analysis

Behavior of pollutants as function of time requires continuous evaluation. The instrumentation (Fig. 18.2) consists of a series of interconnected modules, which automate, stepwise, wet-chemical methodologies. The detection is generally by colorimetric or titrimetric methods.

Oxidants ($O_3$, $Cl_2$ and $NO_x$) are analyzed by measuring the amount of free iodine they liberate from alkaline KI solution. The iodine is proportional to the concentration of the oxidant. The measurement can be done by electrochemical or photometric techniques.

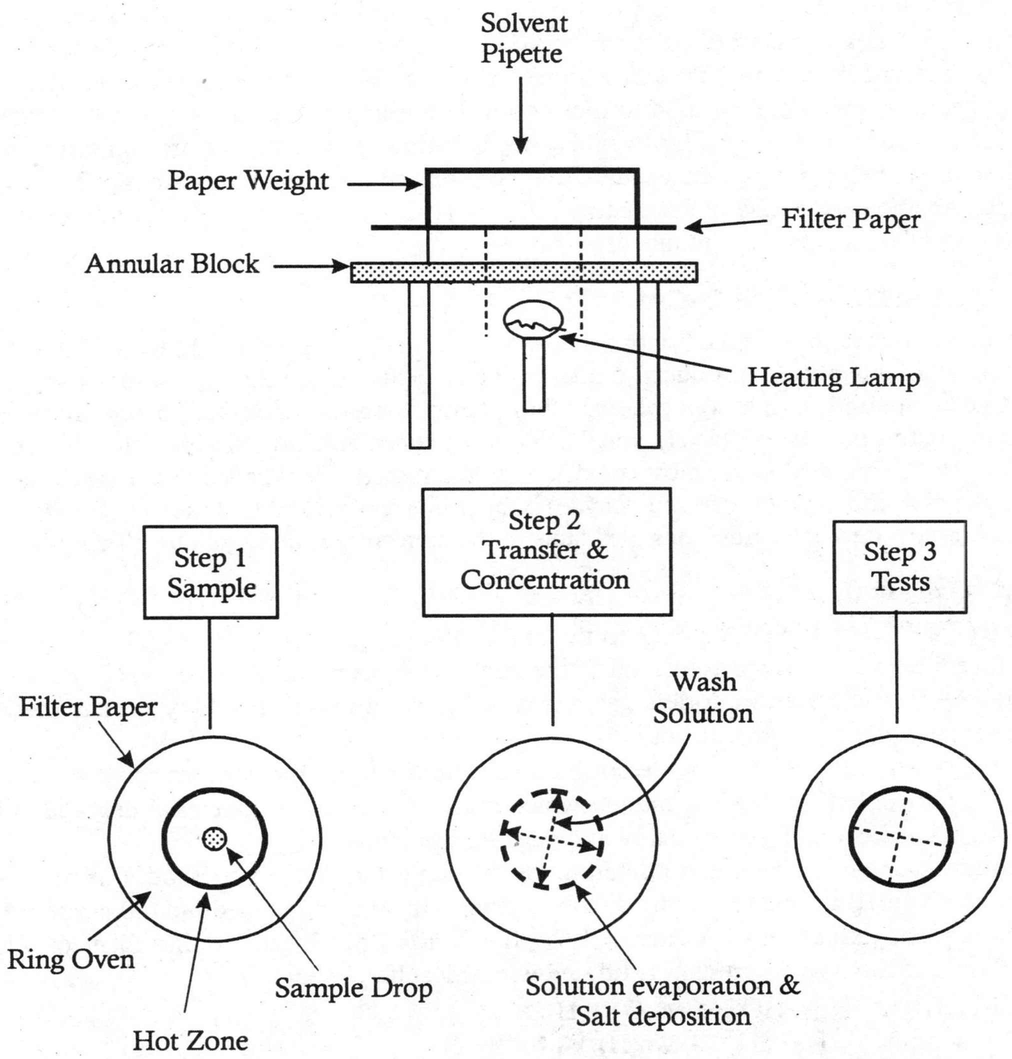


**Fig. 18.1:** Schematic of Ring-oven Colorimetric Setup for Wet-chemical Analysis

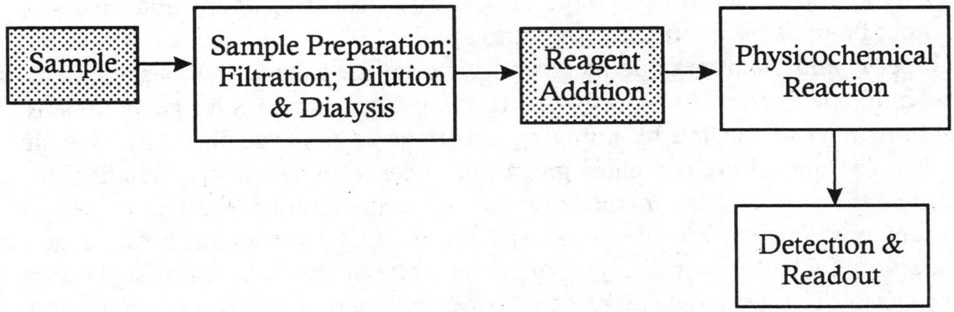


**Fig. 18.2:** Schematic of Continuous Monitoring Setup in Wet-chemical Analysis

An automatic "microsensor" can be used to detect minute quantities of gaseous impurities (0.1 mg/L) by measurement of color produced on a paper sensitized with suitable reagents. The "*Titrilog*" instrument measures the conversion of $SO_3$ to $H_2SO_4$ in an electrolytic cell in the presence of $Br^-$. The instrument can be used to monitor $H_2S$ and mercaptans. *Thomas autometer* is used to measure the electrical conductivity of $H_2SO_4$ solution, produced by the oxidation of $SO_3$. Continuous recording instruments can be used to monitor particulate impurities (>3 μm) by light scattering methods (*Rayleigh* or *Mie* scattering). Radiation counters are (also) used in continuous monitoring of radioactive pollutants.

## 18.2 OXIDANTS AND RADICALS

Oxidants and free radicals in the troposphere occur in both the gaseous and aqueous phases (clouds, fog rain). Photochemically produced oxidants and radicals dictate the fate of many organic and inorganic compounds in the troposphere. Atmospheric oxidants are oxygen, ozone, and hydrogen peroxide, organic peroxides ($RO_2H$), and $Cl_2$. They are considered primary indicators of the quality of urban air. OH radical is the most important of the radicals. It is called "*atmospheric detergent*". Photolysis of polluted air (e.g. formaldehyde) is the primary source of OH radicals. Oxidation of $I^-$ may be employed to determine these oxidants in the concentration range 0.01-10 ppm.

### 18.2.1 Oxygen ($O_2$)

Oxygen ($Z = 8$; $A = 15.999$) belongs to the main subgroup of Group VI, (VIa), of the Periodic Table (See the Periodic Table at the end of the book). It exists in molecular form ($O_2$) in nature. It is a colorless, tasteless and odorless gas, essential to living organisms. Oxygen has the highest electronegativity and electron affinity.

An important parameter to be determined in water analysis is dissolved oxygen (DO). It is related to the content of organic matter, which consumes $O_2$ for its bacterial degradation. DO, therefore, is a measure of water quality and degree of pollution.

A common chemical method for determining oxygen is the Winkler method. It is based on the oxidation of Mn(II) to Mn(IV) by the dissolved oxygen in an alkaline medium (30 per cent KOH), and subsequent oxidation of iodide to $I_2$ by the Mn(IV) produced. Iodine formed, which is proportional to the oxygen present, is titrated with thiosulfate.

$$\begin{aligned} &\text{Mn(II)} + 2OH^- + \tfrac{1}{2}O_2 \rightarrow MnO_2 + H_2O \\ &MnO_2 + 2I^- + 4H^+ \rightarrow \text{Mn(II)} + I_2 + 2H_2O \\ &I_2 + 2S_2O_3^{2-} \rightarrow S_4O_6^{2-} + 2I^- \quad (\textit{Winkler test}) \end{aligned} \quad \text{...(18.1)}$$

The iodine can also be determined colorimetrically by measuring its absorbance at 450 nm. This method cannot be used for continuous monitoring.

Galvanometric and polarographic methods have been used for *in situ* measurement of DO in natural and industrial waters. The electrolytic reaction is spontaneous for galvanometric method, while reaction must be started by applying a voltage to polarize the indicator electrode in polarography. The quenching of yellow-green fluorescence of alkaline adrenaline, by dissolved oxygen, is the basis of one of the methods for the determination of the latter.

Some other physicochemical methods available for the determination of oxygen are—gas chromatography (GC) (Chapter 22), polarography, voltammetry and paramagnetic property of molecular oxygen (Chapter 23), $O_2$-sensing electrode (Clark cell) method (the diffusion current is linearly proportional to molecular oxygen), fluorescence quenching (Chapter 24).

### 18.2.2 Ozone ($O_3$)

Ozone, $O_3$, is formed by an electric discharge through oxygen, and occurs in the stratosphere. Pollutants like $NO_2$ and hydrocarbons favor the formation of ozone at ground level. Ozone is a powerful oxidant. It oxidizes most non-metals and all metals except gold and the platinum metals. It transforms lower oxides into higher ones, and metal sulfides to sulfates.

A simple qualitative test for the detection of ozone is the blue color it gives to a starch-iodide paper. Reverse *West-Gaecke* method is specific for the determination of ozone in air. Wet-chemical as well as physical techniques can be used for determination of ozone.

1. Reaction of ozone with eugenol (4-allyl-2-methoxy phenol) gives formaldehyde as one of the products. It is collected in distilled water and determined spectrophotometrically at $\lambda = 560$ nm after adding Na-dichlorosulfitomercurate(II) and pararosaline.
2. *Indigo colorimetric method*: In acidic solutions, ozone decolorizes indigo and the decrease in absorbance at $\lambda = 600$ nm is proportional to the concentration of ozone. This colorimetric method is quantitative, selective and simple.
3. *Iodine liberation method*: In a buffered potassium iodide solution ozone liberates iodine ($I_2$). Iodine is determined by titration or colorimetrically (iodine-starch blue color).

$$O_3 + 2H^+ + 2I \rightarrow I_2 + H_2O + O_2 \qquad \text{...(18.2)}$$

$$I_2 + I^- \rightarrow I_3^- \text{ (by colorimetry)}$$

4. *Potentiometric method*: The ability of $O_3$ to oxidize iodide ion to iodine in a neutral medium is used in this method. The air sample is split into two streams; one is scrubbed to remove reducing gases and the other to remove reducing gases and $O_3$, before the sample reaches the electrolytic cell. The electrolytic cell contains buffered NaBr, NaI, two Pt-electrodes and one anode. The differential galvanic current produced is directly proportional to the ozone concentration.
5. *SBUV and TOMS methods*: The atmospheric distribution of ozone is determined by satellite based Solar Backscatter Ultraviolet (SBUV) and Total Ozone Mapping Spectrometer (TOMS). The SBUV instrument compares the amount of solar UV with the UV radiation bounced to the satellite by the earth's atmosphere. Since ozone absorbs UV radiation, the decrease in the amount of light that is back-scattered to the satellite relative to that impinging on the atmosphere is a function of the amount of ozone it contains. The TOMS provides daily maps of the global distribution of "total ozone" present between the earth's surface and top of the atmosphere.
6. *By chemiluminescence*: Ethylene reacts with $O_3$ to produce products, which fluoresce. The fluorescence ($\lambda_{max} = 435$ nm) intensity, as measured by a photomultiplier, is dependent on the concentration of $O_3$ in the presence of excess ethylene (sensitivity 0.003-30 ppm). The method can be automated.

$$O_3 + C_2H_4 \rightarrow HCHO^* \qquad \text{...(18.3)}$$

$$HCHO^* \rightarrow HCHO + h\nu \; (300 \le \lambda \le 550 \text{ nm}).$$

7. *Rhodamine-B method*: $O_3$ is determined by treating the sample with rhodamine-B on an activated silica gel. Transition of rhodamine-B from its excited electronic state to ground state yields a characteristic photoluminescence ($\lambda = 590$ nm), the intensity of which is proportional to the concentration of $O_3$ (0.001 ppm).

### 18.2.3 Peroxides

Most common methods for determining peroxides are by electrolytic, chemiluminescence, fluorimetry, and HPLC.

1. *Electrolytic method*: $H_2O_2$ can be determined electrolytically (see Chapter 23). The sample, passed through a liquid scrubber containing HCl-phosphate buffer at pH 7.2, absorbs $H_2O_2$, which is then continuously driven to the three-electrode setup electrolytic detector—the working electrode (a Pt-disc), a reference electrode (saturated calomel electrode) and a counter electrode (a ring-shaped Pt-electrode).
2. *Amperometric method*: Amperometric detection method is based on the utilization of a Prussian Blue bulk modified carbon screen-printed electrode (see Chapter 23).
3. *Fluorescence method*: Non-fluorescent compounds are oxidized to highly fluorescent compounds and assayed (see Chapter 24).
4. *Fluorescence (LIF) method*: Peroxides are separated by HPLC using $H_3PO_4$ at pH 3.5 as eluent, and detected by laser-induced florescence (LIF) or by TDLAS.

### 18.2.4 Radicals

It is extremely difficult to monitor and quantify radicals, because of their reactivity and short-life spans. Chemical amplification method, long-path absorption spectrometry (LPAS), laser-induced fluorescence (LIF), radiochemical tracer methods and electron spin resonance (ESR) methods are some of the techniques employed in their determination.

1. *Chemical amplification method*: This method, applicable for measurement of total $RO_x$ radicals, is based on the chain reaction between the radicals and NO to produce $NO_2$, with CO as the chain propagator. The $NO_2$, thus produced, is detected by chemiluminescence reaction with luminol.
2. *Radiochemical tracer method*: A small amount of radioactive $^{14}CO$ is added to the air sample, and is led into a quartz reaction chamber where it reacts with OH radical to form $^{14}CO_2$. $^{14}CO_2$ is cryogenically enriched and measured by a radiation counter.
3. *Long-path absorption Spectrometry* (LPAS): Long-path absorption spectrometry (LPAS) in the UV range is the only technique that allows the absolute determination of the atmospheric concentration of OH radical. Light from a broadband tunable laser travels a known distance (1-10 km) and is reflected back to a detection system near the light source (Fig. 18.3). The concentration of OH radical is calculated according to the Beer-Lambert law.

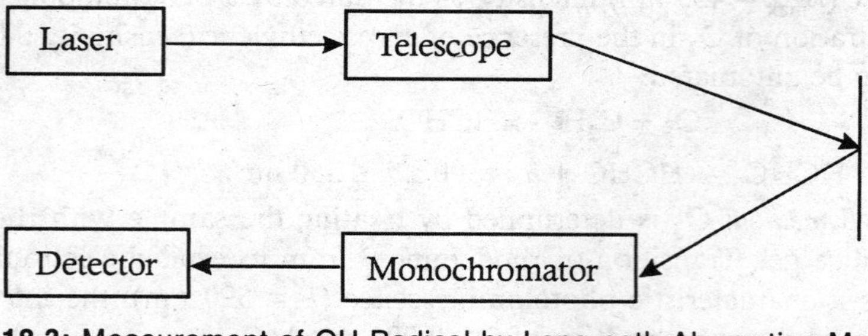


**Fig. 18.3:** Measurement of OH Radical by Long-path Absorption Method

## 18.3 CARBON AND ITS (INORGANIC) COMPOUNDS

Sources of carbon are natural (forest fires) and anthropogenic. Carbon can be in the form of inorganic as well as organic compounds. Carbon that is thermo-evolved between 400-525° C is termed organic carbon, while that is evolved between 700-825° C is designated elemental carbon.

### 18.3.1 Elemental Carbon

Elemental carbon influences the radiative transfer in the atmosphere by its absorbing properties. It, thus, influences regional climate. Of the carbonaceous materials, soot shows the highest absorption for solar radiation of all pollutants.

Determination of elemental carbon in carbonaceous materials is carried out by various methods.

1. *Thermal*: The carbonaceous sample is heated in the presence of oxygen. Elemental carbon (graphite) content is determined by vaporizing the organic carbon first. The carbon ejected at various temperatures is determined in the form of $CH_4$ by flame ionization detector (FID) or as $CO_2$ by electrochemical ($CO_2$-sensitive membrane electrode, Severinghaus electrode), or non-dispersive IR (NDIR) spectrometry.
2. *Chemical*: Organic compounds in the sample are digested in a strong oxidant ($HNO_3$). The remaining residue is determined for its elemental carbon (graphite) content.

   Elemental carbon = (Total carbon-carbon after digestion)
3. *Coulometric method*: The sample is heated to 1,200° C under pure $O_2$ stream, and then acidification of the sample in solution (maintained at pH 10) by $CO_2$ activates a back-electrolysis to regain the initial pH. The amount of coulombic current needed for this operation is directly related to the amount of $CO_2$ that has entered the solution. This method can be automated.
4. *Optical methods*: Optical methods are based on light attenuation. Difference in transmission of light through a clean filter (blank) and through a filter containing particulate matter is correlated to the particulate content (Fig. 18.4).

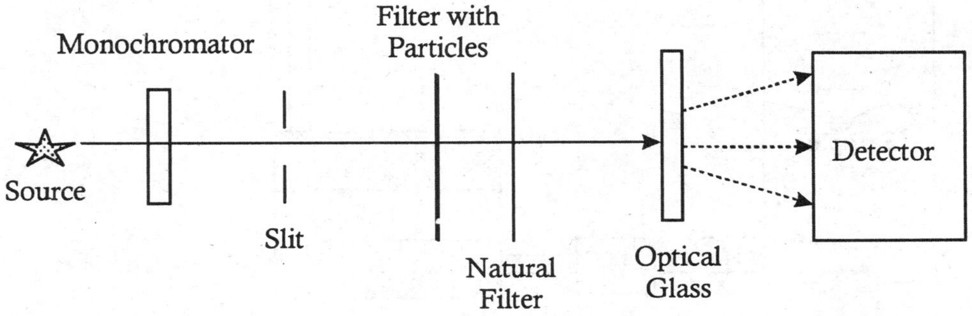


**Fig. 18.4:** Determination of Particulate Carbon Content (Absorption/Reference) by Optical Method

### 18.3.2 Inorganic Carbon (IC)

Inorganic carbon (IC) content can be determined by converting all inorganic carbon to $CO_2$ (organic carbon is not oxidized) by injecting the sample into a chamber packed with phosphoric acid-coated quartz beads. $CO_2$ can be measured by volumetric, conductimetric or coulometric methods.

### 18.3.3 Total Organic Carbon (TOC)

Organic carbon can be biological, chemical or anthropogenic in origin. Total organic carbon

(TOC) is a more convenient way of expressing total organic content in a sample. It is recognized as the best means of assessing the organic content of a water sample. The amount of $CO_2$ generated, after complete oxidation of the organic carbon is taken as a measure of TOC. For the estimation of TOC, the organic molecules are first broken down to single carbon units and molecular forms by thermal incineration at high temperatures or by UV radiation, or oxidation. The $CO_2$ produced can be measured by volumetric, conductimetric, titration or coulometric, non-despersive IR (NDIR) methods and by flame ionization by converting it to methane, $CH_4$.

Automatic analyzers are available to determine TOC. In a TOC analyzer (Fig. 18.5), a sample is packed into a catalyst-packed combustion tube maintained at 1,000° C. Organic carbon is oxidized to $CO_2$ by adding an oxidizing agent (potassium peroxydisulfate, $K_2S_2O_8$), enhanced by UV radiation. Phosphoric acid is added to the sample, which is "sparged" with air or nitrogen to drive off $CO_2$ formed from $HCO_3^-$ and $CO_3^{2-}$ in solution. After sparging, the sample is pumped to a chamber containing a lamp emitting UV radiation of 185 nm. $HO^*$ radical produced reacts with the organic compounds.

$$\text{Organics} + HO^* \xrightarrow{K_2S_2O_8} CO_2\uparrow + H_2O \quad ...(18.4)$$

After the oxidation is complete, the $CO_2$ is sparged from the system and may be determined directly by a non-dispersive IR (NDIR) analyzer or can be reduced to methane ($CH_4$) and measured by a flame ionization detector (FID) (see Chapter 24 for details).

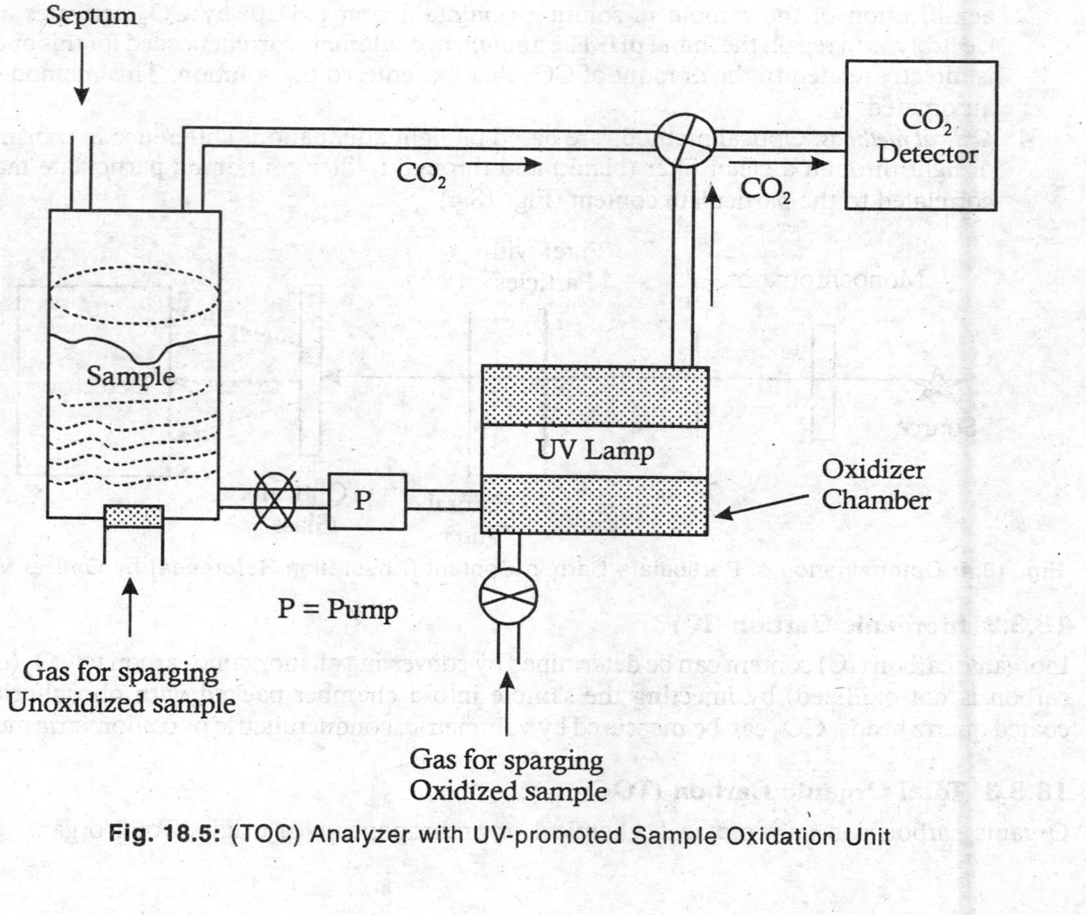


**Fig. 18.5:** (TOC) Analyzer with UV-promoted Sample Oxidation Unit

A dual-channel system can be incorporated to determine the contribution of inorganic carbon. A second sample of equivalent size to the first is injected into the carrier gas stream to a second combustion tube packed with quartz chips wetted with 85 per cent phosphoric acid and heated at low temperature ~150° C. At this low temperature, organic carbon is not oxidized, but $CO_2$ released from inorganic carbon is vaporized. In the pyrolysis process, TOC is determined by subtracting the inorganic carbon (IC) content from the total carbon content. Automatic, on-line TOC monitoring instruments are available. $CO_2$ can also be measured by $CO_2$-sensitive electrodes (*Severinghaus cells*).

TOC can also be determined by the respiratory (spirometric) method. In the respiratory method, the $CO_2$ evolved, due to oxidation of carbon by $O_2$, is absorbed by KOH solution and the resulting pressure drop is monitored continuously by a manometer and correlated to the loss of $O_2$.

### 18.3.4 Carbon Monoxide (CO)

Automobile exhaust is the major anthropogenic source of the pollutant, carbon monoxide. It is also produced in wild fires and fossil fuel burning. The carbon monoxide concentration in the troposphere varies from 10 ppbv to 100 ppbv. CO is determined by electrochemical and non-dispersive IR (NDIR) spectrometric methods. NDIR is a versatile continuous monitoring technique. GC analyzers for CO, based on its prior reduction to $CH_4$, are the most sensitive but they cannot be used for continuous monitoring.

1. **NDIR Spectroscopy:** It is based on the absorption of infrared radiation by carbon monoxide at frequency 2150 $cm^{-1}$ (CO stretching). A double-beam instrument is used and calibrated to null reading with CO-free air. Any CO present in one of the arms disturbs the null reading and thus monitored (Chapter 24 for details). Pre-drying the sample eliminates interference from water. $CO_2$ absorbs at 2350 and 650 $cm^{-1}$ (symmetric stretching and bending vibration, respectively). Therefore, a simultaneous measurement of CO and $CO_2$ is possible with this technique. Other spectroscopic methods are gas-filter correlation spectrometry, and tunable diode laser spectrometry.
2. **GC-FID:** Carbon monoxide can also be measured by flame ionization detector (FID) coupled to a gas chromatograph. From the sample, gases other than CO and methane are selectively absorbed on a pretreatment column. Only CO and methane proceed to the chromatographic column. The eluted gases (CO and $CH_4$) pass, sequentially, through a catalytic (nickel catalyst) reduction column. While $CH_4$ passes through, CO is reduced by hydrogen to methane.

   $$CO + 3H_2 \rightarrow CH_4 + H_2O \qquad \text{...(18.5)}$$

   Two $CH_4$ peaks are detected by FID. The first peak is attributed to $CH_4$ and the second to CO. This method can be used to measure CO, $CH_4$ and hydrocarbons. The method, though sensitive (10 ppbv), cannot be used for continuous monitoring.
3. **Coulometric method:** Air is aspirated from the atmosphere through a filter to retain particulate matter. Interferents are removed by chemical scrubbers, and the gas is passed through a column of $I_2O_5$ (at 160° C) to release free iodine, which is reduced at a graphite cathode. The current produced is proportional to CO present.
4. Conversion of CO to $CO_2$ by atomic oxygen, O, ($CO + O \rightarrow CO_2^*$; $CO_2^* \rightarrow CO_2 + h\nu$), and measurement by chemiluminescence (300-500 nm) (sensitivity ~0.001 ppm).
5. **Thermal catalytic oxidation to $CO_2$:** This method is based on monitoring of heat generated during thermal catalytic oxidation reaction to $CO_2$. Difference in temperature between a cell

where oxidation is occurring, and a reference cell, through which a part of the sample is passing without reaction, can be measured by a thermistor. Catalytic oxidation of hydrocarbons can also be measured by this method.

6. **Electrochemical oxidation method:** Particulate and electroactive impurities are removed by filtration and CO is detected by an electrochemical sensor (Fig. 18.6). The sensor is composed of a sensing electrode (at which CO is oxidized), a counter electrode and a reference electrode setup in sulfuric acid. The sensing electrode is a Teflon-bonded diffusion electrode catalyzed with platinum at which CO is oxidized as follows:

$$CO + H_2O \rightarrow CO_2 + 2H^+ + 2e^- \quad (18.6)$$

Solid-state instrument allows direct, on-the spot measurement of CO concentration in air.

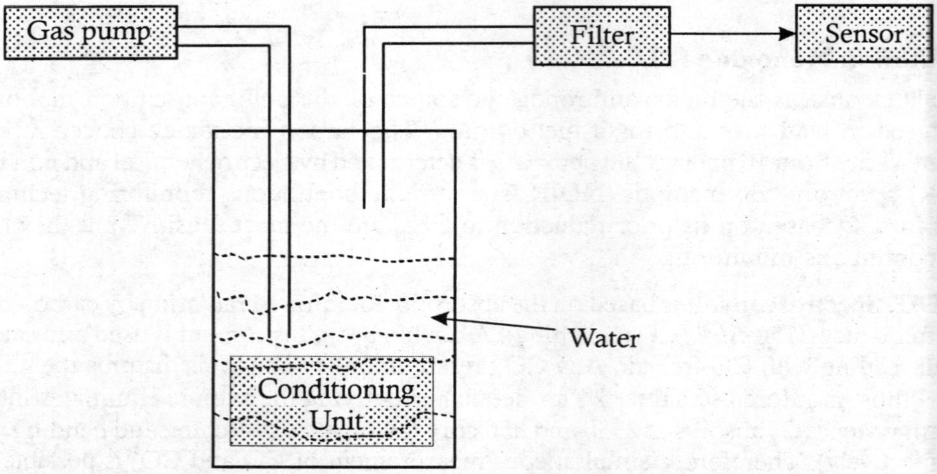


**Fig. 18.6:** Detection of Carbon Monoxide by Electrochemical Oxidation

### 18.3.5 Carbon Dioxide ($CO_2$)

Carbon dioxide ($CO_2$) is released in nature by the oxidation of organic substances. Large amounts of carbon dioxide are produced by the industrial operations also (e.g. calcination of limestone). Non-dispersive IR analysis methods may be employed to determine carbon dioxide.

Wet-chemical methods are based on its acidic character.

1. *Titration method*: After absorbing $CO_2$ in a known excess of alkali (NaOH) to form sodium bicarbonate, the excess alkali is titrated with a standard acid. Completion of the reaction can be determined from the disappearance of pink color of phenolpthalein indicator (pH = 8.3) or potentiometrically.
2. *Zecondroff's method*: $CO_2$ is absorbed in a known excess of $Ba(OH)_2$. $BaCO_3$ is precipitated. Excess of $Ba(OH)_2$ is titrated with HCl.
3. *Electrical conductimetry*: This method can be used to determine atmospheric $CO_2$ as well as $CO_2$ produced in the combustion of hydrocarbons (see Chapter 23).
4. *Non-dispersive IR (NDIR) analysis*: $CO_2$ can be monitored by non-dispersive IR analysis (NDIR) (see Chapter 24 for details).

In view of the potential involvement of $CO_2$ in global temperature rise, monitoring of $CO_2$ is done from several land-based stations (e.g. Alaska, Hawaii, Samoa and the South pole) using continuous infrared analyzers.

### 18.3.6 Methane ($CH_4$)

Methane, $CH_4$, is a colorless and odorless gas. It is a greenhouse gas and is highly explosive and asphyxiating. Methane occurs quite frequently in nature. It is the main constituent of *natural gas* deposits, and *coke gas* (from fuel combustion). It is also released from the bottom of the swamps and stagnant water bodies, where it is formed, called *marsh* gas, or *biogas*, which is a mixture of methane and carbon dioxide produced by bacterial degradation of organic matter and used as fuel.

Methods of determination methane are combustible-gas indicator and volumetric methods.

1. In combustible-gas indicator method, partial pressure of methane is measured. The catalytic combustion of a combustible gas on heated platinum, which is a part of Whetstone bridge. The increase in electrical resistance in the circuit is proportional to the amount of combustible gas.
2. In volumetric method, methane is oxidized to $CO_2$ and $H_2O$ and volume of $CO_2$ is measured.
3. Physical technique involves separation by GC and detection by flame ionization detector (FID).

### 18.3.7 Combustible Gases

Combustible gases can be analyzed by Orsat method, by adsorption of the components in various solvents. A sample of flew gas is passed through a filter to remove particulate matter, and then passed through alkaline pyrogallol (to absorb $SO_2$), NaOH solution (to absorb $CO_2$), and finally through acidic cuprous solution (to absorb CO). Physical methods of determination are GC and detection by thermal conductivity; IR detection for $CO_2$, CO, and $C_2H_4$, and UV detection for $SO_2$ and $NO_2$.

The combustible gas indicators are based on thermal conductivity method. Two platinum wire filaments in a balanced Wheatstone-bridge circuit are heated to relatively low temperature by current from dry cells. One of the filaments in a sample flow stream where combustible gas is present is oxidized or burned catalytically on this filament. The increased temperature resulting from that combustion increases the filament electrical resistance and unbalances the bridge proportionate to the amount of flammable gas or vapor in the sample. The resistance change is measured directly by the meter that is calibrated.

## 18.4 NITROGEN AND NITROGEN COMPOUNDS

Oxides of nitrogen ($NO_x$), ammonia ($NH_3$) and cyanides are some of the important atmospheric and water pollutants. They are among the most important species in water.

Nitrogen, $N_2$, is the major constituent of atmospheric air. Oxides of nitrogen ($NO_x$) form the initiating products in the atmospheric pollution cycle (Fig. 18.7).

Burning of fossil fuels in automobiles, and power plants add to the increase of $NO_x$ pollution.

$$\begin{aligned} NO_2 + h\nu &\xrightarrow{\text{Photochemical}} NO + O \\ O + O_2 &\rightarrow O_3 + M \\ NO + O_3 &\rightarrow NO_2 + O_2 \end{aligned} \qquad \text{...(18.7)}$$

where, $M$ is a third body.

Oxides of nitrogen, especially nitrogen dioxide, $NO_2$, (nitrite) and nitrogen trioxide, $NO_3$, (nitrate) are ubiquitous within environmental, food, industrial and physiological systems. Anthropogenic factors combined with more general mismanagement of our natural resources have created environmental pollution by these compounds. Some examples are: "blue baby" syndrome, due to passage of nitrite into the blood stream resulting in the irreversible conversion of hemoglobin to methemoglobin, thus compromising oxygen uptake and transportation; eutrophication of water bodies; carcinogenic nitrosoamines etc. Nitrogen monoxide (NO) has been shown to play an important role in many metabolic functions (as a neurotransmitter, cytotoxic agent, regulation of vascular tone). As the reaction of nitrogen monoxide (NO) with oxygen leads to the production of nitrite, and it is through the greater stability of the nitrite ion that the action of NO can be detected. The measurement of nitrite can, therefore, provide a reliable measurement of the physiological role NO within the body, and, therefore, can be used as biomarker that enables physicians to gauge the health of humans. Among the ailments for which the monitoring of nitrite and nitrate are useful are sepsis, meningitis, rheumatoid arthritis, and Parkinson's disease. Thus, the presence of nitrite and nitrate within physiological systems is often viewed with justifiable concern. Therefore, accurate and quantitative determination of these compounds is very important in assessing their environmental and health effects and control methods.

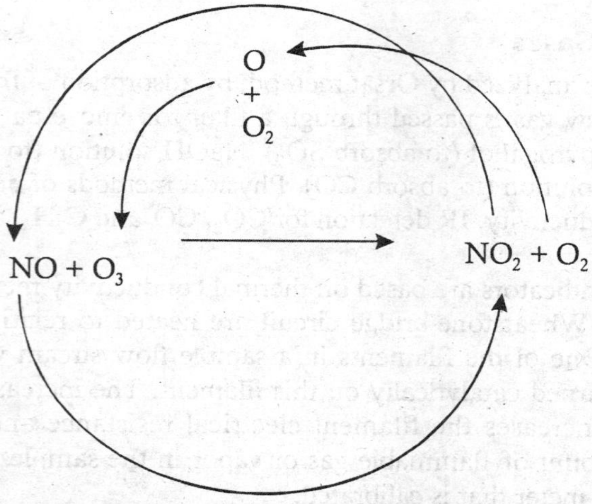


**Fig. 18.7:** Atmospheric Nitrogen Monoxide (NO), Nitrogen Dioxide ($NO_2$) and Ozone ($O_3$) Cycle

There are many quantitative methodologies of determining nitrite and nitrate. Generally, nitrite determination is carried out by chemically reducing the inert nitrate ion to the more reactive nitrite ion before initiating the detection sequence. Spectroscopic methods are by far the most widely used for nitrite/nitrate determination due to the facile assay-type protocols and excellent detection limits. In addition, chemiluminescence, fluorimetric, Raman spectroscopic methods, electrochemical, ion-selective electrode, and chromatographic methods can also be employed for more accurate determination. The increasing demand for rapid on-site analysis makes spectroscopic and electrochemical preferred techniques. Instrumental complexity restricts the operation of other sensitive methods, in spite of their superior performance.

### 18.4.1 Nitrogen ($N_2$)

Nitrogen, $N(Z = 7; A = 14.007)$, is the main component of atmospheric air (78.03 per cent by weight). The major part of atmospheric nitrogen is in the free state.

Often the percentage of nitrogen in gas mixtures is determined by measuring the volume of residual gas after suitable chemical reagents have absorbed all other components. Decomposition of nitrates by sulfuric acid in the presence of mercury liberates NO, which can be measured as a gas. Kjeldahl methods are used in the determination of organic nitrogen and ammonia nitrogen. Alkaline persulfate method may be used to determine total nitrogen.

1. **Kjeldahl Method:** Kjeldahl nitrogen is the sum of amino nitrogen of organic compounds, and inorganic nitrogen (ammonia nitrogen and nitrates). Inorganic nitrogen compounds are converted to ammonium ($NH_4^+$) salt in the presence of concentrated $H_2SO_4$, potassium sulfate and cupric sulfate ($CuSO_4$) as catalyst. C, H, and N are converted to $CO_2$, $H_2O$, and ammonium hydrogen sulfate. After addition of alkali, ammonia is distilled and absorbed in sulfuric acid. The free ammonium ($NH_4^+$) is determined spectrophotometrically by Nessler method or by ammonia-selective electrode. Organic nitrogen can be determined by converting it to $NH_3$ in a stream of $H_2$ gas in the presence of nickel catalyst and the $NH_3$ determined by Indophenol Blue dye method ($\lambda = 630$ nm). Total organic nitrogen is determined by Kjeldhal digestion method, followed by ammonia determination.
2. **Persulfate Method:** Persulfate method gives total nitrogen content of a sample by oxidation of all nitrogenous compounds. At 100° C in alkaline solution persulfate oxidizes organic and inorganic nitrogen to nitrate. Total nitrogen is determined by analyzing the nitrate by cadmium reduction or other methods. "Organic" nitrogen can be estimated from the difference of the individual determinants of total nitrogen and inorganic nitrogen (determination of free ammonium).

### 18.4.2 Dinitrogen Oxide (Nitrous Oxide, $N_2O$)

The primary biogenic sources of nitrous oxide, $N_2O$, are nitrification and denitrification. $N_2O$ is a greenhouse gas. It absorbs IR radiation 200 times as effectively as $CO_2$. It is determined by physical methods, such as long-path IR or GC with thermal conductivity or electron capture detector (ECD). Traces of $N_2O$ can be pre-concentrated on a type 5A molecular sieve and desorbed with helium saturated with water vapor. It is readsorbed on silica gel, cooled by dry ice. Before transferring to the gas chromatograph for detection it is desorbed from silica gel by heating.

### 18.4.3 Nitrogen Monoxide (Nitric Oxide, NO)

Nitrogen forms a number of oxides. All of them can be prepared from nitric acid or its salts. In industry, nitrogen monoxide is an intermediate in the production of nitric acid. Nitrogen monoxide has many physiological functions (neurotransmitter and a cytotoxic agent). NO gets reduced in the presence of strong reducing agents, while strong oxidizing agents can oxidize nitrogen monoxide ($2NO + O_2 \rightarrow 2NO_2$). The concentration of NO is determined through the determination of the more stable $NO_2$.

Wet-chemical and luminescence methods are employed to determine the concentration of nitrogen monoxide. In many cases NO is first oxidized to $NO_2$ (by passing the gas over $CrO_3$ coated on firebrick powder or molecular sieve) and the latter is determined. In that case the result reflects total $NO_x$, that is NO + $NO_2$.

1. **Wet-chemical Methods:** Griess-Saltzmann method (See sec. 18.4.4) can be applied after oxidation of NO to $NO_2$.
2. **Coulometric Method:** Filtered air is aspirated through a chemical scrubber to suppress interfering gases, and NO is oxidized to $NO_2$. $NO_2$ reacts with KI to liberate iodine. The current produced upon reduction of iodine is proportional to NO concentration.

3. **Polarographic Method:** NO is absorbed in NaOH and $O_2$ is freed by bubbling $N_2$. The solution is acidified with HCl; the NO is reduced to $NH_3$ and measured by polarography.
4. **Luminol Method:** For determination of NO by this method, prior oxidation of NO to $NO_2$ is required. $NO_2$ can be detected by the chemiluminescence (at $\lambda = 425$ nm) it produces with luminol in an alkaline solution. It is a highly sensitive method (sensitivity < 30 ppt).
5. **Chemiluminescence Method:** A very sensitive and recognized method for determination of NO is by chemiluminescence reaction between NO and $O_3$ (Fig. 18.8)

$$NO + O_3 \rightarrow NO_2^* + O_2 \qquad \ldots(18.8)$$

$$NO_2^* \rightarrow NO_2 + h\nu \qquad (600 \le \lambda \le 3{,}000 \text{ nm}; \lambda_{max} = 1{,}200 \text{ nm}).$$

Luminescence, due to decay of $NO_2^*$ to ground state is proportional to the concentration of NO when ozone is in excess. This reaction can be used to determine NO directly. Luminescence produced is measured by a photomultiplier tube (PMT). This method is preferred because it avoids wet-chemical treatment and is basically simple and interference-free. Besides, continuous monitoring is possible.
6. $NO_x$ are determined by modern physical techniques such as long-path absorption and differential absorption LIDAR (DIAL), tunable diode laser absorption spectrometry (TDLAS), differential optical absorption spectrometry (DOAS), optoacoustic spectrometry and photo-induced photo-fragmentation/photoionization spectrometry. However, these techniques have low sensitivity and selectivity for determining low concentrations of these compounds.

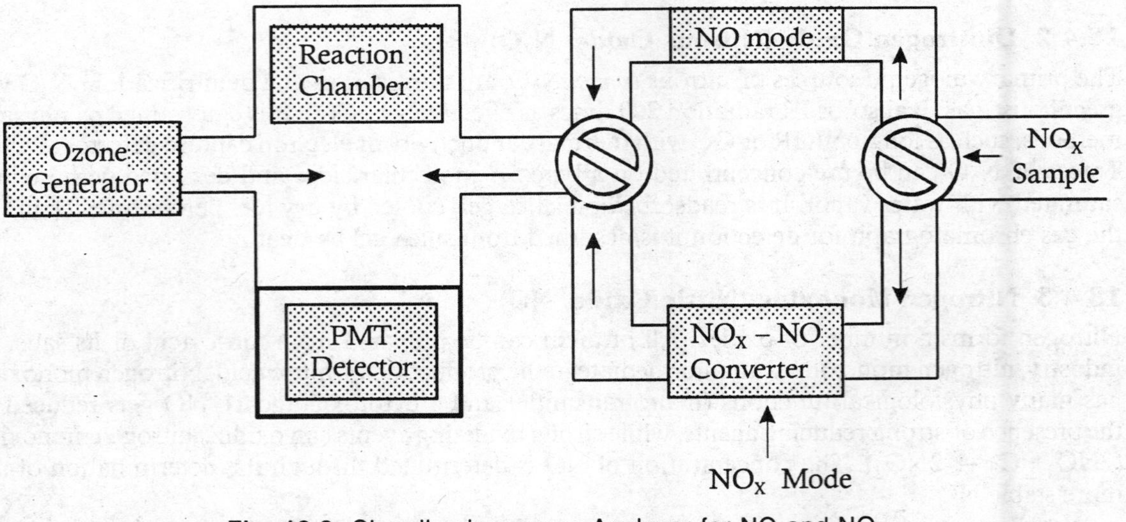


**Fig. 18.8:** Chemiluminescence Analyzer for NO and $NO_x$

### 18.4.4 Nitrogen Dioxide ($NO_2$) and Nitrite ($NO_2^-$)

Nitrogen dioxide, $NO_2$, is reddish-brown poisonous gas with a characteristic odor. It is a very vigorous oxidizing agent. Nitrite ion, ($NO_2^-$), affects hemoglobin. It oxidizes HbFe(II) to methemoglobin, HbFe(III), which is ineffective in carrying oxygen to tissues. This reaction results in oxygen deficiency and ultimately to death. The condition is called *methemoglobinemia.* Nitrite above 0.1 mg/L concentrations is toxic to many organisms. Nitrite may form carcinogenic nitrosoamines within the acidic conditions of the stomach, leading to gastric cancer.

Nitrite ($NO_2^-$) in water is formed either by oxidizing ammonia (by aerobic nitrifying bacteria, e.g. nitrosomonas bacteria) or by reducing nitrates (pseudomonas bacteria).

$$\text{Ammonia} \rightarrow \text{Nitrite } (NO_2^-) \rightarrow \text{Nitrate } (NO_3^-) \qquad ...(18.9)$$

Nitrite ($NO_2^-$) is the most important product in biological oxidation of nitrogenous organic matter of both autochthonous and allochthonous origin. Presence of nitrite in water is indicative of organic pollution. Domestic and agriculture wastes have been the main sources of allochthonous nitrogenous matter. Metabolic wastes of aquatic life and dead organisms add to the autochthonous nitrogenous organic matter. When nitrite is present in excess, it causes excessive growth of algae.

Fluorescence and wet-chemical methods are employed in the determination of $NO_2$ and nitrite ($NO_2^-$).

1. *Fluorescence method*: In this method, both NO and $NO_2$ can be measured by employing dual pathway detector system. In this system, one detector is to measure NO and the other to measure $NO_2$. $NO_2$ pathway has a converter to quantitatively reduce $NO_2$ to NO. The $NO_2$ concentration in the original air stream is the difference of ($NO_2$-NO).
2. *Chemiluminescence method*: The chemiluninescence method given for NO in Section 18.4.3 can be made applicable for determination of $NO_2$, if it is reduced to NO first. The reading after reduction gives NO + $NO_2$ while the reading before reduction is equivalent to NO only.
3. *Luminol method*: Luminol in alkaline solution reacts with $NO_2$ to produce intense chemiluminescence with $\lambda_{max}$ = 425 nm (highly sensitive method, <30 ppt). This method can be adopted for determination of NO, if it is oxidized to $NO_2$ first.
4. *Griess-Saltzmann method*: This method can be applied to determination of $NO_2$ when it is collected as nitrite ($NO_2^-$) (sodium sulfite is well suited for collecting $NO_2$). The method is based on the reaction of $NO_2^-$ with sulfanilamide to form a diazonium salt (at pH = 2-2.5), which on coupling with N-(1-naphthyl)ethylenediamine gives a colored azo dye. The absorbency of the reddish-purple dye is measured spectrophotometrically at $\lambda$ = 545 nm (detection limit = 0.02 → 2 μM).

$$NO_2^- + \text{sulfanilamide} \xrightarrow{CH_3COOH} \text{diazo compound} \qquad ...(18.10)$$

   Diazo compound + N-(1-naphthyl)-ethylenediamine (NEDA) → azo dye

5. *Jacobs-Hochmeister method*: The sample is made alkaline to form nitrite ion ($NO_2^-$). Sodium nitrite is treated with $H_2O_2$ to remove $SO_2$ and then Griess-Saltzmann method is followed.
6. *Phenolic assay*: The chemical reactivity of C-nitroso compounds can be determined by spectroscopic or electrochemical methods. Single analyte species can be determined without significant sample manipulation.
7. Nitrate content in water can be determined by measuring its absorbance at $\lambda$ = 220 nm. Dissolved organic matter also absorbs at this wavelength. However, nitrate does not absorb at $\lambda$ = 275 nm. So, two measurements are made at $\lambda$ = 220 and 275 nm and the difference is taken for the determination of nitrate.
8. *Flow Injection Analysis* (*FIA*): Automatic direct spectrophotometric method for the simultaneous determination of nitrite and nitrate in water samples by flow injection analysis (FIA) is based on the reaction of nitrite and nitrate with the reagent, N-phenylanthranilic acid in sulfuric acid medium (pH ~ 0.5), and the absorbance measured at 410 nm. This method has the advantage of direct determination of nitrite and nitrate and without reducing nitrate to nitrite, as is the case in other methods.

9. *Raman spectroscopy*: Nitrate ion can be determined by monitoring the intensity of Raman bands at 1384 and 2430 $cm^{-1}$.
10. *Electrochemical methods*: Voltammetric and potentiometric methods are in use for determination of nitrite. Ion-selective electrode method is a standard method of test in drinking water. The method can be automated. More selective approach is by employing biological catalysis (e.g. reductase enzymes).
11. *Chromatographic methods*: HPLC and capillary electrophoresis (CE) separation methods, and end column detection by UV, fluorimetric, electron capture, and electrochemical methods are employed for determination of nitrite (and nitrate).

### 18.4.5 Nitrous Acid ($HNO_2$) and Nitric Acid ($HNO_3$)

Nitrous acid ($HNO_2$) and nitric acid ($HNO_2$) are secondary pollutants of high environmental significance. $HNO_2$ is produced by oxidation of NO by OH radical; similarly $HNO_2$ by oxidation of $NO_2$. They are also formed by dissolution of $NO_2$ in water.

$$2NO_2 + H_2O \rightarrow HNO_2 + HNO_2 \quad ...(18.11)$$

The salts of nitrous acid are called nitrites. Alkali nitrites are used in dye industry. Nitrous acid can be determined by several methods. However, there are some inherent difficulties in their determination, caused by the overlapping behavior between the acids and their salts in most of the analytical procedures. Though the salts are mostly present in the aerosols simple filtration procedures do not eliminate their influence. Some of the analytical methods are:

1. Nitrous acid can be trapped as nitrite ions by bubbling through an alkali solution. Nitrite ions are allowed to react with Gries-Saltzmann reagent, resulting in a red-violet azo dye, which is spectrophotometrically measured.
2. *Continuous method*: Differential optical absorption spectroscopy (DOAS) is the widely used method for measurement of nitrous acid (at 354 and 368 nm).
3. Nitrous acid can be selectively separated from $NO_2$ by $Na_2CO_3$ denuder (a glass tube coated internally with a solid sorbent and equipped with a black glass filter column). Following the passage through a debubbler, the scrubbing solution is mixed with ascorbic acid to reduce $NO_2^-$ to NO. A chemiluminescence monitor (Fig. 18.9) determines the NO evolved. It is a continuous monitoring method.
4. *Electrical Condctimetry*: (see Chapter 23).

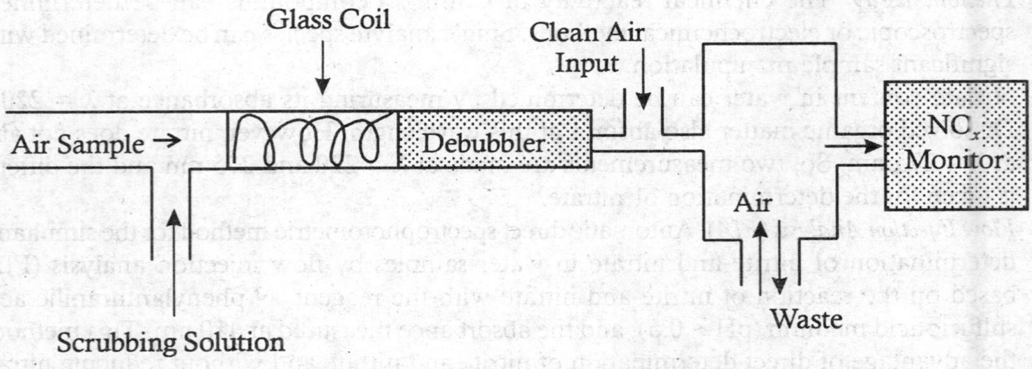


**Fig. 18.9:** Setup for Continuous Measurement of Ambient $HNO_2$

For determination of nitric acid ($HNO_3$), it is first collected in a hallow tube, the inner walls of which have been previously coated with tungstic acid. Following thermal desorption, $HNO_3$ can be determined by ion chromatography (IC) or by a $NO_x$ chemiluminescence detector.

### 18.4.6 Nitrate ($NO_3^-$)

The salts of nitric acid are called nitrates. Inputs of nitrate (and nitrite) to the environment can occur through industrial and domestic combustion processes and gaseous $NO_x$ species converted to $NO_3^-$ through photochemical conversion within the atmosphere. The vast majority, however, arise from agricultural sources. These salts of nitric acid ($NaNO_3$) are used as fertilizers. Of the greatest importance are the nitrates of sodium, potassium, calcium and ammonium. Higher concentrations of nitrate (>10 mg/L) often indicate the effects of nitrogen-containing fertilizers.

Nitrate is estimated by wet-chemical and physical techniques. Generally, the inert nitrate ion is converted to more reactive nitrite, by chemical or photo-induced reduction, before initiating the detection sequence.

1. *Cadmium reduction method*: Nitrate ($NO_3^-$) is reduced to nitrite ($NO_2^-$) by cadmium and the amount of nitrite, $NO_2^-$, formed is determined by Gries-Saltzmann method (at $\lambda$= 545 nm).
2. $NO_3^-$ can be reduced to $NO_3^-$ with hydrazine sulfate and the $NO_3^-$ is determined spectrophotometrically ($\lambda$ = 524 nm).
3. *Fluorescence method*: Reduction of $NO_3^-$ to $NO_2^-$ with hydrazine sulfate and the subsequent determination of the nitrite formed with 2, 3-diaminonaphthalene ($\lambda_{ex}/\lambda_{em}$ = 364/412 nm) (sensitivity 0.01 ppm).
4. $NO_3^-$ can be determined by reducing it to $NH_3$ by titanous chloride and then measuring $NH_3$ potentiometrically by $NH_3$ gas-sensing electrode.
5. *Photo reduction*: Nitrate → nitrite conversion by photo-induced reduction by UV radiation (200 → 300 nm), resulting in the formation of nitrite and oxygen, and nitrite determination.

$$2NO_3^- \xrightarrow{200-300\text{ nm}} 2NO_2^- + O_2 \quad ...(18.12)$$

   This method obviates the use of a chemical reagent for reduction and provides a clean and efficient alternative.
6. *Chromotropic acid method*: Nitrates and nitric acid ($HNO_3$) can be measured by chromotropic acid method. Reaction of nitrate with chromotropic acid gives rise to a yellow colored solution. The absorbance (at 410 nm) is proportional to nitrate concentration. Nitric acid is determined after addition of excess of sodium sulfite (to remove interference due to oxidizing agents), chloride is masked by the addition of antimony (III), (to mask chloride), and addition of urea (to eliminate nitrates as $N_2$ gas).
7. *UV absorbance method*: Absorbance UV radiation ($\lambda$ = 220 nm) by nitrate, measured spectrophotometrically.
8. Nitrate content in water can be determined by measuring its absorbance at $\lambda$ = 220 nm. Dissolved organic matter also absorbs at this wavelength. However, nitrate does not absorb at $\lambda$ = 275 nm. So, two measurements are made at $\lambda$ = 220 and 275 nm and the difference is taken for the determination of nitrate.
9. $NO_3^-$ *ion-selective electrode*: It is a useful method for environmental analysis of nitrates in soil and water; it is a standard method for drinking water analysis.
10. By Ion chromatography (IC).

### 18.4.7 Ammonia ($NH_3$)

Ammonia originates from deamination of nitrogen-containing organic compounds and hydrolysis of urea.

$$CO(NH_2)_2 + H_2O \rightarrow CO_2 + 2NH_3 \quad ...(18.13)$$

Ammonia is quite active chemically. It is a reducing agent. Ammonia and ammonium salts have widespread applications. Ammonium sulfate and nitrate are used as fertilizers and ammonium chloride is used in the dyeing industry, in textile printing and in galvanic cells. Ammonia is a common pollutant of water since it is a primary product of biological decomposition. Its undesirable effects arise from the fact that a number of organisms can utilize ammonia as a source of nitrogen. In addition, $NH_3$ serves as a nutrient for algae and other troublesome aquatic plants, leading to eutrophication of water bodies. In aqueous solutions there is a pH-dependent equilibrium between free $NH_3$ (which is toxic to fish and other aquatic life) and ammonium ions ($NH_4^+$). Concentration of $NH_3$ in water can be used as an indicator of sanitary pollution.

For collection of ammonia from air diffusion denuders, with their inner walls coated with acids, are found useful. Ammonia ($NH_3$) may be determined by spectrophotometric, titrimetric and ion-selective electrode (ISE) methods.

1. *Nessler's reagent method*: Nessler's reagent ($HgI_4^{2-}$), in an alkaline medium, reacts with $NH_3$ and gives rise to an orange-colored compound (ammoniuum dimercury iodide) that is measured at 425 nm.

$$2HgI_4^{2-} + 2NH_3 \rightarrow NH_2Hg_2I_3 + NH_4^+ + 5I^- \quad ...(18.14)$$

2. *Indophenol blue method*: Hypochlorite reacts with ammonium ($NH_4^+$) to form chloramine. This is followed by the reaction of chloramine with phenol, which yields the blue dye, indophenol-blue. In this indirect spectrophotometric method Interference from Ca and Mg is eliminated by the addition of sodium citrate. The color is measured at 630 nm.
3. *Fluorimetric method*: Reaction of $NH_3$ with o-Phthaldialdehyde, a reductant, produces isoindole that is determined fluorimetrically.
4. *Hantsch method*: $NH_4^+$ is treated with formaldehyde and a diketone to give rise to lutidine derivative that is measured by fluorescence method ($\lambda_{ex}/\lambda_{em}$ = 405/510 nm) (sensitivity ~0.2 ppm).

$$\text{Formaldehyde + Diketone + } NH_4^+ \rightarrow \text{Lutidine derivative} \quad ...(18.15)$$

5. *Dimethoxyoxolane method*: The sample is treated with excess of 2, 5-dimethoxyoxolane (at 80° C) in 1, 2-dichloroethane. The resultant pyrrole is subsequently mixed with an excess of (E)-p-dimethylaminoperchloric acid, which forms an intensely blue complex ($\lambda_{max}$ = 630 nm). This method is inexpensive, accurate and reliable. The method is suitable for automation. This method has many advantages over other methods.
6. *Chemiluminescence method*: $NH_3$ is oxidized to NO, which is then determined by chemiluminescence reaction (method of choice). One determination is carried out on the acidic sample, and the other after absorption of $NH_3$ in acidic medium. The difference in measurements corresponds to $NH_3$.
7. Titrimetrically with pH meter.
8. $NH_3$-*selective electrode method*: The $NH_3$-selective electrode setup consists of a gas ($NH_3$) permeable membrane to separate sample solution and ammonium chloride solution (internal solution). Ammonia is converted to aqueous $NH_3$ by raising pH > 11. $NH_3$ diffusing through the membrane changes the pH of the internal solution, which is sensed by a pH electrode. A

chloride ion-sensitive electrode that serves as the reference electrode senses the chloride in the internal solution (sensitivity ~0.03-1.5 ppm).

9. GC coupled to flame thermoionic detector.
10. $NH_3$ is collected on a sorbent (diffusion denuder) impregnated with an acid (oxalic, or tungstic acid) and determined in liquid phase.
11. Tunable diode laser absorption spectroscopy (TDLAS), photoacoustic infrared multi-gas analyzer and Fourier transform IR (FTIR) spectroscopy are more expensive.

### 18.4.8 Photochemical Smog

For photochemical smog formation, sunlight (UV radiation) is the "initiator"("driver"), $NO_x$ the "engine" and volatile hydrocarbons the "fuel".

$$O_3 + h\nu \rightarrow O^* + H_2;\ O^* + H_2O \rightarrow OH + OH$$

$$H_2O_2 + h\nu \rightarrow OH + OH;\ H + O_2 \rightarrow HO_2 \qquad \text{...(18.16)}$$

$$HO_2 + NO \rightarrow OH + NO_2;\ HNO_2 + h\nu \rightarrow OH + NO^*;$$

$$CH_2O + h\nu \rightarrow H + HCO;\ HCO + HO_2 \rightarrow 5CO + H_2O_2$$

The standard method of analysis of photochemical smog is by estimation of oxidant level. Polluted air is bubbled through a solution of potassium iodide and the liberated iodine is measured spectrophotometrically at $\lambda = 352$ nm.

$$O_3 + 3I^- + 2H^+ \rightarrow I_3^- + O_2 + H_2O \qquad \text{...(18.17)}$$

This method gives net oxidant level (total oxidant – total reductant). Peroxyacetyl nitrate (PAN) is determined by IR detector at $\lambda = 5.57$ and $8.62\ \mu m$.

### 18.4.9 Cyanide ($CN^-$)

Cyanides are useful industrial chemicals. Cyanides are used as starting materials in several chemical syntheses (acrylics), metal refining and electroplating. HCN is used as a fumigating agent to kill rodents and other pests in grain storage silos. Free cyanides ($CN^-$; HCN) are toxic pollutants. Cyanide is removed from purified water by alkaline chlorination ($NaCN + Cl_2 \rightarrow CNCl + NaCl$).

Cyanides can be analyzed by spectrophotometric, fluorimetric, titrimetric and cyanide ion-selective electrode methods.

1. *Colorimetric methods*: Hydrogen cyanide (HCN) is liberated from an acidified sample by distillation and purging with air. It is collected by passing it through NaOH solution. Alkaline $CN^-$ is converted to cyanogen chloride (CNCl) by reaction with chloramine-T at pH < 8.0. CNCl forms a red-blue complex on addition of pyridine-barbituric acid. The absorbance is measured spectrophotometrically at $\lambda = 575$, 578 and 582 nm. HCN forms a colored chelate complex with Pd(II)-8-hydroxyquinoline-5-sulfonic acid, which is determined colorimetrically.
2. *Fluorescence method*: Reaction of $CN^-$ with quinones yields a highly fluorescent product that is highly specific for $CN^-$ ($\lambda_{ex}/\lambda_{em}$ = 440/500 nm) (sensitivity ~0.01 ppm).
3. *Titrimetric method*: Alkaline $CN^-$ is titrated with silver nitrate, $AgNO_3$, to form $Ag(CN)_2^-$. A silver-sensitive indicator, such as chromate (*Mohr* method), detects excess $Ag^+$ (at 578 nm), in the presence of an indicator (rhodamine).
4. *Cyanide-selective electrode method*: $CN^-$ in an alkaline medium can be determined potentiometrically by using $CN^-$-selective electrode in combination with a double-junction reference electrode. It is a very sensitive method.

## 18.5 SULFUR AND SULFUR COMPOUNDS

Sulfur, $S$ ($Z = 16$; $A = 32.06$), is a typical non-metal, and occurs in nature in the free state and in various forms, inorganic and organic compounds. It occurs in coal and minerals as sulfides, sulfites, and sulfates and as hydrogen sulfide from organic matter. Compounds of sulfur with various metals are very abundant, and many of them are valuable ores—*iron pyrites* ($FeS_2$), *galenite* (PbS), *zinc blende* (ZnS) and *copper glance* ($Cu_2S$). Roasting of these ores produces atmospheric sulfur dioxide pollution.

The sulfur content of organic compounds can be oxidized to produce $SO_2$ and determined (West-Gaeke, iodometric or fluorescence methods); or can be reduced to $H_2S$ by pyrolysis (at ~1,000° C) in $H_2$ using platinum as catalyst. $H_2S$ is absorbed by zinc acetate and determined (by iodometric, and fluorescence quenching methods).

### 18.5.1 Sulfide ($S^{2-}$)

Sulfide, $S^{2-}$, is produced in hot springs. It is released by decomposition of organic matter, in industrial waste and by bacterial reduction of sulfate.

Sulfide, $S^{2-}$, may be determined by several methods.

1. *Methylene blue colorimetric method*: This most common colorimetric method involves the reaction of aqueous sulfide with dimethylphenyl-1, 4-diamine in the presence of ferric ions. It gives rise to a characteristic blue colored compound (measured at 660 nm).
2. Physical techniques are the ion-selective electrode and fluorescence methods. The potential of Ag/sulfide ion-selective electrode (ISE) is related to the sulfide ion activity in solution. Reaction of $S^{2-}$ with fluorescein mercuriacetate gives a highly fluorescent compcund. The method is very sensitive (0.05 ppb).

### 18.5.2 Hydrogen Sulfide ($H_2S$)

At high temperature, sulfur combines with hydrogen to form hydrogen sulfide ($H_2S$) gas. $H_2S$, is a colorless, highly toxic gas with the characteristic odor of rotten eggs. It is generally produced from the decomposition of organic waste, hot springs and industrial discharge. Hydrogen sulfide is a powerful reducing agent. The presence of $H_2S$ is an indication of organic pollution and low redox potential.

$H_2S$ can be determined by iodometric, colorimetric, coulometric, ion-selective methods.

1. *Lead acetate and silver foil method*: Exposure of lead acetate film to $H_2S$ blackens it due to formation of PbS. Similarly exposure of a silver foil to $H_2S$ turns it black due to the formation of $Ag_2S$.
2. *Iodometric method*: Oxidation of $H_2S \rightarrow S \rightarrow SO_4$ under oxidizing (acidic) condition.
3. *Colorimetric method*: Methylene blue method is based on the reaction of sulfide with ferric chloride and dimethyl-p-phenylenediamine to produce methylene blue, detected colorimetrically at $\lambda = 660$ nm.
4. *Fluorescence quenching*: $H_2S$ quenches the fluorescence of tetraacetoxymercuri-fluorescein. It is a sensitive method and can be automated.

   $H_2S$ + Ferric chloride + Dimethyl-p-phenylenediamine $\rightarrow$ Methylene blue ($\lambda = 660$ nm).
5. Physicochemical methods are micro-coulometric and ion-selective electrode (ISE) techniques.

### 18.5.3 Sulfur Dioxide ($SO_2$)

Combustion of sulfur containing fuels and "roasting" of metal sulfides produce sulfur dioxide ($SO_2$).

$$4FeS_2 + 11O_2 \rightarrow 2Fe_2O_3 + 8SO_2 \quad ...(18.18)$$

$SO_2$ is a colorless, pungent and irritating gas. It is readily soluble in water. Among the sulfur containing pollutants $SO_2$ is the most important. It is one of the criteria pollutants, having been recognized as a major urban pollutant.

There are several qualitative and wet-chemical and physical methods available for the determination of $SO_2$. Wet-chemical methods are mostly based on its reducing properties. Colorimetric, conductimetric, coulometric, flame photometric, dispersive and non-dispersive absorptiometric methods are available for its determination.

1. *Lead peroxide candle method*: Lead peroxide reacts with $SO_2$ forming dark colored lead sulfide (PbS). Other method is conversion of $PbSO_4$ to $BaSO_4$ that can be evaluated gravimetrically or turbidimetrically. This method is simple, qualitative, but time-consuming.

   $$SO_2 + PbO_2 \rightarrow PbSO_4 \quad ...(18.19)$$

2. *The West-Gaeke (Pararosaniline) method*: It is a colorimetric method (Fig. 18.10), based on the absorption and stabilization of $SO_2$ from air by a solution of Na (or K) tetrachloromercurate(II), to form stable dichlorosulfitomercurate(II).

   $$SO_2 + HgCl_4^{2-} + H_2O \rightarrow HgCl_2SO_3^{2-} + 2H^+ + 2Cl^- \quad ...(18.20)$$

   $HgCl_2SO_3^{2-}$ reacts with acid-bleached pararosaniline + formaldehyde to form a purple-red dye, p-rosaniline methylsulfonic acid. The intensity of the color of the dye is measured colorimetrically at $\lambda = 560$ nm. This method is continuous and can be automated (sensitivity 0.005-5 ppm of $SO_2$).

   Average concentration of $SO_2$ in the ambient atmosphere can be monitored from six hours to one week by incorporating permeation and West-Gaeke method. Sampling is accomplished with a penetration device consisting of a glass tube with a silicone membrane across one end. Na-tetrachloromercurate(II) is placed inside the tube in contact with the membrane. $SO_2$, permeating through silicone membrane is stabilized as the mercurate complex, $(HgCl_2SO_2)^-$. The absorbed $SO_2$ is analyzed by West-Gaeke method.

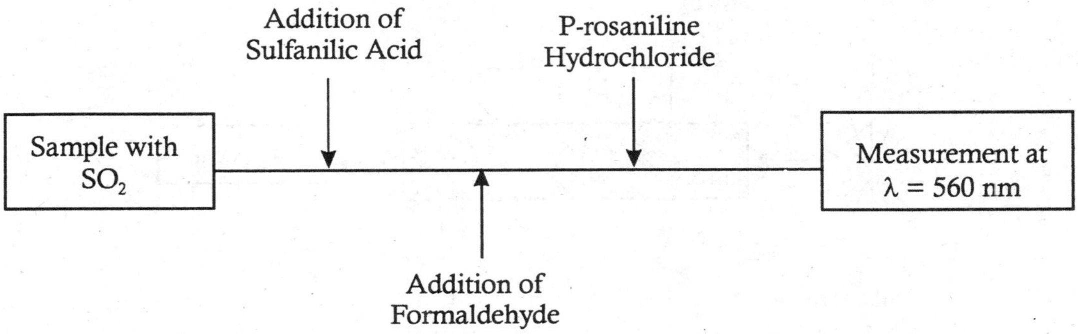


**Fig. 18.10:** $SO_2$ Monitoring based on West-Gaeke Method

3. *Titrimetric method*: Conversion of $SO_2$ to $H_2SO_4$ by $H_2O_2$ can be measured by titration of the acid produced with a standard alkali

   $$SO_2 + H_2O_2 \rightarrow H_2SO_4 \quad ...(18.21)$$

4. *Iodometric method*: The gas sample is bubbled through a solution of $I_2$ + KI. The excess iodine is titrated against sodium thiosulfate.

Chemical methods of $SO_2$ determination are summarized in Table 18.1.

**TABLE 18.1:** CHEMICAL METHODS OF $SO_2$ DETERMINATION

| *Method* | *Procedure* | *Analysis/Remarks* |
|---|---|---|
| Lead peroxide candle | $PbO_2$ converts $SO_2$ to dark colored PbS | Visual; Simple, qualitative, but time-consuming |
| Precipitation | $SO_2$ is absorbed in acidic $H_2O_2$ and sulfate is precipitated as $BaSO_4$ | Gravimetry; Turbidimetry |
| Iodination | $SO_2$ absorbed by NaOH resulting in $Na_2SO_4$ | Iodometry |
| Oxidation | $SO_2$ oxidized by $H_2O_2$ to $H_2SO_4$ | Titrimetriy; Conductimetry |
| Wet-chemical | West-Geake method | Colorimetry |

Physical and physicochemical methods are fluorescence (by UV radiation), polarography, flame photometry, conductimetry, coulometry, and amperometry.

1. *Fluorescence method*: Fluorescence is the basis of $SO_2$ monitors. The air sample containing $SO_2$ is passed continuously through an analyzer, with an UV lamp. $SO_2$ is excited by UV radiation (using Zn (214 nm) or Cd (229 nm) lamp) and the resulting fluorescence (~220-400 nm region) is monitored (sensitivity 0.5-5 ppb).
2. *Flame photometry method*: The flame photometric detector (FPD) relies on the production of sulfur atoms in an $H_2/O_2$ flame. $SO_2$ is passed at controlled flow rate through a fuel-rich $H_2$-air flame. The sulfur chemiluminescence detector (SCD) uses a $H_2/O_2$ flame to convert sulfur species to $SO_2$. $SO_2$ reacts with ozone to produce electronically excited $S_2^*$ molecules ($S + S \rightarrow S_2^*$). $S_2^*$ exhibits ($S_2^* \rightarrow S_2 + h\nu$) chemiluminescence emission bands at 384 and 394 nm, measured by spectrophotometric methods—non-dispersive IR, UV, fluorescence and second-derivative spectrophotometry. In the UV fluorescence method (Fig. 18.11), the air sample is irradiated by a pulsative UV source at 214 nm, and fluorescence is measured at 240-420 nm. This technique affords continuous measurement of $SO_2$ at ppb level.

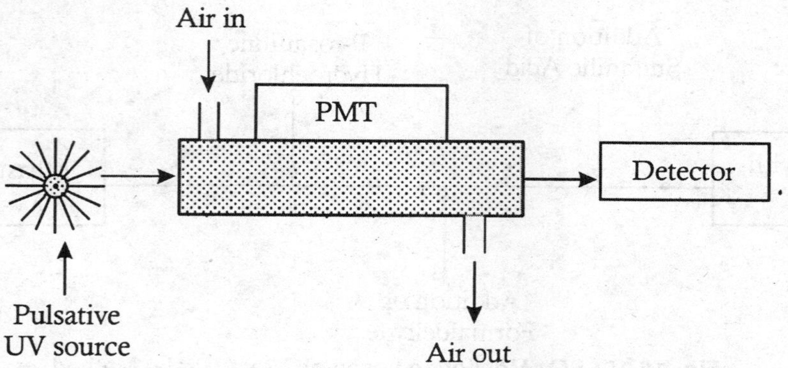


**Fig. 18.11:** Schematic of Pulsative UV Autoanalyzer for $SO_2$ Monitoring

3. *Conductivity method*: $SO_2$ can be determined by measuring the electrical conductivity of the solution due to oxidation of $SO_2$ to sulfuric acid. Continuous measurement is possible (employed in automatic analyzers).

4. *Coulometric method*: $SO_2$ can be measured by coulometric method. The method is based on the ability of $SO_2$ to reduce free $Br_2$ or $(I_2)$. $SO_2 + Br_2(I_2) + 2H_2O \rightarrow SO_4^{2+} + 2Br^- (2I^-) + 2H^+$. The air stream is passed first through a chemical scrubber to retain gases such as $H_2S$, $O_3$ and $Cl_2$, and then to an enclosed coulometric cell. The electrolytic cell consists of two Pt electrodes, an aqueous solution of KBr (KI), $Br_2(I_2)$ and $H_2SO_4$. A third electrode is used to measure changes in anode and cathode potential produced by the absorption of $SO_2$ in solution. The $SO_2$ reacts with the titrant in the cell bridge ($Br_2/Br^-$ couple) causing a decrease in the titrant, and the amount of current required to bring the titrant back to its original level is measured on a recorder.

$$Br_2 + SO_2 + H_2O \rightarrow SO_2 + 2Br^- + 2H^+ \quad \text{...(18.22)}$$

$$2Br^- \rightarrow Br_2 + 2e \text{ (anode reaction)}$$

$$K\left(\frac{Q_s - Q_e}{V \times t}\right) \quad \text{...(18.23)}$$

where, $K$ = calibration curve constant;
$Q_s$ & $Q_e$ = number of coulombs in sample titration and due to evaporation, respectively;
$V$ = flow rate;
$t$ = time

5. *Amperometric (Electrometric) method*: In amperometric method (Fig. 18.12), change in electrode current is correlated to the $SO_2$ concentration. Air sample, after passing through a filter to remove dust, is passed over a heated silver wire to remove $H_2S$, $O_3$ and Cl. The purified air, containing $SO_2$, is bubbled through an electrolytic solution of KBr + $H_2SO_4$. Electrode current indicates decrease in bromide concentration, which is a measure of $SO_2$ concentration. This method is suitable for continuous $SO_2$ monitoring. Portable solid-state instruments are available.

$$SO_2 + Br_2 + 2H_2O \rightarrow H_2SO_4 + HBr \quad \text{...(18.24)}$$

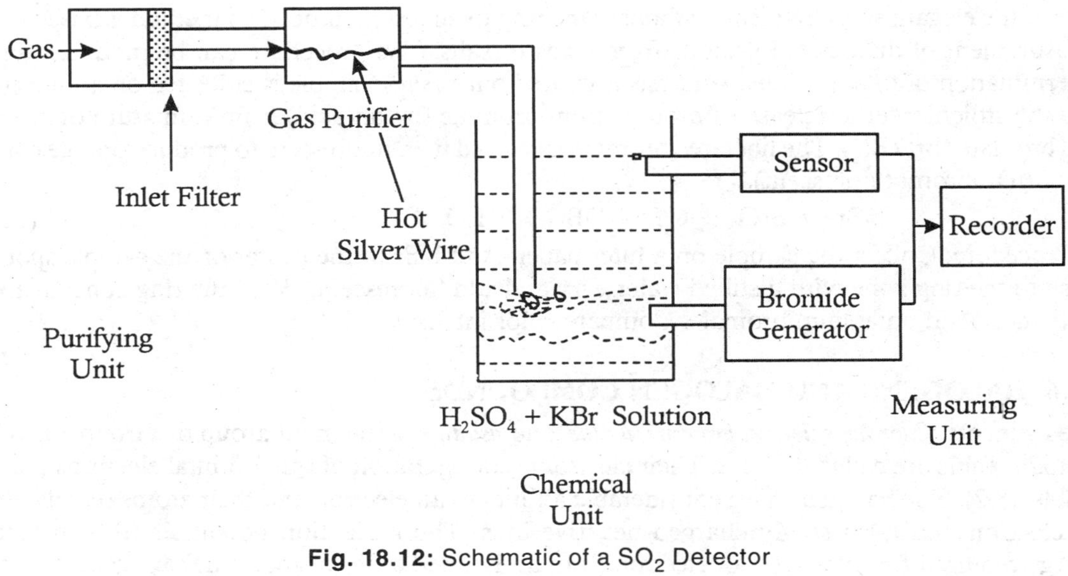


**Fig. 18.12:** Schematic of a $SO_2$ Detector

### 18.5.4 Sulfite $(SO_3)^{2-}$

The source of sulfite ion, $(SO_3)^{2-}$, is industrial sewage. Its high chemical oxygen demand (COD) makes it toxic and therefore a priority pollutant in water monitoring. $(SO_3)^{2-}$ is determined by wet-chemical methods—iodometric and colorimetric methods.

1. *Iodometric method*: Acidic sulfite, $(SO_3)^{2-}$, is titrated with a standard alkaline solution of KI. The liberated iodine is detected by blue color due to its reaction with starch indicator.
2. *Colorimetric method*: Acidified sample is purged with $N_2$ gas and the liberated $SO_2$ gas is absorbed in ferric ion and phenanthrane solution. The color of the complex is measured colorimetrically at $\lambda = 510$ nm.

### 18.5.5 Sulfate $(SO_4)^{2-}$

Sulfate, $(SO_4)^{2-}$, is determined by gravimetric, turbidimetric, colorimetric and ion-chromatography (IC) methods.

1. *Gravimetric method*: Sulfate is precipitated as $BaSO_4\downarrow$ by the addition of $BaCl_2$ to a slightly acidic solution. $BaSO_4$ is insoluble not only in water, but also in dilute acids. The washed and ashed precipitate is weighed as $BaSO_4$.
2. *Turbidimetric method*: Sulfate can be precipitated as $BaSO_4$ of uniform sized crystals in acetic acid medium. $BaCl_2$ can be determined by nephelometer ($\lambda = 420$ nm).
3. $(SO_4)^{2-}$ forms a complex with nitrosulfonic acid and a barium complex forming barium(II)-nitrosulfoazo chelate. The chelate has absorbance peak at 624 nm. Decrease in the peak intensity is a measure of the sulfate, $(SO_4)^{2-}$, present.

### 18.5.6 Sulfuric Acid $(H_2SO_4)$

Sulfuric acid is a strong inorganic acid. It is one of the important products in chemical industry, for manufacture of acids, mineral fertilizers, chemicals, explosives, paints and host of other products.

There are many wet-chemical and conductimetric methods for the estimation of sulfuric acid. One of the elegant wet-chemical methods is the ring-oven test method. The method is based on the measurement of dissociated protons from acids or salts. The procedure can be modified for the determination of total protons. Analysis is carried out *in situ*. Sample is collected on a filter paper and the stoichiometric release of protons from bromine (by an acid) from a mixture of bromide and bromate $(Br^-/Br_2)$. The liberated bromine is reacted with fluorescein to produce orange colored eosin (tetrabromofluorescein).

$$5Br^- + BrO_3^- + 6H^+ \rightarrow 3Br_2 + 3H_2O \qquad ...(18.25)$$

Procedure: Collect the sample on a filter paper. Add KBr to the center of the sample spot and wash to the ring zone with distilled water. Dry and add fluorescein. Wash the ring zone area with methanol. Add potassium bromate. Compare color intensity.

## 18.6 HALOGENS AND HALOGEN COMPOUNDS

The elements, *fluorine, chlorine, bromine, iodine,* and *astatine* in the main group of Group VII of the Periodic Table are called *halogens*. Their electronic configuration of outer orbital electrons is $ns^2np^5$ (Table 18.2). The halogens have considerable affinity to an electron and their atoms readily attach an electron and form singly charged negative ions. The ionization potentials (the amount of energy required for removing an electron from an isolated atom or molecule) are generally high.

As a class halogen elements are nonmetals, and chemically they are all oxidizing agents. Though they display similarities in their chemical properties, there are appreciable differences between them. A growth in the atomic number of the elements in the halogen family from F to At is accompanied by an increase in the atomic radii, diminishing the electronegativity and oxidizing power of the elements. Fluorine is always in the oxidation state of –1 in its compounds. The other halogens exhibit oxidation numbers from –1 to +7.

Halogens ($F^-$, $Cl^-$, $Br^-$, $I^-$) and halides HXs (HF, HCl, HBr and HI) are found in the environment in various forms.

**TABLE 18.2:** PHYSICOCHEMICAL PROPERTIES OF HALOGEN ELEMENTS

| *Parameter* | *F (9)* | *Cl (17)* | *Br (35)* | *I (53)* | *At (85)* |
|---|---|---|---|---|---|
| Electron configuration of outer orbitals | $2s^22p^5$ | $3s^23p^5$ | $4s^24p^5$ | $5s^25p^5$ | $6s^26p^5$ |
| Atomic radius (nm) | 0.064 | 0.010 | 0.114 | 0.133 | ~0.14 |
| Ionic radius (nm) | 0.133 | 0.18 | 0.196 | 0.22 | ~0.23 |
| Ionization potential (eV) | 17.42 | 12.97 | 11.84 | 10.45 | ~9.2 |
| Relative electronegativity | 4.0 | 3.0 | 2.8 | 2.5 | ~2.2 |
| Melting point °C | –219.6 | –101.0 | –7.3 | 113.6 | 230 |
| Boiling point °C | –188.0 | –34.5 | 59.0 | 184.5 | 370 |

### 18.6.1 Fluoride ($F^-$)

Fluorine, $F_2$, ($Z$ = 9; $A$ = 18.999), is a highly reactive and corrosive gas. Fluorides ($F^-$) occur in minerals (*fluorspar*, $CaF_2$). It has the greatest electronegativity among all elements, and fluorine replaces any other halide ion from its compounds. Fluoride in water can be separated from non-volatile constituents by converting it to the acid and distilling it. From the distillate fluoride can be determined by colorimetric, fluorescence and ion-selective methods.

1. *Colorimetric method*: HF produced by treatment with $H_2SO_4$ is trapped with lanthanum-alizarin to form a blue colored complex that is measured colorimetrically at $\lambda$ = 620 nm. This method can be automated. This method is routinely used for fluoride detection in water, foodstuffs and biological materials.
2. Reaction of fluoride with a zirconium-dye complex reduces the intensity of its color. The decrease is proportional to the fluoride concentration. The absorbance is measured colorimetrically at $\lambda$ = 570 nm.
3. *Coulometric method*: Introduction of Zr ions into the solution from Zr electrode: Fluoride combines with Zr ions and the amount of current consumed is proportional to the concentration of fluoride ions.
4. *Fluorescence methods*: Fluoride, $F^-$, may be determined either by the fluorescence of the complex with Zr + Calcein Blue ($\lambda_{ex}/\lambda_{em}$ = 350/410 nm) or by the quenching of the fluorescence of acidic Al-Alizarin Garnet R. The sensitivity is ~0.1-0.001 ppm.
5. *Fluoride selective electrode*: The fluoride electrode is an ion-selective sensor. Doped lanthanum fluoride crystal and calomel reference electrodes are dipped in the sample solution and the fluoride concentration is measured using an ion-meter. This method is practical, sensitive and reliable.

### 18.6.2 Chloride ($Cl^-$)

Chlorine, $Cl_2$ ($Z$ = 17; $A$ = 35.453), is a pungent reactive gas, with high electronegativity and electron affinity. Chlorine displaces the less electronegative halogens (Br, I, and As) from their compounds ($2Br^- + Cl_2 \rightarrow Br_2 + 2Cl^-$). Chloride, $Cl^-$, is one of the major inorganic anions in water and wastewater (NaCl). $Cl_2$ is used in chlorinating water to destroy pathogens. Chlorine has many industrial applications (chlorinated hydrocarbons, chlorine-containing organic solvents). Chlorine is a highly reactive element.

Colorimetric, iodometric, potentiometric, ring-oven and physicochemical methods are in use to determine the chloride concentration.

1. *Colorimetric method*: $Cl^-$ is measured colorimetrically ($\lambda$ = 480 nm).
2. *Dithizone method*: Silver dithizonate releases dithizone in the presence of $Cl^-$, and the dithizone is measured colorimetrically.
3. *Argentometric method*: This method is suitable for estimation of chloride in clear waters. In alkaline medium potassium chromate is used as an indicator in the titration of chloride with silver nitrate. Red-blue colored silver chromate gives the end-point.
4. *Amperometric method*: Chlorine can be determined in water by titration with phenylarsine oxide. At pH 6.5-7.5, bound chlorine reacts very slowly, so that only free chlorine is titrated. On the other hand at pH 3.5-4.5, bound chlorine is also titrated by using an appropriate amount of KI.
5. *Mercuric nitrate method*: $Cl^-$ is titrated with mercuric nitrate, $Hg(NO_3)_2$, to form $HgCl_2$. Excess mercuric ions can be detected from the deep blue color given by diphenylcarbazone, as an indicator (at pH = 2.3-2.8).
6. *Thiocyanate method*: $Cl^-$ reacts with mercuric thiocyanate forming mercuric chloride and liberating thiocyanate. Free thiocyanate reacts with ferric ions to form a red colored ferric-thiocyanate complex. The intensity of the color is proportional to the $Cl^-$ concentration. This method can be automated.
7. *Ortholidine method*: Chlorine can be determined by reaction with ortholidine to yield a colored complex, and the absorbance is measured at $\lambda$ = 453 and 490 nm.
8. *Iodometric method*: Suitable for measuring total chlorine concentration >1 mg/L. Iodine liberated from a KI solution by chlorine at pH < 8 is titrated with sodium thiosulfate ($Na_2S_2O_3$) at pH < 4.0, with starch as the indicator (I-starch blue). In the wastewater analysis, the residual reducing agent, $Na_2S_2O_3$ is titrated with standard iodine.
9. *Potentiometric method*: Useful for the estimation of $Cl^-$ in colored and turbid samples. Potentiometric titration with silver nitrate solution with a glass and Ag : AgCl electrode system.
10. *Amphoteric methods*: These methods are more sensitive. An amphoteric cell contains a polarizable electrode (Pt) and a non-polarizable reference electrode (Cu). When chlorine, a strong oxidizing agent, enters the cell it causes depolarization of Pt electrode and the current generated is proportional to the $Cl_2$ concentration. The pH should be 4.0-4.5, hence buffering is necessary.

#### 18.6.2.1 Chlorine Dioxide ($ClO_2$)

Chlorine dioxide, $ClO_2$, has been used as a bleaching agent in the paper mills and pulp industry. Colorimetric, iodometric and amperometric methods may be employed to determine $ClO_2$. $ClO_2$ releases $I^-$ from an acidic KI solution. The liberated iodine is titrated with a standard solution of sodium thiosulfate ($Na_2S_2O_3$) with starch as the indicator. It is an easy to perform colorimetric method.

### 18.6.3 Bromide ($Br^-$)

Bromine, Br, ($Z$ = 35; $A$ = 79.904), is pungent element that occurs in gaseous or liquid form. Chlorine can liberate bromine from a bromide solution. Bromine dissolves much better in organic solvents such as carbon disulfide, ethyl alcohol, diethyl ether, chloroform, and benzene than in water. This property is taken advantage of for extracting bromine from aqueous solutions. $Br^-$ is determined by gravimetric, colorimetric and ion-chromatography (IC) methods.

1. *Colorimetric methods*: Treatment of $Br^-$ solution with chloramine-T in the presence of phenol red at pH = 4.5 gives rise to a brominated compound whose color varies from reddish to violet, depending on the bromide concentration. This is the basis of a common colorimetric method for its determination.
2. *Fluorescein method*: Bromine can be detected with fluorescein. The yellowish compound yields bright-red colored tetrabromofluorescein (eosin) complex with bromine.
3. *Colorimetric method using iodine*: $Br^-$ is oxidized with calcium hypochlorite to bromate. Then bromate oxidizes iodide and the iodine produced is titrated with sodium thiosulfate.

$$2Br^- + 3Ca(OCl)_2 \rightarrow 2BrO_3^- + CaCl_2 \quad ..(18.26)$$

$$6I^- + 6H^+ + BrO_3^- \rightarrow 3I_2 + Br^- + 3H_2O$$

$$2Na_2S_2O_3^- + I_2 \rightarrow Na_2S_2O_6^{2-} + 2I^-$$

The end-point is determined with starch indicator, which looses blue color when all iodine is titrated.

### 18.6.4 Iodide ($I^-$)

Iodine, I, ($Z$ = 53; $A$ = 126.904), is a violet color solid. Sources of iodine are natural brines and industrial waste. Iodine is used in medicine as antiseptic and in a number of pharmaceutical preparations, and in photochemicals. Higher solubility of iodine in organic solvents is made use of in extracting iodine from aqueous solutions.

Iodide, $I^-$, may be determined by wet-chemical and physicochemical methods. Starch indicator method is a common and routinely employed method.

1. Iodide, $I^-$, is transformed to iodine, $I_2$, by potassium peroxymonosulfate ($KHSO_5$) and is allowed to react with leuco crystal violet and the absorbance of violet dye is measured at $\lambda$ = 592 nm. Iodine reacts with mercuric chloride to form hypoidous acid, which is allowed to react with leuco crystal violet and the absorbance (at pH = 3.5-4.0) is measured at 592 nm.
2. *Titrimetric method*: Argentimetric titration in the presence of dithizone gives a greenish-yellow colored complex.

### 18.6.5 Halogen Compounds

All hydrogen halides (HX, X = F, Cl, Br and I) are colorless gases and are highly soluble in water. Hydrochloric acid (HCl) is one of the most important mineral acids, and large amounts of it are produced as a by-product in the chlorination of organic compounds. Many of its salts find wide industrial and agricultural applications.

Most halogen compounds have to be subjected to dry or wet decomposition procedures to convert bound halogens to halide ions. Decomposition methods involve mineralization treatment with oxidizing agents (e.g. sodium peroxide). Combustion methods are most reliable. Having converted to halide further determination is by any of the methods already given above for halides.

Wet-chemical (colorimeric) methods of analysis of gases and aerosols are listed in Table 18.3.

**TABLE 18.3:** A SUMMARY OF WET-CHEMICAL (COLORIMETRIC) ANALYSIS OF GASES AND AEROSOLS

| *Species* | *Procedure* | $l_{max}$ *(nm)* |
|---|---|---|
| Ammonia ($NH_3$) | Nessler's reagent method: $NH_3 + HgI \rightarrow NH_2Hg_2I_3$ (orange-brown). Indophenol method: alkaline phenol + hypochlorite | 425 |
| Bromide ($Br^-$) | Hypobromite + phenol red → bromophenol blue. Chloramine-T and Iodometric methods | 640 |
| Carbon dioxide ($CO_2$) | Phenopthalein potentiometric method. Sodium bicarbonate and calcium carbonate methods | |
| Carbon monoxide (CO) | Iodine pentoxide ($I_2O_5$) in 10% oleum | |
| Chloride ($Cl^-$) | Argentometric method (for clear waters); Ferric cyanide method; Mercury nitrate method; | |
| | Orthotolidine method | 453; 490 |
| | Iodometric method | 480 |
| Cyanide ($CN^-$) | Chloramine-T method: CNCl + pyridine-pyrozolone + barbituric acid | 575; 578; 582 |
| | Mohr's method: $AgNO_3$ + rhodamine | 578 |
| Fluoride ($F^-$) | Zirconium dye method: $F^-$ + Zr-dye → Zr-fluoride | 570 |
| | Lanthanum-alizarin method | 620 |
| Hydrogen sulfide ($H_2S$) | Methylene blue method | 660 |
| | Iodometric method | |
| Iodide ($I^-$) | K-peroxymonosulfate method; | 592 |
| | Mercuric chloride method | |
| $N_2$ | (Kjeldhal) methods: Digestion with $H_2SO_4 \rightarrow NH_4^+$ | 630 |
| | Persulfate method & Iodophenol method | |
| Nitrite ($NO_2^-$) | Griess-Saltzmann method; Jacobs-Hochmeister method | 545 |
| Nitrate ($NO_3^-$) | Cadmium reduction + Griess-Saltzmann method | 545 |
| | Hydrazine sulfate method | 524 |
| Oxygen ($O_2$) | Winkler method | 450 |
| Ozone ($O_3$) | Reverse West-Gaeke method | 560 |
| | Eugenol-formaldehyde method | 560 |
| | Iodometric method | 600 |
| Sulfide ($S^{2-}$) | Methylene blue method : Iodometric method | 660 |
| Sulfur dioxide ($SO_2$) | West-Gaeke (pararosaniline) method | 560 |
| | Iodometric method | |
| Sulfite ($SO_3^{2-}$) | Fe(III)-phenanthrane method: Iodometric method | 510 |
| Sulfate ($SO_4^{2-}$) | Methylthymol blue method:Nitrosulfonic acid method | 624 |

## BIBLIOGRAPHY

Chen, H., *et al.* (1999), *Int. J. Environ. Anal. Chem.* "Simultaneous spectrophotometric determination of nitrite and nitrate in water samples by flow-injection analysis."

Ciaccio, L. L. (ed.). (1972), Marcel Dekker: New York. "*Water and Water Pollution Handbook.*"

Clark, R. J. H. and Hester, R. E. (1995), Wiley & Sons: New York. "*Spectroscopy in Environmental Science Advances in Spectroscopy Series,*" (Vol. 24).

Davis, J., *et al.* (1999), *Talanta*, **50**(1): 103. "Evaluation of phenolic assays for the detection of nitrite."

Ding, Y., Lee, M. L. and Eatough, D. J. (1997), *Int. J. Environ. Anal. Chem.* "The determination of total nitrite and N-nitroso compounds in atmospheric samples."

Eaton, A. D., *et al.* (eds). (1995), Amer. Public Health Assn: Washington, DC. "*Standard Methods for the Examination of Water and Wastewater.*"

Ewing, G. W. (1977), Academic Press: New York. "*Environmental Analysis.*"

Fellenberg, G. (2000), Wiley: Chichester. "*The Chemistry of Pollution,*" (3rd edn.).

Frei, R. W. and Albarge, J., (eds.). (1986), Gordon & Breach: New York. "*Air and Water Analysis: New Techniques and Data.*"

Gallay, W., *et al.* (1975), Butterworth: London. "*Environmental Pollutants: Selected Analytical Methods.*"

Harrison, R. M. and Perry, R. (1984), Chapman & Hall: London. "*Handbook of Air Pollution Analysis,*" (2nd edn.).

Hitchman, M. L. (1978), Wiley & Sons: New York. "*Measurement of Dissolved Oxygen.*"

Knoll, G. F. (1979), Wiley & Sons: New York. "*Radiation Detection and Measurement.*"

Kouimtzis, T. and Samara, C., (eds.). (1995), Springerverlag: Berlin. "*Airborne Particulate Matter.*"

Lawrebce, N. S., Davis, J. and Compton, R. G. (2000), *Talanta*, **52**(5): 771. "Analytical strategies for the detection of sulfide: A review."

Moorcroft, M. J., Davis, J. and Compton, R. G. (2001), *Talanta*, **52**(5): 785. "Detection and determination of nitrate and nitrite: A review."

Nriagu, N. O., (ed.). (1992), New York: Wiley & Sons. "*Gaseous Pollutants.*"

Peirce, J. J., *et al.* (1998), Butterworth-Heinemann: New Delhi. "*Environmental Pollution and Control,*" (4th edn.).

Perkins, H. C. (1974), McGraw-Hill: New York. "*Air Pollution.*"

Pfafflin, J. R. and Ziegler, E. W. (1976), Gordon & Breach: New York. "*Encyclopedia of Environmental Science and Engineering,*" (2nd edn.).

Quinby-Hunt, M. S., McLaughlin, R. D. and Quintanilha, A. T. (1986), Wiley & Sons: New York. "*Instrumentation for Environmental Monitoring,*" Vols. 1 & 2, (2nd edn.).

Rouessac, F. and Rouessac, A., (eds.). (2000), Wiley: Chichester, MA. "*Chemical Analysis—Modern Instrumental Methods and Techniques.*"

Stern, A. C., *et al.* (1973), Academic Press: New York. "*Fundamentals of Air Pollution.*"

Stevens, R. K. and O'Keffe. (1970), *Annal.Chem.*, **42**: 143A. "*Modern Aspects of Air Pollution Monitoring.*"

Toribara, T. Y., *et al.* (1978), Plenum Press: New York. "*Environmental Pollutants Detection and Measurement.*"

Weisz, H. (1961), Pergamon Press: London. "*Microanalysis by the Ring Oven Technique.*"

Willeke, K. and Baron, P. A. (1993), Van Nostrand-Reinhold: New York. "*Aerosol Measurements: Principles, Techniques and Applications.*"

... (1994), EPA/600/R-93/100: Washington, DC. "*Methods for the Determination of Inorganic Substances in Environmental Samples.*"

... (1999), EPA/625/R-96/010a: Cincinnati, Ohio. "*Compendium Methods for the Determination of Inorganic Compounds in Ambient Air.*"

# 19

# Chemical Methods in Environmental Pollution Analysis—Metals

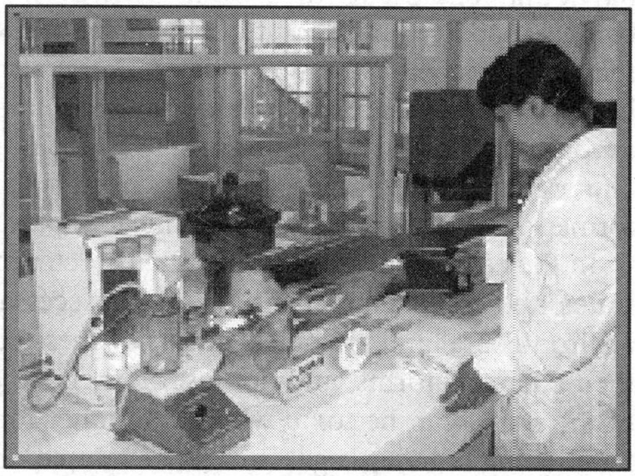

In the case of metals and metalloids, determination of their concentrations in various oxidation states and forms is necessary, because the intensity of toxicity of many metals and their compounds (e.g. As, Cd, Cr, Hg, Pb, Se and V) depends on the oxidation states and the chemical forms. Except for the preliminary treatment and pre-concentration procedures, the methods for their determi-nation do not fundamentally change whether the original sample is from air or water or soil. In the case of air samples, it is a common practice to analyze air particulate for metals. Spectrophotometric methods, based on the color formation between the metals and chelating agents, are among the traditional procedures used for trace metal determination. In modern times these molecular spectroscopic methods have been largely supplanted by AAS and ICP emission spectroscopy, in view of their enhanced sensitivity and precision and multi-element capability and hence fastness. The later, coupled with mass-spectrometric detectors, has proved to be a highly sensitive and precise method of analysis.

There are instances of individual metals being determined by exploiting their special properties. Thus, traces of mercury in air or water can be determined taking advantage of the characteristic UV absorption of its vapor. The elements arsenic, antimony, bismuth, selenium, tellurium, tin, germanium and lead form volatile covalent hydrides. Taking advantage of this property, these elements can be separated from many interfering metals and determined with high sensitivity by AAS, ICP-AES or ICP-MS techniques.

This chapter summarizes the chemical methods for the determination of metals.

Four fifths of all the elements listed in the Periodic Table are metals. Metals occur in widely different physicochemical forms (e.g. oxidation states) in different environmental compartments. They exist in the environment as aerosols and some metals, such as Au, Fe, and Hg, in elemental form as suspensions or colloids. Some compounds of trace elements—As, Hg, Pb and Sn can exist in gaseous form. Soluble trace elements form mostly hydrated, inorganic or organic complexes.

Trace elements, below and above a certain concentration result in metabolic disorders, and some of them, especially, Al, Cd, Cr, Se, V, may constitute serious health hazards. Therefore, determination of metals with high accuracy and sensitivity in various chemical forms (*speciation*) is very important in monitoring the environmental impact, pollution and toxicity of metals. While chemical methods of analysis of metals are dealt in this chapter, passing reference to other techniques (physical) of determination is made, which are dealt separately in other chapters (Chapters 23, 24, 25 and 26).

## 19.1 SPECIATION OF METALS

Maximum permissible levels of pollutants, usually, refer to total concentrations rather than any particular chemical form of it. However, the distribution, accumulation, bioavailability, biodegradability and toxicity of elements depend not only on their concentrations, but also on their physicochemical associations with the surrounding systems. The chemical behavior of a metal is primarily governed by its retention and release reactions. Increase in pH results in increased adsorption of some metals in soils. Metal adsorption also depends on the presence and type of clay and organic matter. These effects are presumably related to partial immobilization of metals due to formation of insoluble metal-organic complexes or increased cationic exchange.

In the case of metals and metalloids determination of their concentration in various oxidation states and molecular forms is, often, necessary. That is, "*speciation*" is necessary because the environmental and biological effects of these pollutants depend on their oxidation states as well as their chemical forms, and not just on their stoichiometric concentration. In some cases, one form of the metal can be toxic whereas another form may be non-toxic or even essential for normal functioning of living organisms. For example, arsenic and chromium are more toxic in their trivalent and hexavalent states, respectively. Hg is more toxic in the organic form ($Hg^+CH_3$). Free (hydrated) metals ions are more toxic to aquatic life.

Metals have a number of common properties. The common physical properties of metals include their high electrical and thermal conductance, luster, ductility (plasticity) and high reflecting power and opacity. All metals are characterized by a comparative readiness to give up their valence electrons and form positive ions. Many metals, chromium, iron, and manganese have different, but always positive, oxidation states in different compounds. That is, metals in free state are reducing agents. The outer electrons in metal atoms are at a considerable distance from the nucleus and are bound to it comparatively weakly. Hence, metal atoms are characterized by low ionization potentials. In general, an increase in oxidation number of an element is attended by an increase in the acid properties. An increase in the ionic radius of an element, at constant charge, is attended by a decrease in the acidic properties.

General experimental protocols for speciation of metals include specimen collection, sizing, concentration (or dilution), separation and detection (Fig. 19.1). Physical methods of determination such as electroanalytical, spectroscopic (AAS, AES), XRF, and NAA are dealt in chapters 23 to 26.

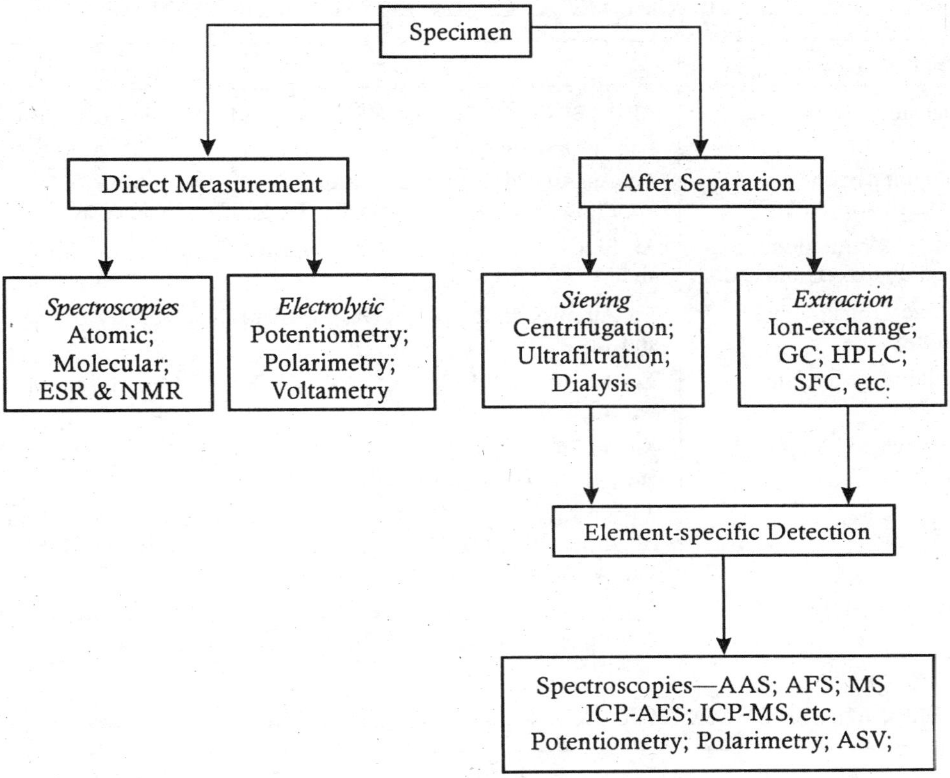


**Fig. 19.1:** Experimental Protocol for Metal Speciation

Methods for determination of metals invariably involve wet-chemical analysis. Chelate-extraction is one of the most extensively used techniques for the analysis of trace metals. Most common chelate-extraction systems are oxines (8-hydroxyquinoline, and dimethylglyoxime), cupferron (ammonium nitrosophenyl-hydroxylamine), dithizone (diphenylthiocarbazone), diethyldithiocarbamate (DDTC), benzidene, and xanthates (Table 19.1). While oxine (8-hydroxyqinoline) is a non-specific complexing reagent, dimethyglyoxime is highly specific reagent, complexing Ni(II) in an alkaline medium and Pd(II) in an acidic medium. Cupferron precipitates large number of heavy metals in acidic media.

Many metals, such as Bi, Cd, Cu, Fe, Ni, and Pb, may be pre-concentrated as their diethyldithiocarbamate (DDTC) chelates using a column filled with Amberlite XAD-4 resin. The retained analytes on the resin are recovered with a small volume of acetone. The metal ions in the effluent may be determined by atomic absorption spectrometry (AAS).

Spectrophotometric or ecotoxicological methods (bioluminescence of luciferin-luciferase system at 465 nm) are simple and rapid, and are suitable for assessing metal contamination in soils and wastes. Selectivity in chelate-extraction systems can be increased by simultaneously adding a complexing agent that preferentially reacts with the interfering metals. This process is called *masking*. Amino polycarboxylic acids, citrate, cyanide, malonate and tartrate are common masking agents used in the selective determination of metals. A large number of elements are extracted into chloroform or carbon tetrachloride as dithizone chelates, and use of masking agents and regulation of pH of the aqueous phase enables selective extraction of some metals (Table 19.2).

**TABLE 19.1:** SOME METAL-CHELATE EXTRACTION SYSTEMS

| *Chelating Agent* | *Metals Extracted* |
|---|---|
| Acetylacetone | Al, Be, Bi, Co, Cr, Cu, Fe, Ga, Hf, Hg, In, Mn, Mo, Pb, Pd, Pu, Ru, Sc, Sn, Th, Ti, Tl, U, V, Zn, and Zr |
| Ammonium pyrrolidine dithiocarbamate (APDC) | Ag, As, Au, Cd, Co, Cr, Cu, Fe, Ga, Ge, Hg, In, Ir, Mn, Mo, Nb, Ni, Os, Pb, Pd, Pt, Re, Rh, Ru, Sb, Se, Sn, Tc, Te, Tl, U, V, W, and Zn |
| Cupferron (Ammonium nitrosophenyl-hydroxylamine) | Al, Bi, Ce, Co, Cu, Fe, Ga, In, Mn, Mo, Nb, Pa, Pb, Pd, Sb, Sn, Th, Ti, U and V |
| Dithizone (Diphenylthiocarbozone) | Ag, Au, Bi, Cd, Co, Cu, Fe, Ga, Hg, In, Mn, Ni, Pb, Pd, Po, Pt, Sn, Te, Ti, Tl and Zn |
| Diethyldithiocarbamate (DDTC) | Ag, As, Au, Bi, Cd, Co, Cr, Cu, Fe, Ga, Hg, In, Mn, Mo, Ni, Pb, Pd, Pu, Sb, Se, Sn, Te, Ti, Tl, U, V, and Zn |
| 1-(2,-pyridylazo)-2-napthol (PAN) | Ag, Bi, Cd, Ce, Co, Cu, Eu, Fe, Ga, Hg, In, Ir, La, Mn, Ni, Pb, Pd, Pt, Rh, Sc, Sn, Th, Ti, U, V, Y, Zn and Zr |
| Oxines | Ag, Al, Am, Ba, Be, Bi, Bk, Ca, Cd, Ce, Cf, Cm, Co, Cr, Cu, Er, Fe, Ga, Hf, Hg, In, La, Mg, Mn, Mo, Nb, Nd, Ni, Np, Pa, Pb, Pd, Pu, Sb, Sc, Sm, Sn, Sr, Th, Ti, Tl, U, V, W, Y, Zn and Zr |
| Theonyltrifluroacetone (TTA) | Ac, Al, Am, Be, Bi, Bk, Ca, Ce, Cf, Cm, Co, Cr, Cs, Cu, Es, Eu, Fe, Fm, Hg, In, Mn, Mo, Pb, Pd, Pu, Ru, Sc, Sn, Th, Ti, Tl, U, V, Zn and Zr |

**TABLE 19.2:** USE OF MASKING AGENTS IN THE DITHIZONE EXTRACTION SYSTEM

| *Masking Agent* | *pH* | *Metals Extracted* |
|---|---|---|
| Bis(2-hydroxyethyl)-dithiocarbamate | Slightly basic | Zn |
| $Br^-$ (or $I^-$) | Acidic | Au, Cu, Pd |
| Cyanide | Acidic | Ag, Au, $Hg^{2+}$, In, $Pd^{2+}$ |
| | Basic | Bi, In, Pb, Po, Sn |
| EDTA | Acidic | Ag, Hg |
| DDTC | Acdic | Bi, Cd, Cu |
| Thiocyanate | Acidic | Au, Cu, Hg |
| Thiosulfate | Acidic | Cd, Cu, In, Ni, Pd, Zn |
| Cyanide + thiocyanate | Acidic | Cu, Hg |
| Cyanide + thiosulfate | Acidic | $Sn^{2+}$, Zn |

Flotation and extraction methods can be employed for separation of Co, Cu, Pb and Ni and elimination of matrix interference of calcium. For separation and concentration of trace metals flotation can be carried out using hexamethylenedithiocarbamate as a complexing and collecting agent. The liquid-liquid extraction of Co, Cu, Pb and Ni is carried out by Na-diethyldithiocarbamate (NaDDTC) as a chelating agent and methylisobutylketone (MIBK) as the extractant. Electrothermal atomic absorption spectrometry (ET-AAS) may be used for determination of extracted analytes.

Flow-injection on-line dilution procedures (Chapter 22) with detection by spectroscopic (Chapter 24) or mass spectrometry (ICP-MS) are available for the determination of various metals, such as As, Cu, Mo, Ni, Pb, Se, and Zn.

## 19.2 ALKALI METALS

The metals of the main subgroup of Group I of the Periodic Table—*lithium, sodium, potassium, rubidium, cesium and francium*—are called *alkali* metals. Alkali metal atoms have one electron in their outermost electron orbital. These elements part with that electron quite readily. This is reflected by low ionization potential for the atoms (Table 19.3). The alkali metals, as a class, exhibit metallic properties; they belong to the most chemically active elements. They combine vigorously with oxygen. Rubidium and cesium self-ignite in the air. Lithium, sodium, and potassium ignite when slightly heated. Since they give up their valence electrons in chemical reactions, alkali metals are the most vigorous reducing agents. Their hydroxides are strong bases.

**TABLE 19.3:** SOME PHYSICAL PROPERTIES OF ALKALI METALS

| *Parameter* | *Li (3)* | *Na (11)* | *K (19)* | *Rb (37)* | *Cs (55)* | *Fr (87)* |
|---|---|---|---|---|---|---|
| Electron configuration in the outer orbital | $2s^1$ | $3s^1$ | $4s^1$ | $5s^1$ | $6s^1$ | $7s^1$ |
| Atomic radius (nm) | 0.155 | 0.186 | 0.233 | 0.245 | 0.266 | 0.28 |
| Ionic radius (nm) | 0.068 | 0.097 | 0.133 | 0.147 | 0.167 | 0.18 |
| Ionization potential (eV) | 5.39 | 5.14 | 4.34 | 4.18 | 3.89 | – |
| Electrode potential (V) | –3.045 | –2.714 | –2.924 | –2.925 | –2.923 | – |
| Density | 0.534 | 0.97 | 0.862 | 1.53 | 1.90 | 2.2 |
| Melting point °C | 180 | 98 | 63.6 | 38.8 | 28.5 | ~20 |
| Boiling point °C | 1,320 | 890 | 755 | 690 | 670 | 620 |
| Flame color | Red | Yellow | Violet | Dark red | Blue | – |
| Emission lines (nm) | 670.8; 610.4 | 589.0; 589.6 | 777.0; 765.5 | 421.6; 420.2 | 459.3; 455.5 | – |

### 19.2.1 Lithium (Li)

Lithium, Li ($Z = 3$; $A = 6.941$), is a silvery white metal. It is manufactured by the electrolysis of a fused mixture of lithium and potassium chloride. Lithium is used in Li-batteries, high-power density nuclear power plants. The isotope $^6$Li is an industrial source for the production of tritium ($^3$H), while the isotope $^7$Li has a low nuclear cross section (~0.003 barn), which permits its use as a heat-carrying agent in uranium reactors in which coolant temperatures above 800° C are required. Lithium-magnesium alloys are used in aerospace and other industries. Lithium and its compounds are also used as rocket fuels, in medicine for treatment of severe depression. Owing to its high chemical reactivity, lithium is used in metallurgy for removing traces of elements (S, N, O and H) from metals and alloys.

Lithium is determined by flame emission photometric method (red color), measured at $\lambda = 670$ nm. Interference from Ba, Ca and Sr can be suppressed by adding a solution of sodium sulfate-sodium carbonate. Preferred techniques are atomic absorption spectrometry (AAS) and inductively coupled plasma (ICP) emission.

### 19.2.2 Sodium (Na)

Sodium, Na ($Z = 11$; $A = 22.99$), is the 6th among the elements in order of abundance, and is a major constituent of seawater. Metallic sodium is a soft silvery metal produced by the electrolysis of molten sodium chloride to which calcium chloride has been added to lower the melting temperature. Sodium is by far the most important alkali metal in terms of industrial use. The most important fields of sodium use are metallurgy, atomic power engineering, and organic synthesis, fertilizer, explosives, detergent and petrochemical industries. In metallurgy, the sodium-thermal method is used to produce a number of refractory metals. It is employed as a catalyst in the production of certain organic polymers. Sodium is an excellent heat-transfer fluid, and because of this property, it has potentially large-scale uses. Sodium-potassium alloy (NaK) is used as liquid metal heat-carrying agent in atomic power engineering (fast-breeder reactors). The mixture has high heat-transfer capacity and does not react with most structural materials. Sodium is important in living matter and is vital in nerve impulse transmission and renal osmosis control of extracellular fluid. It is the most abundant cation in the extracellular fluid. Sodium deficiency (hyponatraemia) leads to Addison's disease, and excess of sodium (hypernatraemia) leads to Conn's syndrome.

Measurement of sodium is generally by flame photometry, AAS and ICP. In flame photometry, the sample is sprayed into a gas flame and the intensity of excitation (yellow color) is measured at $\lambda = 589$ nm.

### 19.2.3 Potassium (K)

Potassium, K ($Z = 19$, $A = 39.10$), ranks 7th among the elements in order of abundance. Major sources of potassium are minerals and fertilizer runoff. It is produced by sodium reduction of molten potassium chloride at 870° C. Potassium is a silver-white metal, very similar to sodium in its appearance and in its physical and chemical properties, but is even more active. It oxidizes rapidly in the air, and vigorously reacts with water to evolve hydrogen.

Potassium compounds are used in chemical manufacturing (fertilizers), bleaching, tanning, detergents, explosives, photography and medicine. Potassium is needed for plant growth (potassium fertilizers) and is an element present in all forms of life. It is the primary inorganic cation within the living cell, and it plays an important role in certain physiological processes (neural signal conduction). Hypokalaemia is usually associated with potassium deficiency, due to decreased intake or external loss. The physiological syndromes are vomiting, diarrhea, metabolic acidosis and alkalosis, leukemia, listlessness, nausea, muscle weakness, cardiac arrhythmia. Hyperkalaemia generally implies that potassium has to be removed from the extracellular compartment. Syndromes are nausea, cardiac arrhythmia and arrest.

Spectrophotometric, AAS and flame-photometric and potentiometric methods are employed for determination of potassium.

1. *Spectrophotometric method*: Cobalt nitrite method, color measured at $\lambda = 425$ nm.
2. *AAS and Flame photometry*: AAS is the preferred method. In flame photometry, the intensity of the characteristic emission line (violet color) is measured at 767 nm.
3. *Potentiometric method*: Potassium is determined by using $K^+$ ion-sensitive membrane electrode and double-junction sleeve-type reference electrode.
4. Ion chromatography (IC), neutron activation analysis (NAA) and X-ray fluorescence (XRF) can be employed to determine both Na and K.

### 19.2.4 Rubidium (Rb) and Cesium (Cs)

Rubidium, Rb ($Z = 37, A = 85.47$) and Cs ($Z = 55, A = 132.905$) are silver-white metals. Cs is the most volatile of the alkali metals (B.P. = 670° C), and it has the second lowest melting point (28.5 °C) among the metallic elements (after mercury). Rb and Cs are used in the production of photoelectric cells. They are generally determined by AAS methods.

## 19.3 COPPER SUBGROUP METALS

The copper subgroup (second subgroup of the group I in the Periodic Table) includes three elementsg—*copper* (Cu), *silver* (Ag), and *gold* (Au). Atoms in these elements have one electron in the outer orbital, but unlike the alkali metal atoms, the preceding electron orbital contains eighteen electrons ($d^{10}ns^1$). The atomic radii of copper subgroup elements are smaller (Table 19.4) than those of main group alkali metals. The small atomic radii also explain the higher ionization energy of their atoms. The elements of the copper subgroup are metals with low activity. Their ions are readily reduced and their hydroxides are weak bases.

**TABLE 19.4:** SOME PHYSICAL PROPERTIES OF COPPER SUBGROUP METALS

| *Parameter* | *Cu* (29) | *Ag* (47) | *Au* (79) |
|---|---|---|---|
| Electron configuration in the outer orbital | $3d^{10}4s^1$ | $4d^{10}5s^1$ | $5d^{10}6s^1$ |
| Atomic radius (nm) | 0.128 | 0.144 | 0.144 |
| Ionization potential (eV) | 7.73 | 7.57 | 9.23 |
| Electrode potential (V) | 0.520 | 0.799 | 1.692 |
| Density | 8.96 | 10.5 | 19.3 |
| Melting point (°C) | 1,083 | 961 | 1,063 |
| Boiling point (°C) | 2,570 | 2,210 | 2,960 |

### 19.3.1 Copper (Cu)

Copper, Cu ($Z = 29, A = 63.546$), is a reddish metal, widely distributed in nature in the form of free metal, sulfides, chlorides and carbonates. The most important copper ores are *copper glance, copper pyrite* and *malachite*. Smelting of ores produces the metal. Pure copper is an excellent conductor of heat and electricity, being inferior in this respect only to silver. Due to its high thermal and electrical conductivity, ductility and high tensile strength, copper is in great demand in electrical and industrial equipment manufacture. Copper forms numerous alloys, the most important being brasses (alloys of copper with zinc) and bronzes (alloys of copper with tin and nickel) that have many industrial applications. Chemically, copper exhibits oxidation numbers +1, +2, and +3 (under special circumstances). Copper is an essential trace element. In plasma, copper is carried by copper-transport protein, caeruloplasmin. Increased serum copper levels lead to acute chronic diseases, such as thyrotoxicosis, malignancy and biliary cirrhosis, and decreased serum copper levels lead to Wilson's disease. Main sources of copper pollution are mining, mine leaching, metal plating and domestic and industrial waste.

Copper is determined by spectrophotometric, AAS and ICP methods.

1. *Neocuproine method*: Cuprous ion ($Cu^+$) reacts with neocuprine (2, 9-dimethyl-1, 10-phenanthroline), in a slightly acidic solution, producing a yellow colored complex. This can be measured measured spectrophotometrically at $\lambda = 457$ nm.

2. *Bathocuproine method*: Water-soluble cuprous ion forms an orange-red chelate in acidic solution, with bathocuproine disulfate. The absorbance is measured at $\lambda$ = 484 nm.
3. *DDTC method*: Diethyldithiocarbamate (DDTC) chelation method; the absorbance is measured at $\lambda$ = 440 nm.
4. *Ring-oven method*: Treat the sample with KCN. Mask interference by adding malonic acid. Determine Cu by spraying saturated solution of dithioxamide in ethanol, producing greenish-blue complex. Ring oven method, though qualitative is useful in field tests.
5. *Fluorescence quenching method*: A very specific and sensitive method (0.1 ppm), based on quenching of the fluorescence of 2-(2'-hydroxyphenyl)benzoxazole in acetone. At pH > 6 the reagent has a green fluorescence, which decreases in the presence of Cu due to formation of non-fluorescent Cu-chelate. This is the basis determination of trace levels of copper. Trace levels Cu(II) can also be determined by quenching of the fluorescence of 4, 5-dihydroxy-1, 3-benzenedisulfonic acid (Tiron). The decrease in fluorescence intensity (at pH = 8) is proportionally to the concentration of Cu(II) ($\lambda_{ex}$ = 294 nm, $\lambda_f$ = 350 nm).
6. *Electrochemical luminescence (ECL) method*: Cu(II) ions in water can be determined by electrochemical luminescence (ECL) method. ECL is generated by reducing $Cu^{2+}$ ions to $Cu^+$ ions with hydroxylamine hydrochloride and then complexing with the chelating agent 2, 9-dimethyl-1, 10-phenanthroline (DMP) to form $Cu(DMP)_2^+$, followed by the oxidation of tr-n-propylamine (TprA). It is a sensitive method (0.1 ppm) for analysis of trace levels of copper in water.
7. *AAS & ICP methods*: AAS with graphite furnace (2,400° C) gives 0.02 μg/mL sensitivity.

### 19.3.2 Silver (Ag)

Silver, Ag ($Z$ = 47, $A$ = 107.868), is found free or combined in minerals and ores of gold, lead, copper, zinc, and nickel. The most important silver (Ag) ore is *silver glance*, or *argentite* ($Ag_2S$). Pure silver is very soft and ductile metal and it is the best conductor of heat and electricity among all the metals. Silver alloys are employed in the production of jewelry and production of domestic articles, coins and laboratory equipment, electric and electronic industry, and in the production of silver-zinc accumulators. Silver compounds are used in silver-plating. Silver salts are used in the production of photographic materials.

Chemically, silver is a metal with low activity. Silver ions suppress the development of bacteria and sterilize potable water even at very low concentration. In medicine, colloidal silver solutions are used for disinfecting the mucous membranes. Silver can cause argyria, a blue-gray discoloration of skin and eyes.

Silver is determined by spectrophotometric, AAS and ICP methods.

1. *Dithizone method*: Many metals can react with dithizone to produce colored coordination compounds. With dithizone silver forms a colored complex, which can be extracted by a suitable solvent and determined spectrophotometrically at $\lambda$ = 462 or 620 nm.
2. *Fluorescence method*: Silver can be determined either by measuring the fluorescence of its chelate with 8-hydroxyquinoline-5-sulfonic acid. It can also be measured by its quenching effect of the fluorescence of eosin + 1,10-phenanthroline complex with Ag. The method is very sensitive (sensitivity ~ 0.01 ppm).
3. AAS (and ET-AAS) and ICP methods are suited for determining trace levels of silver.

### 19.3.3 Gold (Au)

Gold, Au, ($Z$ = 79, $A$ = 196.967), is found in free or combined form in nature. It is the most ductile and malleable of all metals. It is used in coinage, jewelry, dentistry and surgery. It is determined by rhodamine B fluorescence method or by flame AAS method. In rhodamine B fluorescence method, gold forms Au-rhodamine B chelation complex that is fluorescent ($\lambda_{ex}$ = 550 nm, $\lambda_{em}$ = 575 nm). This method is about 20 times more sensitive than atomic absorption method (sensitivity ~ 0.02 ppm).

$$AuCl_4^- + HCl \rightarrow \text{Rhodamine B} \rightarrow \text{Au-Rh B complex} \qquad ...(19.1)$$

## 19.4 ALKALINE-EARTH METALS

The main subgroup of Group II, (IIa), in the Periodic Table includes elements—*beryllium, magnesium, calcium, strontium, barium and radium*. All these elements except beryllium have pronounced metallic properties. In the free state, they are grayish-white substances harder than the alkali metals and have higher melting and boiling points too (Table 19.5).

Chemically, alkaline-earth metals are extremely electropositive and form divalent ionic salts. They can readily part with the two outer orbital electrons, exhibiting an oxidation state of +2 in all their compounds. The oxides of the alkaline-earth elements are basic. The oxides of calcium, strontium, and barium combine directly with water to form hydroxides. Solubility of hydroxides increases from calcium to radium. Unlike the salts of the alkali metals, many salts of the alkaline-earth metals are sparingly soluble in water. These salts include carbonates, sulfates, and phosphates (Table 19.6).

**TABLE 19.5:** SOME PHYSICAL PROPERTIES OF ALKALINE-EARTH METALS

| *Parameter* | *Be (4)* | *Mg (12)* | *Ca (20)* | *Sr (38)* | *Ba (56)* | *Ra (88)* |
|---|---|---|---|---|---|---|
| Electron configuration of the outer orbital | $2s^2$ | $3s^2$ | $4s^2$ | $5s^2$ | $6s^2$ | $7s^2$ |
| Atomic radius (nm) | 0.113 | 0.160 | 0.197 | 0.215 | 0.268 | 0.235 |
| Ionic radius (2+) (nm) | 0.030 | 0.065 | 0.10 | 0.11 | 0.135 | 0.14 |
| Ionization potential (eV) | | | | | | |
| 1st | 9.32 | 7.64 | 6.10 | 5.70 | 5.20 | 5.28 |
| 2nd | 18.21 | 15.03 | 11.87 | 11.03 | 10.0 | 10.15 |
| 3rd | 153.8 | 8.02 | 51.2 | – | – | – |
| Electrode potential (V) | –1.847 | –2.365 | –2.866 | –2.888 | –2.905 | –2.92 |
| Density | 1.85 | 1.74 | 1.54 | 2.57 | 3.7 | ~6 |
| Melting point (°C) | 1,280 | 650 | 850 | 770 | 725 | 700 |
| Boiling point (°C) | 2,900 | 1,105 | 1,480 | 1,380 | 1,600 | 1,140 |

### 19.4.1 Beryllium (Be)

Beryllium, Be ($Z$ = 4, $A$ = 9.012), is a gray metal. The most common mineral of beryllium is beryl, $Be_3Al_2O_3(SiO_2)_6$. Metallic beryllium is prepared by electrolysis of molten chloride. Metallic beryllium has many remarkable properties. Beryllium finds applications in production of corrosion-free alloys and in aircraft engineering due to its lightness, high mechanical strength and melting point. Beryllium is one of the best moderators and reflectors of neutrons in high-temperature nuclear reactors. Thin sheet of beryllium is a good transmitter of X-rays. It is used for making window for X-ray tubes.

**TABLE 19.6:** GENERAL SOLUBILITY RULES FOR SOLUBILITY OF SALTS IN WATER

| *Salts* | *Characteristics* |
|---|---|
| Nitrates | Generally soluble |
| Salts of the alkali metal cations and of the ammonium ion ($NH_4^+$) | Generally soluble |
| Chlorides, bromides and iodides | Generally soluble, with exceptions—the halides of Ag(I), Hg(I), Pb(I), and $HgI_2$ |
| Sulfates and chromates | Generally soluble, with exceptions—Ag(I), Hg(I), Pb(II), Ca(II), Sr(II), and Ba(II), and also $CaCrO_4$ |
| Carbonates, fluorides, hydroxides, oxides, phosphates, sulfide, and sulfites | Generally insoluble, with exceptions—Salts of the alkali metal cations and of the ammonium ion ($NH_4^+$); fluorides of Ag(I), Al(III), Hg(II), and Sn(II); hydroxides and oxides of Ca(II), Sr(II), and Ba(II); and sulfides of Mg(II), Ca(II), Sr(II), Ba(II), Al(III), and Cr(III) |

The chemical properties of beryllium are, to a considerable extent, similar to those of aluminum. Beryllium and its compounds are toxic, causing beryllosis, dermatitis, granulomes in the lung, alveolar wall thickening and lung cancer.

Spectrophotometric, ring-oven, fluorimetric, AAS and ICP methods are employed to determine beryllium.

1. *Spectrophotometric method*: Aluminum reagent is added to a buffered beryllium containing sample solution. A small amount of ethylenediaminetetraacetic acid (EDTA) is added to prevent interference from Al, Co, CU, Fe, Mn, NI, Ti, Zn and Zr. The absorbance of the red colored complex is measured at $\lambda$ = 515 nm.
2. *Ring-oven method*: The measurement of beryllium by the Weisz ring-oven method on paper tape samples is simple and sensitive. Beryllium dust + ammonium acetate mixture is washed to the heated zone of the filter paper with deionized water. Interference from other pollutants is suppressed by adding EDTA solution. Addition of morin reagent (3,5,7,2,4-pentahydroxyflavone) to the ring zone and washing it with ammonia and methanol gives a yellow-green fluorescent product, which is measured colorimetrically. This method is suitable for field studies.
3. *Fluorimetric method*: Beryllium reaction (fused in a bead of sodium fluoride) with morin produces bright yellow fluorescence (sensitivity ~0.01 μg/L). Another method is based on the fluorescence of the chelate formed with 1-Amino-4-hydroxy-anthraquinone (~0.2 ppm) or with 8-hydroxyquinaldine (~0.001 ppm).
4. *AAS & ICP methods*: In AAS method, Be is measured at $\lambda$ = 235 nm in a nitrous oxide-acetylene flame. These two methods are very often chosen for beryllium determination.

### 19.4.2 Magnesium (Mg)

Magnesium, Mg (Z = 12, A = 24.312), a silvery, light metal, is the 8th element in abundance (2.5 per cent of the Earth's crust), after aluminum and iron. It is quite widespread in nature. *Magnesite*, $MgCO_3$, and *dolomite*, $MgCO_3 \cdot CaCO_3$ are the two common minerals of magnesium. It is commercially produced by the electrolysis of molten magnesium chloride. The main uses of magnesium are in the production of variety of alloys (in rocket and aircraft engineering) and in organic synthesis. Its silicates—talcum and asbestos find many applications. Magnesium is an important part of living tissues, especially chlorophyll in green plants.

Magnesium is a powerful reducing agent. It readily dissolves in acids, but does not react with alkalis. Magnesium contributes to hardness of water. Removal of hardness is achieved by chemical softening, reverse osmosis, electrodialysis and ion exchange.

Gravimetric, fluorescence, AAS and ICP methods may be employed to determine magnesium.

1. *Gravimetric method*: After the removal of salts of calcium and other interfering elements, diammonium hydrogen phosphate is added, which precipitates magnesium as magnesium ammonium phosphate. The precipitate is washed and weighed as magnesium pyrophosphate.
2. *Fluorescence method*: An alcoholic solution of Mg-chelate of 8-hydroxyquinoline exhibits fluorescence ($\lambda_{ex}$ = 420 nm, $\lambda_{em}$ = 530 nm) at pH 6.5. This method is now a standard clinical procedure for determination of $Mg^{2+}$ in body fluids (sensitivity ~ 0.01 ppm). Even a more sensitive method is by fluorescence of bis-salicylidene ethylenediamine chelate (~0.0002 ppm).

### 19.4.3 Calcium (Ca)

Calcium, Ca ($Z = 20$, $A = 40.08$), is the 5th most abundant element in the Earth's crust (~3.6 per cent by mass). It is a ductile, hard white metal, and is one of the most widespread elements in nature. It occurs as carbonates (*limestone, marble, calcite and chalk*), sulfate (*gypsum* ($CaSO_4 \cdot 2H_2O$)), fluorides (*fluorspar*), phosphate (*apatite*), and silicate.

It is chemically a very active metal that readily combines with halogens, sulfur and nitrogen. Because of its high chemical activity, it is employed for producing metals such as chromium, cesium, rubidium, uranium, and zirconium from their compounds. Calcium contributes to the total hardness of water. Removal of hardness involves chemical softening, reverse osmosis, electrodialysis and ion exchange.

Calcium is an important constituent of living organisms, found in bones, shells and teeth. Calcium is also involved in neuromuscular excitability and nervous impulse transmission. Some of the calcium-related diseases are Paget's disease (bone deformation and fracture), osteoporosis. Hypocalcaemia results in hyporparathyroidism, excitation of neuromuscular activity (muscle cramps, spasms and convulsion), and renal failure; hypercalcaemia causes hyperparathyroidism, Hodgkin's disease, cardiac arrest, myeloma and leukemia.

Titrimetric, AAS and ICP methods may be employed to determine calcium.

1. *Titrimetric methods*: Calcium can be titrated with EDTA at high pH ~12, using eriochrome black-T as indicator. The color change is from red to blue.
2. *Ca-permanganate method*: Calcium oxalate is precipitated by excess ammonium oxalate from ammonium acetate solution buffered to pH 4.5-5.0. The precipitate is dissolved in dilute $H_2SO_4$ and the liberated oxalic acid is determined by permanganate titration. This method is applicable for wet- or dry-ashed inorganic materials and biological materials.
3. *Fluorescence method*: Fluorescence of $Ca^{2+}$-calcein chelate (sensitivity ~0.2 ppm) is used for its determination.

### 19.4.4 Strontium (Sr) and Barium (Ba)

Strontium, Sr ($Z = 38$, $A = 87.62$), and barium, Ba ($Z = 56$, $A = 137.34$), are silvery-white metals. They occur in nature mainly as carbonates and sulfates. Metallic strontium and barium are very active and react with water. $BaSO_4$ is very sparingly soluble in water and this is taken advantage of in the chemical analysis by gravimetric method. Strontium and barium hydroxides are strong bases. Barium sulfate finds medical applications (barium meal in X-ray diagnosis).

Flame emission, AAS and ICP methods are the preferred methods for determining strontium and barium. Strontium emission is measured at $\lambda = 461$ nm.

All the isotopes of radium are radioactive, of which $^{226}Ra$ (1,600-year half-life) and $^{228}Ra$ (5.8-year half-life) are the important ones. Radium, being chemically similar to calcium, tends to concentrate in bone where α-radiation interferes with red corpuscle production. This can lead to bone cancer. Determination of radium is by methods that come under radiochemistry and nuclear methods (see Chapter 25).

## 19.5 ZINC SUBGROUP METALS

The elements of the second subgroup of Group II of the Periodic Table (IIb), *zinc, cadmium,* and *mercury*. All three are silvery-white metals with relatively low melting points and boiling points. These elements have two electrons in the outer orbital, with configuration $d^{10}ns^2$ (Table 19.7). The $d^{10}$ electrons do not take part in the chemical bonding, but they shield $s^2$ electrons from the nucleus. The zinc group elements exhibit an oxidation state of +2 in their compounds.

**TABLE 19.7:** SOME PHYSICAL PROPERTIES OF ZINC SUBGROUP METALS

| *Parameter* | *Zn* (30) | *Cd* (48) | *Hg* (80) |
|---|---|---|---|
| Electronic configuration | $3d^{10}4s^2$ | $4d^{10}5s^2$ | $5d^{10}6s^2$ |
| Atomic radius (nm) | 0.139 | 0.156 | 0.160 |
| Ionic radius (nm) | 0.075 | 0.1 | 0.11 |
| Ionization potential (eV) | | | |
| 1st | 9.36 | 8.96 | 10.2 |
| 2nd | 17.96 | 16.90 | 18.65 |
| 3rd | 40.0 | 38.0 | – |
| Electrode potential (V) | –0,763 | –0.403 | 0.85 |
| Density (at 20° C) | 7.13 | 8.65 | 13.546 |
| Melting point (°C) | 419.5 | 321 | –38.9 |
| Boiling point (°C) | 906 | 767 | 356.7 |

### 19.5.1 Zinc (Zn)

Zinc, Zn ($Z = 30$, $A = 65.38$), is a white metal with a bluish-gray luster, and is extracted from the minerals such as *sphalerite, calamine, smithsonite* ($ZnCO_3$) and *zinc blende* (ZnS). It has a great variety of applications. Brasses are alloys with zinc and copper as major constituents. Zinc is used in the manufacture of galvanic cells. Zinc sulfide (ZnS) and zinc oxide (ZnO) luminescence, emitting cold light under the action of radiant energy. These compounds, called *luminophors*, are used in cathode-ray tubes, television sets and fluorescent screens. The main sources of zinc pollution in the environment are industrial waste, metal plating and plumbing.

Colorimetric, ring-oven, fluorescence, AAS and ICP methods are employed to determine zinc.

1. *Dithizone method*: Nearly twenty metals can react with dithizone to produce colored coordination compounds. These thizonates are extractable into organic solvents, such as

$CCl_4$. Interference from Au, Ag, Bi, Cd, Co, Cu, Hg, Ni, Pa, Pb, and Sn, in the Zn-thizone reaction, can be overcome by adding sodium thiosulfate and adjusting pH to 4.0-5.5.

2. *Ring-oven method*: Zn in the dust is solubilized by KCN to form zinc tetracyanate complex. The complex is washed to the ring zone of the filter paper by distilled water and when the solvent evaporates, the complex is deposited as a sharply defined ring. A solution of ammonium thiosulfate is also washed to the ring zone. The determination of Zn present is performed colorimetrically by spraying the ring zone with dilute solution of dithizone in $CCl_4$. Absorbance of the intense purple-red complex is measured at $\lambda$ = 520, 540 and 620 nm.
3. *Fluorescence method*: Fluorescence measurement of Zn-chelates is a highly sensitive method. Zn complexed with p-tosyl-8-aminoquinoline (sensitivity ~0.02 ppm) and 2, 2'-methylene-dibenzothiazole (~0.002 ppm) yields strong fluorescence.
4. *AAS & ICP methods*: In AAS method (furnace method is not suitable), Zn is determined at $\lambda$ = 214 nm, with air as oxidant (~0.02 μg/mL).

### 19.5.2 Cadmium (Cd)

Cadmium, Cd ($Z$ = 48, $A$ = 112.40), is a bluish-white rare metal. It is occurs associated with zinc, lead and copper minerals, and resembles zinc in many of its properties. Cadmium has many industrial uses. It is used in electroplating, Ni-Cd batteries and paint and colored glass manufacture, and semiconductors. Cadmium has a high thermal neutron absorption cross-section. This is the reason why cadmium rods are used in nuclear reactors for controlling the rate of the nuclear chain reaction.

Cadmium forms soluble complexes with ammonia and halides (pH > 6.7) and insoluble precipitate with ions, such as carbonate, arsenate, phosphate and sulfide. Sources of cadmium pollution are mining and industrial discharge. Exposure to cadmium can cause renal tubular damage resulting in an aminoaciduria. Cadmium from the soil enters animal kingdom through plants. The major sources of cadmium in food items appear to be grains and seafood.

Spectrophotometric, ring-oven, fluorescence, AAS and ICP methods are employed to determine cadmium.

1. *Dithizone method*: $Cd^{2+}$ ions under suitable conditions react with dithizone to form a pink-red colored complex that is measured at $\lambda$ = 518 nm.
2. *Ring-oven method*: The dust of the sample is processed by treatment with KCN, leading to cadmium cyanide complex. The complex is washed to the heated zone of the filter paper, where it is deposited as a sharply defined ring. Interference due to other pollutants is suppressed by adding sodium thiosulfate. Addition of ferrous dipyridyl iodide gives a colored complex; cadmium can be determined from the absorbance.
3. *Fluorescence method*: Fluorescence of Cd-chelate of p-tosyl-8-aminoquinoline (sensitivity ~0.02 ppm) can be used for determining cadmium.
4. *Electrochemical luminescence (ECL)*: Cd(II) ions in water can be determined by electrochemical luminescence (ECL) method after complexing with 1,10-phenanthroline (50-1,000 ppm).
5. The AAS and ICP methods provide enough sensitivity for determining traces of cadmium. Cd in acidic medium is vaporized by flame (air/acetylene) or graphite oven (graphite furnace is preferred) and measured (at $\lambda$ = 228.8 nm; sensitivity ~0.025 μg/mL).

### 19.5.3 Mercury (Hg)

Mercury, Hg ($Z = 80$, $A = 200.6$), is a silvery metal and is scarce in nature. It occurs chiefly in the form of HgS (*cinnabar*). It is the only metal that is liquid at room temperature (Cs melts at 28.5° C, gallium at 30° C and rubidium at 39° C). It is widely used in chemical (drugs, fungicides, pigments and paints), industrial (textile, paper, leather, ammunition and detonators) and laboratory equipment manufacture (thermometers). Mercury can occur in three oxidation states—0, +1 (mercurous) and +2 (mercuric). In oxygen-containing atmosphere, 0 and +2 states are by far the most stable. $Hg_2Cl_2$ (calomel) is the most important univalent compound.

Major sources of mercury pollution are anthropogenic—mining, industrial effluents and pesticides. Control methods for eliminating mercury include precipitation with $S^{2-}$, ion exchange, filtration, and adsorption on activated carbon or scrubbing. The most common method used is scrubbing by acidic permanganate solution.

Gravimetric, volumetric, spectrophotometric, amperometric, AAS, NAA, XRF and other methods are employed to determine mercury. Organomercury compounds carry C-Hg bonds and are highly volatile. The volatility of mercury species makes gas chromatography (GC) well suited to their speciation (sensitivity ~1 ng/L).

1. *Gravimetric method*: Hg is collected as an amalgam, $HgCl_2$ or HgS, and analyzed gravimetrically (sensitivity 50 μg/L; volumetrically ~10 μg/L).
2. *Dithizone method*: Mercury reacts with dithizone in chloroform to form an orange colored complex, the absorbency of which is measured at $\lambda = 492$ nm (sensitivity ~2 μg/L).
3. *Elemental mercury*: Determination of mercury in gas samples may be carried out by first converting mercury to an amalgam with silver. The amalgam is heated it at 800 °C and Hg is flushed through the cell of a spectrophotometer and absorbance measured at $\lambda = 253.7$ nm (Fig. 19.2). Sensitivity is ~0.01 ng/L.

   Standard technique for measurement of elemental mercury in water samples consists of bubbling vaporous mercury out from water samples using an inert gas, trapping of mercury on the Ag/Au by amalgamation, heating the traps to release the amalgamated Hg and measuring it by cold vapor AAS. The method can be automated and near-real time (temporal resolution ~5 min) continuous measurement of dissolved gaseous mercury (DGM) is possible (sensitivity ~10 pg/L).
4. All forms of mercury can be converted into ionic form by the addition of a strong acid ($HNO_3$ or KI + sulfonic acid) and an oxidizing agent ($K_2Cr_2O_7$). Then an excess of reducing agent ($SnCl_2$) liberates elemental mercury, which is recorded by AAS at $\lambda = 253.6$ nm) (Fig. 19.3). Cold vapor AAS is the method of choice (sensitivity ~0.01 μg/L).
5. *Organic mercury*: Methyl mercury ($HgCl_3$) in fish is extracted with benzene. It is stripped into an aqueous cysteine solution, acidified, back-extracted into benzene and the benzene extract is examined by gas chromatography (GC). Organomercurials ($HgCH_3$) in water are converted into their chlorinated derivatives ($HgCH_3Cl$) to facilitate extraction with an organic solvent and separation and detection by GC-ECD (sensitivity 0.01-0.1 μg/L).
6. *Other methods*: Amperometry, polarography (sensitivity ~20 μg/L), ion selective electrode (~0.02 μg/L), mass spectrometry (MS; ~0.02 μg/L), emission spectrometry (~0.02 μg/L), XRF (~0.01 μg/L) and neutron activation analysis (NAA; ~0.001 μg/L) serve as sensitive methods for determination of mercury. Quenching of the fluorescence of rhodamine B by mercury is a very sensitive method (~0.1 ppm). Microwave induced plasma, as an emission source, is a very sensitive method for total mercury determination (~15 pg/L). The system contains a flow injection (FI) section in which mercury is reduced and deposited on an Au/Pt collector, and carried to the emission source.

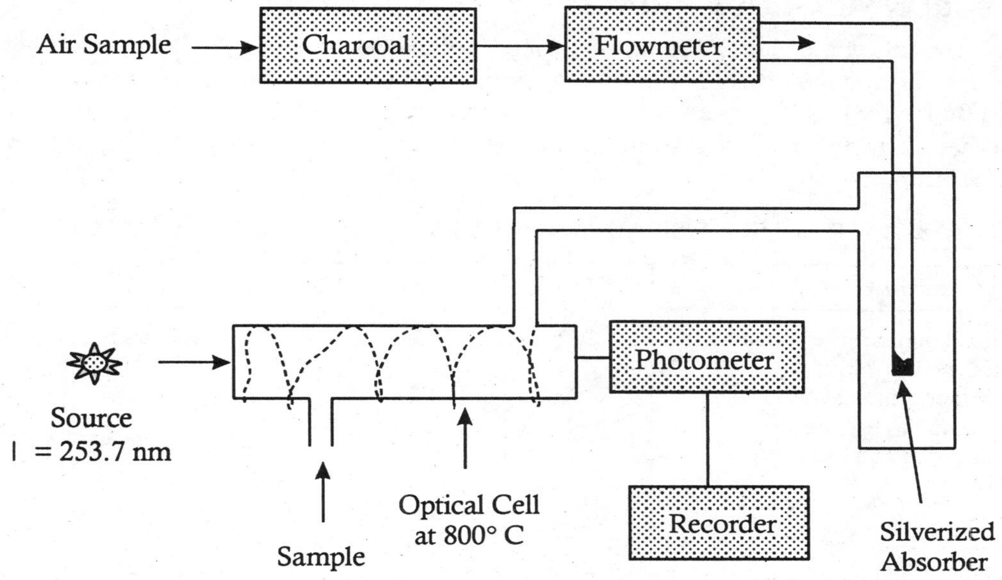


**Fig. 19.2:** Determination of Elemental Mercury in Gas Samples

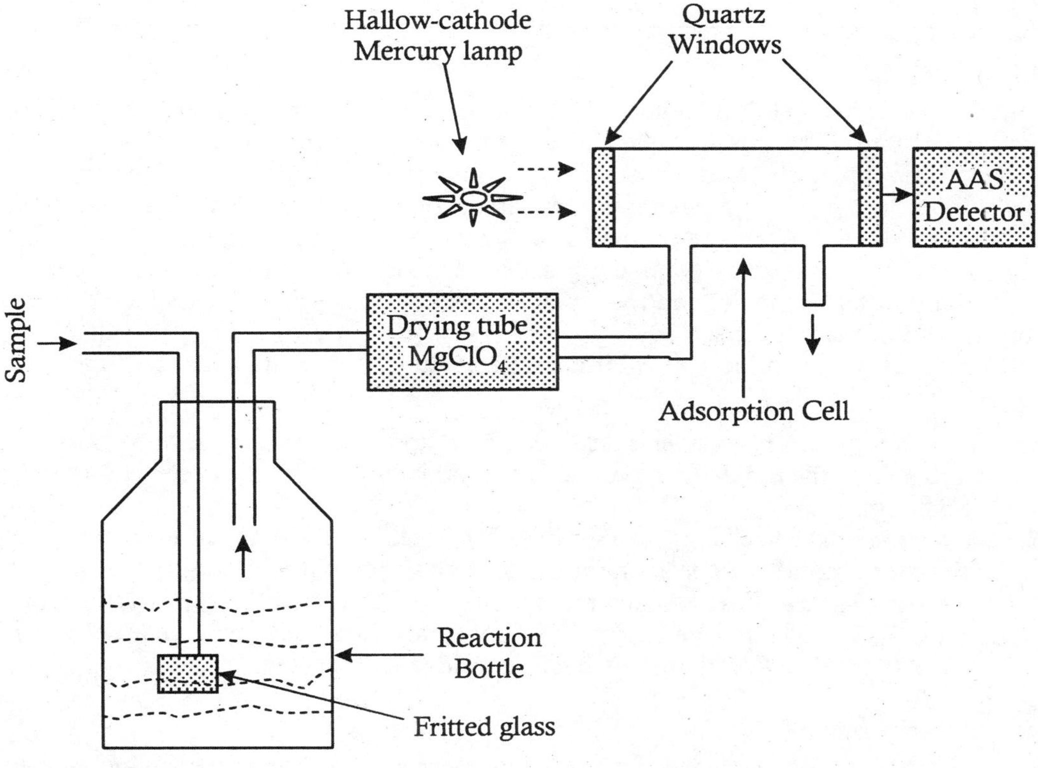


**Fig. 19.3:** Setup for Measuring Mercury by Cold Vapor-AAS Method

## 19.6 BORON SUBGROUP METALS

The five elements of the main subgroup of Group III of the Periodic Table (IIIa) are *boron, aluminum, gallium, indium,* and *thallium*. They are characterized by the presence of three electrons in the outer orbital (Table 19.8), and they exhibit the oxidation states, +1, +2 and +3. Boron is a nonmetal, while other elements are silvery-white metals. Most important elements of this group are boron and aluminum.

**TABLE 19.8:** SOME PHYSICAL PROPERTIES OF BORON SUBGROUP ELEMENTS

| *Parameter* | *B* (5) | *Al* (13) | *Ga* (31) | *In* (49) | *Tl* (81) |
|---|---|---|---|---|---|
| Configuration of electrons in the outer orbital | $2s^22p^1$ | $3s^23p^1$ | $4s^24p^1$ | $5s^25p^1$ | $6s^26p^1$ |
| Atomic radius (nm) | 0.09 | 0.13 | 0.14 | 0.16 | 0.17 |
| Ionic radius (+3 ion) (nm) | 0.02 | 0.05 | 0.06 | 0.08 | 0.10 |
| Ionization potential (eV) | | | | | |
| 1st | 8.3 | 6.0 | 6.0 | 5.8 | 6.0 |
| 2nd | 25.1 | 18.8 | 18.9 | 18.9 | 20.4 |
| 3rd | 37.9 | 28.4 | 30.7 | 28.0 | 29.8 |
| 4th | 259.0 | 120.0 | 64.2 | 57.8 | 50.0 |
| Density | 2.34 | 2.70 | 5.90 | 7.31 | 11.85 |
| Melting point (°C) | 2,200 | 660 | 29.8 | 156.4 | 304 |
| Boiling point (°C) | 2,700 | 2,450 | 2,400 | 1,500 | 1,460 |

### 19.6.1 Boron (B)

Boron, B ($Z$ = 5, $A$ = 10.811), is comparatively scarce in nature. Principal compounds of boron are boric acid and *borax*. The non-metallic characteristics of boron are not akin to those of the other elements of this subgroup, but akin to silicon. At high temperatures boron combines with many metals to form *borides*, which are refractory materials. Boron silicates are used in special grades of glass. Boric acid is used to prepare enamels and glazes, special grades of glass, and as a disinfectant. $^{10}B$ isotope possesses an extremely large capture cross section for thermal neutrons. So this isotope can be used in the fabrication of neutron radiation shields. Boron is an essential trace metal for the healthy growth of many plants.

Boron is usually determined by colorimetric and fluorescence methods. ICP method is also employed.

1. *Curcumin method*: The sample is acidified and evaporated in the presence of curcumin. The absorbance of the red-colored product, rosocyanine, is measured spectrophotometrically at $\lambda$ = 540 nm.
2. *Carmine method*: The change in color from bright red to blue of acidic carmine is a function of the concentration of boron present, measured spectrophotometrically.
3. *Fluorescence method*: Fluorescence measurements of chelates of boron with benzoin ($\lambda_{ex}$ = 365 nm, $\lambda_{em}$ = 480 nm) (sensitivity ~0.01 ppm) and with dibenzoylmethane ($\lambda_{em}$ = 385 nm, $\lambda_{em}$ = 410 nm) (~0.0005 ppm) are highly sensitive and specific.

### 19.6.2 Aluminum (Al)

Aluminum, Al ($Z$ = 13, $A$ = 26.982), is a silvery-gray metal and is the most abundant element in the Earth's crust. It is contained in clays, feldspars, micas, and many other minerals. Its chief ore is bauxite. Aluminum based alloys like duralumin and magnalumin are used in industry. Ruby crystals containing a small admixture of $Cr_2O_3$ are used as quantum generators—lasers. Alum is used in for tanning hides and other industrial purposes.

Aluminum atom has free d-sublevels in the outer electron orbital. Consequently, the coordination number of aluminum and its compounds may be three, four or six (+3, +4 and +6). The oxidation state of aluminum in biological samples is +3.

Spectrophotometric, fluorescence, ion chromatography and AAS methods are employed to determine aluminum.

1. *Erichrome cyanine method*: With erichrome cyanine, aluminum, at pH = 6.0, produces a red-pink colored complex. Its absorbance can be measured at $\lambda$ = 535 nm.
2. *Pyrocatechol violet method*: Pyrocatechol violet reacts with aluminum to produce a colored complex (pH = 6.0), and the intensity of color is measured at $\lambda$ = 580 nm.
3. *Fluorescence method*: Fluorescence measurement of chelates of aluminum with reagents acidic Alizarin Garnet R (sensitivity ~5 ppb), Mordant Blue 9 (~0.5 ppb), Salicyclidene-o-aminophenol (~0.2 ppb), 8-quinolinol and Morin are highly specific and sensitive.
4. *AAS method*: In a nitrous oxide-acetylene flame Al can be determined at $\lambda$ = 309.3 nm.

## 19.7 GALLIUM (GA), INDIUM (IN) AND THALLIUM (TL)

*Gallium*, Ga (Z = 31, A = 69.72), *indium*, In (Z = 49, A = 114.82), and *thallium*, Tl (Z = 81, A = 204.37), are metals belonging to the boron group (Group IIIa) of the Periodic Table. These elements occur in nature in minute concentrations. Gallium exists in liquid state from 30° C to 2,205° C. It is used in high-temperature thermometers (30-1,500° C). These metals are used in semiconductor electronics. Indium is an amphoteric element. Thallium halides are good transmitters of infrared radiation and are used in infrared detectors. Thallium sulfide is employed in photoelectric cells. Thallium and its compounds are very toxic.

Flame AAS and ICP techniques are employed to determine these metals. Fluorescence of their chelates provides sensitive methods for their determination. Rhodamine B (sensitivity ~0.01 ppm), sulfonaphtholazoresorcinal (~0.001 ppm) and 8-hydroxyquinoline (~0.04 ppm) yield fluorescent complexes with gallium and they are exploited for its determination. Thallium can be determined fluorimetrically in HCl solution (sensitivity ~0.01 ppm).

## 19.8 GERMANIUM (GE), TIN (SN) AND LEAD (PB) ELEMENTS

*Germanium* (Ge), *tin* (Sn) and *lead* (Pb) are the metallic elements in the carbon group elements (Group IVa) of the Periodic Table. Carbon and silicon are nonmetals, and though germanium is included among metals it has characteristics of both. The substantial changes in the physical properties from carbon to lead are consequences of increase in atomic size with substantial intervening electronic shells and electronegativity power to attract electrons with increasing atomic number. For example, the attraction between the nucleus of lead and its outermost electrons is less than in carbon; because intervening shells in lead shield the outer electrons. The electronic configuration of this group is $ns^2np^2$ (Table 19.9), and the oxidation states of –2, +2 and +4 are characteristic of the elements of this group.

### 19.8.1 Germanium (Ge) and Tin (Sn)

Germanium, Ge ($Z$ = 32, $A$ = 72.59), exhibits both metallic and non-metallic characteristics (metalloid). Germanium has silvery color and has appearance of a metal. It has semiconductor properties, and this underlies its chief application. $GeO_2$ is used in the composition of glasses having high refractive index in the infrared part of the spectrum. Germanium is determined taking advantage of its fluorescence (greenish-yellow) as benzoin chelate (sensitivity ~2 ppm).

**TABLE 19.9:** SOME PHYSICAL PROPERTIES OF GROUP (IVA) ELEMENTS

| *Parameter* | *C* (6) | *Si* (14) | *Ge* (32) | *Sn* (50) | *Pb* (82) |
|---|---|---|---|---|---|
| Configuration of electron in the outer orbital | $2s^2 2p^2$ | $3s^2 3p^2$ | $4s^2 4p^2$ | $5s^2 5p^2$ | $6s^2 6p^2$ |
| Atomic radius (nm) | 0.077 | 0.124 | 0.13 | 0.15 | 0.175 |
| Ionization potential (eV) | 11.26 | 8.15 | 7.90 | 7.34 | 7.42 |
| Density (at 20° C) | 2.26-3.52 | 2.33 | 5.32 | 5.8-7.3 | 11.34 |
| Melting point (°C) | 3,600 | 1,420 | 950 | 232 | 327.5 |
| Boiling point (°C) | 4,800 | 2,350 | 2,850 | 2,260 | 1,740 |

### 19.8.2 Tin (Sn)

Tin, Sn ($Z$ = 50, $A$ = 118.69), is a silvery-white metal. Alloys of tin, with copper (bronzes) and antimony are used in the manufacture industrial components. Tin exists in two forms. Gray tin (α-tin) is easily crumpled, whereas white tin (β-tin) is malleable and ductile. The spontaneous conversion of the bright metal to a gray powder at low temperatures ("tin pest") seriously hampers the use of tin metal in very cold regions. Organotin (tributyltin) compounds are highly toxic. Trioraganotin compounds are used as fungicides and insecticides.

Tin may be determined by measuring the fluorescence of Sn chelate with 8-hydroxy quinoline-5-sulfonic acid ($\lambda_{ex}/\lambda_{em}$ = 360/515 nm) (sensitivity ~0.01 ppm). Organotin compounds are determined by hydride generation and absorption spectrophotometry. GC or HPLC methods can be employed for separation of organotins, and determined by flame and electrothermal AAS (ET-AAS) and ICP-MS methods. Supercritical fluid chromatography (SFC)-FID technique is a versatile separation method for many metals (see Chapter 22).

### 19.8.3 Lead (Pb)

Lead, Pb ($Z$ = 82, $A$ = 207.19), is a bluish-white, very soft metal. The most important ore of lead is *galenite* (PbS). Lead is used in many engineering and chemical industries—lead batteries, type metals and shot metal, sulfuric acid plants, soldering, automobile industry, radiation shields, pigments and paints, and lead-arsenic pesticides.

All soluble compounds of lead are poisonous. Toxicity increases as solubility increases. Solubility of lead increases with acidity of the medium. Therefore, acidified streams and lakes due to mine leaching and or rain would magnify the lead poisoning of aquatic life. Lead pollution in water may occur due to natural causes as well as domestic and industrial discard (plumbing, soldering waste). Lead in wastewater is removed by chemical precipitation at pH > 6.0 using $CaCO_3$, by activated carbon adsorption, electrodeposition or ion exchange.

Stable isotopes of lead (normal isotopes of lead) are increasingly used in environmental science as tracers and anthropogenic Pb-sources. Due to its extensive uses in industry and its toxicity, accurate determination of it is very essential. Spectrophotometric, ring-oven, fluorescence, polarographic, anode stripping voltammetric (ASV) methods are used for determination of lead. GC or HPLC with an element-selective detector (AAS, AES or MS), and XRF and NAA methods are also used for this purpose. Of these methods, isotopic characterization methods provide high precision.

1. *Spectrophotometric methods*: Acidic lead-containing solution is treated with alkaline citrate-cyanide and extracted with dithizone in chloroform. The red color of the lead thiozonate is measured at $\lambda = 510$ nm.
2. A simple, ultra-sensitive and fairly selective non-extractive spectrophotometric method is with 2, 5-dimercapto-1, 3, 4-thiadiazole (DMTD). Acidic DMTD reacts with lead(II) to give a greenish-yellow chelate, which has an absorption maximum at 375 nm. This method has the required precision and accuracy and is useful for determination of trace amounts of lead in waters, soil samples and biological materials.
3. Another useful method is based on the formation of lead(II) chelate with 2-(5-bromo-2-pyridylazo)-5-dimethylaminophenol in ethanolic medium. Use of HPLC with flow injection (FI) analysis leads to substantial improvement in the analytical performance of the method.
4. *Ring-oven method*: The filtered sample of lead in the dust is digested with $HNO_3$, $H_2SO_4$ and $HClO_4$ to solubilize lead. Acidified $Pb^{2+}$ is made alkaline by treatment with ammonium acetate. The lead acetate is washed to the heated ring-zone of the filter paper where it is deposited as a sharply defined ring. KCN is added to mask interference due to other pollutants. The determination of Pb is performed colorimetrically by spraying dithizone in chloroform, which forms a brick-red complex. The color of Pb-dithizone is compared against standards prepared in a similar manner or measured at $\lambda = 510$ nm.
5. *Fluorescence method*: Lead shows fluorescence in HCl or HBr, measured spectrophotometrically. The method is very sensitive (0.01 ppm).
6. *AAS method*: Sample is wet-oxidized by $HNO_3$ and $H_2SO_4$ or ashed. Lead (and Cd) is selectively complexed by chelating agents, such as ammonium pyrrolidine dithiocarbamate into 4-methylpentan-2-one and extracted. The solution is vaporized by flame (air/acetylene), or graphite furnace (argon/nitrogen at 2,400° C) and the presence of lead is monitored by AAS ($\lambda = 283.3$ nm for Pb, and 228.8 nm for Cd).
7. *Polarography and anode stripping voltametry*: A sensitive method for determination lead(II) with 1,4-bis(proo-2-enyloxy)9-10-anthraquinone (AQ) modified carbon paste electrode is based on non-electrolytic pre-concentration, followed by accumulation period with negative potential, and then a proper anodic stripping. This method can be applied to determination of lead in wastewaters (detection limit ~1 ng/L).
8. XRF and NAA are other methods of analysis of lead.

## 19.9 TITANIUM SUBGROUP METALS

The titanium subgroup—*titanium* (Ti), *zirconium* (Zr) *and hafnium* (Hf), are metals that belong to the secondary subgroup of Group IV, (IVb), of the Periodic Table. The titanium subgroup elements have $(n-1)d^2ns^2$ electronic configuration. The most characteristic oxidation state of this subgroup metals is +4. In the free state, titanium subgroup elements are stable, corrosion resistant, and are refractory with high melting and boiling points (Ti = 1,680° C/3,280° C; Zr = 1,860° C/4,370° C and Hf = 2,150° C/5,400° C).

Titanium ($Z = 22$, $A = 47.88$) is a white metal. The characteristics of titanium, namely resistance to corrosion, high structural strength, low density (half as dense as iron), make it suited for many industrial applications. The metal is used in stainless steel production, internal combustion engineering, aircraft, spacecraft and missiles. Titanium combines with carbon to produce hard titanium carbide. Titanium dioxide ($TiO_2$) is used in the manufacture of refractory glasses, glazes, and enamels. Barium titanate ($BaTiO_3$) is used in high-capacity electrical capacitors, ultrasonic apparatus, and hydroacoustical equipment. Titanium oxide is used as white pigment.

Zirconium ($Z$ = 40, $A$ = 91.22) is found in various igneous rocks, usually as zirconium orthosilicate, zircon. Zirconium does not absorb slow (thermal) neutrons. The high transparency to neutrons, combined with its high resistance to corrosion and high mechanical strength at elevated temperatures place zirconium and its alloys among the principal materials for cladding fuel rods of nuclear power reactors. High affinity of zirconium for gases (oxygen, nitrogen and hydrogen) finds use for it as a "getter" material. $ZrO_2$ is used as abrasive, and refractive material.

Hafnium ($Z$ = 72, $A$ = 178.49) is found in zirconium minerals. It has high neutron absorption cross-section. It is used as control rod material in water-cooled nuclear reactors. It forms alloys with iron, tantalum, titanium and other transition metals. The alloy tantalum-hafnium carbide ($Ta_4HfC_5$) is one of the most refractory materials known (M.P. ~4,300° C).

Flame AAS is the method of choice for the determination of these metals. Zirconium (Zr) may be determined by blue-white fluorescence ($\lambda_{ex}^2/\lambda_{em}$ = 390/465 nm) exhibited when flavanol is added to a slightly acidic solution containing zirconium.

## 19.10 NITROGEN SUBGROUP METALLOIDS

*Arsenic* (As), *antimony* (Sb) and *bismuth* (Bi) belong to the main subgroup of the Group V, (Va), of the Periodic Table. These elements have five electrons in the outer electronic orbital (Table 19.10). Arsenic and antimony are metalloids, exhibiting both metallic and non-metallic properties. However, metallic properties predominate in bismuth. Arsenic and antimony exhibit –3, +3 and +5 oxidation states in their compounds. On the other hand, bismuth forms stable compounds, mainly, in the +3 oxidation state.

**TABLE 19.10:** SOME PHYSICAL CHARACTERISTICS OF GROUP V(A) ELEMENTS

| *Parameter* | *N* (7) | *P* (15) | *As* (33) | *Sb* (51) | *Bi* (83) |
|---|---|---|---|---|---|
| Electronic configuration of outer orbital | $2s^2 2p^3$ | $3s^2 3p^3$ | $4s^2 4p^3$ | $5s^2 5p^3$ | $6s^2 6p^3$ |
| Atomic radius (nm) | 0.071 | 0.13 | 0.148 | 0.161 | 0.182 |
| Ionic (+3) | 0.016 | 0.044 | 0.056 | 0.076 | 0.096 |
| Ionization potential (eV) | 14.53 | 10.49 | 9.82 | 8.64 | 7.3 |
| Density | 1.25/L | 1.83 | 5.72 | 6.68 | 9.80 |
| Melting point (°C) | –209.9 | 44.1 | 815 | 630.5 | 271.3 |
| Boiling point (°C) | –195.8 | 280. | Sublimes at 615 | 1,430 | 1,550 |

### 19.10.1 Arsenic (As)

Arsenic, ($Z$ = 33, $A$ = 74.922), is a dark-gray, brittle element with low thermal and electrical conductivity. It occurs in nature mainly as compounds with metals or sulfur. Arsenic and its compounds are used in metallurgical, agricultural and medicinal purposes (Table 19.11). Arsenic pollution is due to pesticides and chemical waste.

**TABLE 19.11:** USES OF ARSENIC AND ITS COMPOUNDS

| *Compound* | *Uses* |
|---|---|
| As | Not many applications |
| $As_2O_3$ | Glass, enamels, paints and pesticides |
| $As_2O_5$ | Glass industry; wood preservative; pesticides, herbicides and fungicides |
| Organo arsines | Pesticides and herbicides |

Free arsenic and all its compounds are strong poisons. Arsenic can be removed from water by shaking with granulate silicon alloys, precipitation, adsorption on activated carbon (or activated alumina), ion exchange and membrane filtration (see Chapter 28). Oxidation state and pH play a role in the removal efficiency of the process. Arsenic in the pentavalent form As(V) is more readily removed (pH ~10.5) than trivalent arsenic.

Spectrophotometric, fluorimetric, HPLC, electrothermal-AAS (ET-AAS). ICP-AES, ICP-MS methods are employed for the determination of arsenic.

1. *Qualitative method*: Arsenic in the sample is reduced to arsine ($AsH_3$). When it is heated arsine decomposes, and the liberated arsenic forms a characteristic black glittering deposit on the cold parts of the receiver. This is the Marsh Test and the deposit is called 'arsenic mirror'.
2. *Spectrophotometric methods*: Arsenic is reduced selectively by sodium borohydride to gaseous arsine ($AsH_3$) at pH = 6.0. A stream of nitrogen gas sweeps the gaseous product into an absorber tube containing silver diethyldithiocarbamate in pyridine.

   $$AsH_3 + Ag\cdot\text{diethyldithiocarbamate} \rightarrow \text{red-purple complex}$$

   The red color is measured at $\lambda = 520$ nm. Arsenic reacts with silver nitrate + acetic acid forming a reddish-brown complex. Arsenic is decomposed in a closed apparatus in the presence of KBr. Arsenic bromide is distilled and determined colorimetrically using molybdenum blue method.
3. *Fluorescence method*: Arsenic gives fluorescence in HCl or HBr, which is measured spectrophotometrically (sensitivity ~0.2 ppm).
4. *Fluorimetric method*: Arsenic compounds are converted to arsenite, which reacts with thiamine to form thiochrome. The thiochrome is fluorimetrically monitored at 440 nm (with excitation at 375 nm). This method is sensitive, rapid and reproducible.
5. *Electrothermal AAS & ICP-MS methods*: Direct determination of arsenic by atomic absorption spectrometry has a poor detection limit. Inclusion of on-line hydride generation-AAS method increases the sensitivity of detection and reduces interferences from matrix elements. This is the method of choice. Hydride generation can be used as pre- or post-HPLC derivatization. As a pre-column procedure, arsenic compounds are converted to their hydrides, cryogenically trapped and subsequently decomposed in an argon-hydrogen flame. In the post-column derivatization process, hydride generation is used after separation of the arsenic species (by HPLC). Sensitivity of the determination of arsenic is boosted by using hydride generation procedure, and on-line UV photooxidation detector—LC-UV-Hg-ICP-AES tandem. In this method, all arsenic species, pre-concentrated by LC are converted to As(V) either chemically (weakly acidic silver nitrate solution) or by photooxidation. It is then reduced to arsine and determined ($\lambda = 194$ nm). HG-ICP-MS is one of the most sensitive determination methods

of arsenic species, but expensive. Atomic fluorescence spectrometry (AFS), though a less sensitive method, has shorter warm up times and easy handling, and of much lower operational costs.

6. Capillary electrophoresis (CE) coupled to UV, conductometry and ICP-MS (CE-Hg-ICP-MS system) methods are also effective in the determination of arsenic compounds.

### 19.10.2 Antimony (Sb)

Antimony, Sb ($Z$ = 51, $A$ = 121.75), belongs to the main subgroup of the Group V, (Va), of the Periodic Table. It is a silver-white metalloid, with tendency towards metallic nature. It generally occurs as sulfides in the form of *stibnite* or *antimonite*, $Sb_2S_3$. It is also present in trace amounts in natural waters. Antimony is introduced into certain alloys to increase their mechanical strength and hardness (lead alloys in storage batteries). Some antimonides (e.g. GaSb) have semiconductor properties and are used in the electronic industry. Antimony sulfides ($Sb_2S_3$; $Sb_2S_5$) are used in the production of matches and in the rubber industry. Antimony compounds ($SbH_3$) are poisonous.

Antimony may be determined by spectrophotometric, fluorescence, flame AAS and ICP methods.

1. *Spectrophotometric method*: Antimony Rhodamine-B gives a red color complex with antimony(III) iodide, which is measured spectrophotometrically.
2. *Fluorescence method*: Antimony shows fluorescence in HCl or HBr solutions, measured spectrophotometrically (sensitivity ~0.001 ppm).
3. *Electrothermal-AAS* (ET-AAS) *method*: This is the method of choice for determination of antimony.

### 19.10.3 Bismuth (Bi)

The last member of the arsenic subgroup, bismuth, Bi ($Z$ = 83, $A$ = 208.98), is characterized by near metallic properties. It is a lustrous, pinkish-white, brittle metal. Pure bismuth finds application in nuclear power reactors as a heat-carrying agent. Low-melting alloys of bismuth are used in fire-detecting equipment, and as fuses. Compounds of bismuth(III) are used in medicine and in veterinary science. Soluble inorganic bismuth compounds are toxic.

Bismuth is determined by gravimetric, volumetric and spectrophotometric methods. It can be determined spectrophotometric methods by its fluorescence in HCl or HBr solutions (sensitivity ~0.002 ppm). Quinolin-8-ol and Amberlite XAD-7 resin are used for the retention of bismuth, at pH 4.5. The bismuth complex is recovered with nitric acid and analyzed. On-line determination of bismuth involves its pre-concentration; conversion to hydride and determination by flow injection inductively coupled plasma atomic emission spectrometry (HG-ICP-AES).

## 19.11 VANADIUM SUBGROUP METALS

Vanadium subgroup metals, *vanadium* (V), *niobium* (Nb), and *tantalum* (Ta), belong to the secondary subgroup of Group V, (Vb), of the Periodic Table. The metals have high melting and boiling points (V = 1,900° C/3,000° C, Nb = 2,470° C/4,900° C and Ta = 3,000° C/5,500° C) and belong to the class of refractory metals. The oxidation state of their compounds is +5. Vanadium ($Z$ = 23, $A$ = 50.942) is a silvery-white metal, used in alloying steel. Vanadium compounds are used in the chemical and glass industry, in medicine and photography. The compounds are mildly toxic. These elements are released to the atmosphere due to metallurgical operations, by thermal power plants, crude oil and coal extraction. Niobium ($Z$ = 41, $A$ = 92.906) is a soft and ductile metal, used in steel making, tools and dies and supercondctive magnets. It is used, as zirconium alloy, for cladding nuclear reactor fuel. $Nb_3Sn$ is a superconductor below 18.5 K. Tantalum ($Z$ = 73, $A$ = 180.948) is a very hard, silver-gray metal. Its high density, high melting point and resistance to all acids except hydrofluoric acid makes it useful in the manufacture of corrosion-resistant scientific, medical, dental and chemical equipment, cutting tools, aviation and spacecraft parts.

Spectrophotometric, ICP and AAS methods are employed to determine vanadium group metals.

1. *Spectrophotometric method*: Trace amounts of vanadium can be determined by the gallic acid method. The rate of oxidation of gallic acid by persulfate, under acidic conditions, is proportional to the concentration of vanadium. The absorbance is monitored at $\lambda$ = 415 nm.
2. *AAS methods*: AAS with conventional flame mode is not so sensitive. Graphite furnace (2,500° C) method eliminates interference; monitored at $\lambda$ = 318.4 nm (oxidant = nitrous oxide; sensitivity ~2μg/mL).

## 19.12 CHROMIUM SUBGROUP METALS

The chromium subgroup metals, *chromium, molybdenum,* and *tungsten* belong to the secondary subgroup of Group VI, (VIb), of the Periodic Table. The oxidation state of the compounds varies from 0 to +6, the most important being +6. Their properties are similar to the relevant sulfur compounds.

### 19.12.1 Chromium (Cr)

Chromium, Cr ($Z$ = 24, $A$ = 51.996), is a grayish metal, occurring in nature as chromite (FeO $Cr_2O_3$). Chromium is a hard, steely-gray lustrous metal (M.P. = 1,890° C; B.P. 2,480° C). It is used in plating and in manufacture of stainless, acid-resistant and heat-resistant steels (*ferrochromes*). The common oxidation states of chromium are +6, +3 and +2. Chromates are important industrial chemicals. Chromites ($Cr_2O_3$), chromates ($CrO_3$) and dichromates ($Cr_2O_7$) are used in pigments and paints. A common Cr (III) salt is the *chrome alum* ($KCr(SO_4)_2 \cdot 12H_2O$), used in the leather industry for tanning hides and in the textile industry as a mordant in dyeing. Chromates and dichromates are strong oxidizing agents and are widely used for oxidizing various substances. All salts of chromic acid are toxic. Hexavalent (Cr(VI)) compounds are 100 to 1,000 times more toxic than Cr (III) compounds.

Gravimetric, colorimetric, titrimetric, fluorescent quenching, electrothermal-AAS (ET-AAS) and ICP methods are employed to determine chromium. ICP-AES is one of the most sensitive methods (0.001-0.1 ppb). Electronspray mass spectrometry (ES-MS) and isotope dilution mass spectrometry (ID-MS) can provide information about the oxidation states of Cr (III)/Cr (VI) in solution. Most environmental analytical procedures determine Cr (III) and total chromium and calculate the Cr(VI) content by (total Chromium- Cr (III)) difference.

1. *Spectrophotometric method*: Cr(VI) is determined spectrophotometrically by reaction of chromium with acidic diphenylcarbazides. Oxidation of 1, 5-diphenylcarbazole to 1, 5-diphenylcarbazone gives rise to a red colored complex with chromium, which is measured at $\lambda$ = 540 nm.
2. Cr (VI) in water can be determinєd by first filtering the solution and adjusting the pH of filtrate to ~9 to preserve the Cr (VI) oxidation state. Cr (VI) reacts with an azide dye, the color of the product is measured at $\lambda$ = 530 nm.

3. *Fluorescent method*: Chromium may be determined by measuring quenching of fluorescence of triazinylstilbexone (sensitivity ~0.004 ppm).
4. *Electrothermal AAS method*: The sample is heated to 2,400° C; (oxidant = $N_2$; fuel = acetylene), and monitored at $\lambda$ = 358 nm (sensitivity < 0.03 μg/mL).

### 19.12.2 Molybdenum (Mo)

Molybdenum, Mo ($Z$ = 42, $A$ = 95.94), is a silvery-gray refractory metal (M.P. = 2,600° C; B.P. = 5,550° C), obtained from molybdenite ore. It is used in the manufacture of special grades of steel with superior strength. It exhibits oxidation states +2, +3, +4, +5 and +6. Molybdenum is an essential trace element in plants. Molybdenum pollution is mainly due to industrial operations.

Spectrophotometric, fluorescence and AAS methods are employed to determine molybdenum.

1. *Wet-chemical method*: Molybdenum is co-precipitated with $MnO_2$, and solubilized by adding KCN, and determined by spectrophotometry at $\lambda$ = 465 nm.
2. *Fluorescence method*: More sensitive method is fluorescence measurement of Mo-carminic acid chelate (sensitivity ~0.1 ppm).
3. *AAS method*: In AAS method, Mo is monitored at $\lambda$ = 313 nm (flame = nitrous oxide/butylene).

### 19.12.3 Tungsten (W)

Tungsten, W ($Z$ = 74, $A$ = 183.85), is a gray-white heavy metal obtained from wolframite mineral. Among metals it has the highest melting point, 3,400° C (B.P. = 5,900° C), highest tensile strength, and lowest coefficient of linear expansion. Tungsten is used in metallurgy for making high-speed steels (tungsten steels) and alloys. It is the most refractory metal, and is used in a number of heat-resistant alloys. It is used in incandescent lamps, radio electronics, cutting tools, and armor plates for battle tanks and X-ray equipment. Tungsten carbide is next to diamond in its hardness (9.5 in Mohs scale) and is used for hard and tough dies and tools.

Chemically, tungsten is relatively inert, with valence +2 to +6, being most common. Tungsten compounds are very similar to those of molybdenum and same methods of analysis are employed to determine tungsten. Tungsten (W) may be determined by fluorescence measurement of tungsten-carminic acid complex (sensitivity ~0.3 ppm).

## 19.13 MANGANESE SUBGROUP METALS

This group includes the elements of the secondary subgroup of Group VII—*manganese* (Mn), *technetium* (Tc), and *rhenium* (Re). Manganese (Mn) is of great practical importance. Technetium (Tc) is not found in nature. It is produced artificially. Rhenium (Re) has a high melting point, 3,190° C, and rhenium and its alloys with tungsten and molybdenum are used in the production of electric lamps. Alloys of tungsten with rhenium are employed as thermocouples that can be used to measure temperature range from 0 to 2,500° C.

### 19.13.1 Manganese (Mn)

Manganese, Mn ($Z$ = 25, $A$ = 54.938), is a gray-white, hard, brittle metal widely distributed in nature and is 12th in abundance. The most widespread manganese-containing mineral is *pyrolusite* ($Mn_3O_2$). Its main application is in the production of ferro-alloys. Manganese is essential to plant growth and is involved in the reduction of nitrate in green plants and algae. It is an essential trace element for higher animals too. It participates in the action of many enzymes. Manganese pollution (Mn-oxides) is mainly due to metallurgy, ferrous metal industry and industrial waste.

Spectrophotometric, fluorescence, and ICP-AAS methods are employed to determine manganese.

1. *Persulfate method*: Manganese compounds are oxidized to permanganates by persulfate ($Mn \rightarrow MnO_4^-$ by $K_2S_2O_8$) in the presence of silver ions as catalyst. The resulting colored complex is measured at $\lambda$ = 525 and 545 nm.
2. *Formaldoxime method*: Manganese compound is treated with formaldoxime + ferrous ammonium sulfate reagent, and the resulting absorbance is measured at l = 480 nm.
3. *Fluorescence method*: Manganese can be determined by measuring the fluorescence of its chelate, Mn-8-hydroxyquinoline-5-sulfonic acid ($\lambda_{ex}/\lambda_{em}$ = 375/485 nm) (sensitivity ~0.005 ppm).
4. *AAS method*: In an air/acetylene flame (AAS method), manganese has an absorption at $\lambda$ = 280 nm (sensitivity ~0.05 μg/mL).

Technetium, Tc ($Z$ = 43, $A$ = 98.906), does not occur in nature. It is obtained by bombardment of molybdenum with deuteron or neutrons. It is widely used in nuclear medicine (see Chapter 25).

Rhenium, Re ($Z$ = 75; $A$ = 186.2), is a silvery-white and extremely hard metal. It is a rare metal with highest melting point (M.P. = 3,200° C; B.P. = 5,630° C).

## 19.14 IRON FAMILY METALS

The secondary subgroup of Group VIII includes three triads of d-elements. The first of them is formed by the iron family—*iron* (Fe), *cobalt* (Co), and *nickel* (Ni), the second—*ruthenium* (Ru), *rhodium* (Rh), and *palladium* (Pd), and the third—*osmium* (Os), *iridium* (Ir), and *platinum* (Pt). Iron, cobalt and nickel are similar to one another (Table 19.12).

**TABLE 19.12:** SOME PHYSICAL PROPERTIES OF IRON FAMILY METALS

| *Parameter* | *Fe* (26) | *Co* (27) | *Ni* (28) |
|---|---|---|---|
| Electronic configuration | $3d^64s^2$ | $3d^74s^2$ | $3d^84s^2$ |
| Atomic radius | | | |
| $E^{2+}$ (nm) | 0.080 | 0.078 | 0.074 |
| $E^{3+}$ (nm) | 0.067 | 0.064 | – |
| Ionization potential | | | |
| 2nd (eV) | 16.2 | 17.1 | 18.15 |
| 3rd (eV) | 30.6 | 33.5 | 36.16 |
| Electrode potential (V) | –0,440 | –0.277 | –0.250 |
| Density | 7.87 | 8.84 | 8.91 |
| Melting point (°C) | 1,540 | 1,495 | 1,450 |
| Boiling point (°C) | 3,000 | 3,000 | 2,900 |

### 19.14.1 Iron (Fe)

Iron, (Fe; $Z$ = 26, $A$ = 55.847), is the next abundant metal on the globe after aluminum. Important iron ores are *magnetite* ($Fe_3O_4$), *hematite* ($Fe_2O_3$), and *iron pyrites* ($FeS_2$). Iron is a silvery plastic metal, and its main use is in the production of ferrous alloys—irons, steels (iron-carbon alloys containing <2.14 per cent carbon). Iron rusts in humid air and corrodes intensively in water. Special steels are employed to avoid corrosion.

Iron exists in ferrous, +2, and ferric, +3 oxidation states in most of the iron compounds. Iron(II) salts can be readily oxidized to iron(III) salts. Iron(II) salts are often used as reducing agents. Iron salts are used as coagulants in water purification. Iron is an undesirable component of drinking water. Major sources of pollution are mining, acid mine drainage, corroded metal, domestic and industrial waste.

Gravimetric, spectrophotometric, fluorescence, and ICP-AAS methods are employed for the determination of iron.

1. *Gravimetric method*: When alkalis react with solutions of iron(III) salts, red-brown iron(III) hydroxide precipitates. It is insoluble in an excess of the alkali, and is determined gravimetrically.
2. *Phenanthroline method*: Fe(II) gives a red color with 1,10-phenanthroline at pH = 3.2. The color developed is measured at $\lambda$ = 510 and 533 nm. If iron is present in the trivalent state a mild reducing agent like hydroxylamine is added.
3. *Thiocyanate method*: This method is specific for Fe(III). Treatment of Fe(III) salts with potassium thiocyanate, or ammonium thiocyanate gives rise to garnet-red colored complex, which is used for the spectrophotometric determination of iron.
4. *ICP-AAS method*: In AAS method (with air/acetylene), iron is monitored at $\lambda$ = 248 nm.

### 19.14.2 Cobalt (Co)

Cobalt, Co ($Z$ = 27, $A$ = 58.933), is rare in nature and is often combined with arsenic ($CoAs_2$; CoAs). Cobalt is a hard, ductile, lustrous metal resembling iron. It is also ferromagnetic; up to 1,120° C. Cobalt is an industrial metal. Its main uses are in the manufacture of heat-resistant, high-temperature alloys. Radioactive forms are used in medical diagnostics and cancer therapy. Cobalt is a trace element essential in the nutrition (Vitamin B12).

Spectrophotometric, ring-oven, fluorescence quenching and ICP-AAS methods are employed to determine cobalt.

1. *Spectrophotometric method*: Reaction of cobalt with 1-nitroso-2-napthol yields a brown-colored complex that can be measured spectrophotometrically.
2. *Ring-oven method*: The sample containing cobalt is treated with KCN to form cyanocobalt complex. The Co-complex is washed with deionized water to the heated ring zone of the filter paper, where the complex is deposited in a sharply marked zone. Interference due to other pollutants is suppressed by the addition of sodium hydrogen phosphate. Addition of 1-nitroso-2-napthol in acetone to the ring zone gives rise to a brown-colored complex, which may be measured colorimetrically.
3. *AAS methods*: Cobalt is determined by AAS in an air/acetylene flame or ICP-AAS methods. The absorption line at $\lambda$ = 241 nm (sensitivity ~0.2 μg/mL) is suitable for monitoring cobalt.

### 19.14.3 Nickel (Ni)

Nickel, Ni ($Z$ = 28, $A$ = 58.70), occurs in nature chiefly as compounds with arsenic or sulfur. Nickel is a hard (harder than iron), silvery-yellowish metal and is resistant to corrosion and oxidation. It is predominantly used in ferroalloy manufacture (e.g. stainless steels) and coinage. High-temperature nickel alloys are employed in turbines and jet engines, and in electrical heating elements.

Ring-oven, fluorescence quenching and AAS and ICP methods are employed to determine nickel.

1. *Weisz ring-oven method*: Solubilize nickel sample in KCN. Wash it with deionized water to the ring zone of the filter paper, where it is deposited in a sharply divided zone. Add diethylglyoxime and ammonium fumes, to produce a brilliant red colored complex that is measured colorimetrically.
2. *Fluorescence method*: The method is based on the quenching of the fluorescence of Al-1-(2-pyridylazo)-2-naphthol by nickel (sensitivity ~0.003 ppm).
3. *AAS & ICP methods*: Cobalt is determined by AAS using air/acetylene flame or the graphite furnace and monitoring the absorption at $\lambda$ = 232 nm (sensitivity ~0.15 µg/mL). ICP is a more sensitive method (see Chapter 24) and is the preferred method in the case of Co.

## 19.15 PLATINUM SUBGROUP METALS

The platinum group metals include elements of the second and third triads of Group VIII of the Periodic Table—*ruthenium* (Ru), *rhodium* (Rh), *palladium* (Pd), *osmium* (Os), *iridium* (Ir), and *platinum* (Pt). These elements form a group of quite rare metals resembling one another in their properties (Table 19.13). They are uncommon in natural waters.

**TABLE 19.13:** SOME PHYSICAL PROPERTIES OF PLATINUM GROUP METALS

| *Parameter* | *Ru* (44) | *Rh* (45) | *Pd* (46) | *Os* (76) | *Ir* (77) | *Pt* (78) |
|---|---|---|---|---|---|---|
| Outer orbital electron configuration | $4d^7 5s^1$ | $4d^8 5s^1$ | $4d^{10} 5s^0$ | $5d^6 6s^2$ | $5d^7 6s^2$ | $5d^9 6s^1$ |
| Atomic radius (nm) | 0.134 | 0.134 | 0.137 | 0.135 | 0.135 | 0.138 |
| 2nd Ionization potential (eV) | 16.8 | 18.1 | 19.4 | 17 | 17 | 18.6 |
| Density | 12.4 | 12.4 | 12.0 | 22.5 | 22.4 | 21.5 |
| Melting point (°C) | 2,300 | 1,960 | 1,555 | 3,000 | 2,450 | 1,770 |
| Boiling point (°C) | 4,200 | 4,500 | 2,940 | 5,500 | 4,500 | 3,800 |

Ruthenium, Ru ($Z$ = 44, $A$ = 101.07; B.P/M.P = 2,250° C/3,900° C); rhodium, Rh ($Z$ = 45, $A$ = 102.906; M.P/B.P = 1,970° C/3,730° C); osmium, Os ($Z$ = 76, $A$ = 190.2; M.P/B.P = 3,000° C/5,000° C); and iridium, Ir ($Z$ = 77, $A$ = 192.22; M.P/B.P = 2,400° C/4,550° C) are refractory metals. Ruthenium is a silvery-gray metal, used as an alloying agent to harden platinum and palladium. Rhodium is also used an alloying agent to harden platinum and in optical instruments, and thermocouples.

Palladium, Pd ($Z$ = 46; $A$ = 106.40), a gray-white, and extremely ductile metal, is the lightest and lowest-melting (M.P/B.P = 1,550° C/2,930° C) metal of the platinum group. It is used as a catalyst in many reactions. Oxidation states of palladium are +2 and +4. It is determined by spectrophotometric methods. Palladium forms a colored complex with 2,2-dipyridyl-2-pyridylhydrazone (DPPH). The colored complex, Pd(II)-DPPH (2,2-dipyridyl-2-pyridylhydrazone), is extracted with $CHCl_3$ and the absorbance is measured at 560 nm. This method is applicable for monitoring Pd(II) in airborne particulate matter and in automobile exhaust gas converter catalysts.

Osmium is a gray-white, hard and brittle metal and is the densest naturally occurring element. It has the highest melting point. Iridium is one of the denser terrestrial substances. Platinum ($Z$ = 78, $A$ = 195.09) is a white, lustrous, heavy, precious, and malleable metal. It is the most widely used of the six platinum metals. Platinum has high melting point (M.P/B.P = 1,770° C/3,830° C) and extremely resistant to corrosion and chemical attack. It is used in making laboratory instruments, resistance thermometers, thermocouples, and crucibles and in electrochemical apparatus. Its ability to dissolve hydrogen is utilized in the fabrication of hydrogen electrodes. Platinum is an excellent catalyst and is used in oxidation-reduction reactions.

Flame AAS method is employed to determine platinum group elements.

A summary of some wet-chemical methods and spectrophotometric determination of metals are given in Table 19.14.

**TABLE 19.14:** SOME WET-CHEMICAL METHODS FOR DETERMINATION OF METALS.

| *Metal* | *Procedure* | *Masking Agents* | $\lambda_{max}$ *(nm)* |
|---|---|---|---|
| Aluminum (Al) | Erichrome cyanide-R & Pyrocatechol violet methods | Acet, citr, DMP, EDTA, malo, tart. | 535; 580 |
| Arsenic (As) | Silver diethyldithiocarbamate & Ferric ion + Cr(VI) methods | DMP, $S^{2-}$ | 530 |
| Beryllium (Be) | Aluminum-EDTA reagent | $F^-$, SSA, tart. | 515 |
| Bismuth (Bi) | $PbO_2^{2-}$ + $SnO_2^{2-}$ | | |
| Boron (B) (non-metallic) | Curcumin & Caramine methods | – | 540 |
| Cadmium (Cd) | Dithizone method | APCA, citr, $CN^-$, DMP, malo, tart. | 518 |
| Calcium (Ca) | Erichrome black method | APCA, citr, polyphosphates, tart. | |
| Chromium (Cr) | Diphenylcarbazide; $I^-$ + $H_2O_2$; Indigocarmine + $H_2O_2$; $BrO_3^-$ + $I^-$; Luminol + $H_2O_2$ methods | Acet, APCA, citr, SSA, tart. | 540 |
| Cobalt (Co) | Nitroso-napthol; lucigenine + $H_2O_2$; Alazarin + $H_2O_2$; Indigo carmine + H2O2; Luminol + $H_2O_2$; Diphenylcarbazone + $H_2O_2$ methods | APCA, citr, $CN^-$, DMP, malo, tart. | |
| Copper (Cu) | Neocuprine and Bathocuprine methods & hydroquinone + $H_2O_2$; Indigo carmine + $H_2O_2$; Luminol + $H_2O_2$ methods | $NH_3$, $SCN^-$, tart. | 457; 484 |
| Iron (Fe) | Phenanthroline & $I^-$ + $S_2O_8{}^{2-}$; Methyl orange $H_2O_2$; Erichrome black + $H_2O_2$; Luminol + $H_2O_2$ methods | APCA, citr, $CN^-$, DMP, malo, tart. | 533 |
| Lead (Pb) | Gallic acid + $K_2S_2O_8$ & Dithizone methods | Acet, APCA, citr, DMP, tart. | 510 |
| Manganese (Mn) | Persulfate & Diethylaniline; Be-morine fluorescence methods | APCA, citr, $CN^-$, DMP, SSA, tart. | 525 & 545 |
| Mercury (Hg) | $As^{3+}$ + $Ce^{4+}$; Lumiol + $H_2O_2$ & Dithizone methods | APCA, citr, $CN^-$, DMP, tart. | 492 |
| Molybdenum (Mo) | KCN (KI + $H_2O_2$) method | APCA, citr, $CN^-$, DMP, $H_2O_2$, tart. | 465 |
| Potassium (K) | Cobaltnitrite method | .... | 425 |
| Silver (Ag) | Dithizone method; Lucigemine + $H_2O_2$; Sulfanalic acid + $K_2S_2O_8$ | $NH_3$, $SCN^-$, thiourea. | 462 & 620 |
| Vanadium (V) | Gallic acid & Erichrome blue + $KBrO_3$ methods | $CN^-$, EDTA, $H_2O_2$. | 415 |
| Zinc (Zn) | Dithizone method | APCA, citr, $CN^-$, DMP, $NH_3$, tart. | 520 & 540 |

Abbreviations: Acet: = Acetate; APCA = Amino polycarboxylic acids; Citr: = Citrate; CN-: = Cyanides; DMP = 2,3-Dimercapto-1-proponol; EDTA = Ethylenediaminetetraacetic acid; Malo: = Malonate; SSA = Sulfosalicilic acid; Tart: = Tartrate.

## BIBLIOGRAPHY


Adriano, D. C. (1986), Springerverlag: New York. "*Trace Elements in the Terrestrial Environment.*"

Ahmed, M. J. and Al Mamun, M. (2001), *Talanta*, **55**(1): 43. "Spectroscopic determination of lead in industrial, environmental, biological and soil samples using 2,5-dimercapto-1,3,4-thiadiazole."

Albarges, J. B. (ed.) (1980), Pergamon Press: New York. "*Analytical Techniques in Environmental Chemistry.*"

Amyot, M., Auclair, J. C. and Poissant, L. (2001), *Anal. Chim. Acta.*, **447**(1 & 2): 153. "*In situ* high temporal resolution analysis of elemental mercury in natural waters."

Anthemidis, A. N., Themelis, D. G. and Stratis, J. A. (2001), *Talanta*, **54**(1): 37. "Stopped-flow injection liquid-liquid extraction spectrophotometric determination of palladium in airborne particulate matter and automobile catalysts."

Bard, A. J. and Faulkner, L. R. (1980), Wiley & Sons: New York. "*Electrochemical Methods, Fundamentals and Applications.*"

Broekaert, J. A. C., Guecer, S. and Adams, F. (eds.) (1990), Springerverlag: Berlin. "*Metal Speciation in the Environment.*"

Cano-Pavon, J. M., *et al.* (1999), *Intl. J. Environ. Anal. Chem.*, "Analytical methods for mercury speciation in environmental and biological samples—An overview."

Carloli, S. (ed.) (1996), Wiley & Sons: New York. "*Element Speciation in Bioinorganic Chemistry.*"

Colombo, C. and Van den Berg, C. M. G. (1998), *Intl. J. Environ. Anal. Chem.*, "Determination of trace metals (Cu, Pb, Zn and Ni) in soil extracts by flow analysis with voltammetric detection."

Da silva, J. J., Frausto, R. and Williams, R. J. P. (1991), Clarenden Press: Oxford, UK. "*The Biological Chemistry of Elements.*"

Eaton, A. D., *et al.* (eds.) (1995), Amer. Public Health Assn: Washington, DC. "*Standard Methods for the Examination of Water and Wastewater,*" (19th edn.).

Ewing, G. W. (1975), McGraw-Hill: New York. "*Instrumental Methods of Chemical Analysis*", (4th edn.).

Gallay, W., *et al.* (Compilers) (1975), Butterworth: London. "*Environmental Pollutants—Selected Analytical Methods, SCOPE-6.*"

Götzl, A. and Riepe, W. (2001), *Talanta*, **53**(6): 1187. "Field producers to assess soils and waste disposals."

Götzl, A. and Riepe, W. (2001), *Talanta*, **54**(5): 821. "Mercury determination—SPME and colorimetric spot test."

Guerin, T., Astruc, A. and Astruc, M. (1999), *Talanta*, **50**(1): 1. "Speciation of arsenic and selenium compounds by HPLC hyphenated to specific detectors: a review of the main separation techniques."

Guilbault, G. G. (1990), Marcel Dekker: New York. "*Practical Fluorescence.*"

Harrison, R. M. and Wilson S. J. (1982), Pergamon Press: Oxford, UK. "*Analytical Techniques in Environmental Chemistry*", (Vol. 2).

High, B., Bruce, D. & Richter, M. M. (2001), *Anal. Chim. Acta*, **449**(1 & 2): 17. "Determining copper ions in water using electrochemical luminescence."

Hutzinger, O. (ed.) (1990), Springerverlag: Berlin. "*Handbook of Environmental Chemistry.*"

Iyengar, G. V. (1989), CRC Press: Boca Raton, FL. "*Elemental Analysis of Biological Systems,*" (Vol. 1).

Kim, H-S. and Choi, H-S. (2001), *Talanta*, **55**(1): 163. "Spectrofluorimetric determination of copper(II) by its static quenching effect on the fluorescence of 4,5-dihydroxy-1,3-benzenedisulfonic acid."

Krull, I. S. (ed.) (1991), Elsevier: Amsterdam. "*Trace Metal Analysis and Speciation.*"

Leppard, G. G. (ed.) (1983), Plenum Press: New York. "*Trace Element Speciation in Natural Waters and its Ecological Implications.*"

Luconi, M., *et al.* (2001), *Talanta*, **54**(1): 45. "Flow injection spectrophotometric analysis of lead in human saliva for monitoring environmental pollution."

Mousavi, M. F., *et al.* (2001), *Talanta*, **55**(2): 305. "Differential pulse anodic stripping voltammetric determination of lead(II) with a 1,4-bis(prop-2-enyloxy)-9-10-anthraquinone."

Moyano, S., *et al.* (2001), *Talanta*, **54**(2): 211. "On-line preconcentration system for bismuth determination in urine by flow injection hydride generation inductively coupled plasma atomic absorption spectrometry."

Murillo, M., *et al.* (2001), *Talanta*, **54**(2): 389. "Optimization of experimental parameters for the determination of mercury by MIP/AES."

Nriagu, J. O. and Davidson, C. I. (eds.) (1986), Wiley & Sons: New York. "*Toxic Metals in the Atmosphere.*"

Quinby-Hunt, M. S., McLaughlin, R. D. and Quintnilha, A. T. (1986), Wiley & Sons: New York. "*Instrumentation for Environmental Monitoring,*" (Vols I & II).

Perez-Bendito, D. and Rubio, D. (1999), Elsevier: Amsterdam. "*Environmental Analytical Chemistry*: "*Comprehensive Analytical Chemistry Series, Vol. XXXII*", G. Svehla (ed.).

Sanchez-Asensio, J., *et al.* (2001), *Talanta*, **54**(5): 953. "Determination of organotin compounds by hydride generation-gas phase molecular absorption spectrometry."

Sandell, E. B. and Onishi, H. (1978), Wiley & Sons: New York. "*Photometric Determination of Traces of Metals*".

Sanger, D., Outridge, P. and Davis, W. (2000), *Environmental Reviews*, **8**(2): 115. "Stable lead isotopic characteristics of lead ore deposits of environmental significance."

Simon, N. S. (1997), *Intl. J. Environ. Anal. Chem.* "Supercritical fluid carbon dioxide extraction and liquid chromatographic separation with electrochemical detection of methylmercury from biological samples."

Singh, B. and Hoste, L. (2001), *Environmental Reviews*, **9**(2): 81. "*In situ* immobilization of metals in contaminated or naturally metal-rich soils."

Ulrich, S. M., Tanton, T. W. and Abdrashitova, S. A. (2001) *Crit. Rev. Environ. Sci. Tech.*, **31**(3): 241. "Mercury in the aqueous environment: A review of factors affecting methylation."

Uzun, A., Soylak, M. and Elci, L. (2001), *Talanta*, **54**(1): 197. "Preconcentration and separation with Amberlite XAD-4 resin; determination of Cu, Fe, Pb, Ni, Cd and Bi at trace levels in wastewater samples by flame atomic absorption spectrometry."

Van Loon, J. (1985), Wiley & Sons: New York. "*Selected Methods of Trace Metal Analysis. Biological and Environmental Samples.*"

Wang, J., Hansen, E. H. and Gammelgaard, B. (2001), *Talanta*, **55**(1): 117. "Flow injection on-line dilution for multi-element determination in human urine with detection by inductively coupled plasma mass spectrometry."

Weisz, H. (1960), Pergamon Press: London. "Microanalysis by the Ring Oven Technique."

Yang, L., *et al* (2001), *Talanta*, **55**(2): 271. "Determination of Cr(VI) and lead (II) in drinking water by electrokinetic flow analysis systems and graphite furnace atomic absorption spectrometry."

Zendelovska, D., *et al.* (2001), *Talanta*, **54**(1): 139. "Electrothermal atomic absorption spectrophotometric determination of cobalt, copper, lead and nickel in aragonite following floatation and extraction."

—(1995), EPA/600/R-94/111: Washington, DC. "*Methods for the Determination of Metals in Environmental Samples.*"

# 20

# Chemical Methods in Environmental Pollution Analysis—Inorganic, Non-Metallic

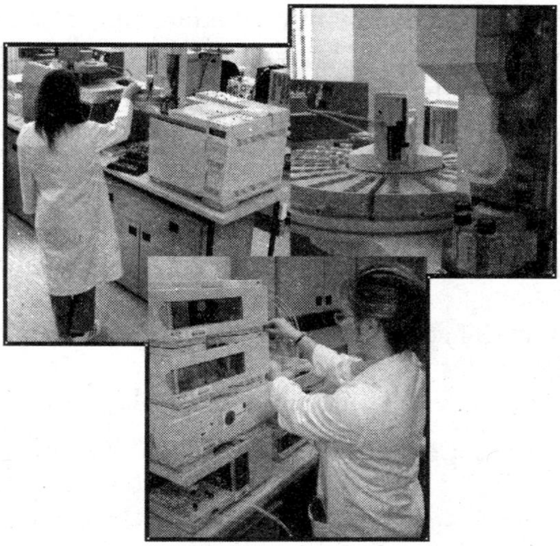

The inorganic, non-metallic particulates such as silica, asbestos, phosphorus and selenium are important environmental pollutants. Solid inorganic particulate and coagulated substances can be determined by gravimetric and volumetric as well as more sophisticated physical techniques. Some of the physical techniques of determination are ion-chromatography (IC), electrical conductimetry, optical and electron microscopy, X-ray fluorescence (XRF) spectroscopy and neutron activation analysis (NAA) and X-ray diffraction and other methods. IC method is suitable for rapid, sequential determination of many common ions, such as $F^-$, $Cl^-$, $Br^-$ nitrites, nitrates, sulfates and phosphates.

This chapter illustrates some of the procedures for determination of some selected inorganic, non-metallic constituents.

Particulate matter, such as asbestos, metal oxides, carbonates, chlorates, nitrates, phosphates, silicates and non-metals such as selenium and tellurium are some of the non-metallic inorganic constituents found in the environment. Some of them are soluble in water, while others are not. The particulate pollutant matter consists of dissolved solids, colloidal inorganic and organic (humic substance) matter. Of these, determination of important constituents of environmental pollution, silicon, phosphorus, and selenium are addressed in this chapter. *Total dissolved solids (TDS)* comprise both organic and inorganic fractions in the residue left upon evaporation of water (103-105° C). Organic fraction can be evaporated on ignition above 600° C, and the residue remaining, as ash is the inorganic fraction.

Solid inorganic pollutants or flocculents coagulated to solids can be determined by gravimetric, volumetric methods as well as more sophisticated physical techniques. Gravimetric methods are simple, but tedious and time-consuming. They also require large quantity of sample. Physical methods are ion chromatography (IC), electrical conductivity, optical and electron microscopy, X-ray diffraction, X-ray fluorescence (XRF) spectroscopy, and neutron activation analysis (NAA). Of many physical techniques, ion chromatography (IC) provides a single instrumental technique that can be used for rapid, sequential measurement of common anions, such as $F^-$, $Cl^-$, $Br^-$, nitrite, nitrate, phosphate, and sulfate.

Inorganic nitrogen in soil occurs predominantly as nitrate and ammonium salts. Total nitrogen (organic + inorganic) can be determined by Kjeldahl method, but the method is tedious. Better and more advanced methods are available for the determination of nitrogen, ammonia and other nitrogen compounds.

## 20.1 SILICON (SI)

Silicon, Si ($Z = 14$, $A = 28.09$), is one of the most abundant elements in the Earth's crust and ranks next to the oxygen in abundance (27 per cent of mass). It is encountered in nature only in compounds, as silica ($SiO_2$) and silicates (feldspars, micas, and kaolin). It is the most important element in the mineral kingdom. Silicon is mainly used in metallurgy, glass, ceramic, cement, and in semiconductor engineering, and as *carborundum* (SiC) in the production of refractory materials.

Gravimetric, colorimetrc, ICP and AAS methods are employed to determine silicon.

1. *Gravimetric method*: Silicates are precipitated as silica ($SiO_2$) by treatment of samples with hydrochloric acid. Dehydrated silica is weighed and volatilized as slicontetrafluoride. The residual impurities are weighed, from which silica ($SiO_2$) content is determined as loss on volatilization.
2. *Colorimetric methods*: Ammonium molybdate at pH ~ 1.2 reacts with silicates and phosphates. Addition of oxalic acid destroys the molybdo-phosphoric acid. The intensity of yellow-colored molybdosilicic acid, measured at $\lambda = 410$ nm, is proportional to the concentration of silica. Silicates react with ammonium molybdate + benzidine giving rise to blue-colored complex (molybdenum blue method). The yellow-colored molybdosilicic acid can be reduced by aminonapthosulfonic acid, leading to an intense blue-colored complex. This method is more sensitive.
3. *Ion-exchange method*: Ion-exchange with a strong base for separation and determination by physical or chemical methods.

### 20.1.1 Asbestos

Asbestos is a magnesium-calcium silicate—$CaO \cdot 3MgO \cdot 4SiO_2$. *Chrysotile* is the most abundant form of asbestos (>95 per cent). Asbestos pollution is due to its mining and milling.

Various physical techniques may be employed to determine asbestos. The use of optical phase-contrast microscopy takes advantage of the refractive index. Electron microscopic studies provide morphological features of asbestos. Powder X-ray diffraction methods, and energy-dispersive X-ray analyses (EDXA) provide structural and morphological features of asbestos.

## 20.2 PHOSPHORUS (P)

Phosphorus, P ($Z = 15$, $A = 30.974$), belongs to the nitrogen Group (Va) of elements in the Periodic Table. Phosphorus is an essential element for plants and animals. In plants, phosphorus is contained chiefly in the seed proteins and nucleic acids, and in animals—in proteins, nucleic acids, milk, blood, tissues, bones and teeth.

Phosphorus has variety of applications in metallurgy, manufacture of matches, semiconductors (GaP and InP), organophosphorus-compounds, pesticides and insecticides, and mineral fertilizers. $^{32}P$ isotope (14-day half-life) is used in tracer studies.

Oxidation states of phosphorus are –3, +3 and +5 (due to the participation of the *d* electrons). The most important of the natural compounds of phosphorus is calcium phosphate, $Ca_3(PO_4)_2$. Inorganic phosphorus is the major pollutant in water bodies. Phosphorus occurs in natural waters as anions of orthophosphoric acid, $H_3PO_4$ ($pK_1 = 2.17$; $pK_2 = 7.31$; and $pK_3 = 12.36$). The anions $H_2PO_4^-$ and $HPO_4^{2-}$ are predominant. Phosphorus, in water, leads to scale formation in boilers. Orthophosphates (fertilizers) are carried into streams by storm run-off. Phosphates are key nutrients stimulating excessive plant growth—weeds and algae, in streams and lakes. Phosphate detergents like sodium tri-polyphosphate ($Na_5P_3O_{10}$) are also major water pollutants. Some of the organic derivatives of phosphorus, termed 'nerve gases,' are among the most toxic substances known to man.

Wet-chemical and physicochemical methods are employed to determine phosphorus.

1. *Vanadomolybdate method*: Ammonium molybdate reacts with orthophosphate in acid medium to form molybdophosphoric acid, which in the presence of vanadium forms yellow-colored vanadomolybdo-phosphoric acid. The intensity of the yellow color of the complex, measured at 400-490 nm (430 nm), is proportional to the phosphate concentration.
2. *Stannous chloride method*: Molybdophosphoric acid formed can be reduced by stannous chloride or ascorbic acid, to form blue-colored complex that is measured spectrophotometrically (640 nm).
3. *ICP method*: Total phosphorus content ($P_T$) of a sample can be determined by the inductively coupled plasma (ICP) detection. Inorganic phosphorus ($P_{in}$) can be determined by extraction and ignition and determination from which the organic phosphorus content ($P_{or}$) can be estimated ($P_{or} = P_T - P_{in}$).

## 20.3 SELENIUM (SE)

Selenium, Se ($Z = 34$, $A = 78.96$), is scarce in nature. Its compounds occur in association with metal sulfides—PbS, $FeS_2$. Selenium is commercially important. It is a semiconductor, and a good photoconductor (a *barrier layer* is formed at the junction of selenium with a metallic conductor that can pass an electric current only in one direction. On illumination, the electrical conductivity of selenium increases more than 1,000-fold). These properties are made use of in semiconductor engineering for the fabrication of rectifiers and photocells with a barrier layer, and alarm devices, mechanical safety devices and in xerography.

Selenium is an essential trace element, but above trace levels it is toxic. Selenium is present in the environment in the oxidation states –2, 0, +4 and +6. In the –2 state, it occurs as hydrogen selenide

($H_2Se$). $H_2Se$ is a toxic gas. In its +4 oxidation state selenium exists as selenite, ($HSeO_3^-$ or $SeO_3^{2-}$), and in its +6 oxidation state as selenate ($SeO_4^{2-}$). Selenate is stable in alkaline and oxidizing solutions. Organoselenium compounds are more toxic than inorganic species owing to their increased fat solubility.

Selenium compounds are highly volatile (Se(IV) forms volatile hydrides). So sampling and determination should be carried out by impingement with water or $Na_2SO_3$, $Na_2S$, or NaOH. Wet-chemical methods, fluorescence, AAS with or without hydride generation, ICP-AES, ICP-MS and AAS methods may be employed to determine selenium.

1. *Diaminonapthalene method*: Selenite(IV) in aqueous acidic medium (pH < 2) can be determined by reaction with 2,3-diaminonapthalene, to produce a strongly fluorescent colored compound. The fluorescence can be excited by 369 nm and its intensity measured at $\lambda = 480$ nm (sensitivity ~0.02 ppm).
2. *Catalytic method*: The method is based on the catalytic effect of selenium in the reaction of methylene blue by sodium sulfide. The method is very sensitive, selective and rapid.
3. Volatile selenium compounds can be oxidized by $H_2O_2$ to Se(VI), which is then determined by digestion with HCl. The analysis can be carried out by spectrophotometric, fluorimetric, hydride AAS, and voltammetric methods.
4. *GC and by AAS, AES or FID-MS methods*: These methods hold much promise. GC-AAS method is both sensitive and specification-specific for the analysis of low boiling organometallic compounds (that include selenium compounds). The technique inherits the separation capability of GC and the specificity of AAS (flame or furnace). Se is detected at 196 nm. LC is more suitable for selenoproteins, as they are thermally unstable and non-volatile.

Wet-chemical methods of analysis for phosphorous, selinium and silicon are summarized in Table 20.1.

**TABLE 20.1:** WET-CHEMICAL ANALYSIS OF INORGANIC, NON-METALLIC COMPOUNDS

| *Species* | *Procedure* | $l_{max}$ *(nm)* |
|---|---|---|
| Phosphorus (P) | Vanadomolybdo-phosphoric acid method | 420 |
| | Stannous chloride (Molybdenum blue) method | 640 |
| Selenium (Se) | Diaminonapthalene method; Methylene blue method | 480 |
| Silicon (Si) | Molybidosilicic acid + oxolic acid method | 410 |

## BIBLIOGRAPHY

Frei, R. W. and Alberge, J. (eds.) (1986), Gordon & Breach: New York. "*Air and Water Analysis: New Techniques and Data.*"

Profumo, A., *et al.* (2001), *Talanta*, **55**(1): 155."Sequential extraction procedure for speciation of inorganic selenium in emissions and working area."

Smith, K. (1991), Marcel Dekker: New York. "*Soil Analysis: Modern Instrumentation Techniques.*"

Sposito, G. (1989), Oxford University Press: New York. "*The Chemistry of Soils.*"

Yaron, B., Calvet, R. and Prost, R. (1996), Springerverlag: Berlin. "*Soil Pollution, Process and Dynamics.*"

—(1986), Wiley Interscience: New York. "*Instrumentation for Environmental Monitoring,*" (21st edn.).

# 21

# Chemical Methods in Environmental Pollution Analysis—Organics

Hundreds of organic compounds, most of them of anthrapogenic origin, are recognized to be pollutants. They are present in air, water and even soil. From the later they reach even groundwater. The number and variety of organic pollutants are staggering.

Organic pollutants can be classified under—(1) conventional pollutants (aldehydes, ketones, alcohols) (2) aromatic hydrocar-bons, (3) polynuclear aromatic hydrocarbons (PAHs), and (4) synthetic organic compounds that include organo-pesticides, fungicides and herbicides. Some of them like the clorofluoro-hydrocarbons are involved in ozone depletion in the stratosphere. Their manufacture and use is being regulated under various International Protocols. Their determination is generally carried out by one or the other variation of GC (like capillary GC) and detection is also by various means, the important one being electron capture. For components, which are nonvolatile HPLC is a frequently employed technique. To tackle the variety of compounds different stationary phase materials are in use. Mass spectrometry is a very frequently used detection system in all these separations.

This chapter outlines the procedures for analysis of organic pollutants.

Organic chemical pollutants comprise a variety of compounds like hydrocarbons, ketones, and aldehydes. They originate from exhaust gases and petrochemical industrial discharge, oils and greases, detergents, pesticides and secondary reaction products. There are four classes of pollutants—(1) conventional pollutants, such as organic aerosols, oils and greases, (2) aromatic hydrocarbons (e.g. phenols, xylenes), (3) polynuclear aromatic hydrocarbons (PAHs), (4) non-conventional synthetic organic compounds, which include organometallic, organophophorous, organosulfur, and halogenated hydrocarbons.

## 21.1 CLASSIFICATION OF ORGANIC POLLUTANTS

Organic pollutants can be classified as volatile organic compounds (VOCs), semi-volatile organic compounds (SVOCs) and non-volatile organic compounds (NVOCs, B.P. > 250° C). VOCs facilitate formation of photochemical oxidants like ozone. They are also responsible for depletion of ozone in the stratosphere and for global warming by trapping IR radiation. Some VOCs are recognized human health hazards. They are present everywhere, outdoors as well as indoors. According to EPA's VOC National Ambient Data Book there are more than 300 compounds, which need attention. Their concentration in ambient atmosphere varies from 0.001 ppbv to 10 ppbv. A similar situation prevails for their indoor distribution.

VOCs are further classified on the basis of their environmental impact.

1. **VOC-OX:** These VOCs are potential agents involved in formation of miscellaneous oxidants like ozone and peroxyacetyl nitrate (PAN). They include alkanes, $C_2$ — $C_8$ alkenes, alkynes, arenes (e.g. xylenes) and terpenes.
2. **VOC-TOX:** These VOCs are toxic to humans, animals and vegetation at trace levels. Nearly 190 of them are also called 'hazardous air pollutants'. They comprise chlorinated solvents, diolefins, benzene compounds and their derivatives containing hydroxyl, nitrate and halogen groups, aldehydes, nitriles and peroxyacetyl nitrates. Many polyaromatic hydrocarbons (PAHs) and organo-pesticides can be included in this category.
3. **VOC-STRAT:** They include freons, halons, and chlorinated hydrocarbons. They contribute for ozone depletion in the stratosphere.
4. **VOC-FORM:** These are secondary pollutants formed as a result of photochemical oxidation of hydrocarbons. They belong to the group of aldehydes, ketones, and free acids. Peroxyacetyl nitrate also belongs to this category.
5. **VOC-CLIM:** VOCs like $CH_4$ and CFCs come in this category. By trapping IR radiation they contribute to greenhouse effect and global warming.

Major sources of VOCs outdoors are automobiles (Table 21.1), chemical and petrochemical industries. Some have secondary origin, resulting from photochemical activity in the atmosphere. The sources of indoor VOCs are building materials, garments, paints, adhesives, and consumer products. Burning and smoking also produce these pollutants. Their effect indoors is magnified by the lack of free ventilation, particularly in the industrially advanced and cold countries. This single factor is responsible for making the concentration of some of the pollutants like formaldehyde, chloroform (from water) and tetrachloroethane (used for dry- cleaning) more indoors than outdoors.

The measurement of organic pollutants in ambient air is often difficult, in part due to the variety of organic substances involved. When determination of a single analyte is required, a direct spectroscopic method is favored. The main aim of analysis varies with the type of VOC to be monitored. For VOC-OX type, which reacts rapidly with the oxidants, prompt quick and frequent analysis and continuous monitoring is preferred. In the case of VOC-TOX analyses the aim is to assess the exposure of humans to the harmful effects of the pollutants. Data on VOC-STRAT and VOC-CLIM pollutants are recorded on a daily or weekly basis at several points on the globe to assess their global effects.

**TABLE 21.1:** POLLUTANTS IN EXHAUST GASES

| *Pollutant* | *Concentration (ppb)* |
|---|---|
| CO | 200-2,000 |
| NO | ~10 |
| $NO_2$ | ~20 |
| $O_3$ | ~10 |
| PAN | ~5 |
| Total Hydrocarbons | 20-50 |

ppb = parts per billion.

For analysis VOCs are initially collected on a cold surface (a cartridge containing adsorbent Tenax or carbon molecular sieve) or cryogenic pre-concentration. Polar VOCs include reactive oxygen-nitrogen containing organic compounds. The determination of hydrocarbons is a three-step process—collection of air samples, pre-concentration of pollutants by a cold trap or by cryogenic methods, separation of hydrocarbon analytes and detection (Fig. 21.1). Non-polar solvents such as n-hexane, n-heptane and dichlorobenzene are used for extraction of non-polar VOCs (hydrocarbons); GC with FID (GC-FID) or ECD (GC-ECD) or GC-MS method (generally) is employed for separation and identification. Polar VOCs (e.g. organic acids) are extracted using polar solvents (ethyl acetate), separated by LC and detected by photometry or conductimetry.

Growing emphasis on environmental monitoring has encouraged the development of more rapid and less expensive methods of analysis of organic air pollutants. Spectroscopic and electrochemical methods have found well suited. FTIR spectroscopy is becoming viable for the remote detection of airborne VOCs. A single instrument can be used for the detection of a variety of compounds. Since the physical collection of a sample is not required, FTIR remote sensors offer possibilities of monitoring large areas, and these could be automated. Infrared differential absorption lidar (IRDIAL) techniques allow remote sensing of gases in the atmosphere from distances up to several kilometers. Differential optical absorption spectroscopy (DOAS) enables *in situ* measurements of airborne pollutants. Electrochemical sensors and biosensors are other types of detectors that are being introduced in environmental studies of pollutants.

## 21.2 CONVENTIONAL ORGANIC POLLUTANTS

Conventional organic pollutants are alcohols, aldehydes, alkanes, alkenes, amines, ketones, organic acids and nitrates.

### 21.2.1 Oxygen-containing Alkyl Compounds

Compounds containing oxygen include alcohols, aldehydes, ketones, organic acids and esters.

R—OH (alcohol); R—CH = O (aldehyde); R-CO-R' (ketone)
R—COOH (acid); R—C—OOR' (ester); (R and R' are alkyl side groups).

Alcohols are industrial solvents and chemical intermediates in the manufacture of various organic compounds. Some of them (methanol) are highly toxic. Phenols are aromatic alcohols. Phenols and other aryl alcohols are more water-soluble and, therefore, are more toxic than aliphatic hydrocarbons. Many are carcinogenic.

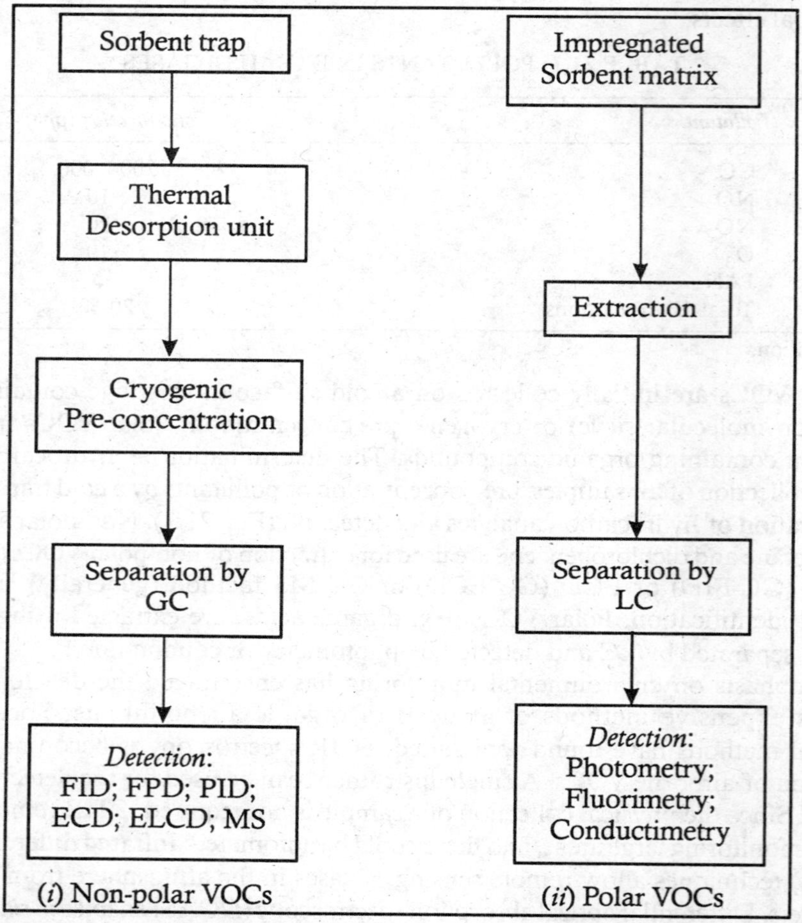


**Fig. 21.1:** Block Diagram of Analysis of (*i*) Non-polar and (*ii*) Polar VOCs

Aldehydes are more toxic than their alcohol analogues. Formaldehyde, or methanal, (*formalin*) is tumorogenic and carcinogenic. Propanal ($H_2C=CHCHO$) is an environmental pollutant.

Many amines are carcinogenic. Acetonitrile ($CH_3-C\equiv N$) is an important commercial nitrile, used in the manufacture of plastics and synthetic rubber. It is highly toxic and carcinogenic.

GC, GC-MS and HPLC methods are the standard methods of determination of these organic compounds.

Aldehydes and ketones are the reaction products of hydrocarbons with OH radical. They are precursors of various oxidants, including ozone, peroxyacyl nitrates and peroxycarboxylic acids. Formaldehyde and acetaldehyde are the most abundant carbonyl compounds of industrial importance. Other $C=O$ compounds are propanal, acetone and benzaldehyde. The concentration of these compounds in clean air is in the low ppbv range. However, in polluted air their level can go up to a few tens of ppbv.

Pre-concentration is necessary before these compounds can be measured in the atmosphere. The carbonyls are known to form well-defined phenylhydrazones and this property is utilized for their concentration. Carbonyls are trapped on solid adsorbents coated with a derivitizing agent, 2,4-dinitrophenylhydrazine (DNPH). The hydrazone formed on the adsorbent is desorbed by a solvent like acetonitrile and determined by HPLC or GC. The most common sorbent used is silica gel. The hydrazones are separated on reversed-phase column and detected by UV spectrometry (at 340 nm).

Carbonyls can be detected with a higher sensitivity by derivatization with a fluorescent reagent and measuring the fluorescence. By converting them into hydrocarbons by reduction with hydrogen in presence of a nickel catalyst, the carbonyls can be determined by GC with a FID detector.

Carboxylic acids are VOC-FORM compounds resulting from the reaction of olefins with ozone. They are also formed by oxidation of alcohols and aldehydes. Though formic and acetic acids are the common acids, the list extends to propionic, butyric, oxalic, and glycolic, lactic and citric acids. Burning of biomass and automobile exhaust are the probable sources of some of these acids. As far as their pre-concentration is concerned, their chemical property makes alkaline coatings the most suitable. KOH coated denuders enable particulate to be separated efficiently from gaseous components. HPLC with ion-exclusion columns is the preferred technique for determining the atmospheric acids. UV spectrophotometry and conductimetry are the two detection devices employed. Ion-exchange columns facilitate the simultaneous determination of organic and inorganic acids. Long-path Fourier transform IR (FTIR) spectroscopy is a versatile technique that does not suffer from the interferences affecting the sampling step.

### 21.2.2 Alkyl Nitrates

It is known for a long time that simple alkyl nitrates occur in the atmosphere. Most of these compounds up to $C_{20}$ are present as gases. Of the total reactive odd nitrogen compounds the $C_3$ to $C_6$ alkyl nitrates account for only 1-2 per cent. Only in the Arctic region this fraction can rise up to 20 per cent. The lighter members of this group ($C_1$-$C_6$) are concentrated on Tenax or Tenax-TA/Carbopack B columns. 5-30 liters of air is sampled this way. Following their thermal desorption they are analyzed by GC. Cryofocusing is included in the procedure. For analyzing higher boiling fractions, the volume of air to be sampled increases to as high as 100-1,000 $m^3$. A mixture of cleaned silica and carbopack C is used as adsorbent. The analytes are separated from the adsorbent by dissolving in a solvent or by Soxhlet extraction and separated and determined by capillary gas chromatography. Non-polar stationary phases are preferred. For determining alkyl nitrates the detection systems found useful are electron capture, massspectrometry and chemiluminescence from interaction with luminol.

OH radical in the presence of NO forms alkyl nitrates in the atmosphere by oxidation of saturated hydrocarbons. The instruments used for the direct determination of the $NO_2$, produced in the thermal decomposition of peroxyacetyl nitrates (e.g. PAN) can be used to measure PANs. The luminol chemiluminescence-based $NO_2$ detector is also very sensitive. Alkyl nitrates can also be determined by GC-ECD, GC-MS and GC-chemiluminescence detection methods

An important group of secondary pollutants are the peroxyacyl nitrates (R—C(O)—O—O—$NO_2$). The peroxyacetyl nitrate (PAN, R = $CH_3$) is the most abundant of this group. In relatively clean atmospheres, as on the oceans, it is formed from the oxidation of hydrocarbons like $CH_4$ and $C_2H_6$. Otherwise, oxidation of isoprene and terpenes lead to its formation.

PAN and related compounds are easily decomposed by heat and hence care has to be taken in the collection and handling of samples. Gas chromatography coupled with an electron capture detector is suitable for measuring these labile pollutants. Glass or Teflon columns packed with a deactivated solid support loaded with Carbowax liquid phase are used in the gas chromatograph. With this combination the limit of detection is about 0.16 ppbv. Another useful method for its determination is based on the indirect chemiluminescence (with luminol) from $NO_2$ formed by the decomposition of PAN.

### 21.2.3 Oils and Greases

Oils and greases (fats) are saturated or unsaturated fatty acids, products produced in petrochemical industrial operations. Oils are liquids at room temperature, while fats are in solid state. Fatty acids are determined, after separating them from hydrocarbons, by adsorption on silica gel. The adsorbed acids are recovered by dissolution in solvents or by Soxhlet extraction methods. The determination is by (1) gravimetric, (2) and IR methods.

1. **Gravimetric Method:** Dissolved or emulsified oils and grease are extracted from water by solvent extraction and determined gravimetrically, after evaporation of the solvent.
2. **IR Method:** Solvent extraction by trichlorotrifluoroethane and detection of C—H (2958 $cm^{-1}$), C—$H_2$ (2924 $cm^{-1}$) bands in the IR region. The intensities of the bands can be correlated to the oil and grease and non-volatile HC content.

### 21.2.4 Phenolic Compounds

Phenols are mononuclear aromatic organic compounds. They are starting materials for the production of dyes, pharmaceuticals, disinfectants, plastics (nylon), resins (adhesives and bakelite) and explosives (e.g. picric acid-trinitrophenol) and pesticides (e.g. dinitrophenol). Commercially important phenols are phenol, catechol, resorcinol, hydroquinone, and cresols (Fig. 21.2). Naturally occurring phenols (e.g. vanillin, thymol, eugenol (Fig. 21.3) are used as flavoring ingredients. Phenols and cresols are still used as biocides and creosote and tar acids as wood protectors. Some phenolic compounds are toxic, and the toxicity depends on the nature and arrangement of functional groups in the molecule.

Phenols are polar (acidic), and, therefore, are more soluble in water than benzene or toluene, due to propensity of hydrogen bond formation on account of the hydroxyl, —OH group.

Phenols can be extracted by: (*i*) pH-dependant solvent extraction, (*ii*) carbon adsorption, (*iii*) stream stripping and (*iv*) biological treatment. They are determined by colorimetric, GC-MS, HPLC, FID and ECD methods.

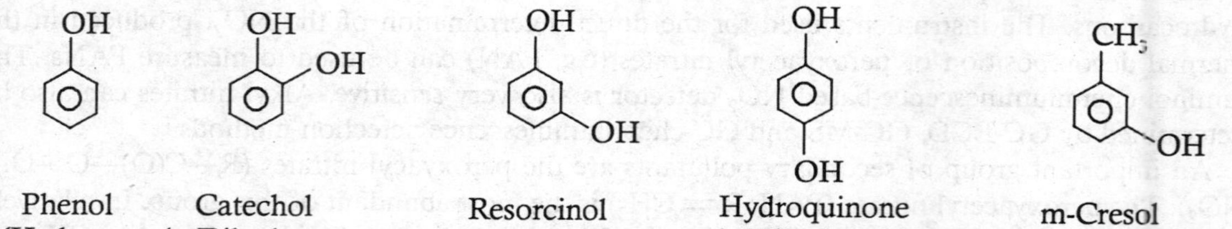

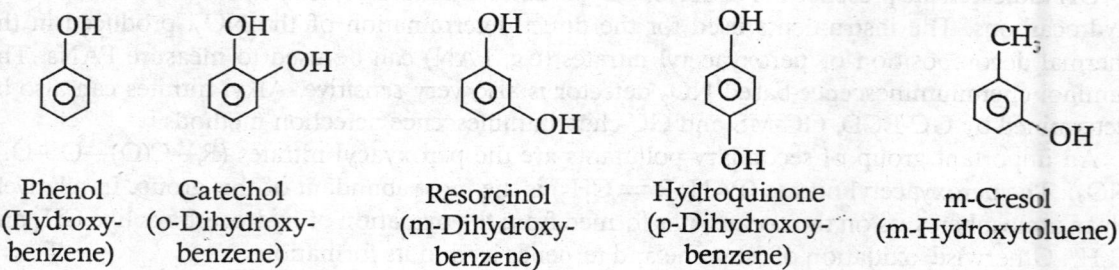

Phenol (Hydroxy-benzene) | Catechol (o-Dihydroxy-benzene) | Resorcinol (m-Dihydroxy-benzene) | Hydroquinone (p-Dihydroxoy-benzene) | m-Cresol (m-Hydroxytoluene)

**Fig. 21.2:** Structural Formulas of some Phenolic Compounds

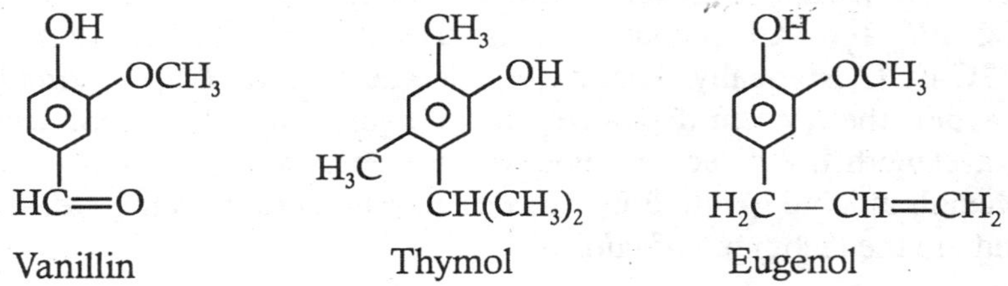


**Fig. 21.3:** Structural Formulas of some Phenol-related Flavoring Compounds

1. **Colorimetric Methods:** The determination of 'phenol index' includes all phenols that undergo coupling reactions. Phenol index determination, without a distillation ('total phenols') is by diazotization of p-nitroaniline followed by coupling with the phenols. The color of the azo dye is measured at $\lambda = 530$ nm.
2. **Steam-distillable Phenol Index:** Reaction of phenols at (pH ~ 8) with 4-aminoantipyrene (4-AAP) gives rise to a characteristic color, measured at $\lambda = 460$ or 550 nm. This method is not very sensitive (~0.2 µg/L). Determination of phenolic compounds in wastewaters is done by a sequential injection analysis (SIA) system, based on oxidative coupling of phenolic compounds with 4-aminoantipyrine (4-AAP) in alkaline solution and spectroscopic analysis at 510 nm. The system can be automated.
3. **Ferricyanide Method:** Reaction of phenols with 4-AAP in the presence of potassium ferricyanide at pH > 8.0, to form a reddish-brown dye. The dye is extracted with chloroform and the absorbance is measured at $\lambda = 460$ or 550 nm.
4. **Fluorescence Method:** Phenols in water can be determined by molecular fluorescence spectroscopy. Phenolate anions (basic form) do not fluorescence in the UV range. Therefore, the sample is made alkaline (pH > 12.0) and compared with the spectrum at neutral or acidic pH condition (pH 4-7.0) (Fig. 21.4).

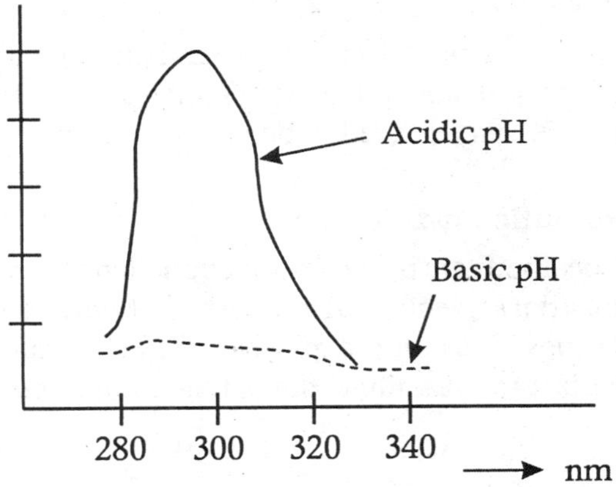


**Fig. 21.4:** Absorption Profiles of Phenols at Acidic and Basic pH

5. **Liquid-liquid Extraction Method:** Methylene chloride ($CH_2Cl_2$) is found to be the most suitable solvent for the extraction of a wide variety of phenolic compounds. The phenolic compounds are adsorbed on a column of XAD-2 resin at an acidic pH and eluted under alkaline conditions. After acidification of the eluate the phenol is extracted into $CH_2Cl_2$,

concentrated and identified by GLC with a flame ionization detector (FID). GC-MS, GC-FID, HPLC, FID and ECD methods are also used. GC-MS is a very sensitive, but expensive method. GC-ECD is equally sensitive, but less expensive. LC, in combination with $C_{18}$ columns, avoids the need to derivatize phenol compounds. Detection can be achieved by variable-wavelength UV detectors (diode-array detectors), for phenols at $\lambda = 280$ nm, and for nitrophenols at $\lambda = 310$ nm. Fluorescence detectors provide far better sensitivity (10 -12g ranges) and are the technique of choice.

## 21.3 HYDROCARBONS

Aliphatic, mono-, and polynuclear aromatic hydrocarbons (PAHS), produced from petrochemical industries are sources of a host of organic pollutants. Non-Methane Hydrocarbons (NMHCs), the sum of all hydrocarbon air pollutants except methane, are significant precursors to ozone formation.

### 21.3.1 Aliphatic Hydrocarbons

Ketones and alkanes (methane, ethane, propane, etc) are saturated aliphatic hydrocarbons, with single C—C bonds, with the general formula, $C_nH_{2n+2}$. Most of the alkanes are gases or liquids at room temperatures and they are the most unreactive of all classes of organic compounds.

Sources of alkanes are petroleum and natural gas with methane (80 per cent), and ethane (10 per cent). They are extremely hydrophobic and they physically interfere with biological and environmental processes. Liquid alkanes form a thin film on water surfaces and choke the underwater aquatic life. Cyclohexanes are cyclic aliphatic hydrocarbons.

Alkenes, or *olefins* are unsaturated hydrocarbons, with the general formula, $C_nH_{2n}$, with at least one C = C double bond (ethylene, propylene). Alkynes are hydrocarbons that contain a triple bond (e.g. acetylene). Alkenes and alkynes are highly reactive. Cycloalkenes are abundant in nature (terpene moieties, $C_{10}H_{16}$). The majority of the reactions of alkenes are characterized by addition to the C = C double bond.

$$—C=C— + YZ \rightarrow Y—C—C—Z \quad \text{(Addition)} \qquad ...(21.1)$$

Infrared (IR) spectroscopy is a useful and a general method for "screening" samples of hydrocarbons prior to performing detailed analysis. Samples are extracted from the matrix with $CCl_4$ and the absorbance at 2930 $cm^{-1}$ is used to determine the concentration of hydrocarbons.

### 21.3.2 Mononuclear Aromatic Hydrocarbons

Mononuclear aromatic hydrocarbons are benzene, toluene, xylene, styrene, etc. (Fig. 21.5). Benzene, toluene, and xylenes are industrial solvents, and are starting materials for the manufacture of a host of chemical compounds—drugs, dyes, and explosives (TNT—trinitrotoluene). Nitrobenzene is used extensively in the manufacture of aniline, the parent compound of many dyes and drugs.

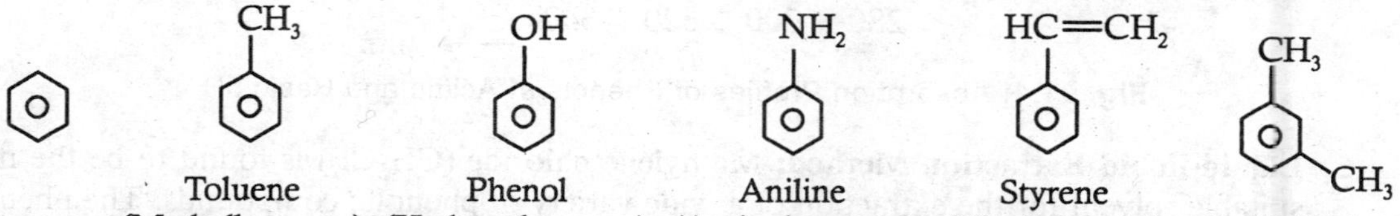


Benzene, Toluene (Methylbenzene), Phenol (Hydroxybenzene), Aniline (Aminobenzene), Styrene (Vinylbenzene), (m-Xylene)

**Fig. 21.5:** Structural Formulas of some Mononuclear Aromatic Organic Compounds

### 21.3.3 Polynuclear Aromatic Hydrocarbons (PAHs)

Polynuclear aromatic hydrocarbons (PAHs) are present in coal tar and fossil fuels. Naphthalene, anthracene, and phenanthrene (Fig. 21.6) are some of them. The anthropogenic PAHs in the environment are formed primarily by pyrolysis of carbonaceous materials. Pyrolysis of halogenated aromatic hydrocarbons generates them. Some of them are carcinogens (e.g. benzpyrene, benzanthracene (Fig. 21.7)). PAHs have very low solubility and hence their toxic effects last long.

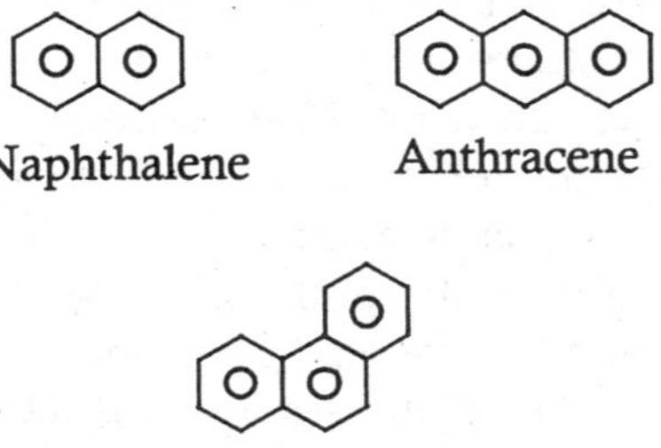

Naphthalene    Anthracene

Phenanthrane

**Fig. 21.6:** Structural Formulas of some Polynuclear Aromatic Hydrocarbons (PAHs)

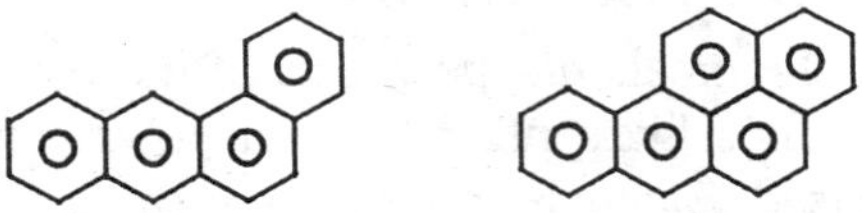

1,2-Benzanthracene    1, 2-Benzpyrene

**Fig. 21.7:** Structural Formulas of Benzanthracene and Benzpyrene

Petroleum products and oil pollution can be characterized by the trace elements like nickel and vanadium that they are associated with. The metallic elements can be determined by AAS and XRF. Some petroleum fractions contain polynuclear aromatic hydrocarbons (PAHs). Hence the fluorescent spectrum can be used as a fingerprint of oil pollution. Simultaneous analysis of multicomponent aromatic hydrocarbons in a mixture of micellar solution can be carried out by fluorimetrc methods.

PAHs (and PCBs) are extracted into organic solvents for isolation from other organics. Individual PAHs in these isolated fractions can be further separated and identified with GC-MS by GC retention and mass spectra. Generally, gas chromatography (GC) and liquid chromatography (LC) are the techniques for the determination PAHs (and PCBs). GC is ideally suited for low molecular mass volatile compounds, while LC is the method of choice for determination of higher molecular mass, low volatile and thermolabile compounds.

1. **GC-FID method:** A fused silica capillary GC column coated with an adsorptive bonded phase (polysiloxane), and detection by FID is a highly sensitive (0.05-0.5 ng) and versatile technique. GC-MS method is also a powerful tool, though expensive, for the identification and quantification of PAHs and PCBs. PCBs can be determined by GC and selective detection by ECD (very sensitive to hologens) or by selective ion monitoring MS (MS-SIM).
2. **Chromatographic methods:** GC is a very sensitive method to determine volatile organic compounds. On the other hand, liquid chromatography (HPLC) is applicable to volatile, scarcely volatile, polar, thermolabile and liquid-soluble analytes (that include chemical pesticides). LC methods are better suited for determining higher molecular mass PAHs (and

PCBs) that are thermally labile and are non-volatile (see Chapter 22). Most common method is HPLC with detection by UV absorption or fluorescence. The use of EI-MS interface in reversed-phase and ion exchange LC has enabled the determination of a large variety of non-volatile organic compounds. PAHs can be determined by UV absorption ($\lambda$ = 255 nm) and fluorescence techniques (excitation $\lambda$ = 280 nm and detection $\lambda$ = 390 and 430 nm). But, GC-MS, HPLC-UV and fluorescence detectors are expensive. They require pre-concentration before analysis. Synchronous fluorescence scan (SFC) method is found to be a method of choice for simultaneous determination of multi-component PAHs in micellar solutions. The method does not require the pre-concentration step. In this method both excitation and emission spectra are scanned simultaneously.

3. Immunoassay methods, such as enzyme-linked immunosorbent assay (ELISA), are gaining ground for screening environmental pollutants. They rely on capillary electrophoresis (immunoaffinity chromatography) for separation and are employed in the determination of PAHs and PCBs, organochlorine pesticides (see Chapters 22 and 24).
4. IR and Raman spectroscopic methods are also used (see Chapter 24).

## 21.4 SYNTHETIC ORGANIC COMPOUNDS

Advent of synthetic organic chemicals has been one of the major causes of ecological pollution. Almost all bulk organic compounds, like synthetic rubber, polystyrene, polyethylene, nylon, polyesters, acrylics, etc. are petroleum-based. Extensive use of chemical pesticides brought along with it health hazards to vegetation, aquatic and animal life. Their high stability, hydrophobicity and persistence result in their gradual accumulation in biological tissues (*bioconcentration*). Chlorodibenzo-p-dioxins are toxic, teratogenic and mutagenic. TCOD is one of the most toxic substances known to mankind.

Synthetic organic compounds (many industrial chemicals, pesticides and PCBs) are generally determined by liquid-liquid extraction (LLE), GC with flame ionization detection (GC-FID), GC-MS, and HPLC with UV absorption and fluorescence detection methods. GC, with packed columns of organosilicon compound mixtures of variable polarity, is the standard method for determining pesticides. Detection is by ECD for organochlorine pesticides. Alkaline flame ionization detector (AFID) or flame photometric detector (FPD) for organophosphrous and organonitrogen pesticides and a FPD for organosulfur pesticides are also used.

### 21.4.1 Halogenated Hydrocarbons

Halogenated hydrocarbons are halomethanes and haloethanes, which are industrial solvents, chlorofluoro carbons, polychlorinated biphenyls (PCBs), and a class of persistent, broad-spectrum insecticides that linger in the environment and accumulate in the food chain. Among them are aldrin, chlordane, DDT, dieldrin, endrin, heptachlor, hexachloride, lindane, mirex, and toxaphene. Some commercially important halogenated hydrocarbons are given in Table 21.2.

The halogenation of a hydrocarbon involves the splitting up of a halogen molecule into two highly energetic halogen atoms. This step is followed attack of the energetic halogen atom upon a hydrocarbon molecule, leading to the formation of a molecule of hydrogen halide and a hydrocarbon *free radical* (free radicals are species having odd numbers of electrons; they are important intermediates in many organic reactions). This newly formed hydrocarbon free radical then attacks an unreacted halogen molecule to form the halogenated hydrocarbon. This reaction (*chain mechanism*) results in the replacement of one or more hydrogen atoms of a hydrocarbon with halogen atoms.

**TABLE 21.2:** SOME COMMERCIALLY IMPORTANT HALOGENATED HYDROCARBONS

| *Compound* | *Formula* | *Uses/Remarks* |
|---|---|---|
| Carbon tetrachloride | $CCl_4$ | Solvent; Poisonous |
| Chloroform | $CHCl_3$ | Solvent; Poisonous |
| Dichlorobenzene |  | Larvicide |
| Ethylchloride | $CH_3CH_2Cl$ | External local anesthetic |
| Freons | $CF_2Cl_2$; $CFCl_3$ | Gaseous refrigerants. Cause for ozone depletion |
| Halothane | $CF_3CHClBr$ | Anesthetic gas |
| Methyl chloride | $CH_3Cl$ | Refrigerant gas |
| Phenacyl chloride | 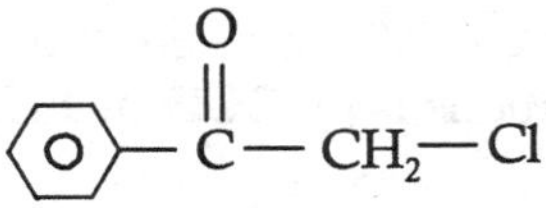 | Teargas used for crowd control |
| Polychlorinated Biphenyls (PCBs) | 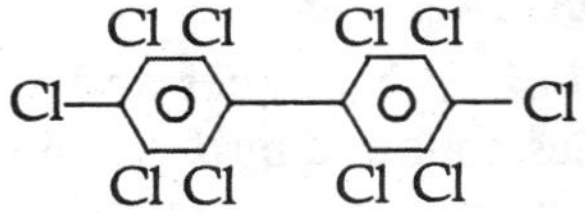 | PCBs are used in lubricants, hydraulic fluids, waxes, paints, insulators and plastics |
| Polyvinylchloride (PVC) | 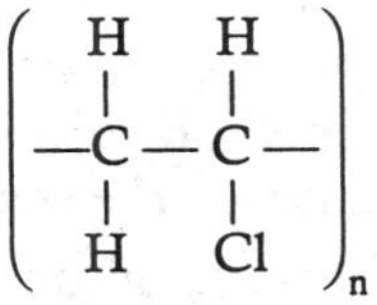 | Plastic polymer |
| Teflon | 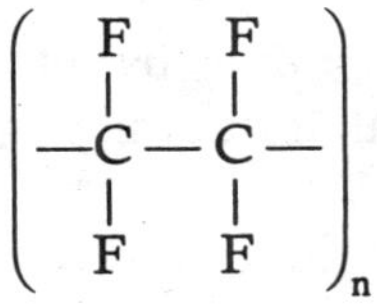 | Plastic material |
| Tetrachloroethylene | 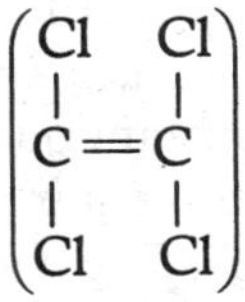 | Dry-cleaning solvent |

Halocarbons (e.g. $CHCl_3$, $CCl_4$, freons) are the target components of VOC-STRAT and VOC-CLIM. They have no known tropospheric sinks, and are known to be "greenhouse" gases on account of their long atmospheric lifetimes, and strong IR absorption bands in the atmospheric window regions. With the large-scale use of these compounds their concentration in the stratosphere has increased and it stands at about 3 ppbv.

Their contribution to ozone destruction in the stratosphere attracts much attention. They undergo photochemical decomposition, giving rise to the two active products, ClO and Cl atom. These secondary products are involved in ozone destruction. CFCs, halons, carbon tetrachloride, and methyl chloroform and HBFCs and ethyl bromide (added by EPA regulations), with ozone depletion potential of 0.2 or more are declared as Class I substances. All HCFCs with ozone depletion potential of less than 0.2 are termed as Class II substances.

Many halogenated hydrocarbons and polychlorinated biphenyls (PCBs) are priority pollutants. Most halogenated hydrocarbons are toxic and are potential carcinogens. Fluorocarbons are less toxic than chlorocarbons. Vinylchloride ($H_2C = CHCl$) is an intermediate for many synthetic polymers, but poses a serious health hazard.

Methods for determination of minor components of this group involve a pre-concentration step. This is achieved either by using cartridges packed with adsorbents like graphitized molecular sieve, or cryogenic traps. Capillary column chromatography, coupled with cryofocusing is employed to resolve the large number of components to be monitored. GC is the method of choice to monitor the most abundant hydro-halocarbons, in the atmosphere, like the freons. Halocarbons, which are abundant, are determination by: (1) total organic halogen (TOX) method, employing microcoulometry or electrolyte conductivity, or by (2) separation (by GC) and detection by microcoulometry, electrolyte conductivity, GC-MS or electron capture (ECD) methods. Preferred detection of volatile halogenated hydrocarbons is by ECD or FID.

#### *21.4.1.1 Total Organic Halogen (TOX) Method*

TOX measurement gives total organic chlorides, bromides and iodides. The method involves the thermal decomposition of the organohalides, followed by determination of the sum of the resulting HCl, HBr and HI produced. Coulometric and conductimetric methods are employed for this determination.

(*i*) TOX by coulometric method: In this method, the sample is decomposed in $O_2$-rich atmosphere and the resultant HX is measured. The steps are—

1. Concentration of the pollutants by adsorption on granular activated carbon (column adsorption module).
2. Removal of inorganic halogens compounds by nitrate wash.
3. Oxidative combustion of organic halides to the corresponding halogen acids (HX) and $CO_2$.
4. Measurement of HX by coulometric titration with silver ions.

After the liquid is passed through the adsorption column, the granular activated carbon is transferred to a combustion boat and heated in a furnace to ~800° C, in a stream of flowing $O_2$. At high temperature, organic halides are converted to HX, $CO_2$ and $H_2O$. The micro-coulometer detector maintains a constant quantity of $Ag^+$ ions, and when $Ag^+$ ions are removed by reaction with HX, the current required for re-establishing the $Ag^+$ ion balance is a measure of the total halogens in the cell. The coulometric instrument for TOX does not consume the carbon adsorbent, so that it can be used again.

(*ii*) TOX by conductimetric detection:

1. Concentration by adsorption on granulated activated carbon.
2. Removal of inorganic halogens by carbonate solution wash,
3. Volatilization of the sample at 800° C by $H_2$-burner, to convert organic halides to HX.
4. Detection of halides by their electrolytic conductivity.

These methods, though not very sensitive, are adequate for many purposes. GC-MS method is very sensitive, but expensive.

#### 21.4.1.2 *Hydrohalocarbons*

A group of hydrofluorocarbons (HFCs) and hydrochlorofluorocarbons (HCFCs) are substituting the CFCs. Since the impact of the new compounds on the environment is not yet adequately established there is a need to monitor them carefully. Since these compounds contain C—H bonds, they are prone to be easily attacked by OH radicals in the troposphere and thus, their migration to the stratosphere is reduced. However, their halogen content makes them harmful to the stratosphere to the extent that they reach that region.

Analysis of hydrohalo carbons is carried out conveniently by GC. At the pptv level they are present, GC-MS is the most preferred technique adopted. Commercially available carbon molecular sieve filled columns are used for their concentration. Following thermal desorption, they are separated in capillary columns and determined through MS.

## 21.5 AGROCHEMICALS

Pesticides, germicides, insecticides, fungicides, herbicides, and defoliants are inorganic, organic and organo-metallic agrochemicals, produced industrially to control insects and pests to promote crop growth. Many of them are phenol-based (e.g. herbicides 2,4-dichlorophenol; 2,4,5-trichlorophenoxy acetic acid), organosulfur, organophosphorus (e.g. malathion, parathion) compounds and halogenated hydrocarbons (e.g. aldrin, dieldrin, DDT, endrin). Almost 70 per cent of all pesticides used by farmers and ranchers are herbicides.

Pesticides enter the atmosphere via aerosol convection and into the aquatic environment by airborne drift (dust and spray), discharge, spillage and percolation. Pesticide residues are separated and concentrated by methods employed for other synthetic organic compounds—by liquid-liquid extraction with a solvent like hexane; and are determined by colorimetric, total halides (TOX), ACh-estarase (AChE) inhibition and bioassay methods. Thin layer chromatography (TLC) and GC are more sensitive and are employed for volatile compounds. For less volatile pesticides, HPLC is employed. Separation and detection methods that are applicable to pesticides are also applicable for determination of herbicides, fungicides and other agrochemicals.

Wet-chemical methods for analysis of some organic compounds are summarized in Table 21.3.

**TABLE 21.3:** WET-CHEMICAL METHODS FOR ANALYSIS OF SOME ORGANIC COMPOUNDS

| *Species* | *Procedure* | $\lambda_{max}$ (nm) |
|---|---|---|
| Phenols | 4-Aminoantipyrine (pH > 8.0) method | 460; 550 |
| | Ferricyanide method | 460; 550 |
| | 3-Methyl-2-benzothiazoline (MBTH) method | 460 & 510 |
| Phosphorus | $PO_4$ + Molybdate ion → phosphomolybdate → Molybdenum blue | |
| Sterols | Aceticanhydride + $H_2SO_4$ + chloroform | 425 |
| Surfactants | Methylene blue to form blue salt | |
| Tannins & lignins | Blue color from tungstophosphoric & molybdophosphoric acids | |

Chemical methods are unsuitable for determination of individual hydrocarbons in mixtures. In the analysis of complex mixtures, it is necessary to combine chemical and physical methods. Many trace organic toxic chemicals (e.g. dioxin-related compounds) can be determined by resonance enhanced multiphoton ionization (REMPI) method. REMPI is an on-line, real-time analytical technique that combines compound-specific selectivity with high sensitivity. The instrument employs a tunable laser for ionization, leading to 2-D spectra (ion-mass and wavelength) and time-of-flight mass spectrometer (see Chapter 22) for detection.

## BIBLIOGRAPHY

Aragon, P., Atienza, J. and Climent, M. D. (2000), *Crit. Rev. Anal. Chem.*, **30**(2 & 3): 121. "Analysis of organic compounds in air: A review."

Barcelo, D. (ed.) (1993), Elsevier: Amsterdam. "*Environmental Analysis: Techniques, Applications and Quality Assurance.*"

Botsford, J. L. (1999), *Environ. Toxicol.*, **14**(2): 285. "*A simple method for determining the toxicity of chemicals using a bacterial indicator organism.*"

Current, R., Meyer, M. J. and Borgerding, A. J. (2001), *Talanta*, **55**(3): 519. "Rapid aqueous sample extraction of VOCs: effect of physical parameters."

Frei, R. W. and Albarge, J. (eds.) (1986), Gordon & Breach: New York. "*Air and Water Analysis: New Techniques and Data.*"

Herber, R. F. M. and Stoeppler, M. (eds.) (1994), Elsevier: Amsterdam. "*Trace Element Analysis in Biological Specimens.*"

Jorgensen, S. E. and Johnsen, I. (1981), Elsevier: Amsterdam. "*Principles of Environmental Science and Technology,*" (Vol. 14).

Lapa, R. A. S., Lima, J. L. F. C. and Pinto, I. V. O .S. (1999), *Intl. J. Environ. Anal. Chem.* "Determination of phenolic compounds in wastewaters by sequential injection analysis and spectrophotometry."

Keith, L. H. (ed.) (1984), Butterworth Pubs: London. "*Identification and Analysis of Pollutants in Air.*"

Minear, R. A. and Keith, L. H. (eds.) (1994), Academic Press: New York. "*Water Analysis: Organic Species.*"

Patra, D. and Mishra, A. K. (2001), *Talanta*, **55**(1): 143. "Investigation on simultaneous analysis of multicomponent polycyclic aromatic hydrocarbon mixtures in water samples: A simple synchronous fluorimetric method."

Pino, V., *et al.* (2001), *Talanta*, **54**(1): 15. "Ultrasonic micellar extraction of polycyclic aromatic hydrocarbons from marine sediment."

Sakellarides, T. M. and Albanis, T. A. (2000), *Intl. J. Environ. Anal. Chem.* "A new organophosphorus insecticides removal process using fly ash."

Thain, W. (1980), Pergamon Press: Oxford, UK. "*Monitoring Toxic Gases in the Atmosphere for Hygiene and Pollution Control.*"

Van Loon, J. C. (1985), John Wiley: New York. "*Selected Methods of Trace Metal Analysis: Biological and Environmental Samples.*"

—(1995), EPA/600/R-95/131: Washington, DC. "*Methods for the Determination of Organic Compounds in Drinking Water.*"

—(1999), EPA/625/R-96/010b: Cincinnati, Ohio. "*Compendium of Methods for the Determination of Toxic Organic Compounds in Ambient Air,*" (2nd edn.).

# SECTION-VII

# Physical Methods in Environmental Pollution Analysis

## OVERVIEW

Chapters 22 to 26 deal with the physical techniques that are useful for the detection and determination of environmental pollutants. Separation methods (Chapter 22) are an integral part of pre-concentration and analysis of pollutants. Their quantification is carried out by several physicochemical and physical techniques, which include electro-analytical methods (Chapter 23), spectroscopic methods (Chapter 24), nuclear analytical techniques (Chapter 25) and other techniques, such as X-ray diffraction and microscopies (Chapter 26).

## SEPARATION METHODS

Separation procedures include pre-concentration of trace level pollutants, and isolation of individual components when necessary. Many of these methods are based on sedimentation, osmosis (and filtration), solvent-extraction, sorption (chromatographic methods) and mass spectrometry (MS).

### Sedimentation Methods

Precipitation and centrifugation, which are among the sedimentation methods, have become an inseparable part of purification and characterization of biological species.

### Osmotic Methods

Dialysis, ultrafiltration, reverse osmosis and electrodialysis are some of the membrane-based methods employed for separation, purification and concentration of analytes. Dialysis is frequently used in chemical, biological, pharmaceutical and environmental studies. Ultra filtration, separation under hydrostatic pressure, is extensively used in separation, purification and concentration of biological macromolecules. Reverse osmosis is ultrafiltration, where the molecular size of the solute is about the same as that of the solvent (~1 nm). Electrodialysis is transport (osmosis) of electrolytes from a solution across a semi-permeable membrane under the influence of an electromotive force.

### Solvent Extraction

Solvent extraction (also known as liquid-liquid extraction, LLE) is a separation as well as a pre-concentration technique. It is based on the differential solubility of solutes in a mixture of two immiscible liquids. One of them is frequently, water. LLE is applicable for both inorganic and organic compounds.

### Sorption Methods

Adsorption, absorption and chemisorption principles are implied in chromatographic and electrophoretic methods. Their extensive use in various fields, that include determination of environmental pollutants, is because of the simplicity in their operation and technical advances. All chromatographic methods are aimed at separation of analytes, present in a liquid or gas sample, based on their differential migration rates in a mobile phase (liquid or gas) and an immiscible stationary phase (solid, liquid or gel). Judicial selection of stationary and mobile phases makes the chromatographic techniques highly versatile and flexible. Thin Layer Chromatography (TLC), Liquid Chromatography (LC) that includes, Size-exclusion (gel filtration/molecular sieve) Chromatography, Affinity Chromatography, Ion-exchange Chromatography (IEC), and Gas

Chromatography (GC) and Supercritical Fluid Chromatography (SFC) are among the variations in the chromatographic technique that can be adopted for successful separations.

In Liquid Chromatography (LC), while the mobile phase is a liquid solvent containing the analytes, the stationary phase depends on the method of separation. In size-exclusion chromatography, molecular (size) separation is achieved on the basis of steric hindrance (molecular sieving) property of porous particles (gels). Affinity chromatography takes advantage for separating the highly specific interactions (enzyme-substrate, antigen-antibody interactions) that exist between biomolecular species. In ion-exchange chromatography, separation is based on the ionic character of the ion-exchange sorbent (the stationary phase). Gas Chromatography (GC) is an extremely sensitive technique that is ideally suited for separation of volatile compounds. Supercritical Fluid Chromatography (SFC) technique exploits the advantages of liquid-like solvent properties of LC methods, and high resolution achievable in GC methods. Chromatographic techniques, combined with mass spectrometry (e.g. GC-MS) are increasingly employed in the trace analysis of environmental pollutants.

Electrophoresis is a separation method based on chemisorption. Separation is achieved based on differential migration (electrophoretic mobility) of ions under the influence of an applied electric field. Gel electrophoresis methods, in conjunction with blotting techniques, are widely used in molecular biology for purification and characterization of biomolecules (e.g. genome project).

### *Detectors*

The success of chromatographic techniques is very much due to the availability of sensitive detection and determination devices for the separated individual components. A wide variety of detectors are in use. Flame ionization (FID), photoionization (PID), electron capture detectors (ECD) and mass spectrometry (MS) are the most frequently used ones. They are useful for analysis of volatile compounds by GC. FID is ideally suited for determination of combustible hydrocarbons (HCs), while ECD is good for determination of electron-rich compounds. Optical absorption, fluorescence and electro-analytical techniques are useful in analyzing eluents from liquid chromatography.

## ANALYTICAL METHODS

### Electro-analytical Methods

Electro-analytical methods (Chapter 23)—Voltammetry, polarography, amperometry, condutimetry, coulometry and potentiometry and their variations, offer sensitive and precise measurement of inorganic and organic trace pollutants in different matrices. Most of these methods rely on the flow of electrons between one or more of the electrodes and the analyte. The names of the methods reflect the measured electric property or its unit. Some of them like voltammetry and polarography have been extensively studied and applied for the determination of inorganic as well as organic pollutants like lead, cadmium, mercury and selenium and pesticides, hazardous chemicals and carcinogens. In voltammetric methods, pre-concentration of analytes is a part of the analysis cycle. Potentiometry, with ion-selective electrodes, has a wide spread use in clinical, biological and environmental fields for carrying out fast, reliable local measurements of ion activities with a remarkable selectivity.

### Spectroscopic Methods

Spectroscopic methods (Chapter 24)—molecular (UV-visible, fluorescence, infrared (IR) and Raman), and atomic (absorption, emission and fluorescence)—are extensively used in quantitative analysis of environmental pollutants (Chapter 24). X-ray photoelectron and Auger electron spectroscopies are dealt under nuclear analytical techniques (Chapter 25).

### UV-Visible Absorption Spectroscopy

UV-visible absorption spectroscopy ($\lambda$ = 150-750 nm range) deals with electronic (vibrational and rotational) transitions of molecules. Most wet-chemical methods of analysis employ UV-visible spectrometry for quantification of an analyte, based on Beer-Lambert law, which correlates the concentration of analytes to the attenuation of the incident radiation by them. Both inorganic and organic pollutants can be determined by this technique. Spectrophotometers are common instruments in all laboratories dealing in chemical analysis.

Turbidimetry and nephelometry are employed to measure the attenuation of light by non-colored particulate and coagulate matter (transmitted by turbidimeters and scattered light by nephelometers). These techniques find application in evaluation and monitoring of particulate pollutants/ contaminants in food, beverage, water and other sources. Diffuse Reflection Spectroscopy (DRS) can be employed in color measurement of films and for studying biological effects of pollutants adsorbed on surfaces, particularly on plants. Resonance-ionization spectroscopy is an extremely sensitive and highly selective analytical tool. Thermal lens spectroscopy (PLS) has wide range of applications in the analysis of contaminants in food and environmental samples.

### Fluorescence Spectroscopy

Fluorescence and phosphorescence are more sensitive than absorption spectroscopic methods ($10^3$-$10^5$ times) and they provide a great degree of sensitivity and specificity. They are widely used in "fingerprinting" analysis of environmental pollutants. They are also employed in enzyme assays, immunoassays, fluorescence-activated cell sorting in biological systems (molecular immunology), clinical pathology and epidemiology. Many substances exhibit either direct or induced fluorescence or fluorescence quenching, and these properties are exploited for their assay. Enzyme-mediated luminescence assays are very sensitive for monitoring cellular events and effects of pollutants on biochemical reactions (e.g. determination of $O_2$-metabolites by luminol assay). Luminescence-based assays are alternate to radioactivity-based immunoassay methods (RIA and IRMA). But, they promise a great potential in environmental monitoring and disease control methods (e.g. in epidemiology, parasitology).

### Infrared (IR) Spectroscopy

Infrared (IR) spectroscopy is a molecular absorption phenomenon. IR spectroscopy is a suitable technique for analysis of environmental pollutants. It can be used in continuous monitoring of atmospheric gases, such as CO, $CO_2$, $SO_2$, and combustible hydrocarbons and contaminants in food and beverages. Non-dispersive IR (NDIR) is commonly used for determination of CO and $CO_2$.

### Raman Spectroscopy

Raman effect manifests due to inelastic scattering of light by molecules. Raman spectroscopy can be employed in the analysis of environmental pollutants. The instrumental setup for Raman studies is simpler and cost effective as compared to that for IR studies. It holds a great promise in environmental pollution analysis.

### *Atomic Spectrometry*

Atomic spectrometry covers a class of spectroscopic methods in which the species examined are in the form of atoms. Absorption (AAS) and emission (AES) methods are the important methods that come under this category. The most common method of introducing a sample is to aspirate it, as a solution, into a high temperature flame (>2,000 °C) or plasma (about 4,000 °C) to atomize it. In AAS, the absorption of characteristic radiation by ground state atoms is measured, and related to the concentration of the analyte in solution, while in AES, the emission of radiation from the excited atoms is measured. Multiple elements can be analyzed by flame resonance AAS as well as by AES methods. Since the techniques require the analyte to be converted to atoms, they are useful only for elemental analysis and for various practical reasons, essentially suitable for analysis of metals. For trace metals these methods have a wide range of application being simple, fast and sensitive. Atomic spectrometry, with all its variations, plays a crucial role in pollution monitoring of air, water and soil samples, after pre-concentration wherever needed. Inductively-coupled plasma (ICP) excitation, coupled with mass-spectrometry is an extremely sensitive and versatile technique for determining trace level pollutants in environmental samples.

### Nuclear Analytical Methods

Nuclear methods for the analysis of environmental pollutants (Chapter 25) involve direct measurement of high-energy radiation ($\alpha$-particles, $\beta$-particles, $\gamma$-rays/X-rays) as well as indirect methods of analysis (e.g. neutron activation analysis, NAA). Radiation counters are the detection devices employed to measure radioactivity. They are gas-filled tubes (proportional, ionization and Geiger-Müller counters) semiconductor (e.g. Si-Li) devices and scintillation counters. Natural radioactivity due to uranium, thorium and their disintegration products (mainly, radon) is all pervasive, apart from any anthropogenic radioactive contamination. Therefore, monitoring of radioactivity is often called for.

X-ray fluorescence (XRF) methods are widely used in qualitative and quantitative analysis of elements above $Z > 5$. Qualitative analysis involves measuring wavelengths of the emitted lines whereas quantitative analysis comprises of measuring the intensities of the lines. X-ray photoelectron spectroscopy (XPS) is used to determine the chemical composition ("fingerprint" analysis) of elements. Auger electron spectroscopy and Rutherford backscattering spectroscopies are used in the analysis of surface characteristics.

### *Nuclear Reaction Analysis (NRA)*

Radioactive tracer techniques provide on-location and on-time information on sediment transport, effluent discharge and physical dilution of pollutants in estuaries, rivers and seashores. They are extensively used in medical diagnostic and therapy, biological and environmental pollution studies.

In Neutron Activation Analysis (NAA), the sample is irradiated with thermal neutrons and the induced radioactivity is measured as a function of time (radioactive decay) and compared with activity of a known standard. One advantage of the NAA method is that pre-concentration before irradiation is not, usually, required. Though NAA methods of analysis provide high sensitivity (ppb level), they are nuclear reactor-based operations and cost-wise more expensive.

### *Radioimmunoassays*

Radionuclide-labeled immunoassays (RIA and IRMA) find place in biomedical, forensic, epidemiological and environmental (food and drug monitoring) studies. Radioimmunoassays combine the high degree of specificity of antibody-antigen (hapten) reaction with high sensitivity and accuracy possible with radioactivity measurements. Radioimmunoassay (RIA) is a competitive immunoassay method, where the analyte (Ag), together with a fixed quantity of radiolabeled anaylte (Ag*) compete with the binding sites of a fixed and limited quantity of antibody (Ab). Immunoradiometric assay (IRMA) is a non-competitive immunoassay method, where excess of radiolabeled antibody (Ab*) is allowed to interact with the analyte (Ag), to be estimated. After the reaction, excess (unbound) Ab* is eluted, and the amount of bound analyte (Ag) in the Ab*-Ag complex is determined from the radioactivity measurement of Ab*.

## Other Techniques

Some of the other techniques employed in the analysis of environmental pollutants are X-ray diffraction and microscopic methods (Chapter 26). These techniques are employed to determine morphological and structural details of pollutants.

### *X-ray Diffraction and Microscopy Methods*

X-ray methods provide structural details at atomic and molecular resolutions. In the analysis of environmental pollutants, powder X-ray diffraction (Debye-Sherer method) is generally used for the determination of composition and morphological details of semi-crystalline (polycrystalline) specimen. Optical microscopy is used for determining the size and number distribution of particulate pollutants, while electron microscopy provides high-resolution ultrastructure details.

## *Noise Pollution*

Noise pollution is increasingly becoming a nuisance and health hazard. Noise is undesirable sound, measured in terms of pressure levels (decibel) by sound level meters. Effects of noise pollution have physical, physiological and psychological dimensions. Noise above 80 decibels (tolerance level) is of concern to human health and prolonged exposure to such noise leads to mental imbalance. General methods of noise pollution control are—(1) removal of the source, (2) reduction of the decibel level, (3) minimizing the vibratory parts of the gadgets and machines by better design, fastening and padding with noise insulators.

# 22

# Physical Methods in Environmental Pollution Analysis—Separation Techniques

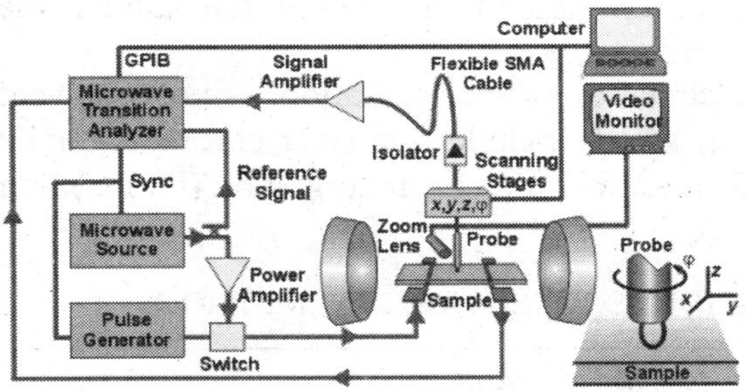


Separation procedures include pre-concentration of trace level pollutants, and isolation of individual components when necessary. Many of these methods are based on sedimentation, osmosis (and filtration), solvent-extraction, sorption (chromatographic methods) and mass spectrometry. Solvent extraction is a separation as well as a pre-concentration technique.

Precipitation and centrifugation have become an inseparable part of purification and characterization of biological species. Dialysis, ultrafiltration, reverse osmosis and electrodialysis are some of the membrane-based methods employed for separation, purification and concentration of analytes.

Adsorption, absorption and chemisorption principles are implied in chromatographic and electrophoretic methods. Their extensive use in various fields, that include determination of environmental pollutants, is because of the simplicity in their operation and technical advances. Electrophoresis is a separation method based on chemisorption. Separation is achieved based on differential migration of ions under the influence of an applied electric field. Gel electrophoresis methods, in conjunction with blotting techniques, are widely used in molecular biology for purification and characterization of biomolecules.

The success of chromatographic techniques is very much due to the availability of sensitive detection and determination devices for the separated individual components. A wide variety of detectors are in use. Flame ionization (FID), photoionization (PID), electron capture detectors (ECD) and mass spectrometry (MS) are the most frequently used ones. They are useful for analysis of volatile compounds by GC. FID is ideally suited for determination of combustible hydrocarbons (HCs), while ECD is good for determination of electron-rich compounds. Optical absorption, fluorescence and electroanalytical techniques are useful in analyzing eluents from liquid chromatography.

This chapter outlines the details of separation techniques used in environmental pollution analysis.

The main instrumental methods in environmental pollution analysis can be classified under: (1) separation/detection methods, (2) electroanalytical methods (3) spectroscopic methods and (4) nuclear analytical methods. Physical methods of separation and purification can be addressed under two categories. The first category is where physical properties such as density, viscosity, conductivity and magnetic properties are determined, for which prior separation of components of a mixture is not necessary, if the components of that mixture differ greatly in their physical characteristics. The second category is where separation methods are employed—particle separation such as sedimentation, filtration, floatation, distillation, sorption, diffusion, and chromatography, and field separation such as ultracentrifugation and electrophoresis, and subsequent analysis by physicochemical and physical techniques. In some cases, like mass spectrometry (MS), separation and detection are integral parts of analysis.

Separation techniques are aimed at resolving individual components of a mixture by exploiting their chemical and physical properties. Separation of components in a high degree of purity is essential for some purposes like identification and their determination. Equilibrium separation and concentration methods depend on the phase-state of the sample mixture and the extracted state of the analyte (Table 22.1). Ease of separation in equilibrium methods is based on the value of the separation factor. If it is large, separation is easy.

Some of the common separation methods are precipitation or sedimentation, osmosis, solvent extraction and sorption (chromatographic) methods, which include thin-layer chromatography (TLC), liquid chromatography (LC), and high performance liquid chromatography (HPLC), ion chromatography (IC) and gas chromatography (GC).

**TABLE 22.1:** CLASSIFICATION OF SEPARATION (CONCENTRATION) METHODS

| *Phase State* | *Extracted State* | *Method of Concentration* |
|---|---|---|
| Gas | Gas | Thermodiffusion |
| Gas-liquid | Gas or liquid; | Evaporation |
| | Liquid; | Wet mineralization |
| | Gas/liquid/solid | Sorption (chromatography) |
| Gas-solid | Solid | Filtration; volatilization; sublimation |
| Liquid | Liquid | Dialysis; chromatography; electrochemical |
| Liquid-liquid | Liquid; | Liquid-liquid extraction; thermodiffusion |
| | Solid | Fire assay |
| Liquid-solid | Solid | Precipitation; evaporation; filtration; sorption; electrodeposition; dialysis; flotation; crystallization |
| Solid-solid | Solid | Crystallization; zone melting |

## 22.1 SEDIMENTATION METHODS

Centrifugation is a sedimentation (precipitation) technique, wherein individual constituents of a mixture are separated by an applied centrifugal force (applied gravitation). The constituents are separated based on their sedimentation rates. The rate of sedimentation of a particle is a function of molecular mass, shape and size of the particle. In a centrifugal field, the rate at which a particle sediments is related to the net centrifugal force, $F_C$, acting on the particle and this is opposed by the buoyant force, $F_b$, and the net sedimentation force $F = (F_C - F_b)$. For a spherical particle

$$F_C = m \cdot \omega^2 \cdot r; \qquad F_b = V \cdot \rho \cdot \omega^2 \cdot r, \quad V = m \cdot V' \qquad \text{...(22.1)}$$

$$F = (F_C - F_b) = m \cdot \omega^2 \cdot r\,(1 - V'\rho) \qquad \text{...(22.2)}$$

Centrifugal force, $F$, is opposed by the frictional drag, $F'$, and at equilibrium $F = F'$.

$$F = f\frac{dr}{dt}; \qquad \text{...(22.3)}$$

Drift velocity $\frac{dr}{dt} = s\omega^2 r$ (by definition)

At equilibrium, $m\omega^2 r(1 - V'r) = f \cdot s\omega^2 r$

But, $$f = \frac{RT}{N_0 D} \qquad \text{...(22.4)}$$

$\therefore$ Molecular mass, $$M_r = \frac{RT}{(1 - V'\rho)} \cdot \frac{s}{D} \qquad \text{...(22.5)}$$

The molecular mass of a particle can be determined without the knowledge of the diffusion coefficient, $D$, by the sedimentation equilibrium method.

$$M_r = \frac{2RT}{(1 - V\rho)\omega^2} \cdot \frac{\ln(C_2 / C_1)}{(r_2^2 - r_1^2)} \qquad \text{...(22.6)}$$

where, $V$ = volume of the particle; $V'$ = partial specific volume of the particle = $(1/\rho_{solute})$; $\rho$ = density of the solution; $m$ = mass of the particle; $\omega$ = angular velocity; $r$ = radius of the particle; $f$ = frictional coefficient; $s$ = sedimentation coefficient; $D$ = diffusion coefficient; $N_0$ = Avogadro's number; $R$ = gas constant; $T$ = absolute temperature; $C_1$ and $C_2$ are the concentrations of analyte at $r_1$ and $r_2$ distances in the centrifuge tube, from the axis of rotation.

The molecular mass, $M_r$, of the particle (molecule) can be evaluated from the slope of the plot of $\ln(C_r)$ versus $r^2$ (Fig. 22.1).

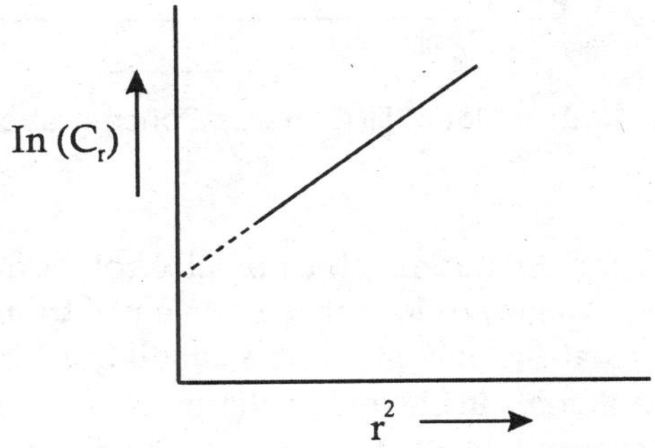


**Fig. 22.1:** Plot of Concentration, $\ln(C_r)$, versus $r^2$ in the Sedimentation-Equilibrium Method

For a non-spherical particle, sedimentation depends not only on its mass, but also on its size and shape. Radius of gyration, $R_g$, translational and rotational diffusion coefficients ($D_t$ & $D_r$) provide information about the size and shape of the particle.

Ultracentrifugation methods have become an integral part of isolation, purification, and physicochemical characterization of biological macromolecules—proteins, nucleic acids, viruses, venoms, toxins, etc.

## 22.2 BARRIER SEPARATION METHODS

Osmotic separation methods are membrane-based (barrier) filtration techniques that can be employed for selective separation of constituents from a mixture. Dialysis, ultra-filtration, reverse osmosis and electrodialysis are some of the methods employed for separation, concentration and purification. Osmosis is a phenomenon in which only the solvent is free to migrate through a barrier (semi-permeable membrane). Most membrane filtration methods employ semi-permeable membranes for selective transport of constituents. In dilute solutions, applicability of the gas law, $PV = nRT$, can be invoked to correlate the hydrostatic pressure with molecular parameters, namely, concentration ($C$) and molecular mass ($M_r$) by osmotic methods.

$$\frac{\Pi}{C} = \frac{RT}{M_r} \quad \text{...(22.7)}$$

where, $\Pi$ = Hydrostatic pressure,
$C$ = solute concentration;
$R$ = gas constant;
$T$ = Kelvin temperature.

Molecular mass, $M_r$ can be evaluated from a plot of $\Pi/C$ versus $C$ (Fig. 22.2).

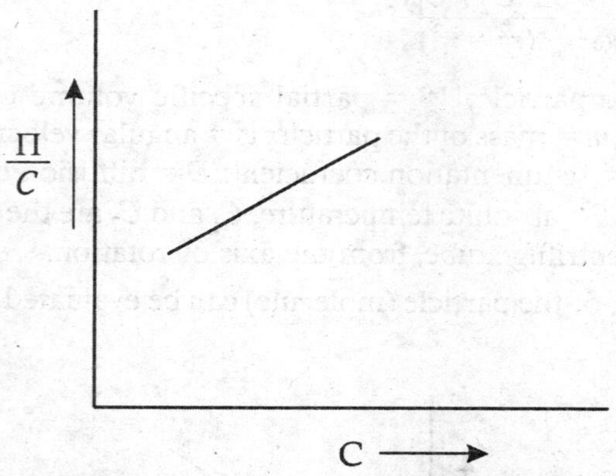


**Fig. 22.2:** A plot of $\Pi/C$ versus Concentration, $C$

### 22.2.1 Dialysis

Normal dialysis is a method for separation of a non-electrolyte from other non-electrolytes or electrolytes present in a mixture. When an electrolyte is separated from other electrolytes, the process is known as Donnan dialysis. Dialysis (including Donnan dialysis) is a frequently used method for recovery and purification of materials in chemical, pharmaceutical, biological, medical (e.g. urine dialysis), and food and beverage industries. Colloids can be effectively separated from dissolved molecules (e.g. purification of biochemical substances such as proteins). Both natural and synthetic membranes are available for dialysis (cellulose acetate, collodion, and gelatin). Cellophane membrane of pore size 0.3-0.5 µm is commonly used.

Permeation is a method that combines the sorption and diffusion properties of the components of a mixture under separation. The mechanisms of permeation consist of sorption of the penetrants on the membrane surface, diffusion through the membrane, and desorption of the penetrants from the other side of the membrane. The method is used for separation of gaseous and liquid pollutants, and also for separation of mixtures with similar chemical structure and close boiling points (e.g. azotropic mixtures).

### 22.2.2 Ultra-filtration

Ultra-filtration is separation under hydrostatic pressure. In ultra-filtration, an inert gas at a constant pressure is passed through a filtration cell supported by a membrane disc. Membranes of cellulose nitrate, cellulose acetate and acrylonitrile polymers, with pore diameters in the range 0.001-0.1 μm are used. This technique is used for separation, purification and concentration of high- and low-molecular mass biological macromolecules—proteins, nucleic acids and their complexes, viruses and polysaccharides. It may also be used to separate trace metals in environmental samples.

### 22.2.3 Reverse Osmosis

Reverse osmosis is an ultra-filtration method in which pressure is applied opposite to and in excess of the osmotic pressure, to force the solvent through the membrane against a concentration gradient. This method is suitable for separation of particulate matter whose molecular size is comparable to that of the solvent (~1 nm). It is employed in the separation of smaller molecular solutes (<500 Dalton), and in the purification of saline water and in wastewater treatment. In this process, application of pressure above the osmotic pressure enables the transport of fresh water from the solute side to the other side of the membrane barrier, leaving more concentrated salt solution behind (see Chapter 28 for details).

### 22.2.4 Electrodialysis

While in normal osmosis solutes are transported due to hydrostatic pressure gradient, in electrodialysis electrolytes are transported from a solution mixture across a semi-permeable membrane barrier under the influence of an applied electromotive force (e.m.f). The apparatus consists of a series of anion and cation exchange membranes, arranged alternately and bound together between a cathode and an anode (Fig. 22.3). The aqueous salt solution is pumped through the membrane stack and a DC potential is applied between the anode and the cathode. The positively charged ions in the solution migrate towards the cathode and *vice versa*. While the positively charged ions (cations) pass easily through the cation exchange membrane, the anion exchange membrane retains them. Similarly, the negatively charged ions in the solution migrate towards the anode and pass easily through the anion exchange membrane.

Electrodialysis is used in the recovery of Cd, Cr, Cu, Ni, and Zn from electroplating solutions, and in stripping pollutants such as Cd, Hg and Pb from industrial waste. Liquid membranes (emulsion membranes, liquid surfactant membranes) are used for removing toxic substances from water, either by diffusion ($NH_3$, $H_2S$ and phenols, etc.) or by metal chelation (Cd, Cr, Cu and Hg) and also for urine dialysis.

## 22.3 SOLVENT EXTRACTION

Solvent extraction is a separation as well as a pre-concentration technique based on difference in solubility of substances in immiscible solvents. In liquid-liquid extraction (LLE), in most cases, water and a water-immiscible solvent (or solvents) are used to create a two-phase system where the analyte is transferred preferentially to the organic phase. Extraction is given by the phase rule

$$\alpha = N + F = Z + 2 \qquad \text{...(22.8)}$$

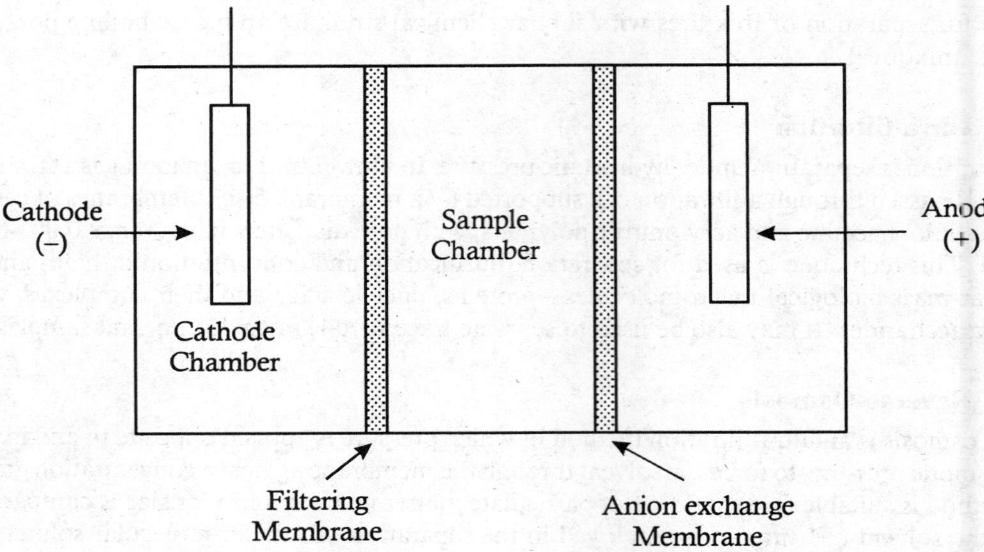


**Fig. 22.3:** Schematic of a Three-chamber Electrodialyser

In the case of distribution of two components *A* and *B* between solvents 1 and 2 their separation factor, α, is given by the equation

$$\alpha = \frac{\left(\frac{N_1}{N_2}\right)^A}{\left(\frac{N_1}{N_2}\right)^B} \qquad \ldots(22.9)$$

where, $N_1$ & $N_2$ = mole fraction of components *A* and *B* in solvents 1 and 2, respectively.

Extraction is normally carried out in a water-organic solvent system. LLE is one of the more frequently used pre-concentration techniques for both organic and inorganic components in water. For metal extraction the procedure involves bringing a water-immiscible organic solvent in contact with an aqueous solution containing a chelating agent. The choice of extractant is dictated by the nature of the target metal. Whereas oxygen-containing solvents more efficiently extract 'hard' metals, S- or P-comtaining solvents better extract 'soft' metals. The low-ionic organometallic compounds are extracted by organic solvents (hexane, pentane, dichloromethane), but inorganic ions require chelating agents to make them pass into the organic phase. Common chelating agents are 8-hydroxy-quinoline (oxine), dimethylglyoxime (DMG), diethydithiocarbamate (DDTC), dithizone (DTz) and acetyl acetone (AcAc).

Solid-phase extraction (SPE) methods can be employed for separation of a variety of organic species. Gas chromatography (GC) is the method of choice for volatile, thermo-stable organic compounds. Solid-state extraction methods (with XAD resins) are capable of isolating ~60 per cent of the water-soluble organic compounds from aerosol samples. The isolated organic matter is nearly free from inorganic ions, which are major constituents of atmospheric aerosols. The content of organic matter can then be estimated from the total organic carbon (TOC), determined by catalytic combustion at 680° C in oxygen. The evolved $CO_2$ can be determined by non-dispersion IR (NDIR). Solid-phase micro-extraction (SPME) followed by GC-MS methods are used for the determination of PAHs, VOCs and pesticides in air, water, and soil samples. Derivatizing analytes can widen the range of application. SPME can be combined with other techniques of sample preparation. SPME provides low-cost evaluation of water properties.

Microwave-assisted extraction method is suitable for thermally labile environmental materials (VOCs) and is superior to Soxhlet extraction. Adsorption by activated carbon is commonly used for collecting trace components from air samples, prior to GC analysis.

Supercritical fluid extraction is a superior extraction method. It uses supercritical fluids (high-purity $CO_2$) and combines the flexibility and versatility of other conventional extraction methods. The method may be automated for trace analyses of metals in natural and sewage waters (Fig. 22.4).

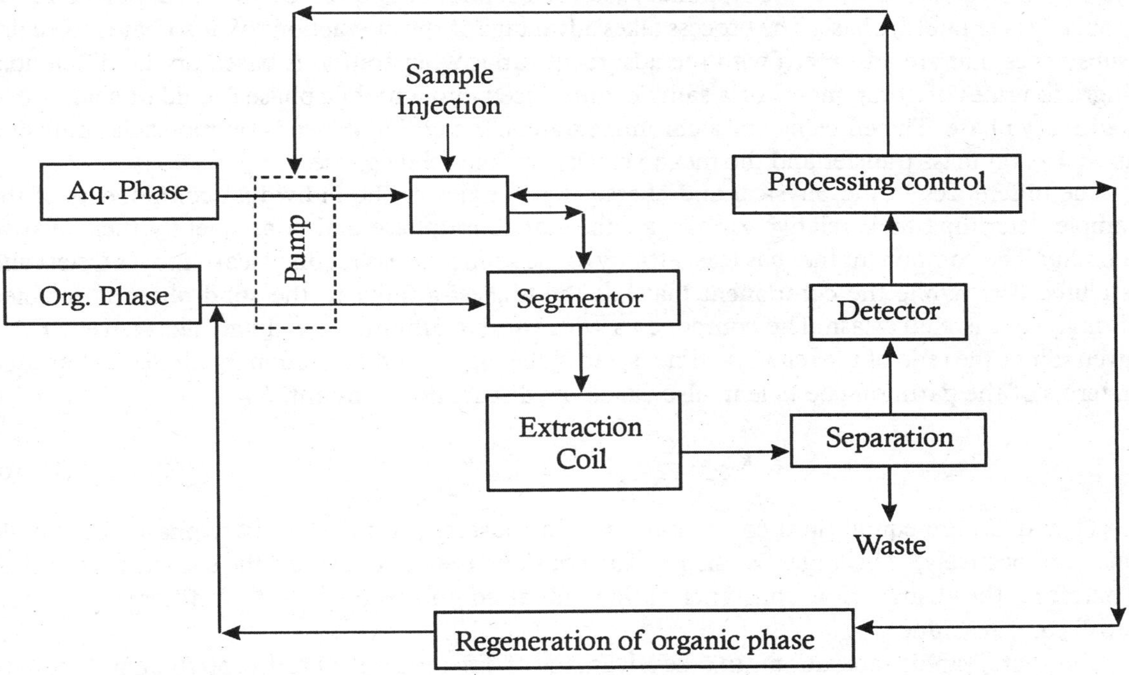


**Fig. 22.4:** Schematic of an Autoanalyzer for Trace Analysis of Metals in Water

## 22.4 SORPTION (CHROMATOGRAPHIC) METHODS

Separation and concentration by sorption methods encompass a variety of techniques by which a group of substances are separated on the basis of their distribution between two phases, one stationary and the other moving. These techniques can be classified (*i*) on the basis of the physicochemical properties involved–absorption, solvent partition, adsorption, and chemisorption, or (*ii*) according to the technique used—paper chromatography, thin-layer chromatography, etc. Although the apparatus used in adsorption chromatography can be the same as that used in partition chromatography, there is an important theoretical difference between the two techniques. The chromatographic technique is either partition or adsorption type depending on the solvent used. While in adsorption chromatography a mixture of components are separated based on the basis of differential adsorption coefficients, in partition chromatography they are separated on the basis of their differential distribution between two phases, one being stationary and the other mobile (e.g. paper-, thin-layer chromatography). Chromatographic and electrophoretic techniques are extensively used for separation, purification and physicochemical characterization of compounds in chemical, biological, pharmaceutical, forensic, medical and environmental sciences. Their extensive use is due mainly to their simplicity of operation and efficiency. As a separation method, chromatography has a number of advantages over solvent extraction, or distillation. It is capable of separating multi-component mixtures, with species ranging from smallest molecules to cells and viruses. Its resolving power is unequalled among separation methods.

Chromatographic techniques are employed for the separation of peptides, proteins, nucleic acids and their constituents, carbohydrates, lipids, drugs, dyes and pollutants. The chromatographic system involves two phases, the stationary (solid or liquid, bound on to a solid support) and the mobile (gas or liquid) phase. The process takes advantage of the interactions of adsorbate molecules (substances that are adsorbed) with the adsorbent surface. Separation is based on the differential migration rates of components of a sample, introduced into a mobile phase (liquid or gas), over a stationary phase. The efficiency of a chromatographic separation depends on molecular diffusion as well as on mass transfer and the mean velocity of the mobile phase.

The differences in the physical and chemical properties of the individual components of the sample determine their relative affinity for the stationary phase and consequently their relative mobility. The component that has least affinity to the solid phase is retained least, moves fastest and is eluted first, while the component that has the highest affinity to the solid phase is retained strongly and is eluted last. The components with varying affinities are eluted successively. For a given solute, the ratio of the retention times spent in the mobile and the stationary phases is expressed in terms of the partition coefficient, also called the distribution constant, $K_D$.

$$K_D = \frac{C_S}{C_M} \qquad \text{...(22.10)}$$

($C_S$ and $C_M$ are equilibrium concentrations of a substance in the stationary phase and mobile phase, respectively). The larger the distribution constant, the more strongly the substance is sorbed. Therefore, the distribution constants of individual components should be different for their satisfactory resolution.

Chromatographic separation can take advantage of differences in (1) adsorption on a stationary phase, (2) solubility, (3) molecular size, (4) charge, (5) binding affinity, and (6) ionic mobility. The sample is dissolved in a mobile phase (liquid or gas) and passed through an immiscible stationary phase. The stationary phase can be a liquid, solid, solid-liquid mixture or gel. Judicial selection of stationary and mobile phases, from a wide range of materials available contributes to the success of the separation.

Chromatographic methods are classified based on the geometry of the system, mode of operation, retention mechanism and phases involved. Choice of a chromatographic method depends on the physicochemical as well as physiological characteristics of eluents (Fig. 22.5). Different techniques are: paper chromatography, thin-layer chromatography (TLC), adsorption chromatography, size-exclusion (gel filtration or molecular sieve) chromatography, ion-exchange chromatography, affinity chromatography, high performance liquid chromatography (HPLC), gas chromatography (GC) (Table 22.2), and supercritical fluid chromatography (SFC).

**TABLE 22.2:** CLASSIFICATION OF CHROMATOGRAPHIC TECHNIQUES

| *Mobile Phase* | *Stationary Phase* | *Classification* |
|---|---|---|
| Gas | Liquid | Gas-liquid (partition) chromatography (GLC) |
| Gas | Solid | Gas-solid (adsorption) chromatography (GSC) |
| Liquid | Liquid | Liquid-liquid (partition) chromatography (LLC) |
| Liquid | Solid | Paper chromatography; Thin-layer chromatography (TLC); Adsorption chromatography; Affinity chromatography; HPLC |
| Liquid | Solid matrix | Size-exclusion (Gel filtration) chromatography |
| Liquid | Ion-exchange resin | Ion-exchange chromatography (IEC) |

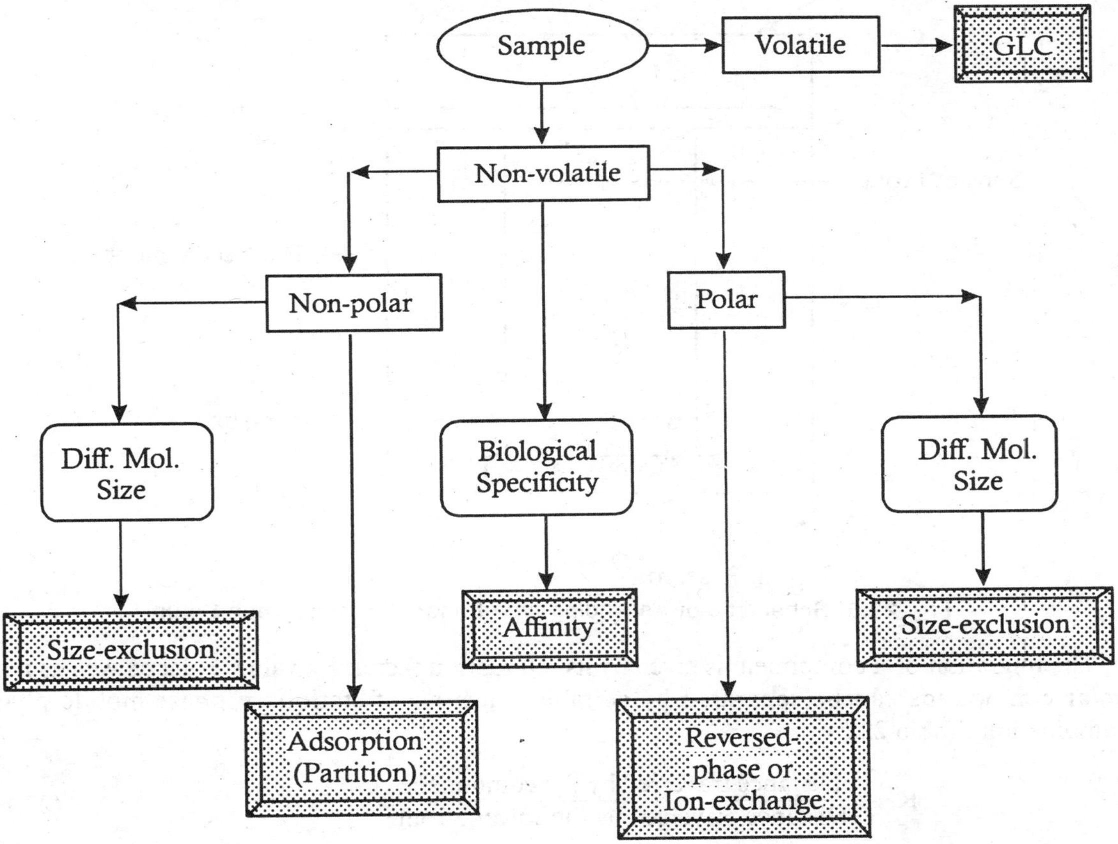


**Fig. 22.5:** Flowchart of Chromatographic Separation Systems

### 22.4.1 Paper Chromatography

Paper chromatography is a plane chromatographic technique, based on partition between water adsorbed on the cellulose molecules acting as the stationary phase and a suitable solvent as the mobile phase. The method is very simple and cost-wise cheap. In the ascending type (Fig. 22.6), the solvent (mobile phase) is placed at the bottom of a closed chamber and allowed to saturate the chamber with its vapor. Spots of sample solution(s) is/are applied over a marked horizontal line on a filter paper and the filter paper is then suspended in the chamber. The setup is such that the bottom edge of the filter paper is in contact with the solvent, but the spots are just above the solvent layer. The mobile phase spreading up the filter paper, drawn by capillary action, dissolves the solutes on the spots and carries them up, to a distance along the paper that is a function of their migration ratios. The filter paper is removed just before the solvent front reaches the top of the paper, solvent front is marked on the paper and it is dried. The positions of the spots on the filter paper are located visually if they are colored, by UV radiation if they fluoresce or by radiochemical detection if they are spiked with radioactive tracers before developing the chromatogram.

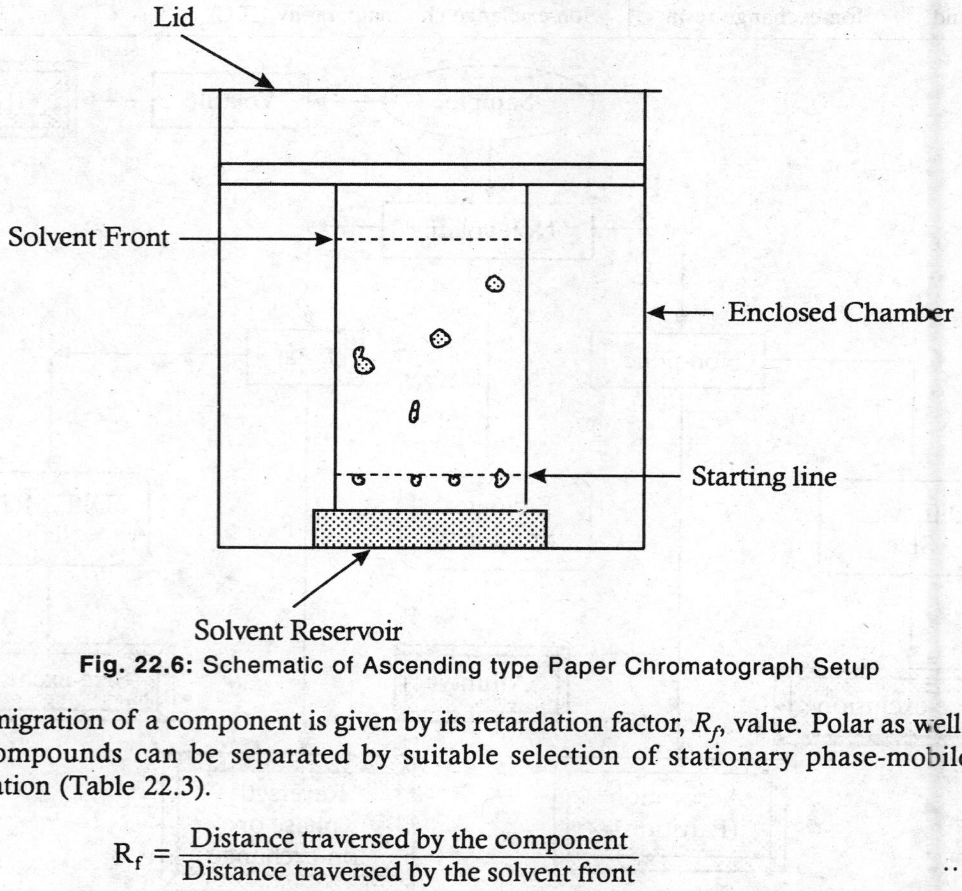


**Fig. 22.6:** Schematic of Ascending type Paper Chromatograph Setup

The migration of a component is given by its retardation factor, $R_f$, value. Polar as well as non-polar compounds can be separated by suitable selection of stationary phase-mobile phase combination (Table 22.3).

$$R_f = \frac{\text{Distance traversed by the component}}{\text{Distance traversed by the solvent front}} \qquad ...(22.11)$$

Two-dimensional paper chromatography is a method that permits better (two-dimensional) resolution of the components. In this method, the sample is spotted at the lower corner of a filter paper, and the chromatogram is developed with a chosen mobile phase. Then the paper is rotated through 90° and using a different solvent, another chromatogram is developed from the spots separated in the first run. The final result is a two-dimensional separation of spots. Two-dimensional chromatograms provided the experimental evidence for the cause of the molecular disease, sickle cell anemia, as due to single amino acid mutation in hemoglobin (Glu → Val in the 6th position in β-chains).

**TABLE 22.3:** MOBILE-STATIONARY PHASE COMBINATION FOR PAPER CHROMATOGRAPHY

| *Stationary Phase* | *Mobile Phase* | *Separation* |
|---|---|---|
| Water | Isopropanol-ammonia | Polar compounds |
| Water | n-Butanol-acetic acid | Polar compounds |
| Water | Phenol | Polar compounds |
| Formamide | Chloroform; benzene; benzene-cyclohexane | Neutral and less polar |
| Phenoxy ethanol | Heptane | Non-polar compounds |
| Liquid paraffin | Dimethyl formamide-methanol | Non-polar compounds |

Cellulose filter paper is the one widely used in paper chromatography. Special types of filter papers are available for the separation of different kinds of compounds (Table 22.4).

**TABLE 22.4:** FILTER PAPERS FOR PAPER CHROMATOGRAPHY

| *Filter Paper Type* | *Separation* |
|---|---|
| Pure cellulose | Polar compounds; and for ion exchange chromatography |
| Carboxylated | Amino acids and protonated amines |
| Acetylated | Reverse-phase chromatography—non-polar compounds and metal cations |
| Alumina and silica coated | Non-polar compounds—fatty acids, steroids, pesticides, vitamins, etc. |

### 22.4.2 Thin-layer Chromatography (TLC)

Thin-layer chromatography (TLC) is also a plane chromatographic method, similar in principle and operational features to paper chromatography. The main difference between paper chromatography and thin layer chromatography lies in the type of support (stationary phase) material. The mobile phase is a solvent or a mixture of solvents. However, the stationary phase is a solid (sorbent). TLC can be made flexible and versatile by employing a variety of sorbents—silica (pH = 3-8), cellulose nitrate, alumina, polyamides and ion exchangers, to separate various kinds of polar, neutral and non-polar compounds. Further, the resolution can be improved by controlling the particle size of the sorbent. Recent developments in TLC (High performance thin layer chromatography, HPTLC), with smaller sorbent plates with uniform particle size and particle distribution, have reduced the operation time to minutes with better resolution and detection limits going down to nanogram and picogram range. These developments, in addition to the fact that its operational features are simple and cost wise cheap, have enhanced its scope of application. It has become the workhorse in the analytical separation and determination methods in pharmaceutical, clinical, forensic and environmental studies. TLC is especially suitable for the separation of lipophilic compounds—fat products (e.g. fat-soluble food additives, flavoring compounds, antioxidants and contaminants), steroids, serum lipids, vitamins, carotenoids and pesticides.

The TLC glass and plastic plates are coated with polar adsorbents for normal-phase. The sample is spotted on the adsorbent layer near the bottom edge of the plate and placed in an airtight developing tank, in which the eluting solvent is placed up to a few millimeters below the sample spots. The operation is similar to that of paper chromatography.

Silica gel, ($SiO_2 \cdot H_2O$), is a commonly used polar sorbent (in the pH range 3-8). Other sorbents are alumina, cellulose and sephadex (Table 22.5). Silica is acidic and is good for the separation of basic molecules. Alumina is basic (pH = 8-11), which is better suited for the separation of weakly acidic compounds from neutral ones. For the reverse-phase method, the sorbents are non-polar materials (e.g. activated carbon and organic polymers). Sorbents with chemically inert surfaces adsorb non-specifically.

**TABLE 22.5:** SORBENTS FOR THIN LAYER CHROMATOGRAPHY (TLC)

| *Sorbent* | *Separation* |
|---|---|
| Silica gel | Alkaloids, amino acids, fatty acids, lipids, oils, steroids, terpenoids |
| Alumina | Alkaloids, amino acids, carotenoids, food coloring, phenols, steroids, vitamins |
| Cellulose | Amino acids, peptides, nucleic acid components |
| Sephadex | Nucleic acid components, peptides, proteins, metal chelates |

Selection of the solvent system depends on the sorbent and the type of analytes to be separated. Alcohols are generally employed for analytes with—OH groups, acetone for carboxyl groups and hexane, heptane and toluene for non-polar analytes. In general, slightly volatile to non-volatile organic compounds are used as solvents. Common protic solvents are water, alcohol and acetic acid. Dipolar aprotic solvents are acetone, acetaldehyde, acetonitrile and nitrobenzene and non-polar solvents are benzene, chloroform and dioxane. Some of the solvent systems from hydrophobic towards hydrophilic are—ether, cyclohexane, toluene, benzene, chloroform, phenol, acetone, ethanol, acetic acid, formamide, and water. Addition of chemically reactive systems (e.g. silver nitrate, silicone oil + $AgNO_3$) to the mobile phase enables special (preparation) recipe.

After development with the mobile phase, the TLC plate is dried and heated to completely evaporate the mobile phase before the zones are detected. Therefore, TLC allows a wide variety of solvents to be used to prepare mobile phase compared with HPLC. The separated spots are generally detected by color, or UV absorption and fluorescence by optical scanning. The identity of TLC spots can be analyzed by combining mass spectrometry (TLC-MS), FTIR (TLC-FTIR), Raman spectrometry, or X-ray fluorescence spectrometry (TLC-XRF).

### 22.4.3 Liquid Chromatography (LC)

Liquid chromatography (LC) is characterized by the use of a liquid mobile phase. Liquid-solid chromatography (LSC) utilizes a solid stationary phase. Silica and alumina are two polar adsorbents commonly used for the stationary phase. The liquid mobile phase moves over the stationary phase in close contact, during which the solutes in the sample are distributed between the two phases, mainly depending on their adsorption properties. The preferred mobile phase is a non-polar or slightly polar solvent. The choice of the stationary phase is dependent on the method of separation—solid matrix for adsorption and ion exchange and a porous gel for gel filtration and affinity binding. In liquid chromatography, the separation is based on the differences in solubility of the components in the coated phase or in their chemical affinity to the coat. The column may be either open or closed (in HPLC).

High performance liquid chromatography (HPLC) has become a standard and routine method for separation and concentration of analytes. An HPLC system consists of a solvent reservoir, high-pressure pumps, sample injection port, separation column, detector, and recorder units (Fig. 22.7). HPLC is the result of attempts to increase the efficiency of liquid chromatography to the same level as that of gas chromatography. This is achieved by using particles of 2-20 μm size to overcome diffusion-related problems in the liquid and solid phase. The resulting high inlet pressure in the column is overcome by resorting to high pressure pumping of solutions. A pump forces one or more mobile phases through the tightly packed high efficiency column. The sample is introduced through an injection system into the entrance to the column. Sample introduction can be improved by flow-injection method, by injecting the sample into a flowing stream, which carries the analyte through a chemical modulator into a detector.

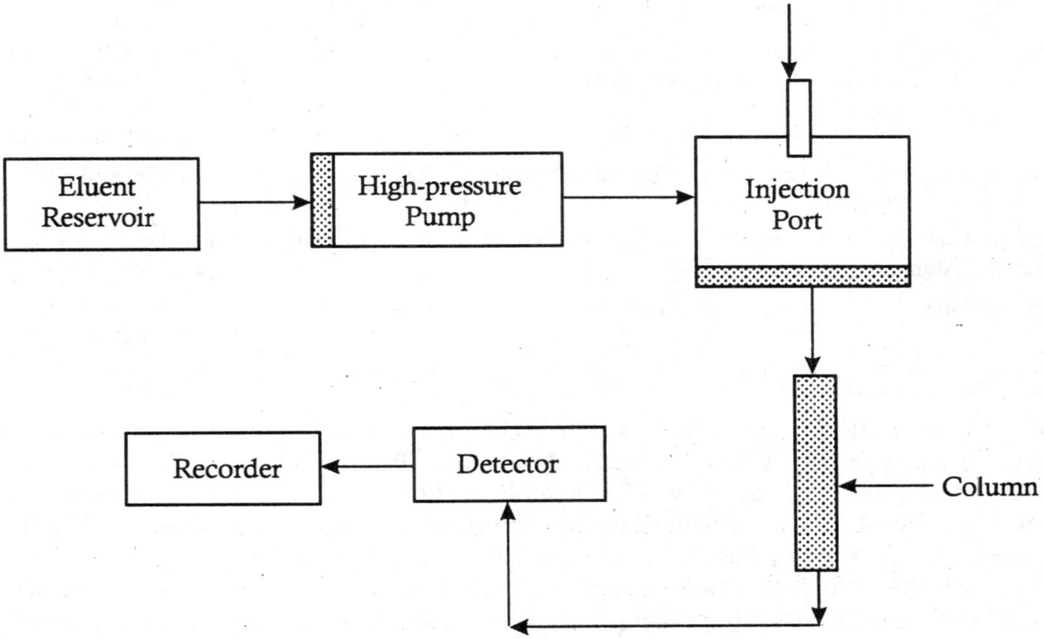


**Fig. 22.7:** Elements of HPLC System

The stationary phase is porous silica on to surface of which are long chain hydrocarbon groups (octyl, hexyl) are chemically bonded by siloxane (Si-O-Si-R) bonds. The mobile phase consists of a solvent or a mixture of solvents—hexane, heptane, ethylacetate for normal-phase, and water, methanol, and acetonitrile for reverse-phase. A stainless steel filter placed before the mixing chamber and the pump protects the high-pressure pump. The sample is introduced into the stream of mobile phase in as short duration as possible by stop-flow or septum injection. A second stainless steel filter placed in front of the column traps the particulates to minimize eddy flow of the mobile phase in the column. Columns are generally stainless steel tubes (~150-500 mm long and ~1-4 mm diameter) that can withstand high pressures. Column packing materials, for the normal-phase, are silica gel, alumina, activated carbon, and porous polymers, and ion-exchange resins. For reverse-phase HPLC, the stationary phase materials are octyl-, or octyldecyl-siloxane (Table 22.6).

There is no universal detector for HPLC. Detectors based on refractive index (RI), optical absorbance, fluorescence, thermal conductivity, and flame ionization are in use. RI detectors have poor sensitivity. Optical absorbance detectors (photodiode array detectors) employ fixed or variable wavelength light sources and monitor the absorbance at the selected wavelengths (sensitivity ~ ng/mL). These detectors are mainly suited for organic compounds but in conjunction with chromogenic reagents can be extended to inorganic materials. Fluorescence detectors are the most sensitive ones (~1000 times more sensitive than absorbance detectors) and are often used. Electrochemical detectors (particularly voltammetry) are used for the detection of metals and a variety of organic compounds (sensitivity pg/mL). Radiochemical detectors, with radiolabeled materials ($^3H$ or $^{14}C$) also can be employed (sensitivity ~ng/mL).

**TABLE 22.6:** SORBENTS AND ELUENTS USED IN SOLID-PHASE LC AND HPLC

| *Sorbent* | *Eluent* | *Nature of Analyte* | *Method* | *Analytes* |
|---|---|---|---|---|
| Octadecyl-bonded silica | Organic | Non-polar & weakly polar | Reverse-phase | PAHs; PCBs; polychloro-phenols; non-polar and organochloro pesticides |
| Styrene-divnylbe-nzene copolymers | Organic | Non-polar & polar | Reverse-phase | Phenols; anilines |
| Activated carbon; Silica gel; Alumina | Organic | Non-polar & polar | Reverse-phase | Alcohols; nitrophenols; polar pesticides |
| Ion-exchange | Water | Ionic organics | Ion-exchange | Phenols; phenoxy acids; aniline |
| Metal-chelates | Solution | Metal complexing | Ligand-exchange | Carboxylic acids |

Reverse-phase HPLC (RP-HPLC) is very well suited for the analysis of non-volatile polar compounds, such as polycyclic aromatic hydrocarbons (PAHs). Adsorbed PAHs on filters are recovered using Soxhlet extraction. The sample is then separated into fractions by column chromatography on silica gel using hexane-chloroform as the mobile phase, and individual components are detected by fluorescence ($\lambda_{ex}$ = 250-280 nm; $\lambda_{fl}$ = 370-390 nm).

Highly selective, efficient separation approach (analogous to those used by chromatography) would be desirable for separation, identification and quantitation of organisms (microbes) also. The classic approach of identification of bacteria is time consuming and applicable for a small percentage of microbes. Immunoassay methods are applicable for a particular microbe only, but ignore broader populations present. Extending the chromatographic approach to isolate and characterize microbes would revolutionize microbiology, medicine, epidemiology and environmental studies.

Advent of HPLC has resulted in tremendous progress in the separation of a wide variety of inorganic and organic compounds, with high sensitivity of detection ($10^{-12}$ g/L). Liquid chromatographic methods, that include HPLC, employ adsorption, partition, size-exclusion, affinity and ion-exchange chromatographic techniques.

### 22.4.3.1 *Adsorption Chromatography*

Adsorption chromatography is based on the principle that certain materials adsorb molecules on their surface by weak Van der Waal forces and the solute molecules partition between the mobile liquid phase (e.g. water) and adsorption sites on the surface of the stationary solid surface (e.g. silica gel). The forces responsible for the interaction between the sample and the adsorbent are (*i*) dispersive forces (non-polar interactions), (*ii*) inductive forces (dipole interactions), (*iii*) hydrogen bonds or (*iv*) ionic forces. Dispersive forces (London forces) are prominent for adsorption of non-polar molecules on a non-polar or on a polar adsorbent. Inductive forces occur due to adsorption of polar molecules on non-polar or polar adsorbents. For polarizable substances (alkenes), induction forces accompany the dispersion forces. Hydrogen bonds occur between electronegative groups and polar parts of a molecule (e.g. ketones, aldehydes, alcohols, amines, and $H_2O$). Competition between the solute and the solvent molecules for the adsorption sites establishes a dynamic process, and the solute molecules are selectively retarded during elution, depending on their physicochemical characteristics, leading to their separation. Paper and thin layer chromatographies are examples of adsorption (partition) chromatography.

In liquid chromatography (that includes TLC), the mobile phase exerts a decisive influence on the separation, unlike in the case of gas chromatography (GC). Therefore, selection of proper eluent is important. Normal eluents are alcohols for analytes with —OH groups, acetone for carboxyl groups and hexane, heptane and toluene for non-polar analytes.

In partition chromatography, if the stationary phase is more polar (e.g. water) than the mobile phase (e.g. an organic liquid), it is called normal-phase chromatography, and if the stationary matrix is less polar than the mobile phase, it is called the reverse-phase. In the normal-phase method, non-polar analytes are eluted first and the most polar analytes the last, and in the reverse-phase method, the polar analytes are eluted first and the non-polar analytes last (Fig. 22.8). Polar and weakly polar compounds are separated on a polar adsorbent and *vice versa*. Reverse-phase technique is widely used to separate amino acids, organic acids, drugs, lipid-soluble compounds, insecticides and pesticides.

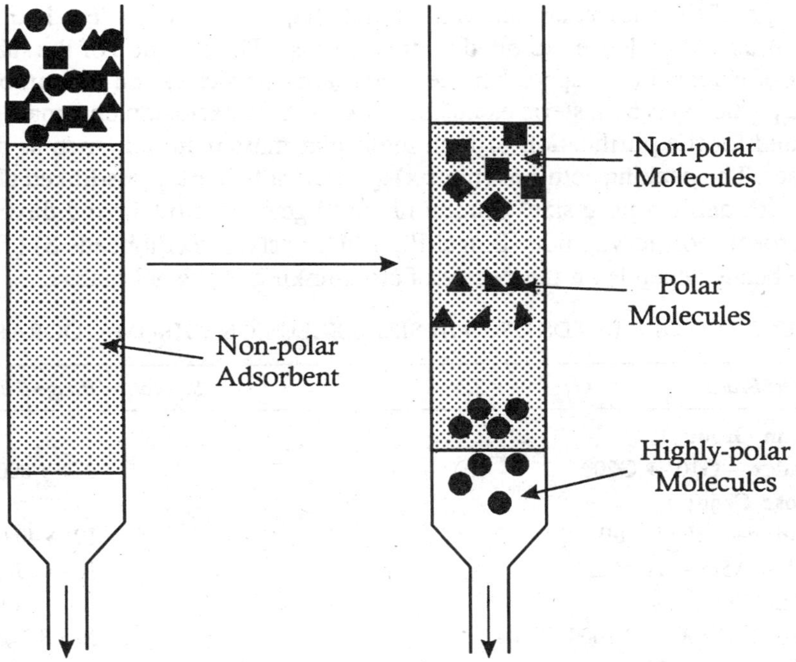


**Fig. 22.8:** Basic Principle of Reverse-phase Liquid Chromatography

Common adsorbents for the normal-phase are polar substances (alkyl amines bonded to silica) (Table 22.7), and the mobile phase is non-polar (hexane, heptane, ethyl acetate). In the reverse-phase technique, the stationary phase is non-polar (alkylsilane groups attached to silica) and the mobile phase is a polar liquid (water, methanol, acetonitrile). The presence of hydroxyl groups in silica gels leads to specific interaction with alcohols, amines, ketones and aldehydes. Silica is acidic and is good for the separation of basic molecules, while alumina is better suited for the separation of acidic compounds.

**TABLE 22.7:** SORBENTS FOR ADSORPTION CHROMATOGRAPHY

| *Sorbent* | *Characteristic* | *Phase* | *Separation* |
|---|---|---|---|
| Alumina | Acidic/basic | Normal-phase | Amino acids, alkaloids, dyestuff, esters, inorganic compounds, sterols and vitamins |
| Silica gel | Acidic | Normal-phase | Amino acids, peptides and sterols |
| Activated carbon | Acidic/neutral | Normal-phase | Amino acids, peptides, carbohydrates |
| Hydroxyapatite | Neutral | Normal-phase | Proteins, nucleic acids, cell extracts, carotenoids |
| Al & Mg-silicates | Neutral | Normal-phase | Alkaloids, esters, glycerides, sterols |
| Polystyrene; Acylic ester polymers | Hydrophobic | Reverse-phase | Insecticides, phenols, steroids, trace metals |

### *22.4.3.2 Size-exclusion (Gel Filtration/Molecular Sieve) Chromatography*

Size-exclusion (gel filtration/molecular sieve) is a separation technique where the separation is based on the basis of the molecular size (mass) and shape of the particles. The molecular sieve properties of porous materials are exploited for this purpose. The partition of the substance between the mobile part of eluent (in the space between the matrix particles) and its immobile part (filling the pores of the particles) is by a steric exclusion process. Size-exclusion chromatography (SEC) is particularly suited for the purification of high molecular mass materials such as proteins, nucleic acids, and viruses. The packing columns (matrix) are generally inert, cross-linked glucose polymers in bead form, with defined pore size. Commonly used gels are cross-linked dextran (Sephadex), agarose (Sepharose), polyacrylamide (Biogel-P), and polystyrene (Biobeads-S) (Table 22.8). The pore size of the beads depends on the degree of cross-linking of the gel matrix.

**TABLE 22.8:** GEL BEADS USED IN SIZE-EXCLUSION CHROMATOGRAPHY

| *Gel Beads* | *Separation Range (kilo Dalton)* |
|---|---|
| **Dextran Group** | |
| Sephadex— G10 → G200 | 0.7 → 600 |
| **Agarose Group** | |
| Sepharose— 2B/4B/6B | 10 → 40,000 |
| Biogel— A5m → A150m | 10 → 150,000 |
| **Acrylamide Group** | |
| Sephacryl S-200 → S-1000 | 5 → 8,000 |
| Biogel P-2 → P-300 | 0.1 → 300 |
| **Polystyrene Group** | |
| Bio-beads SX1/SX2 | 0.4 → 150 |

In gel filtration matrix the solute molecules are retained when they diffuse in and out of the pores. The duration they remain in the pores is a function of their size. If a solution containing molecules of various sizes is passed through a gel matrix column, molecules of size larger than the average diameter of the pores (> "just excluded size") are excluded from entering the beads (size-exclusion). As they suffer no retention in the gel they are eluted first. At the other end of the size spectrum, there is a certain size of molecules for which all spaces in the gel matrix are accessible. They diffuse into the gel beads and are retained by the gel matrix for longer period of time and hence eluted last. Molecules of intermediate sizes show different rates of migration, depending on their size and are eluted in the order of decreasing molecular size (Fig. 22.9). The hydro-dynamical volume of a solute, which determines its elution volume, $V$, is closely related to its molecular mass. $M_r$. The exclusion volume, $V$, of a molecule in size-exclusion process is

$$V = V_0 + KV_i \qquad \ldots(22.12)$$

For small molecules of size < average pore diameter, $K = 1$ and ($V = V_0 + V_i$)

For large molecules of size > average pore diameter, $K = 0$ and ($V = V_0$).

where, $V_0$ = Free volume outside the gel beads, available for all molecular sizes;

$V_i$ = volume inside the pores, available only to small molecules;

$K$ = distribution coefficient.

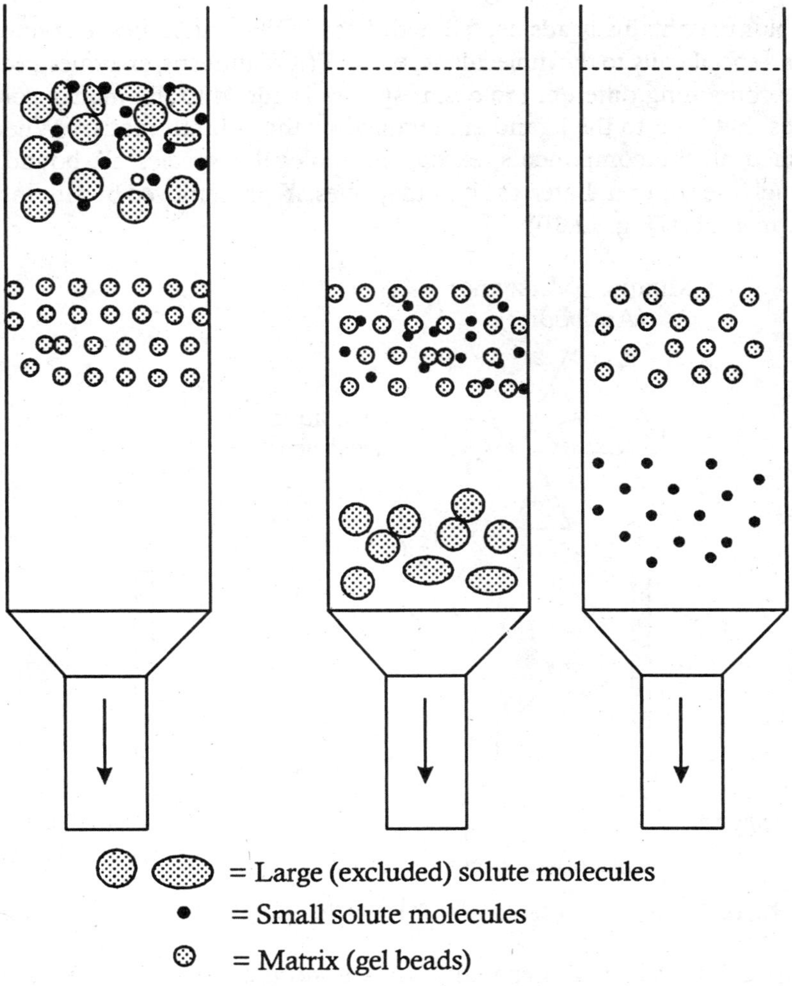


**Fig. 22.9:** Molecular Separation by Size-exclusion (Molecular-sieve) Chromatography

From a plot of elution volume, $V$, against molecular mass, $M_r$, with known molecular mass markers, the elution volume of unknown molecular species can be determined, from which the molecular masses of the unknown molecules can be estimated. Size-exclusion chromatography sorts the solutes in groups of similar $M_r$, which facilitate further separation by other methods. In SEC no change of solvent is needed; so automation of SEC for separation purposes is easy.

### *22.4.3.3 Affinity Chromatography*

Affinity chromatography is a separation method in which molecules can be isolated on the basis of the high specificity of interaction that exists between biomolecules—enzyme-substrate, antigen-antibody, receptor-ligand and the like. The necessary condition is that the material to be purified should be capable of binding reversibly to a specific ligand that is attached to the sorbent matrix.

$$\text{Molecule + Ligand} \Leftrightarrow \text{Molecule-ligand complex} \quad \text{...(22.13)}$$

The matrix column contains beads coated with ligands like antibodies, enzyme substrates and receptors that bind specifically to the molecular species, $X$ ($X$ = antigens, enzymes, receptor proteins). When a mixture containing different molecular species is added to the matrix column, only that molecular species that bind to the ligand are retained by the column matrix. When the sample is passed and washed all the components, except the molecular species, $X$, bound to the affinity matrix pass through the column. Later, the bound species, $X$, are removed by altering the conditions of eluting buffer (e.g. pH) (Fig. 22.10).

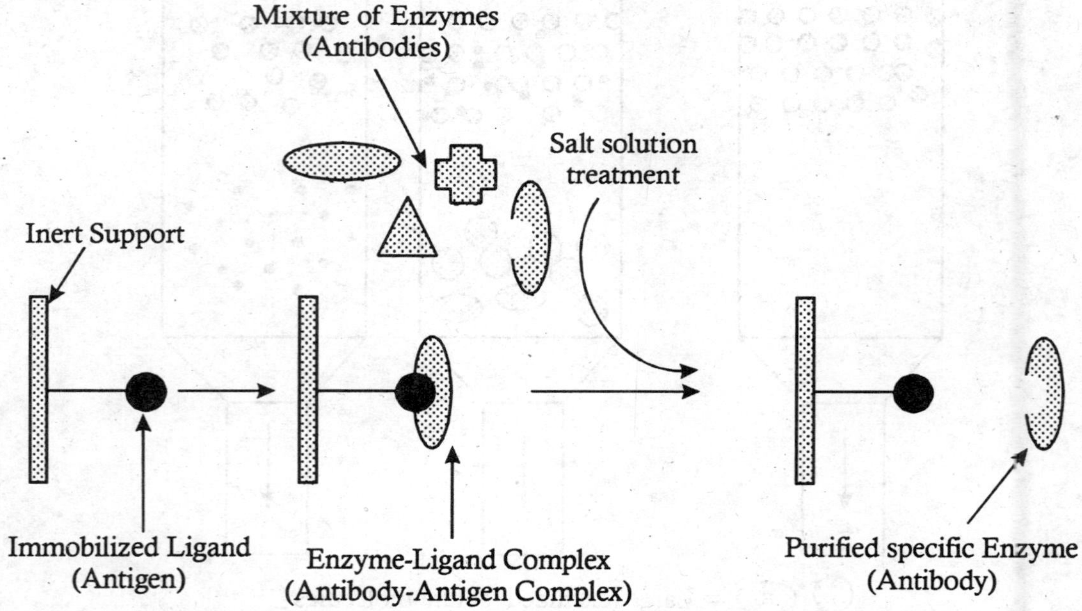


**Fig. 22.10:** Schematic of Molecular Purification by Affinity Chromatography

Suitable affinity columns can be prepared to isolate specific molecular species of interest. This technique has been applied extensively in the purification of antigens, antibodies, enzymes, proteins, hormones, receptors, nucleic acids, carbohydrates, viruses and many other biologically important compounds, and quantified by radiolabeled (RIA and IRMA), enzyme-linked, chemiluminescence and bioluminescence assays. Separation by affinity chromatography and detection by any of the immunoassay methods has a wide range of applications in diverse fields—biology, medicine, clinical chemistry (endocrinology), pharmacology (toxicology), immunology (allergy and oncology), virology, bacteriology and mycology. At present radioactive and non-radioactive immunoassay methods are widely employed in molecular biology, and clinical diagnostics in medicine. Potentially, immunoassay procedures can be applied to non-clinical diagnostics in veterinary and agricultural sciences, food and beverage industries and environmental monitoring studies. The early detection of infectious diseases (e.g. foot and mouth disease, bovine diarrhea, etc) would be of great help in limiting the spread of epidemics in avian, animal and human populations. Detection of trace quantities of hazardous contaminants in food (*Salmonella* and *Listeria*), beverages and water, and bacterial toxins and pathogens (e.g. *E. coli*, and *Clostridium*) and plant and mycotoxins, and pesticides and herbicides in the environment is not only desirable but also necessary. Though immunoassay methods are at present not widely used in environmental monitoring non-radioactive immunoassay methods (chemiluminescence and bioluminescence) have a great potential in determining air, water, food and beverage contaminants.

Metal-chelate chromatography is an extension of affinity chromatography. In this method, immobilized metal ions, $Cu^{2+}$, $Zn^{2+}$, $Mo^{2+}$, $Mn^{2+}$, can be utilized to separate many proteins and peptides that bind to these metals. Ligand-exchange chromatography is another version of affinity chromatography, which is based on selective interactions between the analytes and matrix-bound ligands (ligand-ligand interactions). This method is used in the purification of amines, hydroxy, aldehyde, and keto- compounds, peptides, proteins and nucleic acid constituents.

Chiral chromatographic separation method, with laser-based polarimetric detection, is an emerging technique that has wide applications in pharmacy, forensic science and environmental studies.

#### *22.4.3.4 Ion-exchange Chromatography (IEC)*

Ion-exchange chromatography is a separation method based on ion-exchange process occurring between the mobile phase and the ion-exchange groups bonded to the stationary phase. The stationary phase in the column consists of a polymer with charged groups (ion exchanger) bound to the support material. Mobile counter ions (of opposite charge) are bound to these groups by electrostatic attraction. Ions in the mobile phase, having the same charge as the counter ions can replace them. An 'elution' process can differentially displace the bound ions.

Ion-exchange chromatography is useful for the separation of both inorganic and organic species. There are two types of ion exchangers—cation and anion exchangers. Matrix materials are polystyrene, cellulose or agarose (Table 22.9). Cation exchangers possess negatively charged functional groups and functional groups (sulfonate ($SO_3^-$), carboxylate ($COO^-$)) and attract positively charged analytes. These are also called acidic exchangers because in their simplest state they hold $H^+$ as the counter ion. Anion exchangers have positively charged functional groups (quaternary ammonium ($N^+R_4$)) that attract anionic species. Bio-adsorbers can be used to recover heavy metals in wastewater and to remove Fe, Ca and Mg from hard water. Residual products of cereal processing are used instead of synthetic resins.

**TABLE 22.9:** COMMONLY USED ION-EXCHANGER MATRIX MATERIALS

| *Type of Exchanger* | *Matrix* | *Functional Groups* |
|---|---|---|
| **Anion exchanger (Basic)** | | |
| Weakly basic | Agarose; Cellulose; Dextran; Polystyrene | Aminoethyl |
| Strongly basic | Cellulose; Dextran; Polystyrene | Triethylaminoethyl |
| **Cation exchanger (Acidic)** | | |
| Weakly acidic | Agarose; Cellulose; Dextran; Polyacrylate | Carboxy; carboxymethyl |
| Strongly acidic | Cellulose; Dextran; Polystyrene | Sulfonate; sulfopropyl |

Choice of the exchanger depends on the stability of the sample components, their relative molecular mass and other parameters. Generally if a sample is stable below its isoionic point (pI), giving it a net positive charge, a cation exchanger is useful, whereas if it is stable above its isoionic point, giving it a net negative charge, an anion exchanger is useful. The choice between a strong and weak exchanger also depends on sample stability and the effect of pH on charge of the sample. Weak electrolytes requiring a very low or high pH for ionization are separated advantageously by strong exchangers and *vice versa.* The pH of the buffer should be pH = (pI + 1) of the compounds being separated.

A wide variety of ion-exchange resins are available in the market. Synthetic ion exchangers are usually made of copolymers of styrene and divinylbenzene, having divinylbenzene cross-links. The degree of cross-linking determines the degree of swelling of the gel and the sorption rate. In cation-exchange resins (ion exchangers with acidic side groups), the side groups are sulphonic, carboxyl, hydroxyl or phenolic groups (CM- and SP-Sephadex); while in anion-exchange resins the side groups are aliphatic or aromatic amino groups (DEAE-cellulose). Ion-exchange chromatography has been employed extensively in the separation of polar compounds of biological, pharmacological and medical importance and of environmental concern—metabolites, drugs, foodstuffs, vitamins, amino acids, peptides, nucleic acid components, etc.

Separation by ion exchange chromatography is achieved in two stages—(*i*) first, adsorbing (binding) the charged species on the ion-exchange column, then (*ii*) eluting the spatially resolved components (Fig. 22.11).

Ion chromatography (IC) is a modification of the ion-exchange process, with an anion exchange column (separator column), a cation exchange column (suppressor column), and a conductivity detector for rapid, sequential measurement of common anions such as $F^-$, $Cl^-$, $Br^-$, ($CN^-$), ($NO_2^-$), ($NO_3^-$), ($PO_4^{3-}$), and ($SO_4^{2-}$). It is an effective method for pre-concentration of hydrated ions, charged complexes and metals.

Detection of anions can be carried out by various methods. UV spectrometric and AAS detectors are selective. Spectrophotometry can be used as a universal detection method by post-column reaction of the sample ions with chromogenic reagents. Electrochemical detection is most widely used detector in ion chromatography. Conductimetric detector is universal as same detector can be used for detecting inorganic and organic ion species. Membrane eluent suppressor achieves this, which was the key to the success of ion chromatographic method for anion measurement. Membrane suppressor is a cation-exchange membrane capable of continuously converting eluent and separated anions to their acid forms. To suppress the conductivity of the eluent, the solution is passed through a suppressor column that contains strong acid resin and the anions are detected by the conductivities against the background of low conductivity eluent, measured by conductimetry. The output ion current is a plot of conductance versus time.

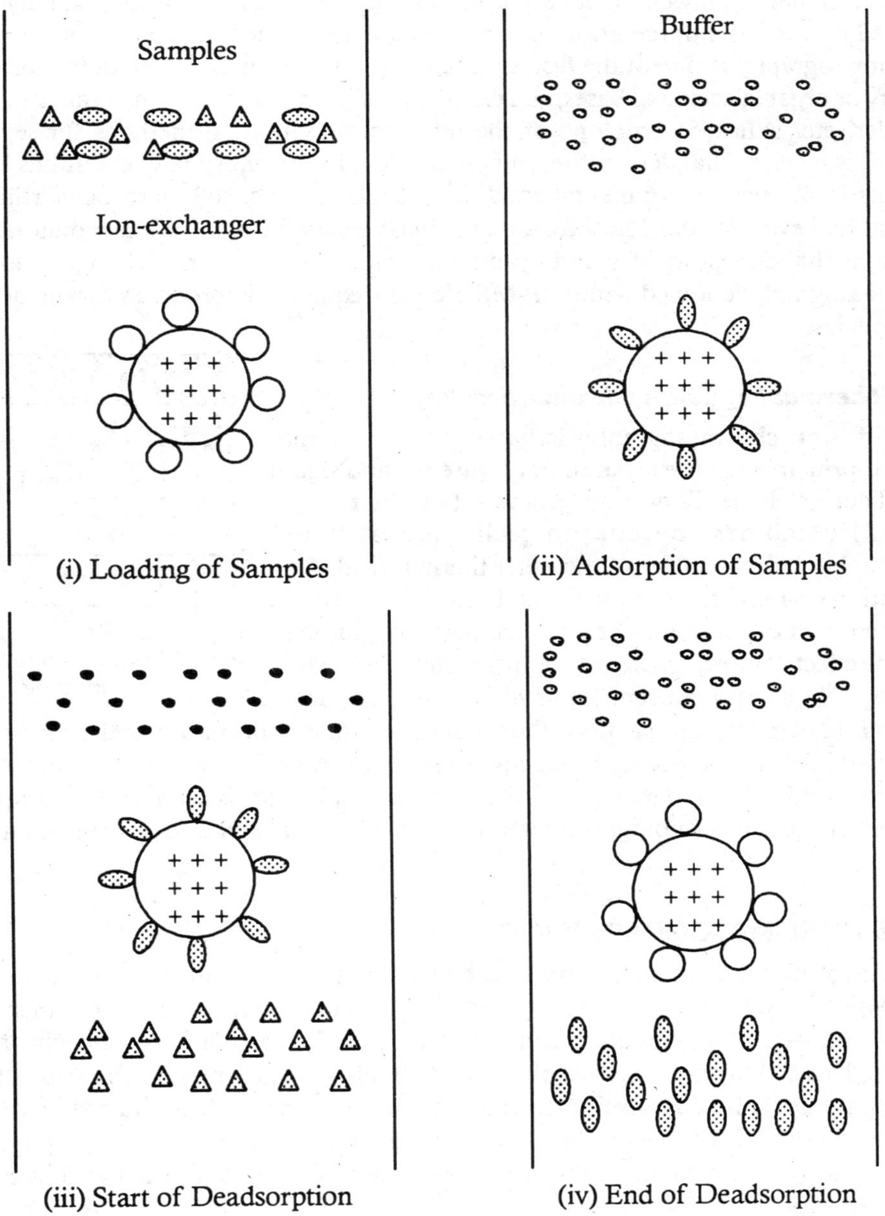


**Fig. 22.11:** Schematic Features of Ion-exchange Chromatographic Separation

Ion chromatograph setup consists of an injection valve, sample loop, guard column, separator column, membrane suppressor, conductivity detector and output signal recorder. The general procedure is—A water sample is injected into a stream of carbonate-bicarbonate eluent and passed through a series of ion exchangers. The anions of interest are separated on the basis of their relative affinities for strongly basic exchangers (strongly basic separator columns). The separated anions are converted to their highly conductive acid forms and are measured by conductivity. The anions are identified on the basis of retention time as compared to standards.

Ion chromatography is one of the fastest separating methods used in the industrial purification and recovery of organic acids and bases, and extraction of precious metals, and uranium, lanthanides, and actinides, etc. It has become one of the most commonly used methods for separating and identifying ion species—halides, sulfite, sulfate, nitrite, nitrate, phosphate, etc. It has been used to monitor ions in diverse environmental conditions, such as brine, soil, aerosols, fertilizers, blood, urine, foodstuffs, beverages, etc. Ion chromatography is very useful as an on-line monitoring technique for major ions, halides, pesticides and cyanide in water samples. Ion-exchange, ion-pair reverse phase chromatography coupled with ICP-MS are the frequently approaches for the determination of arsenic species.

### *22.4.3.5 Thermal Diffusion Chromatography*

Thermal diffusion chromatography is based on the thermo-gravitational principle for the separation of gaseous and liquid mixtures. Thermal Field Flow Fractionation (TFFF) method (Fig. 22.12) combines the elution technique of liquid chromatography with the effect of the outer thermal field.

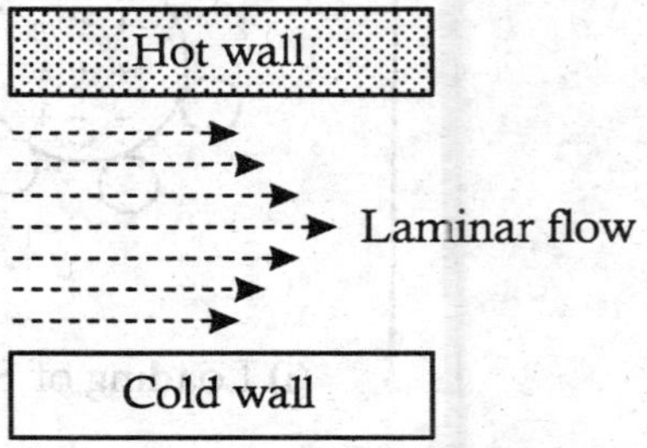


**Fig. 22.12:** Principle of Thermal Field Flow Fractionation (TFFF) Method

The solution carrying the sample flows through a horizontal column where it is exposed to a vertical temperature gradient. (1) Under the effect of transverse temperature gradient, individual components of the sample concentrate at various distances from the cold wall. (2) Under laminar flow, the components that concentrate at the center are carried more rapidly while the components near the walls flow slower. The separated components can be detected as in liquid chromatography. TFFF method is suitable for separation of azotropic mixtures, thermo labile compounds (biological materials) and macromolecules (e.g. organic and biological polymers).

## 22.5 GAS CHROMATOGRAPHY (GC)

Gas chromatography is a separation method by which molecules can be resolved on the basis of their distribution between a gas and ether a liquid or a solid. The method can be considered under two subdivisions—(1) Gas-liquid chromatography (GLC) and (2) Gas-solid chromatography (GSC), named according to the mobile phase used. While gas is the mobile phase in both cases, the stationary phase is a solid adsorbent matrix in GSC and a nonvolatile liquid coated on to an inert solid support in GLC, respectively. The GSC is a form of adsorption chromatography, while GLC is a form of partition chromatography. The selectivity in GC, in both the cases, is based on the stationary phase used.

Gas chromatography (GC) is a very sensitive and selective technique, widely used for separation and determination of volatile compounds. All organic compounds that can be volatilized can be determined qualitatively and quantitatively (sensitivity $10^{-12}$-$10^{-14}$ g) by GC. Adsorption onto porous non-polar organic polymers in conjunction with thermal desorption is the most widely used method for determining organic compounds of low volatility. The most commonly used adsorbents are activated carbon, silica gel, alumina, molecular sieves, and porous organic polymers like Tenax, Porapak and XAD resin). Van der Waal forces, active over small distances, cause adsorption on activated carbon. It is recommended as a universal sorbent for sampling of all non-polar and semi-polar compounds. Oxidized activated carbon is a selective multifunctional cation exchanger. Electrostatic forces are responsible for adsorption on silica, alumina and zeolite gels.

Though inorganic compounds are generally non-volatile, those that can be volatilized can also be studied by GC (Table 22.10). Many inorganic compounds that are non-volatile (boranes, silanes, germanes, organotin, organolead, organomercury, chromium, etc) can be converted to volatile compounds by derivatization and can be resolved by GC. Inorganic gas chromatography (IGC) has been developed as a method of analysis of volatile compounds like hydrides or chlorides.

**TABLE 22.10:** SOME ELEMENTS THAT CAN BE VOLATILIZED

| *Elemental Species* | *Volatilized Compound* |
|---|---|
| Elemental species | H, Hg, $N_2$ and halogens |
| As, Bi, Ge, Pb, S, Sb, Se, Sn, Te | Hydrides |
| B, Mo, Nb, Si, Ta, Tl, V, W | Fluorides |
| Al, As, Cd, Cr, Ga, Ge, Hg, Mo, Sb, Sn, Ta, Tl, V, W, Zn, Zr | Chlorides |
| As, Bi, Hg, Sb, Se, Sn | Bromides |
| As, Sb, Sn, Te | Iodides |
| As, C, H, Os, Re, Ru, S, Se | Oxides |

## 22.5.1 Principles of Operation

The basic principle of GC separation technique is—when a mixture of volatile materials, transported by a carrier gas/mobile phase (argon, helium or nitrogen), is passed through a column containing a stationary medium (column matrix) each volatile component will be partitioned between the carrier gas and the adsorbent phase according to its physicochemical properties. In GLC, the stationary phase is a liquid of high boiling point (e.g. silicone grease) coated on an inert granular solid (e.g. diatomaceous earth, poly-tetrafluoro ethylene, (PTFE)). In GSC the columns used are carbon, silica and alumina (for the separation of hydrocarbons, $H_2$ and $O_2$), alumina, molecular sieves, and porous polymers (for aqueous solutions, acids and other polar materials).

GC columns can be either packed column GC, or open tubular column (OTC) GC. In packed column GC the stationary phase is packed in narrow tubes of suitable dimensions, while in open tubular column (OTC) GC the stationary phase is supported as thin film on the inner surface of the column wall ((Wall-coated open tubular columns (WCOT) and surface coated open tubular columns (SCOT)). Porous layer OTC (PLOT) columns are used as molecular sieves, while SCOT columns are used for partition.

The main components of any GC setup (sample injection port, the column and the detector) must be temperature controlled, and each must be independently controllable. The temperature programming is ideally suited for the analysis of wide boiling range compounds. The column is maintained at a regulated temperature (temperature-programmed) and when a sample solution is introduced into the column, the components are vaporized and are flushed through the column by the carrier gas at different rates. Each separated analyte appears as a peak on the chromatogram, and the intensity of each recorded signal peak gives a measure of the quantity of a certain analyte.

There are three essential steps in GC analysis (Fig. 22.13): (1) sample introduction, (2) selective sample transport and (3) detection.

1. The contaminated air (or sample dissolved in volatile solvent and flash vaporized) is passed through a cartridge containing the adsorbent (Tenax or carbon molecular sieve). The carrier gas from a container is led into the GC column. The sample (cartridge) containing the analytes is transferred to the heating block maintained at ~400° C, where the analytes are vaporized and injected into the stream of carrier gas on to a trap and subsequently on to the front of the GC column. The analytes are flushed as a plug of vapor by the carrier gas stream into the column inlet. The column is first held at low temperature, and the column temperature is uniformly increased (temperature programmed).
2. The solutes are adsorbed at the head of the column by the stationary phase and then desorbed by the fresh carrier gas ($N_2$, He, Ar). The analytes in gaseous form are transported through the stationary column by the carrier gas stream and emerge through the column in reverse order of their retentivity and are measured by a detector. Thus, each analyte is transported through the column to the detector at a specific rate, depending on its volatility, thus resulting in resolution of analytes (bands).
3. Detection is carried out by any one of the systems—thermal conductivity (TCD), thermoionic (TID), flame ionization (FID), electron-capture (ECD), electrolytic, coulometric, and conductimetric detectors, or by mass spectrometry (MS).

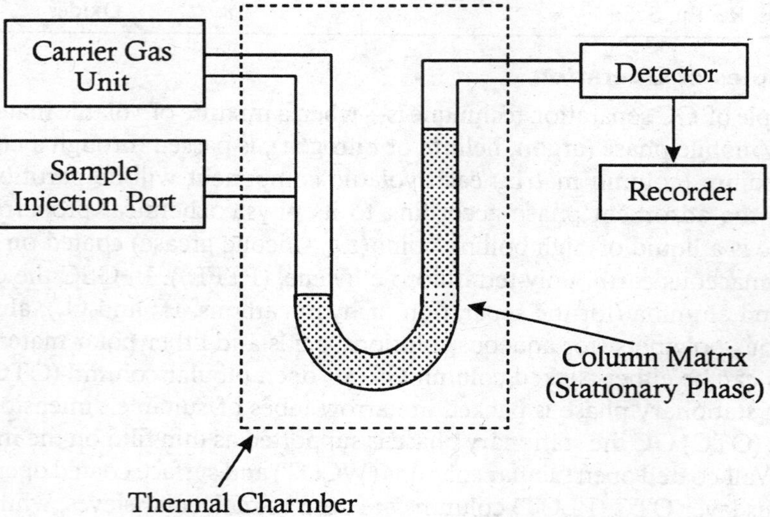


**Fig. 22.13:** Basic Modules in Gas Chromatography (GC) Setup

In GSC, the distribution between the two heterogeneous phases depends on adsorption, while dissolution is involved in GLC. In gas chromatography, accurate measurement of relative retention data, retention time and retention volume, is utilized for quantitative analysis of complex mixtures. Retention time, $t_R$, is the time elapsed from the injection of the solute into the column up to the moment when the maximum concentration of the solute is eluted from the column.

Retention time $t_R = t_M + t_S$ ...(22.14)

Retention volume $V_R = V_M + V_S$

where $t_M$ and $t_S$, = time the component spends in the mobile and stationary phase, respectively; $V_M$ and $V_S$ = retention volumes. The simplest way to identify a solute is to compare the retention time ($t_R$) with that of a known solute.

The time required for detection is dependent on the specific compound and corresponds to its distribution constant, $K_D$. The peak of elution (retention) curve is a part of the chromatogram, recording the response of the detector to a component. The quantitative composition of the mixture can be found by either measuring the areas under the peaks (internal normalization), or standard addition, or external standardization or internal standardization. This can be quantified further by mass spectrometric (MS) methods. Combination of GC-MS provides a rapid characterization ("fingerprinting") and the identification of the sources of environmental pollution.

### 22.5.2 Detector Systems

The success of gas chromatography is dependent to a large extent upon the efficiency of the detection system and its suitability for its intended purpose. Detector systems are based on the physicochemical properties of the analytes and the degree of sensitivity required.

As solutes elute from the column, they interact with the detector. The detector converts this interaction into an electronic signal that is sent to the data system. The magnitude of the signal is plotted versus time (from the time of injection) and a chromatogram is generated.

The selection of a particular detector type depends on several parameters—such as detector noise level; detector sensitivity; minimum detectable amount (MDA, which is the amount of analyte that produces a signal that is above the noise level); linear dynamic range (LDR, which is the sample concentration over which the detector response is linear). MDA and LDR for a specific analyte may be determined by plotting peak area against concentration in the carrier gas or in the sample analyzed.

There are a variety of detectors that are used in gas chromatography, to provide selective responses to particular groups of compounds to simplify the chromatograms from complex samples. Each detector has its own characteristics (selectivity, sensitivity, linear range, cost, etc.). The detectors can broadly classified, based on selectivity.

1. General (or universal) detectors that respond to all compounds (e.g. TCD), which differ from the carrier gas.
2. Selective detectors (e.g. ECD; FPD; NPD) that respond to range of compounds, which have some common chemical or physical property.
3. Specific detectors that respond to a single chemical species.

Besides selectivity, detectors can also be classed according to how they respond to the amount of analyte passing through. Concentration-dependent detectors (ECD; FID; TCD) produce a signal that is related to the concentration of solute in the gas stream present in the detector at any particular time. Mass-flow detectors (FID; FPD; NPD) produce a signal that is related to the rate at which solute molecules enter the detector.

While, thermal conductivity detector (TCD) is a general-purpose detector responding to most compounds to be analyzed by GC, electron capture detector (ECD), flame ionization detector (FID), and photoionization detector (PID) are the most frequently used detectors for determining volatile organic compounds (VOCs) in atmospheric samples (Table 22.11). The ECD and PID are especially attractive because they are selective and mask interfering peaks. The FID is very sensitive to most VOCs (except formaldehyde), but is not sensitive to non-combustible compounds. Combination of detectors like PID-FID is useful for general-purpose monitoring of a wide variety of VOCs. The FID-ELCD (electrical conductivity detector) and PID-ELCD (halogen mode) detector systems enable simultaneous determination of halogenated, aromatic and unsaturated species.

The use of an IR spectrometer as a detector for GC has been of great value. As carrier gases (He, $H_2$, $N_2$) do not absorb IR radiation and a better signal sensitivity is obtained by using FTIR spectrometry.

**TABLE 22.11:** DETECTORS FOR MONITORING VOLATILE ORGANIC COMPOUNDS (VOCs)

| *Detector System* | *VOC Group* | *Applicability* | *MDA(g/mL)* | *LDR* |
|---|---|---|---|---|
| Atomic emission Detector (AED) | Many compounds | Wider applicability | $10^{-12}$-$10^{-14}$ | |
| Electron capture Detector (ECD) | VOC-FORM; VOC-STRAT; VOC-TOX | Highly sensitive for compounds containing electronegative atoms | $10^{-12}$-$10^{-14}$ | $10^4$ |
| Electrical conductivity Detector (ELCD) | VOC-TOX | Specific to Cl and Br; S and N | $10^{-12}$-$10^{-14}$ | $10^5$ |
| Flame ionization Detector (FID) | VOCO-OX; VOC-TOX; VOC-CLIM | Good for TOX and CLIM | $10^{-10}$-$10^{-12}$ | $10^7$ |
| Flame photometric Detector (FPD) | VOC-CLIM | P and S selective; and halogen-specific | $10^{-10}$-$10^{-12}$ | $10^4$ |
| Photoionization Detector (PID) | VOC-OX; VOC-TOX | Wider applicability | $10^{-10}$-$10^{-12}$ | $10^5$ |
| Thermal conductivity detector (TCD) | All VOCs | Universal | $10^{-7}$-$10^{-9}$ | $10^5$ |
| Mass spectrometry (MS) | VOC-OX; VOC-TOX; VOC-FORM | High sensitivity; expensive | $10^{-10}$-$10^{-12}$ | $10^5$ |
| FTIR spectrometry | VOCs and absorbing gases | Better signal sensitivity | – | – |

VOC = Volatile organic compound; VOC-OX = VOC oxidants; VOC-TOX = VOC toxicants; VOC-STRAT = VOC stratospheric; VOC-FORM = VOC photochemical oxidants; VOC-CLIM = VOC greenhouse (see Chapter 21); MDA = Minimum detectable amount (sensitivity); LDR = Linear dynamic range.

### *22.5.2.1 Atomic Emission Detector (AED)*

Atomic emission detector (AED) is based upon atomic emission to determine (simultaneously) the atomic emissions of many of the elements in solutes that elute from a GC capillary column. The eluent from a GC capillary column end is introduced into microwave-energized helium plasma. The plasma is sufficiently energetic to atomize all of the elements in a sample and to excite their characteristic atomic emission spectra. These spectra are then scanned with a spectrometer with flat diode array detector, capable of detecting emitted radiation from about 170 to 780 nm (Fig. 22.14).

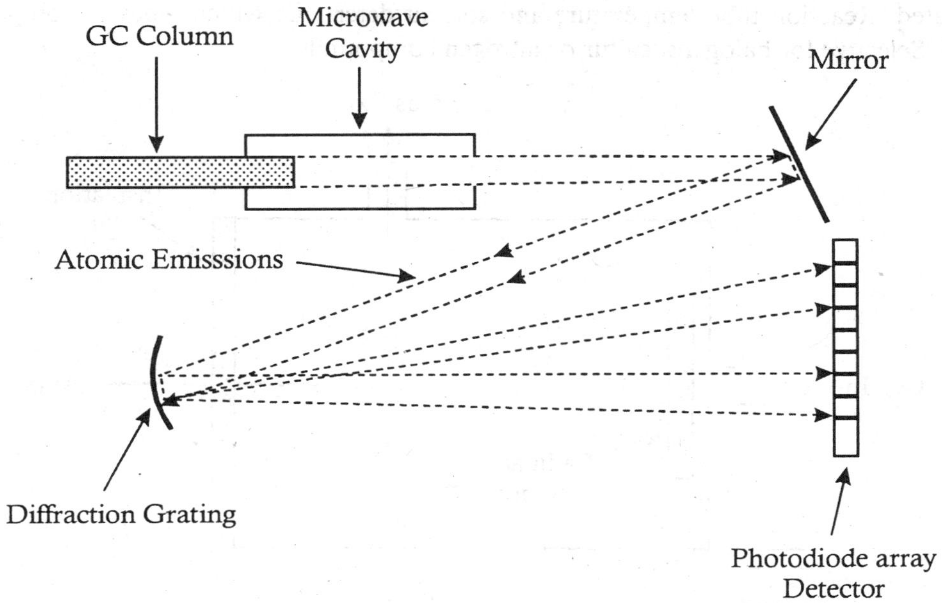


**Fig. 22.14:** Schematic of Atomic Emission Detector (AED)

### *22.5.2.2 Electron Capture Detector (ECD)*

Electron capture detector (ECD) is ideally suited for the detection of high electron-affinity compounds—halogens, peroxides and quinones and nitro group-containing organic compounds. It is very sensitive for halogenated compounds ($10^{-12}$ g) and is particularly used for the detection of halogenated pesticides, insecticides and herbicides, and halocarbon residues in environmental samples. But it is insensitive toward functional groups such as amines, alcohols and hydrocarbons.

The detection is based on the changes in electrical conductivity of gases in an ionization chamber, caused by the presence of electron-acceptor molecules. The device (Fig. 22.15) consists of a cylindrical cell with two electrodes—an outer source electrode and a central collector electrode. The source electrode is a β-particle emitter ($^3H$ or $^{63}Ni$). The high-energy electrons emitted by the radioactive source ionize the carrier gas (Ar, He or $N_2$) producing secondary electrons, which are collected by the anode, maintaining a standing detector current (due to ionization process) between the pair of electrodes. When organic molecules that contain electronegative functional groups, such as halogens, phosphorous, and nitro groups pass by the detector, they capture some of the electrons and reduce the current measured between the electrodes. The electronics of the detector system is maintained such that the pulse rate when a sample enters the ECD and captures secondary electron(s) is linearly related to the sample concentration. Element selective detection (ESD) can be combined with GC for speciation of elements in ionic form (Fig. 22.16).

### *22.5.2.3 Electrical Conductivity Detector (ELCD)*

Compounds are mixed with a reaction gas and passed through a high temperature reaction tube. Specific reaction products are created and they along with a solvent are passed through an electrolytic cell (Fig. 22.17). The change in the electrolytic conductivity of the solvent is measured and a signal is generated. Reaction tube temperature and solvent determine which types of compounds are detected. Selective for halogens, sulfur or nitrogen compounds.

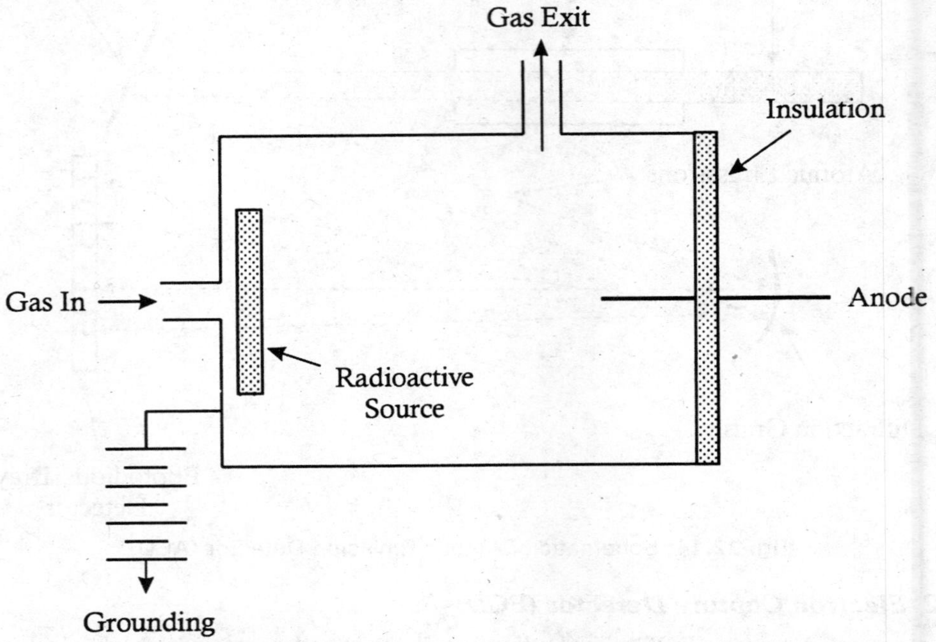


**Fig. 22.15:** Schematic of Electron Capture Detector (ECD)

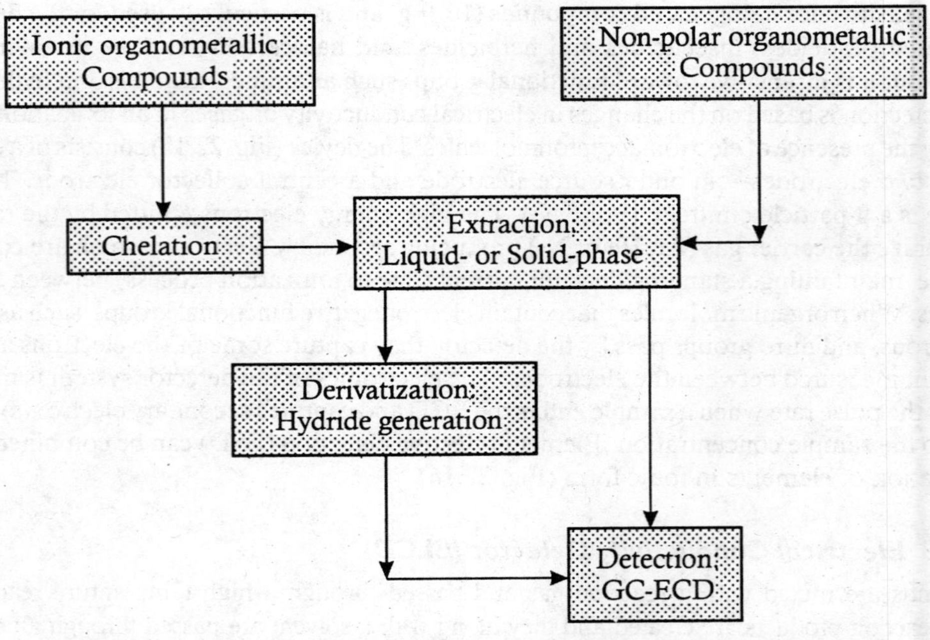


**Fig. 22.16:** Schematic of Metal Speciation Analysis by GC-ECD Method

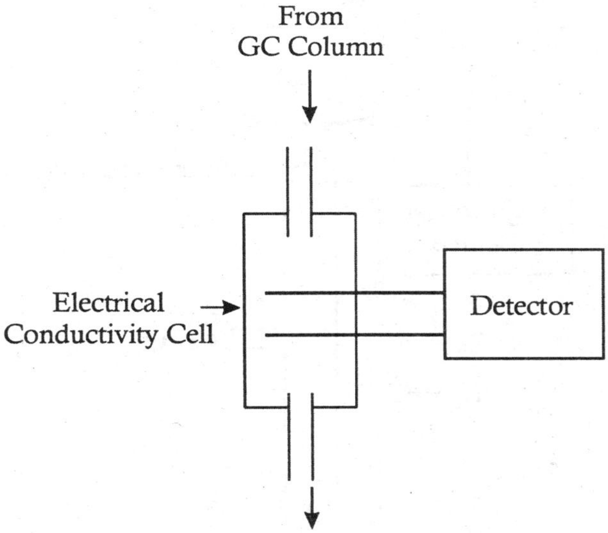


**Fig. 22.17:** Schematic of Electrical Conductivity Detector (ELCD)

### *22.5.2.4 Flame Ionization Detector (FID)*

Ionization detectors are specific property detectors, based on electrical conductivity of gases. The source of ionization energy is either a flame or nuclear radiation. The flame ionization detector (FID) is ideally suited for the detection of hydrocarbons and is widely used in GC analysis. This is because it responds to all organic compounds containing C—H bonds and has high sensitivity (~$10^{-12}$ g/L) and large linear response (~107). It is reliable, rugged and easy to operate. With it direct and continuous monitoring is possible. The response of the FID is proportional to the number of carbon atoms present in the compound. FID detects only the compounds that contain C—H bonds and does not respond to pollutants that do not contain C—H bonds, such as $H_2$, $O_2$, $CO_x$, $H_2O$, $SO_x$, $NO_x$, etc. Functional groups, such as C$=$O; O—H, halogens and amines do not yield in the flame. Therefore, FID is not useful for analysis of inorganic gases and organic compounds in water.

A FID detector consists of an air/hydrogen diffusion flame and collector plate (Fig. 22.18). The effluent from the column is mixed with hydrogen and air and then ignited electrically (2,100° C), at a small metal jet. A high polarizing voltage is applied between the two electrodes around the burner nozzle and produces an electrostatic field. Most organic compounds produce ions and electrons that can conduct electricity through the flame. Negative ions migrate to the collector electrode above the flame and positive ions migrate to the high voltage electrode. The so generated ionization current between the two electrodes is directly proportional to the hydrocarbon concentration in the sample that is burned by the flame.

Current flow is a measure of the amount of carbon burned, and hence of the quantity of hydrocarbons entering the detector per unit time, and are counted electrically. He-detector is more sensitive compared to $H_2$-detector. Helium atoms are excited above its ionization potential (24.5 eV) by a β-emitting source. The excited $He^*$ atoms, in turn, excite molecules that have less ionization potential than 24.5 eV. For example, the ionization potential of $CH_4$ is 14.5 eV, about the same as that of $H_2$ (15.6 eV), and, therefore, the He-detector is better suited to detect $CH_4$.

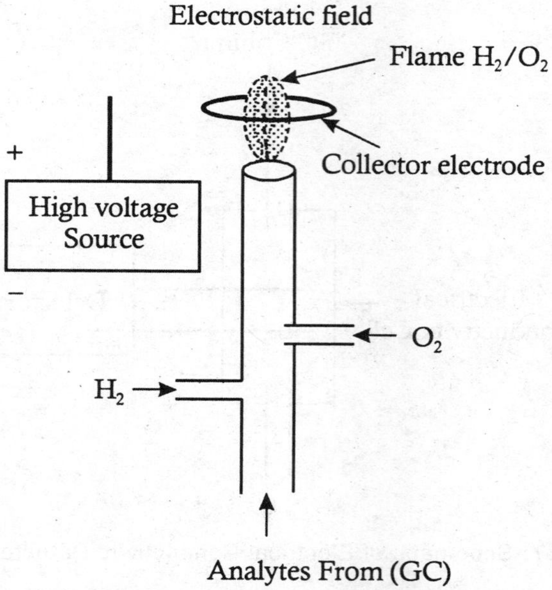


**Fig. 22.18:** Schematic of Flame Ionization Detector (FID)

### 22.5.2.5 *Specialized Flame Ionization Detectors*

The FID can be modified to be species-specific. Modified FID, specific to sulfur and phosphorus, is called flame photometric detector (FPD). In a FID setup, if a sodium salt is positioned above the flame, the ion current can be made highly specific to phosphorus. Similarly, high selectivity can be achieved for nitrogen and phosphorus, if a rubidium salt is used. Such a modified FID is called a thermoionic detector (TID) or alkali flame ionization detector (AFID).

FPD device (Fig. 22.19) uses the chemiluminescent reactions of the compounds in a hydrogen/air flame as a source of analytical information that is relatively specific for substances containing sulfur and phosphorus atoms. The flame photometric detector (FPD), chemiluminescence detector (CLD), and thermal energy analyzer (TEA), act by measuring the chemiluminescence of these elements. In this detector, the effluent is passed into a low-temperature hydrogen/air flame, which converts part of the phosphorus to an HPO species that emits bands of radiation centered about 510 and 526 nm. Sulfur in the sample is simultaneously converted to $S_2$, which emits a band centered at 395 nm. CLD uses chemiluminescence reaction of NO with $O_3$ (nitrosoamines and nitrogen-compounds).

Thermoionic detector (TID or NPD) is the most common specific-specific (N and P) GC detector based on the ionization of analyte in the presence of heated alkali source. The NPD is a highly sensitive but specific detector. It gives a strong response to organic compounds containing nitrogen and/or phosphorus. Compared with the flame ionization detector, the NPD is approximately 500 times more sensitive for compounds containing phosphorus and 50 times more sensitive for nitrogen-bearing species. These properties make nitrogen-phosphorus detector particularly useful for detecting and determining the many pesticides that contain phosphorus- determination of fungicides, herbicides in drinking water, and nitrophenols in groundwater.

This detector is similar to the design to the FID but with an important difference: an electrically heated silicate bead doped with an alkali salt mounted between the jet and the collector

(Fig. 22.20). The column effluent is mixed with hydrogen, passes through the flame tip assembly, and is ignited. The hot gas then flows around an electrically heated rubidium silicate bead. The heated bead forms plasma (600-800° C). The combustion products of nitrogen and phosphorus compounds interact with the alkali metal ions and produce thermionic electrons, which are attracted to the collector and give rise to the increase in current. The number of ions hitting the collector is measured and a signal is generated.

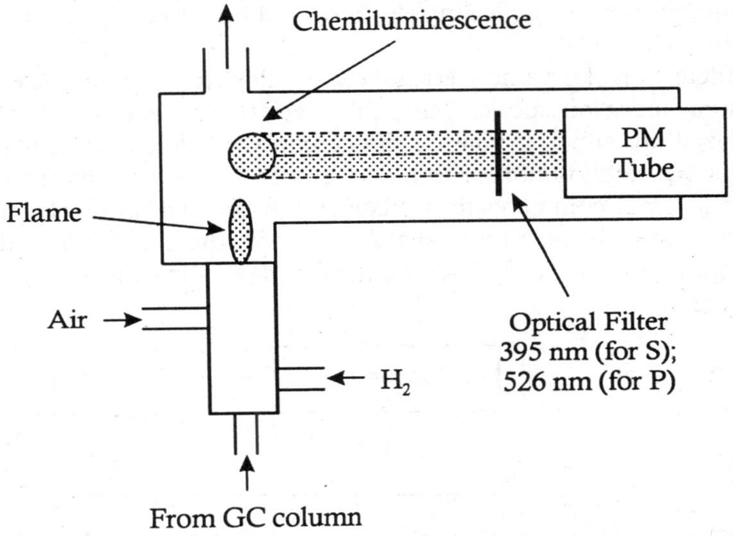


**Fig. 22.19:** Schematic of Flame Photometric Detector (FPD)

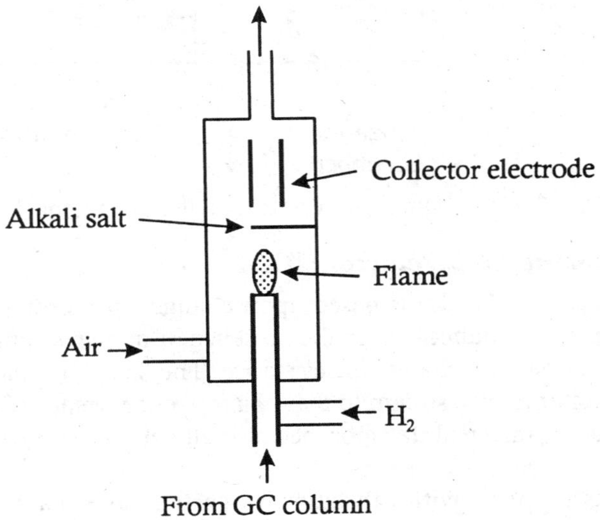


**Fig. 22.20:** Schematic of Nitrogen, Phosphorus (Thermoionic) Detector (NPD or TID)

### *22.5.2.6 Photoionization Detector (PID)*

The selective determination of aromatic hydrocarbons or organo-heteroatom species is performed by the photoionization detector (PID). In the photoionization detector, the column eluent is irradiated with an intense bean of ultraviolet radiation, which causes ionization of the molecules ($R + h\nu \rightarrow R^+ + e^-$).

PID is equipped with an UV source that emits photons through an optically transparent widow (LiF and NaF crystals), into an ionization chamber (Fig. 22.21). A signal is generated when a compound in the gas stream is ionized by photons. Photoionization occurs in the case of molecules whose ionization potential is less than that of the photons. A positively biased high-voltage electrode accelerates the resulting ions to a collecting electrode and the resulting current is proportional to the concentration of the analyte.

It is used for the detection of organic— (HCs, ketones, aldehydes, amines, esters, organo-metallic compounds) and some inorganic species (purgeable species). The reason why the compounds that are routinely analyzed are either aromatic hydrocarbons or heteroatom containing compounds (like organosulfur or organophosphorus species) is because these species have ionization potentials (IP) that are within reach of commercially available UV lamps. The available lamp energies range from 8.3 to 11.7 eV, that is, lambda max ranging from 150 nm to 106 nm. Although most PIDs have only one lamp, lamps in the PID are exchanged depending on the compound selectivity required in the analysis.

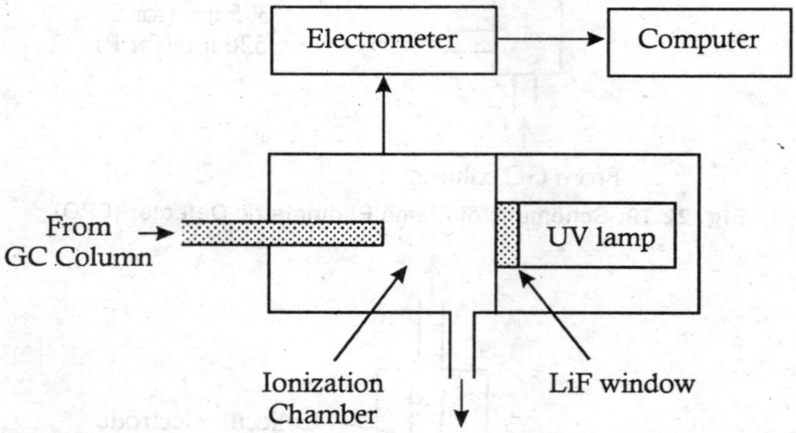


**Fig. 22.21:** Schematic of Photoionization Detector (PID)

### *22.5.2.7 Thermal Conductivity Detector (TCD)*

Thermal conductivity detector (TCD) is based upon changes in thermal conductivity of the gas flowing. Changes in thermal conductivity by the presence of analytes, cause a temperature rise in the element, which is sensed as a change in resistance. The TCD compares the heat-conducting ability of the exit gas (contaminant) stream to that of a reference stream of pure carrier gas (helium or hydrogen). It is called a universal detector, because all substances give a signal and it is non-destructive.

The detector consists of two cavities through which the gas streams flow (Fig. 22.22). The heating elements are thermistors or resistance wire (Pt wire) thermometers, which form two resistor arms of a balanced Wheatstone bridge circuit. In the absence of any analyte, the sample gas and the reference gas streams are adjusted to cool respective resistors equally and the Whetstone bridge remains in electrical balance (null deflection). When an analyte enters the sample cavity of the detector, the thermal conductivity of the mixture goes down and less heat is transferred away from the transducer. Consequently, the temperature goes up, resulting in a change in the resistance of the heating element that is converted into an electrical signal. The current signal is a measure of the concentration of the analyte. TCD is not very sensitive (~$10^{-5}$ g/L), but it detects a variety of organic and inorganic compounds including those not detected by FID.

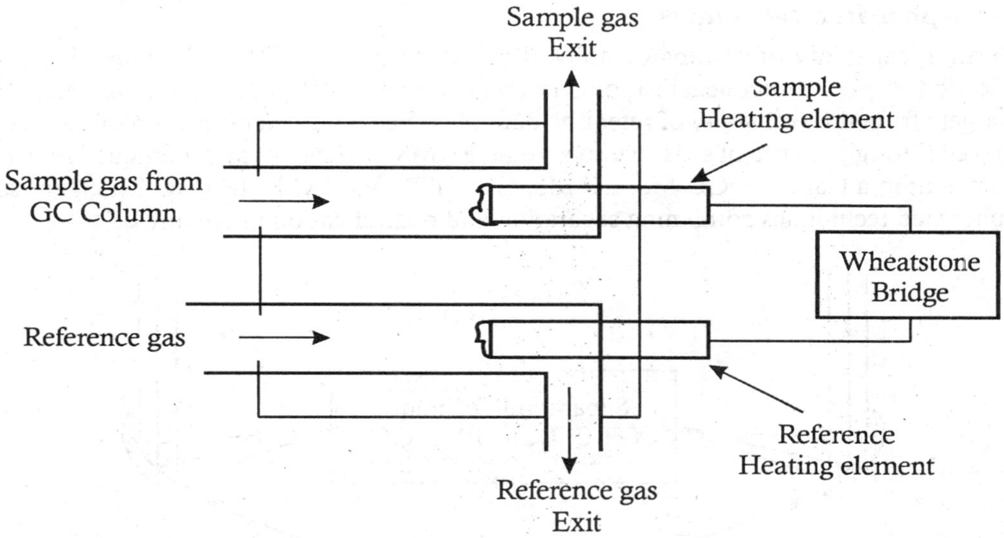


**Fig. 22.22:** Schematic of Basic Modules in Thermal Conductivity Detector (TCD)

#### *22.5.2.8 Other Detectors*

Conductimetry, coulometry, and amperometry detectors are some of the electrolyte-based detectors, used for the detection of various analytes, either general or species-specific (e.g. microcoulometric detectors are halogen-specific) (see Chapter 23 for details).

### 22.5.3 Supercritical Fluid Chromatography (SFC)

Gases above their critical temperature and under high pressures are supercritical fluids with better mobility (which is the case with gases) and solvating properties (which case with liquids). The SFC technique exploits this liquid-like solvating property of supercritical fluids to separate polar, thermally unstable, chemically reactive, and non-volatile compounds. In GC, rapid molecular diffusion leads to mass transfer of the solute between the mobile and stationary phases, resulting in better resolution and shorter analysis times. However, GC methods can only be used to separate compounds that can be volatilized. SFC combines the better mobility of gases and higher solvating power of liquids to achieve efficient separation. In the case of some substances, which are thermally unstable or cannot be resolved by GC this technique works better than HPLC. GC-SFC-LC can be combined in a single uninterrupted chromatogram.

$CO_2$ is the commonly used supercritical fluid, because of its availability in pure form and its easy detection by flame ionization detector. $N_2O$, and $NH_3$ are more polar fluids. n-pentane is used for the separation of polycyclic aromatics. SFC finds application in the separation of pesticides and their metabolites.

Supercritical fluid extraction method may be employed in large-scale purification of polluted streams and soils. SFC with heated and pressurized $CO_2$ and added metal-binding chemical compound is used to clean up radioactive contaminants in soil. Under these conditions $CO_2$ flows like a gas, dissolves like a liquid, but behaves with chemical properties unlike gases or liquids. The method is non-destructive, leaving the soil intact, and is industrially safe and environment friendly.

### 22.5.4 Hyphenated Techniques

Identification capability of chromatography alone is generally insufficient for identification and quantification of every component in a complex mixture. The limitation in chromatographic methods arises largely from the ambiguity of retention data. A more certain identification of each peak can be achieved through the use of sensitive spectrometer (MS or IR) to obtain a unique "finger-print" of each eluent in a teal time. GC-MS, LC-MS, GC-FTIR, and LC-FTIR (Fig. 22.23) are some of the hyphenated techniques combining separation and quantification procedures.

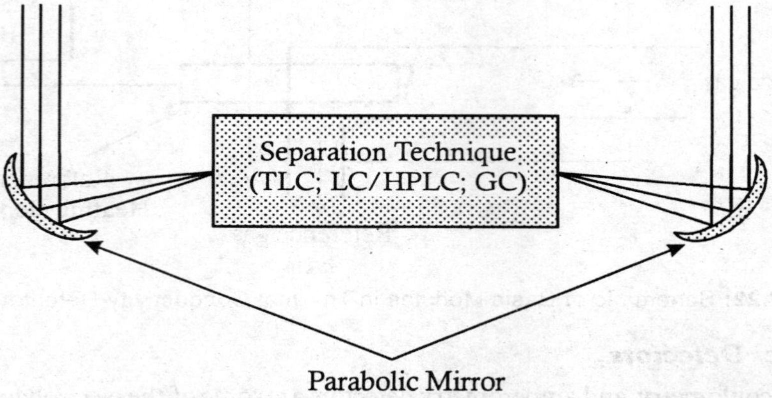


**Fig. 22.23:** Schematic of FTIR Combined with Chromatographic Methods

GC-MS is one of the most sensitive analytical systems in the separation, identification and quantification of volatile organic and inorganic compounds. The system can be used in the study of drugs, secondary metabolites, toxins, in pollution monitoring and ecological studies. In environmental monitoring trace level organic pollutants can be detected only after concentration.

(*i*) In closed-loop stripping analysis (CLSA) method, concentration of trace levels of volatile organic compounds is realized by recirculating stream of and flushing out the organic compounds which are adsorbed by an activated carbon filter and are subsequently extracted from the filter with carbon disulfide ($CS_2$). A portion of the extract is injected into a capillary column of GC-MS setup. This method is suitable for analysis of broad spectrum of organic compounds in water.

(*ii*) In purge-and-trap method, volatile organic compounds are transferred from the aqueous to vapor phase by bubbling an inert gas through the liquid and a sorbent trap collects the vapors. After purging is completed, the sorbent trap is heated to desorb the compounds on to a GC column and the separated components are analyzed by MS. This method is well suited to determine purgeable organic compounds in municipal and industrial wastewater systems, and a large number of organic contaminants at very low concentrations otherwise not detected by other techniques.

Growing emphasis on environmental monitoring has encouraged the development of more rapid and less expensive methods of analysis. These are spectroscopic and electrochemical sensor arrays, which provide good sensitivity and selectivity at relatively low cost. Some of the spectroscopic methods, such as infrared differential absorption (IRDIA), and differential optical absorption spectroscopy (DOAS) are capable of carrying out remote sensing measurements of airborne pollutants.

### 22.5.5 Applications

Applications of GC techniques are wide and varied. GC is the most widely used technique for determining volatile organic compounds (VOCs) with suitable stationary phases (polysiloxanes, polyethylene glycol (PEG) etc), dairy and food products for volatile components (aldehydes, esters, and ketones) to determine age and rancidity. The volatile nature of the flavors and essential oils are particularly suited for GC analysis. Pyrolysis GC is employed for separation and identification of materials such as plastics, natural and synthetic polymers, paints, polynuclear aromatic hydrocarbons (PAHs), halogenated pesticides, organophosphorus, organosulfur, organolead and organomercury compounds. Selective retention of straight-chain molecules on the molecular sieves can be utilized for separation of aldehydes from ketones, and acids from esters. Chemical conversion gas chromatography and derivatization methods lead to an ever-expanding area of application of GC in biochemistry, pharmacology, medicine and environmental monitoring. Its success is due to its inherent efficiency as to the variety of detection systems available.

HPLC and GC systems are highly versatile separation methods. While GC provides high resolution and speed of separation it is applicable only to volatile compounds. It is not appropriate to many polar, non-volatile and thermo-labile compounds. On the other hand, HPLC is applicable to many such compounds. Supercritical fluid extraction with enzyme immunoassay allows faster extraction and more sensitive and selective analysis than conventional (Soxhlet) methods. A comparison of their suitability is given in Table 22.12.

**TABLE 22.12:** COMPARISON OF HPLC AND GC SEPARATION METHODS

| *HPLC Separation* | *GC Separation* |
|---|---|
| Organic and inorganic volatile and non-volatile and thermolabile compounds | Generally organic, volatile and thermo-stable compounds |
| Operational parameter are flexible and separation wide range of analytes possible | Operational parameters are less flexible and separation of only volatile analytes. |
| Moderate resolution and efficiency | High resolution and high efficiency |
| Mobile phase (liquid) is forced through the stationary columns under pressure | Mobile phase (gas) flushes vaporized analytes through columns |
| Better separation by varying composition of mobile phase while the sample is flushed | Better separation with column temperature control |
| The stationary phase must be chemically bonded to the support matrix | Stationary phase is adsorbed (coated) on the supported matrix |
| Detection (generally) by UV absorbance | Wide variety of detectors |

## 22.6 MASS SPECTROMETRY (MS)

Spectroscopic methods of characterization (UV, IR etc) are based on excitation of molecules. Mass spectrometry is based upon the ionization of solute molecules in the ion source and the separation of the ions generated on the basis of their mass/charge (*m/Ze*) ratio by an analyzer unit. The (*m/Ze*) ratio is characteristic of the target molecule. The analyzer may be a magnetic sector analyzer, a quadruple mass filter, or an ion trap. Ions are detected by a dynode electron multiplier. For the formation of mass spectrum the molecules of the substance must be ionized. The resulting ions are focused and accelerated into a mass filter. The mass filter selectively allows all ions of a specific mass to pass through to the electron multiplier. All of the ions of the specific mass are detected.

All mass spectrometers have three basic units—(1) an ionization chamber for generating the molecular ion species under vacuum, (2) a mass analyzer for separating ion species according to (*m/Ze*) ratio, and (3) a detector system to identify ionic species. If GC is combined with MS, then an interfacing module is required. Electron impact (EI), electrostatic or chemical ionization, thermal ionization or any other suitable method may be employed to produce ionized particles. MS can be employed to determine the molecular mass as well as structural features of molecules, and has the ability to detect a wide variety of compounds.

### 22.6.1 Ionization Chamber

Ionization chamber is the most important part of a mass spectrometer. Electron impact ionization (EI) (Fig. 22.24) is the most widely employed method of ionization. In this setup, sample (from GC or LC) is led into the ionization chamber maintained at high vacuum through a 'molecular leak' (orifice). Electrons, emitted from an electron gun (a tungsten filament coated with thorium oxide), are driven from the filament at ~70 V to anode (trap), and they interact with the sample, present in vaporous state. The interaction results in either loss of an electron from the substance, that would produce a cation, or electron capture that would produce an anion. The positive ions produced in the ionization chamber are gently drawn out by a mild electrostatic field, maintained between the repeller (positive slit placed after the 'molecular leak') and the first accelerating slit (negative slit placed after the electron gun). A strong electrostatic field (500 $\rightarrow$ 4000 V), maintained between the first and the last accelerating slit, accelerates the ions of masses, $m_1$, $m_2$, and $m_3$, to the mass analyzer. EI ionization methods are better suited for gases and organic molecules; but lack selectivity.

Other ionization methods are tailor-made to specific substances. In chemical ionization (CI) method, a reacting gas (e.g. methane) is introduced into the ionization chamber (at ~100 Pa). The spectra are relatively simple. Chemical ionization (CI) is based on the EI source, but little fragmentation occurs. CI spectrum provides better information. The fragmentions method operates with stable ions of mass lower than that of original molecule. It can differentiate the components of various structural groups that molecular ions of which have same (*m/Ze*) values. In field-ionization method (FI), the sample is exposed to the effect of an electric field gradient, leading to the formation of a molecular ion that is not excited. Only slight fragmentation occurs and the spectrum is simple and enables easier determination of molecular formulae. Field desorption method is a special case of FI, characterized by desorption of ions formed from molecules adsorbed on the emitter surface (ion probe). This method is suitable for analysis of thermally labile substances (like biological materials). Ionization by inductively coupled plasma method is an electrode-less technique that avoids impurities from source electrodes.

When energetic particles (argon ions) strike the surface of a solid, neutral atoms and secondary charged particles are ejected from the target in a process called sputtering. In the secondary ion mass spectrometry (SIMS) method, these secondary ions are used to give information about the target material. Photo-ionization and resonance ionization methods (laser probes) achieve higher elemental selectivity, because they take advantage of much more neutral atoms emitted in the sputtering process. In the pyrolysis or thermal degradation method, volatile substances are ejected from the material, under vacuum and ionized by EI. This method is well suited for environmental studies and monitoring and for the identification of certain micro-organisms.

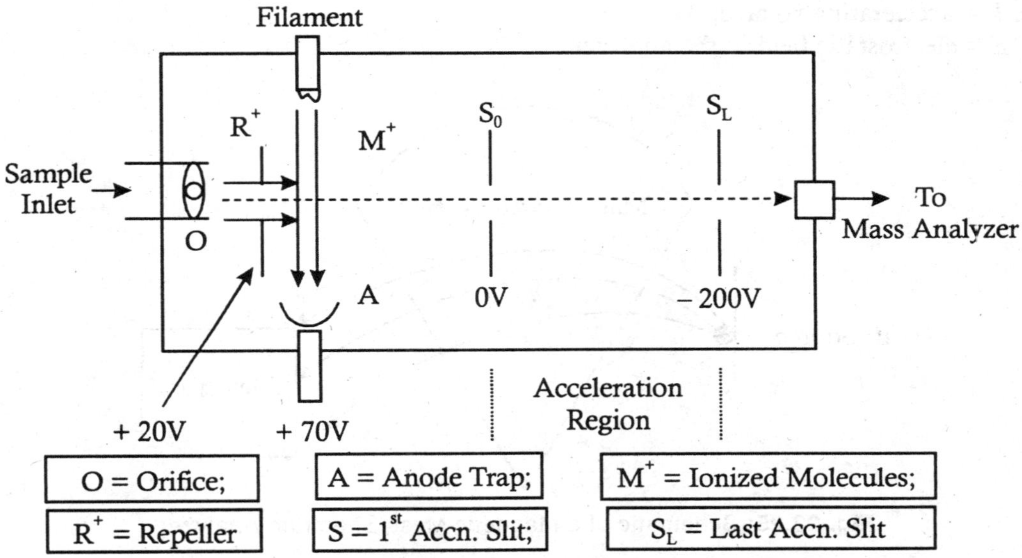


**Fig. 22.24:** Schematic of Electron-Impact (EI) Ionization Chamber

### 22.6.2 The Mass Analyzer

The function of the mass analyzer or ion-beam analyzer is to separate the ionic components according to their $(m/Ze)$ ratio. This can be accomplished with static magnetic fields, time-varying electric fields, and time-of-flight methods. The resolving power $(m/\Delta m)$ of an MS is a measure of its ability to separate adjacent masses that are displayed as peaks on the detector. Magnetic deflection analyzers (Fig. 22.25) are well suited for high-resolution mass spectrometry (momentum spectrometry). These operate by deflection of charged particles in vacuum through a uniform magnetic field and the $(m/Ze)$ ratio is given by

$$\left(\frac{m}{Ze}\right) = \frac{B^2R^2}{2V} \qquad \ldots(22.15)$$

where, $B$ = magnetic flux density;
$R$ = radius of the trajectory;
$V$ = acceleration voltage.

By varying $B$, ions of different mass (but of the same velocity) can be made to have the same trajectory ($R$ = constant). This is the principle of magnetic scanning and is commonly used in environmental analysis. The resolution can be improved by a combination of electric and magnetic analyzers (double focusing mass spectrometer).

Secondary ion mass spectrometry (SIMS) is a double-focusing mass spectrometry technique, where electric and magnetic fields are arranged in tandem in such a way (first by energy and then by mass) that ion beams that emerge from the source slits in divergent directions and different velocities are refocused. Ions from the ionization chamber emerge with varying kinetic energies. These ions follow a circular trajectory in the electrostatic analyzer, whose radius, $R_e$, is given by

$$R_e = \frac{2V}{E} \qquad \text{...(22.16)}$$

where, $V$ = accelerating voltage;
$E$ = electrostatic field in the analyzer.

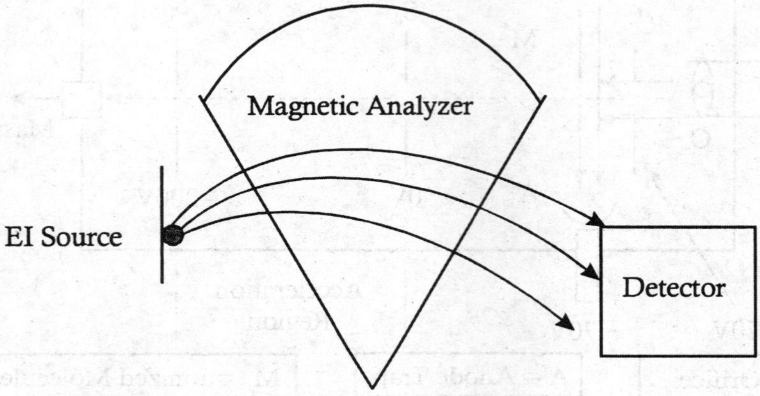


**Fig. 22.25:** Schematic of a Magnetic Mass Spectrum Analyzer

Ions emerge from the electrostatic analyzer with varying masses but with the same velocity and enter the magnetic sector analyzer to undergo $m/Ze$ analysis. SIMS provides very high accuracy mass analysis (1 ppm), and is used for the detection of nuclides. The instrument has micro-focus ion sources, cesium and $O_2$ or argon. The Cs source enables a probe size of 200 nm and the $O_2$ (argon) source 500 nm.

Non-magnetic analyzers are also in use to separate ions. Time-of-flight mass analyzer does not have magnetic or electric fields. Analysis relies on the differing speeds (arrival times) of ions with the same energy but different masses, and the mass spectrum is a function of time.

$$v = \frac{2ZV}{m} \qquad \text{...(22.17)}$$

Quadruple analyzer, also called RF-DC mass filter, is a non-magnetic mass analyzer (Fig. 22.26) that is preferred for GC-MS, and water analysis. The analyzer consists of four cylindrical rods, symmetrically placed (two along $X$-axis and two along $Y$-axis) around a central axis ($Z$-axis). The electrodes are connected so that opposing rods have the same potential ($X$-rods (+) and $Y$-rods (–)). A radio frequency (RF) potential is applied to the electrodes (with $V_Y = -V_X$) of quadrupoles with a low voltage of direct current. The fixed DC and oscillating RF fields cause the ions to undergo complicated trajectories through the quadruple filter. The ions of a certain ($m/Ze$) value entering the field either pass through the analyzer (along $Z$-axis) or hit some of the electrodes, depending on the frequency and applied current onto the quadruple filter. This way, ions of different masses can be filtered and transmitted. Knowledge of the potentials and frequency specifies the mass of the analyte.

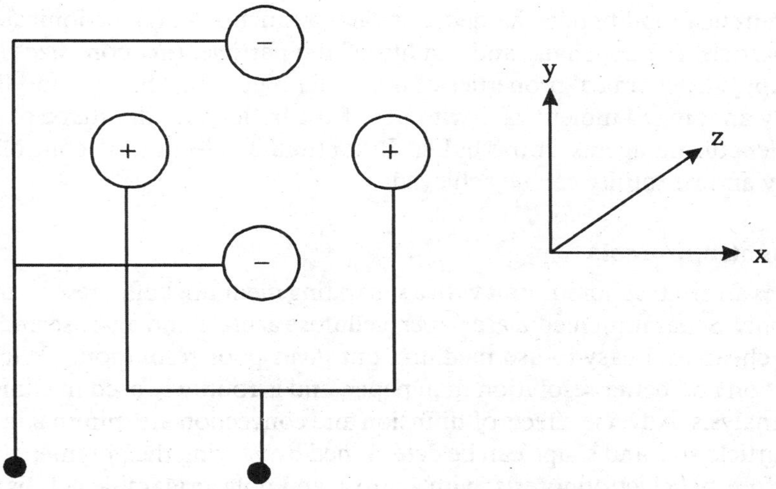


**Fig. 22.26:** Principle of Quadrupole Mass Analyzer

## 22.7 ELECTROPHORESIS

Electrophoresis is a separation method based on the differential migration of charged particles in a solution towards the electrodes of opposite polarities under the influence of an applied electric field. The extent of such a separation is time dependent. The technique is extensively used in the separation of biological molecules—peptides, proteins and nucleic acid fragments. It has become a standard separation technique in molecular biology and is identified by various blotting techniques (e.g. the genome project).

The rate of migration of a charged particle under the influence of an electric field is dependent on (*i*) the electrophoretic mobility of the charged particle and (*ii*) the frictional drag in the solution, and other physicochemical parameters. By optimal control of these parameters, the electro-phoretic method can be made versatile for the separation of polar molecules such as amino acids, peptides, proteins, (enzymes, hormones, antibodies, etc), carbohydrates, and nucleic acids (DNA, RNA and their components).

The rate of migration is directly proportional to the net charge of the particle, and inversely to its size, and the viscosity of the medium. The force, $F$, acting on a particle of charge, $q$, in an electric field of strength, $E$, is opposed by the frictional drag, $F'$.

$$F = qE;\ F' = fv \qquad \text{...(22.18)}$$

$$v = \frac{qE}{f} = \frac{qE}{6\pi\eta r} \qquad \text{...(22.19)}$$

The electrophoretic mobility, $\mu = v / E = \dfrac{q}{6\pi\eta r}$ (Stokes equation) ...(22.20)

If $\mu$ is positive (pH > the isoelectric point), the particle moves towards the cathode.

If $\mu$ is negative (pH < the isoelectric point), the particle moves towards the anode.

where $v$ = terminal velocity;

$f$ = frictional coefficient,

$\eta$ = viscosity coefficient;

$r$ = radius of the particle.

The electrophoretic mobility of a charged particle is a function of (*i*) the ionizable groups on the surface of the particle, (*ii*) the shape and rigidity of the particle, (*iii*) pore size of the separation matrix, and (*iv*) physicochemical properties of the buffer medium. The sign and the magnitude of charge carried by an ionized moiety vary with pH of the buffer, and the shape of the particle may be affected by denaturing agents in the buffer. By optimum selection of controlling parameters, greater flexibility and versatility can be achieved.

### 22.7.1 Gel Electrophoresis

A setup comprises an electrophoretic unit with a separating medium, buffer reservoir and accessories, and a power supply. Separating media are paper, cellulose acetate and agarose and polyacrylamide gels. Paper is a cheap and easy-to-use medium, but gives poor resolution. Cellulose acetate has minimal adsorption and better resolution than paper, and is routinely used in clinical, forensic and environmental analysis. Adverse effects of diffusion and convection are minimal in gels. In addition to separation, particle size and shape can be determined by varying the polymer concentration and pore size. Therefore, gel electrophoresis, with agarose and polyacrylamide gels, is extensively used in the separation and physicochemical characterization of peptides, proteins and nucleic acids.

Both horizontal and vertical slab gel electrophoresis systems are in use. Slab gels have the advantage of loading several samples simultaneously. Horizontal setup (Fig. 22.27) is used for the separation of large-size macromolecules—DNA, plasmids, restriction fragments, with agarose gels, which have large pore sizes. Vertical polyacrylamide slab gels are used for purification of peptides, proteins, and other biological molecules. Polyacrylamide gel electrophoresis (PAGE) has become a standard method for protein purification and characterization in molecular biology and clinical medicine.

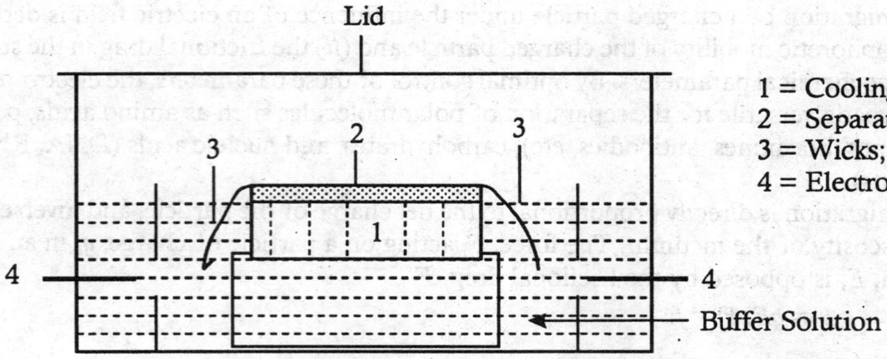


**Fig. 22.27:** Schematic of Horizontal Gel Electrophoresis Setup

*Immunoelectrophoresis* is a gel elctrophoretic technique for the determination of specific proteins in biological fluids. The technique is based on the specific interaction between immunoglobulins and their antigens. It consists initially of separation of the proteins of the serum, or fluid, by electrophoresis. After electrophoresis, antiserum is placed in a trough, cut adjacent and parallel to the direction of electrophoresis and the proteins and antibodies are allowed to diffuse towards each other. Interaction between the antibodies and protein antigens results in the precipitation of immune complexes within the gel, which takes the form of a series of arcs, each arc resulting from the precipitation of each serum protein with its own specific antibody.

### 22.7.2 Capillary Zone Electrophoresis (CZE)

HPLC, coupled with element-selective detectors is well suited for speciation analysis than for total elemental determination. However, HPLC procedures are time consuming and expensive. Capillary zone electrophoresis (CZE) is an alternative to liquid chromatography (LC) on account of its ability to separate ionizable compounds. CZE separations occur entirely in a liquid phase. As no mass transport between mobile and stationary phases is involved scope for peak broadening is negligible. Separation of analytes in CZE is based on differences in their electrophoretic mobilities, which can be modified by adding suitable soluble reagents. Thus, CZE is more efficient, expeditious and selective than LC. It is used in the determination of organic and biological compounds in environmental samples, polar pesticides (phenoxyacid pesticides, paraquat and diaquat), organophsophorus pesticides and herbicides (sensitivity $10^{-15}$ M/L).

## 22.8 OTHER METHODS

Magnetic particle separation method is gaining much importance in recent years. It is solid-phase separation method, akin to affinity chromatography, but it can be used in batch separation also. The method is simple, inexpensive and easy to operate. It combines high surface capacity with fast, efficient separation without the need for centrifugation. Thus, it provides a solid-phase separation methodology that avoids many of the disadvantages of other phase separation techniques while retaining their advantages. The technique relies on immobilization of analytes (antibodies in immunoassays) and separating them using simple operation equipment. Compounds in trace levels can be separated and detected. Cell sorting by this method, using antigen-coated magnetic particles, is a much cheaper alternate method to fluorescence activated cell sorter (FACS) method.

## BIBLIOGRAPHY


Ahuja, S. (ed.) (1997), Amer. Chem. Soc: Washington, DC. "*Chiral Separation: Applications and Technology,*" Washington.

Albarges, J. B. (ed.) (1980), Pergamon Press: Oxford: , UK. "*Analytical Techniques in Environmental Chemistry.*"

Alonso, V. E., de Torres, G. and Cano Pavon, J. M. (2001), *Talanta,* **52**(2): 219. "*Flow Injection on-line Electrothermal Atomic Absorption Spectrothermetry.*"

Alfussi, Z. B. and Wai, C. M. (eds.) (1992), CRC Press: Boca Raton, FL. "*Preconcentration Techniques for Trace Elements.*"

Aragon, P., Atienza, J. and Climent, M. D. (2000), *Crit. Rev. Anal. Chem.*, **30**(2&3): 121. "Analysis of Organic Compounds in Air: A Review."

Barcelo, D. (ed.) (1993), Elsevier: Amsterdam. "*Environmental Analysis: Techniques, Applications and Quality Assurance.*"

Barcelo, D. (ed.) (1996), Elsevier: Amsterdam. "*Applications of LC-MS in Environmental Chemistry.*"

Barker, J. (1999), Wiley & Sons: Chichester, UK. "*Mass Spectrometry.*"

Baugh, P. J. (ed.) (1993), IRL Press: Oxford, UK. "*Gas Chromatography: A Practical Approach.*"

Braithwaite, A. and Smith, F. J. (1996), Chapman & Hall: London. "*Chromatographic Methods,*" (5th edn.).

Brown, P. R. and Hartwick, R. A. (1988), John Wiley: New York. "*Practical HPLC.*"

Chu, T-Y., *et al.* (2001), *Talanta* **54**(6): 1163."Microwave-accelerated derivatization processes for the determination of phenolic acids by gas chromatography—mass spectrometry."

Clement, R. E. (ed.) (1990), Wiley Interscience: Toronto. "*Gas Chromatography- Biochemical, Biomedical and Clinical Applications.*"

Colume, A., *et al.* (2001), *Talanta,* **54**(5): 943. "Evaluation of an Automated Solid-Phase Extraction System for the Enrichment of Organochlorine Pesticides from Waters."

Davies, R. and Pearson, M. (1988), Wiley & Sons: New York. "*Mass Spectrometry.*"

Dean, P. D. G., Johnson, W. S. and Middle, F. A. (1985), IRL Press: Oxford, UK. "*Affinity Chromatography: A Practical Approach.*"

Eaton, D. A., *et al.* (eds.) (1995), Amer. Public Health Assn: Washington, DC. "*Standard Methods for the Determination of Water and Wastewater*," (19th edn.).

Fang, L. and Xiobai, X. (2000), *Intl. J. Environ. Anal. Chem.*, "Capillary electrophoretic separation and determination of chlorophenolic pollutants in industrial wastewaters."

Fifield, F. W. and Haines, P. J. (2000), Blackwell Science: Oxford, UK. "*Environmental Analytical Chemistry*," (2nd edn.).

Forrest, G. C. and Rattle, S. J. (1983), Churchill-Livingstone: Edinburgh, UK. "*Immunoassays for Clinical Chemistry*," (2nd edn.).

Frei, R. W. and Albarge, J. (eds.) (1986), Gordon & Breach: New York. "*Air and Water Analysis: New Techniques and Data.*"

Fried, B. and Sherma, J. (1999), Marcel Dekker: New York. "*Thin-layer Chromatography*," (4th edn.).

Gehrke, C. W., Wixom, R. L. and Bayer, E. (eds.) (2001), Elsevier: Amsterdam. "*Chromatography—A Century of Discovery (1900-2000)—The Bridge to the Science/Technology.*"

Gübitz, G., *et al.* (2001), *Crit. Rev. Anal. Chem.*, **31**(2): 141. "Chemiluminescence Flow-Injection Immunoassays."

Guerin, T., Astruc, A. and Astruc, M. (1999), *Talanta*, **50**(1): 1. "Speciation of Arsenic and Selenium Compounds by HPLC Hyphenated to Specific Detectors: A Review."

Haddad, P. R. and Jackson, P. E. (1990), Elsevier: Amsterdam. "*Ion Chromatography: Principles and Applications.*"

Hagberg, J., *et al.* (2000), *Intl. J. Environ. Anal. Chem.* "Application of Capillary Zone Electrophoresis for the Analysis of Low Molecular Weight Organic Acids in Environmental Samples."

Hancock, W. S. (ed.) (1990), Wiley Interscience: New York. "*HPLC in Biotechnology*."

Havenga, W. J. and Rohwer, E. R. (2000), *Intl. J. Environ. Anal. Chem.* "The use of SPME and GC-MS for the chemical characterization and assessment of PAH pollution in aqueous environmental samples "

Hearn, M. T. W. (ed.) (1991), VCH Pubs; New York. "*HPLC of Proteins, Peptides and Polynucleotides.*"

Heftman, E. (ed.) (1992), Elsevier: Amsterdam. "*Chromatography.*"

Herber, R. F. M. and Stoeppler, M. (eds.) (1994), Elsevier: Amsterdam. "*Trace Element Analysis in Biological Specimens.*"

Inczedy, J. (1986), Ellis Harwood: Chichester, UK. "*Ion-exchangers and their Applications.*"

Katz, E., *et al.* (1999), Marcel Dekker: New York. "*Handbook of HPLC.*"

Keith, L. M. (ed.) (1991), Lewis Pubs: Chelsea, MI. "*Environmental Sampling for Trace Analysis: A Practical Guide.*"

Keith, L. M. (ed.) (1992), Ann Arbor Science Pubs: Ann Arbor, MI. "*Advances in Identification and Analysis of Organic Pollutants.*"

Kline, T. (ed.) (1993), Marcel Dekker: New York. "*Handbook of Affinity Chromatography*."

Lambropoulou, D. A., Konstantinou, I. And Albanis, T. A. (2000), *Intl. J. Environ. Anal. Chem.* "Determination of herbicides in natural waters using solid phase microextraction (SPME) and gas chromatography coupled with flame thermoionic and mass spectrometric detection."

Landers, J. P. (ed.) (1997), CRC Press: Boca Raton, FL. "*Handbook of Capillary Electrophoresis*," (2nd edn.).

Lee, M. L. & Markides, K. E. (1987), *Science*, **235**: 1342. "*Chromatography with Supercritical Fluids.*"

Li, S. F. Y. (1992), Elsevier: Amsterdam. "*Capillary Electrophoresis: Principles, Practice and Applications.*"

Lombas-Garcia, E., *et al.* (1998), *Intl. J. Environ. Anal. Chem.* "Supercritical fluid extraction versus ultrasonic extraction for the analysis of polycyclic aromatic hydrocarbons from reference sediments."

Mar Gonzalez, M., Gallego, M. and Valcarcel, M. (2001), *Talanta*, **55**(1): 135. "Determination of arsenic in wheat flour by electrothermal atomic absorption spectrometry using a continuous precipitation-dissolution flow system."

Meloan, C. E. (1999), Wiley Interscience: New York. "*Chemical Separations—Principles, Techniques and Experiments.*"

Melvin, M. (1987), John Wiley: New York. "*Electrophoresis.*"

Mori, S. and Barth, H. G. (1999), Springerverlag: Heidelberg. "*Size-exclusion Chromatography*."

Narayanan, P. (2001), Bhalani Pubs: Mumbai. "*Clinical Biophysics: Principles and Techniques.*"

Narayanan, P. (2003), New Age Intl Pubs: New Delhi. "*Essentials of Biophysics*,"(2nd Print).

Niessen, W. M. A. and Voyksnev, R. D. (eds.) (1999), Marcel Dekker: New York. "*Liquid Chromatography—Mass Spectrometry*," (2nd edn.).

Nilsson, T., *et al.* (1997), *Intl. J. Environ. Anal. Chem.* "Analysis of volatile organic compounds in environmental water samples and soil gas by solid-phase microextraction."

Quinby-Hunt, M. S., McLaughlin, R. D. and Quintinilha, A. T. (1986), Wiley & Sons: New York. "*Instrumentation for Environmental Monitoring*," (Vols 1 & 2).

Perez-Bendito, D. and Rubio, S. (1999), Elsevier: Amsterdam. "*Environmental Analytical Chemistry*"—"*Comprehensive Analytical Chemistry, Vol XXXII*", G. Svehla (ed.).
Rickwood, D. (ed.) (1984), IRL Press: Oxford, UK. "*Centrifugation: A Practical Approach.*"
Rubinson, J-F. and Rubinson, K. A. (1998), Prentice & Hall: London. "*Contemporary Chemical Analysis.*"
Scott, R. P. W. (1995), Marcel Dekker: New York. "*Techniques and Practice of Chromatography.*"
Sewell, P. and Clark, B. (1988), John Wiley: New York. "*Chromatographic Separations.*"
Shibamoto, T. (ed.) (1998), Marcel Dekker: New York. "*Chromatographic Analysis of Environmental and Food Toxicants.*"
Sievers, R. E. (ed.) (1995), Wiley & Sons: New York. "*Selective Detectors: Environmental, Industrial and Biochemical Applications.*"
Simon, N. S. (1997), *Intl. J. Environ. Anal. Chem.* "Supercritical fluid carbon dioxide extraction and liquid chromatographic separation with electrochemical detection of methyl mercury from biological samples."
Skoog, D. A., West, D. M. and Holler, J. (1988), Saunders: Philadelphia, PA. "*Fundamentals of Analytical Chemistry,*" (5th edn.).
Soriano, J. M., *et al.* (2001), *Review, Crit. Rev. Anal. Chem.*, **31**(1): 19. "Analysis of carbamate pesticides and their metabolites in water by solid phase extraction and liquid chromatography: A."
Subramanian, G. (ed.) (1994), VCH Pubs: New York. "*A Practical Approach to Chiral Separation by Liquid Chromatography.*"
Taylor, L. T. (1995), Wiley-VCH Pubs: New York. "*Supercritical Fluid Extraction.*"
Thomson, S. and Budzinski, H. (1999), *Intl. J. Environ. Chem.* "Determination of polychlorinated biphenyls and chlorinated pesticides in environmental biological samples using focused microwave-assisted extraction."
Timerbaev, A. R. (2000), *Talanta*, **52**(3): 573. "Element speciation analysis by capillary electrophoresis."
Touchstone, J. C. and Sharma, J. (eds.) (1985), Wiley & Sons: New York. "*Techniques and Application of Thin Layer Chromatography.*"
Turrio-Baldassarri, L., *et al.* (1999), *Intl. J. Environ. Chem.* "Supercritical fluid extraction of bivalve samples for simultaneous GC-MS determination of polychlorobiphenyls and polycyclic aromatic hydrocarbons."
Van Loon, J. C. (1982), CRC Press: Boca Raton, FL. "*Chemical Analysis of Inorganic Constituents of Water.*"
Van Loon, J. C. (1985), John Wiley: New York. "*Selected Methods of Trace Metal Analysis: Biological and Environmental Samples.*"
Varga, B. *et al.* (2001), *Talanta*, **55**(3): 561. "Isolation of water-soluble organic matter from atmospheric aerosols."
Weeks, I. (1992), Elsevier: Amsterdam. "*Chemiluminescence Immunoassay*"—"*Comprehensive Analytical Chemistry,*" Vol. XXIX, G. Svehla.(ed.).
Weiss, J. (1995), VCH Pubs: New York. "*Ion Chromatography,*" (2nd edn.).
Wilson, K. and Walker, J. M. (1994), Foundation Books: New Delhi. "*Principles and Techniques of Practical Biochemistry,*" (4th edn.).
Wu, C. S. (ed.) (1995), "*Handbook of Size-exclusion Chromatography.*"
Yost, R. S., Ettre, L. S. and Conlon, R. D. (1980), Perkin-Elmer: Norwalk, CT. "*Practical Liquid Chromatography: An Introduction.*"
Zlotorzynski, A. (1995), *Crit. Rev. Anal. Chem.*, **25**(1): 43. "The application of microwave radiation to analytical and environmental chemistry."
Zygmunt, B., Jastrzebska, A. and Namienik, J. (2001), *Crit. Rev. Anal. Chem.*, **31**(1): 1. "Solid phase microextraction—a convenient tool for the determination of organic pollutants in environmental matrices."
—(1999), EPA/625/R-96/010b: Cincinnati, Ohio. "*Compendium Methods for the Determination of Toxic Organic Compounds in Ambient Air,*" (2nd edn.).

# 23

# Physical Methods in Environmental Pollution Analysis—Electroanalytical Techniques

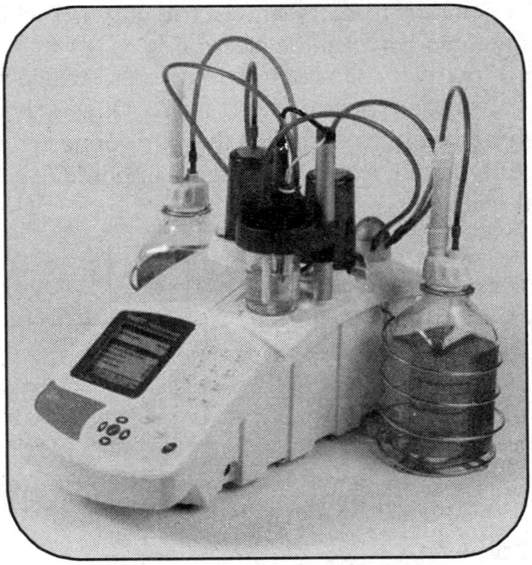

Various electroanalytical techniques such as voltammetry, polaro-graphy, amperometry, conductimetry, coulometry and potentio-metry and their variations, offer sensitive and precise measurement of inorganic and organic trace pollutants in different matrices. Most of these methods rely on the flow of electrons between one or more of the electrodes and the analyte. The names of the methods reflect the measured electric property or its unit. Some of them like voltammetry and polarography have been extensively studied and applied for the determination of inorganic as well as organic pollutants like lead, cadmium, mercury and selenium and pesticides, hazardous chemicals and carcinogens. In voltammetric methods, pre-concentration of analytes is a part of the analysis cycle. Potentiometry, with ion-selective electrodes, has a wide spread use in clinical, biological and environmental fields for carrying out fast, reliable local measurements of ion activities with a remarkable selectivity.

This chapter outlines the electroanalytical techniques.

Electroanalytical techniques use electrically-conductive probes, called electrodes, to make contact with the (ionic) analyte solution and measure voltage, current, conductivity or some other electrical parameter. The measured parameter is related to the identity of the analyte or to its concentration in solution. Most of the electroanalytical methods rely on the flow of electrons between one or more of the electrodes and the analyte. The analyte must be capable of either accepting one or more electrons (undergo reduction) from the electrode or donating one or more electrons (undergo oxidation) to the electrode.

The major electroanalytical methods are voltammetry (polarometry), amperometry, coulometry, conductimetry, and potentiometry (ion-selective electrodes). Voltammetry and polarography are among the very few techniques that are equally suitable for analyzing inorganic, organic and organo-metallic compounds. Voltammetry is a technique in which the potential is varied in a regular manner while the current is monitored. Polarography is a sub-type of voltammetry that utilizes a liquid metal (usually mercury) electrode. Potentiometry measures electric potential (voltage) while maintaining a constant electric current (nearly zero) between two electrodes. Amperometry monitors electric current (amperes) while keeping the potential constant. Conductimetry measures conductance (the ability of a solution to carry an electric current) while a constant alternating current (AC) potential is maintained between the electrodes.

An electrolytic cell consists of two electrode-electrolyte systems, termed half-cells, joined internally. An electrolytic cell can be galvanic or electrolytic (Fig. 23.1). A galvanic cell is one in which reactions occur spontaneously at the electrodes when they are connected externally by a conductor (e.g. battery cells). In an electrolytic cell, external voltage is imposed (e.g. charging of rechargeable batteries) to drive a chemical reaction.

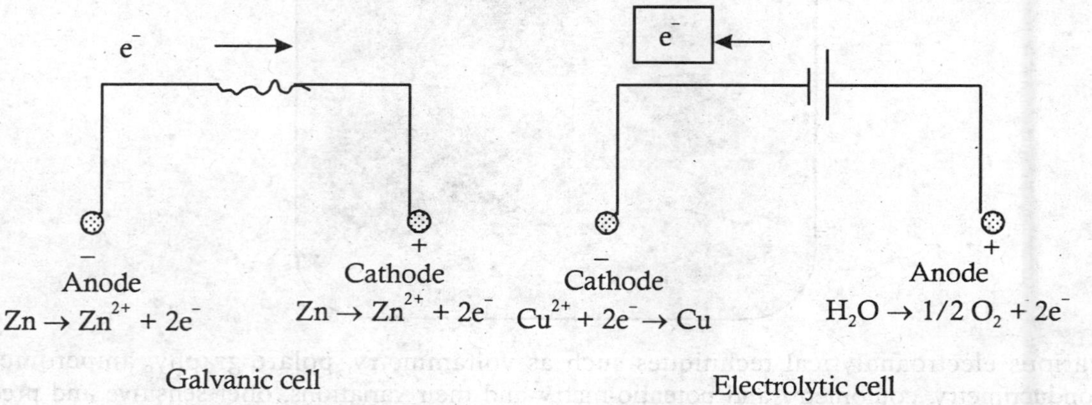


**Fig. 23.1:** Chemical Reactions at an Electrolytic Cell

The mobility (flux) of electroactive (reducible and oxidizable) species in solution is a function of concentration gradient (diffusion), electrical potential gradient (charge migration) and hydrodynamic velocity (convection) in the solution. At equilibrium, the flux can be determined by the Nernst-Planck equation.

$$J(x,t) = -\frac{\partial C(x,t)}{\partial x} - \frac{z\mathfrak{F}}{RT} DC(x,t)\frac{\partial \phi(x,t)}{\partial x} + v(x,t)C(x,t) \qquad \text{...(23.1)}$$

Under controlled non-diffusive and non-connective situations, the electrode potential at equilibrium is given by the Nernst equation.

$$E = E^0 + \frac{RT}{n\Im} In\left(\frac{C_o}{C_X}\right) \qquad \ldots(23.2)$$

where $J$ = flux;

$D$ = diffusion coefficient;

$C$ = concentration;

$z$ = charge;

$\Im$ = Faraday constant;

$n$ = number of electrons;

$\phi$ = electric potential;

$v$ = hydrodynamic velocity of the solution.

## 23.1 VOLTAMMETRIC METHODS

Voltammetry is useful for qualitative and quantitative analysis of a wide variety of molecular and ionic materials. It is a dynamic electroanalytical technique in which the potential is varied in a regular manner while the current is monitored. Three-electrode cells are normally used (Fig. 23.2). A controlled potential is applied to a working electrode (WE), and the parameter under control is the potential difference between the working electrode and the reference electrode (RE). The potential at which reaction occurs is a characteristic of the analyte, and the amount of current that is measured is proportional to its concentration, according to Nernst equation. The WE is the site at which the redox reaction and charge-transfer take place. The analyte reacts electrochemically at the working electrode. The reference electrode is such that its potential is constant regardless of the solution into which it is dipped. Two of the most commonly employed reference electrode systems are Ag/AgCl and the calomel electrode. The purpose of the auxiliary electrode (AE) is to carry most of the current. The potential is controlled between the working electrode and the reference electrode, while current flows through the solution between the working electrode and the auxiliary electrode (AE), and measurements are made at predetermined time intervals. Platinum and carbon are commonly used as auxiliary electrodes. The solution is purged with an inert gas to eliminate $O_2$. Voltammetric technique involves a reaction due to an imposed external voltage greater than the spontaneous potential of the cell, to obtain a response (current) that is related to the concentration of the analyte in the solution.

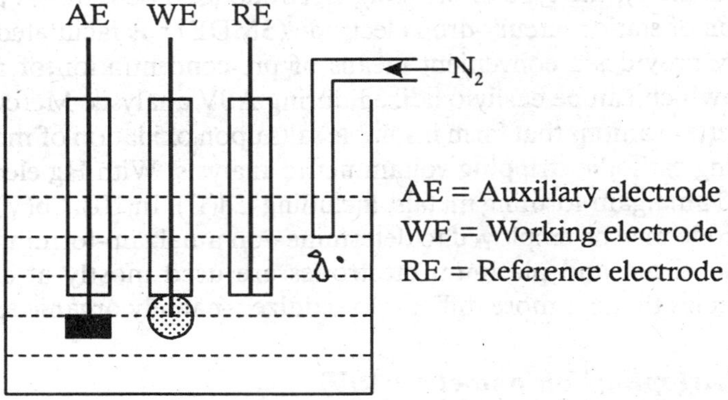


**Fig. 23.2:** Schematic of a Voltammetric Cell

The manner in which the potential of the working electrode varies with time is called a voltage ramp. In dealing with a multi-component solution, the potential is initially adjusted to a value at which no electrochemical reaction occurs at the working electrode. Then the potential is scanned in a direction that makes an electrochemical reaction more favorable. If a reduction reaction is to be studied the voltage applied to the cell is more negative (for a cathodic process) and voltage is made more positive for oxidation processes (for anodic process) than the values of the equilibrium potential of the corresponding oxidation-reduction systems, calculated using Nernst equation. As a result of the shift in the electrode potential from the equilibrium value (polarization), certain physicochemical processes take place at the electrode. A voltagram is a plot of the current as a function of the applied voltage. In voltammetric methods, pre-concentration of analyte can be a part of the analysis cycle. Voltammetric techniques are classified depending on the potential waveform applied—linear-scan voltammetry (LSV), differential pulse voltammetry (DPV) and stripping voltammetry.

### 23.1.1 Stripping Voltammetry

Voltammetric methods are employed to study redox reactions, and the different forms of voltammetry differ in the type of varying potential that is applied to the indicator electrode. Stripping voltammetric method (anodic stripping and cathodic stripping) is a two-step process:

1. In pre-electrolysis step (pre-concentration step), the analyte species (metal ions) are deposited (reduced) either as an amalgam or as mercury complex on a stationary electrode such as a hanging mercury drop (HMDE) or a thin mercury film electrode (TMFE) or by adsorbing the analyte on the solid electrode (glassy carbon or gold) surface.
2. In the anodic-stripping step, the deposited species are stripped from the electrode back into solution. The metal is *stripped* (oxidized) from the electrode at an oxidizing potential. To carry out the stripping or reoxidation step, the potential is gradually increased and the accompanying current, ($i_p$) is characteristic of the metal and proportional to metal ion concentration. An even better strategy is to strip the metal using a differential pulse signal.

The magnitude of stripping is proportional to the concentration of the analyte species. The most commonly used waveforms in the stripping voltammetry are the linear sweep and the differential pulse (DP) modes. The DP waveform has been widely used because it is more effective at discriminating charging currents and permits lower limits of detection.

In stripping voltammetry, the type of working electrode used determines the sensitivity of the method. Introduction of static mercury drop electrode (SMDE) has facilitated the use of stripping techniques. Mercury provides a convenient means of pre-concentration of many metal ions by forming amalgams, which can be easily oxidized during ASV analysis. Mercury can also be used to pre-concentrate certain anions that form insoluble salts upon oxidation of mercury. These species can be reduced during cathodic-stripping voltammetric analysis. With Hg electrodes, ASV can be used to determine 25 amalgam-forming metals, including Cd(II), In(III), Sb(V), Sn(IV) and Zn(II). Glassy carbon electrode may be employed to determine non-amalgam-forming metals, such as Ag, As, Cu, Hg, and Se. Gold and platinum electrodes are used mostly at anodic potential for determination of species that are more difficult to oxidize—namely organic species.

#### *23.1.1.1 Anodic-stripping Voltammetry (ASV)*

When the potential is held at a negative value (Fig. 23.3), followed by scanning in a positive direction, the technique is known as Anodic-stripping voltammetry (ASV). This is one of the most sensitive electrochemical techniques (10-100 times more sensitive than AAS) for detection of metals in solution. The method does not require sample extraction or pre-concentration. It is also a non-destructive technique and four to six elements can be analyzed simultaneously.

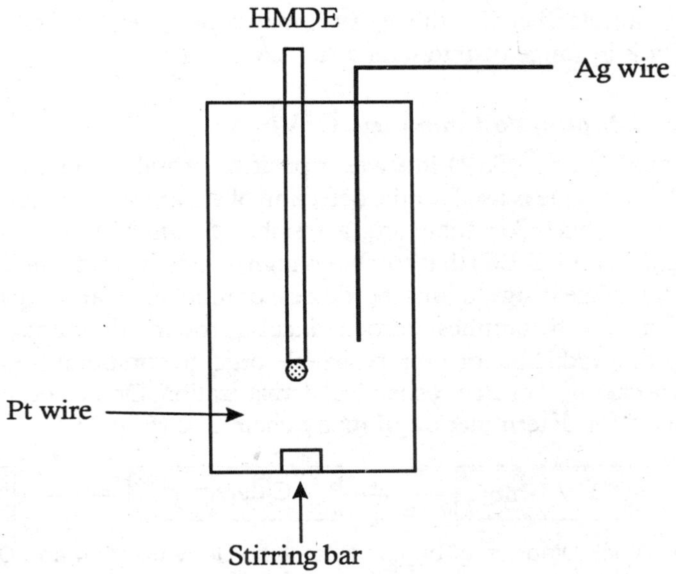


HMDE = Hanging Mercury Drop electrode

**Fig. 23.3:** Schematic of Electrochemical Cell for Anodic-stripping Voltammetry (ASV)

The potential that should be applied to the electrode is according to the Nernst equation,

$$E = E^0 + \frac{RT}{n\Im} . \ln\left(\frac{\gamma_{ox}}{\gamma_{red}} . \frac{C_{ox}}{C_{red}}\right) \qquad ...(23.3)$$

where $E$ = Potential applied to the electrode;
$E^0$ = standard potential;
$\gamma$ = activity coefficient;
$C$ = concentration of the analyte;
ox & red = oxidation and reduction states.

The ASV cell consists of a three-electrode potentiostat (with plating/stripping electrode, a reference and a counter electrode), a voltage ramp generator, a current measuring device, an inlet for deoxidizing gas (a gas to expel oxygen) and a stirring mechanism. The potentiostat maintains a constant negative potential at the working electrode relative to the reference electrode.

Differential pulse or square waveforms are used for stripping. They provide enhanced sensitivity and resolution. The differential pulse type waveform consists of a series of pulses superimposed on a linear voltage ramp, while the square waveform consists of a series of pulses superimposed on a staircase potential waveform. The current is measured just prior to application of the pulse and after the applied pulse. The difference between the two currents is plotted as function of linear sweep voltage ramp potential. Square wave stripping is faster and is about ten times more sensitive.

Anodic-stripping voltammetry, and potentiometric stripping analysis (PSA) are multi-element analysis methods, and are used in the determination of heavy metals (As, Cd, Co, Cu, Hg, Ni, Pb, Sn, Zn) in water and biological specimens. Cd, Cu, Pb and Zn can be determined simultaneously in real time. ASV method is preferable to atomic absorption spectrometry (AAS), which only yields the total concentration. ASV is especially suited for the determination of metals in seawater and other saline water samples, because the high salt matrix provides excellent conductivity. On the other hand the salinity is a source of interference in AAS.

#### *23.1.1.2 Cathodic-stripping Voltammetry (CSV)*

Cathodic-stripping voltammetry (CSV) involves a positive or reduction current passing through the circuitry. The CSV technique is used for the detection of organic compounds that form insoluble complexes on the Hg surface. For analyzing a number of analytes in a sample, HPLC with electrochemical detection (HPLC-EC) that combines high selectivity and sensitivity, can be adopted. Environmental pollutants like drugs, toxins, pesticides, oranochloro- and organo-phosphorus, and organo-sulfur compounds, nitrosamines, agrochemicals, growth stimulants, can be analyzed by this technique. Many non-reducible or non-oxidizable organic compounds can be converted into electroactive species via chemical or electrochemical derivatization. On-line derivatization procedures (HPLC-hv-EC) are used for determination of many chemicals (Fig. 23.4).

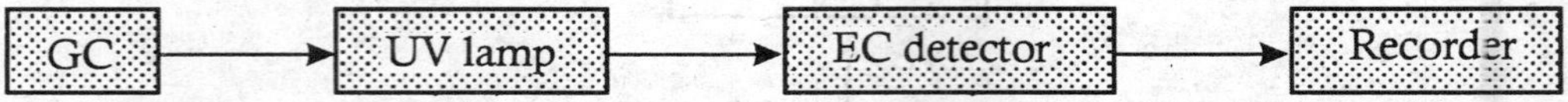


**Fig. 23.4:** Block Diagram of on-line Photolytic Derivatization and Detection

#### *23.1.1.3 Adsorptive-stripping Voltammetry (AdSV)*

Several metals do not form amalgams with mercury, and consequently cannot be determined by conventional stripping voltammetry. Their measurement is accomplished by adsorptive-stripping voltammetry (AdSV). In this method pre-concentration is achieved by adsorbing a metal chelate on the surface of the electrode. The reaction current of either the metal ion or the ligand in the adsorbed complex is used to quantify the surface-bound species that is directly related to the metal concentration in the bulk of the solution. Adsorptive voltammetric technique, with its inherent sensitivity, offers a new approach to speciation. It provides a direct determination of total Co, Cu, Ni and Ti, and also of a number of organic compounds.

While adsorptive stripping voltammetry (AdSV) is commonly used for monitoring trace metals, the reverse process (adsorptive-catalytic stripping process) provides a sensitive procedure for the voltammetric determination of catalysts.

### 23.1.2 Polarography

Polarography is a voltammetric method of analysis in which the working electrode is a falling mercury drop. This technique is used almost exclusively for the analysis of reducible species, usually metal cations (As, Cd, Cu, Pb, Ti, Zn). In a polarographic cell the cathode is the mercury drop released at regular intervals from a capillary tube. The anode is either mercury or calomel. A reservoir regulates the height of mercury in the cathode tube and the rate of dropping from the capillary. Initially the potential is adjusted to a value at which no electrochemical reaction takes place at the indicator electrode. Then the electrode is made more cathodic for a reduction reaction to take place. The measured current is small before the reaction starts. At a certain electrode potential the reduction reaction starts and electrons are withdrawn from the electrode. This leads to increase in the current flow. The polarogram involves a plot of the current as a function of the applied potential. The height (current) of the wave or the peak, as measured by extrapolating the linear portion of the curve prior to the wave or peak, is directly proportional the concentration of the analyte. For quantifying the concentration the measured height is compared with a working curve.

Maximization of the Faradaic over the charging current has been accomplished by pulse polarography. One of the widely used pulse modes in trace analysis is differential pulse (DP) polarography. In this mode small potential pulses are superimposed on a conventional DC voltage ramp and applied to the DME near the end of the drop time. The current is sampled just before application of the pulse and again at the end of the pulse when charging current has decayed. It is the difference between these two current measurements that is displayed. DP polarography is an extremely sensitive method, and it is applicable to both inorganic and organic analysis.

## 23.2 AMPEROMETRIC METHODS

Amperometric titration is a special adaptation of the polarographic principle. Amperometric detection relies on the measurement of current passing while keeping the potential constant. The current that flows between the working electrode and the reference electrode is related to the concentration of the analyte. Amperometric detectors are the most commonly used electroanalytical tools for flowing systems (in flow injection analysis), and are suitable for monitoring oxidant gases ($Cl_2$, $O_2$, $O_3$, etc). Amperometric titration curve is a plot of current as a function of titrant volume. The oxidizing sample (e.g. chlorine-containing sample) is added in increments and the current generated due to the depolarization of the electrode is measured. Residual $ClO^-$ is determined by titration in the presence of acidic KI solution (pH 3.5-4.5).

$$ClO^- + H_2O + 2I^- \rightarrow Cl^- + 2OH^- + 2I_2 \quad \ldots(23.4)$$

The apparatus (Fig. 23.5) consists of a cell connected to a microammeter with necessary electrical accessories. It includes a platinum electrode, a salt-bridge to provide electrical contact and a reference electrode (Ag-AgCl in a saturated NaCl) connected by means of salt-bridges. Electrodes are encased in plastic jackets (to keep them free of deposits and foreign matter), and attached to a 1.35-volt DC power source. Amperometric detectors are by far the most commonly employed electrochemical (EC) detectors in conjunction with HPLC. EC detection is possible in normal-phase chromatography where non-conducting eluents are employed.

Modern voltammetric and amperometric detection methods when coupled with liquid chromatography (LC) or flow-injection analysis (FIA) have several advantages over spectrophotometric techniques:

1. Pre-concentration is a part of the analysis, so better sensitivity is achieved.
2. Detection limit is improved.
3. Measurements are possible in turbid media.
4. Easy to operate and inexpensive.
5. Portable to remote areas (therefore, useful for environmental monitoring).

#### *23.2.1.1 Coulometry*

Coulometry is a method that monitors the analyte by measuring the quantity of electricity (coulombs) consumed during its electrochemical reaction. It involves the quantitative electrochemical conversion of a constituent in solution from one initial oxidation state to another and calculating its amount from the quantity of current passed ($I \times t$), according to Faraday law. In a constant-potential coulometric option the potential of the working electrode is controlled at a value at which only a single electrochemical reaction can occur (at a time). Direct coulometric determination is employed for the continuous determination of $SO_2$ and $H_2S$, based on the oxidation of sulfur compounds by generated bromine.

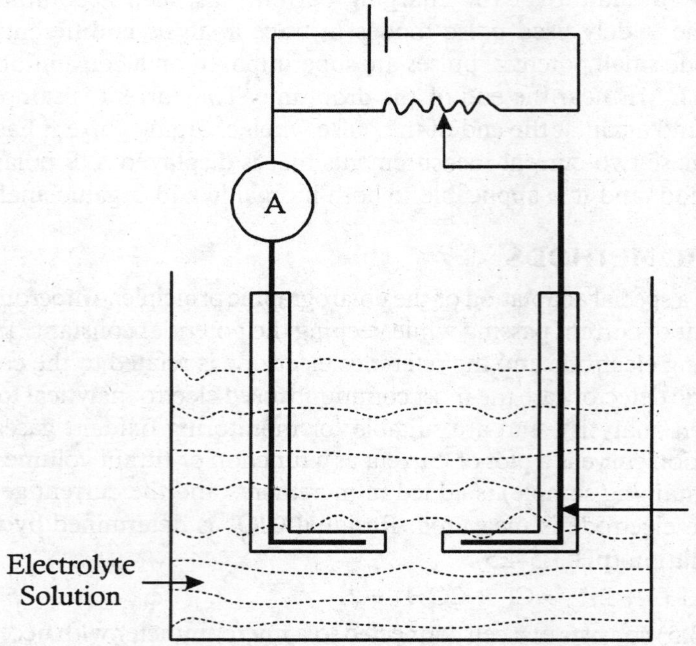


**Fig. 23.5:** Schematic of Amperometric Titration Apparatus

## 23.3 CONDUCTIMETRIC METHODS

Conductimetry is a method in which the ability of an analyte to conduct an electrical current is monitored. Since all ions contribute to the conductivity of a solution, the method is not selective; and it is also not suitable for analysis of undissociated molecules.

The two major uses of conductimetry are in monitoring the total conductance of a solution and to determine the end point of a titration that involves ions. As a method of testing, conductimetry can be employed to perform titrations both in concentrated and dilute solutions, colored, turbid and contaminated solutions and also in the presence of redox agents, which limit the use of indicators. Oscillometry is high-frequency alternate current conductimetry. In this technique both electrode and electrode-less cells can be used. Electric conductivity measurements rely on determining changes in electric resistance or capacitance or some other electrical parameter. The specific electrical conductivity of ionic conductors can be given as

$$\gamma = \frac{A}{R} \qquad \text{...(23.5)}$$

where, $\gamma$ = specific electrical conductivity;
$A$ = effective surface area;
$R$ = electric resistance.

In conductimetry analysis, the test gas comes in contact with the absorbing reagent in the reaction space. The gas without the absorbed component leaves the apparatus and the absorption solution passes into the conductivity cell. The determination is carried out by comparative measurement of the values for the absorption solution prior to and after the reaction.

Conductimetric methods are simple and are used in titration of acids and bases by neutralization, and analysis of protic and non-protic solutions and solvents. Conductivity method is frequently used in the analysis of gaseous atmospheric pollutants, to monitor water purity, particularly the salt content and in the quantification of photosynthesis, respiration, fermentation and analysis of other biological processes. They are also used in determination of relative humidity, moisture content in soils, cereals and agriculture feeds.

### 23.3.1 Gasometry

Conductimetric methods are widely used in environmental monitoring of gaseous pollutants such as $H_2S$, $NH_3$, $SO_2$, $CO_2$, and $NO_2$. The gas is passed through deionized water of measured conductivity and the solution then reaches the overflow, into which a second conductometric cell is built. The difference of the two conductivities, which is proportional to the quantity of absorbed gas, is recorded. For the analysis of polluted gases, a conductimetric gas-liquid absorption method can be automated for continuous determination of $H_2S$, $NH_3$, $SO_2$, $CO_2$, and $NO_2$. The electrical conductivity of the absorbing solution can be measured before and after the reaction. The method is simple and appropriate.

### 23.3.2 Water Quality Monitoring

Conductivity measurement is well suited for monitoring water quality. Electrolyte content of rainwater samples by conductimetry, together with pH and other physicochemical data are helpful in monitoring acid rain. The method can also be used for the determination of the mineral content of water and effluents (e.g. nitrates) and the degree of impurity and acidity of sewage water. In this method, the electrolytic conductivity is measured by placing two electrodes into the same sample and measuring the resistance between the inert electrodes. A Whetstone Bridge setup is used to carry out this (Fig. 23.6). In Whetstone Bridge, the variable resistance, RS, is adjusted until zero is recorded in the null detector (inductor salinometers are more commonly used).

Coulson conductivity detector can be used in the analysis of nitrogenous organics, halogenated hydrocarbons, pesticides, herbicides, pharmaceuticals and nitrosamines (halogens, N, S and NO). The detector contains a reference and analytical electrodes, a gas-liquid contactor and a separator. Solvent enters the cell and flows by the reference electrode. The electrometer measures the difference in conductivity at the reference electrode (solvent) and the analytic electrode (solvent + carrier + reaction products). Compounds eluted from a GC column enter a rector tube maintained at ~900° C. They are mixed with a reaction gas—hydrogen for estimating X, and NO-containing compounds and air for S-containing compounds, and analyzed by the Coulson detector.

### 23.3.3 Bioorganic Analysis

The conductimetric gas-liquid absorption method is also applicable for determining carbon and sulfur content of organic compounds. The gas, formed during the combustion of the sample, is absorbed in an alkaline solution and change in electrical conductivity of solution is measured.

Conductivity methods are used in food and beverage industries for quality control of meat, fruit juices, wines and other food products.

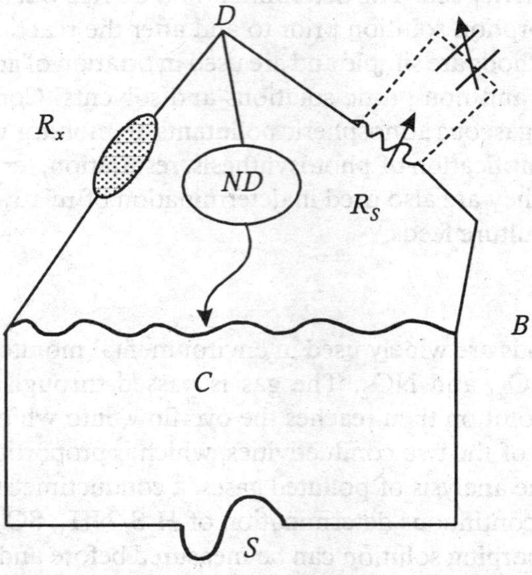


$R_s$ = Resistance of the Sample; $R_x$ = Variable Resistance;
$ND$ = Null Deflector

**Fig. 23.6:** Electrolytic Conductivity Detector Setup

Conductimetry can be used to determine hydrocarbon content in protein synthesis and biomass concentration studies. The method is based on measuring the electrical conductivity and the temperature coefficient of the sample, and the hydrocarbon concentration is determined from the time needed for an abrupt decrease in the temperature coefficient of the conductivity of the sample. The electrodes are cooled to ~10° C, lower than the freezing point of the liquid hydrocarbons. During measurement, the electrodes dipping in the fermentor become fully covered within a few minutes by a layer of solid hydrocarbons. The formation of this hydrocarbon layer, which is a good insulator, results in an abrupt decrease in the electrical conductivity. The temperature coefficient, α, is determined by

$$\alpha = \frac{1\Delta\gamma}{\gamma\Delta T} \qquad \ldots(23.6)$$

where, $\gamma$ = conductivity measured with clean electrodes; $\Delta\gamma$ = difference between the conductivity measured before and after the electrodes have been coated with the hydrocarbons; $\Delta T$ = difference between the temperature of the sample and the cooled electrodes.

By measuring the time elapsed from the beginning of the measurement to the abrupt reduction of conductivity and the appropriate temperature coefficient, the hydrocarbon content of the sample can be determined graphically. Biomass concentration can be obtained from a graph of temperature coefficient versus biomass weight.

## 23.4 POTENTIOMETRIC METHODS

In potentiometric methods, the potential of a working electrode with respect to a reference electrode at null current is determined. The potential of the working electrode varies with the concentration of the analyte, while the potential of the reference electrode is constant. In some cases the indicator electrode is a metal (Pt, Au, Ag) or other conductive material (graphite, glassy carbon). In such a case the electrode indicates the potential of the bulk solution in which it is dipped. More selective are the pH electrode, ion-selective electrodes (for $F^-$, $Cl^-$, $NO_3^-$ etc.), enzyme electrodes, and gas-sensing electrodes (for $O_2$, $NH_3$ etc). The potential difference between an indicator electrode and a standard electrode is a log function of the activity of an ion in solution according to the Nernst equation

$$\Delta E = \frac{2.3RT}{Z\Im}\log(a_z) \quad \text{...(23.7)}$$

where, $\Im$ = Faraday constant;
$Z$ = charge (+ for cation; – for anion).

### 23.4.1 pH Electrode

The pH electrode is an ion-selective electrode specific to $H^+$ ion (see Chapter 17 also). Ag/AgCl or a calomel electrode is used as an internal reference electrode. It consists of a solution of KCl in contact with Ag/AgCl or solid mercurous chloride (calomel) and mercury (Hg | $Hg_2Cl_2$ || KCl | test solution). Changes in potential can be measured by making contact with the internal filling solution using a reference electrode and the contact with the sample is achieved with the second reference electrode via a salt-bridge. The external reference electrode can either be a separate probe or a combination electrode. Reference electrodes must be stable and non-polarizable. The internal reference electrode and reference solution are separated from the analyte solution by a glass membrane that is chosen to respond to a specific analyte, namely $H^+$.

The relationship between the potential and the activity of a specific ion in solution is based on the Nernst equation

$$\Delta E = E^0 - \frac{RT}{n\Im}.\ln a_i - E_{\text{ref}} \quad \text{...(23.8)}$$

where, $\Delta E$ = Potential difference;
$E^0$ = standard indicator potential at $a = 1$;
$T$ = absolute temperature;
$R$ = gas constant;
$n$ = number of ions;
$\acute{A}$ = Faraday constant;
$a_i$ = activity of the ion species, $i$.

The activity is related to concentration and is calibrated by comparison with a series of standards.

The pH is related to redox potential. The redox pH of a system is when redox pair is in equilibrium. The pH of water is a very important parameter that controls the species and many redox reactions play a central role in the environmental behavior of many elements—oxidation states of metals in solution and their reactions. The mobility of a large number of elements relevant to biological processes is strongly dependent on the pH of the medium.

### 23.4.2 Gas-sensing and Ion-selective Electrodes

The selectivity of the ion-selective electrodes results from the nature of the interaction between the membrane and the analyte. Most of the ion-selective electrodes can be classified under one of the major groups.

(*i*) Fixed charge ion-exchanger (e.g. glass electrodes for $H^+$, $Na^+$, $K^+$).

(*ii*) Liquid phase with dissolved ion-exchanger - porous hydrophobic membrane, which supports an anion or cation-exchanger, which is insoluble in the analyte solution (e.g. for $Ca_2$).

(*iii*) Solid-state materials (e.g. $LaF_3$ for $F^-$; AgCl for $Cl^-$, $Br^-$, and $I^-$) use a sparingly soluble solid, ionically conducting, either alone or suspended in an organic polymeric material, as the membrane.

(*iv*) Neutral-carrier ion-selective electrodes are similar in design to the liquid-ion-exchanger electrodes.

The ideal ion-selective electrode is one that responds only to one analyte ion. Electrode selectivity is achieved by interposing an ion-specific material between the sample and an internal reference system. Many indicator electrodes such as solid (glass), liquid, solid-state, polymer membranes and non-conventional electrodes are available for determination of gases, anions and cations (Table 23.1). Solid-state electrodes are ion-selective electrodes in which the sensor is a thin layer of a solid, which is sparingly soluble and is a conductor of that specific ion. In ion-selective field-effect transistor (ISFET), the intrinsic gate insulator is also used as the electroactive (chemical sensing) surface. Composite ISFETs incorporate a solid-state membrane (e.g. AgCl) or a polymeric membrane doped with an ion-exchanger or ionophore over the gate of a field-effect transistor, and the current flow through the transistor is monitored. The current flow is controlled by the charge applied to the gate, which is determined by the concentration of analyte in the membrane on the gate. These electrodes are rugged and simple and are useful for fieldwork.

**TABLE 23.1: GAS-SENSING AND ION-SELECTIVE ELECTRODES**

| *Electrode/Membrane* | *Gas/Ion Detected* | *Applications* |
|---|---|---|
| Glass | $H^+$ | Redox and pH measurements |
| Glass | $Li^+$, $Na^+$, $K^+$ & Ag | Analysis of seawater, soil, serum etc. |
| Glass | $NH_3$ & $NH_4^+$ | Ammonia and water pollution |
| Glass | $CO_2$ | Blood and environment |
| Glass | $SO_2$, & $NO_2$ | Air pollution monitoring |
| Glass | $NO_3^-$ | Water treatment and microbial growth |
| Solid ($Ag_2S$) | $F^-$, $Br^-$, $I^-$, $CN^-$ & $SO_3^-$ | Environmental studies |
| Polyethylene | $O_2$ | Redox, and dissolved oxygen determination |
| Membrane | $CO_2$, $SO_2$, $NO_x$ & $H_2S$ | Environmental studies |
| Liquid membrane | Ca, $NO_3^-$ & $Cl^-$ | Environmental studies |
| Lanthanum fluoride | $F^-$ | Environmental studies |
| Silver sulfide | $Cd^{2+}$, $Cu^{2+}$ & $Pb^{2+}$ | Environmental pollution studies |

Gas-sensing electrodes are designed to monitor dissolved gases. They are prepared by wrapping a gas-permeable (polymer) membrane over a pH electrode. Between the membrane and the pH sensing surface is a thin layer of an electrolyte. A flat-ended glass electrode covered with a layer of bicarbonate solution and a gas permeable membrane has been in use for a long time for monitoring carbon dioxide. When the electrode is placed in an atmosphere of $CO_2$, the gas diffuses through the polymer membrane and on reaching the bicarbonate solution reacts with it and establishes equilibrium with respect to hydrogen ions. The glass electrode will sense this. The pH change is related to the concentration of gas that diffuses through the membrane. That, in turn, is related to the partial pressure of the gas outside. $O_2$, $CO_2$, $NO_2$, $SO_2$, $NH_3$, $H_2S$, HCN and HF can be measured by suitably choosing the electrolyte.

Several nutrient species, such as $NH_3$, nitrite, nitrate, phosphate, etc, which are important in environmental pollution studies, may be determined by using gas-sensing or ion-selective electrodes. Field-effect transistor devices (ISFET-based Severinghaus type $pCO_2$ electrodes) are suitable for the direct and on-line measurements of $pCO_2$ in blood and other body fluids and for measurement of changes in surface charge density.

### 23.4.3 Enzyme Electrodes

Biomembrane electrodes (enzyme electrodes) are similar in design to gas-sensing electrodes. Enzyme electrodes are biosensor devices (biological recognition elements, such as enzyme, antibody, receptor, DNA, or micro-organism). The result is an electrode, which is useful for the assay in a manner as simple as a pH measurement with a glass electrode. The outer permeable membrane is used to hold a gel between the two membranes. The gel contains an enzyme that selectively catalyzes the reaction of the analyte. The principle of enzyme electrode is: an immobilized enzyme (immobilized by adsorption, protein cross-linking, and entrapment with a gel matrix or covalent bonding to a matrix) is used to react with the analyte to be assayed with a high degree of specificity. The immobilized enzyme is placed over an ion-selective electrode, which measures either the decrease of one of the ($O_2$ in Clark's cell) reactants or a product ($CO_2$ in Severinghaus electrode). In enzyme-modified ion-selective field-effect transistor (EnFET) a gel film containing an enzyme is covered on the electroactive gate film. Enzyme-substrate interactions create pH changes within the gel film, which are measured.

In an amperometric enzyme electrode a platinum electrode is covered with a plastic membrane to which an enzyme gel layer adheres. The measured current is proportional to the amount of electroactive substance generated in the membrane. For example, in determination of glucose, the measured current is proportional to the loss of $O_2$ inside the immobilized glucose oxidase.

### 23.4.4 Applications of Potentiometric Methods

Gas-sensing and ion-selective electrodes are employed in instrumental analytical procedures because of their speed, sensitivity and reliability. They find wide clinical and biological application. They find extensive use in monitoring environmental pollution and in the assay of pesticides, drugs and enzymes. Though enzyme electrodes are used primarily in clinical work (estimation of glucose, urea, and amino acids), they are of potential value in environmental monitoring and control studies too.

#### *23.4.1.1 Determination of Ammonia*

A gas-sensing probe can be employed to determine $NH_3$. It is a gas electrode with a pH-sensitive tip (Fig. 23.7). The e.m.f. of such an electrochemical cell is a function of the gas of interest. The ammonia probe is a glass electrode on which a gas permeable polymer membrane is fitted. Between the membrane and the electrode a thin layer of ammonium chloride is introduced. When the electrode is placed in an ammonia-containing atmosphere $NH_3$ diffuses through the membrane, changing the pH of the internal solution. The glass electrode senses the pH change in the internal solution. This change is related to the partial pressure of ammonia in the external atmosphere.

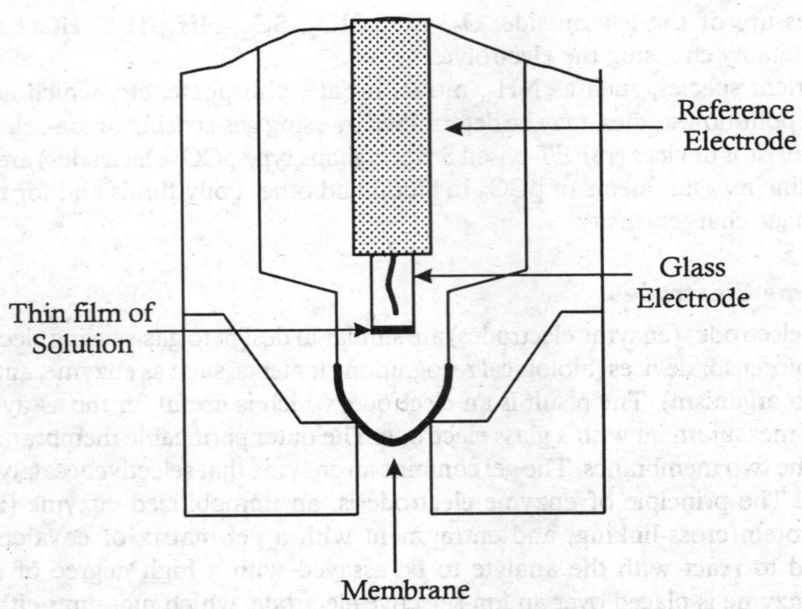


**Fig. 23.7:** Gas-sensing (pH) Probe

### *23.4.4.2 Determination of Nitrate*

Ion-selective electrode (ISE) can be used for determining nitrate concentration. The setup consists of a liquid ion-exchange electrode with an inert membrane support. The ion-exchanger is a long-chain alkyl ammonium nitrate. Nitrate ($NO_3^-$) can be determined by other methods too. ($NO_3^-$) is separated from nitrite ($NO_2^-$) by HPLC and measured by a conductivity detector or by a UV detector.

### *23.4.4.3 Assay of Enzymes*

Ion-selective electrodes are employed to determine activities of many enzymes, which are of considerable importance in monitoring of food and beverage products. For example, the activity of the enzyme rhodanase may be determined by a cyanide-selective electrode that follows the decrease in $CN^-$ ion during the reaction, which is catalyzed by rhodanase. Similarly, sulfide-selective electrode is used to monitor the amount of thiocholine released during the catalysis by cholinesterase.

$$CN^- + S_2O_3^{2-} \xrightarrow{\text{Rhodanase}} SCN^- + SO_3^{2-} \quad \text{...(23.9)}$$

$$\text{Acetylthiocholine} + H_2O \xrightarrow{\text{Acetylcholinesterase}} \text{Thiocholine} + CH_3COOH$$

## 23.5 DETERMINATION OF OXYGEN

Oxygen is involved in many redox reactions, fermentation, respiration, photosynthesis, and metal corrosion. Accurate determination of oxygen is very important in fermentation systems, food processing, enzyme assay, clinical medicine, sewage and industrial discharge and water treatment processes and environmental monitoring. Gas-sensing electrode (Clark electrode) method, paramagnetic detection and fluorescence are some of the methods employed to determine oxygen.

### 23.5.1 Ion-sensitive Electrode Method

Determination of oxygen can be carried out by an oxygen-selective membrane electrode (Clark oxygen electrode). The system consists of a platinum cathode and a silver anode, both immersed in a saturated KCl solution, and separated from the test solution by an $O_2$-permeable membrane. When a potential difference of –0.6 V is applied across the electrodes, electrons are generated at the Ag-anode, which are used to reduce oxygen. The oxygen tension at the Pt-cathode drops and more oxygen diffuses towards it. The current thus produced is proportional to oxygen tension.

At silver anode: $4Ag + 4Cl^- \rightarrow 4AgCl + 4e^-$ ...(23.10)

At platinum cathode: $O_2 + 4H^+ + 4e^- \rightarrow 2H_2O$

### 23.5.2 Magnetic Detection

Paramagnetic gases, such as $O_2$, can be detected, based on the measurement of their magnetic susceptibility. Thermomagnetic, magnetomechanical, and magnetopneumatic methods can be employed to determine oxygen. In thermomagnetic method (magneto-oxygen analyzer, Fig. 23.8), the air sample is introduced into an electrically heated cross-tube of an annular chamber, half of which is exposed to a non-homogeneous magnetic field. When a gas stream containing oxygen enters the cell, oxygen, being a paramagnetic substance, is attracted towards the magnetic field and tends to flow through the cross-tube. As $O_2$ is heated, it loses its paramagnetism and is replaced by the cooler paramagnetic $O_2$. The cooling caused by the flow of oxygen creates a change in electrical resistance in the probe wire, and the resistance is proportional to total oxygen in the carrier gas stream. Another method is by null method (Fig. 23.9). In this method, the gas stream is fed into measuring and reference chambers with two heated filaments, connected in a Whetstone bridge. The filament in the measuring cell is placed between the poles of a magnet. If the gas contains $O_2$, the magnetic field in the measuring cell leads to an increased flow that is proportional to the oxygen concentration.

Magnetomechanical methods measure the mechanical force exerted on a body located in a non-homogenous magnetic field. A rod-shaped test body is attached to a platinum filament and suspended between the polar extensions (containing diamagnetic gas) of a magnet. The torsion of the filament is determined by the reflection of a light beam from the mirror attached to the filament. The scale is calibrated to measure partial pressure of $O_2$. The magnetopneumatic methods are based on measuring the potential difference created in two branches by drawing of oxygen into the magnetic field.

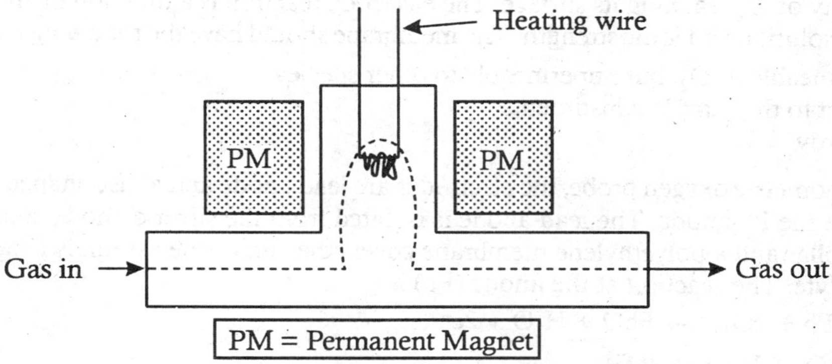


**Fig. 23.8:** Schematic of Magneto-Oxygen Analyzer

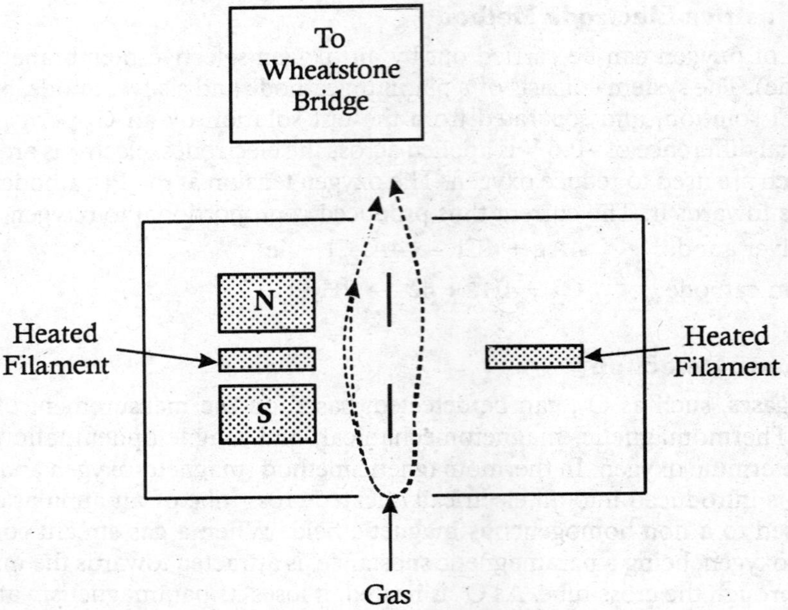


**Fig. 23.9:** Schematic of Thermomagnetic Oxygen Analyzer

### 23.5.3 Measurement of Dissolved Oxygen (DO)

Another important parameter to be considered in environmental pollution monitoring is the determination of dissolved oxygen (DO) in water. Without the DO, water bodies become inhospitable for gill-breathing aquatic organisms.

The amount of oxygen dissolved in water can be determined by amperometric, conductimetric, coulometric, potentiometric and voltammetric methods. A frequently used method takes advantage of gas-sensing probes. The following steps are involved:

1. $O_2$ from the sample passes through the $O_2$-permeable membrane (polyethylene or polypropylene).
2. $O_2$ transfers through the electrolyte between the membrane and an inert $O_2$-sensing electrode.
3. When $O_2$ reaches the cathode, it is reduced, thereby generating a current equivalent to the quantity of $O_2$ reaching its surface. The electrode reaction is a function of the electrode, pH of the solution and ionic strength. The membrane should have the following characteristics—

(*i*) Permeable to $O_2$, but impermeable to other species
(*ii*) Inert to the sample constituents
(*iii*) Sturdy.

In a galvanometric oxygen probe, the electrodes are lead anode and a disc-shaped silver cathode, located inside the Pb-anode. The lead anode is isolated from the silver cathode, and both are held in a plastic collar and a polyethylene membrane covers the end of the assembly (Fig. 23.10). KOH is the electrolyte. The reaction at the anode (Pb) is

$$Pb + 2OH^- \rightarrow PbO + H_2O + 2e^- \quad \ldots(23.11)$$

$$2e^- + \tfrac{1}{2}O_2 + H_2O \rightarrow 2OH$$

$$Pb + \tfrac{1}{2}O_2 \rightarrow PbO$$

Unless free DO is available, the reaction does not take place, and the detector does not detect any current. Intensity of current is proportional to $O_2$ level in the solution. The probe can be employed to determine $O_2$ activity and concentration. The probe is suitable for measurement of DO *in situ* and for continuous monitoring.

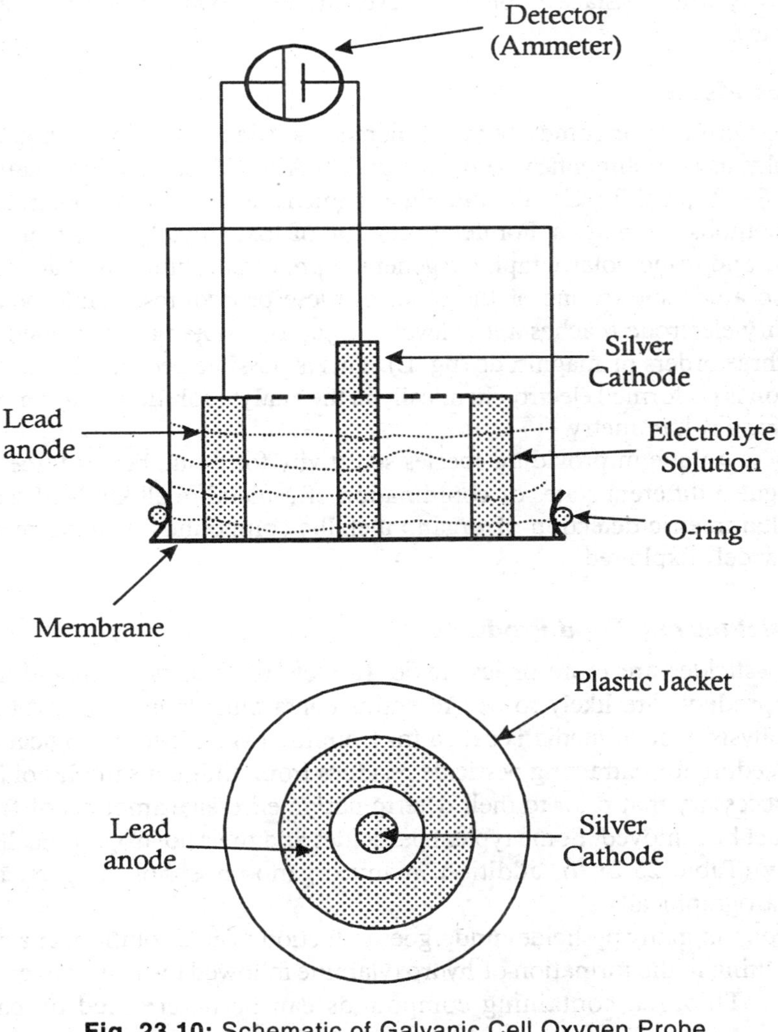


**Fig. 23.10:** Schematic of Galvanic Cell Oxygen Probe

## 23.6 ELECTROANALYTICAL METHODS IN POLLUTION MONITORING

There is no analytical technique, which can be automatically considered to be 'the best' for all problems in pollution monitoring. It is evident that for all pollution determination in modern analytical systems instrumental methods are employed. This chapter has outlined a number of techniques based on electrometric phenomena. Whether it is a case of toxic metal(s) determination or of other hazardous chemicals, methods permitting simultaneous determination of several trace components are a *priori* superior to single element techniques. In this context some of the electroanalytical procedures like voltammetry have an advantage over many other processes.

### 23.6.1 Voltammetry and Polarography

Voltammetry and polarography provide a very promising approach to environmental pollution monitoring since they can measure a wide range of toxic species, from heavy metals to organics at levels down to and below ppm-level. If the respective suitable mode is applied, they combine the excellent sensitivity with satisfactory precision even at the determination limits, and inherent high degree of accuracy.

#### *23.6.1.1 Trace Metals*

For trace metal determination some analytical chemists consider the twin techniques—voltammetry and polarography, as complimentary to AAS and ICP-AES. Most of the inorganic compounds of Cr, Ni, Co, Fe, Se, Zn, As, Be, Cd, Pb and Hg are amenable to sensitive polarographic and other voltammetric methods of analysis. For determination of toxic metals like Pb and Cd in biological fluids like blood and urine polarography is generally preferred. Similarly, trace metals in air-filter samples are also amenable to one or the other of these procedures. While polarography at the dropping mercury electrode reaches a low level of 1 μg/L, stripping voltammetry can extend this limit down by three orders of magnitude (ng/L). This is possible because, in the latter technique a pre-concentration is performed electro-chemically in the analyte solution. Pulse mode is the superior version of stripping voltammetry.

Voltammetry is unique in providing species sensitivity for a number of trace metals due to its ability to distinguish different states of coordination of a metal in all kinds of water. Coupling of HPLC with voltammetric detection of organo-metallic compounds of toxic metals, e.g. organo-mercurials are widely exploited.

#### *23.6.1.2 Pesticides and Food Products*

Practically all pesticides are more or less toxic. In view of their wide spread use, they or their decomposition products are likely to be present as contaminants in all environmental matrices. Hence, their analysis in such media is called for from the hygienic-toxicological aspects. There is no uniform procedure for extracting pesticide residues from different samples of biological origin. However, it is necessary that prior to their determination excessive amounts of fat, wax and color in the final extract be removed. Some typical pesticides and their polarographically active group(s) are given below (Table 23.2). In addition, many organo-phosphorous pesticides can also be determined polarographically.

The $-NO_2$ group in many pesticides undergoes reduction in acid solution, at a mercury cathode, as a $4e^-$ step resulting in the formation of hydroxylamine followed by a $2e^-$ step giving a substituted ammonium ion. Thiourea containing compounds can be determined by cathode-stripping voltammetry.

#### *26.6.1.3 Environmental Carcinogens*

Electroanalytical methods provide sensitive, accurate and convenient procedures for the determination of a variety of 'chemical carcinogens' in water, air and food.

Polycyclic hydrocarbons with four or more fused rings, which are recognized carcinogens, can be determined by anodic differential pulse voltammetry at a glassy carbon electrode with good sensitivity. When sulfolane is the solvent the concentrations that can be determined are about $5 \times 10^{-7}$ M but if acetonitrile is used as solvent the limit can go down to $2 \times 10^{-8}$ M. For analyzing a number of compounds HPLC separations, coupled with electro-chemical detection is a very useful method. Aromatic amines and related nitro-compounds can also be determined by similar techniques. Polarography of azo-compounds has been extensively studied. Concentration in the range $10^{-6}$ M can be determined by differential pulse polarography.

**TABLE 23.2:** SOME PESTICIDES AND THEIR ACTIVE GROUP(S)

| *Pesticide* | *Active Group(s)* |
|---|---|
| DDT | $-CCl_3$ |
| Trichlorfon | $-CCl_3$ |
| Fenitrothion | $-NO_2$ |
| p-Nitrophenol | $-NO_2$ |
| Thiron | —S—S— |
| Parathion | $-NO_2$ |
| Gutathion | 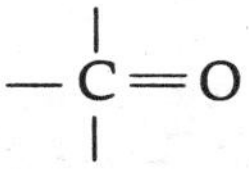 |
| Morestan | 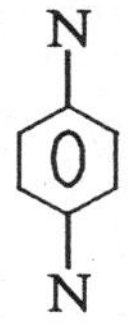 |
| Diquat | 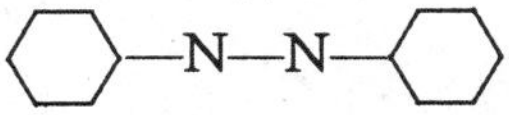 |

One group of compounds that has attracted much attention as carcinogens consists of N-nitroso amines. These compounds are present to varying degree in nearly every section of the environment, e.g. in air, water, food, cigarette smoke, synthetic cutting oils, cosmetics and even in the popular beer. These compounds undergo an irreversible $4e^-$ reduction in acid solution, forming the corresponding unsymmetrical hydrazine and an irreversible $2e^-$ reduction in basic solutions to give nitrous oxide and the precursor amine. Differential pulse techniques in acid solution provide highly sensitive methods for their determination. HPLC with polarographic detection is a suitable technique for determining a number of compounds of this type.

## BIBLIOGRAPHY


Bard, A. J. and Faulkner, I. R. (1980), Wiley & Sons: New York. "*Electrochemical Methods, Fundamentals and Applications.*"

Barek, J., *et al.* (2000), *Int. Rev. Anal. Chem.*, **30**(1): 37. "Polarographic and Voltammetric Determination of Chemical carcinogens."

Bond, A. M. (1980), Marcel Dekker: New York. "*Modern Polarographic Methods in Analytical Chemistry.*"

Craig, L. J. (1986), Longmans: Birmingham, UK. "*Organometallic Compounds in Environment: Principles and Reactions.*"

Covington, A. K. (ed.) (1979), CRC Press: Boca Raton, FL. "*Ion-Selective Electrode Methodology.*"

Estela, J. M. *et al.* (1995), *Crit. Rev. Anal. Chem.*, **25**(2): 91. "Potentiometric stripping analysis."

Ewing, G. W. (1975), McGraw-Hill: New York. "*Instrumental Methods of Chemical Analysis,*" (4th edn.).

Fleet, B. (1980), Pergamon Press: Oxford, UK. "*Recent Advances in Electrochemical Techniques for Environmental Pollution Monitoring and Control,*" (Vol.3).

Guilbault, G. G. (1976), Marcel Dekker: New York. "*Handbook of Enzymic Methods of Analysis.*"

Harrison, R. M. and Wilson, S. J. (1982), Pergamon Press: Oxford, UK. "*Analytical Techniques in Environmental Chemistry*," (Vol. 2).

Kissenger, P. T. and Heineman, W. R. (eds.) (1984), Marcel Dekker: New York. "*Laboratory Techniques in Electroanalytical Chemistry*."

Leppard, G. G. (ed.) (1983), Plenum Press: New York. "*Trace Element Speciation in Natural Waters and its Ecological Implications*."

Ryan, T. (ed.) (1984), Plenum Press: New York. "*Electrochemical Detectors*."

Skoog, D. A., Holley, J. F. and Nieman, T. A. (1998), Harcourt & Bruce: Philadelphia, PA. "*Principles of Instrumental Analysis*," (5th edn.).

Skoog, D. A., West, D. M. and Holley, J. F. (1994), Saunders College: Philadelphia, PA. "*Fundamentals of Analytical Chemistry*," (6th edn.).

Smolkova-Keulemansova, E. and Fetl, L. (1991), Elsevier: Amsterdam. "*Analysis of Substances in the Gaseous Phase*": "*Comparative Analytical Chemistry*", Vol. XXVIII, G. Svehla (ed.).

Smyth, M. R. and Vos, J. G. (eds.) (1992), Elsevier: Amsterdam. "*Analytical Voltammetry*"; "*Comprehensive Analytical Chemistry*", Vol. XXVII, G. Svehla, (ed.).

Smyth, W. F. (ed.) (1980), Elsevier: Amsterdam. "*Electroanalysis in Hygiene, Environmental, Clinical and Pharmaceutical Chemistry*."

Wang, J. (1985), Verlagchemie: Weinheim. "*Stripping Analysis: Principles, Instrumentation and Applications*."

Wang, J. *et al.* (2001), *Talanta*, **54**(1): 147. "Determination of micromolar bromate concentrations by adsorptive catalytic stripping voltammetry of the molybdenum-3-methoxy-4-hydroxymandelic acid complex."

# 24

# Physical Methods in Environmental Pollution Analysis—Spectroscopic Techniques

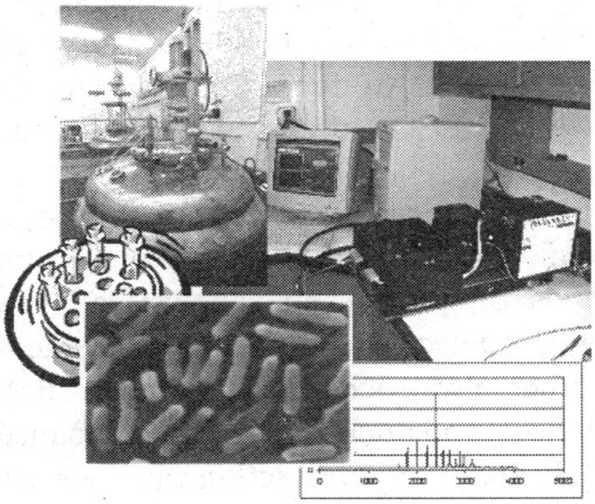

Spectroscopic methods, molecular (UV-visible, fluorescence, infrared (IR) and Raman), and atomic (absorption, emission and fluorescence) are extensively used in quantitative analysis of environmental pollutants.

UV-visible absorption spectroscopy (l ≈ 150-750 nm range) deals with the electronic transitions of molecules. Most wet-chemical methods of analysis employ UV-visible spectrometry for quantification of an analytes, based on Beer-Lambert law, which correlates the concentration of analytes to the attenuation of the incident radiation by them. Both inorganic and organic pollutants can be determined by this technique. Spectrophotometers are common instruments in all laboratories dealing in chemical analysis.

Turbidimetry and nephelometry are employed to measure the attenuation of light by non-colored particulate and coagulate matter. These techniques find application in evaluation and monitoring of particulate pollutants/contaminants in food, beverage, water and other sources. Diffuse Reflection Spectroscopy (DRS) can be employed in color measurement of films and for studying biological effects of pollutants adsorbed on surfaces, particularly on plants. Resonance-ionization spectroscopy is an extremely sensitive and highly selective analytical tool. Thermal Lens Spectroscopy (TLS) has wide range of applications in the analysis of contaminants in food and environmental samples.

Fluorescence and phosphorescence are more sensitive than absorption spectroscopic methods and they provide a great degree of sensitivity and specificity. Infrared (IR) spectroscopy is a molecular absorption phenomenon. Raman effect manifests due to inelastic scattering of light by molecules. Raman spectroscopy can be employed in the analysis of environmental pollutants.

Atomic spectrometry covers a class of spectroscopic methods in which the species examined are in the form of atoms. Absorption (AAS) and emission (AES) methods are the important ones that come under this category.

This chapter deals with the quanlitative and quantitative spectroscopic analysis of environmental pollutants.

Spectroscopic analysis is based on a measure of the interaction of electromagnetic radiation with matter, as a function of wavelength or frequency (energy). The technique provides a set of versatile analytical tools for determining the compositional characterization of constituents in a matrix, by identifying wavelengths and intensities of absorption or emission lines of the constituents in a sample.

Atoms and molecules are excited or ionized by interaction with radiation, thermal or electromagnetic, depending on the energy of the incident radiation. High-energy radiation (X-rays) has sufficient energy to ionize matter. X-ray analysis is dealt under nuclear methods (Chapter 25). Radiation in the ultraviolet (UV), visible and infrared (IR) regions does not have sufficient energy to remove the orbital electrons of molecules to ionize them. The energies in these wavelength regions result in electronic, vibrational and rotational excitation of molecules. Transitions from excited states to ground states of a species (molecular or atomic) give rise to spectral lines, characteristic of that species. Absorption occurs when radiation quanta possesses the energy corresponding to an orbital transition. Emission occurs when an excited molecule or atom returns to lower energy or ground state. Widely employed molecular spectroscopic methods are—UV-visible and infrared (IR) (absorption), fluorescence (emission) and Raman (scattering). Atomic absorption (AAS), atomic emission (AES) and atomic fluorescence (AFS) are frequently used for the determination of trace metals in different matrices (Fig. 24.1).

Various types of spectroscopic methods are employed in environmental analysis. The type of spectroscopy suitable for analysis depends on the energy region of the electromagnetic spectrum and its interaction with matter (Table 24.1). A great advantage of spectroscopic analysis is that a large number of elements can be determined simultaneously. The detection limit of a given element depends upon its excitation potential, atomic mass and other physicochemical characteristics. Magnetic resonance spectroscopy—electron spin resonance (ESR) and nuclear magnetic resonance (NMR), are not commonly used in pollution analysis. NMR and X-ray diffraction methods are powerful and versatile techniques employed not just to detect and quantify molecular species, but also to elucidate complete three-dimensional architecture of molecular species.

**TABLE 24.1:** SPECTROSCOPIC METHODS BASED ON INTERACTION OF RADIATION WITH MATTER

| *Spectroscopy* | *Wavelength* ($\lambda$) | *Type of Interactions* |
|---|---|---|
| $\gamma$- ray (Mössbauer) | <0.15 nm | Nuclear |
| X-ray | 0.01-100 nm | Core electrons |
| Ultraviolet (UV) | 100-350 nm | Bonding electrons |
| Visible | 350-750 nm | Bonding electrons |
| Infrared (IR) | 0.75-100 μm | Molecular vibrations/rotations |
| Microwave | 0.75-10 mm | Molecular rotations/orientations |
| Electron Spin Resonance (ESR) | ~30 mm | Electron spin |
| Nuclear Magnetic Resonance (NMR) | 0.5-10 m | Nuclear spin |

## 24.1 UV-VISIBLE SPECTROSCOPY

Absorption spectroscopy in the UV-visible region is the most often used spectral method of analysis. The basic approach is to measure the ratio of the absorbed electromagnetic radiation by the analyte to that of the incident radiation. Energy of a molecule, $E$, is sum total of electronic transitions, bond vibrations and rotations and molecular translations.

$$E = E_{elec} + E_{vib} + E_{rot} + E_{trans} + \ldots \qquad \ldots(24.1)$$

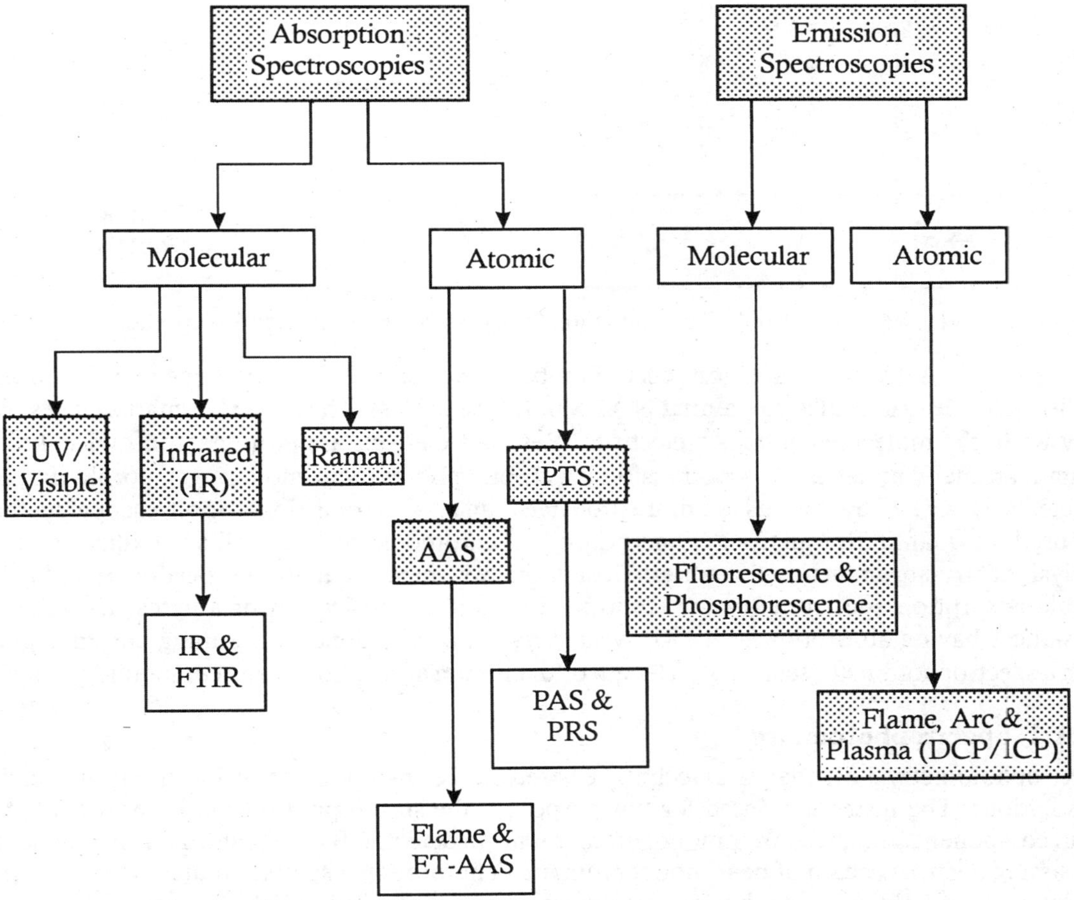


ET-AAS = Electro-thermal AAS
PTS = Photo-thermal spectroscopy
PAS = Photo-acoustic spectroscopy
PRS = Photo-refraction spectroscopy
ICP = Inductively-coupled plasma

**Fig. 24.1:** Flowchart of Optical Spectroscopic Techniques

In molecular spectroscopy, the UV-visible spectroscopy encompasses 100-750 nm wavelength region. UV spectroscopy deals with electronic transitions that arise due to inductive, resonance and solvent effects, while visible spectroscopy deals with the electronic and vibrational transitions of molecules. The electronic and vibrational energy levels and hence the spectra are characteristic of a given system. Electrons in $n$, $\sigma$, and $\pi$-orbitals are excited to antibonding levels ($\sigma^*$, and $\pi^*$). $\sigma \rightarrow \sigma^*$, $n \rightarrow \sigma^*$, $n \rightarrow \pi^*$, and $\pi \rightarrow \pi^*$ transitions are possible (Fig. 24.2).

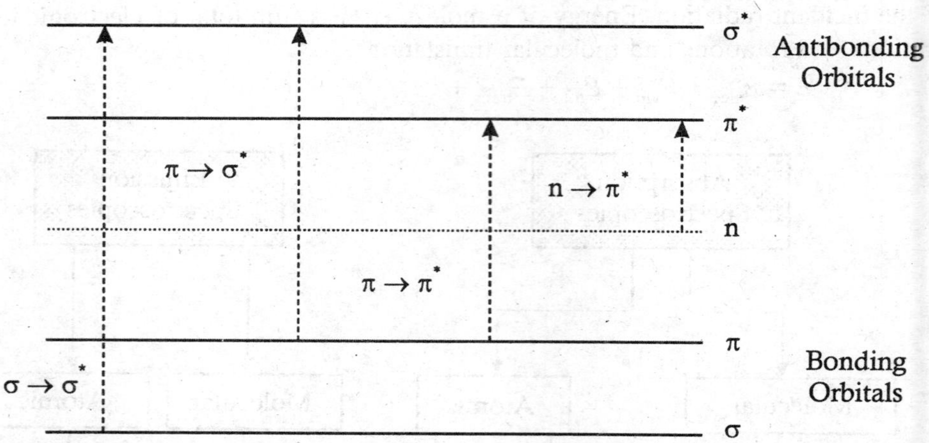


**Fig. 24.2:** Electronic Transitions in Molecules between σ, π and *n*-Orbitals

The $n \rightarrow \sigma^*$ and $n \rightarrow \pi^*$ transitions occur in substances, which have at least one free electron pair in the molecule (molecules containing N, O, S or halide atoms). The $n \rightarrow \pi^*$ excitation takes place only when the molecule contains π-electrons. Resonance effects are prominent in molecules that contain aromatic moieties ($\pi \rightarrow \pi^*$ transitions), chromophores (benzenoid rings), porphyrins, etc., (Table 24.2). $\pi \rightarrow \pi^*$ transitions are more probable than $n \rightarrow \pi^*$ transitions. Spectroscopic methods are used for quantitative analysis (fingerprinting of chromophores), as well as auxiliary tools for analysis of structure-function dynamics. Most of the wet-chemical analysis procedures employ UV-visible absorption spectrometry for identification and quantification of anlytes. Detection of substances having absorption in the UV-visible region can be done by scanning chromatograms with a spectrophotometer set at a wavelength of maximum absorption of the compound of interest.

### 24.1.1 Spectrophotometry

Spectrophotometry is an analytical technique based on the absorption of visible or ultra-violet light by solutions. The instrument used for this purpose is the spectrophotometer. It consists of three main components, a light source, monochromator and a detector. Basically the instrument operates by passing a narrow beam of near-monochromatic light through a sample solution and measuring the intensity of light reaching the detector. In practice, the intensity of light ($I$) passing through the sample is compared to the intensity of light ($I_0$) passing through a 'blank'.

Beer-Lambert law defines the degree of absorption of monochromatic light by a homogeneous absorbing medium (Fig. 24.3). It gives the relationship between absorption of radiation in a solution, where chromophores (solutes) are distributed homogeneously, and the solute concentration.

$$I = I_0 \exp(-abC) \qquad \ldots(24.2)$$

$$\log \frac{I_0}{I} = abC = A \text{(absorbance)} \qquad \ldots(24.3)$$

where, $I_0$ = intensity of the incident radiation;
$I$ = intensity of the transmitted radiation;
$a$ = absorption coefficient;
$b$ = absorption path length;
$C$ = concentration of the absorbing species.

Concentration of a solute can be estimated from a plot of $\log\left(\frac{I_0}{I}\right)$ versus concentration, $C$ (Fig. 24.4).

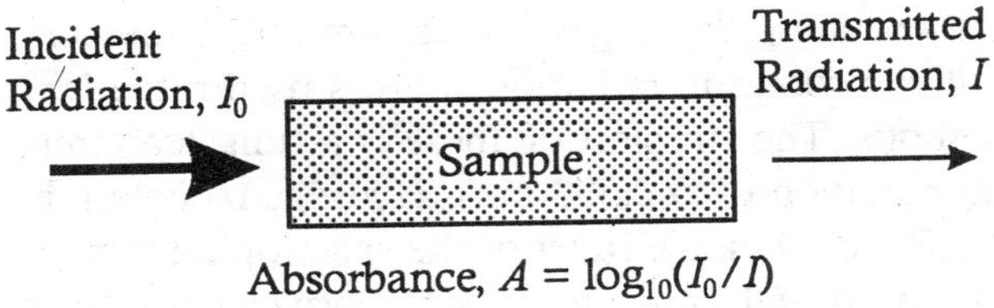


**Fig. 24.3:** Principle of Beer-Lambert Law

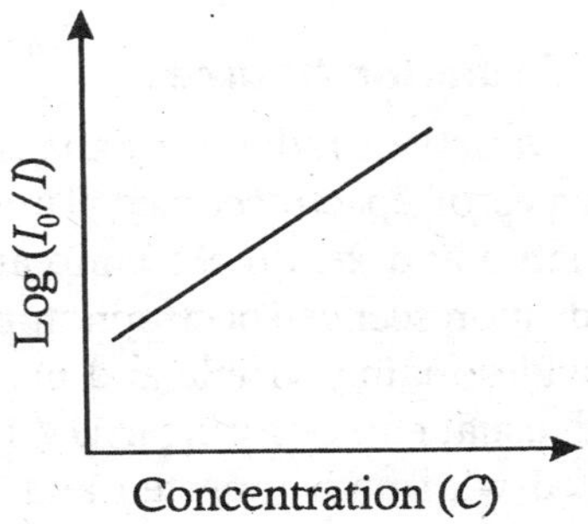


**Fig. 24.4:** Absorbance-Concentration Relationship according to the Beer-Lambert Law

**TABLE 24.2:** SOME CHROMOPHORE MOIETIES AND THEIR ABSORPTION MAXIMA

| *Chromophore* | *Moiety* | $\lambda_{max}$ *(nm)* | *Transition* |
|---|---|---|---|
| Alkene | —(C=C)— | 180 | $\pi \rightarrow \pi^*$ |
| | $—(C{=}C)_2—$ | 220 | |
| | $—(C{=}C)_3—$ | 260 | |
| Alkyne | —(C≡C)— | 180 | $\pi \rightarrow \pi^*$ |
| Carbonyl | C=O | 150 | $\pi \rightarrow \pi^*$ |
| | | 190 | $n \rightarrow \sigma^*$ |
| | | 280 | $n \rightarrow \pi^*$ |
| Thio | C=S | 240 | $\pi \rightarrow \pi^*$ |
| | | 500 | $n \rightarrow \pi^*$ |
| Nitro | $NO_2$ | 280 | $n \rightarrow \pi^*$ |
| Azo | —(N=N)— | 190 | $\pi \rightarrow \pi^*$ |
| | | 340 | $n \rightarrow \pi^*$ |
| | —(N≡N)— | 170 | $\pi \rightarrow \pi^*$ |
| Carboxyl | COOH | 180 | $n \rightarrow \sigma^*$ |
| | | 290 | $n \rightarrow \pi^*$ |
| Amide | CONH | 215 | $n \rightarrow \pi^*$ |
| Peptide bond | —(C—N)— | 190 | Small π–electron system |
| Nucleic acid bases | | 265 | Small π–electron system |
| Phenyl | | 270 | Small π–electron system |
| Naphthalene | | 310 | Large π–electron system |
| Anthracene | | 380 | Large π–electron system |
| Porphyrins | | >400 | Extended π– electron system |

Practically all spectrophotometers are fitted with a scale that reads both optical density (*A*), which is a logarithmic scale and percent transmittance, which is an arithmetic scale. Optical density scale is the most often used one.

#### 24.1.1.1 Radiation Sources

There are a variety of radiation sources to cater at different wavelength regions of radiation and to different types of spectroscopies. Tungsten filament lamp, hydrogen/deuterium lamp, mercury discharge lamp and xenon arc lamp are some of the common radiation sources used n the UV-visible and fluorescence spectrophotometric methods. The simplest of these is a tungsten lamp, which provides mainly visible and near infrared radiation (320-2,500 nm). Hydrogen/deuterium lamp is the usual choice for near UV radiation (180 to 400 nm). Most of the spectrophotometers are provided with both tungsten and deuterium lamp and depending on the wavelength range needed and one of them can be chosen at any given time. Xenon arc provides continuous spectrum in UV-visible region and is mostly used in spectrofluorimeters. Mercury arc provides number of spectral lines superimposed on continuum background in the UV region (200-400 nm).

#### 24.1.1.2 Monochromator

In modern spectrophotometers a monochromator is used to select the incident wavelength. The light emitted by the source is collimated by means of the 'entrance slit' to direct a narrow beam to the dispersing medium. The critical component of the monochromator is the 'dispersing element,' which disperses the light beam from the source according to wavelength. A prism or a diffraction grating is used for this purpose. By suitably rotating the mounting of the dispersing element a narrow beam of radiation of specified wavelength can be made to pass through the 'exit slit' and then through the optical cell in which the sample or blank solution is kept. Cells (usually 1.00 cm) are available in glass, plastic or quartz. Quartz cells are transparent to UV-visual light.

#### 24.1.1.3 Detector Systems

The spectrophotometer is fitted with a radiation detector, which produces a current or voltage when struck by radiation. The signal is a function of the wavelength and intensity of the radiation. All electromagnetic radiation detectors are optical transducers, either photon, or thermal radiation detection instruments that convert radiation signals to electrical pulses that can be amplified and modulated and recorded. Photon detectors are employed for measurement of radiation from *X*-rays to visible regions, and thermal detectors are employed for measurement in the infrared and thermographic measurements. Photovoltaic (barrier-layer) cells, photoconductive tubes and photomultiplier tubes (PMTs) are some of the photon detectors.

Photovoltaic (barrier-layer) cells are commonly used photo transducers for measuring radiation in the visible region (400-800 nm). They do not need any external power supply; they serve to convert radiant energy into electrical energy (they operate as current sources). Photovoltaic cells are rugged and simple, and they are used in cameras and solar cells. Photovoltaic cells have a potential for environment-friendly electricity production (basis of solar cells).

The device (Fig. 24.5) consists of a thin layer of semiconductor material (selenium) is deposited on a dielectric substance. Metallic electrodes and suitable enclosure with a window are added to complete the device. A potential barrier develops between the semiconductor layer and the metal surface. Incidence of light on the surface of the cell causes the flow electrons from the semiconductor to the metal and the current generated is proportional to the intensity of the incident radiation.

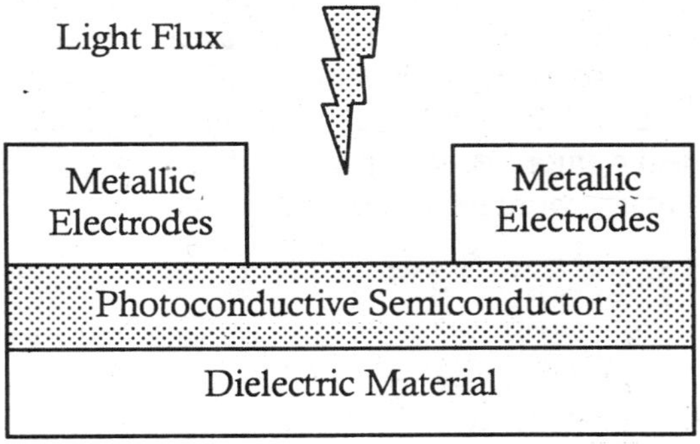


**Fig. 24.5:** Schematic of a Photovoltaic Device

Photoemissive tubes are vacuum tubes (vacuum photocells) (Fig. 24.6), with cathodes coated with light-sensitive materials (bialkali materials—$SbCs_3$, $SbCs_3O$, BiAgOCs), capable of emitting photoelectric electrons. The anode picks up the electrons that are released.

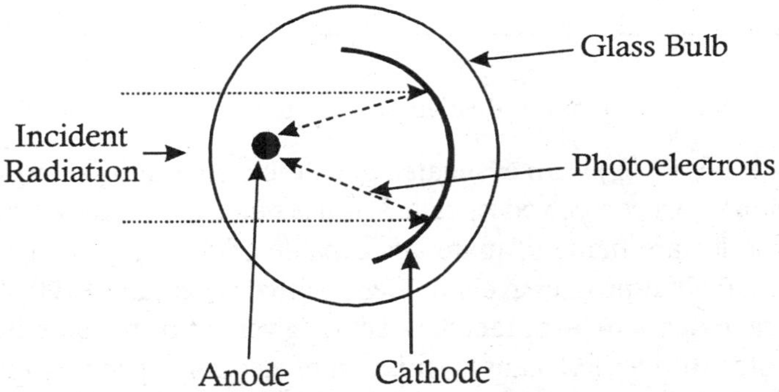


**Fig. 24.6:** Components of a Photoemissive Cell (Vacuum Photodiode)

Another system of detectors is photodiodes and photodiode arrays. A photodiode is a semiconductor diode, which depends for its operation on the inner photoelectric effect. In a p-n junction photodiode (Fig. 24.7), current flows only if the positive and negative terminals of the batter are connected to the p-side and n-side of the diode, respectively. Reversal of the battery terminal changes a *p-n junction* diode from a conductor to an insulator. Radiation impinging upon such a reverse-biased p-n junction diode surface creates electron-hole pairs, and the current produced is proportional to the intensity of the impinging radiation. A one-dimensional array of such diodes is called diode array detector. The advantage of this system lies in its ability to simultaneously measure the power at essentially all wavelengths. When such an array is used, the wavelength-dispersing unit is located after the sample cell. Photodiode assemblies are not used for very high-resolution measurements. Because of the fastness with which the spectrum can be scanned, they are used for making kinetic measurements and in chromatographic detectors.

The most widely used photon detection system is the photomultiplier tube (PMT). An electron energy (>100 eV) falling on a suitable photosensitive electrode material can eject more electrons, the number depending on the energy of the first electron and the nature of the electrode material.

This phenomenon is known as *secondary electron emission*. The number of emitted electrons due to one incident electron is the secondary emission coefficient (amplification). Photomultiplier tube (PMT) is essentially a vacuum phototube supplemented by a device to amplify a photocurrent due to secondary electron emission. The PMT system transduces the light signals into electrical signals with amplification. PMTs are extremely sensitive photodetectors, with extremely low dark current (the D.C. thermoionic current that flows without any light) and are used in all types of radiation measuring detectors—UV-visible, fluorescence, and atomic absorption/emission spectrophotometers, cameras, dosimeters, scintillation counters, tomographies, etc.

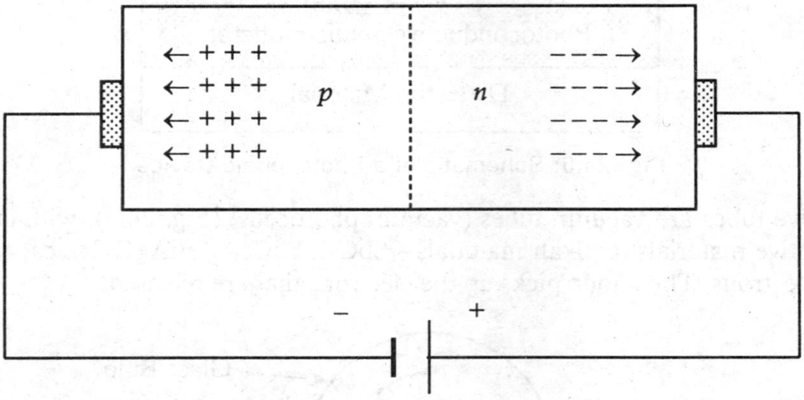


**Fig. 24.7:** Schematic of Reverse-bias p-n Junction Photodiode

A photomultiplier tube (PMT) is an evacuated glass tube, consisting of a photoemissive surface (*photocathode*), additional target electrodes, called *dynodes* and an anode, resealed in an evacuated tube (Fig. 24.8). Dynodes are made of materials capable of strong secondary electron emission. Each dynode is maintained at progressively higher positive potential (+100 V) than the preceding one so that electrons each dynode can attract the electron given off by the previous dynode. Radiation impinging on the photocathode (PC) causes emission of electrons (primary electrons), the number (photocurrent = $I_{ph}$) proportional to the intensity of incident radiation. These electrons are attracted towards the 1st dynode, which is positive with respect to the photocathode; liberating in the process more electrons (secondary electrons; $I_1 = \sigma I_{ph}$). These secondary electrons created at the 1st dynode are attracted towards the 2nd dynode, which is more positive than the 1st dynode, and produce more secondary electrons ($I_2 = \sigma^2 I_{ph}$). In this way a cascade of electrons are produced at each dynode, and the secondary electrons created at the last dynode are attracted to the anode with an electron amplification ($I_n = \sigma^n I_{ph}$) of $\sim 10^6$.

### 24.1.2 Turbidimetry/Nephelometry

Turbidimetry (absorption method) and nephelometry (scattering method) are employed to measure the attenuation of the intensity of transmitted or scattered light, due to the presence of suspended particulate matter. Beer-Lambert law does not apply in these cases. Transmitted light is measured along the axis of the incident light for turbidimetry, and scattered light is measured at 90° to the incident light for nephelometry (Tyndall scattering). The intensity ratio, $(I_0/I)$ of the sample is compared with standard suspension(s) of known concentration(s). The instrumental components are similar to those used in spectrophotometry. Mercury vapor lamp with green filter or He-Ne laser would provide monochromatic radiation. Photodiodes or photomultiplier tube (PMT) would serve as detectors.

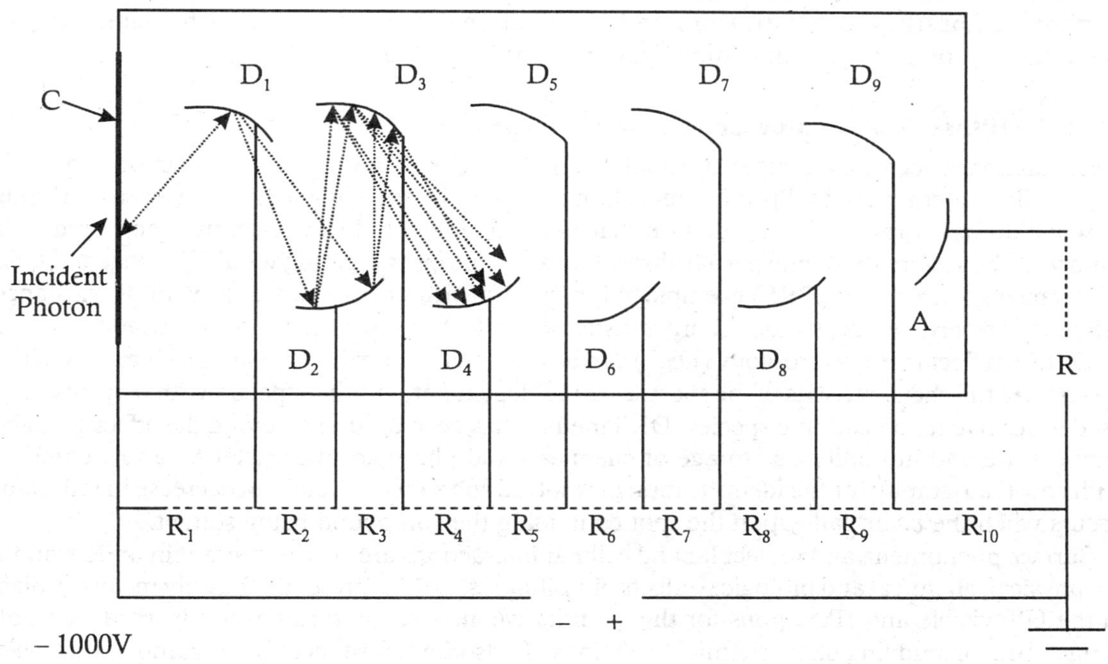


C = Photosensitive cathode; $D_1$, $D_2$,....$D_9$ = Dynodes (cathodes);
A = Anode; R, $R_1$, $R_2$.... = Resistances

**Fig. 24.8:** Schematic of a Photomultiplier Tube (PMT)

Turbidimetry and nephelometry are widely used in biological, clinical, pharmaceutical, food, beverage, dairy, wastewater treatment, and environmental pollution monitoring studies to analyze suspended particulate matter, contaminants and pollutants. Dilute suspensions of micro-organisms and colloidal matter may be measured by nephelometry, while turbidimetry is reserved for denser suspensions. Dual-beam nephelometer is used in smoke detection. In this instrument, one beam of light is passed through a cell containing air drawn from within the building, and another beam of light is passed through a cell containing air drawn from outside the building. The difference in the intensity of scattered light is proportional to the smoke particles generated within the building.

## 24.2 ABSORPTION AND REFLECTANCE SPECTROSCOPIES

Several spectroscopic methods are available to study structural and dynamic characteristics of molecular complexes.

### 24.2.1 Action Spectroscopy

Action spectroscopy is used to characterize photo pigments spectroscopically and kinetically. Action spectrum is representation of the photo response of chromophores as a function of wavelength, λ. Action spectroscopy is based on the Einstein-Stark equivalence—the amount of transformed pigment is proportional to the number of photons absorbed by the pigment. The Beer-Lambert law does not hold good to the pigment molecules, but theory postulated by Mie is more useful in interpreting the action spectra. Action spectroscopy is employed to study the function of photo pigment molecules in photosynthesis, photo activation and inactivation (rhodopsins, phytochromes, etc.) and photodimerization (e.g. thymine dimerization in nucleic acids).

### 24.2.2 Diffuse Reflectance Spectroscopy (DRS)

The radiation reflected by a finely dived solid consists of a regular or specular reflection component and a diffusion component. Specular reflection occurs at the surface with no transmission through it, while diffusion results from radiation that has penetrated and subsequently reappeared at the surface of the system following partial absorption and multiple scattering within the system. Diffuse reflectance spectroscopy (DRS) is employed in color measurement studies of various substances, influence of particle size, surface characteristics and effect of moisture on color variations.

Diffuse reflectance spectroscopy (DRS) can be used in identification of cations by means of their spectra. As the characteristics of the spectra are pH-dependant, multiple spectra with reagents can be used to resolve many cationic species. DRS method has been used in studying the effects of aging, temperature and humidity on storage of chemicals and pharmaceuticals. DRS can be combined with chromatography for the identification of resolved substances. A sudden decrease in reflectance occurs when the beam falls upon the spot containing the compound being sought.

Surface phenomena and molecular and cellular interactions are very important in understanding the physical, chemical and biological effects of pollutants. DRS is presently the only means available in the UV, visible and IR regions for the quantitative analysis of substances adsorbed on a solid surface and for studying photochemical reactions of adsorbed substances. For example, anthracene dimerizes to dianthracene under UV radiation, and it is oxidized to anthraquinone in the presence of oxygen. In either case the rate of conversion is dependent on the nature of the adsorbent. Thus, by making use of color changes resulting from adsorbent-adsorbate interactions, it is possible to study the surface phenomena of adsorbents and equilibrium constants for the association of molecular species adsorbed on solid surfaces. DRS may be employed in the quantitative study of interaction of particulate matter, pollen and aerosols on leaves, foliage and biological tissues. DRS analysis of color and color change in food products due to aging, spoilage and additives and dyes is more quantitative. DRS can be used to monitor process color changes resulting from smoking and pickling of meats, browning of milk on heating, darkening effect due to roasting process and also due to irradiation.

Diffuse reflectance method may be used to study tanning of skin and the effects on other biological tissues (e.g. bone and collagen) as a result of UV radiation and application of skin creams and ointments. Reflection oximetry has been developed for the measurement of the oxygen saturation, (OS), of human blood.

$$OS = A + \left[\frac{B}{I_r(\lambda)}\right] \quad \ldots(24.5)$$

where, $A$ & $B$ are constants, and $I_r(\lambda)$ is the intensity of the reflected light.

### 24.2.3 pH-dependent UV Absorption Spectroscopy

pH-dependent UV absorption method (in the pH range 5-12) can be employed to study the absorption of pollutants. Most phenols are in acid form at pH $<$ 5 (except picric acid) and they are in their dissociated phenolate form at pH = 12. The absorption profile, the displacement of $\lambda_{max}$ towards longer wavelength region (bathochromic shifts) at alkaline pH is measured to monitor the presence of phenols. The difference in absorption profiles (at pH = 5 and 12) is due to conversion of free phenol to the phenolate anion.

The apparatus (Fig. 24.9) has the standard components essential in photometry—an UV radiation source, sample holder, color filters, optical system, detectors (PMTs) and recorder. UV radiation from a mercury vapor lamp is passed through the sample holder and then split into two beams. One beam passes through a filter that transmits $\lambda$ = 290 nm (measuring wavelength), and the other beam through the second filter that transmits reference wavelength, $\lambda$ = 365 nm. The intensities are recorded by a photomultiplier tube (PMT) and recorded. The differential output of PM tube is linearly proportional to the phenol concentration.

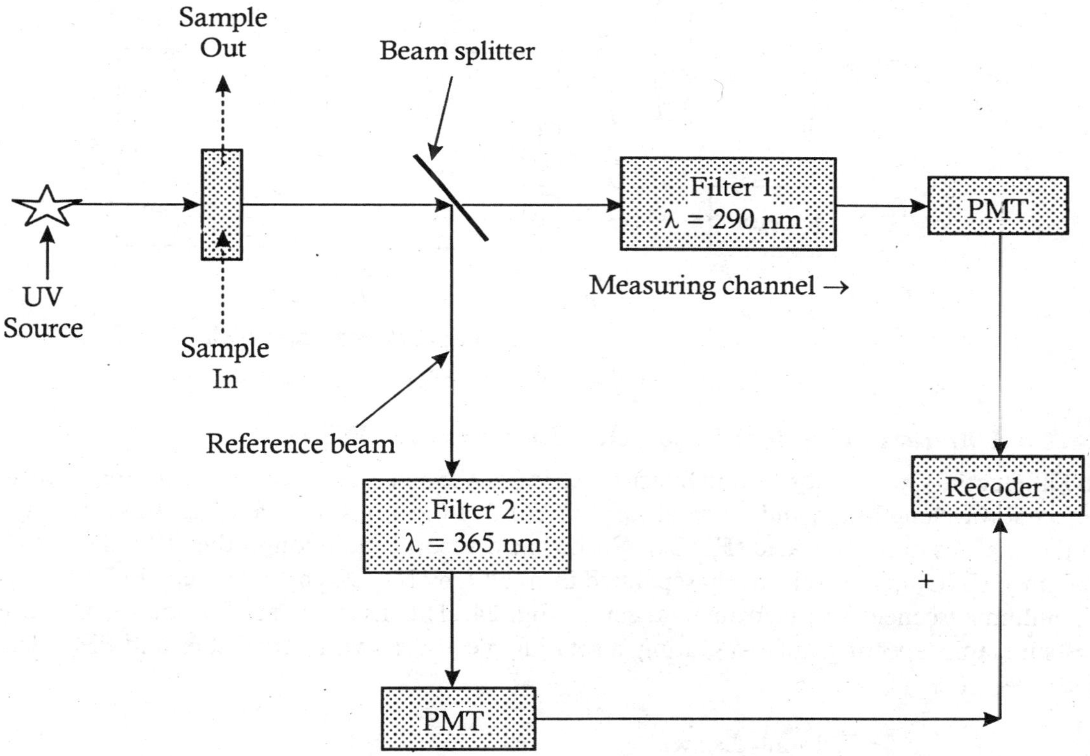


**Fig. 24.9:** Schematic of Photometric Analyzer Setup

### 24.2.4 Correlation Spectrometry

Correlation spectrometry is a technique that yields increased specificity and sensitivity by comparing the spectrum of the air pollutants with a standard replica spectrum of that substance. In this method, the light beam passing through the sample is modulated at frequencies corresponding to the absorption bands of the target gas. Changes in the intensity of the light beam are measured only at the wavelengths where the target absorbs. This allows measurements to be made in the presence of other absorbing gases. The spectrometer (Fig. 24.10) has a wheel with two similar cells, one containing the gas to be analyzed, and the other containing a non-absorbing gas. When the wheel is rotated, the two gas cells are placed alternately in the light beam. The rotating filter wheel acts as a chopper that modulates the radiation received by the detector only at wavelengths corresponding to the absorption spectrum of the sample gas. The sample gas is passed through a sample cell. The sample cell causes additional absorption of the radiation received by the detector. Correlation spectrometry is used for measurement of gaseous pollutants such as $NO_2$ and $SO_2$.

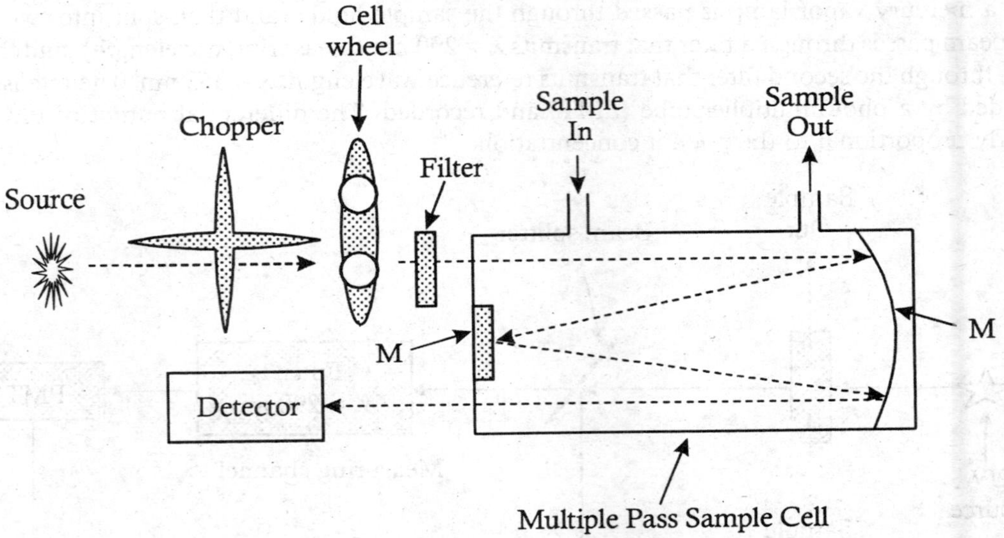


**Fig. 24.10:** Schematic of Correlation Spectrometry

### 24.2.5 Differential Optical Absorption Spectrometry (DOAS)

In Differential Optical Absorption Spectrometric (DOAS) method, measurements are made at two adjacent wavelengths, on and off an absorption band of the target molecules. DOAS is widely used in the analysis of nitrous acid ($HNO_2$). Nitrous acid absorbs radiation in the UV region (354 and 360 nm). $HNO_2$ can be selectively separated from $NO_2$ by $Na_2CO_3$ by a denuder and measured by chemiluminescence. The instrumental setup (Fig. 24.11) consists of an UV source, xenon high-pressure lamp, spectrograph—a grating, a rotating metal discs with radial slits, and PMT detector and a recorder.

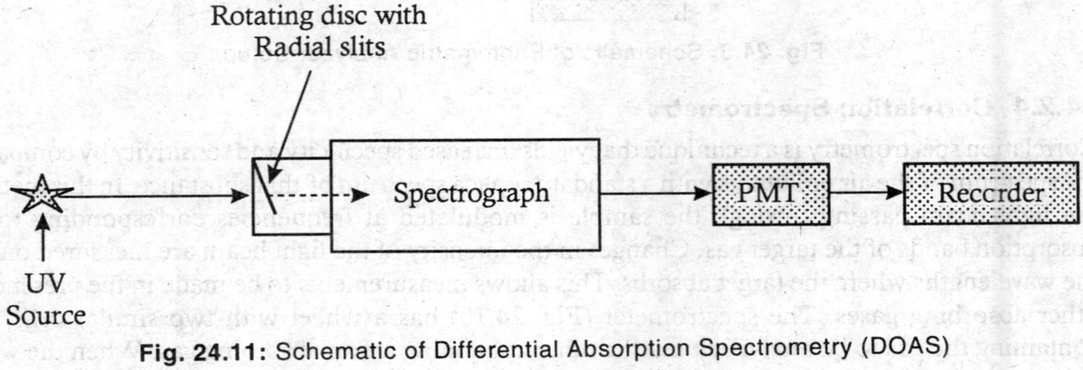


**Fig. 24.11:** Schematic of Differential Absorption Spectrometry (DOAS)

### 24.2.6 Second-derivative Spectroscopy

In conventional spectroscopy, attenuation of radiation through a sample may be due to more than one substance absorbing at the same wavelength. Second-derivative spectroscopy provides change in intensity with wavelength for individual absorbers (Fig. 24.12). It measures the second derivative of light intensity $\left(\frac{d^2I}{d\lambda^2}\right)$ as a function of wavelength, λ, thereby providing the presence of specific absorption bands. The detection limit in second-derivative spectroscopy is quite high (Table 24.3) and it is used for trace analysis of gases in the environment.

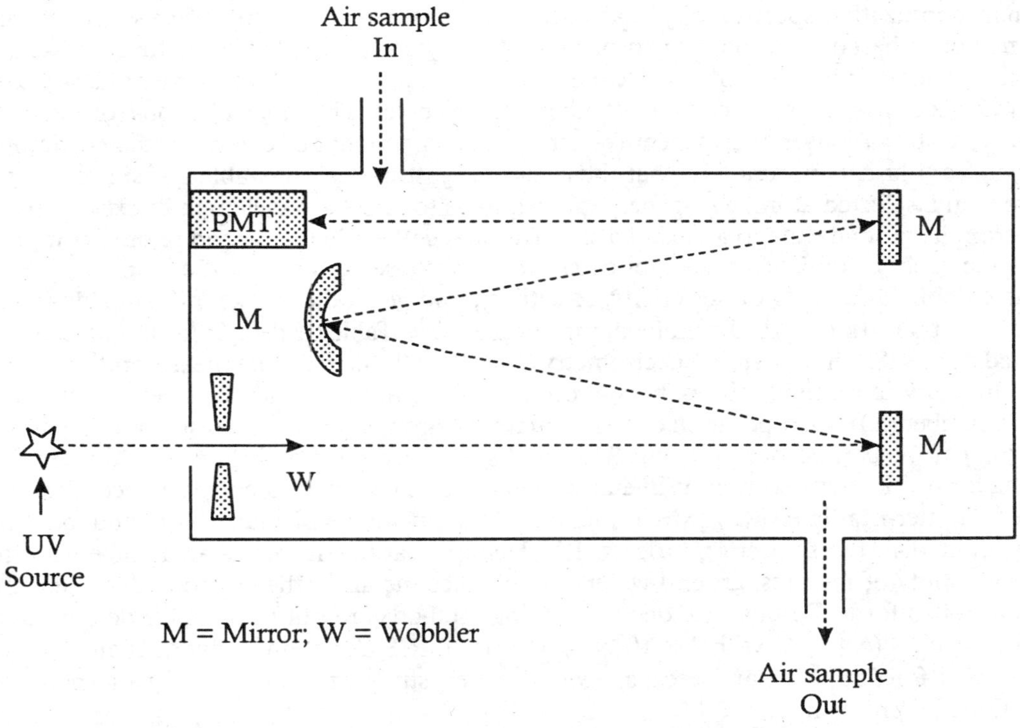


**Fig. 24.12:** Schematic of Second-derivative Spectrometer Setup

**TABLE 24.3:** DETECTION LIMIT IN SECOND-DERIVATIVE SPECTROSCOPY

| *Pollutant* | *Detection Limit (ppm)* |
|---|---|
| Ammonia | 0.001 |
| Benzene | 0.025 |
| Nitrogen monoxide (NO) | 0.005 |
| Nitrogen dioxide ($NO_2$) | 0.04 |
| Ozone ($O_3$) | 0.04 |
| Sulfur dioxide ($SO_2$) | 0.001 |

### 24.2.7 Doppler-limited Spectroscopy

Doppler effect limits the line width of a molecular absorption transition. This can be minimized by use of a laser as a source of light. Tunable laser sources with extremely narrow bandwidths achieve a resolution on the order of Doppler line width (0.001-0.05 nm). Laser spectrometers are inherently more sensitive than conventional broadband source types. The extremely narrow line width of a laser beam permits it to undergo multiple reflections through a sample without spatial spreading and interference, thus providing a long absorption path length. High-intensity laser source reduces the detection duration and coupled with a high sensitivity detector, laser spectrometer is an ideal instrument for detecting and characterizing short-lived intermediate species in chemical reactions.

### 24.2.8 Resonance-Ionization Spectroscopy (RIS)

Resonance-ionization spectroscopy is an extremely sensitive and highly selective analytical tool. The method is based on resonance absorption of analyte atoms followed by further excitation to ionization continuum. It exploits high coherence and monochromatic properties of lasers to ionize and eject electrons from selected types of atoms or molecules. The molecular ions (or electrons) are then detected by various means to identify elements or compounds and determine their concentration. Photons from lasers are tuned so that their energies just match allowable transition energies for electrons in a selected atom. The general procedure is to select a wavelength to excite a particular atom from its ground state to an excited state (resonance), while the second photon completes the ionization process. Ionization will not occur unless the laser is tuned to the atom.

The combined features of selectivity, sensitivity, and generality make RIS suitable for a wide variety of applications. All the elements in the Periodic Table, except helium and neon can be detected with RIS. Though mass spectrometry (MS) is well suited for the measurement of isotopes, it has difficulty in resolving the isobars (atomic nuclei with same mass number (A), but different atomic number (Z)). Incorporation of RIS in mass spectrometry solves the isobar problem. Isotope counting can be carried out by using RIS for Z-selectivity sorting, followed by quadruple mass filtering for A-selectivity sorting. While in secondary ion mass spectrometry the secondary ions are utilized, sputter-irradiated RIS (SIRS) method takes advantage of much more abundant neutral atoms emitted in the sputtering process. RIS has applications in many areas of environmental concern—study of ice caps, ground water, role of trace metals in the environment, medicine and biology and so forth. Radioactive decay counting methods are not suitable for determining long half-life isotopes (e.g. $^{81}Kr$ with $2 \times 10^5$ yrs). However, RIS can count individual atoms and thus can be used for dating polar ice caps, ground water, study of hydrological patterns and other environmental problems.

### 24.2.9 Photoacoustic Spectroscopy (PAS)

Photoacoustic spectroscopy (PAS) gas analyzer provides a system to analyze a mixture of gases with high sensitivity. In PAS, the gas sample is contained in a resonant, sealed cell (Fig. 24.13), and is irradiated with pulsed $CO_2$ laser. A rotating chopper modulates the radiation. Absorption of radiation by the sample gas causes its temperature to rise resulting in a series of pressure pulses, which are detected by microphones and IR detectors placed in the walls of the cell. Detection of a particular gas is achieved by irradiation of the sample with light of specific wavelength at which that gas absorbs. PAS can be used for point measurements in a gas sample, and is suitable for a wide range of gases and volatile organic molecules.

### 24.2.10 Thermal Lens Spectrometry (TLS)

Thermal Lens Spectrometry (TLS) is based on generation of heat during the non-radiative relaxation of the absorbed optical radiation. The increase in temperature of the sample is determined by changes in its physical properties, such as density and refractive index. Measurement of absorbance, through change of temperature-dependent refractive index of the sample, which is exploited in TLS, offers detection sensitivity, as it is less subject to errors arising from light scattering in the sample. High sensitivity of TLS technique can be combined with the high selectivity of separation techniques, such as HPLC (HPLC-TLS), and IC (IC-TLS), which enables determination of environmental pollutants at ppb levels.

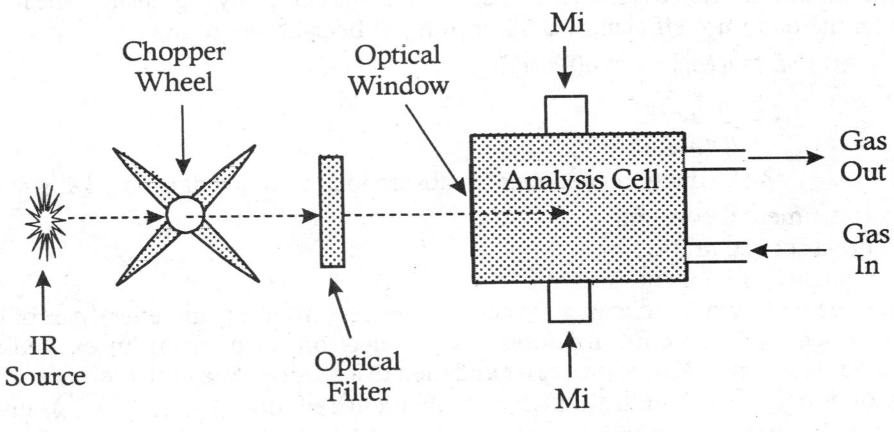


**Fig. 24.13:** Schematic of Photoacoustic Spectrometry (PAS) Setup

Both IR and microscopic TLS spectrometers are available for analysis. Differential TLS spectrometers can be used to circumvent the problem of relatively large absorbance of solvents in the IR spectral region. They operate on the basis of positional dependence of thermal lens signal. The possibility of focusing laser beams to points smaller than 1 mm in diameter enables absorbance on the microscopic scale and measurements on living cells and tissues becomes possible.

TLS techniques have found wide-ranging application in the analysis of trace concentration of contaminants in many environmental and food samples. This technique offers better detection sensitivity in the analysis of heavy metals (Cd, Cr, Fe, Mn, Ni, Pb) in water samples. TLS is well suited for determination of environmentally relevant organic compounds like pesticides. Enzymatic processes can further increase selectivity-e.g. selectivity of detection of organo-phosphorous- and carbamate pesticides can be enhanced by adding specific enzyme (AChE) into TLS flow-injection system.

Sensitive, fast and cost-effective analytical methods are needed for detection of undesirable contaminants, pollutants and toxicants at trace level in foods and beverages. TLS is well suited for the characterization of liquid beverages such as fruit juices, wines and other liquids. Content of free fatty acids (FFA) and trans-unsaturated fatty acids (TFA) should be low in edible oils and fats. While these compounds can be determined by IR spectroscopy, the sensitivity is low. These and other chemicals can be determined with high sensitivity (ppb level) by TLS techniques.

## 24.3 FLUORESCENCE SPECTROSCOPY

Molecules absorb electromagnetic radiation and are excited to higher energy states. Luminescence is the phenomenon of emission of electromagnetic radiation upon transition from excited electronic states to the ground state.

$$MX_n + h\nu \rightarrow MX_n^* \rightarrow MX_n + h\nu \quad ...(24.6)$$

When there is no energy loss during these electronic transitions ($\lambda_{inci} = \lambda_{emit}$), it is a simple case of luminescence, called photo-luminescence. If the energy for excitation is obtained from a chemical reaction, the process is termed chemiluminescence and if it is by thermal energy it is called thermo-luminescence. Bioluminescence is release of electromagnetic energy by organisms. The luminescence, $L$, depends on the quantum efficiency, $\varphi$. According to Beer-Lambert law,

$$L = K\varphi I_0[1 - \exp(-\varepsilon bc)] \quad ...(24.7)$$

$$\varphi = \frac{h\nu_{emitted}}{h\nu_{absorbed}} \quad ...(24.8)$$

$$\varepsilon = \text{Extinction coefficient} = \text{absorbance} \times \text{molecular mass } (A \times M) \quad ...(24.9)$$

where, $K$ = instrumental constant;
$c$ = molar concentration;
$I_0$ = intensity of incident radiation.

However, if the molecular species contain chromospheres (absorbing moieties), part of the incident energy is absorbed by these chromophores and an electron drops from an excited state to an intermediate state by non-radiative processes and then to a ground state with emission. The emitted radiation is of longer wavelength than that of incident radiation, ($\lambda_{emit} > \lambda_{inci}$), giving rise to electronic effects—fluorescence and phosphorescence (Fig. 24.14). The quantum efficiency, $\varphi$, is

$$\varphi = \frac{\text{No. of photons emitted}}{\text{No. of photons absorbed}} \quad ..(24.10)$$

Greater the quantum efficiency, $\varphi$, the greater will be the fluorescence or phosphorescence.

Fluorescence is the emission process without any change in the spin orientation. It is the singlet → singlet state transition (paired electrons transition) process, and the lifetime is ~$10^{-8}$-$10^{-4}$ s (Fig. 24.15). Phosphorescence is the emission process with a change in the spin orientation during transition. It is the triplet → singlet state transition (unpaired electrons transition) process with much longer lifetime ~$10^{-6}$-$10^{2}$ s. In phosphorescence the excited state is paramagnetic. In either case, the emitted light is, often, of a longer wavelength than the absorbed radiation.

$$\lambda_{phos} > \lambda_{fluo} > \lambda_{inci} \quad ...(24.11)$$

$$\nu_{phos} < \nu_{fluo} < \nu_{inci}$$

A fluorimeter or a spectrofluorimeter is used to carry out fluorescence measurements. Components of these instruments are the same as those of spectrophotometers with additional elements. A fluorescence spectrophotometer has two optical elements, one for wavelength selection for excitation of chromophores and the other for detection of fluorescence wavelength(s), characteristic to specific chromophores. The optical elements are arranged transverse to each other (Fig. 24.16). Essential components are: (1) excitation source—tungsten lamp, Hg-vapor lamp (365, 405, 436 and 546 nm), Xe-arc lamp (200-800 nm) or tunable dye laser, (2) excitation monochromator (or filter) for selection of excitation wavelength, (3) emission monochromator (or filter) for selection of fluorescence emission wavelength and (4) detector (photomultiplier or array photodiodes).

The desired wavelength of excitation ($\lambda_{ex}$) is selected with the help of a monochromator or an optical filter (primary filter) placed between the radiation source and the sample. The wavelength for the fluorescent emission ($\lambda_{em}$) to be measured is selected with the help of an additional monochromator or a second optical filter (secondary filter) placed between the sample and a photodetector located at 90° angle from the incident light path. Spectrofluorimeters permit the recording of two types of spectra: (*i*) excitation spectrum, (*ii*) emission spectrum. The excitation spectrum is a plot of fluorescence intensity versus excitation wavelength ($\lambda_{ex}$) with the emission wavelength held constant ($\lambda_{em}$ = constant). The spectrum is identical with UV absorption spectrum. The emission spectrum is a plot of the fluorescence intensity versus $\lambda_{em}$, with the excitation wavelength held constant, ($\lambda_{ex}$ = constant).

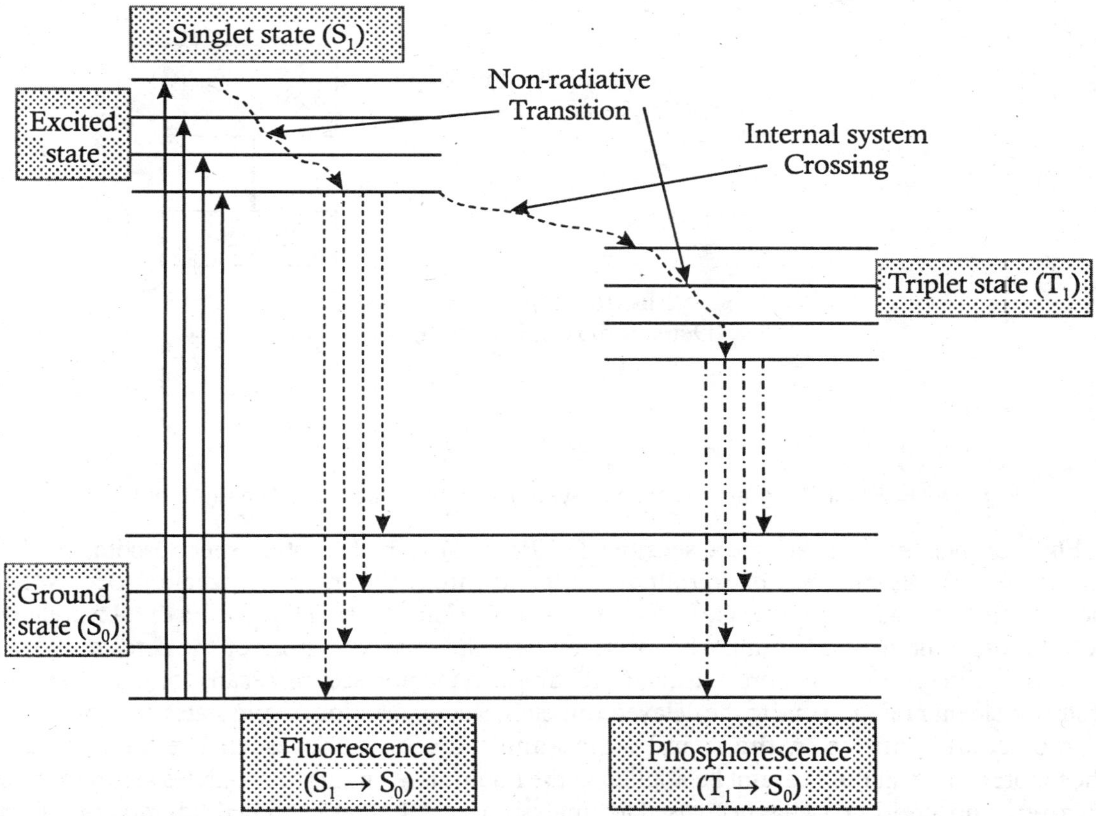


**Fig. 24.14:** Electronic Transitions (Jablonski Diagram) depicting Fluorescence and Phosphorescence

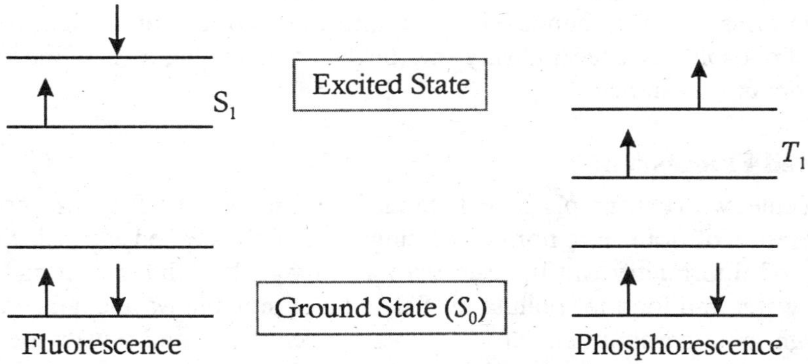


**Fig. 24.15:** Orientation of Spins in Fluorescence and Phosphorescence

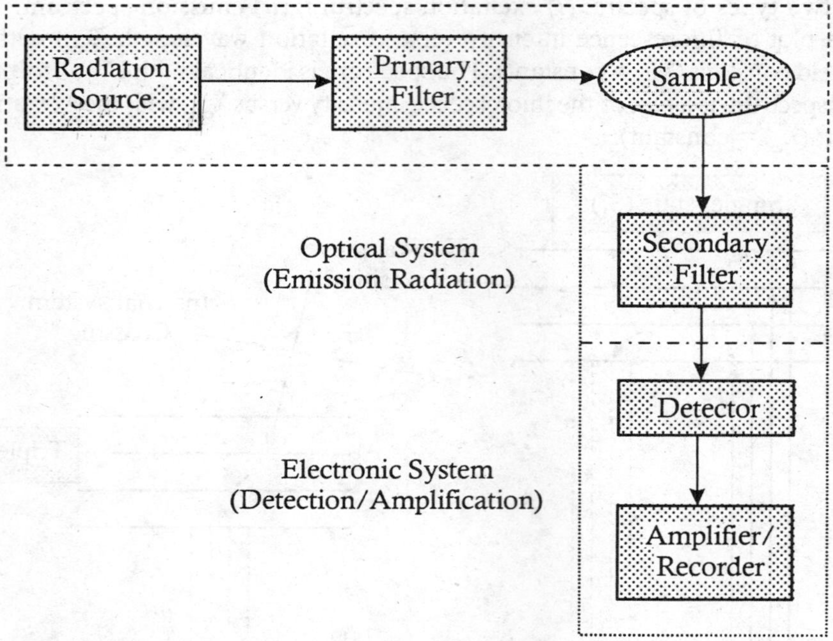


**Fig. 24.16:** Block Diagram of Optical Systems in Fluorescence Measurement Setup

Fluorescence methods are more sensitive (~$10^3$ times) than absorption spectroscopic methods. This is because (*i*) minimal background interference from the incident radiation occurs as the measurement is made at right angle to the incident radiation, and (*ii*) elimination of all wavelengths except those emitted by the sample by the secondary filter placed in front of the detector.

Phosphorescence is even more sensitive (~$10^3$ times) than fluorescence, because of non-interference of stray incident radiation due to the delayed emission. Reorganization of spin states as a consequence of molecular motions in the liquid state annuls the phosphorescence effect, therefore, phosphorescence method cannot be applied in the liquid state, unless the analytes are immobilized or frozen. Phosphorimetry has been used for analysis of metabolites in body fluids, proteins, nucleic acid components, pharmaceuticals, and many environmental pollutants. Phosphorescence is more sensitive and selective than fluorescence in the analysis of carbamate pesticides and trace analysis of polycyclic aromatics, such as benzo[e]pyrene and anthraquinone, which are carcinogenic. Measurement of phosphorescence intensity and the data of mean lifetimes of the substances can be used to identify many substances.

### 24.3.1 Induced Fluorescence

Generally, molecules with chromophores exhibit intrinsic fluorescence. Fluorescence due to external agents (incorporation of flours into non-fluorescing molecules) is called extrinsic fluorescence. The high sensitivity of fluorimetry can be used with advantage, by "fingerprinting" method, in the analysis of air, water and food for pollutants. Selection of suitable wavelengths for excitation and emission through monochromators helps in increasing selectivity and sensitivity.

Fluorescence activated cell sorting (FACS) is one of the applications of extrinsic fluorescence that is widely used in molecular biology and clinical diagnostics. In FACS, a fluorescent, labeled antigen, that is complementary to the required antibody specificity, is introduced into the culture prior to sorting. The labeled antigen binds only those hybridomas secreting the required antibody and the cell sorter can be programmed to dispense into the clone plate only those cells carrying the fluorescent label.

Fluorescence is the basis of $SO_2$ monitoring. $SO_2$ is excited using Zn (214 nm) or Cd (229) lamp and the fluorescence (220-400 nm) is monitored. Laser-induced fluorescence technique is used to detect various atomic, free radical and molecular species (Table 24.4). OH radical in the atmosphere is monitored by Laser-induced excitation of OH at 282 nm and fluorescence measurement at 309 nm.

**TABLE 24.4:** LASER-INDUCED FLUORESCENCE DETECTION

| *Species* | $\lambda_{max}$ *(nm)* |
|---|---|
| H (radical) | 122 |
| O (radical) | 130.5 |
| OH (radical) | 306.5 |
| Cl | 139 |
| $CH_3O$ | 304 |
| $NO_3$ | 663 |

## 24.4 LUMINESCENCE-BASED ASSAYS

Fluorescence techniques find extensive application in molecular biology (enzyme assays, immunoassay), clinical pathology (determination of ions, elements, lipids, amino acids, proteins, steroids, metabolites, blood typing) and biomedical studies (scintillation counting, organ imaging, cell sorting) and environmental pollution monitoring. The major use of fluorimetry in environmental assays is to detect materials in concentrations too low for absorption spectrometry. It is helpful in the assay of air and water pollutants and pathogens and additives, vitamins, steroids, hormones, and drugs, carcinogens, pesticides in foods and grains and beverages.

Fluorescence is less likely in compounds with $n \rightarrow \pi^*$ transitions. Only relatively few aliphatic and saturated cyclic compounds fluoresce and phosphoresce. Heterocyclic compounds (e.g. indoles, coumarins), usually, do not fluoresce in non-polar liquids, but may do so in polar solvents. On the other hand, molecules with planar moieties (aromatic rings) exhibit strong fluorescence, because of $\pi^* \rightarrow \pi$ transitions. Conjugated, fused rings enhance fluorescence. The presence of metal ions can influence the luminescence characteristics of organic molecules. Paramagnetic transition metal ions generally produce large quenching effects. Diamagnetic, non-transition metal ions are usually poor quenchers and hence form good fluorescent chelates.

Many inorganic substances (in the solid state) either fluoresce or phosphoresce directly, or after addition of inorganic reagents (HCl or HBr). This behavior is used in the assay of many inorganic ions—As, Bi, Fe(II), $I^-$, Os, Pb, Sb, Se, Te and Tl. The formation of a highly fluorescent metal chelate, with an organic legend, can lead to a sensitive and highly specific method for the determination of many elements—Al, As, Au, B, Be, Ca, Cd, Cu, Ga, Ge, Hg, Mg, Pd, Rh, Ru, S, Sb, Se, Sn, Te, Tl, Zn and Zr (see Chapters 20, 21and 22 for details). The ligand should be an organic molecule, with O or N, which is itself non-fluorescent. Upon complex formation with a metal ion, the *n*-orbital electrons are utilized in binding the metal and they become less accessible to excitation. Strong fluoresce occurs due to co-planarity and rigidity of the conjugated structure of the metal-chelate (Fig. 24.17).

**Fig. 24.17:** Example of Non-fluorescent and Fluorescent Chelate Moieties

In addition, some ions and Cu, Fe, Ni, $O_2$ fluoride and cyanide can be determined indirectly by measuring the amount of quenching they produce in the fluorescence of a chelate or by causing the release of a ligand which can then react to form a fluorescent product (e.g. determination of cyanide).

$$\text{Non-fluorescent chelate} + CN^- \rightarrow \text{Fluorescent } (CN^-) - \text{Chelate} \quad \ldots(24.12)$$

Very sensitive and highly selective fluorescence procedures have been developed for the assay of organic compounds—amino acids, proteins, enzymes, nucleic acid components, vitamins, steroids, drugs and environmental pollutants. Some organic compounds are naturally fluorescent (intrinsic flours), possessing planar moieties and delocalized π-bonding (phenylalanine, tryphtophan, tyrosine, nucleic acid bases, ATP, FAD, NAD, vitamin A and aromatic hydrocarbons), which can be used for their assay. Fluorescence method has been a valuable tool in identifying the mechanism of photosynthesis, because different chlorophylls, pheophytins and other photosynthetic precursors and products are fluorescent at different wavelengths. Other organic compounds, themselves non-fluorescent or weakly fluorescent, can be converted to good fluors by simple chemical reaction or change of environment. This feature is used to study phenols in water. Phenols do not fluoresce in the basic medium (anionic form). However, they do fluoresce in the acidic medium (Fig. 24.18). The spectra of phenols in the basic and acidic media are compared to quantify the presence of phenols in water.

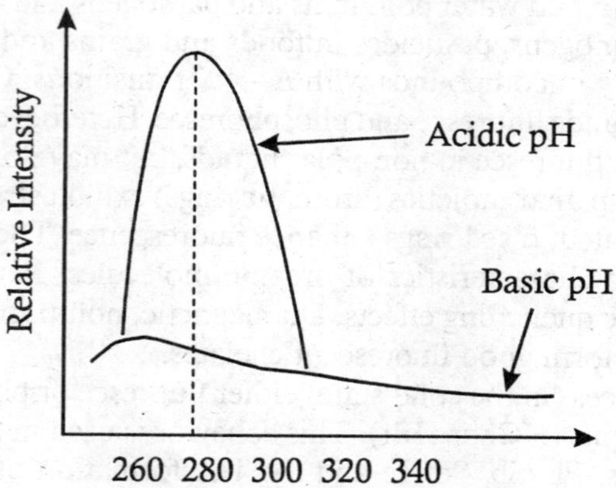


**Fig. 24.18:** Fluorescence Emission Spectra of Phenols as a Function of pH

### 24.4.1 Enzyme-mediated Luminescence Assays

Enzyme-mediated luminescence assays are based on chemiluminescence, bioluminescence, and fluorescence. Enzyme-catalyzed peroxide decomposition reactions are made use of in chemi- and bioluminescence reactions. Luminol (5-Amino-2,3-dihydrophthalazine-1,4-dione) and its derivatives are commonly used in chemiluminescence assays. Luminol is a particularly useful tool for monitoring active $O_2$ concentration. It is used as indicator of cellular events involving $O_2$-metabolites. Horseradish peroxidase is frequently used as catalyst for the luminol reaction. Certain metal ions can exert a profound effect on luminol chemiluminescence. Iron is also one of the important catalysts. So, iron-containing biological complexes (e.g. cytochrome c, myoglobin, hemoglobin, haematin) can be determined with great sensitivity by employing chemiluminescence. Another enzyme-catalyzed chemiluminescence system makes use of dioxetanes (phosphates). A large number of thermally stable dioxetanes undergo chemiluminescence reaction on elevation of temperature (thermoluminescence).

Bioluminescence is chemiluminescence in biological systems. High quantum yields are possible in bioluminescence. Bioluminescence is used in many non-immunoassay applications (metabolic pathways, cellular redox reactions, and cell activation). Luciferin (benzothiozole), colentrazine (imidazolopyrazine), and proteins, pholasin, aequorin, obelin, and photoproteins (green fluorescent protein, lumazine protein) are some of the bioluminescent compounds. Any change in cellular respiration or disruption of cell structure results in a change in respiration with a concurrent change in the rate of bioluminescence. Thus, bacterial bioluminescence test provides a rapid, reliable and convenient means of determining the toxicity of waste material.

### 24.4.2 Chemiluminescence Immunoassays

The chemiluminescence immunoassay methods combine high specificity and sensitivity achievable from immunochemical methods with photochemical luminescence detection methods. These are alternate choices for radioactivity-based methods (RIA and IRMA; Chapter 25), with the difference being that the antigen is labeled with a fluorescent material instead of a radioactive material, and are better suited for environmental monitoring and quality assurance because of several advantages. Some of these are:

1. Radioactive materials are not involved. Therefore, there is no need for stringent safety.
2. No need for prior separation of the immune complexes (homogeneous immunoassay and fluorescence-quenching methods).
3. Equipment can be set up in laboratories, diagnostic centers and field stations.
4. Long shelf-life of chemicals.
5. Chemiluminescent labels can be incorporated into the analyte, whereas this is not possible with radioactive labeling methods due to catastrophic radiation damage.
6. Accuracy is comparable to radioactivity based assay methods like radioimmunoassay (RIA) and immunoradiometric assay (IRMA).

Analogous to radioactivity-based immunoassays, chemiluminescence assays are named depending upon the assay method—chemiluminescence immunoassay (CIA), competitive-binding immuno-chemiluminometric assay (CB-ICMA) and two-site immuno-chemiluminometric assay (ICMA).

Generally fluorescence immunoassays use lanthanide (Erbium) chelates as fluorophores and the measurement is based on time-resolved fluorescence. In contrast to conventional fluorophores, lanthanide element cations have a much longer fluorescence lifetime and thus suppress the background fluorescence, the technique known as dissociation-enhanced lanthanide fluoro-immunoassay (DELFIA). Further, the difference between the absorption (excitation) maximum and emission maximum (Stokes-shift) is greater than for conventional fluorophores, thus increasing the sensitivity of the measurements.

$$\text{Stokes shift} = \left(\frac{1}{\lambda_{ex}} - \frac{1}{\lambda_{em}}\right)x10^7 \qquad ...(24.13)$$

Chemiluminescence immunoassay (CIA) is analogous to conventional radioimmunoassay (RIA) in that labeled antigen competes with unknown labeled antigen for biding to an antibody. The free and bound fractions of the antigen are isolated and the enzymic activity in one or the other is measured and related to the concentration of the unlabeled antigen.

Competitive-binding immuno-chemiluminometric assay (CB-ICMA) is analogous to immunoradiometric assay (IMRA) in that the antigen reacts with excess of labeled antibody. After removal of the excess antibody, the enzyme activity in the bound fraction is measured and related to the concentration of antigen.

In homogeneous immunoassays (enzyme-labeled or fluor-labeled immunoassays), separation of the free and bound labeled material is not required, unlike the other types of assay. Monitoring is carried out, based on "energy transfer" effects, by measuring the change of chemiluminescence intensity upon immune complex formation. Depending on whether the enzyme is activated or inhibited, the assay can be known as enzyme-multiplied immunoassay (EMIA) technique or enzyme-inhibition immunoassay.

In the enzyme-labeled immunoassay, the enzyme-labeled antigen is allowed to compete with the analyte for the limited binding sites on the antibody molecules. Enzyme activity is suppressed by binding of the labeled antigen ($Ag^*$). The presence of analyte causes displacement of the enzyme-labeled antigen ($Ag^*E$) and enzyme activity and substrate turnover. The intensity of color formed is directly proportional to the concentration of the analyte. Another method of assay is by fluorescence polarization, using plane-polarized light. Loss of polarization is a measure of the analyte concentration.

In fluor-labeled immunoassay, the chemiluminescence labeled antigen is allowed to react with an antibody that has been labeled with an "energy acceptor." Thus, upon immune complex formation and triggering of the chemiluminescence reaction, the emission is quenched due to the close proximity of the chemiluminescent label to the quencher. The intensity of light emission in a given assay tube is proportional to the amount of competing analyte antigen introduced into the system. Monitoring of changes in wavelength of emitted light can also be used for assaying. Energy transfer to a fluorescent acceptor molecule, which emits at a longer wavelength than the chemiluminescent molecule itself (non-radiative, intermolecular energy transfer), can achieve a change in wavelength of light emission from chemiluminescent system.

Most common detectors for chemiluminescence are photomultiplier tubes (PMTs) and photodiodes. A more sophisticated approach is single-photon-counting imaging. Two-dimensional array diode arrangement (charge-coupled device) is used to measure discrete photonic events within a plane of the photocathode, enabling a positional element to be introduced into the measurement of intensity of sample emission.

To-date, major applications of chemiluminescence assays have been in the field of clinical medicine and diagnostics. But, luminescence assays have potential in non-clinical fields like environmental monitoring and control studies, testing of contaminants in water, food and beverages, detection of toxins, infectious vectors, pathogens and parasites (parasitology). With the availability of camera luminometers to measure chemiluminescence, analysis can be carried out without sophisticated instrumentation, in diagnostic centers, laboratories and field stations.

## 24.5 INFRARED (IR) SPECTROSCOPY

The vibrational and rotational frequencies (eq. 24.1) are in the visible, infrared (IR) and far infrared regions of the electromagnetic spectrum. Infrared spectroscopy is divided in three regions for operational convenience- near infrared (0.7 → 1.5 μm; 4,000 → 12,500 $cm^{-1}$), mid infrared (1.5 → 10 μm; 4,00 → 4,000 $cm^{-1}$), and far infrared (10 → 100 μm; 40 → 400 $cm^{-1}$).

IR spectroscopy is an absorption phenomenon. The molecule should exhibit a change in the electric dipole moment for it to absorb infrared radiation.

$$\frac{\partial \mu}{\partial x} \neq 0 \text{ for IR-active} \qquad \text{...(24.14)}$$

Therefore, IR spectroscopy is used to study polar interactions, proton conduction, hydration, and denaturation, metal coordination, etc., in the 2.5-25 μm (4000 → 400 $cm^{-1}$) range. IR spectroscopy is very useful for "screening" samples prior to performing other analyses and identifying source of the pollutants. Stretching frequencies in the "finger print" region (1500 → 650 $cm^{-1}$) provide useful markers for chemical groups (Table 24.2).

IR spectrometers are employed in the analysis of trace pollutants in air ($CO_2$, CO, $SO_2$, and $NO_x$). Hydrocarbons are detected in the 1250 → 700 $cm^{-1}$ (2.5-14.5 μm), O—H, N—H and C—H stretching at 3600 → 2800 $cm^{-1}$, double bonds and N—H deformation at 1850 → 1400 $cm^{-1}$, $CH_3Cl$ at 733 $cm^{-1}$, $CH_3OH$ at 1020 $cm^{-1}$, esters at 1740 and 3100 $cm^{-1}$, C—C at 1500 → 600 $cm^{-1}$, C—O at 1240 → 1100 $cm^{-1}$, C=C, and C=O at 1700 $cm^{-1}$, C≡C, and C≡N at 2200 $cm^{-1}$, $SO_2$ at 1250 → 1000 $cm^{-1}$. The 2950 → 3700 $cm^{-1}$ region is not used, to avoid absorption by water or $CO_2$.

Different types of radiation sources, optical systems and detectors are needed to scan different infrared regions. Radiation sources must be continuous and they must glow at high temperatures. For the near-IR region, a tungsten filament would serve as a source. For the mid-IR region the sources are generally inert solids (Nernst glower = mixture of oxides of rare earth refractory materials; globar = silicon carbide (SiC); carbon rod) heated to ~2000 K and lasers such as $CO_2$ laser (at $\lambda$ = 10 mm) and tunable semiconductor diode lasers. Radiation from a mercury-vapor lamp is employed in the far-IR region. Tunable diode lasers are useful for monitoring atmospheric pollutants (NO, CO, $O_3$, $SO_2$) because their emission wavelengths can be tailored to coincide with IR absorption lines of these gases and their output can be sufficiently collimated for transmission over distances of several kilometers for field applications. Diode lasers are used in low-pressure sampling, *in situ* source monitoring, long-path atmospheric monitoring, and passive IR heterodyne detection (see Chapter 26). Optical systems include either quartz prisms or diffraction gratings.

Detectors are photodiodes, thermocouples (or thermopiles), bolometers, Golay cells or pyroelectric (barium titanate) detectors. Photodiodes (0.3-4.0 mm) are intrinsic photon detectors (electric conduction by radiation). Bolometer is an electrically conducting material with a high temperature coefficient of resistance. The change of resistance due to radiation is measured electrically in a Wheatstone bridge circuit. Golay cells are pneumatic detectors that depend on the expansion of a heated gas. Energy absorbed by a small volume of enclosed gas expands and the resulting pressure change distorts a diaphragm carrying a mirror. A beam of light deflected by the mirror on to a photocell is monitored. A small angular deflection of the mirror is sufficient to change the light intensity falling on the photocell from zero to maximum. Multiple-reflection (echelon-type) IR spectrometers (Fig. 24.19) increase the effective beam path length and thereby sensitivity.

### 24.5.1 Applications

Infrared spectroscopy has several advantages, which make it suitable for monitoring and analyzing environmental pollutants. IR spectroscopy is used in almost all types of chromatography to identify the separated components because:

1. The systems can be investigated without changing its composition.
2. The IR spectrum is a specific property of the given substance—geometrical isomers and molecules containing different atomic isotopes have each a different spectrum.
3. The method is sensitive and selective for the analysis of complex mixtures.
4. Continuous analysis can be performed, which lends itself to kinetic investigation.
5. Results are quick.
6. A very small amount of substance is required.

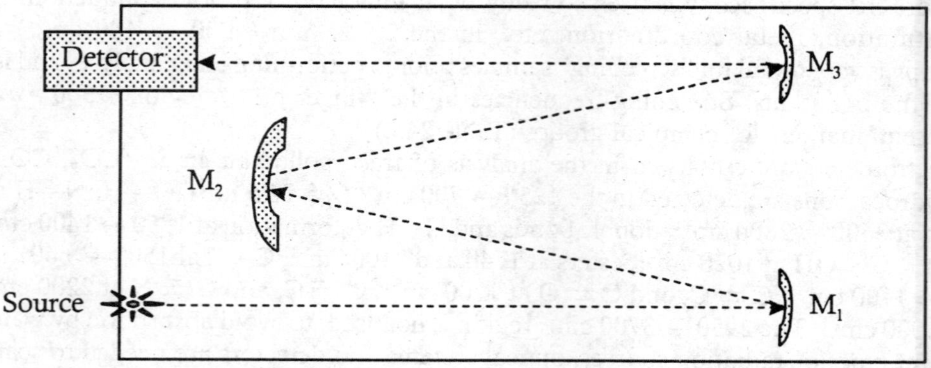


**Fig. 24.19:** Schematic of a Multiple-reflection (Echelon-type) IR-Spectrometer

### 24.5.1.1 *Test for Purity*

The IR spectrum of the sample is compared with that of the pure (standard) substance, and the bands arising from the impurities are identified. If the sample contains C≡O, C≡C, C≡N or other groups and they are absent in the matrix component, then the presence of the groups can be identified from "fingerprint" comparison. When a double-beam instrument is used, recording the difference spectrum of the sample containing the impurity and the pure sample can easily identify the bands due to the impurity. The pure component need not be used as a reference; instead a previously prepared mixture containing a known amount of the component can be used. In this case, the difference spectrum also indicates the quantity of purity.

### 24.5.1.2 *Analysis of Gases*

Analysis of gases like CO, $CO_2$, $SO_2$, $SO_3$, combustible gases (e.g. $CH_4$) and other impurities in air can be done by IR spectroscopy. Non-dispersive IR (NDIR) spectrometry is also suitable for analysis of CO, $CO_2$, and $CH_4$.

Analysis can be done by computer-controlled IR spectroscopy. At constant path length, the absorbance, *A*, of radiation is proportional to the concentration, *C*, of the pollutant. So, unknown concentrations of pollutants can be calculated if the absorbance is first measured for some standard sample. For a single component system the absorbance is

$$A = KC \quad \ldots(24.15)$$

For a two-component system

$$A_1 = K_{11}C_1 + K_{12}C_2 \text{ (for } \lambda_1) \quad \ldots(24.16)$$
$$A_2 = K_{21}C_1 + K_{22}C_2 \text{ (for } \lambda_2)$$

The procedure for a multi-component system is involves the following steps.

1. Preparation of standard sample.
2. Selection of analytical λ for each component, which is accomplished with a short digital scan routine. The starting λ for each component, number of steps and step size are stored in the computer memory. The λ selection is based on minimizing the interference and maximizing sensitivity.
3. Comparison of all absorbance readings to a reference λ, which is selected in a region of non-absorbance.
4. Constants used for calculating unknown concentrations are determined from data obtained from the standard samples and stored in the computer.

#### *24.5.1.3 Water Analysis*

IR spectroscopy can be employed to determine carbon content of water, water content and moisture in foods, grains and other agricultural products, and soil contaminants. Reflectance spectra in the visible-near infrared (VNIR) region can be correlated to spectrally active solid contaminants in soil (organic matter and clay content) by multivariate calibration procedure using partial least squares (PLS) regression method.

#### *24.5.1.4 Industrial Chemicals*

IR spectroscopy is also used in testing of contaminants in commercial products. Some of the inorganic contaminants are fluoride, chloride, chlorate, nitrite, nitrate, sulfate, oxide, peroxides, and phosphates. Organic contaminants include organic salts, aldehydes, ketones, esters, amines, amino acids, azides, alkaloids, steroids, vitamins, sugars, cellulose and its derivatives, organometallic compounds. In this category can also be included hydrocarbons, petrochemical products, pigments and dyes, fertilizers, organophosphorus-, organochloro-, and organosulfur pesticides, herbicides and many other industrial chemicals found in water and soil samples (by vapor-phase FTIR).

#### *24.5.1.5 Food Additives*

IR analysis is useful for the determination of moisture content in foodstuffs as well as in detection of additives and adulterants in foods and edible oils, beverages. Thermal lens spectrometry (TLS) is applied in food and environmental research. Selective reagents, biosensors and chromatographic separation methods (IC, HPLC) are combined with TLS. The main advantages of these analytical methods include improved sensitivity and selectivity, simplicity, minimized need for sample preparation and handling and rapid analysis.

### 24.5.2 Non-dispersive (Fourier-transform) Spectrometry

Both absorption and emission can be used for gas analysis. But, most gas analyzers are of non-dispersive type (NDIR or NDUV). Fourier-transform spectrometry is commonly applied in the IR region (FTIR). In FTIR spectroscopy the entire frequency range is measured simultaneously. Several components of an atmosphere can, therefore, be measured together. As a whole range of wavelengths are passed through the sample simultaneously, the number of elements, $M$, measured per unit time is increased, and the signal to noise ratio ($S/N$) is improved by $\sqrt{M}$, and the time can be reduced by $1/M$. FTIR spectroscopy is suitable for study of reaction kinetics. It is also used as a GC detector (GC-IR) since the carrier gases He, $H_2$, $N_2$ don not absorb IR radiation.

A widely used device for determination of atmospheric CO and $CO_2$ is non-dispersive infrared (NDIR) spectrometry. In this method the entire emission from the source passes through the absorption cell (no dispersion). The NDIR instruments can either have selective detectors (with positive filtration), or non-selective detectors (with negative filtration). In the positive filtration mode (Fig. 24.20*a*) radiation from the source passes simultaneously through a cell that has the test mixture that absorbs IR, and through the reference cell that does not absorb IR. The difference in the signals received by the two detectors is measured. The measurement is rendered selective by filling the detector with the same gas that is in the cell.

In an analyzer with the negative radiation filters, all the radiation from the source passes through the measuring cell (Fig. 24.20*b*) filled with the test gas mixture and is then split into two beams. One beam passes through the reference cell where IR is not absorbed and the other beam through the test cell filled with gas whose concentration is to be determined. Selectivity is achieved at the cells and non-selective detectors carry out the detection.

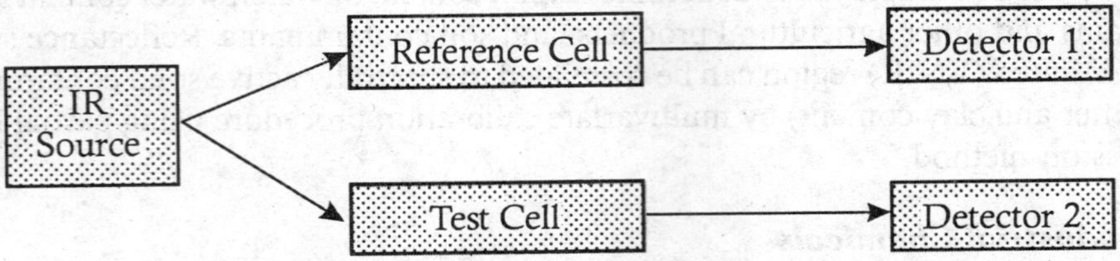


**Fig. 24.20(*a*):** Schematic of Positive Filtration Non-dispersive IR (NDIR) Measurement

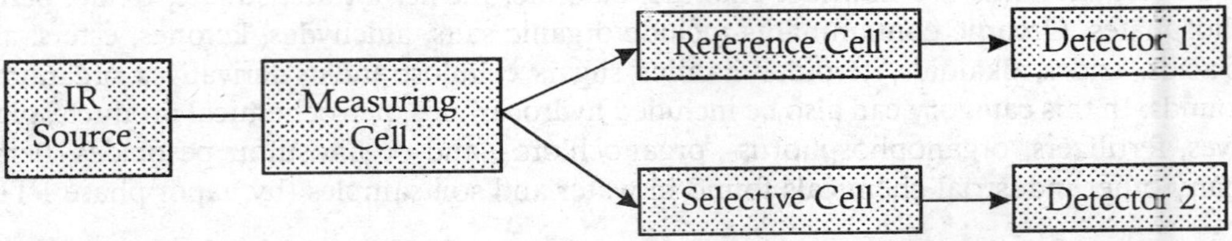


**Fig. 24.20(*b*):** Schematic of Negative Filtration Non-dispersive IR (NDIR) Measurement

CO and $CO_2$ can be monitored by rotating shutter based on the preferential absorption of IR radiation by these gases (Fig. 24.21). Water vapor interferes in this method. Therefore, the gases should be dry.

The instrument consists of an IR source, a beam chopper, sample and reference cells and an IR detector. The reference cell is filled with a gas ($N_2$) that does not absorb IR radiation. The sample cell is continuously flushed with air containing the pollutant (CO or $CO_2$) for detection. The detector has two sealed absorption chambers of equal volume, separated by a flexible membrane diaphragm. Both the chambers of the detector are filled with the gas that is to be detected so that *energy, characteristic of the gas to be measured is selectively absorbed.* The chopper passes IR radiation through the reference cell and the sample cell alternatively, and the two beams finally reach the detector. If there are no IR absorbents (CO or $CO_2$) in the sample cell the IR signals from both the compartments of the detector balance electronically and the diaphragm stays in the null position. If the sample that is flushed with air contains IR absorbers (CO or $CO_2$), they absorb radiation proportional to the amount present. Since more IR radiation passes through the reference cell, the temperature of the detector compartment under the reference cell is higher. The unequal amounts of energy received by the two compartments of the detector cause the membrane diaphragm to move (bend), and the curvature of the diaphragm is a measure of the IR absorbents present in the sample air.

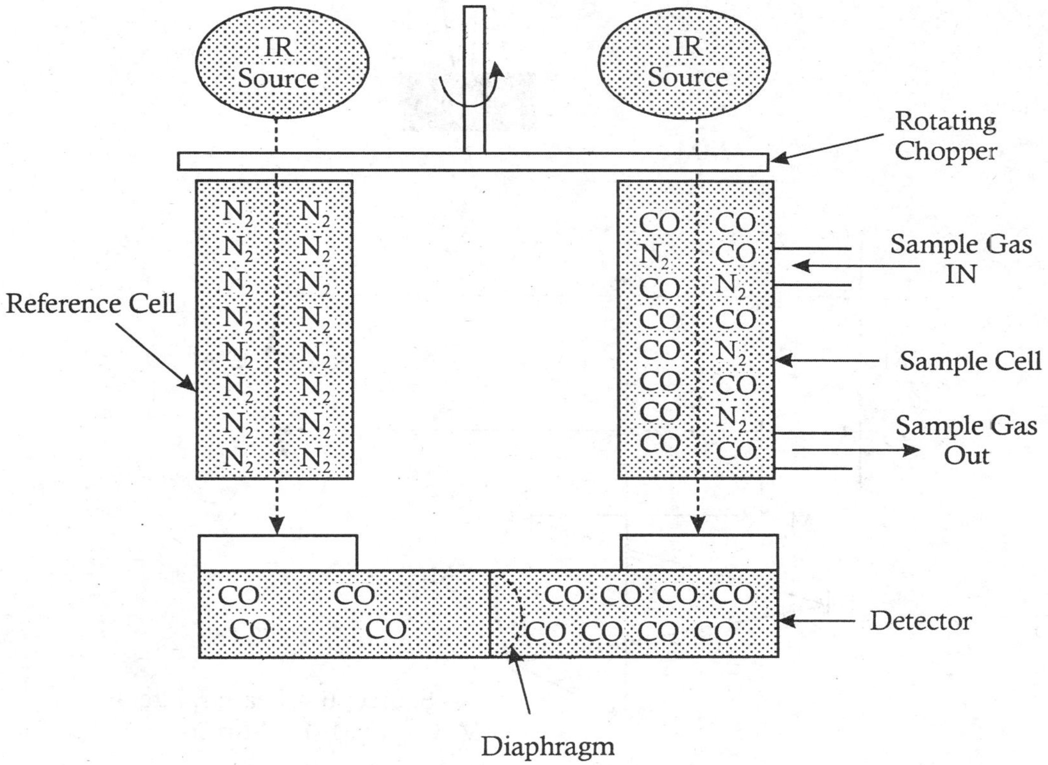


**Fig. 24.21:** Schematic of a Non-dispersive Infrared (NDIR) Analyzer

### 24.5.3 Non-dispersive Interferometry

Mathematical computation of intensities measured on a non-dispersive interferometer (Fig. 24.22) to arrive at an absorption spectrum is extremely useful in far-IR spectroscopy. In the far-IR region, the low output power of the source reduces the signal to nearly the noise level of the detector. With the interferometer, the full output of the source over the range covered is allowed to fall on the detector so that the noise is no longer a matter for concern. A further advantage over grating instruments is the relatively large spectral range handled without the need for changing filters or other optical parts.

The optical elements consist of a light source (S), convergent mirrors, beam-splitter (B), fixed plane mirror ($M_3$) and movable mirror (MM), sample and a detector unit (D). Light from the source (high pressure Hg-vapor lamp) is collimated to fall on the beam-splitter (B). The beam is reflected to the fixed plane mirror ($M_3$) or transmitted to the movable mirror and eventually to the condenser mirror ($CM_2$). The reflected beam is focused on to the sample and the signal is recorded by the detector (D). The intensity is a function of the path difference ($\Delta$) between the light beams. The intensity of the source at any frequency, after absorption by the sample, can be expressed as a function of the intensity measured by the detector and ($\Delta$), the extra distance moved by the mirror.

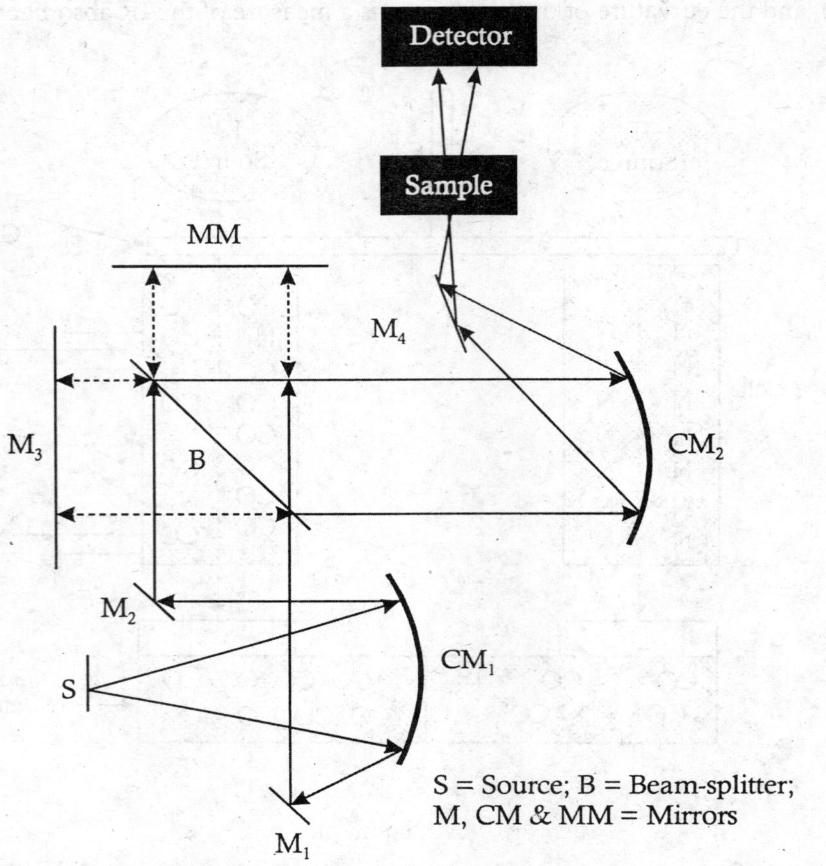


**Fig. 24.22:** Schematic of Interferometric Spectrometer

## 24.6 RAMAN SPECTROSCOPY

Scattering of radiation, in the UV-visible region, by molecules gives rise to elastic scattering (Rayleigh scattering), wherein there is no change in the energy (frequency/wavelength) of the radiation, and inelastic scattering (Raman scattering), wherein there is a change in the energy (frequency/wavelength). Raman effect is an inelastic molecular scattering of light, with lower frequency (Stokes-Raman) and higher frequency (Anti-Stokes-Raman) lines.

$$\nu_0 = \nu_{scatter}\ (\lambda_0 = \lambda_{scatter})\quad \text{(Rayleigh scattering)}$$

$$\nu_0 < \nu_{scatter}\ (\lambda_0 > \lambda_{scatter})\quad \text{(Stokes-Raman spectrum)} \qquad ...(24.17)$$

$$\nu_0 < \nu_{scatter}\ (\lambda_0 > \lambda_{scatter})\quad \text{(Anti-Stokes-Raman spectrum)}$$

Though Raman spectroscopic techniques are mostly confined to research areas, they can find place in monitoring environmental pollutants in air and water. Raman spectroscopy is vibrational spectroscopy, and is complementary to IR spectroscopy in its information content. However, it has several advantages over IR spectroscopy.

1. Raman effect is a scattering phenomenon in contrast to IR. Therefore, the sample need not be transparent.

2. Raman spectrum can be obtained for compounds in any state (gaseous, liquid, solid, gel etc.).
3. Water is not an impediment (as is the case in IR) and studies can be conducted in aqueous (natural) media, without much preparation.
4. It is applicable to polar as well as non-polar compounds (unlike IR). It is ideal for "fingerprint" type analysis of organic compounds (e.g. hydrocarbons).
5. It has a wider spectral range (10-6000 $cm^{-1}$).
6. Resonance Raman studies (there is no IR analogy) enable analysis of molecular features of chromophores.
7. Antistokes-Raman studies can be used to monitor intermediate molecular species.
8. Radiation sources and instrumental setup need not be expensive.

Raman effect is a scattering phenomenon, and to be Raman-active, the molecule must exhibit change in polarizability.

$$\frac{\partial \alpha}{\partial x} \neq 0 \text{ (Raman-active molecule)} \qquad ...(24.18)$$

Both IR and Raman spectroscopy provide vibration frequencies (stretching, bending and skeletal) of chemical groups, like C—C, C=C, C=N, C≡C, C≡N, C—S, S—S (Table 24.2). While C≡C, C=C, C—C and S – H vibrations are weak or absent in IR spectra they are, on the other hand, strong in Raman spectra. A common light bulb can be a radiation source for Raman studies and specimen holders can be of glass.

Determination of parameters of individual particles (as against bulk) can give precise information on chemical compounds present, their sources, and history of the environment. The spectral content of Raman lines provides information on the chemical identity and molecular state of the samples under study. With the availability of high-intensity monochromatic laser sources (He-Ne = 632.8 nm, Ar = 488 nm and 514.5 nm, Kr = 568.0 nm), sensitivity and selectivity of analysis and the potential of Raman methods have increased tremendously.

One of the chief drawbacks of Raman spectroscopy has been the weak intensities of Raman lines and the interference from fluorescence background. In fact, fluorescing impurities have to be removed from every sample. These problems have been circumvented by the availability of suitable diode lasers in the near infrared (NIR) region to excite Raman lines (e.g. 1064 mm) that are not affected by fluorescence. In that case Raman spectra can be recorded with minimal sample preparation.

Another technique of great promise is the Coherent Anti-Stokes Raman Spectroscopy (CARS). In the CARS method (wave-mixing phenomenon), two strong collinear laser beams of different frequencies ($\nu_1 > \nu_2$) irradiate the sample. If the frequency difference, $\nu_1 - \nu_2$, is equal to the frequency of a Raman-active transition line, $\nu_R$, then the efficiency of the wave mixing is enhanced and Stokes ($\nu_R = 2\nu_2 - \nu_1$) and anti-Stokes lines ($\nu_R = 2\nu_1 - \nu_2$) are produced by wave mixing due to non-linear polarization of the medium. CARS produces intensities ~$10^4$-$10^5$ times great. This enhanced signal level can greatly reduce the time necessary to record Raman spectra. Owing to the coherence of the general signals, the divergence of the output beam is small, and good spatial discrimination against the background signal is achievable.

Though Raman spectroscopy is not used extensively in environmental pollution analysis, recent advances in instrumentation do hold a great potential for their introduction in the future.

## 24.7 ELECTRON SPIN RESONANCE (ESR) SPECTROSCOPY

Electron spin resonance (ESR) or electron paramagnetic resonance (EPR) spectroscopy is a branch of absorption spectroscopy that deals with the transitions between magnetic energy levels of electrons of unpaired spins. The technique is used to study (monitor) paramagnetic species, ions of transition elements, Cu- and non-heme metalloproteins and free radicals. For paramagnetic atoms, the value of '$\gamma$' may differ from the free-electron value of 2.00232 (varying from $1 \rightarrow 6$ or more). Different paramagnetic spectra can usually be distinguished quite easily by their $\gamma$-values. It is also used to study the effect of UV and X-rays on biological materials, bacterial spores, and radical intermediates as a function of time (decay of ESR signals in time), and to detect residual free radicals in irradiated food products. Use of ESR method in environmental monitoring studies is limited as the sample should be in solid state and very low temperature setup is needed for detecting ESR signals.

## 24.8 ATOMIC SPECTROMETRY

Light absorption and emission in atoms is a process that depends on excitation of an atom from one state to another. Since the energy levels of the different states (ground and excited states) are fixed by the electronic configuration of the atom, the energy difference between them is discreet (quantized). This implies, atom will absorb or emit light of discreet wavelengths, corresponding to the energy difference of these states. The wavelengths of light absorbed or emitted are known as 'characteristic wavelengths' because they are representative of a specific atom going through these changes. There are relatively few, narrow characteristic wavelengths of significance. They are called lines because, ordinarily, they are less than 1 nm in width. Realizing the complexity of arc and spark emission spectra of elements and the associated problems of using them for analytical purposes several investigators explored their relatively simple atomic spectra for this purpose, and met with great success.

The important methods based on spectroscopy of atomic species are: Atomic Absorption Spectrometry (AAS), Atomic Emission Spectrometry (AES) and Atomic Fluorescence Spectrometry (AFS).

### 24.8.1 Atomic Absorption Spectrometry (AAS)

This method for the determination of trace metals was developed in 1950s by Alan Walsh and is now universally popular and widely practiced. Atoms in the ground are excited by absorption light, and they return to the ground state with emission of continuous as well as line spectra. The AAS method relies on the absorption line spectra of dissociated free metallic atoms. The line spectrum (transition lines) is characteristic for each element present in the sample and the intensity of the transition lines is correlated to the metal atomized by the thermal device (flame or furnace). In the AAS what is measured is the percentage fraction of radiation absorbed ($A_T$), which is proportional to the analyte metal.

$$A_T = \frac{\text{Intensity of absorbed radiation}}{\text{Intensity of incident radiation}} = (\Delta I / I_0) \times 100 \qquad \ldots(24.19)$$

Essential components of an AAS setup consist of—(1) a light source for generating characteristic radiation, (2) an atomizing unit for generation of atomic vapor of the analyte in a hot flame or a furnace, (3) optical system (monochromator and dispersion unit), detector and recorder system for detection and measurement of absorption signals (Fig. 24.23).

#### *24.8.1.1 Radiation Sources*

Lines of both emission and absorption spectra for atoms involve transitions to or from ground state and occur at identical energies. However, the emission lines are usually more in number than the absorption lines. The lines common to both are called 'resonance lines.' For measuring absorption one of the prominent resonance lines is chosen from the radiation source. A hollow cathode lamp is one device, which has served as a reliable radiation source for AAS.

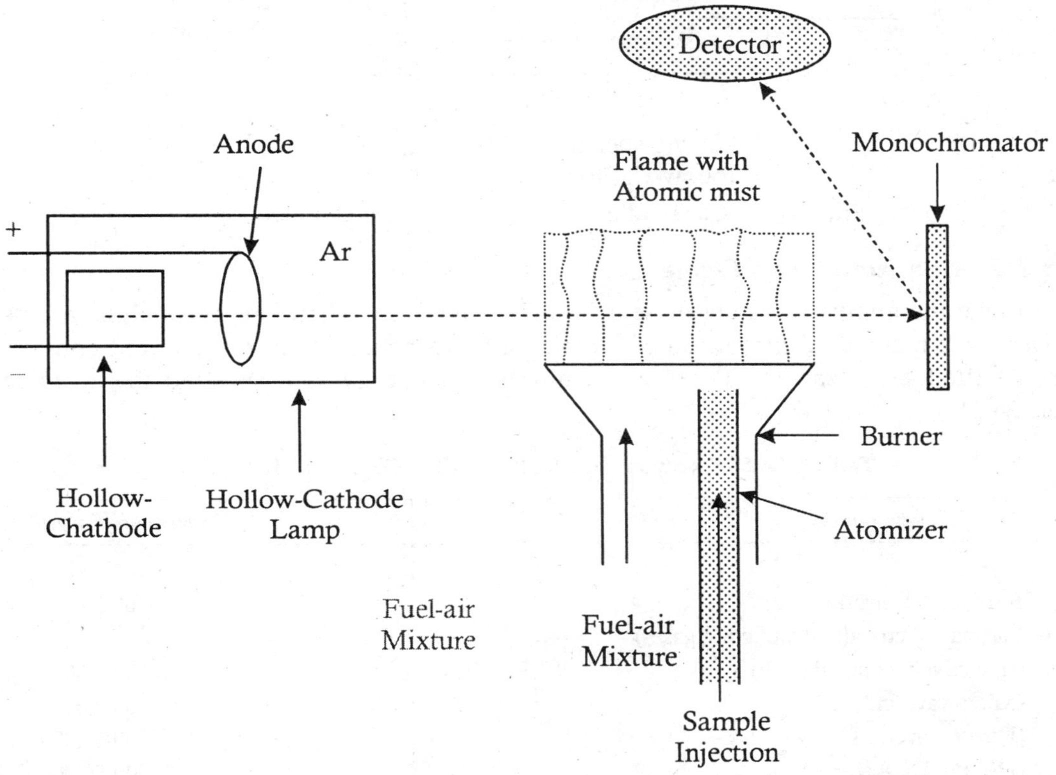


**Fig. 24.23:** Schematic of Flame Atomic Absorption Spectrometry (Flame-AAS)

A hollow-cathode lamp is essentially a discharge tube filled with an inert gas such as Ar, He or Ne at low pressure (1 to 5 torr). The anode is of tungsten and as each metal has its own characteristic absorption wavelength, the cathode in the lamp is of the same metal as the analyte metal (Fig. 24.24). Passing of current spurts metal atoms from the cathode and collisions of these atoms with inert gas lift the metal atoms to the excited state and their transitions to their ground states (sputtering) cause emission of characteristic line spectra.

Vapor-discharge lamps have also been used as sources of line spectra for volatile elements such as, Cd, Cs, Hg, K, Na, Ru, Tl and Zn. Electrodeless discharge lamps (EDL) (arcs) are high-intensity line sources, used in place of hollow-cathode lamps for many elements (Ag, Al, As, Au, B, Be, Bi, Ca, Co, Cr, Cu, Fe, Ga, Ge, Hg, K, Mg, Mn, Mo, Ni, Pb, Pd, Pt, Rb, Sb, Se, Sn, Tl, W, Zn and Zr). In discharge lamps, a small quantity of metal salt whose spectrum is desired is introduced into a quartz tube containing inert gas and sealed. The discharge lamp is placed in a microwave oven and the plasma excites the metal atoms, leading to the line spectrum of the metal.

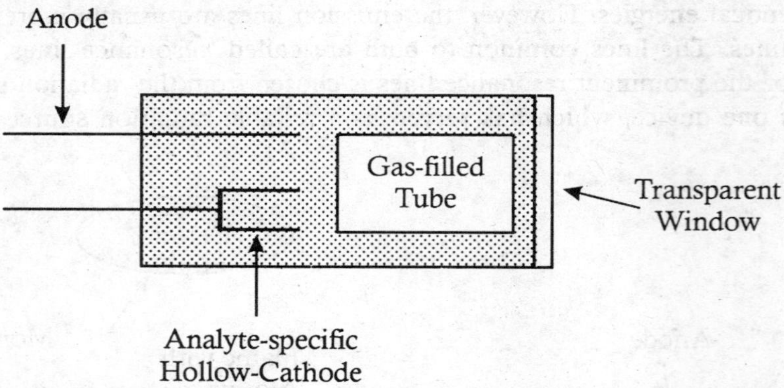


**Fig. 24.24:** Schematic Diagram of a Hollow-Cathode Lamp

### *24.8.1.2 Atomization in a Flame*

In a sample the analyte is rarely present in an atomic form. Therefore, its analysis necessarily involves some method of conversion of sample (or at least a part of it) to its atomic constituents. This is called atomization. There are a number of energy sources used for atomization (Table 24.5).

**TABLE 24.5:** ENERGY SOURCES USED FOR ATOMIZATION

| *Energy Source* | *Temperature Range* (° C) |
|---|---|
| Flame | 1,700-3,100 |
| Furnace (electrothermal) | 1,200-3,000 |
| Plasma (electrically conducting gases) | |
| (Inductively coupled, ICP) | 6,000-8,000 |
| (Microwave induced, MIP) | 5,000-7,000 |
| (Direct current, DCP) | 6,000-10,000 |
| (DC arc, DCA) | 4,000-5,000 |

Atomization by flame method is usually achieved by preparing the sample as a dilute solution and aspirating this with the oxidant or fuel-oxidant mixture before it passes into the combustion area. The heat of the flame and the action of the reducing gases convert the constituents of the sample to atomic state. AAS measurement is based on the concentration of the ground state atoms in the flame as only they are measured by the spectrophotometric principle—absorption of light from a beam passing through the flame. Conversion of sample to its neutral atoms is never complete. The degree of atomization is dependent upon various factors like energy levels of electrons in the atom, flame temperature, retention time of the components in the flame, composition of the flame gases (fuel rich or fuel deficient) and various types of matrix effects. At very low flame temperature (<1,500 K) atomization may not take place to a significant extent. On the other hand, at high temperatures (>4,000 K) some of the atoms may be excited to higher energy states, causing emission of characteristic radiation as they relax to ground or lower energy states. Therefore, for AAS measurement the flame composition and temperature are chosen to keep the population of free atoms as high as possible in the flame.

#### 24.8.1.3 Flame Techniques

The common fuel gases used for the flame and the temperature are given in Table 24.6.

**TABLE 24.6:** FLAME GASES USED IN AAS

| *Flame* | *Temperature (° C)* | *Flame* | *Temperature (° C)* |
|---|---|---|---|
| Argon-hydrogen | 1,580 | Air-propane | 1,725 |
| Air-hydrogen | 2,200 | Air-acetylene | 2,600 |
| Oxygen-hydrogen | 2,800 | Oxygen-propane | 2,900 |
| $N_2O$-acetylene | 3,000 | Oxygen-acetylene | 3,300 |

In combustion flames the temperature is 2,500 to 3,500 K. The burner system consists of a gas control unit, nebulizer, mixing chamber and the burner head. The nebulizer aspirates the sample solution through a capillary, forming a fine aerosol mist. It is mixed with fuel and oxidant gas in a mixing chamber before it is led into the burner head, where the sample drops are desolvated, the solid vaporized and the analyte metals atomized. Nebulizers are either ultrasonic type or of venturi type. The burners used for the flame are elongated to give a light path of about 10 cm to enhance the absorption. Modulation of the light beam upstream of flame, by a rotating chopper, allows detector to reject emission generated within the flame.

Fuel-rich flames provide a reducing atmosphere that helps to break up more refractory materials like metal oxides into atoms. Only about 35 elements can be detected by AAS with low temperature flames (<2500° C). It is not possible to detect Al, B, Be, Ge, Mo, Si, Tl, Ti, V, W and Zr. Premixed hydrogen-air flame is suitable for AAS determination of elements which are easily atomized. Many metals can be determined by direct air-acetylene flame method (Ag, Au, Bi, Ca, Cd, Ce, Co, Cr, Cu, Fe, Ir, K, Li, Mg, Mn, Na, Ni, Pa, Pb, Pt, Rh, Ru, Sb, Sn, Sr, Th and Zn), while higher temperatures ($N_2O$-acetylene flame) are needed for atomization of refractory oxide-forming elements (Al, Ba, Fe, Mo, Os, Rh, Si, Sn, V and W) (Table 24.6). $N_2O$-acetylene provides a strong reducing atmosphere and permits the detection of low concentration of non-dissociable compounds that remain undetected in air-acetylene flame—Al, B, Be, Er, Eu, Gd, Hf, Ho, Ir, La, Lu, Nb, Nd, Pr, Re, Sc, Si, Sn, Ta, Tb, Tm, U, V, W, Y, Yb, and Zr. Argon-hydrogen and air-hydrogen flames are preferred for the analysis of As, Cd, Pb, Se and Zn. As and Se are determined after their conversion to volatile hydrides through reduction by sodium borohydride reagent in acidic medium. These hydrides are flushed continuously by an inert gas (argon/nitrogen) (refer to Chapter 15 for details).

Solvent extraction prior to AAS determination by air-acetylene flame method is suitable for the determination of low concentrations of Ag, Cd, Co, Cr, Cu, Fe, Mn, Ni, Pb and Zn. The method consists of sample concentration by chelation and solvent extraction. Ammonium pyrrolidine dithiocarbamate (APDC), sodium diethyldithiocarbamate (Na-DDTC) and dithizone (DZ) are often used as chelating agents. These reagents form insoluble chelates that are extracted by methylisobutyl ketone (MIBK) or n-butylacetate (Table 24.7) followed by aspiration into an air-acetylene flame. The organic phase is either directly injected or back-extracted to a water phase and then injected into the flame.

Water phase → organic phase → aspiration

Water phase → organic phase → water phase → aspiration

**TABLE 24.7:** COMPARISON OF APDC, DDTC AND DZ AS EXTRACTING AGENTS

| *APDC* | *DDTC* | *DZ* |
|---|---|---|
| pH > 4 | pH > 5 | pH 5-6 and > 9 |
| Many elements can be extracted | Less number of elements than with APDC | Less number of elements than with DDTC |
| High sensitivity | High sensitivity | Back-extraction to acidic solution is needed |
| Leads to precipitation | Leads to precipitation | Perfect distribution |
| Dissociates at high temperature | Stable | Stable |
| Low purity | High purity | High purity |

Solvent extraction followed by AAS with $N_2O$-acetylene flame is suitable for determination of low concentrations of Al and Be. The method consists of chelation with 8-hydroxyquinoline, extraction with methylisobutyl ketone (MIBK) and aspiration in an $N_2O$-acetylene flame.

Cold vapor method is a flameless AAS method, particularly suitable for the determination of mercury. It is based on the existence of mercury vapor in the atomic state at ambient temperature. Mercury in air or water is trapped in (KI + sulfonic acid) solution, and reduced to atomic vapor state at room temperature by adding stannous chloride. Flushing mercury vapor into an absorption cell with air follows this step. Quantification is done by measuring the absorbance at $\lambda = 253.7$ nm (Chapter 19).

Amalgam-AAS method is specific for the determination of mercury. Mercury-containing aerosol, collected on a nitrocellulose filter paper is vaporized in a combustion boat. The emerging vapor is successively passed through a heated quartz tube that contains CuO and Ag to remove interfering substances, an anhydrous $CaSO_4$ column to remove water, and finally through a quartz tube containing Ag chips to trap Hg as silver amalgam. The Ag-Hg amalgam is vaporized and flushed into a flameless AAS (264 nm) (Fig. 24.25).

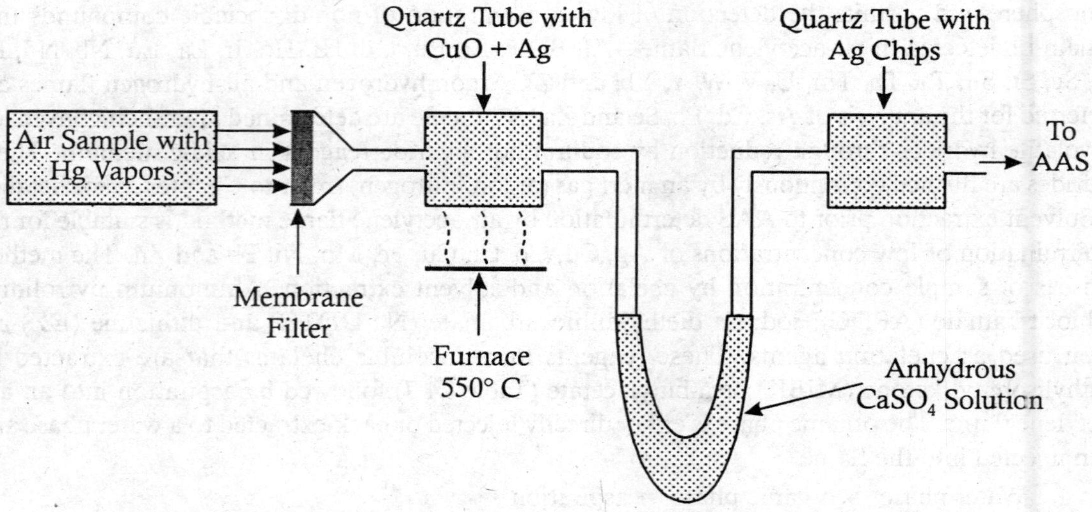


**Fig. 24.25:** Determination of Hg by Ag-Hg Amalgam-AAS Method

### *24.8.1.4 Limitations of Flame Atomization*

The temperature of the flames is not high enough to excite significant fractions of refractory metals or elements, which readily form refractory oxides in the flame. Sensitivity is poor for such metals like B, Nb, Ta, Ti and W. To overcome this type of problems energy sources other than the flame have been put to use.

### *24.8.1.5 Non-flame Atomization (ET-AAS)*

AAS lacks sensitivity in the flame mode. One frequently used alternative is electrothermal method (ET-AAS)—electrothermally heated quartz furnace (QF-AAS) and graphite furnace (GF-AAS). The commonest form of a flameless atomizer is the carbon rod system (Fig. 24.26). Graphite furnace at high temperature (3,000° C) results in increased atomization. In this method, the sample is placed in a cup of conductive carbon. Three heating cycles usually follow each with variable current and time settings. Initially a small current is passed to heat and evaporate the solvent (drying stage). Then current is increased in char cycle (ashing stage) to burn off organic and other matrix compounds. In the final stage (atomization cycle), the temperature is raised to 2,000-3,000 °C to atomize the analyte. The atomized analyte absorbs characteristic incident radiation passing through the graphite tube.

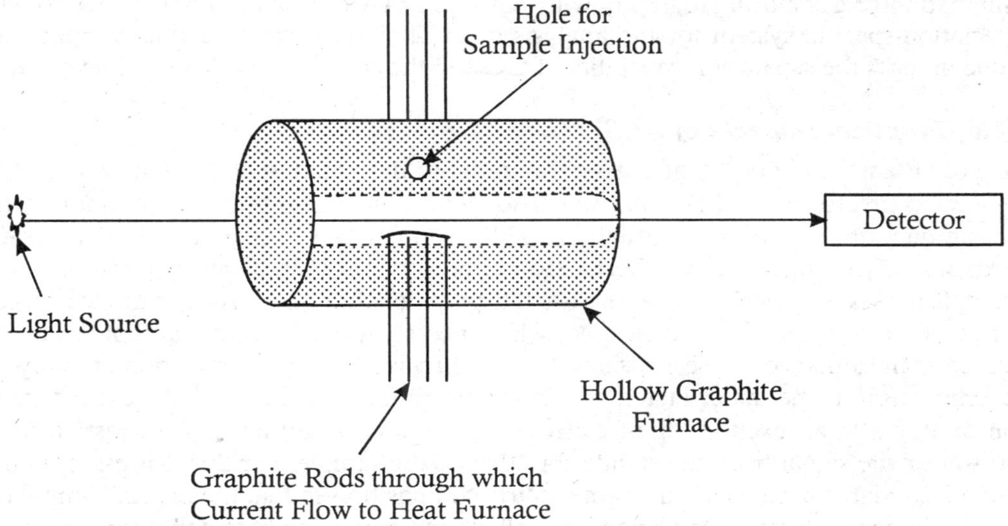


**Fig. 24.26:** Schematic of Graphite Furnace Setup for Electrothermal-AAS (ET-AAS)

The main advantage of the furnace mode of atomization lies in higher sensitivity (about 100 times) and much smaller sample required than for flame-AAS. Hence, in terms of absolute scale it is more sensitive (Table 24.8). While the flame measurement requires about $10^{-9}$ g of material, the furnace technique may require as little as $10^{-14}$ g. The detection limit is pushed down to sub-ppb level. Other advantages of the method are possibility of using solid samples, avoiding chemistry problems associated with flame and possibility of reducing oxide formation by the carbon of the furnace acting as a reductant. ET-AAS is suitable for determination of microquantities of Ag, Al, As, Ba, Be, Cd, Co, Cr, Cu, Fe, Mn, Mo, Ni, Pb, Sb, Se and Sn. At the same time the procedure is not without some disadvantages. They are—(1) it is not as convenient as the flame; (2) instrumentation is more costly and (3) precision and accuracy are poorer compared with the flame. The other methods of atomization are not so commonly used.

**TABLE 24.8:** COMPARISON OF DETECTION LIMITS FOR VARIOUS AAS METHODS

| *Element* | *Flame* (μg/L) | *Graphite Furnace* (μg/L) |
|---|---|---|
| Antimony (Sb) | 100 | 1 |
| Arsenic (As) | 100 | 3 |
| Beryllium (Be) | 2 | 0.02 |
| Boron (B) | 700 | – |
| Cadmium (Cd) | 1 | 0.003 |
| Chromium (Cr) | 3 | 0.1 |
| Copper (Cu) | 2 | 0.02 |
| Iron (Fe) | 5 | 0.1 |
| Lead (Pb) | 1 | 0.003 |
| Manganese (Mn) | 2 | 0.02 |
| Mercury (Hg) | 500 | 2 |
| Tin (Sn) | 20 | 20 |

Combination of gas chromatography with AAS (GC-AAS) provides both a sensitive-element and speciation-specific system for the analysis of low-boiling organometallic compounds. The technique inherits the separation capability of GC and the specificity of AAS (flame or furnace).

### *24.8.1.6 Practical Aspects of AAS*

AAS is not an absolute method of analysis. In order to determine specific elements by AAS the instrument has first to be calibrated. A comparison with standards is the most common method of performing quantitative analysis. Usually a calibration curve is prepared which is made up of standards of known concentration. Alternately, the method of standard additions is used, especially if the sample matrix is not well known. It is usual to prepare solutions for AAS in dilute hydrochloric acid. However, nitric acid can also be used, where necessary. Sulfuric acid and phosphoric acids are avoided as the anions of these acids interfere in volatilization and atomization of many metals. Flame temperature is the major factor, which sets the number of atoms in the excited state. The fraction of atoms in an excited state (relative to the ground state) has to be constant to enable comparison of the signal from the standard to that of the sample. For this it is essential that the sample and standards have nearly the same matrix composition so that the matrix components do not alter the flame parameters. Only under controlled conditions of operation the atom concentration in the lamp beam is proportional to the total (formula) concentration of the particular element in the sample.

Apart from flame conditions the factors affecting atomization are chemical and physical parameters. For example, presence of anions, like phosphate, inhibits free atom formation of some metals like calcium. It is possible to overcome such effects by using flame of higher temperature or adding a releasing agent like lanthanum, which binds the phosphate and sets calcium free to atomize. In some cases addition of a chelating agent like EDTA also helps in a similar way.

Among matrix effects one has also to consider the presence of solvents other than water in the sample or standards. Presence of organic solvents has some beneficial effects too. This aspect becomes important when trace metals in samples are separated from matrix components by solvent extraction, for concentration or for avoiding interference. In such a case, two possibilities exist: (1) aspirating the organic phase into the flame and (2) stripping the analytes from the organic phase and aspirating the aqueous solution. It is advantageous to use the organic phase for analysis by AAS. For example, a solution of nickel (10 μg/mL) in different solvents has given absorbances 0.0177 in water, 0.0630 in acetone, 0.0780 in heptane and 0.0362 in carbon tetrachloride. The results also show the inflence of solvent type on sensitivity of determination.

The most widely used background correction technique is use of continuum source deuterium arc) alternately with hollow-cathode lamp. Both the element being determined and the background absorb the light from hollow-cathode lamp. Only the background, not the element of interest, essentially absorbs the light from the continuum source. Application of Zeemann or Smith-Hieftje correction can minimize background interface. Zeemann background correction is based on the principle that magnetic field splits the spectral line into two linearly polarized lines. One component ($\sigma$-component) is absorbed only by the background, whereas the other ($\pi$-component) is absorbed by both the atoms of element of interest and the background. Smith-Hieftje background correction is based on the principle that absorbance measured for a specific element is reduced as the current to the hollow-cathode lamp is increased, while absorption of nonspecific absorbing substances remains the same at all current levels.

#### *24.8.1.7 Flame Resonance Spectrometry (FR-AAS)*

Conventional AAS is reasonably sensitive for trace analysis. However, in some cases, like determination of pollutants the sensitivity might not be sufficient. In such cases either pre-concentration has to be introduced or some way must be found to improve sensitivity. Extraction of metals with solvents after chelation offers both possibilities. Separation of required elements and concentration is achieved. Moreover, the presence of solvent in the aspirated solution some times improves the sensitivity. Another limitation is only one element can be determined at a time. This limitation is circumvented by Flame resonance spectrometry (FRS), which is a multi-element AAS. The setup consists of a light source, a furnace, a flame and a detector system (Fig. 24.27).

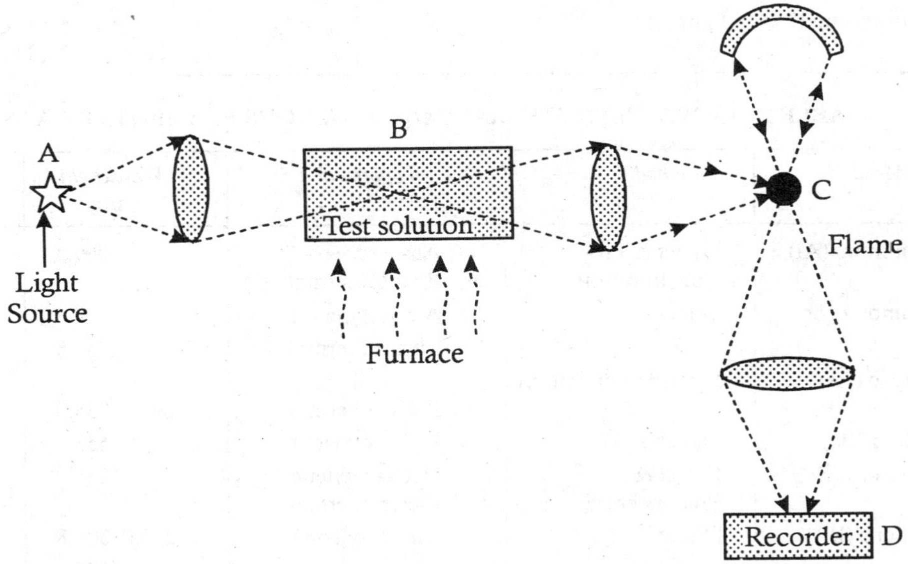


**Fig. 24.27:** Schematic of Flame Resonance Spectrometer-AAS Setup

Light from the source (A) passes through the furnace (B), containing the test sample, and focussed on to the flame (C). The furnace atomizes the sample. A pure solution of the required element is sprayed into the flame (C). Resonance occurs if the wavelength of the radiation of the light source (A) matches with the absorption wavelength of that particular element (resonance condition). Under the resonance condition, there is greater absorption by the particular element and the decrease in intensity at the flame (C) is recorded by a detector (D). Sequential multi-element AAS can be accomplished by using multi-element hollow-cathode, a set of single-element hollow-cathode tubes or a continuous source (xenon arc).

Various AAS methods and their characteristics are summarized in Table 24.9, and absorption maxima for various metals by AAS are given in Table 24.10.

**TABLE 24.9:** SUMMARY OF ATOMIZATION METHODS IN AAS

| *Flame* | *Atmosphere* | *Method* | *Temperature* °C | *Metals Determined* |
|---|---|---|---|---|
| Air-acetylene | Oxidative | Direct | 2,300 | Ag, Au, Bi, Ca, Cd, Cs, Co, Cr, Cu, Fe, Ir, K, Li, Mg, Mo, Mn, Na, Ni, Os, Pa, Pb, Pt, Rh, Ru, Sn, Sr, Th, Zn |
| Air-acetylene | Oxidative | Preconcentration | 2,300 | Ag, Cd, Co, Cr, Cu, Fe, Mn, Mo, Ni, Pb, Zn |
| $N_2O$-acetylene | Reducing | Direct | 2,900 | Al, Ba, Be, Ca, Mo, Os, Rh, Si, Th, Ti, V |
| $N_2O$-acetylene | Reducing | Preconcentration | 2,900 | Al, Be |
| Ar-$H_2$ or $N_2$-$H_2$ | – | Hydride | 1,580 | As, Se |
| Flameless | – | Cold vapor | – | Hg |
| Flameless | – | Amalgam | – | Hg |
| Electrothermal | Flameless | – | >2,000 | Ag, Al, As, Be, Cd, Co, Cr, Cu, Fe, Mn, Mo, Ni, Pb, Sb, Se, Sn |

**TABLE 24.10:** WAVELENGTH MAXIMA OF VARIOUS ELEMENTS BY AAS

| *Metal* | *Method(s)* | *Vaporization* | *Wavelength* (nm) | *Detection* (ppm) |
|---|---|---|---|---|
| Aluminum (Al) | Direct & Pre-concentration | Air-acetylene & Electrothermal | 309.3 | 0.05 |
| Antimony (Sb) | Direct | Air-acetylene & Electrothermal | 217.5; 259.8 | 0.2 |
| Arsenic (As) | Hydride generation | Argon-$N_2$ & Electrothermal | 193.7; 235.0 | 0.2 |
| Barium (Ba) | Direct | $N_2O$-acetylene | 553.6 | 0.05 |
| Beryllium (Be) | Direct & Preconcentration | $N_2O$-acetylene & Electrothermal | 235.0 | 0.002 |
| Bismuth (Bi) | Direct | Air-acetylene | 223.0; 306.8 | 0.05 |
| Boron (B) | Direct | Air-acetylene | 249.8 | 0.01 |

| *Metal* | *Method(s)* | *Vaporization* | *Wavelength* (nm) | *Detection* (ppm) |
|---|---|---|---|---|
| Cadmium (Cd) | Direct & Pre-concentration | Air-acetylene & Electrothermal | 228.8; 326.0 | 0.005 |
| Calcium (Ca) | Direct | Air-acetylene & $N_2O$-acetylene | 422.7; 393.4 | 0.005 |
| Cesium (Cs) | Direct | Air-acetylene | 852.7 | 0.05 |
| Chromium (Cr) | Direct & Pre-concentration | Air-acetylene & Electrothermal | 358.0; 360.5 | 0.005 |
| Cobalt (Co) | Direct & Pre-concentration | Air-acetylene & Electrothermal | 240.7; 243.4 | 0.005 |
| Copper (Cu) | Direct & Pre-concentration | Air-acetylene & Electrothermal | 218.0; 324.7; 349.2 | 0.005 |
| Gold (Au) | Direct | Air-acetylene | 242.8 | 0.02 |
| Iridium (Ir) | Direct | Air-acetylene | 264.0 | 2. |
| Iron (Fe) | Direct & Preconcentration | Air-acetylene & Electrothermal | 248.3; 260.0; 372.0 | 0.005 |
| Lead (Pb) | Direct & Pre-concentration | Air-acetylene & Electrothermal | 217.0; 283.3; 405.8 | 0.01 |
| Lithium (Li) | Direct | Air-acetylene | 670.8 | 0.005 |
| Magnesium (Mg) | Direct | Air-acetylene | 285.2; 383.8 | 0.005 |
| Manganese (Mn) | Direct & Pre-concentration | Air-acetylene & Electrothermal | 279.5; 403.0 | 0.003 |
| Mercury (Hg) | Ag-Hg amalgam | Flameless | 253.7 | 0.5 |
| Molybdenum (Mo) | Direct & Pre-concentration | Air-acetylene; $N_2O$-acetylene & Electrothermal | 313.2; 379.8 | 0.05 |
| Nickel (Ni) | Direct & Pre-concentration | Air-acetylene & Electrothermal | 262.0; 341.5; 441.5 | 0.005 |
| Osmium (Os) | Direct | Air-acetylene & $N_2O$-acetylene | 291.0 | 1. |
| Palladium (Pd) | Direct | Air-acetylene | 247.6 | 0.02 |
| Platinum (Pt) | Direct | Air-acetylene | 266.0 | 0.1 |
| Potassium (K) | Direct | Air-acetylene | 766.5 | 0.005 |
| Rhodium (Rh) | Direct | Air-acetylene & $N_2O$-acetylene | 344 | |
| Rubidium (Rb) | Direct | Air-acetylene | 780.0 | 0.005 |
| Ruthenium (Ru) | Direct | Air-acetylene | 350.0 | 0.3 |
| Selenium (Se) | Hydride generation | Argon-$N_2$ & Electrothermal | 196.0; 251.6 | 0.2 |
| Silicon (Si) | Direct | $N_2O$-acetylene | 251.6 | 0.1 |
| Silver (Ag) | Direct & Pre-concentration | Air-acetylene & Electrothermal | 328.0 | 0.005 |
| Sodium (Na) | Direct | Air-acetylene | 589.0 | 0.005 |
| Strontium (Sr) | Direct | Air-acetylene | 460.7 | 0.01 |

| *Metal* | *Method(s)* | *Vaporization* | *Wavelength* (nm) | *Detection* (ppm) |
|---|---|---|---|---|
| Thallium (Tl) | Direct | Air-acetylene & $N_2O$-acetylene | 276.8 | 0.8 |
| Tin (Sn) | Direct | Air-acetylene & Electrothermal | 224.6; 303.4 | 0.05 |
| Titanium (Ti) | Direct | Air-acetylene & Electrothermal | 364.3 | 0.2 |
| Vanadium (V) | Direct | $N_2O$-acetylene | 318.4 | 0.05 |
| Zinc (Zn) | Direct & Pre-concentration | Air-acetylene & $N_2O$-acetylene | 213.8; 307.6; 481.0 | 0.002 |

### 24.8.2 Atomic Emission Spectrometry (AES)

In atomic emission spectrometry (AES) the analyte atoms are raised to an excited electronic state through energetic collision with flame constituents. When the atoms relax to the ground state, radiation is emitted that is characteristic of the element. The predetermined wavelength is isolated using a monochromator and detected by a photomultiplier tube. The intensity of emission is proportional to the concentration of the analyte in the sample solution.

AES is a versatile, rapid and sensitive method for the analysis of trace metal constituents in a variety of matrices. For most AES analysis rather dilute solutions (0.1 to 100 mg/mL) are required. The detection limits are 0.1 to 5 ppm. The experimental setup for AES is similar to the one used for AAS. It consists of—(1) atomizer/excitation system (e.g. ICP), (2) dispersive optics (for wavelength selection) and detector system (e.g. PMT, CCD) (Fig. 24.28). Emission spectrometer can be sequential type, wherein one wavelength is scanned at a time, or direct-reading type, wherein placing many PMTs behind the exit carries out simultaneously multi-elemental analysis slits on the periphery of a Rowland circle.

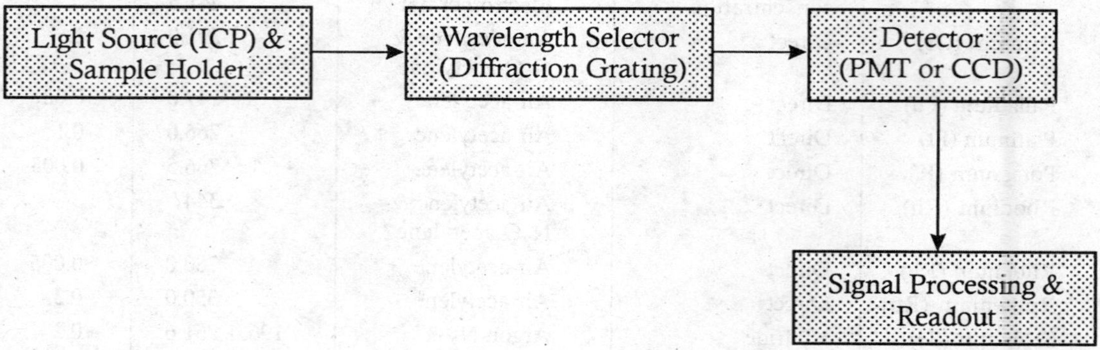


**Fig. 24.28:** Schematic of Atomic Emission Spectrometric (AES) Setup

Unlike AAS, the AES method does not require a separate lamp for each element. It produces emission for all atomized elements simultaneously. With the availability of inductively coupled plasma (ICP) source, many of the practical problems of AES have been alleviated. Therefore, AES is a convenient method for multi-element analysis and is superior to AAS.

#### 24.8.2.1 Atomization/Excitation

For AES, higher temperatures are used so that even those elements with resonance lines below 300 nm are excited. The proportion of excited atoms, out of the total atoms of an element present in a flame, depends both on the energy difference from the ground state to the excited state and the temperature of the flame. These effects are illustrated by a hypothetical pair of elements: If the energy of the excited state is 2.1 eV (600 nm) the fraction of excited atoms in the flame will be $5 \times 10^{-4}$ at 3,000 K and $1 \times 10^{-3}$ at 3,500 K (increase by a factor of 2). If, on the other hand, the energy of excited state is 4.2 eV (300 nm), at the same temperatures the fractions of excited atoms are $2 \times 10^{-7}$ and $2 \times 10^{-6}$ respectively (factor of 10). For AES a higher temperature, compared to AAS, is desirable. If excitation temperatures greater than 4,000 K are required sources such as the direct current plasma (DCP), microwave-induced plasma (MIP) and inductively-coupled plasma (ICP) modes are used.

Direct current plasmas (DCP) are of two types: (*i*) current-carrying DC plasmas, and (*ii*) current-free DC plasmas. In current-free DC plasma sources, a DC arc burns in a chamber between two ring-shaped carbon electrodes. The stabilizing gas (Ar or He) is introduced tangentially into the space between the two electrodes, and the plasma is transformed through the ring cathode by the vortex coolant flow.

In microwave induced plasma (MIP) method, the solution is fed into the injection system. Heating to 370-400 K evaporates the solvent. The residue, vaporized at higher temperature (1,000-3,000 K), then reaches, along with the working gas, the plasma zone where it is atomized and excited. Coupling of MIP to a gas chromatography system is suitable for determination of metal chelates, mercury and other elements. For example, it is possible with MIP-GC to determine mercury in its various organic forms. This is essential for monitoring environmental and food pollutants. In such a setup (Fig. 24.29), the outlet of the GC column is connected to MIP. The carrier gas (Ar or He) in the GC column is also the working gas for MIP.

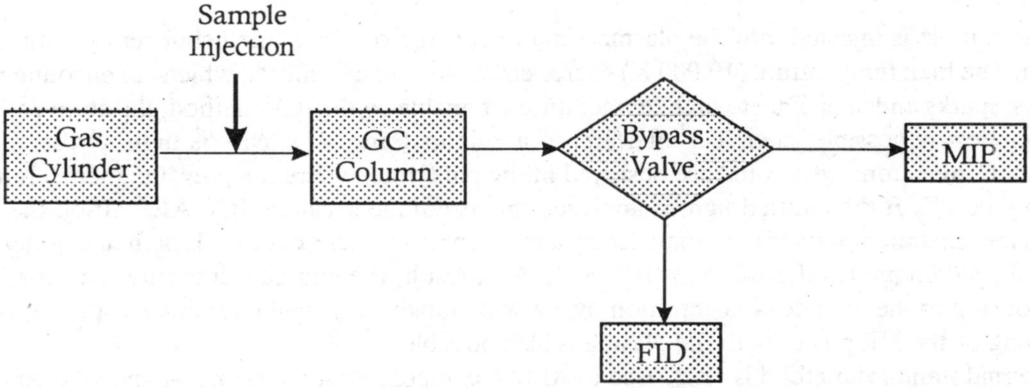


**Fig. 24.29:** Schematic of a GC Column with MIP/FID Detector

The principle of energy transmission in an inductively coupled plasma (ICP) setup (Fig. 24.30) is that of the transformer, in which the primary winding is the HF coil and the secondary winding the plasma.

An ICP system consists of a flowing stream of argon gas, ionized (plasma) by an applied RF field. The RF field is inductively coupled to the ionized gas by a water-cooled coil surrounding quartz "torch" that supports and confines the plasma. The magnetic fields generated by the RF (4-50 MHz) induction currents accelerate argon ions in one direction and electrons in the other direction, leading to ionization and heating.

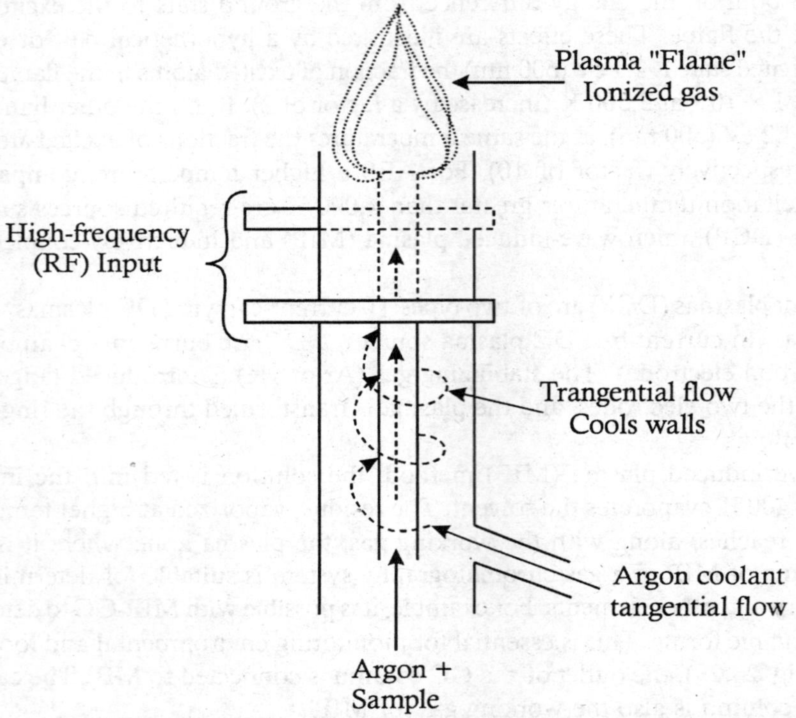


**Fig. 24.30:** Schematic of Inductively-coupled Plasma (ICP) Setup for AES

The sample is injected into the plasma via a pneumatic or ultrasonic nebulizer operating with argon. The high temperature (10,000 K) source eliminates matrix effects, which are encountered in flames, sparks and arcs. Due to high temperature attainable by the ICP method, the atomization of all elements is possible, chemical interference is minimal and sensitivity is high. Ordinarily the spectra of free atoms generated are observed in the tail flame, where the prevailing temperature is about 4,000 K. If the emitted light is analyzed the technique is called 'ICP-AES'. If on the other hand, the generated particles are analyzed by a mass-spectrometer on the basis of their charge/mass ratio the technique is referred to as 'ICP-MS'. As a result, the emission intensity retains a linear relationship to the analyte concentration over a wide range. A combined inductive and capacitive coupling of the HF power with the plasma is also possible.

Internal standard method is often used in AES to compensate some errors. A known quantity of internal standard (S) is added to the sample solution (e.g. Li for the analysis of Na and K). A set of standards of analyte is prepared with the same amount of S added. The calibration graph is prepared by plotting the ratio of emission $S_E$ to that of the analyte, $A_E$, against the concentration of analyte, $A_C$. The sample concentration is obtained from the calibration plot using the ratio $S_E / A_E$ for the sample. Characteristic wavelengths and detection limits for metals by the AES method are given in Table 24.11.

**TABLE 24.11:** CHARACTERISTIC WAVELENGTHS AND DETECTION LIMITS FOR METALS BY AES METHOD

| *Element* | *Wavelength (nm)* | *Detection (ppm)* |
|---|---|---|
| Aluminum (Al) | 308.0 | 0.01 |
| Antimony (Sb) | 206.0 | 0.5 |
| Arsenic (As) | 194.0 | 0.2 |
| Barium (Ba) | 455.0 | 0.05 |
| Beryllium (Be) | 313.8 | 0.01 |
| Bismuth (Bi) | 250.0 | 0.02 |
| Cadmium (Cd) | 227.0 | $10^{-6}$ |
| Calcium (Ca) | 318.0 | 0.02 |
| Chromium (Cr) | 268.0 | 1.0 |
| Cobalt (Co) | 228.7 | 0.005 |
| Copper (Cu) | 324.7 | 0.001 |
| Gallium (Ga) | 403.3 | 0.5 |
| Germanium (Ge) | 265.0 | 0.1 |
| Gold (Au) | 242.8 | 0.005 |
| Indium (In) | 410.5 | 0.2 |
| Iron (Fe) | 260.3 | 0.01 |
| Lead (Pb) | 220.0 | 0.02 |
| Magnesium (Mg) | 279.2 | 0.001 |
| Manganese (Mn) | 258.5 | 0.001 |
| Molybdenum (Mo) | 202.0 | 0.003 |
| Palladium (Pd) | 340.5 | 2.0 |
| Platinum (Pt) | 266.0 | 50.0 |
| Selenium (Se) | 196.0 | 0.2 |
| Silicon (Si) | 212.6 | 0.5 |
| Silver (Ag) | 228.0 | 0.0001 |
| Strontium (Sr) | 460.7 | 0.03 |
| Tellurium (Te) | 408.3 | 0.01 |
| Thallium (Tl) | 191.0 | 0.01 |
| Tin (Sn) | 303.4 | 0.1 |
| Vanadium (V) | 292.4 | 0.05 |
| Zinc (Zn) | 214.0 | 0.0005 |

Several instrumental methods are available for the determination of metals. Of these, atomic spectroscopy and anode-stripping voltammetry (ASV) are the preferred ones for trace analysis. The choice of the method depends on the physicochemical characteristics of the analyte and sensitivity and accuracy of the technique, availability of the equipment and the cost factor. Though ASV is a very sensitive method, it is applicable to only a small number of metals (Table 24.12).

choice of the method depends on the physicochemical characteristics of the analyte and sensitivity and accuracy of the technique, availability of the equipment and the cost factor. Though ASV is a very sensitive method, it is applicable to only a small number of metals (Table 24.12).

**TABLE 24.12:** SENSITIVITY OF VARIOUS METHODS FOR METAL ANALYSIS

| *Method* | *Sensitivity (Moles)* | *Comments* |
|---|---|---|
| UV-Visual Spectrophotometry | $10^{-5}$-$10^{-9}$ | Versatile and routinely used |
| Flame-AAS | $10^{-6}$-$10^{-7}$ | Simple and versatile |
| Flameless AAS | $10^{-7}$-$10^{-9}$ | |
| AES | $10^{-5}$-$10^{-8}$ | Versatile |
| Anode-stripping voltammetry (ASV) metals | $10^{-9}$-$10^{-10}$ | Very sensitive, but applicable to only a few metals |
| Polarography | $10^{-5}$-$10^{-8}$ | For electro-active metals and compounds only |
| Ion-selective electrodes (ISE) | $10^{-5}$-$10^{-6}$ | Ion selective but moderate sensitivity |

## BIBLIOGRAPHY


Alonso, V. E., de Torres, G. A. and Pavon, C. J. M. (2001), *Talanta*, **55**(2): 219. "Flow Injection On-Line Electrothermal Atomic Absorption Spectrometry."

Anderson, P., *et al.* (1998), *Int. J. Environ. Anal. Chem*. "Comparison of Techniques for the Analysis of Industrial Sols by Atomic Spectrometry."

Campbell, A. K. (1988), Ellis Harwood: Chichester, UK. "*Chemiluminescence—Principles and Applications in Biology and Medicine.*"

Cassella, A., *et al.* (2001), *Talanta*, **54**(6): 1087. "Fourier transform infrared spectrometric determination of ziram."

Clark, R. J. H. and Hester, R. E. (1995), Wiley & Sons: New York. "Spectroscopy in Environmental Science: Advances in Spectroscopy Series," (Vol. 24).

Colthup, N. B., Daly, L. H. and Wiberely, S. E. (1990), Academic Press: San Diego, CA. "*Introduction to Infrared and Raman Spectroscopy*," (3rd edn.).

Dodeigne, C., Thunus, L. and Lejeune, R. (2000), *Talanta*, **51**(3): 415. "*Chemiluminescence as a Diagnostic Tool: A Review.*"

Eaton, A.D., *et al.* (eds.) (1995), Amer. Public Health Assn: Washington, DC. "*Standard Methods for Examination of Water and Wastewater*," (19th edn.).

Ferro, J. R. and Nakamoto, K. (1994), Academic Press: New York. "*Introduction to Raman Spectroscopy.*"

Fifield, F. W. and Haines, P. J. (2000), Blackwell Science: Oxford, UK. "*Environmental Analytical Chemistry*," (2nd edn.).

Franko, M. (2001), *Talanta*, **54**(1): 1. "Recent Applications of Thermal Lens Spectrometry in Food Analysis and Environmental Research."

Gally, W., *et al.* (Compilers). (1975), Butterworth: London. "*Environment Pollutants: Selected Analytical Methods.*"

Gübitz, G., *et al.* (2001), *Crit. Rev. Anal. Chem.*, **31**(2): 141. "*Chemiluminescence Flow-Injection Immunoassays.*"

Guilbault, G. G. (1990), Marcel Dekker: New York. "*Practical Fluorescence.*"

Jenkins, R. (1999), Wiley Interscience: New York "*X-ray Fluorescence Spectrometry*," (2nd edn.).

Lakowicz, J. R. (1999), Kluwer-Academic Press: New York. "*Principles of Fluorescence Spectroscopy*," (2nd edn.).

Mandelis, A. (ed.) (1992), Elsevier: New York. "*Principles and Perspectives of Photo-thermal and Photo-acoustic Phenomena.*"

Narayanan, P. (2001), Bhalani Pubs: Mumbai. "*Clinical Biophysics: Principles and Techniques.*"

Narayanan, P. (2003), New Age Intl. Pubs: New Delhi. "*Essentials of Biophysics*," (2nd Print).

Quinby-Hunt, M. S., Mc Laughlin, R. D. and Quinitinilha, A. T. (1986), Wiley & Sons: New York. "*Instrumentation in Environment Monitoring.*"

Rubinson, J. F. and Rubinson, K. A. (1998), Prentice & Hall: London. "*Contemporary Chemical Analysis.*"

Skoog, D. A., Holler, F. J. and Nieman, T. A. (1998), Harcourt & Bruce: Philadelphia, PA. "*Principles of Instrumental Analysis*," (5th edn.).

Underfriend, S. (1970), Academic Press: New York. "*Fluorescent Assay in Biology and Medicine*."

Weeks, I. (1992), Elsevier: Amsterdam. "Chemiluminescence Immunoassay". *In*: "*Comprehensive Analytical Chemistry*, Vol. XXIX', G. Svehla (ed.).

Weiz, B. (1985), VCH Pubs: New York. "*Atomic Absorption Spectroscopy*," (2nd edn.).

White, C. and Argauer, R. (1970), Marcel Dekker: New York. "*Fluorescence Analysis: A Practical Approach*."

White, R. (1990), Marcel Dekker: New York. "*Chromatography/FTIR and its Applications*."

# 25 Physical Methods in Environmental Pollution Analysis—Nuclear Analytical Techniques

Nuclear methods for the analysis of environmental pollutants involve direct measurement of high-energy radiation ($\alpha$-particles, $\beta$-particles, $\gamma$-rays/X-rays) as well as indirect methods of analysis (e.g. neutron activation analysis, NAA). Radiation counters are the detection devices employed to measure radioactivity. They are gas-filled tubes (proportional, ionization and Geiger-Müller counters), semiconductor (e.g. Si-Li) devices and scintillation counters. Natural radioactivity due to uranium, thorium and their disintegration products (mainly radon) is all pervasive, apart from any anthropogenic radioactive contamination.

X-ray fluorescence (XRF) methods are widely used in qualitative and quantitative analysis of elements above $Z > 5$. Qualitative analysis involves measuring wavelengths of the emitted lines whereas quantitative analysis comprises of measuring the intensities of the lines. X-ray photoelectron spectroscopy (XPS) is used to determine the chemical composition ("fingerprint" analysis) of elements. Auger electron spectroscopy and Rutherford backscattering spectroscopies are used in the analysis of surface characteristics.

This chapter deals with some of the nuclear analytical methods employed in the analysis of environmental pollutants.

Nuclear analytical methods for analysis of environmental pollutants involve direct measurement of high-energy radiations (α-rays, β-rays or γ-rays/X-rays) from the unstable nuclei of the analytes by various radiation counters, as well as indirect methods of analysis (Fig. 25.1). The energy of the emission is used for qualitative analysis, while the intensity of the emitted particle/radiation is used for quantitative analysis.

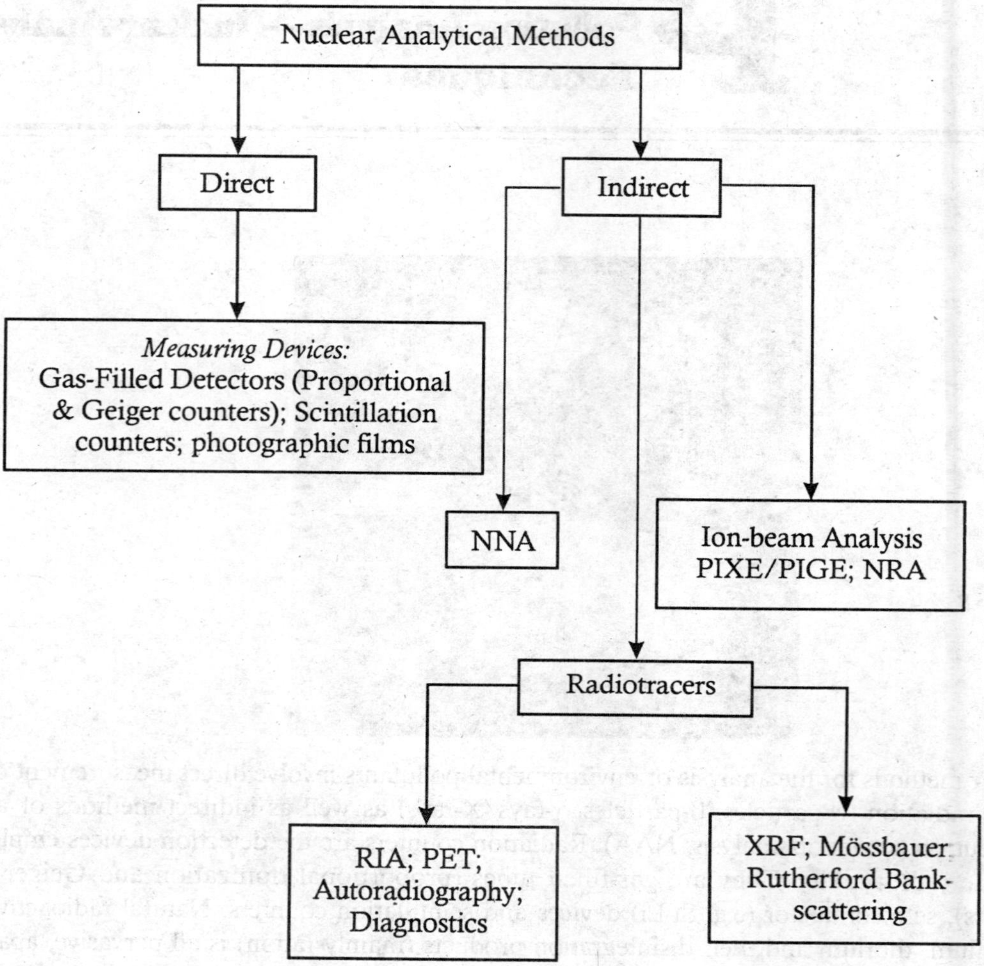


PIXE/PIGE: Particle induced X-ray/gray emission;
NAA: Neutron activation analysis; PET: Positron emission tomography;
RIA: Radioimmunoassay : XRF: X-ray fluorescence

**Fig. 25.1:** Flowchart of Nuclear Analytical Methods

## 25.1 RADIATION MEASURING INSTRUMENTS

Measurement of high-energy radiation relies on its interaction with matter, to produce physical or chemical effects that are proportional to the absorbed radiation. There are a number of radiation measuring devices, which include gas-filled radiation counters (ionization chambers, proportional counters and Geiger-Müller tubes), semiconductor detectors, and scintillation (solid and liquid) counters. γ-ray spectrometry is employed to determine γ-ray emitting radionuclides, particularly in soil samples. α- and β-ray autoradiography is also an excellent screening technique to identify radioactive particles in soil. The simplicity of operation and relative cheapness make autoradiographic techniques (e.g. differential autoradiographic imaging) very useful in large-scale screening of environmental and biological samples.

### 25.1.1 Gas-filled Radiation Counters

Gas-filled detectors have a sealed metal chamber (cathode) and an axial metal wire (anode) and are filled at low pressure with an inert gas like argon or krypton (Fig. 25.2). When a charged particle passes through the gas it transfers its energy to the atoms resulting in their excitation or ionization. When the excited atom spontaneously returns to its ground state a part or whole of its energy is given out in the form of an electromagnetic radiation (photon). For typical gases the emitted radiation is in the UV region of the spectrum. A charged particle creates thousands of excitations in its path and there will be a flash of UV photons. Some detectors (scintillation and Cherenkov detectors) are based on counting these flashes directly.

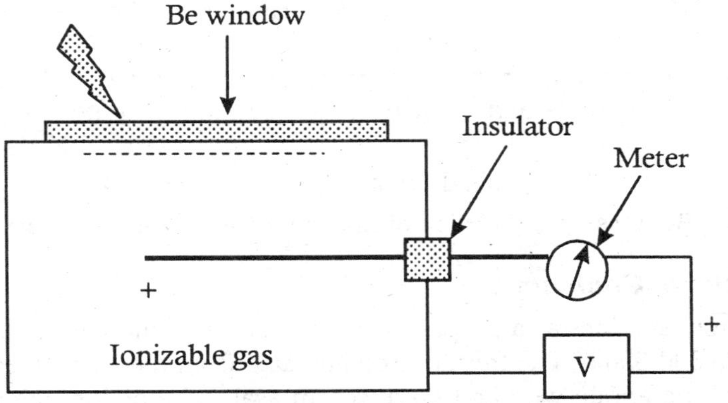


**Fig. 25.2:** Schematic of a Gas-filled Radiation Counter

When ionization takes place in a gas, an ion pair is produced, consisting of an electron and a relatively heavy positive ion. Under the influence of an applied electric field the positive ions and the electrons drift toward the oppositely charged electrodes. The functioning characteristics gas-filled radiation counters (ion chambers, proportional counters and Geiger-Müller detectors) are based on the movement of these charges, which are influenced by the applied voltage (Fig. 25.3).

#### *25.1.1.1 Ion Chambers*

An ion chamber is a device in which two electrodes are arranged on opposite sides of a gas filled container. The movement of ion pairs formed in the gas by incident radiation, under an applied electric field, constitutes an electric current that can be measured in an external circuit. Above a certain voltage the current no longer depends on the voltage and this is the region (100-300 V) normally chosen for the operation of an ion chamber. Under these conditions the current measured in the external circuit is equal to the formation of charges in the gas by incident radiation. Air-filled ion chambers, operated in current mode, are a common type of survey meters used to monitor personal exposure to γ-rays.

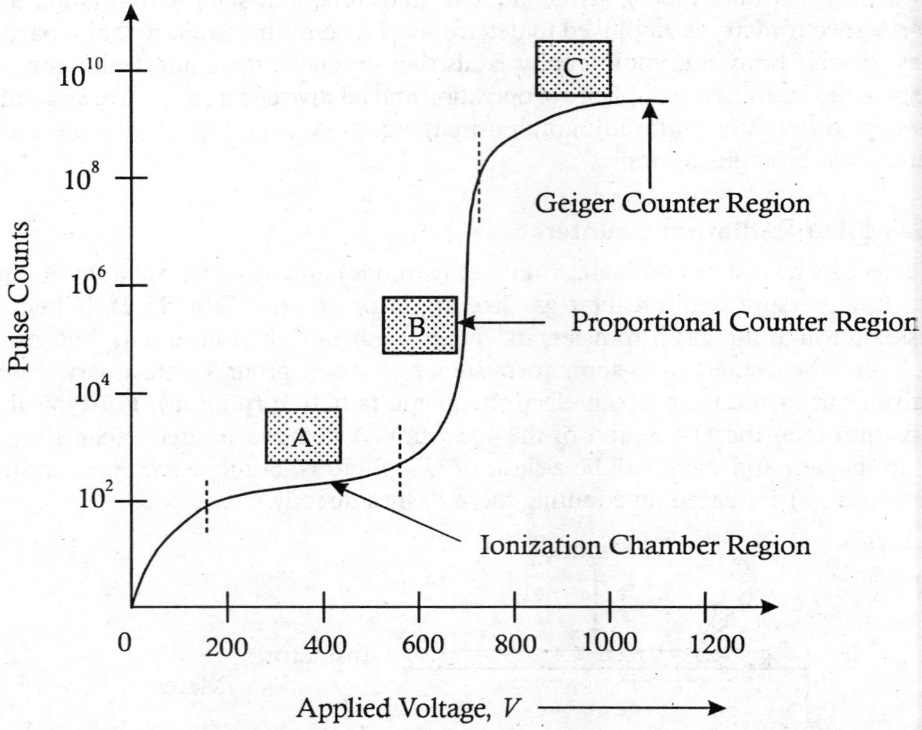


**Fig. 25.3:** Ionization Regions as a Function of Applied Voltage in a Gas-filled Radiation Counter

### *25.1.1.2 Proportional Counters*

The gas filled detector functions as a proportional counter when the voltage applied between the electrodes is between 300-800 V. The free electron liberated from the atom in excitation by the UV photon drifts toward the anode wire and produces an avalanche of ion pairs. In a proportional counter the spread of avalanches is inhibited through the addition of a second gas (e.g. methane) that absorbs the UV photons without liberating electrons. In this region of operation the amplitude of the electric pulse is proportional to the initial number of ion pairs produced due to the primary ionization process. This region is energy-sensitive and is suited for detecting particles of high ionizing power (α-particle). Proportional counters not only count the particles but also indicate their type and energy.

### *25.1.1.3 Geiger-Müller (GM) Counters*

In a GM counter, the chamber is filled with a mixture of argon and alcohol, and the insulated anode is maintained at 1,000-1,500 V (Fig. 25.4). A thin mica or glass window acts as an entry surface for incident radiation. Ionizing particles (α and β) entering the chamber initiate ion pair production. When the electrons are accelerated toward the anode they collide with, and ionize more atoms of the gas. A cascade process ensues and the GM tube produces pulses as large as one volt. Because of such a large pulse the electronics attached to the counter are simple. GM tubes cannot discriminate between different types of radiation. The counter simply registers an ionization event as a single electric pulse (heard as 'click' in a loud-speaker). The detection efficiency of γ-rays by GM counters is low because of the penetrating power of such radiation. However, as a detector of radiation (α, β or γ) a GM tube is still a useful instrument. The output of a portable GM survey meter can be displayed as a rate meter giving the average rate of pulse production.

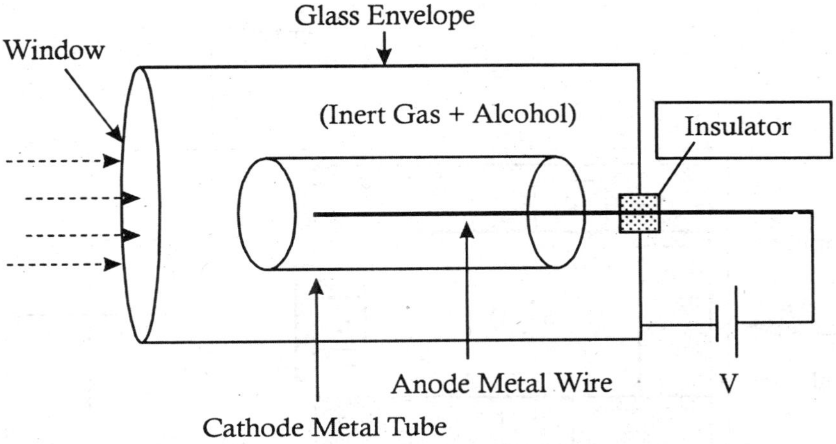


**Fig. 25.4:** Basic Features of Geiger-Müller (GM) Counter

### 25.1.2 Semiconductor Detectors

Semiconductor transducers are of major importance as detectors of γ-rays/X-rays. In semiconductor detectors an electric field is present throughout the active volume. The passage of an energetic charged particle through a semiconductor transfers energy to electrons in the valence band and promotes them into the conduction band, resulting in an electron-hole pair. The subsequent drift of electrons and holes toward electrodes on the surface of the semiconductor material generates current pulses in much the same way as motion of ion pairs in a gas-filled chamber.

Semiconductor detectors offer the best energy resolution provided by common detectors. A 1 MeV charged particle loosing its energy in a semiconductor creates about 300,000 electron-hole pairs. This number is about 10 times larger than the number of ion pairs that would be formed by the same particle in a gas. As a consequence, the charge packet for equivalent energy loss by the incident particle is 10 times larger, improving the signal to noise ratio as compared with a pulse type ion chamber. Another advantage is, the medium being a solid, full energy of the ionizing particle can be absorbed in a relatively thin detector. Moreover, it is practical to fully absorb fast electrons such as β-particles. Therefore, spectroscopic methods can be employed to measure the energies of fast electron radiation. Generally, α-ray spectrometry is carried out using silicon detectors and γ/X-ray spectrometry with Ge or Si detectors (Fig. 25.5).

### 25.1.3 Scintillation Counters

When energetic particles traverse through certain transparent materials they create excited atomic or molecular states, which decay through emission of visible or UV fluorescence. Such materials are known as scintillators. This property is the basis for detection of these high-energy particles. The scintillations are converted to electric pulses through the use of photomultiplier tubes (PMTs) or photodiodes and counted instrumentally (Fig. 25.6).

The desirable characteristics of scintillators are high sensitivity (fraction of the original particle energy that is converted to light energy), short decay time, linear dependence of the light output on the energy deposited, good optical quality and availability in large sizes at moderate cost. Since no single material possesses all these attributes a choice of scintillator has to be made keeping the application in view.

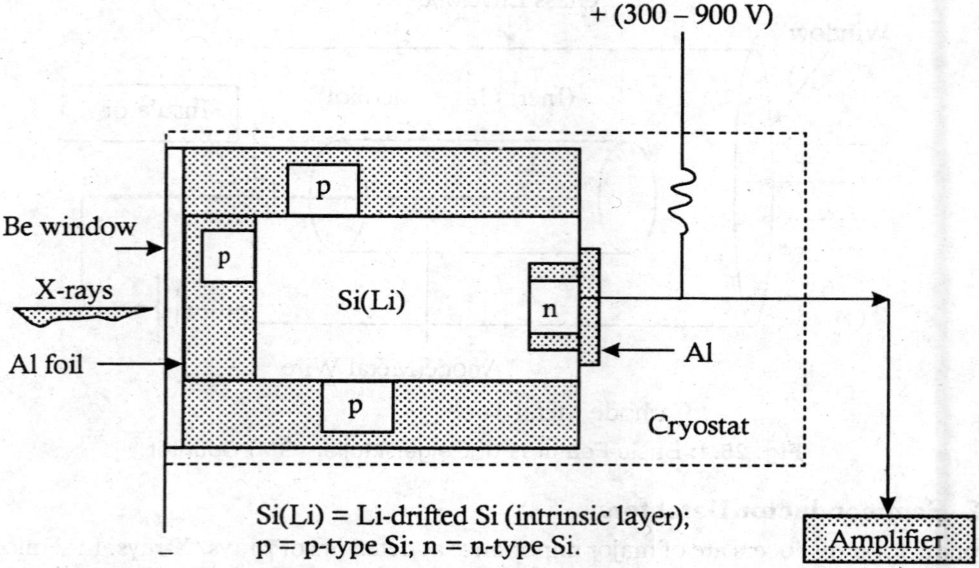


**Fig. 25.5:** Schematic of Lithium-drifted Silicon (Si(Li)) X-ray Photon Detector

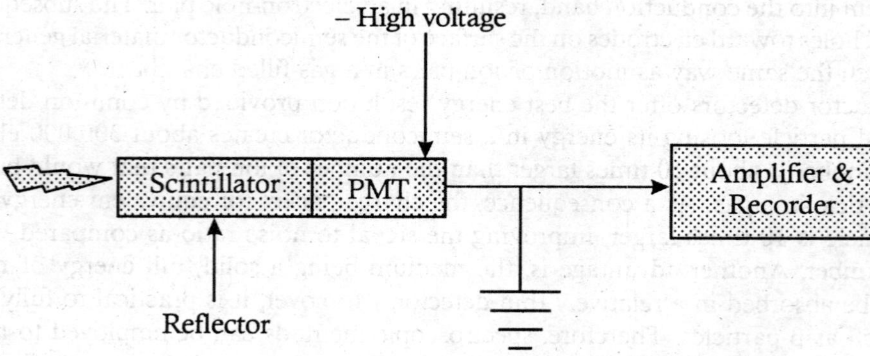


**Fig. 25.6:** Schematic of Solid Scintillation Counter Setup

### *25.1.3.1 Solid Scintillation Counting*

In solid scintillation counting, the radiation comes from a source external to the scintillator. When the radiation strikes the scintillator, electrons are excited and when they reach to the ground state, they emit energy as light. Solid scintillators are most suitable for counting $\gamma$-radiation.

Among the inorganic scintillators thallium, doped NaI and CsI are used in the form of transparent single crystals. Scintillation efficiencies are of the order of 13 per cent and decay times are 0.23 and 0.08-3.3 microsecond, respectively, for NaI and CsI, which are not good enough for high counting rate. While the emission frequency of NaI suits PMT; that of CsI is good for photodiodes. ZnS(Ag) is one of the oldest scintillation materials. However, this material cannot be obtained in single crystal form and is used as 'phosphor screens', made up of layers of painted material.

A number of organic molecules exhibit fluorescence when excited by ionizing radiation. Anthracene and stilbene (polynuclear aromatic hydrocarbons) are among the earliest known materials. An attractive feature in their case is the small decay time of the order of nanoseconds. However, their scintillation efficiency is poorer than that of inorganic materials.

#### *25.1.3.2 Liquid Scintillation Counting*

Liquid scintillation counting is used for counting β-, and γ-rays and rarely α-rays. The radioactive sample is dissolved in a vial with a primary and secondary scintillants. The secondary scintillant enables light to be emitted of a wavelength to which the photomultiplier tube (PMT) is most sensitive.

Organic scintillators (flours) are sometimes used in the form of solution. The flour is dissolved in an organic solvent like toluene and a dispersing agent. Solutions of this type are called 'scintillation cocktails' (Fig. 25.7). They are used for counting β. The counter is equipped with a multi-channel analyzer, which can display sample results as a spectrum and computes count rate by the method of applied spectrum analysis. The sample to be counted is directly dissolved or dispersed in the cocktail, thus avoiding attenuation problem. These counters are particularly suitable for low energy β-emitters like tritium ($^3H$) and carbon ($^{14}C$).

Organic scintillators can be incorporated into a transparent plastic matrix and the detector can be cast in various forms. For this purpose the fluor is added to a monomer, which is subsequently polymerized. A third substance is usually added to modify light emission frequency to match the response of PMT or photodiode.

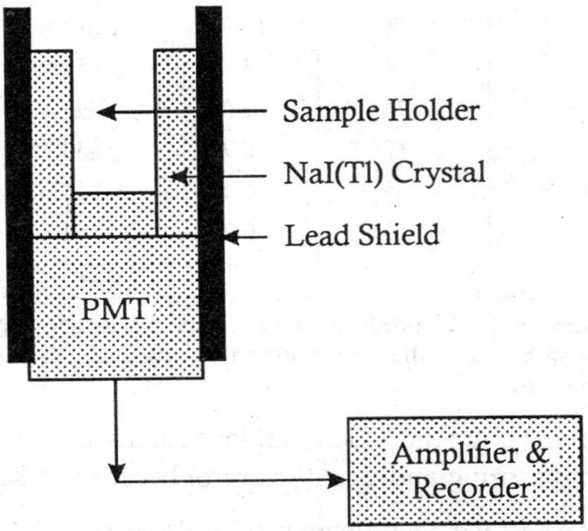


**Fig. 25.7:** Schematic of Liquid Scintillation Counter Setup

## 25.2 FEATURES OF X-RAY SPECTROSCOPY

X-ray spectroscopy, like optical spectroscopy, is based upon measurement of absorption, emission, fluorescence, scattering and diffraction. X-ray absorption and fluorescence methods are widely used for qualitative and quantitative determination of all elements in the Periodic Table above the atomic number $Z > 5$ (after beryllium). X-ray fluorescence (XRF) methods are extensively used in sedimentation flow analysis, air quality monitoring and speciation of metals in water. Major elements that can be determined in the sediments by this method are $_{12}Mg \rightarrow {}_{17}Cl$, $_{20}Ca$, $_{22}Ti \rightarrow {}_{35}Br$, $_{37}Rb \rightarrow {}_{42}Mo$, $_{47}Ag$, $_{48}Cd$, $_{50}Sn$, $_{52}Te$, $_{53}I$, $_{80}Hg$, $_{82}Pb$ and $_{83}Bi$.

*Note:* X-ray diffraction comes under a separate category. It is a powerful and versatile method for structure elucidation. Complete three-dimensional architecture of compounds can be determined, at atomic and molecular levels, depending on the state of the substance. Principles and operational methodology are briefly mentioned under other methods (Chapter 26).

The atomic processes involved in X-ray spectroscopy are quite simple. When a high-energy particle/radiation interacts with an atom, the atom may eject specific core electron (K-, or L-electron) creating a vacancy, which is filled by transition of an electron from the higher energy shells. The energy equal to the difference between these two energy states is emitted as an X-ray photon. K-series involve electron transitions from the L, M, N, ... shells to the K-level. Similarly, L-series arise from electron transitions from M, N, ... shells to the L-level (Fig. 25.8). The elements in a specimen are excited by absorption of primary X-rays and emit their own characteristic fluorescent X-rays. Thus, the X-ray line spectrum emitted by an ionized atom is characteristic of the element involved (Table 25.1) and the intensity of the spectral lines is proportional to the concentration of that element.

**TABLE 25.1:** X-RAY LINE SPECTRA OF SOME ELEMENTS

| *Element* | *Atomic Number* (*Z*) | *K-series* (Å) | | *L-series* (Å) | |
|---|---|---|---|---|---|
| | | *α-line* | *β-line* | *α-line* | *β-line* |
| Sodium (Na) | 11 | 11.910 | 11.617 | –** | –** |
| Potassium (K) | 19 | 3.742 | 3.454 | –** | –** |
| Chromium (Cr) | 24 | 2.290 | 2.095 | 21.714 | 21.323 |
| Rubidium (Rb) | 37 | 0.926 | 0.830 | 7.318 | 7.075 |
| Cesium (Cs) | 55 | 0.40 | 0.355 | 2.892 | 2.683 |
| Tungsten (W) | 74 | 0.21 | 0.184 | 1.476 | 1.282 |
| Uranium (U) | 92 | 0.126 | 0.11 | 0.91 | 0.72 |

(1 Å = 0.1 nm; ** = Elements below Z < 23 produce only K-series. For heavier elements, additional series (M, N, ...) are found at longer wavelength, but as their intensities are very weak, they are not of much importance in conventional spectroscopic analyses).

X-ray spectroscopic techniques are the widely employed analytical methods in elemental analyses, environmental pollution monitoring and quality control because of several advantages:

1. It offers high precision and accuracy. Better precision than wet-chemical and other instrumental methods.
2. Sensitivity and detection limits do not vary greatly across the Periodic Table.
3. Spectra are simple to interpret as compared to the UV-visible spectra.
4. Spectral lines are generally independent (for elements of high atomic number) of the physical and chemical state in which the element exists. For example, the $K_{\alpha}$ lines for elements of higher atomic number are the same whether the target is a pure metal, oxide or sulfide.
5. Rapid and quantitative determination of elements of atomic number $Z > 5$.

6. Minimal sample quantity and sample preparation is needed.
7. It is a non-destructive method of analysis.

Instrumental factors are important in elemental analysis by X-ray spectrometry. Some of these factors are: source of X-rays and their characteristics, monochromator and the detector and recording system.

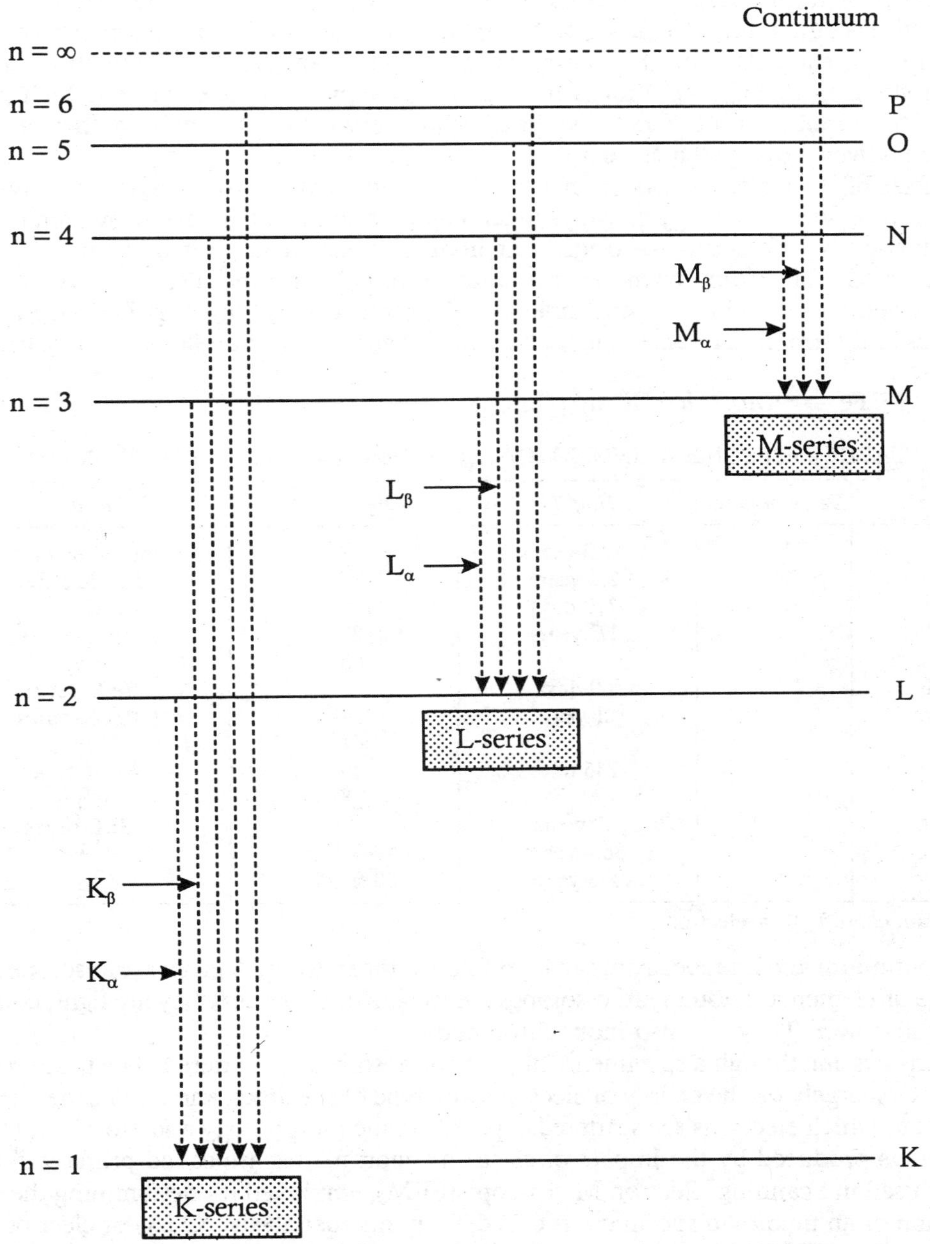


**Fig. 25.8:** Schematic Representation of Characteristic X-ray Line Spectra

### 25.2.1 Sources of X-rays

Sources of X-rays are: (*i*) X-ray tubes (Coolidge tubes), (*ii*) radioisotopes, (*iii*) electron microscope, (*iv*) primary X-rays when X-ray fluorescence is to be studied, and (*v*) synchrotron radiation.

The most common source of X-rays for analytical work is the X-ray tube. Coolidge tube is a sealed vacuum tube where a metal anode (anticathode), such as Cr, Cu, Mo, Ni or Ag is bombarded with a beam of high-energy electrons to produce X-ray continuum (*Bremsstrahlung*), and characteristic spectral lines, depending on the applied voltage. For the excitation of a characteristic line spectrum of each element, the minimum threshold voltage, $V_0$, should be above the 'critical excitation voltage,' $V_K$ ($V_0 > V_K$) for that element. Though X-ray tubes provide high power and versatility, the production of X-rays by electron bombardment is a highly inefficient process as most of the power is dissipated as heat.

The effects of X-rays and γ-rays on matter are synonymous, as both X-rays and γ-rays are high-energy photons, with the energy of γ-rays much higher than that of X-rays. However, the mode of their production is different; γ-rays originating in the nucleus are generated by nuclear transitions, while X-rays basically come from electronic transitions. X-rays are also produced as a part of radioactive decay of certain isotopes (Table 25.2) This process also involves K-electron capture by the nucleus and formation of an element of the next lower atomic number ($Z - 1$), releasing an X-ray photon.

$$^{55}\text{Fe} \rightarrow {}^{54}\text{Mo} + h\nu \text{ (X-ray photon)} \qquad \ldots(25.1)$$

**TABLE 25.2:** SOME RADIOISOTOPE SOURCES FOR X-RAY SPECTROSCOPY

| *Isotope* | *Decay process* | *Half-life* | *Energy* (keV) | *Radiation* |
|---|---|---|---|---|
| Ti | $\beta^-$ | 12.3 years | | Continuum & Ti K-series |
| $_{55}$Fe | EC | 2.7 years | 5.9 | Mn K-series |
| $_{57}$Co | EC | 270 days | | Fe K-series |
| $_{109}$Cd | EC | 1.3 years | 22 | Ag K-series |
| | | | 88 | γ |
| $_{125}$I | EC | 60 days | | Te-K series |
| $_{145}$Sm | | 1 year | 39 | Pm K-series |
| | | | 61 | γ |
| $_{153}$Gd | | 236 days | 40 | Eu K-series |
| | | | 100 | γ |
| $_{210}$Pb | $\beta^-$ | 22 years | | Bi L-series |
| $_{238}$Pu | | 86.4 years | 99 & 150 | γ |
| $_{241}$Am | | 458 years | 60 & 100 | γ |

EC = electron capture; $\beta^-$ = electron.

Both continuum and line spectrum can be employed for analysis. A given radioisotope is suitable for a range of elements. Though radioisotopes are weak X-ray sources, they are light, compact and consume no power. They are also monochromatic.

Electrons passing though a specimen will either be absorbed or scattered. The process that takes place depends largely on the energy of electrons and type of specimen and its thickness (Fig. 25.9). The extent to which electrons are scattered depends on the morphology and atomic number of the target. X-rays produced by the impact of electrons upon matter are called primary X-rays. The technique used in Scanning Electron Microscopy (SEM), employed for determining the elemental composition of an unknown specimen, is called X-ray microanalysis (XMA) or electron probe X-ray microanalysis (EPXMA). Electron microscopy and XMA are based on the various types of interactions between the electron beam and specimen.

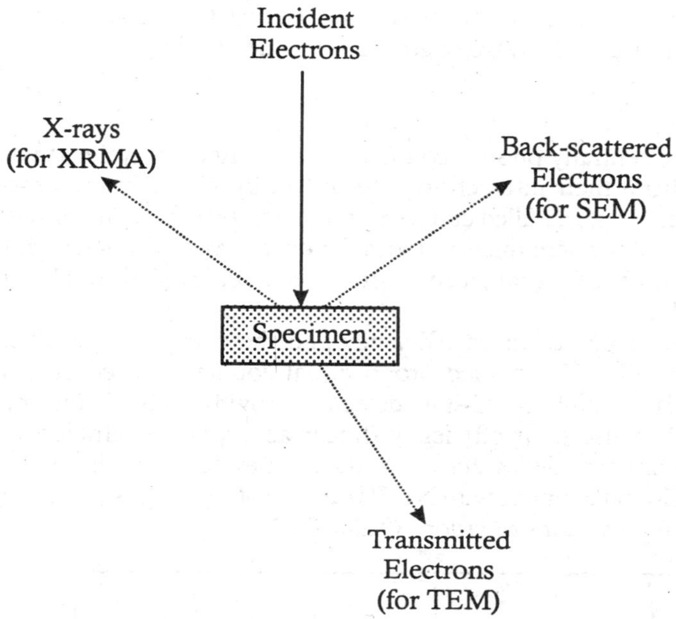


**Fig. 25.9:** Types of Interactions of High-energy Electrons with Matter

In EPXMA and SEM, an electron beam is focused on to a nanometer-sized probe to generate several signals upon interaction with the sample. Morphological studies are based on the detection of secondary and back-scattered and *Auger* electrons (section 25.4.2). The wavelength or energy-dispersion of emitted X-rays can provide compositional information of the specimen. EPXMA and SEM have been successfully used for the identification, classification and source characterization of aerosol particles.

When primary X-rays are used for excitation, the scattered radiation is called the secondary (or fluorescent) X-rays. Every material when subjected to irradiation by γ- or X-rays becomes a source of secondary radiation. This arrangement has the advantage of eliminating the X-ray continuum from the primary source. Wherever it is possible, the synchrotron sources provide all the advantages of desired X-ray intensities and wavelength (energy) selection. Special lines (M, N,... series) (which are very weak from the other sources) can be employed in the analysis.

### 25.2.2 Filters and Monochromators

Filters and monochromators are generally used wherever monochromatic radiation is required for analysis. Thin strips of suitable metals (target-filter combination) can be used to obtain monochromatic radiation, which is generally used in X-ray diffraction methods. A thin metal foil with an absorption edge between two spectral lines will eliminate the background radiation and attenuate one of the spectral lines (e.g. Ni filter attenuates Cu-$K_\beta$ line (1.392 Å) allowing Cu-$K_\alpha$ line (1.542 Å)). 'Balanced filter' method employs two separate measurements of the counting rates with two filter foils made of elements whose photoelectric absorption edges are on either side of the $K_\alpha$ line of the element to be determined (e.g. Ni weakly absorbs both Cu- and Fe-X-rays, whereas Co strongly reduces the intensity of copper X-rays). Use of secondary radiation (fluorescence) and radioactive isotopes is another way of eliminating the X-ray continuum. Crystal monochromators are employed in X-ray spectrometers. Most advantageous is the use of synchrotron radiation source, as it allows selection of desired wavelength from a wide continuous spectrum of wavelengths.

### 25.2.3 Detectors

X-ray detectors are essentially photon counters. The energies (wavelengths) of X-ray photons are detected primarily from their interaction with matter by photoelectric absorption. Photographic films, scintillation counters, gas-filled counters (in the proportional counter region with amplification ~$10^3$-$10^5$ counts/s) and semiconductor diodes (Li-drifted Si or Ge-detectors) are used to detect X-ray photons. Separation of signal from noise can be achieved through pulse-height or energy-discrimination methods.

Detector selection depends on its efficiency and the energy range of particles involved. For detecting radiation <34 keV, Xe-filled proportional counters are commonly used if the energy resolution required is not high. Solid-state detectors provide high resolution, but they are good for <20 keV energies, because their efficiency (linear absorption coefficients) falls of rapidly with increasing energy, being practically zero at >80 keV. Besides, they should be operated at very low temperatures. Scintillation counters with NaI(Tl) are used for energies >35 keV (Fig. 25.10). Detectors are selected based on these considerations (Table 25.3).

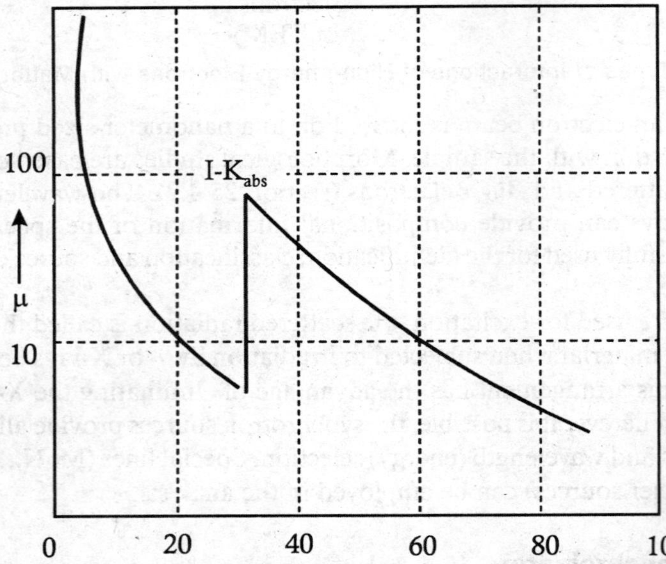


**Fig. 25.10:** Mass Absorption Coefficient, μ, (for NaI) as a Function of X-ray Energy

**TABLE 25.3: CHOICE OF DETECTORS FOR MEASURING X-RAYS**

| *Atomic Number* (*Z*) | *X-ray Energy* (keV) *K* | *L* | *Detector System* |
|---|---|---|---|
| 10-20 | 1.0-3.5 | – | Proportional counters (Ne or Ar) & Si(Li) |
| 20-35 | 3.5-13.5 | – | (Proportional counters (Ar or Xe) & Si(Li) |
| 35-55 | 13.5-33.5 | – | Proportional counters (Xe or Kr) & Si(Li) |
| 55-62 | 33.5-40 | – | NaI:Tl |
| 62-92 | – | 5.6-13.6 | Proportional counters (Ar or Xe) & Si(Li) |

### 25.2.4 Pulse-Height Analyzers

Pulse-height analyzers or discriminators are electronic sieves that reject pulses below and above certain preset bandwidths. A single circuit is frequently used to reject pulses below a minimum value (such as suppression of noise pulses). A pair of discriminators set to different voltages $V_L$ and $V_U$ and used in an anti-coincidence configuration becomes a single-channel analyzer (Fig. 25.11). All the pulses below $V < V_L$ are stopped by the lower discriminator, and the pulse $V > V_U$ pass through both the discriminators, but are rejected by the anti-coincidence circuit. There is no output if neither discriminator is activated nor both of them are activated. That is, there is an output only if the lower discriminator, but not the upper discriminator, is activated. The difference in voltage, $\Delta V = (V_U - V_L)$ defines the energy widow (bandwidth) in which pulses are passed to a scalar counter and display unit. Multichannel pulse-height analyzers can determine simultaneously several elements in the material under investigation.

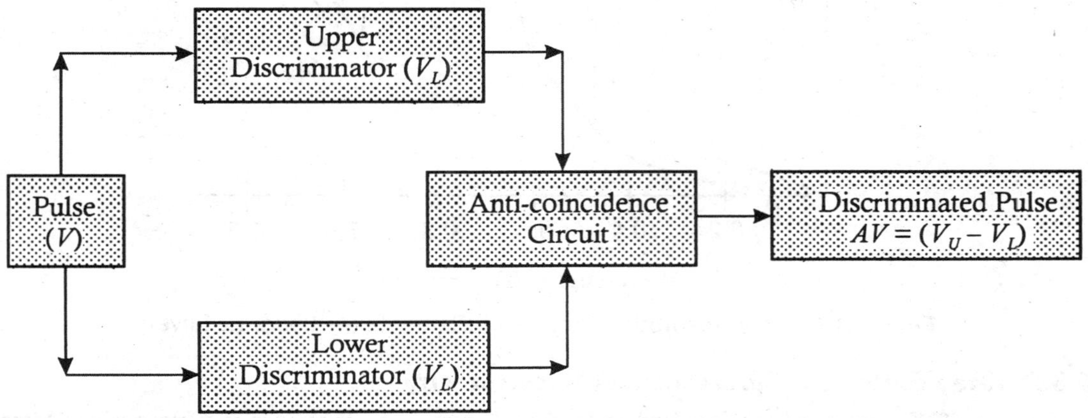


**Fig. 25.11:** Schematic of Anti-coincidence Pulse-height Analyzer for Energy Discrimination

## 25.3 X-RAY ABSORPTION AND EMISSION (FLUORESCENCE)

Both X-ray absorption and X-ray fluorescence (XRF) are available for environmental monitoring. X-ray fluorescence studies of materials are of considerable practical importance.

### 25.3.1 X-ray Absorption Spectrometry

The X-ray absorption spectra of elements are simple and consist of well-defined absorption peaks (Fig. 25.12). Photoelectric absorption is more prominent. X-ray absorption does not occur at a particular wavelength, but changes gradually with wavelength until it reaches an absorption edge. The absorption coefficient changes sharply at X-ray wavelength corresponding to the energy just required to remove an electron from a specific inner shell to form an ion. The wavelengths of particular absorption peaks are characteristic of the element. In absorption studies, the attenuation of an X-ray line serves as the analytical variable for the elemental composition of solids and liquids. Analyte atoms best absorb X-rays that are just shorter in wavelength than their relevant absorption edges and the mass attenuation coefficients are the highest. Wavelength selection is accomplished either by filters or monochromators.

From a plot of mass absorption coefficient versus wavelength, the concentration of the analyte can be determined by the equation

$$I = I_0 \exp(-\mu_m \cdot x) \qquad ...(25.2)$$

where $I_0$ and $I$ = incident and transmitted intensities,
$\mu_m$ = mass absorption coefficient and
$x$ = thickness of the specimen

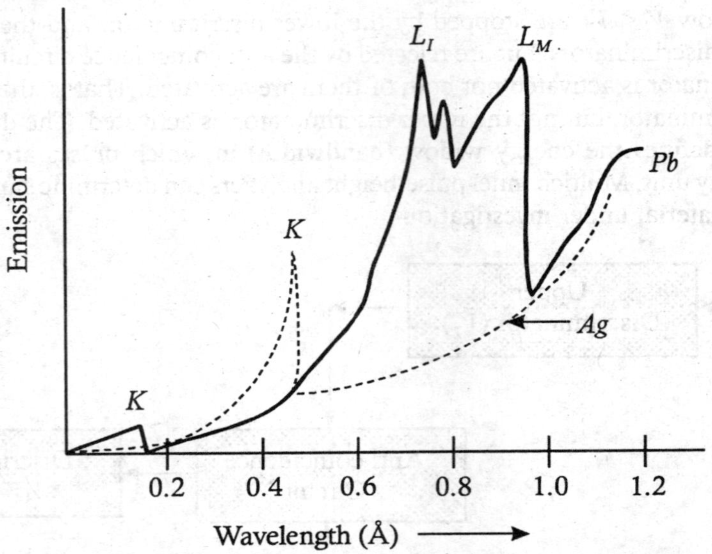


**Fig. 25.12:** X-ray Absorption Profiles of Elements (— Lead; ... Silver)

### 25.3.2 X-ray Emission (Fluorescence) Spectrometry

Absorption of X-rays produces excited ions, which return to their ground state via electronic transitions, emitting characteristic fluorescence lines. The efficiency of production of fluorescent X-rays depends on the energy of the incident X-rays and the absorption edge. Photoelectric absorption is highest at incident energy slightly greater than the binding energy of the shell involved. The probability of fluorescence decreases at incident energies greater than the binding energy of the element of interest. Thompson and Compton scattering can be minimized if mono-energetic radiation with photon energies slightly higher than the fluorescence peaks of interest. So, it is important to optimize the energies of the incident radiation.

X-ray fluorescence and location of absorption edges can be used to identify qualitatively the elements present in a sample. The isotopic abundance of a particular element can be determined by measuring the difference in the X-ray absorption just above and just below an absorption edge of that element and the spectral location of a particular element can be obtained from the position of the absorption edge.

### 25.3.3 Total Reflection X-ray Fluorescence (TXRF)

High background and severe matrix effects limit the efficiency of XRF. These factors are taken care in total reflection X-ray fluorescence (TXRF) spectrometry. TXRF is a variation of ED-XRF technology, and is based on total reflection of X-rays at glancing angle < critical angle. The impinging X-ray beam is highly collimated and strikes the sample below the critical angle of incidence, leading to total reflection, hence the scattered X-ray background is almost eliminated, thereby enhancing the detection accuracy (nanogram range). TXRF has multi-elemental analysis capability, needs small sample volume, and can be used to analyze surface phenomena with lower limit of detection,

### 25.3.4 X-ray Spectrometers

Three types of X-ray spectrometers are available—the wavelength-dispersion, energy-dispersion and nondispersive types.

#### 25.3.4.1 *Wavelength-dispersive X-ray Spectrometer*

In wavelength-dispersive spectrometers, (WD-XRF) (Fig. 25.13), the specimen is irradiated by primary X-rays and the secondary X-rays emitted from the sample, reflected by a wavelength-dispersive element, are received by a detector. Essential components of a monochromator are a set of collimators and a dispersive element (crystal analyzer mounted on a goniometer head). Rotation of the goniometer at an angle, θ, is coupled to the rotation of the detector by an angle, 2θ, to satisfy Bragg's diffraction condition

$$2d\sin\theta = n\lambda \quad \text{...(25.3)}$$

where, $d$ = interplanar spacing of the crystal analyzer;
$n$ = order of diffraction

The detector scans the wavelengths and the areas under the peaks are related to analyte concentrations.

Both single-channel and multi-channel spectrometers are available. Single-channel instruments are provided with X-ray sources, chromium for longer wavelength and tungsten for shorter wavelength regions. In a multi-channel setup, wavelengths are measured by placing analyzer-detector pairs at appropriate locations around the circumference of the goniometer. Each analyzer crystal (Table 25.4) must be at an angle, θ, relative to the sample holder that corresponds to a desired analyte line. About 10-30 elements can be analyzed simultaneously by multi-channel spectrometers.

**TABLE 25.4:** CRYSTAL ANALYZERS EMPLOYED IN WAVELENGTH-DISPERSION X-RAY SPECTROMETERS

| *Analyzer* | *Reflecting Plane* | *Lattice Spacing* (Å) | *Wavelength Range* (Å) | *Resolution* |
|---|---|---|---|---|
| LiF | (220) | 1.42 | 0.25-2.70 | Very high |
| | (200) | 2.01 | 0.35-3.90 | Very high |
| Topaz | (303) | 1.36 | 0.24-2.60 | High |
| Silicon | (111) | 3.14 | 0.55-5.98 | High |
| PET | (002) | 4.40 | 0.77-8.50 | High |
| Gypsum | (020) | 7.60 | 1.32-14.5 | Medium |
| KAP | (1010) | 13.32 | 2.32-24.4 | Medium |

PET = pentaerythritol; KAP = Potassium acid phthalate.

#### 25.3.4.2 *Energy-dispersive X-ray Spectrometer*

X-ray spectra can also be presented as a function of photon energy (instead of wavelength). This is achieved by an energy-dispersive spectrometer, (ED-XRF), which consists of an X-ray tube or a radioactive source, a sample holder, semiconductor detector and electronic arrangement for energy discrimination and measurement (Fig. 25.14).

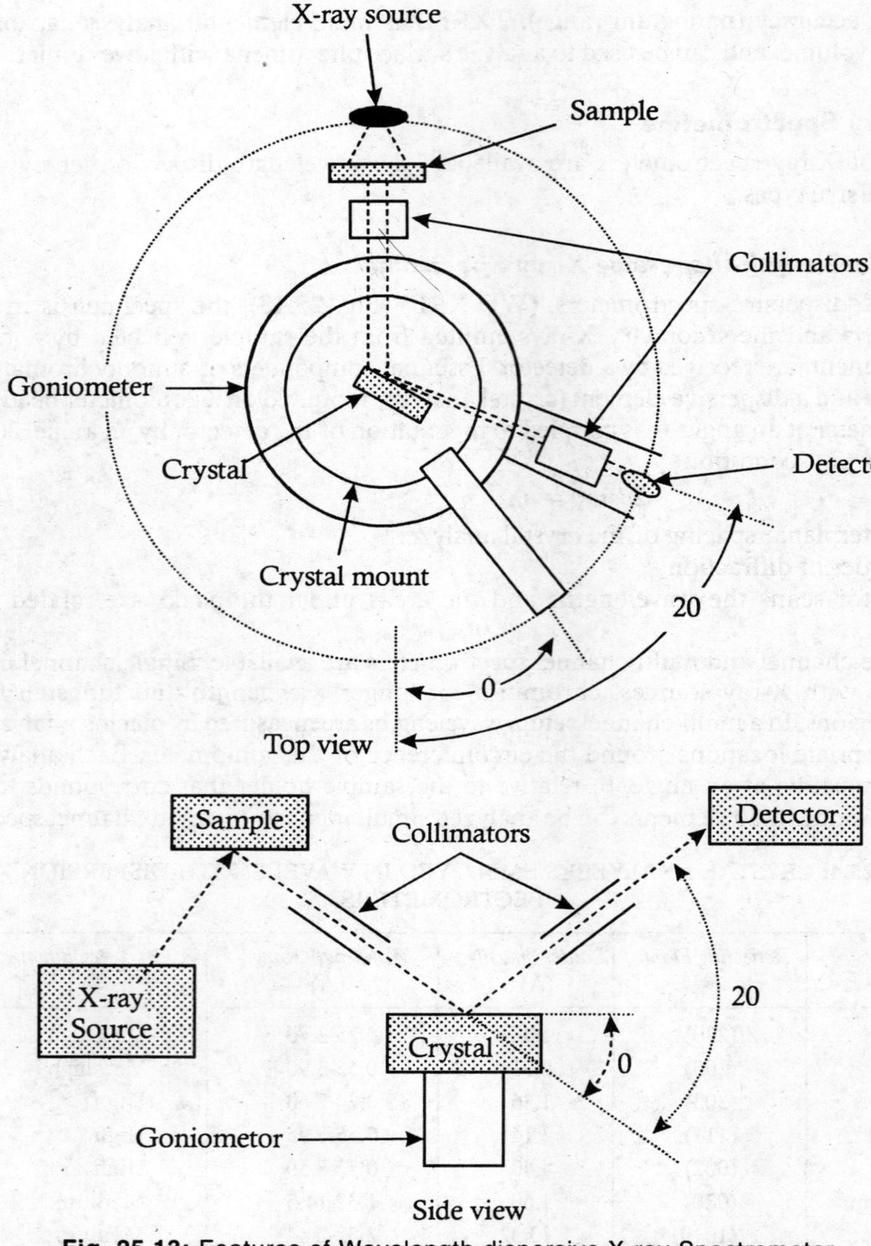


**Fig. 25.13:** Features of Wavelength-dispersive X-ray Spectrometer

In this system, the detector generates a pulse the amplitude of which is proportional to the photon energy. Pulse-height analyzer circuitry distributes the pulses according to their amplitudes and records the number of each energy band. Thus the analyzer generates an energy spectrum and the position of the peaks can be used to identify the elements in the sample.

Energy-dispersive instrumentation is simple and absence of collimators, diffraction elements, and moving parts greatly increase its sensitivity. These features also permit the use of weak radiation sources (radioactive isotopes). Energy-dispersive systems are superior in shorter-wavelength regions. Both primary and secondary excitation modes can be used (Fig. 25.15). By choosing one or the other, it is possible to have control over the distribution and level of excitation radiation. With direct (primary) excitation mode, one can use high power available in the 'bremsstrahlung' radiation. It provides efficient excitation at different channels. Secondary excitation mode is used to furnish a nearly monochromatic exciting radiation to the sample, by directing bremsstrahlung radiation on a target (secondary fluorescer). One may select desired wavelength by changing targets. This mode is used in trace element determination. ED-XRF is very useful for heavy metal aerosols.

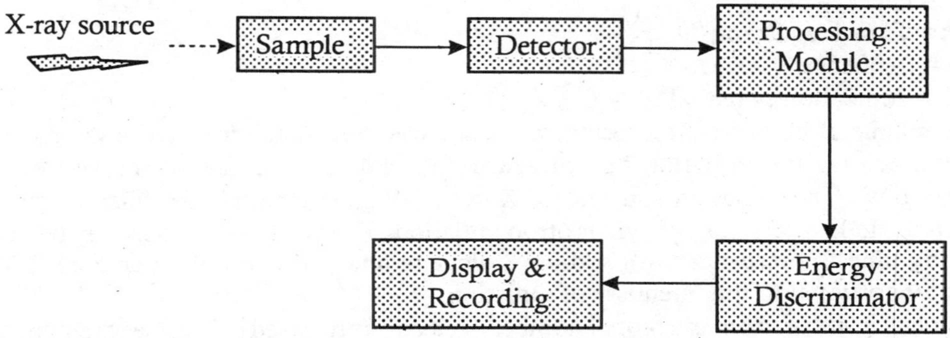


**Fig. 25.14:** Modules in Energy-dispersive X-ray Spectrometer Setup

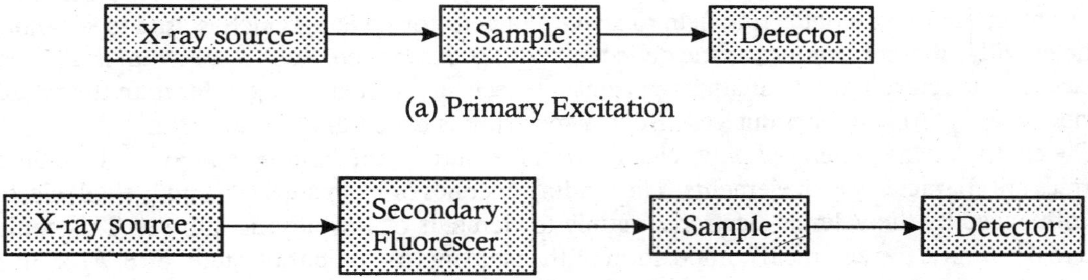


(a) Primary Excitation

(b) Secondary Excitation

**Fig. 25.15:** Excitation Modes of Samples in Energy-dispersive X-ray Spectrometry

## 25.4 ION SPECTROSCOPIES

Photoelectron (XPS) and Auger electron spectroscopy, ion scattering spectroscopy (ISS), secondary ion mass spectroscopy (SIMS), and atom probe field ion microscopy (AP-FIM) are some of the physical techniques that come under ion spectroscopic methods. In the photoionization process, the molecular ion formed may be unstable and dissociate rather than fluoresce. The ionic species may be investigated qualitatively and quantitatively using a mass spectrometer (photoionization mass spectrometry).

### 25.4.1 X-ray Photoelectron Spectroscopy (XPS)

Principles of the photoelectron spectroscopy are based on the photoelectric effect. When a beam of high-energy photons (vacuum ultraviolet and X-ray) passes through a substance it will be attenuated by a number of processes. The photoelectric effect is the phenomenon (for which *Albert Einstein* was awarded the Nobel Prize) in which a photon interacts with a target atom and ejects a bound electron. When the energy of the incident radiation exceeds the ionization energy, part of its energy is spent in overcoming the binding energy of the electron, dislodging it from its shell. The rest of the energy is transferred to the ejected electron in the form of kinetic energy. Photoionization process can be represented by

$$M_i + h\nu \rightarrow M_f^+ + e(\text{E}) \quad \text{...(25.4)}$$

$$E = h\nu - I_{\text{if}} - \Delta E_{\text{vid}} - \Delta E_{\text{rot}} \quad \text{...(25.5)}$$

where, $M$ = molecular species;

$E$ = energy;

$h$ = Planck constant;

$\nu$ = frequency;

$I_{\text{if}}$ = ionization potential, $i \rightarrow f$.

The technique of photoelectron spectroscopy invloves the bombardment of a sample with high-energy monochromatic radiation and measuring the kinetic energies of the ejected electrons (photoelectrons). The radiation sources are X-rays, UV sources from He- filled discharge tubes (He-I = 58.4, He-II = 30.4 nm) or synchrotron radiation. The UV fluorescence may be detected by monotoring blue fluorescence with sodium salicylate and a photomultiplier tube (PMT). The output is proportional to the intentisty of radiation.

A photoelectron spectrum is a plot of number of electrons released ($N$), verses their energies. The energies of electrons released are analyzed either by: (*i*) electrostatic deflection type analyzers or (*ii*) retarding analyzers. In deflection type analyzers, particles from a source follow a curved path under the influence of a deflecting field to arrive at a detector system, which is used to separate the particles with different energies as the deflection voltage is scanned. Such a spectrum is differential spectrum. The retarding type analyzer permits all electrons with energy greater than the retarding voltage ($E > V_R$) to reach the detector. Such a spectrum is called integral spectrum.

The energy of the generated photoelectron is a measure of the binding energy, and the binding energies are characteristic of elements. The binding energies of inner shell (K- and L-shell) electrons are influenced by the valence electrons, mainly in elements of low atomic number. This could be used to determine the chemical composition of the element (e.g. oxidation states of S, Al & Si, etc).

### 25.4.2 Auger Electron Spectroscopy

The ejection of an inner core electron (say from the K-shell) creates a vacancy that could be occupied by an outer electron (from L-, M-, N-shells) with emission of characteristic X-rays. The energy can also be transferred to any other outer orbital electron, ejecting it from the atom, instead of producing X-ray. This ejected electron is called 'Auger electron' (Fig. 25.16), and the spectroscopy is called Auger electron spectroscopy, which is similar to XPS, except that monoenergetic photons are used for excitation in XPS.

Emission of an Auger electron may occur only if the liberated energy is greater than the binding energy of an electron in a given level. Auger effect can take place if

$$E_K - E_L \geq E_L;\ E_L - E_M \geq E_M \text{ etc.} \quad \text{...(25.6)}$$

Auger effect generally occurs in light elements, and K-lines are used to determine elements of $Z$ = 20 – 50 and L-lines for $Z > 50$. Auger electron spectroscopy is employed to study surface analysis. High-energy primary electrons from an electron gun, from a scanning electron microscope setup or some other source, impinge upon the target and an electron energy analyzer analyzes back-scattered electrons that comprise secondary electrons and the Auger electrons. The analyzer allows the sampling of the secondary electrons and the Auger electrons, and detection of their energy distribution function.

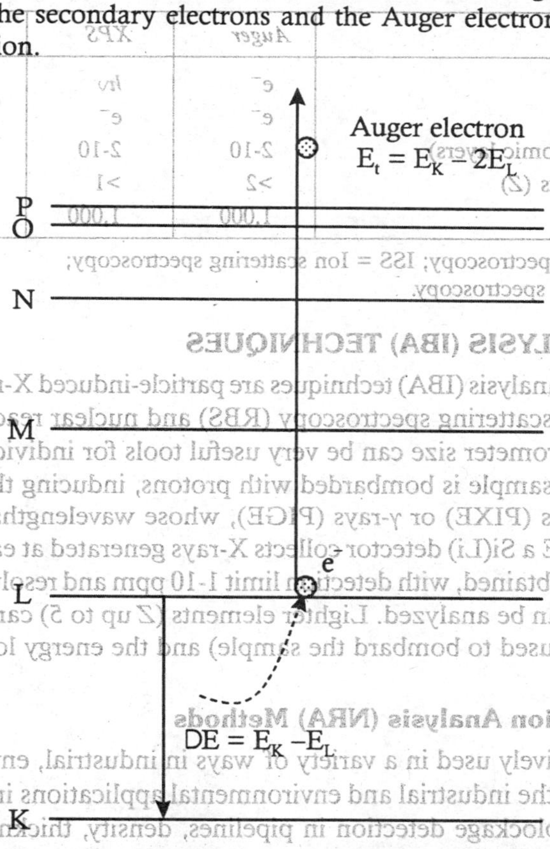


**Fig. 25.16:** Emission of an Auger Electron from L-shell

The Auger spectrum is a plot of number of Auger electrons ($dN(E)/dE$) versus the electron energy *E*. Qualitative analysis is straightforward, involving comparison of the prominent negative peak energies with tabulated energies from standard spectra of the elements. Besides the energies, the shape of the Auger spectrum is characteristic for the element analyzed. Standard "fingerprint" spectra of the elements, in most cases, allow a straightforward identification of the elements present and the composition of the bombarded surface.

Electron spectroscopies are commonly employed to study surface constituents and thin films. Among these, ISS is the most surface specific (Table 25.5). In this method, a monoenergetic beam (0.1-10 keV) strikes the surface of a target and the energy distribution of ions scattered in a particular angle is measured. The energy spectrum contains information about the mass and number of atoms in the very first layer of the solid. The related Rutherford back-scattering spectroscopy (RBS) with high-energy primary ions (~100 keV) is less suited for surface studies but is a powerful tool for non-destructive analysis of thin films (~10 nm). SIMS is useful for both surface and bulk analysis of solids. The ultimate possibility in microanalysis seems to be the atom probe field ion microscopy (AP-FIM). Optical spectroscopy can also be applied to surface analysis, if the emission is restricted to surface constituents.

**TABLE 25.5: COMPARISON OF SOME ELECTRON SPECTROSCOPIES**

| *Parameter* | *Auger* | *XPS* | *ISS* | *SIMS* |
|---|---|---|---|---|
| Excitation | $e^-$ | $h\nu$ | $I_{prim}$ | $I_{prim}$ |
| Emission | $e^-$ | $e^-$ | $I_{sec}$ | $I_{sec}$ |
| Depth of analysis (atomic layers) | 2-10 | 2-10 | 1 | 1-3 |
| Detection of elements (*Z*) | >2 | >1 | >1 | All |
| Sensitivity (ppm) | 1,000 | 1,000 | 1,000 | 1 |

XPS = X-ray photoelectron spectroscopy; ISS = Ion scattering spectroscopy;
SIMS = Secondary ion mass spectroscopy.

## 25.5 ION-BEAM ANALYSIS (IBA) TECHNIQUES

Most common ion-beam analysis (IBA) techniques are particle-induced X-ray/$\gamma$-ray emission (PIXE/PIGE), Rutherford back-scattering spectroscopy (RBS) and nuclear reaction analysis (NRA). Ion beams collimated to micrometer size can be very useful tools for individual particle analysis.

In these methods, the sample is bombarded with protons, inducing the excited elements in the sample to produce X-rays (PIXE) or $\gamma$-rays (PIGE), whose wavelengths are characteristic of the elements present. In PIXE a Si(Li) detector collects X-rays generated at each beam position. A real-time X-ray image is thus obtained, with detection limit 1-10 ppm and resolution 0.5-10 μm. Elements above sodium ($Z > 11$) can be analyzed. Lighter elements ($Z$ up to 5) can be identified by the back scattering of $\alpha$-particles (used to bombard the sample) and the energy lost in the nuclear recoil.

### 25.5.1 Nuclear Reaction Analysis (NRA) Methods

Radionuclides are extensively used in a variety of ways in industrial, environmental, medical and scientific fields. Some of the industrial and environmental applications include oil exploration and recovery, material flow, blockage detection in pipelines, density, thickness and moisture content measurements. Environment related applications are determination of sedimentation discharge in estuaries, monitoring of effluent discharge and dispersion of pollutants into rivers and water bodies. In all radiometric methods the analysis depends on the detection and measurement of radionuclides, either present or added.

The component to be determined is made to react quantitatively with a radioactive reagent of known specific activity ($S_0$). The reaction product is separated and its radioactivity ($\chi$) is determined. The mass of the reaction compound ($M_r$) is

$$M_r \propto (\chi / S_0) \qquad \ldots(25.7)$$

When a quantitative separation of the pure component is difficult, recourse may often be taken, with advantage, to radioactive tracer or isotope dilution methods. Tracer analysis is simple in principle. A known quantity (in terms of activity) of radioactive tracer of the constituent to be measured is added to the sample at the start of the analysis; the constituent is separated and determined. The tracer can be used to follow the distribution of the element in different streams by following the distribution of the isotope in terms of radioactivity.

Isotope dilution method is a simple, sensitive and a practical method of analysis. The unique advantage of the isotope dilution method is that quantitative isolation of the compound from the sample is not necessary. The mass of the tracer, $M_t$, is related to the mass of the sample, $M_0$, and specific activities according to the relation

$$M_t = M_0\left(\frac{S_0}{S_0 - 1}\right) \qquad ...(25.8)$$

Isotope dilution method, and measurement of mean residence time are used for measurement of mass and volume, respectively. If a known aliquot of radiotracer, $M_t$, with known specific concentration, $C_t$, is introduced into a system and a sample of known mass, $M_s$, is taken for a tracer concentration, $C_s$, the mass and volume are given by the relationships.

$$M_s = M_t\left(\frac{C_t}{C_s} - 1\right) \qquad ...(25.9)$$

$$V_s = V_t\left(\frac{C_t}{C_s} - 1\right)$$

$$M_s \propto M_t\left(\frac{N_t}{N_s}\right) \qquad ...(25.10)$$

where, $N_t$ and $N_s$ are count rates in an aliquot of the injection solution after dilution with a tracer material and sample after homogenization, respectively.

### 25.5.2 Measurement of Natural Radioactivity

Natural radioactive nuclides are present in soil and water. U-238, U-235, Th-232, Ra-226, K-40 and Be-7 are the isotopes generally assayed in soils. In water, the radioactive nuclides of concern are Ra-221, Pb-214, Bi-214, K-40 and Na-20. Water can also carry weak β-emitters like $^3$H (tritium) and $^{14}$C. The γ-ray emitters are usually analyzed by γ-ray spectrometry, after chemical treatment of the sample. In the case of water samples, usually, the measurement is carried out after evaporation, or direct introduction of sample in the 'scintillation cocktail.'

Measurement of density, thickness and material flow (ρ . *l*) are based on the γ-ray attenuation by materials. Nucleonic density gauges, based on γ-ray transmission, can be used for determining soil density and soil compactness. Nucleonic thickness gauges are based on γ-ray or β-ray back-scattering by materials. Radiotracer techniques provide on-location and on-time data for sediment transport and effluent discharge into water bodies and physical dilution and dispersion patterns of pollutants in water bodies.

Nucleonic gauges can be used for continuous monitoring of moisture. The method is based on neutron scattering. The neutron source ($^{241}$Am-Be) emits fast neutrons, which interact with the medium. If the medium contains moisture, the fast neutrons would be thermalized and scattered. The amount of scattered neutrons is proportional to the moisture content.

### 25.5.3 Neutron Activation Analysis (NAA)

Neutron activation analysis (NAA) is an induced radioactive elemental analysis technique. NAA is a sensitive technique useful for performing both qualitative and quantitative multi-element analysis of major, minor and trace elements in a variety of matrices. It is based on the measurement of characteristic radiation from radionuclides formed directly or indirectly from irradiation of the material of interest by atomic particles. The technique has found application in environmental analysis, nutritional and health related studies, geological as well as material sciences.

The basic requirements to undertake NAA are a source of neutrons and instrumentation suitable for measuring γ-rays. Although a number of nuclear reactions are known for producing radioactive product nuclei, the most common one that is exploited for NAA is the neutron capture or (*n*, γ) reaction. Of the different neutron sources (reactors, accelerators and radio-isotopic neutron emitters) the most reliable and universally opted one is a research nuclear reactor. Thermal neutron zone in a reactor (energy below 0.5 eV) is the most suitable one for irradiation.

Since most of the samples subject to analysis carry many elements, a detailed knowledge of the reactions that occur when neutrons interact with the target is necessary to arrive at reliable analytical values.

The sequence of events that take place in neutron irradiation is as follows: when a target nucleus captures a neutron a compound nucleus forms in an excited state. This primary product emits prompt γ-rays and attains a less energetic state. In many cases the new nucleus is radioactive and decays with a characteristic half-life, emitting γ-ray(s) of well-defined energy. Most of the NAA procedures depend on the measurement of delayed γ-ray(s).

The protocol generally followed to calculate concentration of an element in the unknown sample comprises—irradiation of the sample and a standard containing a known amount of the element of interest together with thermal neutrons (in a nuclear reactor) and subsequent measurement of the induced radioactivity as a function of time by a γ-ray spectrometer. The activation of the element will be via the $(n, \gamma)$ reaction to produce an isotope of the target with one mass number higher (e.g. $^{127}I\ (n, \gamma)\ ^{128}I$). About 70 elements can be made radioactive by neutron irradiation.

$$^{55}_{25}\text{Mn} + ^{1}_{0}\text{n} \xrightarrow{X(n,\gamma)Y} {}^{56}_{25}\text{Mn (radioactive)} \rightarrow \quad \ldots(25.11)$$

The energy and the intensity of the resultant radiation are characteristic of the isotope and its quantity.

Theoretically, the activity produced by a given element can be calculated from

$$A = N\sigma f S \quad \ldots(25.12)$$

$$\lambda = (\ln 2/\text{half-life}) \quad \ldots(25.13)$$

where, $A$ = observed radioactivity (disintegrations/s = Bq);
$N$ = number of target atoms;
$\sigma$ = atomic activation cross-section ($10^{-24}$ cm$^2$);
$f$ = nuclear particle flux (neutrons/cm$^2$/s);
$S$ = saturation coefficient;
$\lambda$ = decay constant;
$t$ = irradiation time

The cross-section expresses the tendency of the inactive nuclei to absorb neutrons and become activated. In practice, for analysis one uses exclusively the relative method by irradiating known standards simultaneously with, and under the same conditions, as the sample.

Depending on the complexity of the matrix and the concentration at which a component has to be determined there are two broad approaches to NAA. With modern day facilities like automated sample handling, γ-ray measurement with solid state detectors (Ge/Li, operating at liquid nitrogen temperature, 77 K) and computerized data processing it is often possible to determine simultaneously 30 or more elements in most samples without chemical processing. This type of purely instrumental procedure is commonly called "Instrumental Neutron Activation Analysis, (INAA)". However, if the circumstances necessitate chemical separations after irradiation, for elimination of interferences or to concentrate the radioactive isotope of interest, the procedure is called "Radiochemical Neutron Activation Analysis (RNAA)".

### 25.5.3.1 Molecular Activation Analysis (MAA)

In many cases of pollution investigation, for judging biological and environmental effects of element(s), a straightforward total concentration of element(s) is not sufficient but a more detailed analysis of the species of the element present is called for. MAA is an adaptation of NAA to obtain information about the chemical species of elements in systems of interest. In this technique the high degree of sensitivity of NAA is intelligently coupled with some of the effective physical, chemical and biological separation procedures. The first step in MAA is the selective separation of various species. This is followed by identification and determination by NAA.

A simple example of the application of MAA is the determination of mercury in human hair. It is known that methyl mercury has a strong toxic effect on human embryo. To study the transfer mechanism of mercury, mainly methyl mercury, from pregnant women to infants a procedure has been developed to determine longitudinal variation of total, inorganic and organic mercury in their hair. The procedure for methyl mercury takes advantage of the volatility of methyl-mercury cyanide (Me-Hg-CN). Hair sample is mixed with potassium hexacyano-ferrate and sulfuric acid. The resulting Me-Hg-CN is volatilized and absorbed on cystine paper, which is irradiated and directly counted.

#### *25.5.3.2 Sensitivities Available for NAA*

There are many factors, which determine the sensitivities attainable in NAA. They are the irradiation parameters (neutron flux and irradiation and decay times), measurement conditions (detector efficiency, counting time) and nuclear parameters (isotopic abundance, neutron activation cross section, and half-life and γ-ray abundance of the product nucleus). Detection limits of selected elements in typical environmental samples are given in Table 25.6.

**TABLE 25.6: DETECTION LIMITS OF SELECTED ELEMENTS IN TYPICAL ENVIRONMENTAL MATRICES BY NAA**

(ASSUMED IRRADIATION FLUX: $10^{13}/cm^2/sec$)

| *Element* | *Coal (mg/kg)* | *Coal Fly-ash (mg/kg)* | *Atmospheric Particulates ($ng/m^3$)* | *Soil Sediments (mg/kg)* |
|---|---|---|---|---|
| Al | 20 | 35 | 5 | 30 |
| As | 0.05 | 0.25 | 0.015 | 0.45 |
| Br | 0.05 | 0.3 | 0.1 | 0.4 |
| Cd | 0.5 | 4 | 0.1 | 5 |
| Co | 0.015 | 0.1 | 0.03 | 0.05 |
| Cr | 0.5 | 3.5 | 1 | 1.5 |
| Cu | 1 | 5 | 1 | 1.5 |
| Hg | 0.05 | 1 | 0.02 | 1 |
| Mn | 0.1 | 2.5 | 0.5 | 8 |
| Mo | 0.1 | 2.5 | 0.35 | 2.5 |
| Ni | 10 | 40 | 2.5 | 60 |
| Sb | 0.03 | 0.1 | 0.015 | 0.075 |
| Se | 1 | 4 | 0.1 | 3 |
| U | 0.1 | 0.5 | 0.03 | 0.05 |
| V | 0.3 | 0.5 | 0.03 | 1.5 |
| Zn | 2 | 10 | 1 | 5 |

#### *25.5.3.3 Some Environmental Applications of NAA*

A wide variety of environmental samples are amenable for NAA for trace elements. They include samples of aerosols, dust, fossil fuels and their ashes, soils and sediments, sewage sludge, and surface and ground water. For assessing environmental pollution 'biomonitoring' is one of the techniques used. One approach in this line is the direct monitoring, which is based on measuring the quantity of pollutants in suitable organisms or plant matter, rather than in samples from the environment. For this purpose NAA can be applied to any of the following: animals, birds, insects, fishes, aquatic and marine biota, seaweed, algae, lichens, mosses and plant materials. For example, a great deal of attention has been given on the feasibility of human hair as a material easily accessible for non-invasive sampling in individuals and population groups to determine occupational or environmental exposure. Thus, a correlation is established between arsenic content of hair of children living in various distances from a pollution source and the expected degree of arsenic contamination of the air.

## 25.6 CHOICE OF ANALYTICAL TECHNIQUES

There is a variety of instrumental techniques available for the analysis of environmental pollutants, and the choice of an analytical technique depends on various factors—suitability, availability, sensitivity, and cost effectiveness. The AAS, ICP-AES and ICP-MS are the primary choices for metal speciation. By contrast, non-metals such as H, C, N, O, P and S and the halogens cannot be determined at their typical concentration levels in airborne particles. NAA is an extremely sensitive and versatile method for determination of trace elements. However, since it requires access to a nuclear research reactor, it is less widely applied for elemental analysis than techniques like AAS, ICP, ICP-MS and XRF. NAA is preferred for rare earth and some other elements. However, limits of detection offered by NAA, AAS and ICP-AES exhibit large fluctuations among the elements. By contrast, limits of detection vary very little from element to element in XRF and PIXE methods. Once sample is prepared XRF is more rapid than NAA. Non-destructive techniques are not sensitive enough to detect Cd and Mo. Comparison of detection characteristics of trace elements and detection limits of analytical techniques are given in Tables 25.7 and 25.8.

**TABLE 25.7:** COMPARISON OF SOME ANALYTICAL TECHNIQUES FOR ANALYSIS OF TRACE ELEMENTS

| *Technique* | *Accuracy* | *Sensitivity* |
|---|---|---|
| AAS | Poor | Good |
| ICP-AES | Average | Good |
| ICP-MS | Average | Good |
| ICP-IDMS | Good | Average |
| XRF | Good | Good |
| TXRF | Average | Good |
| PIXE | Poor | Average |
| NAA | Good | Good |

## 25.7 RADIOIMMUNOASSAYS

Radioimmunoassays are techniques for the estimation of particular compounds, based on competition for binding to antibody between the analyte in the sample and radioactively labeled analyte. The more of the analyte there is in the sample, the less of radioactively labeled analyte will bind to the antibody. These techniques combine the high degree of specificity of the antibody-antigen reaction with the extreme sensitivity of radioactivity measurement. These methods were first used to measure insulin (*Yalow & Berson,* 1960) and the thyroid hormones (*Ekins*, 1960) in human plasma. Since then, they have been extended to practically all areas of biology and medicine, pharmacology, forensic medicine and environmental and food sciences. Radioimmunoassays are used to detect and estimate a wide variety of analytes, such as peptides, proteins, tumor markers, nucleic acid components, steroids, vitamins, bacterial and viral antigens, drugs, toxins, antibiotics, food additives, carcinogens and other hazardous chemicals, pesticides, herbicides and other pollutants).

**TABLE 25.8:** DETECTION LIMITS OF SOME ANALYTICAL METHODS FOR TRACE ELEMENTS

| *Element* | *AAS (μg)* | *ICP-AES (μg)* | *ICP-MS (μg)* | *XRF (μg)* | *PIXE (μg)* | *NAA (μg)* |
|---|---|---|---|---|---|---|
| As | 0.2 | 3 | 0.04 | 4 | 0.4 | 0.03 |
| Cd | 0.003 | 2 | 0.06 | 6 | 10 | 0.6 |
| Cr | 0.01 | 4 | 0.6 | 15 | 1 | 0.03 |
| Cu | 0.02 | 4 | 0.3 | 6 | 0.3 | 0.03 |
| Fe | 0.02 | 0.5 | – | 10 | 0.5 | 6 |
| Hg | 2 | 20 | 0.02 | 7 | 1 | 0.03 |
| Ir | – | 40 | 0.07 | – | 15 | 0.006 |
| Mn | 0.01 | 1 | 0.1 | 10 | 0.6 | 0.001 |
| Mo | 0.02 | 8 | 0.04 | 5 | 2 | 0.3 |
| Ni | 0.2 | 6 | 0.1 | 5 | 0.4 | 0.3 |
| Sb | 0.1 | 20 | 0.05 | 8 | 15 | 0.001 |
| Sc | 0.5 | 50 | 0.8 | 2 | 0.4 | 0.03 |
| Sn | 0.1 | 17 | 0.06 | 8 | 16 | 1 |
| V | 0.2 | 4 | 0.03 | 20 | 1.5 | 0.03 |
| Zn | 0.001 | 1.5 | 0.2 | 5 | 0.3 | 0.3 |

*More accurate methods are underlined*

**AAS:** Atomic absorption spectrometry;

**ICP-AES:** Inductively coupled plasma-atomic emission spectrometry;

**ICP-MS:** Inductively coupled plasma-mass spectrometry;

**ICP-IDMS:** Inductively coupled plasma-isotope dilution mass spectrometry;

**XRF:** X-ray fluorescence;

**TXRF:** Total reflection XRF.

**PIXE:** Particle induced X-ray emission;

**NAA:** Neutron activation analysis.

## 25.7.1 Radioimmunoassay (RIA)

The general principle of RIA is the reaction between the analyte (antigen, Ag) in the sample or standard, a fixed amount of the radiolabeled antigen ($Ag^*$) and a limited amount of a specific antibody (Ab). This is followed by the separation of the immune complex and measurement of the radioactivity associated with it. In order to allow the competitive-binding principle to operate, the antibody must be present in a limited amount relative to the total quantity of the antigen present. As the labeled antigen ($Ag^*$) and unlabeled antigen (Ag) molecules compete for the binding sites of the antibody (Ab), the amount of radiolabeled antigen (Ag*) that can bind to the antibody (Ab) is related inversely to the amount of unlabeled antigen (Ag) present. That is, the radioactivity associated with the immune complex is inversely related to the analyte concentration. The immune complex is separated from the reaction mixture and the radioactivity of the complex is measured. A measure of the amount of analyte present in a sample is obtained by constructing a standard curve (Fig. 25.17).

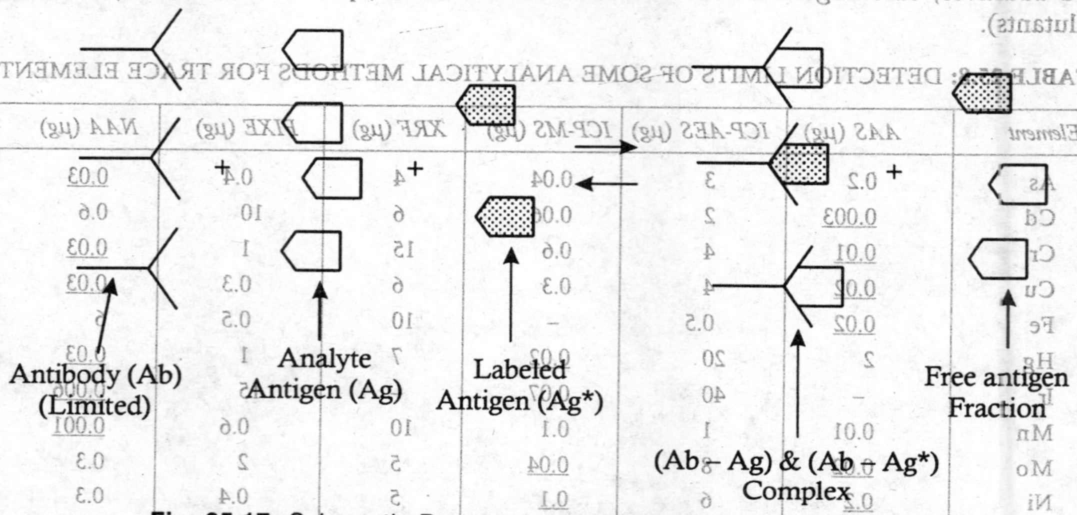


**Fig. 25.17:** Schematic Representation of Radioimmunoassay (RIA)

A typical radioimmunoassay procedure will consist of the following steps:

1. A fixed, limited quantity of the antibody (Ab) is incubated with a fixed quantity of labeled antigen (Ag*) and varying quantities of cold antigen (in the form of standards or patient's serum).
2. At the end of the reaction, the antibody-bound antigen (Ag-Ab and Ag*-Ab) is physically separated from the free antigen (by adsorption, precipitation, ion-exchange).
3. The radioactivity of the antigen-antibody complex is measured using radiation counters.
4. A calibration curve is obtained by plotting a graph relating the standard antigen concentrations and the radioactivity associated with the corresponding antigen-antibody complex (Fig. 25.18). This curve is used to read the unknown sample concentrations.

### 25.7.2 Immunoradiometric Assay (IRMA)

Immunoradiometric assay (IRMA) is a non-competitive immunoassay that utilizes radiolabeled antibody (Ab*) instead of a radiolabeled antigen, Ag* as the tracer. In this type of assay, the analyte (in the standard or in the sample) binds simultaneously to two different antibodies. One antibody is labeled with a radioisotope and serves as the tracer, while the other antibody is immobilized on a solid surface and facilitates the separation of the immune complex at the end of the reaction. Both the antibodies are in excess. At the end of the reaction, the antigen-antibody complex bound to the solid surface is separated and the radioactivity associated with it is measured. In IRMA, the radioactivity of the antigen-antibody complex is directly related to the analyte concentration. As in the case of RIA, a calibration curve is obtained for IRMA and the concentration of the sample is read from it (Fig. 25.19).

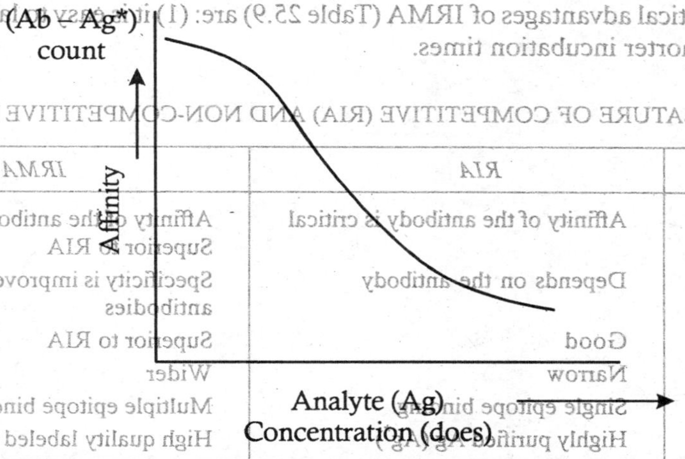


**Fig. 25.18:** Dose-response Relationship (Ag*-Ab) versus Analyte (Ag) Concentration

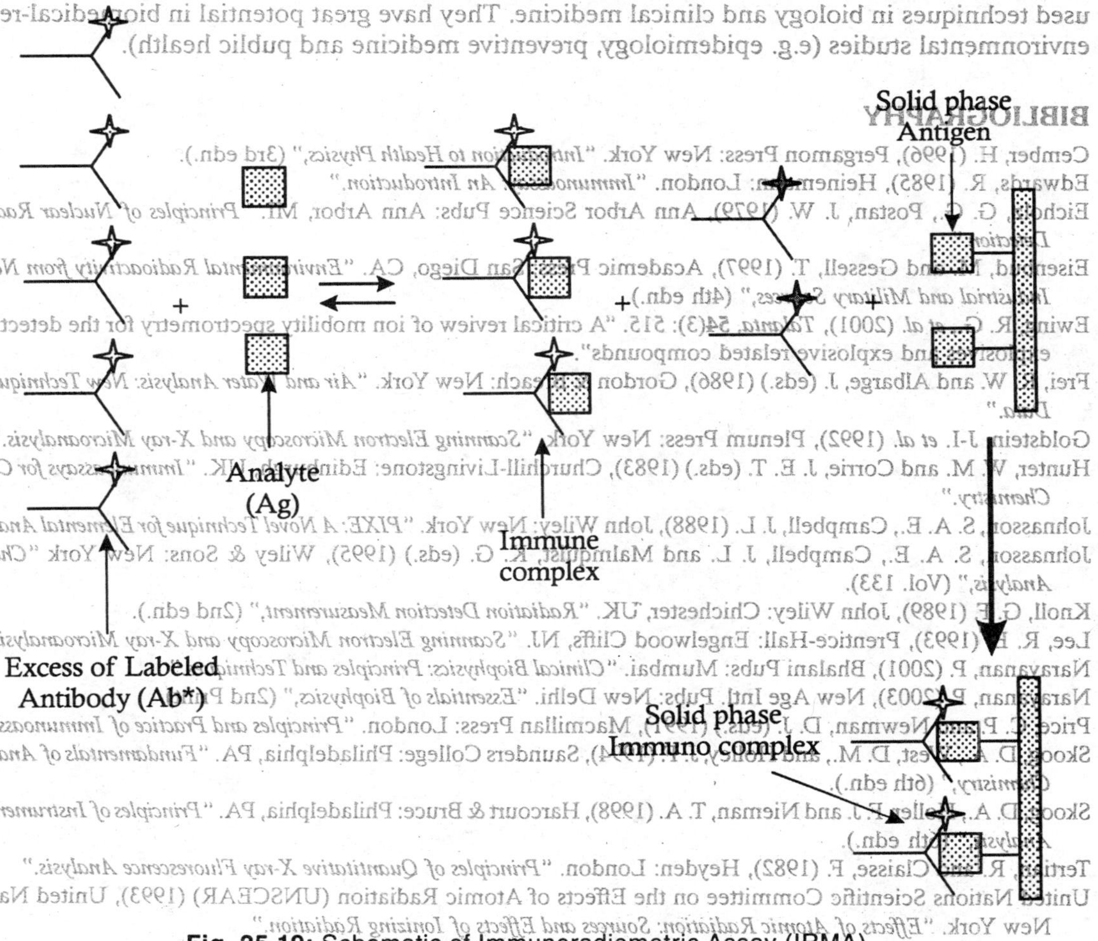


**Fig. 25.19:** Schematic of Immunoradiometric Assay (IRMA)

Some of the practical advantages of IRMA (Table 25.9) are: (1) it is easy to label macromolecules than haptens, (2) shorter incubation times.

**TABLE 25.9:** FEATURE OF COMPETITIVE (RIA) AND NON-COMPETITIVE (IRMA) ASSAYS

| *Parameter* | *RIA* | *IRMA* |
|---|---|---|
| Sensitivity | Affinity of the antibody is critical | Affinity of the antibody is less critical; Superior to RIA |
| Specificity | Depends on the antibody | Specificity is improved due to the use of two antibodies |
| Precision | Good | Superior to RIA |
| Dynamic range | Narrow | Wider |
| Binding | Single epitope binding | Multiple epitope binding |
| Reagents | Highly purified Ag ($Ag^*$) | High quality labeled reagent ($Ab^*$) |

Immunoassays—enzyme-based (Chapter 24) and radioactivity-based are routine and widely used techniques in biology and clinical medicine. They have great potential in biomedical-related environmental studies (e.g. epidemiology, preventive medicine and public health).

## BIBLIOGRAPHY

Cember, H. (1996), Pergamon Press: New York. "*Introduction to Health Physics*," (3rd edn.).

Edwards, R. (1985), Heinemann: London. "*Immunoassay: An Introduction.*"

Eicholz, G. G., Postan, J. W. (1979), Ann Arbor Science Pubs: Ann Arbor, MI. "*Principles of Nuclear Radiation Detection.*"

Eisenbud, M. and Gessell, T. (1997), Academic Press: San Diego, CA. "*Environmental Radioactivity from Natural, Industrial and Military Sources*," (4th edn.).

Ewing, R. G., *et al.* (2001), *Talanta*, **54**(3): 515. "A critical review of ion mobility spectrometry for the detection of explosives and explosive related compounds".

Frei, R. W. and Albarge, J. (eds.) (1986), Gordon & Breach: New York. "*Air and Water Analysis: New Techniques and Data.*"

Goldstein, J-I. *et al.* (1992), Plenum Press: New York. "*Scanning Electron Microscopy and X-ray Microanalysis.*"

Hunter, W. M. and Corrie, J. E. T. (eds.) (1983), Churchill-Livingstone: Edinburgh, UK. "*Immunoassays for Clinical Chemistry.*"

Johnasson, S. A. E., Campbell, J. L. (1988), John Wiley: New York. "*PIXE: A Novel Technique for Elemental Analysis.*"

Johnasson, S. A. E., Campbell, J. L. and Malmquist, K. G. (eds.) (1995), Wiley & Sons: New York "*Chemical Analysis*," (Vol. 133).

Knoll, G. F. (1989), John Wiley: Chichester, UK. "*Radiation Detection Measurement*," (2nd edn.).

Lee, R. E. (1993), Prentice-Hall: Engelwood Cliffs, NJ. "*Scanning Electron Microscopy and X-ray Microanalysis.*"

Narayanan, P. (2001), Bhalani Pubs: Mumbai. "*Clinical Biophysics: Principles and Techniques.*"

Narayanan, P. (2003), New Age Intl. Pubs: New Delhi. "*Essentials of Biophysics*," (2nd Print).

Price, C. P. and Newman, D. J. (eds.) (1991), Macmillan Press: London. "*Principles and Practice of Immunoassays.*"

Skoog, D. A., West, D. M., and Holley, J. F. (1994), Saunders College: Philadelphia, PA. "*Fundamentals of Analytical Chemistry*," (6th edn.).

Skoog, D. A., Holler, F. J. and Nieman, T. A. (1998), Harcourt & Bruce: Philadelphia, PA. "*Principles of Instrumentation Analysis*," (6th edn.).

Tertian, R. and Claisse, F. (1982), Heyden: London. "*Principles of Quantitative X-ray Fluorescence Analysis.*"

United Nations Scientific Committee on the Effects of Atomic Radiation (UNSCEAR) (1993), United Nations: New York. "*Effects of Atomic Radiation: Sources and Effects of Ionizing Radiation.*"

Vourvopoulos, G. and Womble, P. C. (2001), *Talanta*, **54**(3): 459. "*Pulsed Fast/Thermal Neutron Analysis: a Technique for Explosive Detection.*"

Wagner, H. N. Jr. and Szabo, Z. (eds.) (1995), Saunders: Orlando, FL. "*Principles of Nuclear Medicine.*"

Willeke, K. and Baron, P. A. (eds.) (1993), Van Nostrand-Reinhold: New York. "*Aerosol Measurement—Principles, Techniques and Applications.*"

Yalow, R. S. and Berson, S. A. (1960), *J. Clin. Invest.*, **39**: 1157.

... (1983), Wiley Interscience: New York. "*Instrumentation for Environmental Monitoring, (Vol I: Radiation),*" (2nd edn.).

# 26

# Physical Methods in Environmental Pollution Analysis—Other Techniques

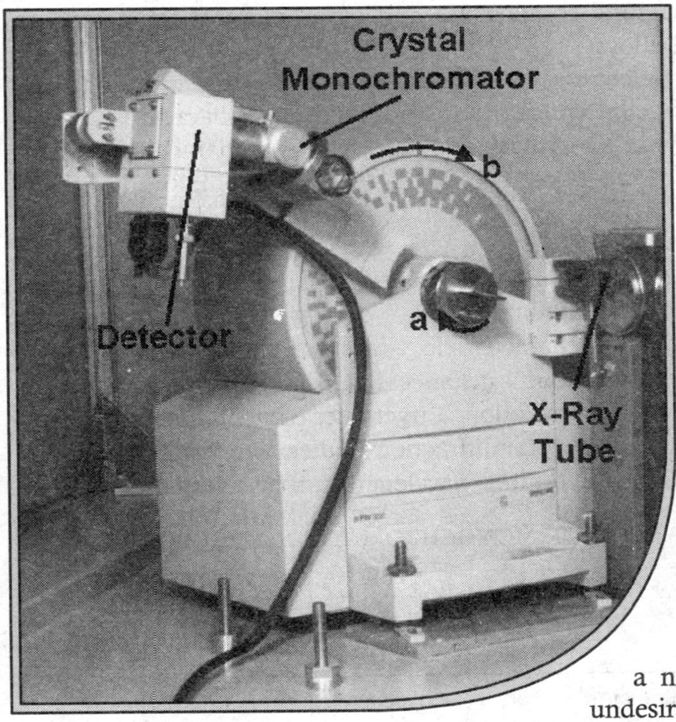


Some of the other techniques employed in the analysis of environmental pollutants are X-ray diffraction and microscopic methods. These techniques are employed to determine morphological and structural details of pollutants.

X-ray methods provide structural details at atomic and molecular resolutions. In the analysis of environmental pollutants, powder X-ray diffraction (Debye-Sherer method) is generally used for the determination of composition and morphological details of semicrystalline (and polycrystalline) specimen. Optical microscopy is used for determining the size and number distribution of particulate pollutants, while electron microscopy provides high-resolution ultrastructure details.

Noise pollution is increasingly becoming a nuisance and health hazard. Noise is undesirable sound, measured in terms of pressure levels (decibel) by sound level meters. Effects of noise pollution have physical, physiological and psychological dimensions. Noise above 80 decibels (tolerance level) is of concern to human health and prolonged exposure to such noise leads to mental imbalance. General methods of noise pollution control are—(1) removal of the source, (2) reduction of the decibel level, (3) minimizing the vibratory parts of the gadgets and machines by better design, fastening and padding with noise insulators.

This chapter outlines the details of X-ray diffraction and microscopic methods and noise pollution and its effects on human population.

Some of the other techniques of analysis of pollutants are—ultrastructure and morphological studies by X-ray diffraction, and electron diffraction and electron microscopy, and analysis of noise (mechanical) pollution.

## 26.1 X-RAY DIFFRACTION METHODS

Diffraction (optical, electron, neutron and X-ray) methods are aimed at obtaining structural details of objects at various resolutions. Of these, X-ray diffraction techniques are the most versatile and to-date the only methods that can be employed to obtain three-dimensional architecture of molecules at atomic (~1 Å), and molecular (>1 Å) resolutions, depending on the state of the matter. The discovery of X-rays by Konrad Röntgen, and X-ray diffraction by Max von Laue and Paul Ewald, were both fortuitous, but they are recognized as landmark discoveries that led to the advancement of science in many fields. Structure determination by X-ray diffraction is highly complex and involved, and therefore, this technique is employed in the analysis of environmental pollutants where their structure elucidation is required. "Powder diffraction" X-ray analysis is the method, generally employed, for determining structural composition and morphological characteristics, and stoichiometry of clays, minerals and alloys.

X-rays are produced when high-energy electrons are decelerated by a target (anticathode- Cu, Mo, Ni, Ag, Fe, etc), giving rise to X-ray continuum (Bremsstrahlung) and characteristic X-ray lines (spectra) if the applied voltage is above the critical voltage, which is a characteristic of the target. That is, X-ray spectral lines are element-specific and this feature is employed in X-ray microanalysis (XMA) to determine the elemental composition of materials. Characteristic X-ray line spectra are used in structure determination.

Diffraction effects come into fore if the dimensions of the object to be probed and that of the probe (wavelength) are of the same order of magnitude. X-rays belong to the electromagnetic spectrum with short wavelengths. Wavelengths of characteristic X-rays (e.g. Cu-$K_\alpha$ = 1.542 Å) used in X-ray diffraction studies are of the order of interatomic distances (e.g. C – C ~ 1.54 Å). Therefore, only X-rays are suitable to probe matter at atomic resolution. Target-filter combination (Table 26.1) is used to obtain monochromatic radiation (Fig. 26.1) for diffraction studies. Synchrotron radiation provides a wavelength continuum from which the desired wavelength can be selected.

**TABLE 26.1:** TARGET-FILTER COMBINATION FOR MONOCHROMATIC X-RAY GENERATION

| *Target (Anticathode)* | *Filter* | *Wavelength of K-Absorption Edge* (Å) | *Wavelength* (Å) |
|---|---|---|---|
| Molybdenum (Mo) (42) | Zirconium (40) | 0.6888 | $K_\alpha$ = 0.711<br>$K_{\alpha1}$ = 0.70926<br>$K_{\alpha2}$ = 0.71354<br>$K_\beta$ = 0.63225 |
| Copper (Cu) (29) | Nickel (Ni) (28) | 1.4869 | $K_\alpha$ = 1.54178<br>$K_{\alpha1}$ = 1.5405<br>$K_{\alpha2}$ = 1.54434<br>$K_\beta$ = 1.39217 |

| *Target (Anticathode)* | *Filter* | *Wavelength of K-Absorption Edge* (Å) | *Wavelength* (Å) |
|---|---|---|---|
| Cobalt (Co) (27) | Iron (Fe) (26) | 1.7429 | $K_\alpha = 1.789$<br>$K_{\alpha 1} = 1.7889$<br>$K_{\alpha 2} = 1.79279$<br>$K_\beta = 1.62073$ |
| Iron (Fe) (26) | Manganese (Mn) (25) | 1.8954 | $K_\alpha = 1.937$<br>$K_{\alpha 1} = 1.93597$<br>$K_{\alpha 2} = 1.93991$<br>$K_\beta = 1.75654$ |
| Chromium (Cr) (24) | Vanadium (V) (23) | 2.2676 | $K_\alpha = 2.290$<br>$K_{\alpha 1} = 2.28962$<br>$K_{\alpha 2} = 2.29352$<br>$K_\beta = 2.08479$ |

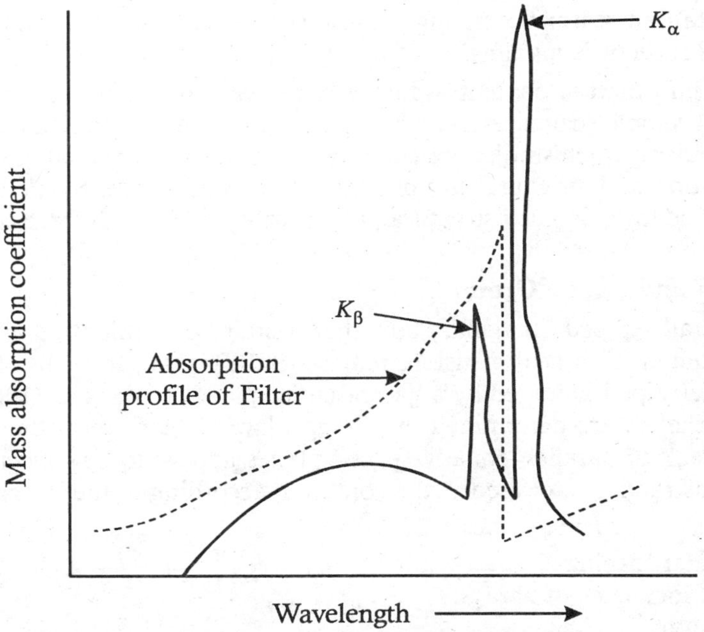


**Fig. 26.1:** Target-Filter Combination to obtain Monochromatic X-rays

### 26.1.1 Principles of X-ray Diffraction

The basic principles of X-ray diffraction (scattering) are conceptually similar to those of optical and electron microscopies. A beam of electromagnetic radiation (or electrons) is focussed on to an object by a lens system and the scattered rays are recorded and analyzed. There exists to-date no proper lens system for focussing X-rays (refractive index of X-rays = 1 in all media). What are recorded (by film or electronic methods) in any X-ray diffraction method are the scattered X-rays (diffraction patterns). In the absence of a lens system, image of the object is not formed but is reconstructed by mathematical image-reconstruction processes from the diffraction data. The operational procedures involved in the transformation of the diffraction patterns to an image are highly intricate and involved.

Matter in solid state can exist as amorphous (e.g. glass, clay), semi-crystalline (e.g. metals, minerals) or single-crystalline state (crystalline sugar, NaCl, etc). Though the amount of diffracted energy is the same for a substance in whatever forms it exists, the information content obtainable is not the same. Completely amorphous solids are those with random distribution of internal structure (liquid-like). Amorphous materials give rise to diffuse scattering. Semi-crystalline, or polycrystalline solids are those with random distribution of crystallites (e.g. powdered composition of crystallite materials or alloys). They give rise to diffraction lines and arcs (e.g. minerals, alloys and fibers such as DNA, collagen, cellulose, etc). A single-crystalline material is the one with periodic arrangement of its internal structure (atoms, molecules or clusters of molecules) in three-dimension (e.g. individual crystalline NaCl, ammonium sulfate, globular proteins and viruses, etc). Due to the periodic arrangement and internal spatial organization the scattering is coherent and the diffracted rays are channeled in certain directions only, giving rise to discrete diffraction spots.

1. **Amorphous state:** Scattering is random and diffuse. Information content is minimal.
2. **Semi-crystalline state:** Scattering is semi-random. Gives rise to diffraction lines and arcs.
3. **Single-crystalline state:** Scattering is coherent. Gives rise to discrete diffraction spots. Information content is maximal.

Due to limited information content available in the case of amorphous and semi-crystalline materials, only very simple structures in higher space groups can be studied by powder diffraction (Debye-Sherer) methods. Such studies are common in metallurgy, mineralogy and environmental studies of composition and structural morphology of clays and soils. Single-crystalline state is a prerequisite for elucidating detailed structural information by X-ray diffraction.

### 26.1.2 Unit Cell and Space Group

A 'unit cell' is a parallelepiped (imaginary box) that contains one unit of pattern and translational repeat of the unit cell in all three dimensions represents the crystal. In terms of Miller indices, the unit cell is a parallelepiped bounded by adjacent lattice planes of the sets (100), (010) and (001). The unit cell parameters are determined by 'Bragg's law', which states that a beam of X-rays incident upon a stack of parallel, equally spaced planes appear to be reflected by these planes (Fig. 26.2). For constructive interference, the condition according to the Bragg's equation is

$$2d\sin\theta = n\lambda \qquad \text{...(26.1)}$$

where, $d$ = interplanar spacing;
$\theta$ = angle of incidence of X-rays;
$\lambda$ = wavelength;
$n$ = order of reflection, 1, 2, 3, ...

Bragg's equation enables determination of unit cell parameters from the diffraction patterns. The biggest $d$-spacing corresponds to the smallest Bragg angle and hence the unit cell parameters can be calculated from the smallest Bragg angles (from the diffraction photographs).

### 26.1.3 Structure Determination

Bragg's equation is useful in determining the unit cell parameters and geometry (external morphology) only. It is inadequate for structure determination, that is, the determination of positions of atoms and molecules in the unit cell (in the asymmetric unit to be precise). Structure determination requires an understanding of the wave properties, as X-ray diffraction is a wave phenomenon.

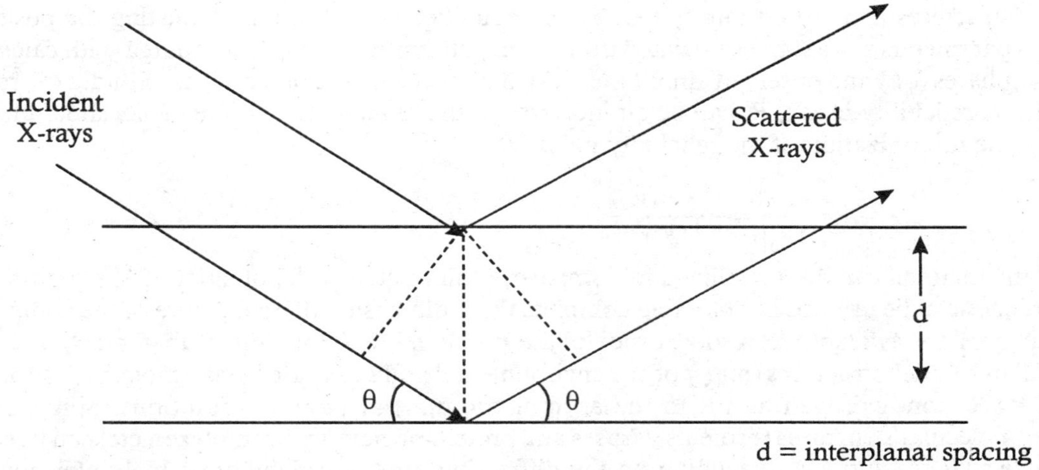


**Fig. 26.2:** Bragg's Interpretation of X-ray Scattering ($2d\sin\theta = n\lambda$)

A wave is characterized by: (*i*) amplitude and (*ii*) phase. These two parameters must be available for each Bragg reflection, *hkl*, (diffraction spot) for image reconstruction by Fourier transformation. Amplitudes for Bragg reflections, *hkl*, can be obtained from the intensity data of the diffraction patterns, recorded photographically, electronically or by any other means. Amplitude of a reflection, *hkl*, is the square root of its intensity. For structure determination the information pertaining to both the amplitudes (intensities) and the phases of diffracted reflections is required. All X-ray diffraction techniques record intensity data only, that is, the square of the amplitudes of reflections, *hkl*; diffraction data cannot provide the 'phase' information. As the phases of the diffracted spots cannot be obtained experimentally, mathematical methods have to be invoked to obtain phase information, corresponding to each diffracted spot. This is the classic "phase problem" encountered in X-ray crystallography, and resolving the "phase problem" (mostly by mathematical procedures) is central and crucial to the structure determination by any X-ray diffraction techniques.

In the case of amorphous, semi-crystalline and fibrous materials, which are incidentally of interest in the analysis of environmental pollutants, the information content obtainable is meager. Generally, powder X-ray diffraction (Debye-Sherrer) methods are used to elucidate structure of substances that crystallize in higher space groups (cubic, tetragonal and orthorhombic). Fiber X-ray diffraction methods are employed to determine structures of fibrous molecules like nucleic acids, collagen, fibrous proteins and polymers. Debye-Sherer camera or powder diffractometer is universally used in powder diffraction studies.

In the analysis of environmental pollutants by powder X-ray diffraction methods, the general protocol followed is as follows:

1. The specimen is ground to a fine powder and packed in a glass capillary (a thin wire if it is a metal or alloy) and the diffraction pattern are recorded on photographic films or electronically (diffractometer). The diffraction pattern consists of concentric arcs (rings). The unit cell parameters and the contents (formula mass) in the unit cell are computed applying the Bragg equation.

2. For determining the structure, the intensity profiles of the diffracted arcs are compared with those of similar solved structures to characterize the structural features of the test specimen.
3. Refinements, improvements and structural variations of the test specimen from the solved structures are carried out by iterative computational methods, by adjusting the positional parameters so that the calculated diffraction patterns (FC-maps), computed with calculated phases ($\phi C$) and observed amplitudes ($F_0$), matched with the observed amplitudes ($F_0$-maps).
4. A reliability index, $R$, can be an indicator of the refinements. Refinements should lead to the minimization of the reliability index, $R$.

$$R = \frac{|\Delta F|}{|F_0|} = \frac{\left||F_0| - |F_C|\right|}{|F_0|} \qquad \text{...(26.2)}$$

If the material can be crystallized in a single-crystalline state, full potential of X-ray diffraction techniques can be utilized in obtaining complete three-dimensional architecture of the compound. Procedures are available to resolved the "phase problem" (e.g. the "direct methods") and solve "small molecule" structures (many of the environmental pollutants are "small molecule" structures with <100 non-hydrogen atoms in the asymmetric unit) at atomic resolutions. Single-crystal macromolecular structures (proteins, viruses and protein-nucleic acid complexes, etc) can be solved at molecular resolutions, depending on the diffraction limit. Availability of high-intensity, high degree collimation, and multi-wavelength radiation sources (synchrotron radiation sources), and versatile X-ray detectors with increased sensitivity, increased dynamic range, and increased counting rate, have made it possible fast and accurate data-collection. With these developments in technology, the potential of X-ray diffraction techniques has increased tremendously, not just in elucidating the three-dimensional structures, but also in real-time analyses of structural dynamics at molecular level (time-resolved X-ray crystallography).

## 26.2 MICROSCOPIC METHODS

All microscopies (optical, electron, etc) are aimed at obtaining finer structural and morphological details with "meaningful magnification." The structural details obtainable from any microscopy are set by the resolution limit of the lens system. The resolution is defined as the minimum distance below which two point objects are not distinguishable as distinct and separate entities. Magnification is simply the number of times the image is larger than the object. For optical microscopes, the resolution limit, $d$, is set by the diffraction effects of the lens system, and according to the Abbe's criterion

$$d = \frac{0.612\lambda}{n \cdot \sin\alpha} \approx 0.5\lambda \qquad \text{...(26.3)}$$

where, $\lambda$ = wavelength of light;
$\alpha$ = aperture angle;
$n$ = refractive index of the medium.

The resolution in optical microscopy cannot be better than 200 nm (0.2 μm). Optical microscopic methods are commonly employed in environmental monitoring for particle size and number distribution determination. Electron microscopy provides higher resolution to obtain structural details at cellular level. The resolution limit in electronic microscopy is set by lens aberrations (spherical, chromatic and astigmatic). The outer most electrons being refracted more strongly than those near the axis cause spherical aberration. Chromatic aberration is caused by fluctuations in the accelerating voltage of the electron gun.

Scanning electron microscopy (SEM) is used in environmental studies to obtain morphological features at high resolution. In SEM, an ultra-thin electron beam scans the specimen surface in raster pattern (as in a TV picture scan), and the back-scattered electrons are collected as function of position by a detector system. The data are stored in a computer and the image reconstruction is carried out from the stored data. SEM provides surface topographic information to about 50 Å (~5 nm) resolutions. X-rays produced in SEM setup are used for elemental analysis by XRMA technique (dealt in Chapter 25). SEM provides a large depth of focus and causes minimum damage to the specimen. Specimen preparation is easier than in transmission electron microscopy (TEM). Therefore, SEM is used for studying biological specimen like pollen, tissues, and cells and for surface studies to assess the degree of damage caused by pollutants. SEM can also be used for determination of thickness (size) of particulate matter by shadow casting. This is carried out by shadowing particles with carbon or a metal from a known angle and measuring the length of a shadow on a photographic plate from which the thickness can be calculated.

$$d = l \tan\alpha \qquad \ldots(26.4)$$

where, $d$ = thickness of the particle;
$l$ = length of the shadow;
$\alpha$ = shadow angle;
$d = l$, for
$\alpha = 45°$

### 26.2.1 Electron Diffraction

Elastic scattering in crystalline specimen gives rise to electron diffraction, a process very similar to X-ray diffraction. Electron diffraction is suitable for qualitative and quantitative study of crystalline solid surfaces (clay, coal or ore samples) and thin (crystalline) films (<1 μm) for which X-ray diffraction is not suitable. Single crystalline materials give rise to discrete diffraction spots, while polycrystalline specimens give rise to arcs and circles centered round the incident electron beam (Debye-Scherer rings). Data on unit cell and space group may be obtained by comparing the diffraction patterns with ASTM or X-ray diffraction patterns.

## 26.3 NOISE POLLUTION

Noise is undesirable sound. Sound is mechanical transfer of energy without transfer of mass. Propagation of sound waves (pressure fluctuations) requires a medium. The velocity, $C$, of sound in a medium is given by

$$C = 344\sqrt{\frac{T}{T_0}}\ \text{m / s}^1 \qquad \ldots(26.5)$$

where, $T$ = temperature in absolute scale;
$T_0$ = 293 K.

### 26.3.1 Intensity of Sound

The intensity of sound ($I$) is measured in terms of pressure levels. The response of human ear to sound pressures is quite remarkable, extending over a wide range of pressure, $2 \times 10^{-5}$ to 100 $N{\cdot}m^{-2}$. That is, the response is in the $1 : 10^7$ range. The intensity ($I$) response is logarithmic and is represented by decibels, dB.

$$L_p = 20 \log_{10} \frac{P}{P_{ref}} \, .dB \qquad \text{...(26.6)}$$

$$dB = 10 \log\left(\frac{A}{A_0}\right) \qquad \text{...(26.7)}$$

Pressure level, *LP*, is the logarithmic ratio of measured sound pressure, *P*, and a reference sound pressure, $P_{ref}$, ($P_{ref} = 2 \times 10^{-5}$ N·m$^{-2}$ ($10^{-12}$ W·m$^{-2}$)), which is the lowest sound pressure some people can detect. Some of the sound pressure levels are given in Table 26.2.

**TABLE 26.2:** SOUND PRESSURE LEVELS

| *Sound Pressure* ($N/m^2$) | *Sound Pressure Level* (*dB*) | *Source/ Effect* |
|---|---|---|
| $2 \times 10^{-5}$ | 0 | Threshold of hearing |
| – | 10 | Rustling of leaves |
| $2 \times 10^{-4}$ | 20 | Whisper at 1 m distance |
| – | 30 | Solitary place |
| $2 \times 10^{-3}$ | 40 | Quite room |
| – | 50 | Business office |
| $2 \times 10^{-2}$ | 60 | Normal conversation |
| – | 70 | Regular noise |
| $2 \times 10^{-1}$ | 80 | Traffic noise |
| – | 90 | Heavy vehicular noise |
| $2 \times 10^{0}$ | 100 | Pneumatic drill/aircraft noise |
| – | 110 | Roc-band music |
| $2 \times 10^{1}$ | 120 | Threshold of pain |
| – | 130 | Rocket launch |

### 26.3.2 Sound Levels

Sound levels are frequency-weighted, and sound level meters are provided with a set of frequency-weighted networks, with A-, B- and C-scales (Fig. 26.3). Sound levels are marked according to the weighting scheme scales, dB-(A), dB-(B) and dB-(C). A-scale has a very prominent frequency-dependence for frequencies below 1000 Hz. A-scale is the most useful one, as it approximates the response of the human ear to sound levels below 55 dB. The B-scale is useful for measuring steady exposure to continuous noise (drone). C-scale is practically linear, with little dependence on frequency in the greater part of the audible frequency range.

### 26.3.3 Noise Pollution Measurement

Sound is due to pressure fluctuations. The pressure, *P*, is related to the density, ρ

$$P = C^2 \rho \qquad \text{...(26.8)}$$

$$V^2 P - \frac{1}{C^2} \frac{\partial^2 P}{\partial t^2} = S(x,t) \qquad \text{...(26.9)}$$

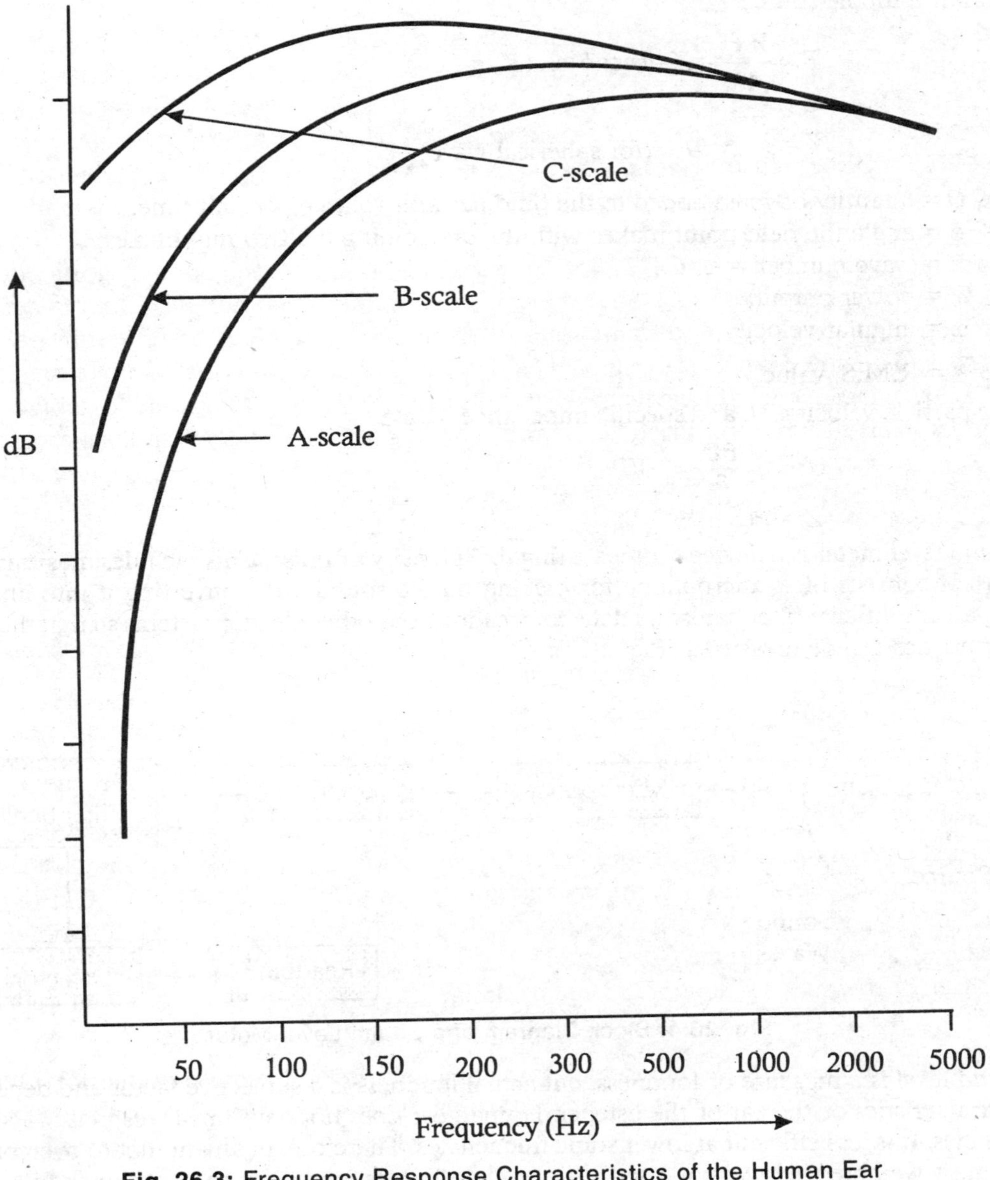


**Fig. 26.3**: Frequency Response Characteristics of the Human Ear

1. If there is no source, $S(x, t) = 0$, and

$$\frac{\partial^2 P}{\partial x^2} - \frac{1}{C^2}\frac{\partial^2 P}{\partial t^2} = 0 \text{ (with no source term)} \qquad \text{...(26.10)}$$

2. For a monopole,

$$S(x,t) = -\frac{\partial Q}{\partial t} \qquad \text{...(26.11)}$$

3. For a dipole source

$$P = \frac{WQ_0}{4\pi r}(K \cdot d\cos\phi)\sin(K \cdot r - \omega t) \quad \text{...(26.12)}$$

$$W = \frac{P^{-2}}{\rho C} \cdot 4\pi^2 \text{ (for spherical wave)} \quad \text{...(26.13)}$$

where, $Q$ = quantity of mass added to the fluid per unit volume per unit time;
$\phi$ = angle the field point makes with the line joining the two monopoles;
$K$ = wave number = $\omega / C$;
$W$ = power output;
$\omega$ = angular velocity;
$\sqrt{P^{-2}}$ = R.M.S. value.

The particle velocity, $U$, and specific impedance, $Z$, are

$$U = \rho_0 \frac{\partial U}{\partial t} = -VP \quad \text{...(26.14)}$$

$$Z = PC \quad \text{...(26.15)}$$

Sound level meter is a device for measuring the intensity of noise. This includes music and other sounds. It consists of a microphone for picking up the sound and converting it into an electric signal, an amplifier, a filter bank and detector/readout and other electric systems so that the desired characteristics can be measured (Fig. 26.4).

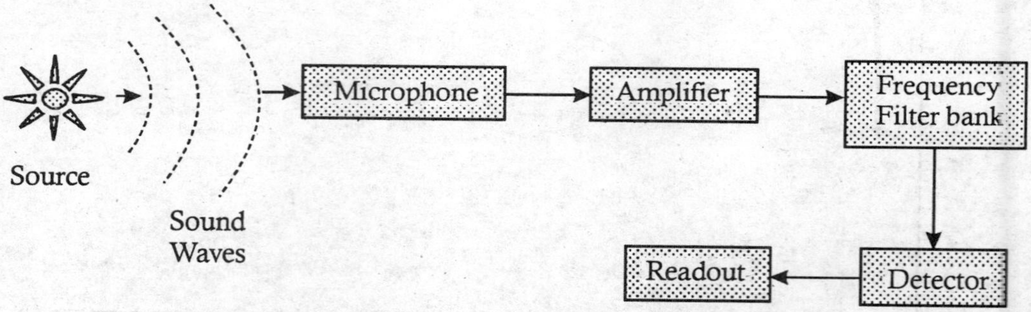


**Fig. 26.4:** Block Diagram of a Sound Level Meter

Sound level is a measure of loudness, but actual loudness is a subjective factor and depends on the characteristics of the ear of the listener. Human ear does not have linear response to all sonic frequencies. It is less efficient at lower sonic frequencies. Therefore, in an attempt to overcome this problem, a weighted network is used, to correlate human response to noise. It is logarithmic measurement for sound power level. ($P_{ref} = 2 \times 10^{-5}$ N·m$^{-2}$).

Microphones are the instruments used to measure sonic vibrations (noise). They convert (transduce) acoustic energy to electrical signals. There are three types of microphones—condenser, piezoelectric and dynamic type.

### *26.3.1.1 Condenser Microphone*

Condenser microphones operate on the principle of a capacitor. A condenser microphone consists of an electrically charged diaphragm and a back-plate (Fig. 26.5).

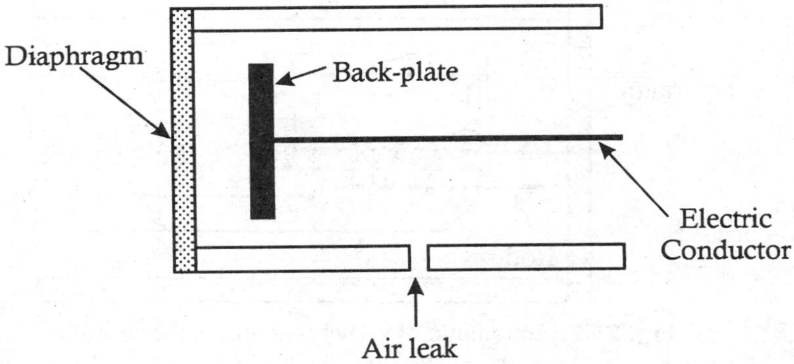


**Fig. 26.5:** Schematic of a Condenser Microphone

The diaphragm and back-plate acts as electrode plates of a capacitor. Deflections of the diaphragm due to sound pressure fluctuations translate to capacitance changes and the electrical signals generated are directly proportional to the deflections of the diaphragm. That is, to the sound pressure fluctuations.

Condenser microphone is a well-suited device for measuring sound pressure in field because of its flat linear response over a wide frequency range, high sensitivity and long-term operational stability.

### 26.3.3.2 Piezoelectric Microphone

Piezoelectric microphones operate on transduction of mechanical strain on a piezoelectric material to generate electric signal. Such a device (Fig. 26.6) utilizes piezoelectric materials (quartz, Rochelle salt, tourmaline, ammonium dihydrogen phosphate, etc) to convert acoustic energy to electric signals. The movement of the diaphragm under the action of sound waves produces mechanical strain on the piezoelectric material that is transduced into an electrical signal that is proportional to the strain.

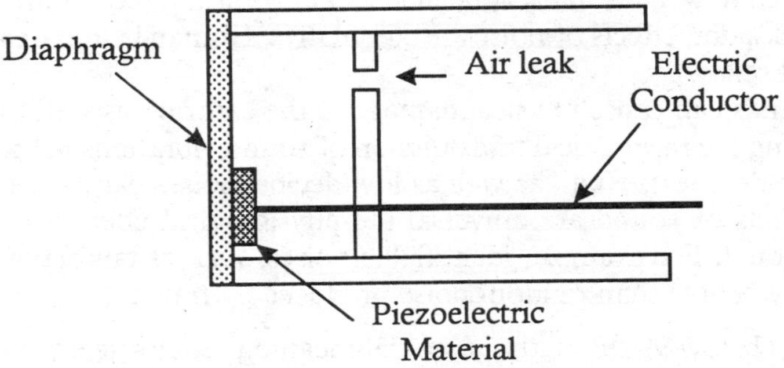


**Fig. 26.6:** Schematic of a Piezoelectric Microphone

### 26.3.3.3 Dynamic Microphone

In a dynamic microphone (Fig. 26.7), electrical signals are generated by a moving coil in a magnetic field due to the mechanical vibrations of the diaphragm. The deflection of the diaphragm is proportional to the sound pressure.

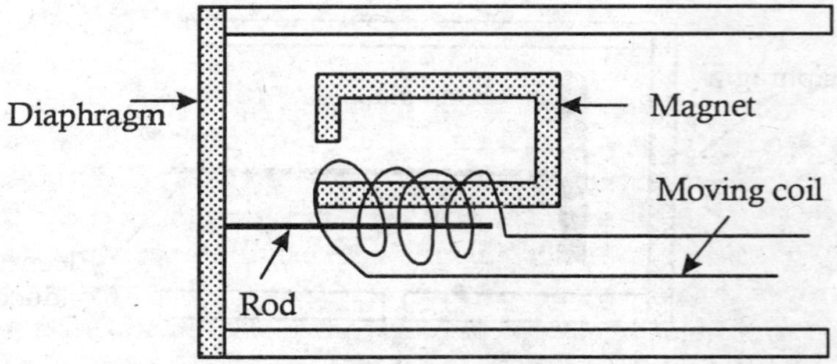


**Fig. 26.7:** Schematic of a Dynamic Microphone

### 26.3.4 Effects of Noise Pollution

Noise pollution is a redundant term. Noise itself is a pollutant that is undesirable for human well-being and for aesthetic life. It is difficult to conceive modern life without some kind of noise or other. Urban noise is a factor that is increasingly becoming a nuisance and a health hazard and is thus an environmental pollutant. Effects of noise pollution are felt at individual and local level, and have physical, physiological and psychological effects. Noise may affect humans physiologically (damage to hearing organs) and psychologically (irritation, lack of concentration and other responses). Damages can be long lasting and irreversible.

#### *26.3.4.1 Physical and Physiological Effects*

Loudness of the sound represents the amount (amplitudes) of the sound wave vibrations. In addition to loss of hearing due to aging (*presbyousis*), environmental noise can also cause discomfort and loss of hearing. Prolonged exposure to noise of certain frequencies can cause damage to the inner ear. It is called 'sensorineural' or 'perceptive hearing loss.' This could be permanent in nature. Noise of any kind, at any level, affects concentration and efficiency of an individual. Noise alters the rhythm of the heartbeat, dilates blood vessels and causes headache and irritability. While, infrasounds disrupt mental concentration and discomfort, ultrasounds (>20,000 Hz from jet engines, high speed drills) lead to fatigue, nausea, headache and vomiting. Both infra- and ultrasounds can be health hazards. Besides, effects of noise are cumulative. Human beings do not 'get used' to noise in the physiological sense.

High-decibel sounds can cause physical damage to the eardrum as well as to the ossicles in the middle ear, disrupting the mechanical transmission of sound vibrations in the hearing organs, thus leading to complete deafness. High- as well as low-decibel noises can cause physiological effects. While, physical effects of sound are universal the physiological effects of sound can vary from individual to individual. For example, identifiable noises, such as talking or music, may be more distracting for many people than random noise produced by traffic.

1. Steady noises below 90 dB, without special meaning, do not seem to interfere with human performance.
2. Noises at higher decibels, >100 dB, can cause auditory damage, nervous fatigue and loss of balance.
3. Irregular bursts of noises are more disruptive.
4. High frequency component of noise, >1000 Hz, may produce more interference.
5. All types of noise reduce efficiency of work, leading to annoyance, irritation and mental fatigue.

6. Infrasounds disrupt mental concentration and discomfort.
7. Ultra-, and infrasounds can be health hazards.

Urban ecology is beset with perpetual noise pollution of various kinds and levels. Noise above 80 dB is concern for human health. Noise level of 130 dB is the extremity of human tolerance. High volume traffic, construction and supersonic aircraft noises present a special problem for urban noise pollution control. Sonic boom leads to property damage as well.

It is not just decibel noise levels, but also its nature. The noise level and its duration have also marked effect on the behavioral pattern of humans and animals. Prolonged exposure to sound levels, above 130 dB, leads to mental imbalance. Audio fatigue by permanent noise produces hearing loss.

In acute acoustic trauma, resulting from short, intense exposures to noise, the effects are due to mechanical damage caused on account of excessive vibration of organ of Corti and damage to outer ear hair cells in the basal membrane, close to the oval window. Longer exposures lead to complete damage of organ of Corti, rupture of Reissner's membrane, resulting in perforation of the tympanic membrane.

### 26.3.5 Noise Pollution Control

Noise has become ubiquitous and is encountered everywhere in modern urban life—in homes, on roads, work places and recreation centers, just about every place. There are various recommended procedures for control of noise pollution. Eliminating the source of noise, though desirable, is not always possible, because any mechanical vibration can create noise. General steps in reduction of noise pollution at work places are:

1. Reduction in sound levels (if feasible).
2. Minimizing the vibratory parts of the equipment and gadgets by fastening and mechanical support.
3. Interrupting the path of the sound (by screens and absorbers).
4. Preventing the recipient from unwanted sounds (by screens, absorbers, earmuffs, and soundproof work places).

## BIBLIOGRAPHY

Agarwal, S. K. (1991), Himanshu: Udaipur. "*Pollution Ecology*."

Bragdon, C. R. (1971), University of Pennsylvania Press: New York. "*Noise Pollution*."

Brown, M. E. (ed.) (1988), Elsevier: Amsterdam. "*Handbook of Thermal Analysis and Calorimetry*," (Vol. 1).

Cuniff, P. F. (1977), Wiley & Sons: New York. "*Environmental Noise Pollution*."

Glusker, J. P. and Trueblood, K. N. (eds.) (1985), Oxford University Press: New York. "*Crystal Structure Analysis: A Primer*," (2nd edn.).

Gondal, M. A. and Mastromarino, J. (2000), *Talanta*, **53**(1): 147. "Lidar system for Remote Environmental Studies."

Hinkely, E. D. (ed.) (1976), Springerverlag: Heidelberg. "*Laser Monitoring of the Atmosphere*."

Kryter, K. D. (1970), Academic Press: New York. "*The Effect of Noise on Man*."

Ladd, M. F. C. and Palmer, R. A. (eds.) (1978), Plenum Press: New York. "*Structure Determination by X-ray Crystallography*."

Landsbert, H. E. (1969), Doubleday: New York." *Weather and Health*."

Lazzarotto, B. *et al.* (1998), *Intl. J. Environ. Anal. Chem.* "A Raman differential absorption Lidar for Ozone and Water Vapor Measurement in the Lower Troposphere."

Lee, R. E. (1993), Prentice-Hall: Engelwood Cliffs, NJ. "*Scanning Electron Microscopy and X-ray Microanalysis*."

Monteith, J. L. (1973), Edward Arnold: London. "*Principles of Environmental Physics.*"
Monteith, J. L. (1976), Academic Press: London. "*Vegetation and Atmosphere,*" (2nd edn.).
Moran, J. M., Morgan, M. D. and Weinberg, J. H. (1973), Little Brown: Boston, MA. "*An Introduction to Environmental Science.*"
Narayanan, P. (1989), *Physics Education*, (6): 217. "X-ray structure determination of biomolecules."
Narayanan, P. (2003), New Age Intl. Pubs: New Delhi. "*Essentials of Biophysics,*" (2nd Print).
Oort, A. H. (1970), *Sci. Amer.*, **223**: 54. "The energy cycle of earth."
Peirce, J. J . Weiner, R. F. and Vesiland, P. A. (1998), Butterworth-Heinemann: New Delhi. "*Environmental Pollution and Control,*" (4th edn.).
Priest, J. (1973), Addison-Wesley: New York. "*Problems of Our Physical Environment.*"
Purdom, P. W. (ed.) (1980), Academic Press: New York "*Environmental Health,*" (2nd edn.).
Turiel, I. (1975), Prentice-Hall: Engelwood Cliffs, NJ. "*Physics: The Environment and Man.*"
Van Wijk, W. R. (ed.) (1963), North-Holland: Amsterdam. "*Physics of Plant Environment.*"
Waldbott, G. L. (1973), Mosby: St. Luis, MS. "*Health Effects of Environmental Pollution.*"
Welch, B. L. and Welch, A. S. (1970), Plenum Press: New York. "*Physiological Effects of Noise.*"
Woodward, F.I. and Seehy, J. E. (1983), Butterworth: London. "*Principles and Measurements in Environmental Biology.*"

# SECTION-VIII

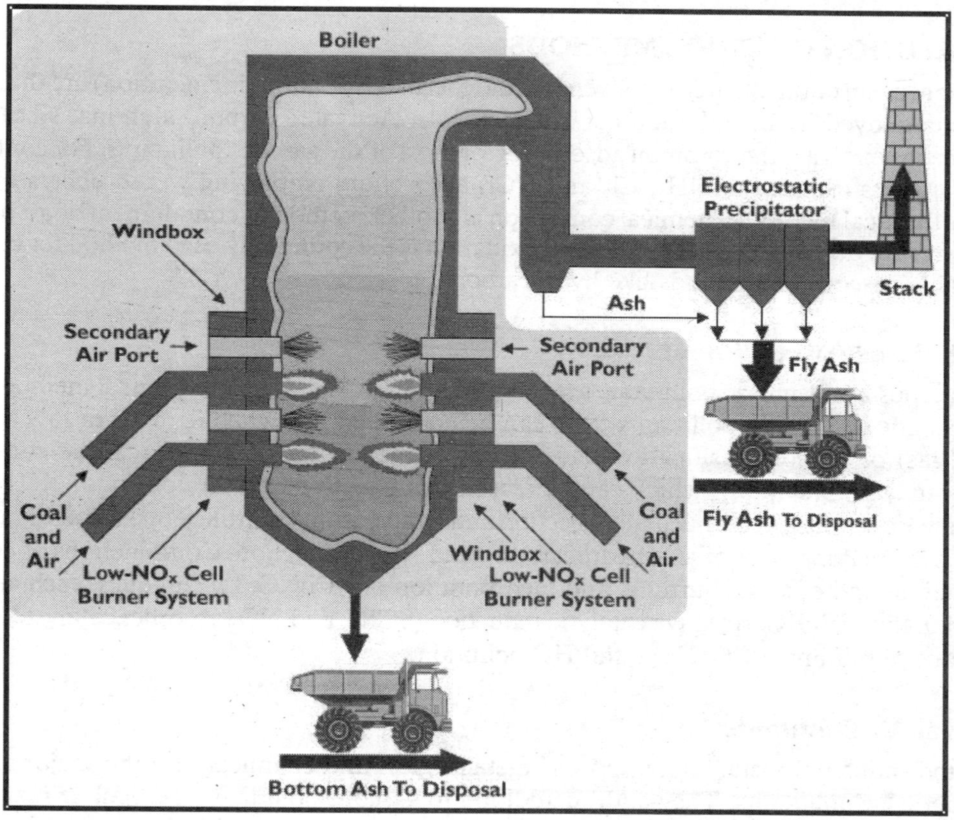


# Environmental Pollution Control

## OVERVIEW

Chapters 27 to 30 deal with the environmental pollution control methods—air (Chapter 27), water (Chapter 28) and soil pollution control (Chapter 29) methods and issues and parameters concerned with environmental health (Chapter 30).

Pollution control depends on the state of the pollutants, their sources, physicochemical characteristics and socioeconomic considerations for their containment. The general approach towards abatement/containment is: (1) elimination or change of operation procedures in the sources, (2) better recycling/utility schedules and (3) eco-friendly dispersion/disposal methods.

## AIR POLLUTION CONTROL METHODS

Adsorption, absorption, chemical conversion and thermal oxidation (incineration) are the standard methods employed in air pollution control protocols. Activated carbon, alumina, silica gel and diatomaceous earth are the common adsorbents for most of the gaseous pollutants. For containment of soluble gases ($SO_2$, $H_2S$, $NH_3$, $Cl_2$ and HCl) absorption, employing wet-scrubbers or packed towers, is the ideal method. Chemical conversion is another method of containment of air pollutants (e.g. conversion of $SO_2$ to $H_2SO_4$). Thermal oxidation is the commonly used method for converting combustible gaseous compounds like hydrocarbons.

### Control of Gaseous Pollutants

Major gaseous atmospheric pollutants are $SO_2$, $H_2S$, $NO_x$, $NH_3$ and CO. Control methods have to be tailor-made to specific pollutants. $SO_2$ can be controlled by switching over to low-sulfur fuel (natural gas) or by coal gasification and solvent extraction methods. $SO_2$ can be converted to $H_2SO_4$ (catalytic oxidation), which can be reused. An effective way for control of $NO_x$ is by off-stoichiometric burning, controlled fuel injection and temperature-controlled operations. Automobile exhaust is the major source of CO, hydrocarbons, and $NO_x$ production. Combinations of controlled fuel injection, and efficient burning, and incorporation of "$DeNO_x$" techniques, such as exhaust gas recirculation (EGR), selective catalytic reduction (SCR), and $NO_x$ absorber catalyst (NAC) are the options to minimize CO, $NO_x$, and HC pollution.

### Industrial Air Pollutants

Urban and industrial installations (power, metallurgical and chemical) are the major sources of atmospheric air pollutants. They add, in addition to gaseous pollutants, particulate matter, metal fumes and a variety of chemical pollutants.

Power generation is the major source of air pollution that includes thermal pollution. Electric power generation, to-date, relies heavily on fossil fuels and nuclear fuels. Both types are beset with specific pollutants and accompanying environmental hazards. Coal burning adds to the "greenhouse" effect and global warming. Nuclear power plants pose inherent radiation hazards (e.g. three-mile island and Chernobyl disasters), and long-term storage problems of spent fuels. Development of alternate clean-energy sources (wind, geothermal and solar energy) seems to be the only solution for the future eco-friendly world.

Mining and metallurgical operations generate large quantities of particulate pollutants, gaseous and metal fumes. General methods of control are judicious use/combination of collection methods—filters, bag houses, wet-scrubbers and electrostatic precipitators.

Chemical and petrochemical industrial operations produce inorganic (acids and alkalies), organic (solvents, synthetic products, fertilizers), and biologically hazardous chemical products (agrochemical products). General methods of control are—collection by bag-houses and wet-scrubbers, thermal incineration, catalytic conversion and after-burner oxidation.

## WATER POLLUTION CONTROL METHODS

Water is the ultimate source and conduit for accumulation and dispersal of environmental pollutants in the lithosphere. Therefore, the well-being and quality of vegetation and living organisms (ecosystems) on this planet are intimately connected with the quality of water bodies. Contamination of water sources arises from discharge of sewage, organic, biological and industrial and hazardous waste products. Control of water pollution may be achieved by organizing—(1) separate water supply sources for different purposes, (2) wastewater treatment and storage and (3) avoiding direct discharge of untreated effluents to water bodies and strict adherence to domestic, urban and industrial norms of water management protocols.

### Sewage Water Treatment

Sewage water contains sediments, scum, oil, inorganic, metal and organic pollutants, pathogens and disease-causing vectors. The water purification methods follow stepwise treatment protocols that consist of several levels—(1) primary (physical), (2) secondary (biological), and (3) tertiary (chemical/biological). The degree of purification depends on the type of wastewater and the purpose of usage of treated water.

#### *Primary Treatment*

The primary wastewater treatment protocol is a simple physical process to remove suspended matter, grit and scum by filtration.

#### *Secondary Treatment*

The secondary wastewater treatment protocol incorporates measures for further purification to remove oxygen-demanding organic pollutants (carbonaceous pollutants). It is a biological process that allows microorganisms to degrade organic matter present in the effluents from the primary treatment process in an aerated atmosphere until the BOD of wastewater has been reduced to acceptable levels.

#### *Tertiary Treatment*

Tertiary wastewater treatment protocol includes steps to remove dissolved organic and inorganic compounds that include algal nutrients (phosphorus and nitrogen compounds). The process involves removal of nitrogenous and phosphorus compounds and toxic metals by chemical methods.

Removal of organic compounds in wastewater is accomplished by adsorption, reverse osmosis, vacuum distillation and freezing. Ion-exchange, reverse osmosis and electrodialysis are some of the other techniques employed to remove inorganic pollutants in wastewater.

Nitrogenous organic compounds can be transformed to simple non-nitrogenous compounds by nitrification/denitrification processes. Ammonia can be removed by air stripping. Phosphorous can be removed by addition of coagulants, sorption, ion-exchange, electrodialysis and reverse osmosis methods.

### Industrial Wastewater Treatment

Industrial wastewater disposal poses unique problems because of the presence of hazardous chemicals (acids, alkalis, solvents and organic compounds) and trace level metals and effluent hot water. Treatment protocols depend on the type of industrial operations. Floatation and sedimentation can remove grit and debris. Salts are removed by ion-exchange, reverse osmosis. Metal pollutants can be removed by various physicochemical methods. Lime or lime with sulfide treatment removes most heavy metals (e.g. Cd, Cr, Cu, Hg, Ni Pb, Zn, etc) as insoluble hydroxides and basic salts. Combustible residues may be incinerated and the final waste is compacted and disposed off as landfills, sand beds and fertilizers.

## LAND POLLUTION CONTROL METHODS

Land pollution is primarily due to mining, agriculture and animal husbandry and forest degradation. Strict control of land degradation operations is one of the primary steps needed to avoid or minimize atmospheric and hydrospheric pollution and degradation of ecosystems.

Solid waste consists of garbage, rubbish, trash, industrial and agriculture runoff, animal and human waste and sewage sludge. Waste disposal protocols involve collection, storage, processing, transport and dispersal. Processing includes separation, screening, size reduction, compaction and treatment and incineration/pyrolisis, etc.

The principal disposal operations are land filling, composting (biological degradation), incineration (volume reduction) and recycling methods. Incineration is not a total solution as it contributes to atmospheric as well as thermal pollution.

## ENVIRONMENTAL HEALTH

Environmental health refers to environmental conditions that determine the quality of living organisms that inhabit various ecosystems. Maintaining environmental health requires knowledge of environmental science and technology (environmental engineering). It involves evaluation and optimization of various solutions to minimize the environmental pollution level.

### Industrial Ecology

Urbanization and industrialization have ushered in a host of consumer products, synthetic hazardous chemicals. Their production, usage and disposal norms have created complex environmental pollution problems. While it might be possible to find remedies to point-source pollutants, it is rather difficult to monitor and find suitable remedies for end-product disposal problems.

### Environmental Pollution and Health

There is an inverse correlation between environmental pollution and health & wellbeing of the populations. Of all the environmental media, hydrosphere is the most important one for accumulation and transport of different types of pollutants—hazardous chemicals, metals, pathogens.

### Pathogens

Pathogenic microorganisms thrive in humid atmosphere and water bodies. Vectors like mosquitoes require water for hatching. Man-made mosquito breeding grounds thrive due to water stagnation, poor sanitation and proliferation of urban slums. Mosquitoes are vectors for many diseases—

malaria, encephalitis and yellow and viral fever. Proliferation of flies is also due to breakdown of sanitation and unhygienic practices of food storage and consumption. Flies are vectors for typhoid fever and enteric diseases. Rodents, which breed and proliferate amidst slum dwelling human habitats, are responsible for various types of plagues and diseases.

### Toxins and Poisons

Host of chemicals (e.g. alkaloids and man-made chemicals) and metals are toxic and poisonous. Many synthetic chemicals used in agriculture and food preservation are chronic poisons and carcinogens. In addition, microbial action in food products produces various toxins (e.g. botulism) and poisons. Therefore, food preservation is an important part of environmental pollution control and environmental hygiene.

### Food Preservation

Inefficient methods of preservation of agriculture produce and food products and their spillage are responsible for proliferation of pests and pathogenic microorganisms. In addition, food preservation has global importance from the economic, hygienic, social and environmental pollution control aspects.

There are various kinds of food-preservation techniques and technologies, from conventional drying, pickling and curing to modern methods of vacuum packing, lyophilization, freeze-drying and radiation treatment by $\gamma$-rays. Radiation treatment seems to hold a great promise, in spite of apprehensive common public. It greatly extends the shelf life of agriculture produce, vegetables, cereals, grains, spices and food products. Radiation treatment eliminates the necessity of fumigants. Bioengineering methods auger a great promise in enhanced food production, quality improvement and preservation.

Additives and coloring substances are still used in food products. While some of the natural additives are necessary for preservation as well as for flavor, taste and aesthetic sense, proliferation of synthetic (artificial) additives, flavors and coloring materials is of concern from the point of view of pollution of foods and beverages. Some of these chemicals are known or potential carcinogens. In addition some artificially enriched animal feeds, nutrients and food products may be sources of epidemics (e.g. mad-cow disease outbreak in Britain). There is an urgent need for comprehensive analysis and monitoring of environmental pollutants in food and drug industry to evolve and implement guidelines for better and healthy condition for living organisms on this planet.

### Biodegradation and Bioremediation

Biodegradation and bioremediation are part of the strategy to keep the concentration of environmental pollutants at very low levels. "Biodegradation" refers to any biologically mediated structural alteration of a compound. Microbial biodegradation is biologically catalyzed reduction of complex chemicals and some of them may be environmental toxicants or pollutants. Microbial biodegradation is carried out by various chemical transformations—hydrolysis, dehalogenation, dealkylation, etc. Hydrolysis is the common means by which microorganisms inactivate toxicants—esters, anhydrides, amides and acyl derivatives. Controlled application of microorganisms for destruction of chemical pollutants is common in wastewater treatment, toxic waste disposal by bioreactors and biofilters.

Genetic engineering promises to provide organisms with novel and presently non-existing biodegrading abilities. A variety of problems can, perhaps, be tackled in future through the use of genetically engineered microorganisms.

# 27

# Air Pollution Control Methods

Mthods such as adsorption, absorption, chemical conversion and thermal oxidation are the standard methods employed in air pollution control protocols. Activated carbon, alumina, silica gel and diatomaceous earth are the common adsorbents for most of the gaseous pollutants. For containment of soluble gases ($SO_2$, $H_2S$, $NH_3$, $Cl_2$ and HCl) absorption, employing wet-scrubbers or packed towers, is the ideal method. Chemical conversion is another method of containment of air pollutants (e.g. conversion of $SO_2$ to $H_2SO_4$). Thermal oxidation is the commonly used method for converting combustible gaseous compounds like hydrocarbons.

Major gaseous atmospheric pollutants are $SO_2$, $H_2S$, $NO_x$, $NH_3$ and CO. Control methods have to be tailor-made to specific pollutants. $SO_2$ can be controlled by switching over to low-sulfur fuel or by coal gasification and solvent extraction methods. An effective way for control of $NO_X$ is by off-stoichiometric burning, controlled fuel injection and temperature-controlled operations. Combinations of controlled fuel injection, and efficient burning, and incorporation of "$DeNO_x$" techniques, such as exhaust gas recirculation (EGR), selective catalytic reduction (SCR), and $NO_x$ absorber catalyst (NAC) are the options to minimize CO, $NO_x$, and HC pollution.

Urban and industrial installations (power, metallurgical and chemical) are the major sources of atmospheric air pollutants. Power generation is the major source of thermal pollution. Electric power generation relies heavily on fossil and nuclear fuels. Both types are beset with specific pollutants and accompanying environmental hazards. Coal burning adds to the "greenhouse" effect and global warming. Mining and metallurgical operations generate large quantities of particulate pollutants, gaseous and metal fumes. Chemical and petrochemical industrial operations produce inorganic (acids and alkalis), organic (solvents, synthetic products, fertilizers), and biologically hazardous chemical products (agrochemical products). General methods of control are—collection by bag-houses and wet-scrubbers, thermal incineration, catalytic conversion and after-burner oxidation.

This chapter deals with some of the technologies in control/abatment of air pollutants.

Air pollution control methods depend on the state of the pollutants, type of their sources (point source or random source), their physicochemical characteristics and prevalence, and their level of impact on the environment and socioeconomic considerations for their containment. Pollution control aims at containment of air, water, land and industrial pollution to safeguard the ecology, environment and wellbeing of the living systems. Control methods are inter-related and several. Optimal methods should take into account ecological, aesthetic and socioeconomic considerations.

## 27.1 CONTROL STRATEGY

Air pollution control takes into account the source, the type of pollution, meteorological conditions including wind speed, and geographical terrain around the place. Pollution control methods include collection, conversion, minimization and dispersal protocols. Control from point sources (industrial installations) is relatively easy compared to random and mobile sources (transport, agricultural and domestic operations). For point sources, control operations include identifying the emission point, collection, treatment and dispersion or disposal of contaminants (Fig. 27.1).

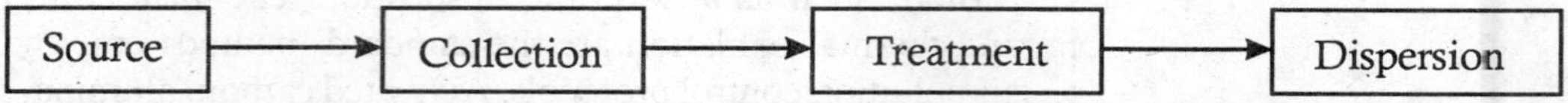


**Fig. 27.1:** Block Diagram of Pollution Control from Point Sources

The general strategy is:

1. Elimination or change of the source or the process parameters, generating the pollutants, is the desired method. For example, to reduce sulfur emissions from a power plant the path lies in desulfurizing coal, use of low-sulfur coal or other fuel or switching over to alternate less-pollutant energy sources like wind, geothermal and solar energy sources. The last mentioned one is not practical in the present state of development of the technologies.
2. Collecting and recycling of pollutants is another possibility. Gaseous pollutants like $SO_2$ are amenable for collection and disposal. Collection methods are, generally, based on absorption, adsorption, precipitation, scrubbing and conversion to less toxic compounds (e.g. conversion of CO to $CO_2$).
3. Combustible pollutants can be disposed by incineration or conversion to liquid or solid form and dispersed.

## 27.2 COLLECTION METHODS

Various collection procedures are in use, depending on the type of pollutant and the level of its undesirability. General and routinely employed methods are precipitation, absorption, adsorption, chemical conversion and thermal oxidation.

### 27.2.1 Precipitation Methods

Gravitational settling, centrifugal separation, fabric filtration, wet collection and electric precipitation, and chemical methods of precipitation are some of the standard precipitation methods in use.

### 27.2.2 Absorption Methods

Absorption, also called scrubbing, is a diffusion process, carried out by selective wet-scrubbers or packed towers for soluble gases (e.g. $SO_2$, $H_2S$, HCl, $Cl_2$, NO and $NH_3$). Solubility is the most important factor in the selection of a solvent for absorption. In addition, the solvent should be relatively non-volatile, non-flammable, and chemically stable and is of low toxicity and pollution potential. Water is the most commonly used solvent in absorption systems.

Counter-current flow of scrub solution and the gas stream is effective in trapping the pollutants and in optimum utilization of reagents. The gas stream moves upwards through a packed bed and the absorbing solution is injected at the top of the packing. The most contaminated gas is at the bottom in contact with the most contaminated solution and the clean gas is at the top of the tower with relatively clean solution.

### 27.2.3 Adsorption Methods

Contaminant gases are adsorbed by porous solid materials (matrix), and released later by elution, or thermal treatment. Common adsorbent materials are activated carbon, alumina, silica gel, and diatomaceous earth. Adsorption is a viable technique for many odor control systems. Activated carbon is a highly adsorbent form of carbon used to remove odors and toxic substances from liquid or gaseous emissions. In waste treatment, it is used to remove dissolved organic matter from waste drinking water. It is also used in motor vehicle evaporative control systems. Activated carbon or alumina with a chemical reactant can convert the adsorbed contaminants to less odorless compounds.

### 27.2.4 Chemical Conversion Methods

Chemical conversion processes are often employed for collection and control of gaseous pollutants. Under favorable conditions the resultant products can find commercial applications. As an example, $SO_2$ can be trapped by reaction with CuO or $CaCO_3$.

$$\begin{aligned} CuO + \tfrac{1}{2}O_2 + SO_2 &\rightarrow CuSO_4 \\ CaCO_3 + \tfrac{1}{2}O_2 + SO_2 &\rightarrow CaSO_4 \\ SO_2 + Na_2SO_3 + CuO + H_2O &\rightarrow 2NaHSO_3 \end{aligned} \quad \ldots(27.1)$$

### 27.2.5 Thermal Oxidation Methods

Thermal oxidation is also a practical method (combustion, incineration and flaring) for converting combustible gaseous compounds to $CO_2$ and $H_2O$, halogens to HX, and S to $SO_2$. The main variables controlling the efficiency of a combustion process are temperature ($T$), time ($t$) and turbulence. At a constant combustion temperature the destruction and removal efficiency (DRE) is

$$DRE\% = \left(\frac{VOC_{in} - VOC_{out}}{VOC_{in}}\right) \times 100 \quad \ldots(27.2)$$

Direct-flame afterburners (refractory-lined oxidizers operated at temperatures 750-850° C) are the most commonly used devices in industry. Catalytic afterburners are used in industry for the control of solvent and organic vapor emissions, including those associated with automobiles and transport vehicles. In catalytic oxidation (catalytic incineration) oxidizes volatile organic compounds (VOCs) by using solid catalysts (metals or metal oxides) to promote the combustion process. Catalytic incinerators require lower temperatures (350-600° C) than conventional thermal incinerators, thus saving fuel and other costs. First, the contaminated air is preheated to reach a temperature necessary to initiate catalytic oxidation. Then the preheated VOC-laden air is passed through a bed of solid catalyst for its rapid oxidation. Off-gas scrubbing is required to control acid vapors of halogenated hydrocarbons.

Incineration/pyrolysis (destructive distillation in the absence of oxygen) is not a total solution because such a process causes thermal pollution and gives rise to residual waste that has to be disposed off.

## 27.3 CONTROL OF GASEOUS POLLUTANTS

Major gaseous pollutants in the atmosphere are $SO_2$, $H_2S$, $NO_x$, $NH_3$ and CO. Control methods are general or tailor-made to specific pollutant.

### 27.3.1 Control of $SO_2$

Sulfur is present in coal and sulfur-containing minerals (iron pyrites, $FeS_2$; lead sulfide, PbS, etc). Crushing of coal and flotation of sulfides can reduce the sulfur bearing minerals but any such process cannot eliminate sulfur associated with coal itself, in the form of organic compounds. Sulfur in fuel oils can be removed in high-pressure catalytic reactors, in which hydrogen combines with sulfur to form $H_2S$. $H_2S$, being toxic is adsorbed by activated carbon and is converted to $SO_2$ for further disposal. By switching over to low-sulfur fuel (natural gas) or by an integrated coal gasification cycle, or fluidized bed combustion of coal mixed with an adsorbent such as lime or limestone (flue gas desulfurization) $SO_2$ emission can be controlled.

Flue gas desulfurization is achieved by catalytic oxidation using vanadium pentoxide. This catalyst promotes oxidation of $SO_2$ to $SO_3$, which can be absorbed in water to form $H_2SO_4$. The latter can be disposed conveniently (Fig. 27.2).

$$SO_2 + \frac{1}{2}O_2 \rightarrow SO_3 \qquad \ldots(27.3)$$

$$SO_3 + H_2O \rightarrow H_2SO_4$$

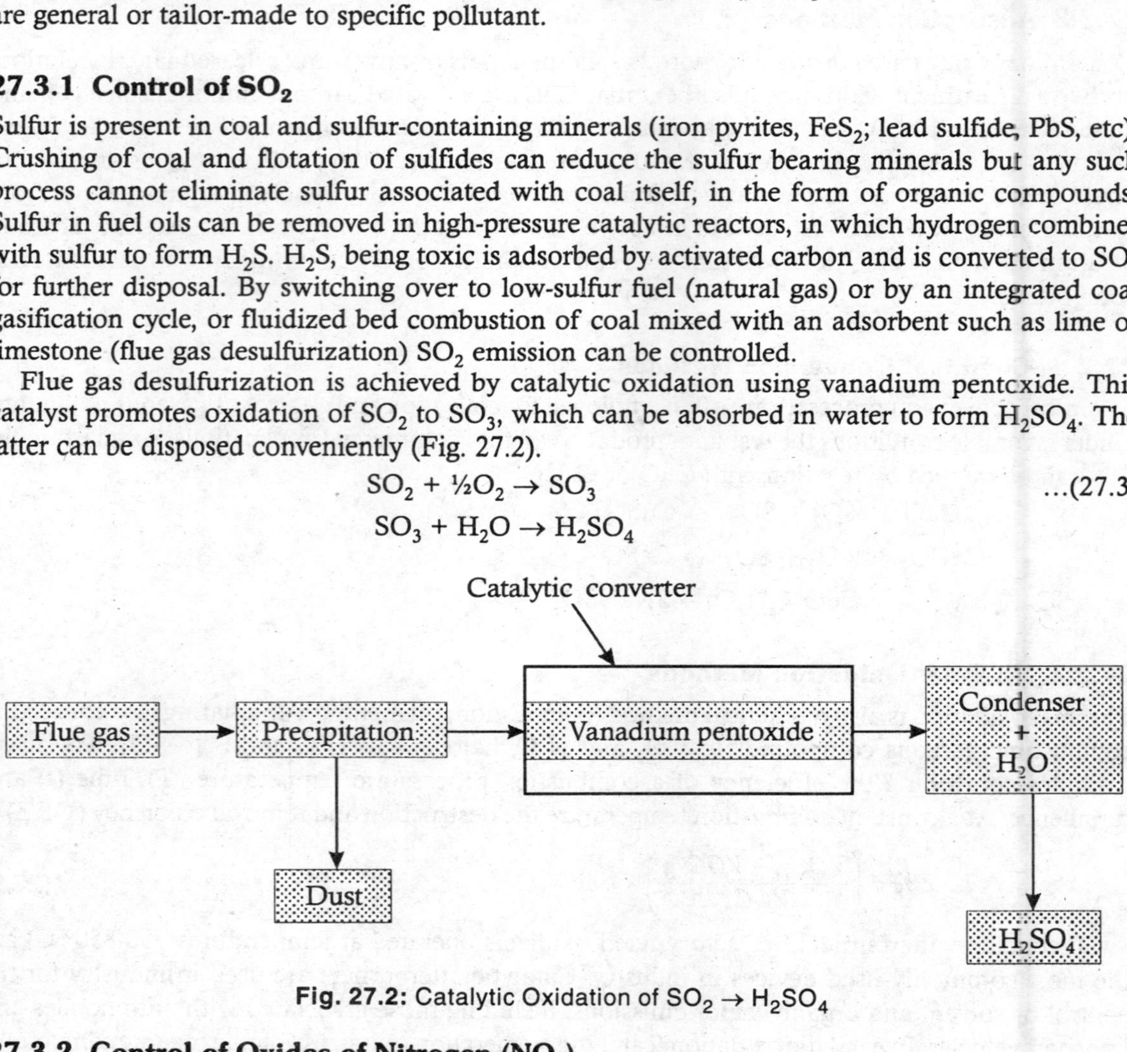


**Fig. 27.2:** Catalytic Oxidation of $SO_2 \rightarrow H_2SO_4$

### 27.3.2 Control of Oxides of Nitrogen ($NO_x$)

Electric power generators and automobiles, which use fossil fuels, are the major sources for $NO_x$ and hydrocarbon (HC) pollutants due to incomplete combustion. In petrol-based vehicles, the major pollutant parts are fuel tank (HCs), carburetor (HCs), crankcase (HCs), and exhaust (HCs,

CO, NO/$NO_2$). Emissions from petrol-based vehicles contain large quantities of carbon monoxide, hydrocarbons and $NO_x$, and lesser quantities of particulate matter. On the other hand, emissions from diesel-based vehicles contain minimum quantities of CO, and HCs, but contain worrisome quantities of $NO_x$, and particulate mater.

The thermal fixation of atmospheric $N_2$ and $O_2$ produces thermal $NO_x$, while the combustion of chemically bound $N_2$ in the fuel produces fuel $NO_x$ (in the two cases the composition of $NO_x$ with respect to different oxides of nitrogen is different). The formation rates of $NO_x$ from air as well as from fuels are kinetically limited, with the amount of $NO_x$ formed being less than the equilibrium value. Both nitrogen monoxide (NO) and nitrogen dioxide ($NO_2$) are formed in higher temperature combustion of air. Oxidation of NO in the atmosphere enhances its $NO_2$ content.

There are three major routes for $NO_x$ formation in combustion processes. One mechanism involves thermal dissociation of $N_2$ and $O_2$ and subsequent reaction of nitrogen and oxygen atoms to form oxides of nitrogen. The extent of thermal $NO_x$ formation depends on combustion characteristics such as temperature, oxygen and nitrogen concentration in the flame zone, and residence time.

Since the combustion conditions control the formation rates of thermal $NO_x$, modifying that process, for example by use of staged-air burners or staged-fuel burners, its formation can be suppressed. The peak or maximum flame temperature is the most important parameter that determines the potential for $NO_x$ formation. The strategies for controlling the formation of thermal $NO_x$ are:

1. Reduction of peak temperature.
2. Reduction of oxygen concentration in the high temperature zone.

Reducing the combustion temperature effectively reduces thermal $NO_x$, but not fuel $NO_x$. Fuel $NO_x$ depends mainly on the air/fuel ratio. There are several effective methods for controlling the fuel $NO_x$ production and release ("de$NO_x$") technology: (*i*) low-excess-air operation, (*ii*) off-stoichiometric combustion (source emission reduction) in multi-stages, the primary flame zone being fuel-rich and the secondary zone being fuel-lean, and (*iii*) after-treatment of tail gas (i.e. removal of $NO_x$ after combustion). Low $NO_x$ burners ensure that initial fuel combustion occurs within fuel-rich conditions (low $O_2$ conditions). Burner modification, combined with an after-treatment process is a viable method to achieve reduced $NO_x$ emissions. Flue or exhaust gas recirculation and inlet water injection reduce flame temperature and $NO_x$ formation in the flame zone.

1. $NO_2$ formation is limited by restricting the excess amount of $O_2$ in the combustion process, over that needed to burn the hydrocarbon fuel. This situation minimizes the available $O_2$ for reaction with $N_2$:

$$CH_4 + O_2 \rightarrow 2O_2 + CO_2 \qquad \text{...(27.4)}$$

$$N_2 + O_2 \rightarrow NO$$

2. "Lean $NO_x$" technology relies on the use o hydrocarbons present in the engine exhaust to serve as reductants.
3. Exhaust gas recirculation (EGR) method reduces $NO_x$ both by reducing $O_2$ concentration and the temperatures in the burning flame.
4. Selective catalytic reduction (SCR) or selective non-catalytic reduction methods may be employed to carry out tail gas after-treatment. Selective catalytic reduction (SCR) is a chemical process that converts NO into $N_2$ and $H_2O$. The process involves mixing the gas with a

reductant (ammonia or urea) and passing the mixture over a bed of SCR catalyst ($Ti_2$ + $V_2O_5$ + $WO_3$). The catalyst allows the reaction process to take place at low temperatures (300-400° C). Oxidation of NO to $NO_2$ upstream of the SCR catalyst substantially increases the performance of the catalyst. The SCR process is likely to be a technique of choice for compression-ignited (diesel) engines. Zero ammonia technology (ZAT) or $NO_x$ trap does not require the injection of ammonia or urea. In this process $NO_x$ is converted into $NO_2$ and adsorbed on to the catalyst. Use of solid reductants other than urea (or ammonia) has many practical advantages. For example, solid ammonium carbamate can be decomposed to give $CO_2$ and $NH_3$ above 60° C. Selective non-catalytic reaction method requires higher temperatures for operation.

5. Non-selective treatment of contaminated gas, which is a wet removal process, may be carried out first to remove dust and poisonous chemicals, and then $NO_x$ and $SO_x$. The scrubber technology utilizes a counter-current packed tower design with an alkali scrub.
6. $NO_x$ adsorber catalyst (NAC) relies on the chemical adsorption of $NO_x$ (by alkaline earth oxides) under "fuel-lean" exhaust conditions and periodic regeneration of the catalyst under brief "fuel-rich" excursions.

### 27.3.3 Control of Hydrocarbons and Carbon Monoxide (CO)

Connecting fuel tank and carburetor systems to an activated carbon canister and burning them in the engine can stop evaporation of hydrocarbons (HCs) in fuel tank and carburetor. The crankcase can be closed off with safety valve ventilation and the gases recycled into the manifold.

Carbon monoxide (CO) is due to incomplete burning of carbonaceous fuel. Automobiles are the major sources of carbon monoxide pollution. In automobiles, during expansion, the combustion process becomes kinetically linked and CO concentration is much higher than the equilibrium value (Fig. 27.3). Non-equilibrium behavior is most evident for fuel-rich mixture and therefore, increasing air/fuel ratio is the required change.

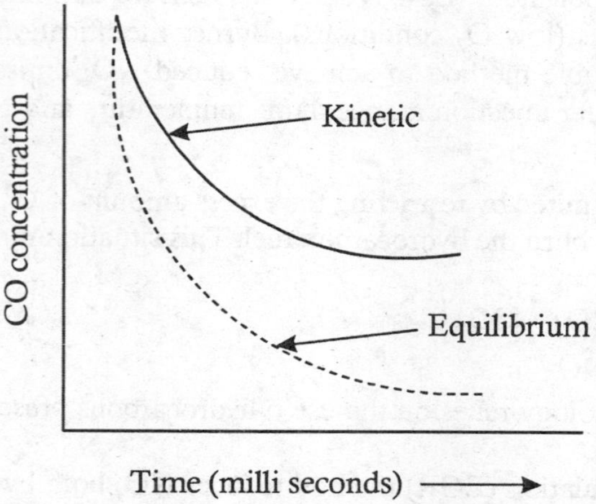


**Fig. 27.3:** CO Composition during Expansion in Combustion Process

Automobile exhaust is a major source of both NO and CO pollution. Stratification of the engine with two compartments, the first for pre-ignition of air/fuel mixture and the second for providing a broad flame for efficient burning, helps in better fuel burning and CO reduction. NO reduction is accomplished in the first stage by burning a fuel-rich mixture and then depleting the $O_2$ at the catalyst (Pt-Rh). Air is introduced in the second stage and CO is oxidized at lower temperature. Incorporation of the three-way catalytic converter in the exhaust system (with unleaded petrol) provides better oxidation of CO and HCs to $CO_2$ and $H_2O$ and reduction of NO to $N_2$.

## 27.4 CONTROL OF INDUSTRIAL AIR POLLUTION

Atmospheric air pollution from urban and industrial operations is mainly due to power generation plants, transportation, mining, mineral processing and metallurgical and chemical industries.

### 27.4.1 Power Generation Plants

Power is needed for practically all activities of technological societies. (Fig. 27.4). Because our world depends so much on energy, we need to find sources of energy that will last a long time. We should also use energy sources that produce as little pollution as possible.

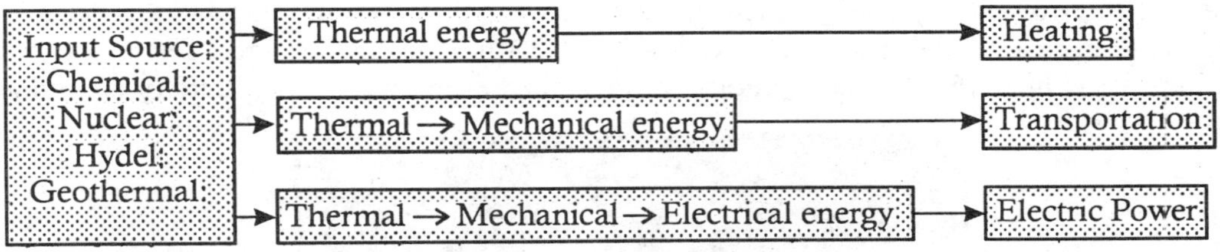


**Fig. 27.4:** Power Generation and Utilities

Types of pollutants in energy-producing industries depend on the type of fuel used, combustion process and other parameters. Power generation is connected not only with the technical aspects, but also on social, economic and environmental factors. Electric power generation to-date relies heavily on fossil fuel burning, hydropower stations, and nuclear reactors. Fossil fuel combustion systems use coal and oil-based products. The particulate size, shape, flammability, corrosiveness, toxicity and other characteristics of aerosols have to be taken into consideration while designing control systems. Many fossil fuels are contaminated with sulfur, heavy metals, and other dangerous pollutants. Suitable chemical and combustion processes (Fig. 27.5) can achieve reduction of $SO_2$ and $NO_x$.

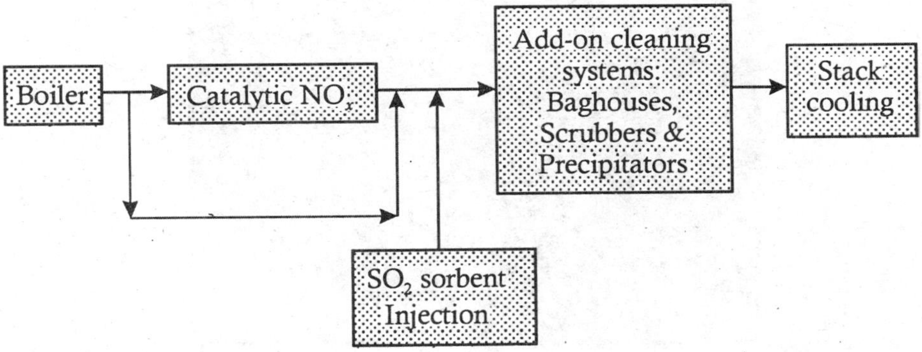


**Fig. 27.5:** Block Diagram of $SO_2$ and $NO_x$ Reduction in an Energy-generation Plant

### 27.4.2 Non-conventional Energy Sources

The power generation from coal and other fossil fuels has several limitations—"greenhouse" reactions leading to global climate change; increase in average temperature that would result in melting of icecaps, rise in ocean levels and change in climate patterns. Some of the non-conventional energy sources—hydel, geothermal, ocean currents, wind, solar, and biomass have also been exploited though to a limited extent. Large-scale hydel resources are costly to tap both economically and environmentally, and they disrupt human and animal habitats and biodiversity.

Geothermal and ocean currents, wind and solar power plants seem to be clean energy sources in the long-term perspective. Small-scale hydro, biomass, wind and solar power may become major sources of renewable energy in the future. Sun is a huge reservoir of energy ($\sim 10^{15}$ kW) and could provide all the energy requirements, if harnessed properly (by properly designed and cost-effective photovoltaic systems). Biomass gasification and bio-methanation are possible environment-friendly sources of energy. The thermochemical system of "*pyrolysis*" and gasification of the biomass includes a combination of *pyrolysis* and *steam reforming* process.

Wind energy could become competitive and cost-effective with emerging technologies. It is one of the cleanest forms of power, which can alleviate environmental degradation. Wind energy causes neither harmful pollution (no $SO_x$, $NO_x$, $CO_x$) nor hazardous radioactive waste. Small wind turbines (Fig. 27.6) can be assembled and installed even on rooftops. Battery charging wind

**Fig. 27.6:** A Wind Turbine

turbines are particularly useful for remote homes, sail boats, telecommunication systems and water pumping. If it is combined with solar array panels, the power supply becomes more consistent under varied weather conditions. Using of turbines of larger dimension increases the power potential. Many countries in the world are exploring the feasibility of wind as one of the environmental-friendly, renewable energy source (Table 27.1).

**TABLE 27.1:** GLOBAL WIND POWER UTILIZATION

| *Country* | *Power (MW)* | *Country* | *Power (MW)* |
|---|---|---|---|
| Germany | 4,990 | USA | 2,700 |
| Spain | 2,050 | India | 1,100 |
| Britain | 360 | China | 300 |
| Italy | 300 | Canada | 130 |
| Japan | 80 | France | 40 |
| New Zealand | 35 | Norway | 15 |
| Australia | 10 | Russia | 5 |

Hydrogen as an energy source combines the virtues of electricity and fossil fuels without their disadvantages. Hydrogen has a number of attractive qualities. It is a non-polluting, non-toxic fuel, which, when burned, discharges only water, and possibly a small amount of nitrogen oxide (if it is used with air at high temperature). Since hydrogen is a carbon-free fuel, it provides $CO_2$-free combustion in internal combustion (IC) engines. The combination of hydrogen with fossil fuels such as gasoline, natural gas, ethanol, methanol for fueling IC engines provides less emission, and increase performance. Adding 5 % hydrogen to the gasoline provides 30% to 40% less $NO_x$ emission. Although there are many advantages hydrogen as fuel, there are some impediments in using hydrogen in vehicles. The major challenge is to provide satisfactory answers to the four main problems—its production, its use as fuel, its storage and its distribution.

All non-conventional energy sources have some limitations because of which they have not been fully accepted at present anywhere in the world. They can only be considered as possible energy sources in the future at local, regional and global levels.

### 27.4.3 Metallurgical Industries

Pollutants from metallurgical operations are predominantly dust, fumes, metal oxides, sulfur dioxide and carbon monoxide. Hydrogen fluoride (HF), hydrogen sulfide ($H_2S$), Hg vapor are other gaseous pollutants.

Iron and steel industries are major industries in all countries of the world. Industrialization started and was sustained by this industry. Industrial operations incorporate integrated processes of coke making, sintering, smelting, rolling and alloy making. Pollutants in various processes are:

1. *Coke making*: Particulate matter, coal dust, HCs and $H_2S$.
2. *Sintering*: The major pollutants are CO, halides, ammonia, HCs and $SO_2$.
3. *Iron & steel making*: Iron ore, limestone and coke mixture is heated to reduce iron oxide to metallic iron. Major pollutants are particulate matter, metallic fumes, iron oxide, CO and $H_2S$.
4. *Ferrous alloys*: Pollutants are metallic fumes and particulate matter, and oxides of other metals in galvanizing etc.

Air pollutants and control methods in metallurgical operations are given in Table 27.2.

**TABLE 27.2:** AIR POLLUTANTS AND CONTROL METHODS IN METALLURGICAL OPERATIONS

| *Metal* | *Process* | *Pollutants* | *Control* |
|---|---|---|---|
| Aluminum (Al) | Electric | HF, CO, $NO_2$, $SO_2$, $H_2S$, & $CS_2$, | Coated filter baghouses; Fluid bed scrubbers & Electrostatic precipitators. |
| Arsenic (As) | Byproduct of Cu & Pb smelting | $AS_2O_3$, | Baghouses. |
| Beryllium (Be) | Ore processing | Beryllium dust | Baghouses; Cyclones; Filters. |
| Cadmium (Cd) | Byproduct of Pb & Zn smelting | Cd and oxides of Cd | Baghouses; Filters; Cyclones; |
| Copper (Cu) | Smelting and refining processes | Metal fumes & $SO_X$, | Electrostatic precipitators. |
| Lead (Pb) | Smelting & refining | Dust, fumes, trace metals, $PbO_2$, $SO_2$ & CO | Sintering; Baghouses & Electrostatic precipitators. |
| Mercury (Hg) | Thermal decomposition | Fumes & $SO_2$ | Scrubbers; Baghouses & Electrostatic precipitators. |
| Zinc (Zn) | Smelting & refining | Trace metals; CO; $SO_2$ | Scrubbers; Baghouses & Electrostatic precipitators. |
| | | | Scrubbers; Baghouses & Electrostatic precipitators. |
| Refractory metals (Mo, Sn, Ti & Zr) | Many processes | Particulates; fumes; trace metals; Cl & $NH_3$ | Conversion & Baghouses. |
| Iron & Alloys | Sintering, metal making, galvanizing and alloy making | Particulates, coal dust, metal fumes, metal oxides, halides, CO, $NH_2$; $NO_x$, halides, HCs, $H_3S$ & $SO_2$ | Baghouses; Electrostatic precipitators; Cyclones and flaming of CO. |

**TABLE 27.3:** POLLUTANTS DUE TO INORGANIC CHEMICAL PROCESSES

| *Chemical* | *Process* | *Pollutant (s)* | *Control* |
|---|---|---|---|
| **Acids** | | | |
| Hydrochloric (HCl) | Byproduct of chlorination | HCl | Absorption |
| Hydrofluoric (HF) | Fluorospar + $H_2SO_4$. | HF, $SiFO_4$, $CO_2$ & $SO_4$ | Scrubbers |
| Nitric ($HNO_3$) | Ammonia oxidation of air + NO | $NO_x$ | Absorption, adsorption & catalytic reduction |
| Sulfuric ($H_2SO_4$) | Oxidation of $SO_3$ (contact process) | Fluorides, $SO_x$ & $H_2SO_4$ mist. | Wet scrubbers |
| Phosphoric | Phosphate rocks + $H_2SO_4$ | $H_3PO_4$, $H_2S$, fluorides & particulate | Scrubbers & Baghouses |
| **Alkalis** | | | |
| Caustic soda (NaOH) | Electrolytic chlor-alkali process | Particulates, Cl, $NH_3$, $CO_2$, CO, $H_2$ & Hg | Wet scrubbers |
| Soda ash ($Na_2CO_3$) | Solvay process | Particulates & NH3 | Scrubbers |
| Ammonia ($NH_3$) | Catalytic process of $H_2$ & $N_2$ | CO, HCs and $NH_3$ | Wet scrubbers |
| **Other** | | | |
| Chlorine (Cl) | Chlor-alkali process | Particulates & $NH_3$ | Wet scrubbers |

### 27.4.4 Chemical Industries

The atmospheric pollutants from chemical industries are inorganic, organic and biological in nature (from food and beverage production).

1. *Inorganic pollutants*: Inorganic chemical processes are mostly concerned with the production of mineral acids, bases, fertilizers and halides. Major acids are HCl, $H_2SO_4$, $HNO_3$ and phosphoric acid. Major bases produced are lime (calcium oxide), soda ash (sodium carbonate) and caustic soda (NaOH). Fertilizers are phosphates (from phosphate rocks) and nitrates (due to $HNO_3 + NH_3$). Chlorine is the byproduct in the electrolytic processes of brine. Air pollution by inorganic compounds and control methods are given in Table 27. 3.
2. *Organic pollutants*: Organic pollutants are mainly from petroleum-based operations (drilling, storing and transporting) and production of petrochemicals. Oxidation and combustion of organic matter releases large amounts of gaseous pollutants to the atmosphere. Biological treatment of polluted air and odors is possible by degradation of hydrocarbon containing waste-air, by a combination of biofilters with downstream adsorbers and chemical scrubbers. Air pollution by organic compounds and control mechanisms are given in Table 27.4.
3. *Biological pollution*: Biological pollutants are mainly from food, animal husbandry and beverage industries. Expected ambient air quality standards are given in Table 27.5.

**TABLE 27.4:** ORGANIC AIR POLLUTANTS AND CONTROL METHODS.

| *Chemical* | *Process of Production* | *Pollutant(s)* | *Control* |
|---|---|---|---|
| Acrylonitirle (Vinyl cyanide) $CH_2CHCN$ | Chemical process | Acetonitrile, CO, propane, propylene | Thermal incineration |
| Adipic acid $COOH(CH_2)_4COOH$ | Oxidation of cyclohexane by air (in the manufacture of synthetic fibers) | $NO_x$ | Catalytic conversion |
| Carbon | Combustion of oil | $H_2$, $C_2H_2$, CO, $H_2S$ & $SO_2$ | Flaring |
| Carbon monoxide (CO) | Oil refining, cocking, incineration | CO | Afterburner |
| Ethylene dichloride | Chemical process | CO, $CH_4$, $C_2H_4$, & $C_2H_6$ | Heat boiler |
| Ethylene oxide | Chemical process | $CH_4$, $C_2H_4$, & ethylene oxide | Catalytic afterburner |
| Formaldehyde | Chemical process | CO, methanol & formaldehyde | Wet scrubbers for liquids and afterburners for vent gases |
| Hydrocarbons (HCs) | Boilers, flares, heaters, waste effluents | Many | Many methods |
| $NO_x$ | Petrochemical boilers & flares | $NO_x$ | Catalytic converters |
| Pesticides | Chemical processes | Pesticides | Controlled use |
| Plastics | Polymerization of organic compounds | Particulates, chemicals and vapors | Precipitation, dry scrubbing and cyclones |
| $O_x$ | Boilers, flares & acid sludge | $SO_x$ | Wet scrubbers |
| Soaps & detergents | Fatty acids + alkali | Particulates and surfactants | Floatation, coagulation, etc. |

**TABLE 27.5:** AMBIENT AIR QUALITY STANDARDS

| *Pollutant* | *Accepted Concentration* | *Detection Method(s)* | *Control Method(s)* |
|---|---|---|---|
| Carbon monoxide (CO) | 10 ppm (8 hr mean) (10 $mg.m^{-3}$) | Non-dispersive IR (NDIR) | Afterburner oxidation |
| Nitrogen dioxide ($NO_2$) | 0.25 ppm(470 $\mu g.m^{-3}$) | Saltzman & Colorimetric (with NaOH) | Catalytic conversion |
| Sulfur dioxide ($SO_2$) | 0.05 ppm (24 hr mean) (150 $\mu g.m^{-3}$) | Pararosaniline & Coductometric | Wet scrubber |
| Hydrogen sulfide ($H_2S$) | 0.03 ppm (40 $\mu g.m^{-3}$) | Cadmium hydroxide & Staractan | Adsorption; conversion |
| Lead (Pb) | 1.5 $\mu g.m^{-3}$ (annual mean) | Dithizone | Scrubbers & baghouses |
| Hydrocarbons (HCs) | 0.25 ppm (160 $\mu g.m^{-3}$) | Flame ionization (FID-GC) | Scrubbers and baghouses |
| Petrochemical oxidants | 0.10 ppm (200 $\mu g.m^{-3}$) | Neutral buffered KI; Chemiluminescence | Scrubbers and baghouses |
| Suspended particulate matter | 100 $\mu.m^{-3}$ | High-volume sampling | Precipitation; cyclones; conversion |

Conversion formula for mass to ppm.

$$\text{Concentration}(mg / m^3) = \text{Concentration}(ppmV) \times \frac{M \times (12.187)}{(273.15 + ^\circ C)}$$

where, $M$ = molecular mass.

## BIBLIOGRAPHY


Adamson, R. G. (1973), Bellhaven House: Ontario. "*Pollution: An Ecological Approach.*"
Andrews, W. (ed.) (1972), Prentice-Hall: Engelwood Cliffs, NJ. "*Environmental Pollution.*"
Bockris, J. O. M. (1977), Plenum Press: New York. "*Environmental Chemistry.*"
Buffle, J. and Van Leeuwen, H. P. (eds.) (1992), Lewis Pubs: Chelsea, MI. "*Environmental Particles,*" (Vol. 1).
Butler, J. D. (1979), Academic Press: New York. "*Air Pollution Chemistry.*"
Carless, J. (1993), "*Renewable Energy: A Concise Guide to Green Alternatives.*"
Cunningham, W. P. (ed.) (1994), Gale Pubs: Detroit, IL. "*Environmental Encyclopedia.*"
Ewing, G. W. (1977), Academic Press: New York. "*Environmental Analysis.*"
Faith, W. C. and Arthur, A. H. Jr. (1992), John Wiley: New York. "*Air Pollution.*"
Harrison, R. M. and Perry, R. (1984), Chapman & Hall: London. "*Handbook of Air Pollution Analysis.*"
Hutzinger, O. (ed.) (1990), Springerverlag: Berlin. "*Handbook of Environmental Chemistry.*"
Jorgensen, S. E. and Johnson, I. (1981), Elsevier: Amsterdam. "*Principles of Environmental Science and Technology,*" (Vol., 14).
Keith, L. H. (ed.) (1984), Butterworth Pubs: London. "*Identification and Analysis of Pollutants in Air.*"
Kouimitzis, T. and Samara, C. (eds.) (1995), Springerverlag: Berlin. "*Airborne Particulate Matter.*"
Liu, D. H. F. and Liptak, B. G. (eds.) (2000), Lewis Pubs: Boca Raton, FL. "*Air Pollution.*"
Markert, B. (ed.) (1994), VCH Pubs: New York. "*Environmental Sampling for Trace Analysis.*"
McGrath, J. J. and Barnes, C. D. (1982), Academic Press: New York. "*Air Pollution: Physiological Effects.*"
Nriagu, N. O. (ed.) (1992), Wiley & Sons: New York. "*Gaseous Pollutants.*"
Parker, H. W. (1977), Prentice-Hall: Engelwood Cliffs, NJ. "*Air Pollution.*"
Perkins, H. C. (1977), McGraw-Hill: New York. "*Air Pollution.*"
Perry, R. and Young, R. J. (eds.) (1979), Chapman & Hall: London. "*Handbook of Air Pollution Analysis.*"

Stern, A. C. (ed.) (1984), Academic Press: Oxford, UK. "*Air Pollution*," (Vol. 1).
Stern, A. C., *et al.* (1984), Academic Press: New York. "*Fundamentals of Air Pollution*," (2nd edn.).
Thain W. (1980), Oxford: Pergamon Press. "*Monitoring Toxic Gases in the Atmosphere for Hygiene and Pollution Control*,"
Turk, A., *et al.* (1978), Saunders: Philadelphia, PA. "*Environmental Science*," (2nd edn.).
Wark, K. and Warner, C. F. (1976), IEPA: New York. "*Air Pollution: Its Origin and Control*."
...(1990), Natl. Inst. Standards & Technology: Gaithersburg, MD. "*Standard Materials for Environmental Research, Analysis and Control*."
... (1993), McGraw-Hill: New York. "*Environmental Science and Engineering*."

# 28

# Water Pollution Control Methods

Water is the ultimate source and conduit for accumulation and dispersal of environmental pollutants in the lithosphere. Therefore, the well-being and quality of vegetation and living organisms (ecosystems) on this planet are intimately connected with the quality of water bodies. Contamination of water sources arises from discharge of sewage, organic, biological, industrial and hazardous waste products.

Control of water pollution may be achieved by organizing—(1) separate water supply sources for different purposes, (2) wastewater treatment and storage and (3) avoiding direct discharge of untreated effluents to water bodies and strict adherence to domestic, urban and industrial norms of water management protocols.

Sewage water contains sediments, scum, oil, inorganic, metal and organic pollutants, pathogens and disease-causing vectors. The water purification methods follow stepwise treatment protocols that consist of several levels—(1) primary (physical), (2) secondary (biological), and (3) tertiary (chemical/biological). The degree of purification depends on the type of wastewater and the purpose of usage of treated water. The primary wastewater treatment protocol is a simple physical process to remove suspended matter, grit and scum by filtration. The secondary wastewater treatment protocol incorporates measures for further purification to remove oxygen-demanding organic pollutants. Tertiary wastewater treatment protocol includes steps to remove dissolved organic and inorganic compounds that include algal nutrients.

Industrial wastewater disposal poses unique problems because of the presence of hazardous chemicals (acids, alkalis, solvents and organic compounds) and trace level metals and effluent hot water. Treatment protocols depend on the type of industrial operations. Floatation and sedimentation can remove grit and debris. Salts are removed by ion-exchange, reverse osmosis. Metal pollutants can be removed by various physicochemical methods. Lime or lime with sulfide treatment removes most heavy metals as insoluble hydroxides and basic salts. Combustible residues may be incinerated and the final waste is compacted and disposed off as landfills, sand beds and fertilizers.

This chapter deals with some of the methods controlling water pollutants.

Water is essential for plants, aquatic animals, and land-based organisms. Water is used for many purposes—drinking, bathing, washing, cleaning, domestic and industrial needs and transport. Clean drinking water is an essential need of mankind. In many parts of the world, streams, rivers, ponds and lakes are used as drinking water sources. To prevent these resources from pollution effective treatment of wastewater is a necessity. BOD, pH, ammonium, nitrite, nitrate and phosphorus are the most important parameters, the optimization of which is essential to ensure proper functioning of wastewater treatment plants.

## 28.1 WATER AND WASTEWATER

There are many factors that influence the quality of water. Clay minerals are among the most common suspended matter found in natural waters. Because of their structure and high surface area per unit mass, these minerals have a strong tendency to adsorb chemical species from water. Wastewater is the spent or used water from a home, community, farm, or industry that contains dissolved or suspended matter. Domestic, gardening, farming, and industrial usage of water commonly adds contaminants and chemicals to the discharge. Domestic pollution is mainly due to food, household and sanitary activities. Agriculture-based pollution is due to crop wastes, fertilizers and pesticide and herbicides. Industrial pollution is due to food, beverages, wood, paper mills, and power and industrial operations and waste discharge.

When organic waste is present in water, the nutrients (nitrogen and phosphorus) are utilized by various organisms present in the aquatic environment. In that process they extract dissolved oxygen (DO) for their respiration, creating biochemical oxygen demand (BOD). Thus, a number of nutrient species such as ammonia, nitrates, phosphates and phosphoric acid are considered hazardous or serious water pollutants. Ammonia is also toxic and its toxicity increases with increase in pH and temperature. A water body with a pH value 6.0-9.0 can hinder the activity of the micro-organisms needed for the natural cleaning process.

Water pollution is caused primarily by the drainage of contaminated waters into the surface water or ground waters (aquifers). For drinking purposes water must be free from pathogenic organisms and hazardous chemicals. Wastewater can pollute water bodies through chemicals, metals, plant nutrients, waste heat and pathogens (Table 28.1). Contamination through human activities can spread infectious diseases. Therefore, water supply and water disposal should be optimized to maintain water standards for drinking, domestic and other purposes. This is achieved by: (*i*) selection of suitable water sources for different purposes, (*ii*) wastewater treatment before discharge into water sources and (*iii*) storage and treatment facilities for industrial effluents.

**TABLE 28.1:** CHARACTERISTICS OF WATER AND STANDARDS

| *Parameter* | *Effects* | *Standard for Drinking Water* |
|---|---|---|
| **Physical** | | |
| Temperature | Affects density, viscosity, surface tension, and dissolved oxygen (DO) and hence biological activities of aquatic life | Should not be >40° C |
| Dissolved Oxygen (DO) | Low dissolved oxygen (DO) affects biological activities of aquatic life | 6-7 ppm |
| Turbidity | Suspended matter excludes light and affects aquatic photosynthesis | 5 ppm |

| *Parameter* | *Effects* | *Standard for Drinking Water* |
|---|---|---|
| Color | Due to dissolved matter and chemicals | 15 ppm |
| Odor | Volatile gases such as $H_2S$ and organic decay | 3 ppm |
| Taste | Due to dissolved matter, salts and ions<br>Impairs quality of water | – |
| **Chemical** | | |
| Acidity | – | pH = 6-7.5 |
| Total dissolved solids (TDS) | – | 500 ppm |
| Arsenic | Disinfectant and oxidant | 0.01-0.05 ppm |
| Chloride | Toxic | 250 ppm |
| Chromium | – | 0.05 ppm |
| Copper | – | 1 ppm |
| Cyanide | – | 0.02 ppm |
| Fluoride | – | 1 ppm |
| Iron | – | 0.3 ppm |
| Lead | – | 0.05 ppm |
| Nitrate | – | 50 ppm |
| Sulfate | – | 250 ppm |
| Zinc | – | 5 ppm |
| **Biological** | | |
| Coliform | Health hazard | 1 coliform/100 mL |

## 28.2 WASTEWATER TREATMENT PROCESSES

Wastewater from home, community, farm or industrial discharge, contains dissolved or suspended matter and is unfit for domestic use and/or hazardous to life and wellbeing. Pollutants alter the physical, chemical and biological characteristics of water (Table 28.2).

**TABLE 28.2:** SUBSTANCES THAT IMPAIR WATER QUALITY

| *Substance* | *Undesirable Effects* | *Maximum Allowed Level (mg/L)* |
|---|---|---|
| Chloride | Taste and smell | 200 |
| Fluoride | Fluorosis | 1 |
| Phenols | Taste | 0.001 |
| Detergents | Taste and foaming | 0.2 |
| Hydrogen sulfide | Taste and odor | 0.05 |
| Calcium | Water hardness | 50 |
| Magnesium | Water hardness | 30 |
| Copper | Taste | 0.05 |
| Iron | Taste and color | 0.1 |
| Manganese | Taste and color | 0.05 |
| Zinc | Taste and color | 5 |
| Ammonia | Growth of organism & pH | 0.05 |

Municipal sewage contains sediments, scum, oil, grease, organic algal nutrients, oxygen-demanding compounds, pathogens, etc. Sediments constitute a mixture of clay, silt, sand, minerals and organic matter. The purpose of water treatment is to correct the deficiencies in water quality. The purification is usually based on stepwise cleaning. General wastewater treatment processes are listed in Table 28.3.

**TABLE 28.3:** GENERAL WASTEWATER TREATMENT PROCESSES

| *Process* | *Application* | *Quality Improvement* |
|---|---|---|
| Coagulation and sedimentation | Primary treatment (physical method)<br>Removal of phosphorus.Lime-soda application | Reduction in turbidity<br>Reduces nutrient phosphorus<br>Reduces hardness (removes $Ca^{2+}$ and $Mg^{2+}$) |
| Gravity filtration | Primary treatment | Reduction in suspended particles; low turbidity |
| Filtration and chlorination | Municipal water supplies treatment | Reduction in turbidity and bacteria |
| Fine screening | Removal of raw sewage | Reduction in BOD |
| Microstraining | Clarification prior to filtering | Reduction in microscopic particles |
| Carbon filters | Adsorption | Removal of taste and odor |
| Ion exchange | Demineralization | Removal of ions |
| Electrochemical desalting | Reverse osmosis | Reclamation of water from saline and brackish water |

Water-pollution control primarily involves the removal of impurities from wastewater before it reaches natural water bodies or aquifers. Wastewater must be purified or treated to some degree in order to protect public health and to prevent deterioration of existing water quality. In developed countries specifications are laid out for permitting treated water to be let into surface or underground water sources. Some of the parameters that have to be controlled are turbidity, dissolved oxygen (DO), BOD, coliforms, acidity and toxic substances. To meet these requirements wastewater, in particular sewage, treatment processes consist of several levels of purification.

1. *Primary treatment*: At this stage the process is aimed to remove about 60 per cent of suspended solids and 35 per cent of BOD. Dissolved impurities are not removed.
2. *Secondary treatment*: About 85 per cent of both suspended solids and BOD is removed. In many countries a minimum level of secondary treatment is obligatory.
3. *Tertiary Treatment*: This is put into practice when a high degree of purification is desired. In addition to total solids and BOD, dissolved impurities like nitrate and phosphate are also eliminated. Nearly 99 per cent of all impurities from the sewage are removed by an elaborate treatment. The effluent from the tertiary treatment is almost of drinking water quality. For the same reason it is very expensive and is undertaken only when high quality water is in demand.

Whatever might be the level of water treatment in each case, the last step, before discharging it into a body of surface water, is disinfection. This is achieved by chlorination. In modern water treatment plants ultraviolet irradiation and ozone treatment are often chosen for this purpose.

### 28.2.1 Primary Wastewater Treatment Processes

Primary wastewater treatment is a simple physical process to remove materials that can either float or can readily settle under gravity. Primary wastewater treatment process usually involves screening, grit removal and sedimentation, pre-treatment for coagulation, flocculation, and settling and sludge removal. Finally, a simple post-treatment with chlorine (Fig. 28.1) is also carried out.

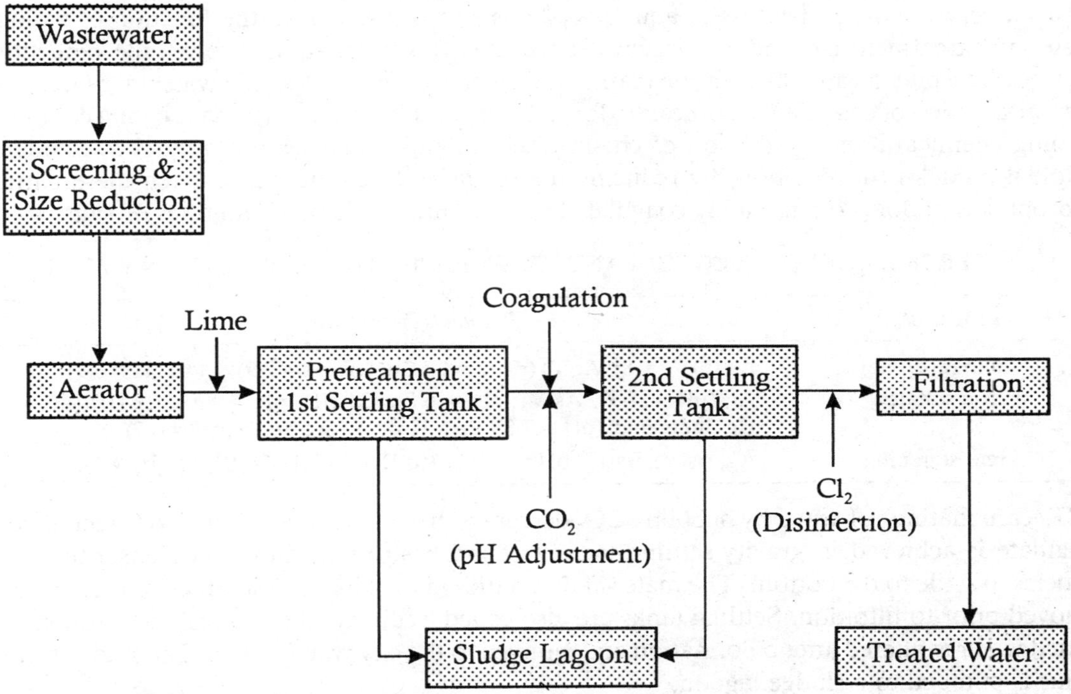


**Fig. 28.1:** Block Diagram of a Primary Wastewater Treatment Plant

#### *28.2.1.1 Screening*

The first step in the removal of floating material in wastewater is by bar screens (grit removers) consisting of long steel bars spaced ~2.5 cm apart. Grit chambers are designed to slow down the flow so that suspended matter (silt, sand and gravel) can settle down. The second step is comminution, accomplished with a circular grinder designed to grind solids coming through the screens into pieces of size abut <0.3 mm. Floatation is a process to remove grease and oil.

#### *28.2.1.2 Pretreatment and Filtration*

The next step is aeration. The wastewater, free from debris and suspended particles, is led into an aeration chamber. Addition of oxidizing agents (chlorine or potassium permanganate) at alkaline pH oxidizes Fe(II) to Fe(III), which hydrolyzes and gets precipitated. $Ca^{2+}$ and $Mg^{2+}$ bicarbonates and sulfates cause water hardness. Addition of lime after aeration raises the pH of water (~11) and precipitates $Ca^{2+}$ and $Mg^{2+}$. The precipitate formed is allowed to settle in the primary basin. Water softening by cation exchange is also used ($Na^+$ ion is exchanged for $Ca^{2+}$ ion in solution). Coagulants such as alum (ferric and aluminum sulfates) are added to coagulate colloidal suspensions (Table 28.4).

Coagulation is a physicochemical process involving the agglomeration of colloidal particles (0.001-1 μm) to facilitate their subsequent removal by sedimentation and filtration. Colloids are hydrophobic (petrochemicals), hydrophilic (proteins and nucleic acids, etc) and association colloids (micelles and soaps and detergents). The surface charge on a colloid may prevent aggregation, but this is pH dependent. In the neutral pH range (~7), most colloidal particles in natural waters have a negative charge (proteins, algal and bacterial cells). The aggregation and settling of micro-organism cells is a very important process in aquatic systems and is essential to the function of biological wastewater treatment (secondary wastewater treatment). Coagulation involves the reduction of electrostatic repulsion and flocculation (gathering together of fine particles in water by gentle mixing after the addition of coagulant chemicals to form larger particles) depends upon bridging compounds forming chemical bonds. Addition of chemical coagulants enhances the mutual attachment of colloidal particles (flocculation) by reducing the repulsive forces acting between these particles. Hydrophobic colloids often readily coagulate by the addition of small quantities of salts.

**TABLE 28.4:** CHEMICAL COAGULANTS TO PRECIPITATE POLLUTANTS IN WATER

| *Coagulant* | *Pollutants Precipitated* |
|---|---|
| Alum (alkaline) | Ag (pH = 6-8); As(V) (pH = 6.7); Cr(III) (pH = 7-9) & Pb (pH = 6-9) |
| Ferric sulfate | Ag (pH = 7-9); As(V) (pH = 6-8); Cd(III) (pH > 8.0); Cr(III) (pH = 6-9); Hg (inorganic) (pH = 7-8); Pb (pH = 6-9) & Se (IV) (pH = 6-7) |
| Lime softening | Ag; As(V); Ba(II) (pH = 10-11); Cd(III); Cr(III); Fe(II); Mn(II) & Pb |

"Recarbonating" of water by bubbling $CO_2$ decreases the pH of water. Settling of precipitate and coagulate is achieved in gravity-settling tanks (second basin) that allow the denser-than-water particles to settle to the bottom. The material that settles in the tanks is known as 'sludge' which is removed prior to filtration. Settling tanks are also called 'sedimentation tanks' or 'clarifiers.' The treated water is filtered after chlorination and pumped to the city water mains. The sludge from the basins is pumped to a sludge lagoon.

### 28.2.2 Secondary Wastewater Treatment Processes

The objective of secondary wastewater treatment by biological processes is to remove soluble organic pollutants (carbonaceous materials). The most harmful effect of organic matter in wastewater lies in removal of oxygen due to the intervention of micro-organisms. The characteristics of wastewater are measured in terms of its BOD and COD. Streams receiving organic wastes are capable to some extent, of self-purification through micro-organism-mediated oxidation of organic matter. Micro-organisms consume the organic impurities as food, converting it into $CO_2$, $H_2O$, and energy for their growth and reproduction. The same phenomenon is utilized in biological waste treatment. Secondary wastewater treatment by biological processes consists mainly of allowing micro-organisms to degrade organic matter, in solution or in suspension, in the presence of added $O_2$ until the BOD of the wastewater has been adjusted to acceptable levels. The treatment plants are designed to reduce the high-energy compounds to low-energy chemicals.

The yield of a secondary wastewater treatment is expressed as mass of secondary sludge produced per mass of BOD removed

$$\text{Yield} = \frac{\text{Mass of Suspended Solids}}{\text{Mass of BOD}} \qquad \text{..(28.1)}$$

Biological processes depend on pH, temperature, type of substrates, and the availability of nutrients and minerals. After the organic pollutants are nearly eliminated the micro-organisms are separated and recycled while the purified water is collected for further treatment or pumping to utilities after disinfection.

The principal accomplishment of secondary wastewater treatment lies in the removal of suspended matter by settling, flotation, biochemical degradation and adsorption on the sludge by the primary wastewater treatment processes followed by biological degradation processes. The biochemical degradation reactions (redox reactions) may proceed either in the presence of free oxygen (aerobically), or in its absence (anaerobically). Bacterial respiration is a biochemical process in living protoplasm (cell), whereby energy is made available for endothermic life processes.

Organic matter oxidation (respiration)

$$C_xH_yO_z + O_2 \rightarrow CO_2 + H_2O + \text{energy} \quad \ldots(28.2)$$

Cell material synthesis

$$C_xH_yO_z + NH_3 + O_2 + \text{energy} \rightarrow C_5H_7NO_2 + CO_2 + H_2O$$

Cell material oxidation

$$C_5H_7NO_2 + 5O_2 \rightarrow 5CO_2 + 2H_2O + NH_3 + \text{energy}$$

### 28.2.3 Different Wastewater Treatment Systems

Biological wastewater treatment systems employ several methods for purification. Both aerobic and anaerobic treatment systems are in use.

#### *28.2.3.1 Aerobic Wastewater Treatment Systems*

Aerobic treatment is process by which microbes decompose complex organic compounds in the presence of oxygen and use the liberated energy for reproduction and growth. Various kinds of aerobic wastewater treatment processes are in use, and the most common are:

1. *Aerobic stationary-contact systems*: Irrigation beds, sand filters and trickling filters.
2. *Aerobic suspended-contact systems*: Aerobic lagoons (Fig. 28.2).

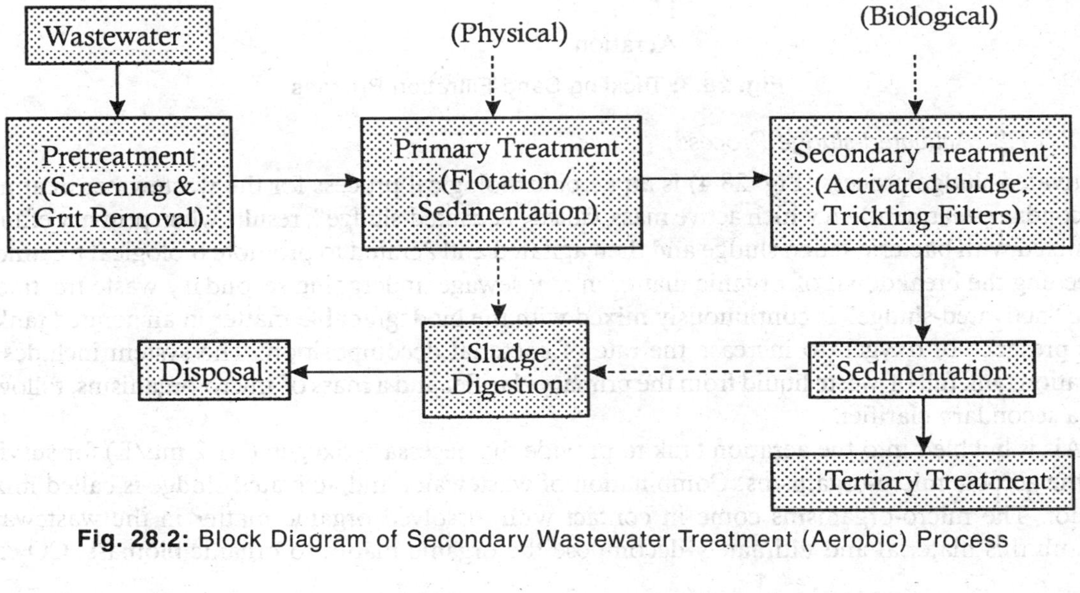


**Fig. 28.2:** Block Diagram of Secondary Wastewater Treatment (Aerobic) Process

### *28.2.3.1.1 Sand Filtration*

Filtration is intended to remove solids from wastewater by: (*i*) microscreens, (*ii*) sand filters or (*iii*) diatomaceous earth filters. Microscreens are mechanical filters. Sand filters are gravity filters, allowing water to trickle through. Diatomaceous earth filters are also packed beds.

"Air stripping" can be done to remove certain ions. For example, at neutral pH, ammonium ions, $NH_4^+$, exist in equilibrium as

$$NH_4^+ \Leftrightarrow NH_3 + H^+ \qquad \ldots(28.3)$$

Above pH > 10, the species is $NH_3$, which can be removed by blowing air and the expelled gas is trapped. Sufficient oxygen is provided through recirculated water to maintain aerobic condition.

One of the commonly used biological wastewater treatment processes is the trickling filter method (Fig. 28.3) in which sewage water is sprayed continuously over the top of a deep bed of stones or other solid support covered with deposits of micro-organisms in the form of biofilm or immobilized cells. As the wastewater trickles down, the microbes absorb the dissolved oganic compounds, thus lowering the BOD of the sewage. The bed is periodically (or continuously) aerated by an air bubbling device. The recycling of effluent to the location upstream of trickling filters as an influent to provide multi-pass filtering improves efficiency in removal of organic pollutants.

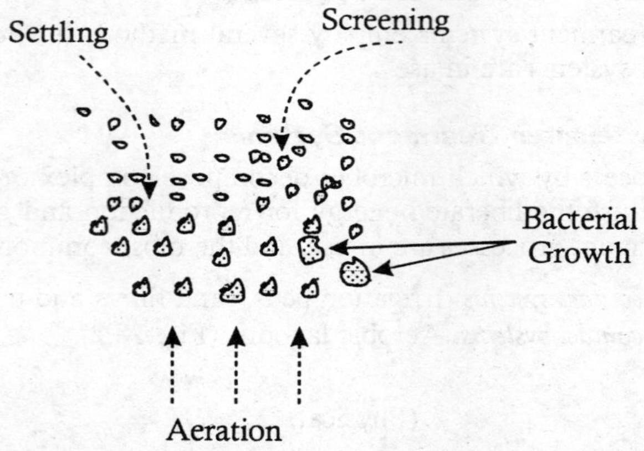


**Fig. 28.3:** Trickling Sand Filtration Process

### *28.2.3.1.2 Activated-sludge Process*

Activated-sludge process (Fig. 28.4) is an aerobic biological process for the wastewater treatment (secondary treatment), in which active mass, called "activated-sludge", results when primary effluent is mixed with bacteria-laden sludge and then agitated and aerated to promote biological treatment, speeding the breakdown of organic matter in raw sewage undergoing secondary waste treatment. The "activated-sludge" is continuously mixed with the biodegradable matter in an aerated tank in the presence of oxygen to increase the rate of bacterial decomposition. The system includes an aeration tank full of waste liquid from the primary clarifier and a mass of micro-organisms, followed by a secondary clarifier.

Air is bubbled into the aeration tank to provide the necessary oxygen (~1-2 mg/L) for survival of the aerobic micro-organisms. Combination of wastewater and activated-sludge is called mixed liquor. The micro-organisms come in contact with dissolved organic matter in the wastewater, adsorb this material and ultimately decompose the organic matter to organic biomass, $CO_2$ and

$H_2O$ and convert nitrogen to $NH_4^+$ and phosphorus to $PO_4^{3-}$ (orthophosphates). The micro-organisms are separated from the mixed liquor in a settling tank when most of the organic matter (i.e., food for the micro-organisms) has been used up. The micro-organisms remaining in the settling tank have no food available, and are thus "hungry" (activated-sludge). Part of the settled micro-organisms, called the 'return activated-sludge,' from the second clarifier is pumped back into the aeration tank, where they find more food in the organic contaminants in the fresh sewage liquid entering there from the primary clarifier. The process is repeated. The combination of a high concentration of "hungry" cells from the return sludge and a rich food source from the influent sewage under oxygenated conditions provides optimum conditions for the rapid degradation of organic matter. This recirculation protocol is a key feature of the activated-sludge process.

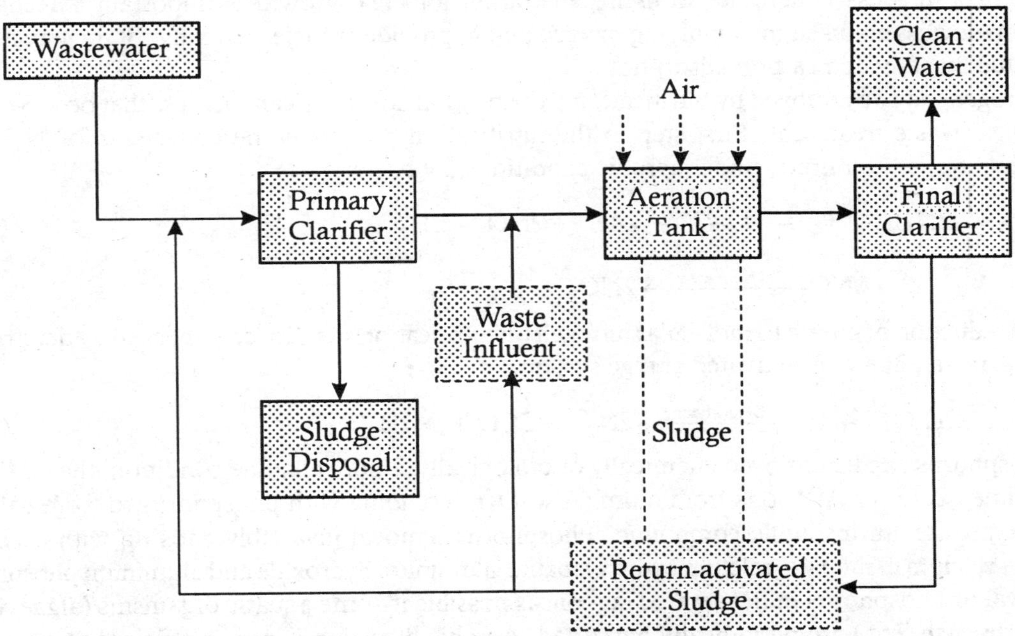


**Fig. 28.4:** Block Diagram of an Activated-sludge Treatment System

The activated-sludge process provides two ways for the removal of BOD. BOD may be removed by: (*i*) oxidation of organic matter to provide energy for the metabolic process of the microorganisms, and (*ii*) incorporation of the organic matter into cell mass (biomass) (Fig. 28.5). In the first pathway carbon is removed as $CO_2$, and in the second pathway carbon is removed as solid mass.

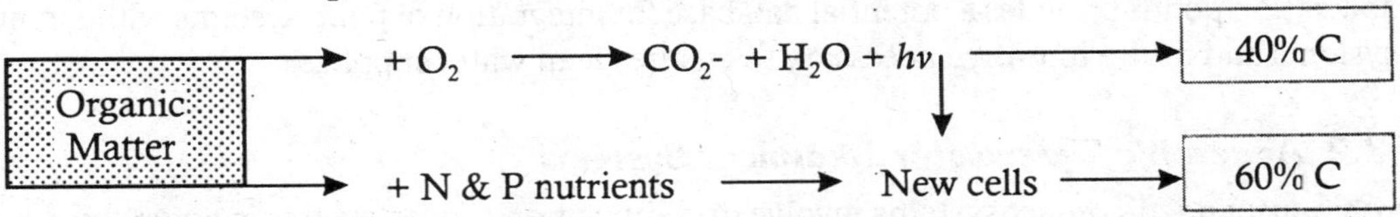


**Fig. 28.5:** Metabolic Pathways for Aerobic Microbial Degradation of Organic Wastes

Activated-sludge treatment produces more micro-organisms than necessary and, therefore, some of the micro-organisms are wasted. The disposal of part of the activated-sludge is necessary. Aggregation and settling of micro-organisms is a very important step in the activated-sludge (biological) wastewater treatment process. About 70 per cent of the secondary sludge must be treated and disposed. A significant fraction of aggregated microbial cells is removed as 'bacterial-floc.'

*28.2.3.1.3 Aerated Lagoons*

A popular secondary wastewater treatment setup is the aerated lagoon (sewage lagoon), also called oxidation pond or stabilization pond. These are shallow ponds designed to treat wastewater through interaction of sunlight, bacteria and algae. They speed up the natural process of biological decomposition of organic waste by stimulating the growth and activity of bacteria that degrade organic waste. The process is similar to the activated-sludge system, but without recirculation of sludge. The process is aerobic; so light penetration for algal growth is important. Mechanical aeration devices are used for supplying oxygen and to provide sufficient mixing. BOD may also be corrected by activated carbon adsorption.

Nitrogen may be removed by nitrification. Nitrification is a significant process that occurs during biological waste treatment. First step in the nitrification process is the conversion of $NH_3$ and organic nitrogen to nitrate, under aerobic conditions, by bacteria.

$$2NH_4^+ + 3O_2 \xrightarrow{\textit{Nitrosomonas}} 4H^+ + 2NO_2^- + 2H_2O \quad \text{...(28.4)}$$

$$2NO_2^- \xrightarrow{\textit{Nitrobacter}} 2NO_3^-$$

The reduction of nitrate to molecular nitrogen (denitrification step) is accomplished by denitrifying bacteria in an anaerobic activated-sludge system.

$$4NO_3^- + 5\{CH_2O\} + 4H^+ \xrightarrow{\text{Denitrifying}} 2N_2\uparrow + 5CO_2\uparrow + 7H_2O \quad \text{...(28.5)}$$

Phosphorus can be removed chemically or biologically. Under alkaline condition, the $Ca^{2+}$ ions from lime (CaO) or $Al^{3+}$ ions from alum ($Al_2(SO_4)_3$) combine with phosphorus to form calcium hydroxyapatite, an insoluble compound. Phosphorus removal invariably ends up with excess of solids—calcium carbonate, calcium hydroxyapatite, aluminum hydroxide and aluminum phosphates. Removal of phosphorus in the aerated lagoons is possible if some aquatic organisms (algae, water hyacinths, fish, etc.) are periodically harvested. Aerobic digestion is not as efficient as anaerobic digestion.

The stabilization pond is an ecosystem governed by the nature of the communities that it supports, and the environmental conditions in which it is maintained. Monitoring rates of change using biological and chemical techniques provides useful information for adjusting retention times or creating additional pond capacity to enhance the treatment process. Information used to manage the stabilization ponds provides a potential database for integration of pond systems with membrane filter systems and future initiative necessary to ensure clean water supplies.

### 28.2.3.2 Anaerobic Wastewater Treatment Systems

Anaerobic suspended-contact systems involve decomposition of wastewater in anaerobic lagoons. In an anaerobic wastewater treatment process (Fig. 28.6), anaerobic bacteria stabilize the organic matter in the absence of free oxygen (oxygen is poisonous to anaerobic bacteria). This process is widely used for stabilization of sludge collected from primary and secondary settling tanks and for soluble wastes in anaerobic lagoons.

In the 1st stage, organic materials are converted to simple organic acids (acetic, propionic, etc.) by acid-forming bacteria. In the 2nd stage, the fatty acids are converted to $CO_2$ and methane, thereby stabilizing BOD and COD. Anaerobic bacterial treatment processes enable the conversion of organic pollutants to methane gas, which can be burned out (or used as an energy source). Anaerobic treatment processes are very sensitive to operating parameters and their environments—pH, temperature and the presence of toxins.

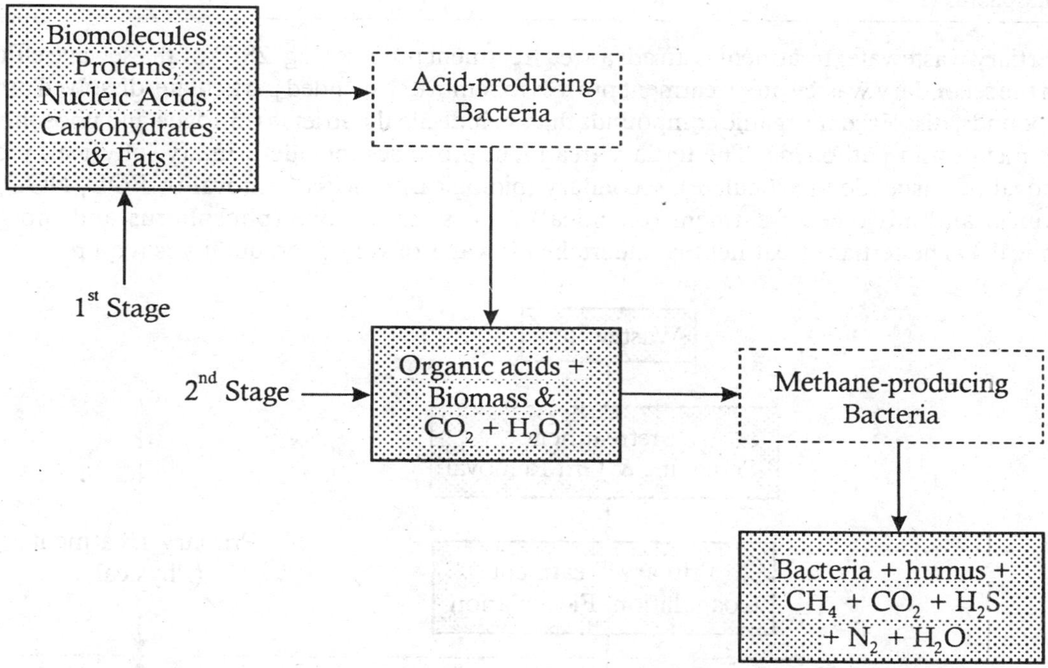


**Fig. 28.6:** Anaerobic Wastewater Treatment Process

The general performance of anaerobic digesters has been improving as a result of better reactor designs, operating conditions, and use of specialized microbial consortia. In the case of domestic sewage, the overall process is improved by the adoption of an anaerobic pretreatment step (e.g. upflow sludge blanket (USB) reactor), and coupling of new methods of nutrient removal, such as anaerobic ammonium oxidation and chemical precipitation of phosphorus. In the case of industrial wastewater, new additives are being developed that help the anaerobic sludge deal with unbalanced or toxic wastewater. Another development is engineering of anaerobic sludge granules, which improve catabolic activity of the sludge and shorten the adaptation period of the microbial consortia to xenobacteria. New biotechnologies (e.g. high-solids fermentation technology) are also being marketed for the anaerobic removal of sulfate and heavy metals.

### 28.2.4 Tertiary Wastewater Treatment Processes

Secondary wastewater treatment reduces suspended matter and BOD, but not nitrogen and phosphorus (Table 28.5). The removal of nitrogen and phosphorus and dissolved inorganic salts is relatively expensive and is used only in advanced wastewater treatment (tertiary wastewater treatment).

**TABLE 28.5:** PERFORMANCE OF WASTEWATER TREATMENT PROCESSES

| *Parameter* | *Raw Wastewater (mg/L)* | *After Primary Treatment (mg/L)* | *After Secondary Treatment (mg/L)* |
|---|---|---|---|
| BOD | 250 | 175 | 15 |
| Suspended solids | 200 | 60 | 15 |
| Phosphorus (P) | 8 | 7 | 6 |

Tertiary wastewater treatment is an advanced treatment process (Fig. 28.7) of the sewage effluents from the secondary wastewater treatment process to remove suspended particulate, dissolved organic compounds, dissolved inorganic compounds that include algal nutrients (nitrogen and phosphorus), toxic metals and pathogens. The tertiary treatment processes include primary (physical) process (removal of suspended particulate), secondary (biological) process (removal of BOD, NOD, and organics) and advanced treatment (chemical) process to remove (phosphorus and inorganic chemicals). The tertiary treatment is undertaken if water of very good quality is required.

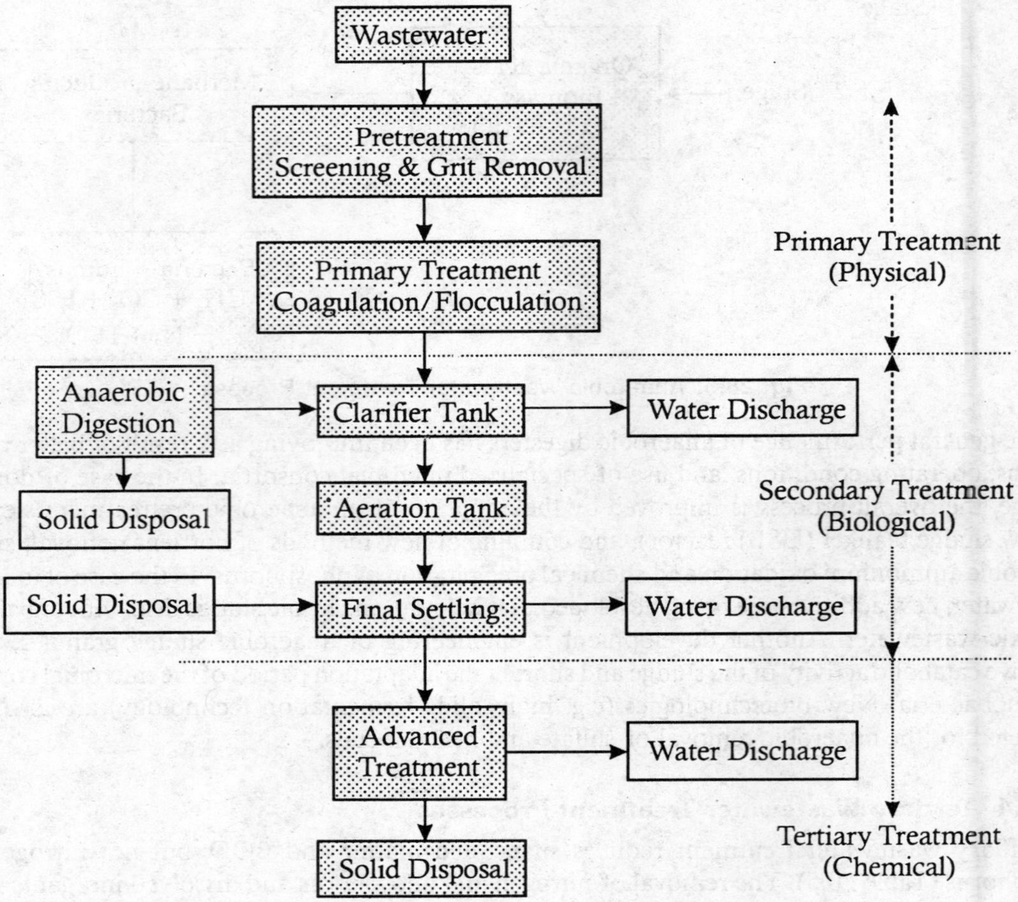


**Fig. 28.7:** Block Diagram of Advanced Wastewater Treatment Protocol

Precipitation and coagulation in the primary treatment process remove suspended particles. In the secondary treatment process, methanol is metabolized under aerobic conditions, utilizing $O_2$ from nitrates and releasing nitrogen to the atmosphere. Removal of BOD is accomplished in the biological unit (Fig. 28.8). Nitrogen is present in the form of nitrates and ammonia. Biological nitrogen removal from wastewater is accomplished by successive microbial transformation, the nitrification-denitrification processes. Removal of $NH_3$ (removal of NOD) can be achieved by adding lime to the water to raise its pH (>10), followed by aeration to strip the gas from the liquid ("air stripping"). A physicochemical process called "ammonia stripping" may be used to remove ammonia from sewage. But, in practice, $NH_3$ is removed by nitrification. It is a two-step process in which ammonia is first converted into nitrates by a class of aerobic micro-organisms. The nitrates are further metabolized by another class of anaerobic bacteria that convert it to nitrogen, which escapes from the water because it is sparingly soluble in water.

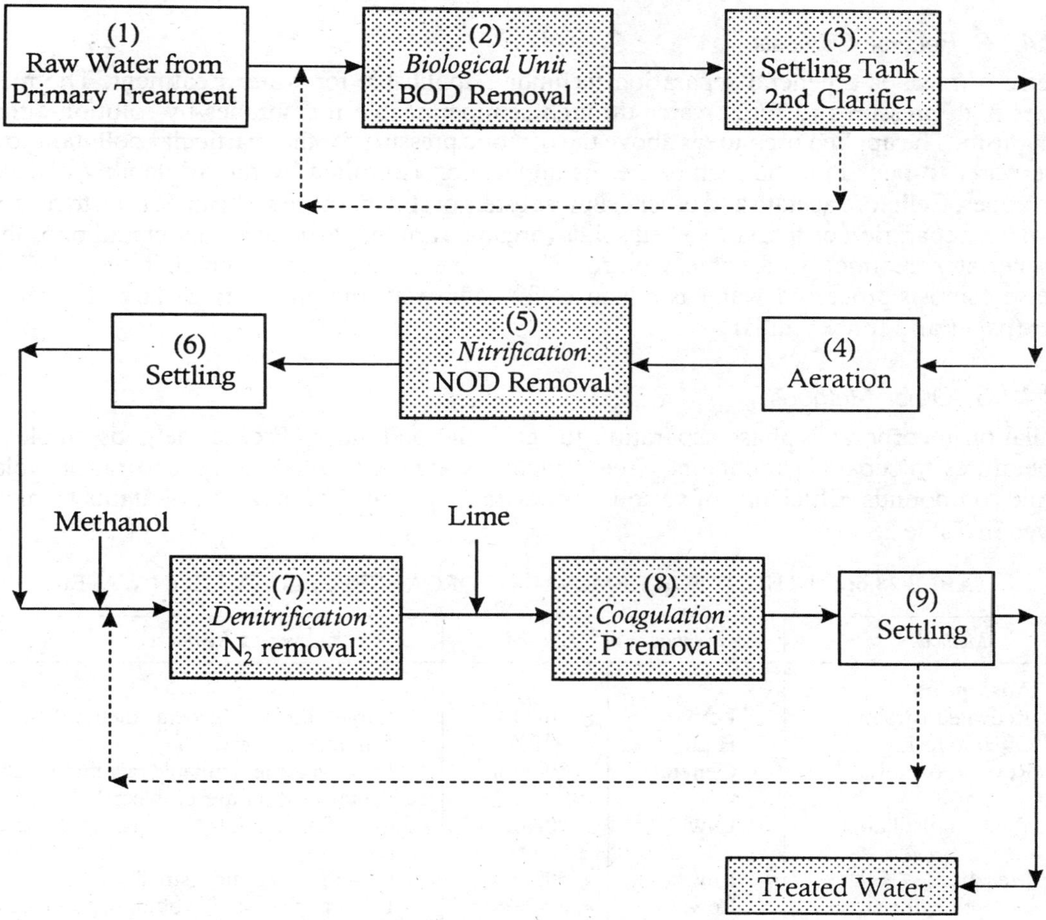


**Fig. 28.8:** Advanced Wastewater Treatment to remove Inorganic Compounds

Removal of organic and inorganic compounds can be accomplished by several methods. Organic pollutants can be removed by: (*i*) adsorption on activated carbon or organic resins (polyurethane), (*ii*) reverse osmosis, (*iii*) vacuum distillation and (*iv*) freezing. Inorganic substances can be removed by: (*i*) ion exchange, (*ii*) reverse osmosis and (*iii*) distillation.

### 28.2.4.1 Removal of Organic Matter

Several methods, most common being adsorption on activated carbon, freeze-drying and distillation accomplish removal of organic matter.

#### 28.2.4.1.1 Adsorption

Adsorption of organic compounds on activated carbon is an important step in the tertiary wastewater treatment process. The exhausted carbon can be discarded or regenerated by burning off the organics in a furnace. Foam separation achieves selective removal of soluble and colloidal pollutants like alkyl-benzenesulfonate (ABS), surfactants and phenols.

#### 28.2.4.1.2 Reverse Osmosis

Reverse osmosis is a general separation technique applicable for water treatment. The method utilizes high pressures to force water through semipermeable membranes by solution-diffusion mechanism. The applied pressure is above the osmotic pressure for that particular pollutant so as to cause water to penetrate the membrane. Its application is limited by the availability of suitable membranes. Cellulose acetate and other cellulose esters of suitable pore size can be used to desalinate salt water. Separation of total dissolved solids (organic carbon, particulate matter and phosphates) in wastewater treatment is feasible. Control of pH is necessary to avoid precipitation of $CaCO_3$. Reverse osmosis-processed water is adequate for industrial units that utilize large quantities of water (paper and timber mills).

#### 28.2.4.1.3 Other Methods

Distillation incorporates phase separation to reject the pollutants. Freeze methods employ low temperatures to separate pollutants. Freeze methods are well suited for separation of (volatile) organic compounds. Efficiency of various processes for removal of organic pollutants from water is given in Table 28.6.

**TABLE 28.6:** METHODS FOR REMOVAL OF ORGANIC POLLUTANTS IN WATER

| *Method* | *Selectivity* | *Efficiency* | *Recovery & Remarks* |
|---|---|---|---|
| **Adsorption** | | | |
| Activated carbon | Low | 10-20% | Large volumes (general method) |
| Organic resins | High | ~10% | Limited recovery |
| Reverse osmosis | General | 80-90% | Large volumes (suitable membranes and salt contamination are problems) |
| Vacuum distillation | Low | 80-90% | Only for non-volatile organics (good recovery) |
| Freezing | Low | 50% | Good for organics (small volumes) |
| Freeze-drying | Low | 80-90% | For organics (small volumes; good recovery) |

### 28.2.4.2 Removal of Inorganic Compounds

Coagulation, ion exchange, and reverse osmosis and electrodialysis are the commonly employed methods to remove inorganic water pollutants. The ion-exchange process used for removal of inorganic compounds consists of passing the water successively over a solid cation-exchanger and a solid anion-exchanger. The cation-exchanger is regenerated with strong acid and anion-exchanger with strong alkali.

### *28.2.4.2.1 Removal of Nitrogen and Phosphorus*

Nitrogen is algal nutrient. Major processes for the removal of nitrogen from wastewater are:

(*i*) Ion exchange
(*ii*) Reverse osmosis
(*iii*) Electrodialysis
(*iv*) Air stripping of ammonia: pH is raised to 11 with lime and ammonia is stripped from water by air in a stripping tank.
(*v*) *Bacterial treatment*: Nitrogen-containing organic compounds are transformed to simple non-nitrogenous compounds, by nitrification-denitrification (Fig. 28.9) treatment processes.

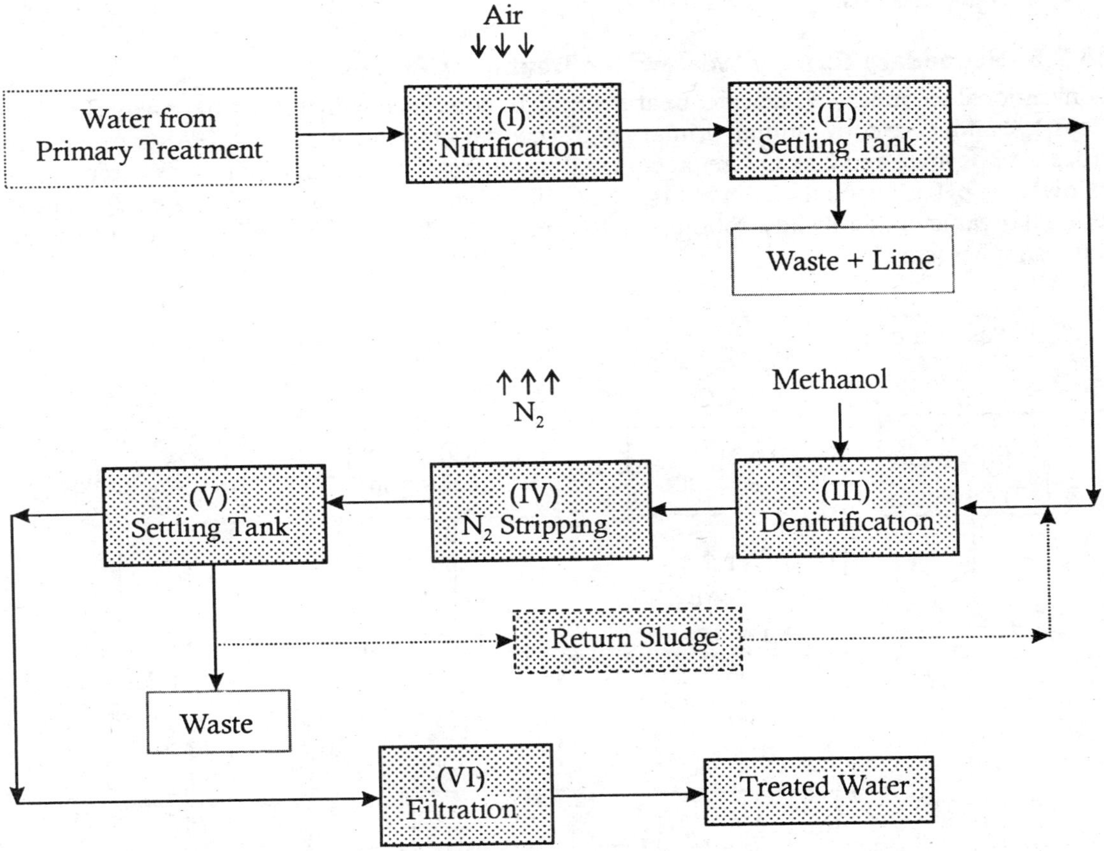


**Fig. 28.9:** Nitrification-Denitrification in Wastewater Treatment Process

The secondary effluent is saturated with air until all nitrogen is converted to nitrate ion through biological oxidation (nitrification) (Chamber I). Waste is removed from the settling tank and lime is added in Chamber (II) to remove phosphate. Methanol is added in the denitrifying tank (III) to serve as carbon and energy source for the denitrifying bacteria growing anaerobically. $N_2$ is stripped off in the next step (IV) and particulate and biomass are settled out in the next chamber (V). Return active-sludge is recycled (III → V) or treated effluent is diverted to the final filtration step (VI) to remove residual particulate. The value of total N after the treatment should be <18 mg/L.

Phosphorus is removed from water to reduce algal growth. This can be achieved by:

(*i*) *Addition of coagulants*: Lime ($Ca(OH)_2$), alum ($Al_2(SO_4)_3$, magnesium ($MgSO_4$) and ferric salts.

(*ii*) *Sorption*: Percolating water through a column of activated alumina removes phosphates. The column is regenerated by back washing with NaOH followed by acidification with $HNO_3$.

(*iii*) *Ion exchange*: Water is passed through an anion exchanger bed. Regeneration of ion exchanger is accomplished by NaCl wash.

(*iv*) *Electrodialysis*: By applying a direct current across a body of water separated into layers by membranes alternately permeable to cations and anions.

(*v*) Reverse osmosis.

### 28.2.5 Secondary-Tertiary Wastewater Treatment Process

Conventional secondary wastewater treatment can be combined with tertiary treatment (Fig. 28.10). Steps 1, 2 and 3 are conventional primary and secondary treatment processes. After the secondary process, phosphate and $CaCO_3$ are precipitated by adding lime, $Ca(OH)_2$, and $CO_2$ (steps 4 and 5), followed by pH adjustment (by bubbling of $CO_2$ in step 6), ammonia stripping (step 7), removal of residual organic matter by adsorption (step 8). Finally purified effluent is diverted to the municipal water supplies (step 9).

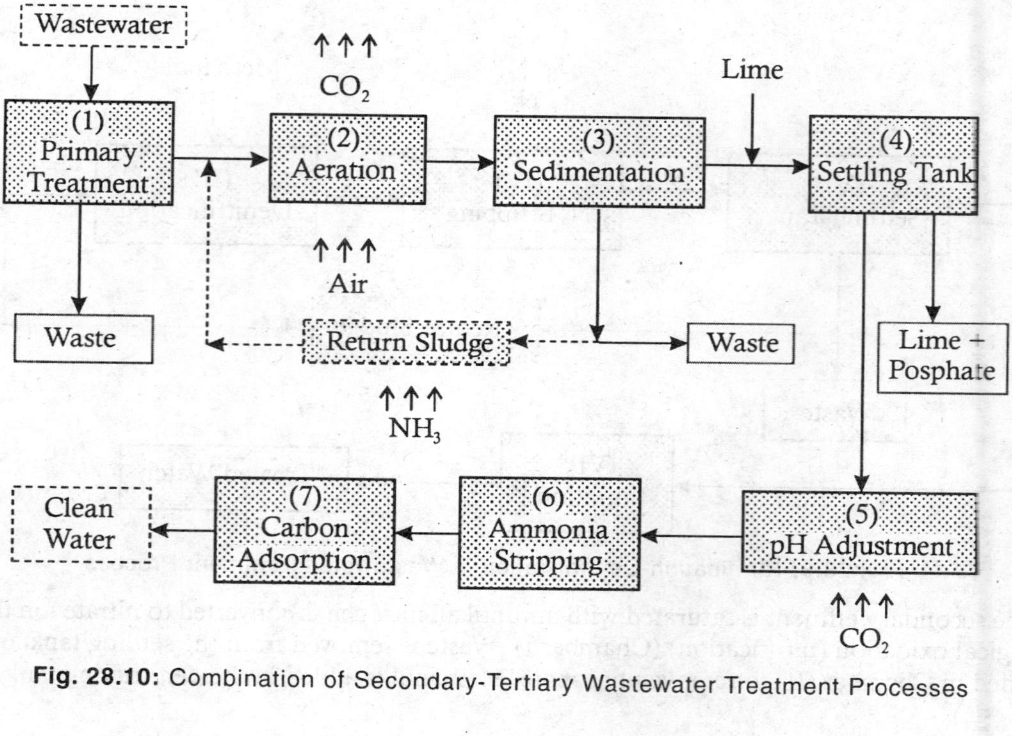


**Fig. 28.10:** Combination of Secondary-Tertiary Wastewater Treatment Processes

### 28.2.5.1 Physical-Chemical Treatment of Wastewater

Physical-chemical wastewater treatment system operates without recourse to biological treatment processes (Fig. 28.11). This system may be adequate for many industrial wastewater treatment requirements. Major steps are:

1. Grinding and screening and grit removal
2. Coagulation and removal of phosphate
3. Flocculation and sludge removal
4. Lowering of pH
5. Adsorption on activated carbon and filtration
6. Disinfection.

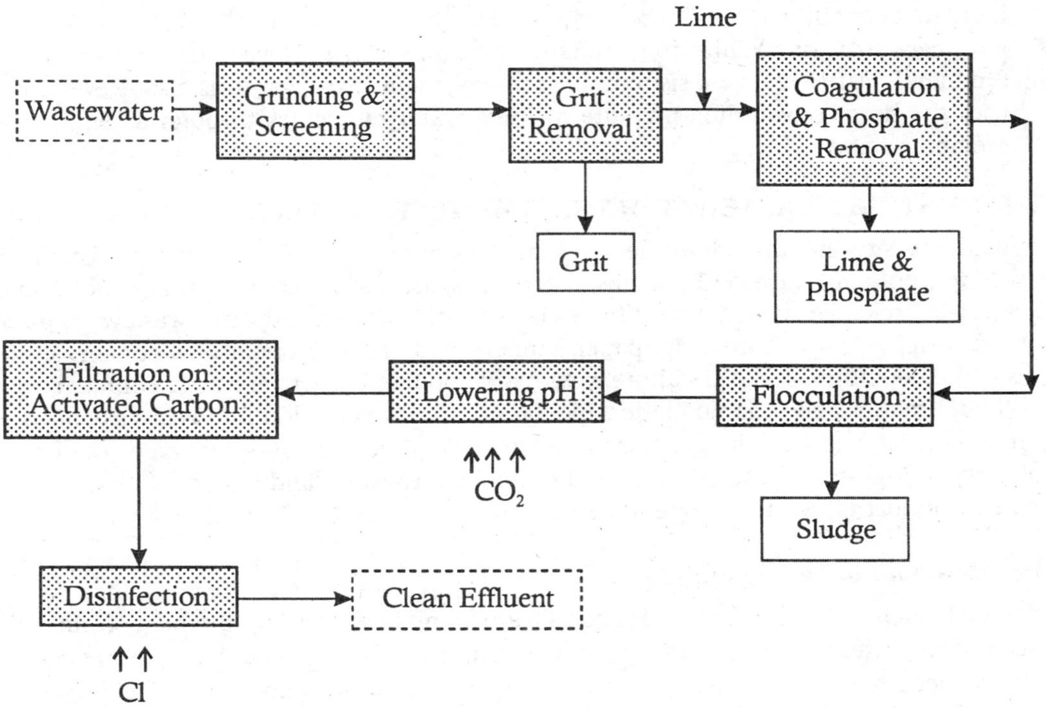


**Fig. 28.11:** Physical-Chemical Treatment of Wastewater

### 28.2.5.2 Disinfection

Water for human consumption and bathing is disinfected to destroy disease-causing micro-organisms (pathogens). Currently, the primary disinfecting agent is chlorine (and chloro-lime). However, chlorine produces some problems in aqueous ecosystems. Alternative disinfecting agents are UV radiation and ozone. UV radiation destroys pathogens effectively and its use as disinfecting agent eliminates the need to handle chemicals. Besides, it does not cause taste and odor problems. Ozone inactivates a wider range of micro-organisms than chlorine. In addition, $O_3$ has a high disinfection kill power, releases limited by-products, is non-reactive with $NH_3$, and has an excellent ability for removing undesirable odor and color. But, UV and ozone treatment technologies are expensive, and, therefore, are not in wide use.

#### 28.2.5.3 *Point-of-use (POU) Treatment of Wastewater*

Point-of-use (POU) treatment of wastewater is preferred in such cases where centralized treatment is not always a feasible option or for water faucet treatment for drinking and cooking purposes. A POU device uses a sequential combination of filtration, adsorption, ion exchange or reverse osmosis steps:

1. **Water "softeners":** They remove ions with positive charge (cations). Water softeners exchange "hard" ions (Mg, Ca) for "soft" ions (Na, K) that will not form precipitates and deposits. Water softeners have practically no ability to remove organic pollutants.
2. **Filtration:** Filtration, following coagulation by alum, ferric chloride or ferric sulfate, is used for removing sediment and gross particulate and suspended matter.
3. **Adsorption:** Granular activated carbon can remove most organic chemicals (but not inorganic contaminants—"minerals") from water by "adsorbing" them.
4. **Reverse Osmosis:** While ion exchange columns can remove inorganic contaminants ("minerals"), reverse osmosis removes some of the micro-organisms also. Reverse osmosis uses the home's waterline pressure to force water molecules through a semi-permeable membrane filter.

## 28.3 INDUSTRIAL WASTEWATER TREATMENT PROCESSES

All industrial wastewater discharge affect in some way or other the normal characteristics of water bodies. Major pollutants are acids, alkalis, inorganic salts, suspended matter, organic compounds and heat. Heat exchange via water constitutes a large fraction of total industrial wastewater discharge. Many industrial wastes—from pulp, paper, tannery and cannery operations contain suspended and dissolved contaminants. Agriculture, veterinary, food and beverage wastes contain suspended and colloidal matter and organic matter. Floatation or sedimentation removes suspended solids. Salts may be removed by ion exchange or reverse osmosis. Sludge is removed by filtration or percolation through sand beds and disposed off by incineration, landfill, etc.
Some of the industrial waste treatment methods are given in Table 28.7.

### 28.3.1 Removal of Heavy Metals

Heavy metals such as As, Cd, Cr, Cu, Hg and Pb are found in wastewater discharge from a number of industrial processes as well as from contaminated aquifers. Because of toxicity their concentration must be reduced to very low levels prior to release of the wastewater. Various physicochemical treatment processes effectively remove trace metals from wastewater (Table 28.8). Lime or lime with sulfide treatment (most heavy metals are sulfide seekers) removes most heavy metals as insoluble hydroxides and basic salts.

Removal of toxic metals in potable water includes processes of precipitation/coagulation, adsorption, ion exchange, membrane filtration (reverse osmosis) and point-of-use technologies. POU methods are preferred when centralized treatment is not feasible (such as in remote areas and single houses) and water is needed for drinking and cooking purposes. These systems offer ease of installation, simplify operation and maintenance and generally have lower capital costs. POU treatment usually involves a single-step.

Coagulation technology can be successfully employed to achieve removal of toxic metals. For example, coagulation of As(V) with alum, ferric chloride or ferric sulfate can remove >90 per cent. Softening technology can be implemented by water systems to achieve >90 per cent removal.

Alumina filtration is an adsorption process for removal of toxic metals. Oxidation state of the metal plays a role in its removal/treatment processes. For example, arsenic in pentavalent state is more readily removed than the trivalent arsenite form.

**TABLE 28.7:** REMOVAL OF WASTE IN INDUSTRIAL WASTEWATER TREATMENT

| *Industrial Wastes* | *Characteristics* | *Treatment Methods* |
|---|---|---|
| **Energy** | | |
| Coal-based | Particulate & $H_2SO_4$; corrosive | Precipitation & scrubbing |
| Steam-based | Inorganics, dissolved solids & heat | Neutralization, cooling by aeration of ashes |
| Nuclear power-based | Radioactive particulates | Concentration & dispersion |
| **Chemical** | | |
| Acids/alkalis | Corrosive and pH | Neutralization |
| Detergents | High BOD | Floatation and precipitation |
| Hydrocarbons (HCs) & pesticides | High organic matter & high BOD | Dilution, adsorption & controlled discharge |
| Phosphates & nitrates | Suspended particulates and nutrients | Coagulation and settling |
| **Petrochemical** | | |
| Oils | Emulsions | Floatation |
| Refining | Sulfur; dissolved salts & high BOD | Diversion, recovery, acidification & incineration |
| Rubber & plastics | Halides & high BOD | Aeration, chlorination, sulfonation & biological treatment |
| Glass & ceramics | Suspended particulates | Precipitation & discharge |
| Food & beverage | Suspended solids, organic matter, fermentable products, oils and fats, detergents & high BOD | Screening, floatation, coagulation, settling, filtration, aeration & biological treatment |
| Wood & paper | Suspended particulates, variable pH | Settling, aeration, biological treatment & recovery of by-products |
| Textile & leather | Suspended matter & high BOD | Neutralization, coagulation, aeration and biological treatment |

## 28.4 SLUDGE REMOVAL/DISPOSAL

Sludge discharged after secondary treatment contains suspended solids, organics, pesticides, traces of heavy metals (Cd, Cr, Cu, Hg, Ni, Pb and Zn) and pathogenic micro-organisms. It is malodorous, putrefying and hazardous to health. Sludge treatment is needed before its final disposal. Many times it is treated under aerobic conditions.

### 28.4.1 Sludge Treatment

In the sludge digestion process (a biological process), organic solids are decomposed into stable substances. At the same time the volume of the sludge is reduced by bacterial degradation (by methane-producing bacteria). Digestion, thus, reduces total mass of solids and destroys pathogens. Bacteria metabolize organics anaerobically.

$$2\{CH_2O\} \xrightarrow{\text{Methanobacteria}} CO_4\uparrow + CO_2\uparrow \quad \ldots(28.6)$$

Methane and $CO_2$ are recycled. Digested sewage sludge is dewatered before disposal. Sludge-drying beds provide the simplest method of dewatering.

**TABLE 28.8:** REMOVAL OF TRACE METALS IN INDUSTRIAL WASTEWATER TREATMENT

| *Metal* | *Biological* | *Al + Activated Carbon* | *$FeCl_2$ + Activated Carbon* | *Lime + Activated Carbon* |
|---|---|---|---|---|
| Arsenic (As) | – | ~90% | 95% | – |
| Cadmium (Cd) | 20-40% | ~55% | ~98% | ~99.5% |
| Chromium (Cr(III)) | 40-80 | 99 | 99 | 98 |
| Copper (Cu) | 50 | 98 | 95 | 90 |
| Lead (Pb) | 50-90 | 96 | 99 | 99 |
| Mercury (Hg) | 20-70 | 98 | 99 | 90 |
| Nickel (Ni) | 10-40 | 30 | 30 | 99.5 |
| Zinc (Zn) | 30-80 | 30 | 95 | 70 |

### 28.4.2 Landfill

The final destination of treated sewage sludge usually is the land. But, proper planning and care are required in its disposal. Dry sludge is disposed off as fertilizer. Sanitary landfills are disposal sites for non-hazardous solid wastes spread in layers, compacted to the smallest practical volume, and covered by material applied at the end of each operating day. Dewatered sludge can be buried underground in a sanitary landfill. If dewatering by sand beds is considered impractical due to lack of land, mechanical dewatering techniques, such as pressure filter, belt filter and centrifugation may be used. Secure chemical landfills are disposal sites for hazardous waste, selected and designed to minimize the chance of release of hazardous substances into the environment.

### 28.4.3 Incineration

Incineration (combustion) is a burning, or rapid oxidation process, accompanied by release of energy in the form of heat and light. It also refers to controlled burning of waste, in which heat chemically alters organic compounds, converting into stable inorganic such as carbon dioxide and water.

Sludge can be reduced in volume by incineration. Any organic compound can be oxidized in an aqueous solution if sufficient energy (heat and pressure) is supplied. Thus, wet air oxidation (incineration) is to convert organic compounds to $H_2O$, $CO_2$ and ash residue, based on Zimmerman process. The combustion occurs in two stages. In primary combustion, moisture is driven off, and the waste is ignited and volatilized. The residue and fly ash are taken out. In the secondary combustion, the remaining residue is oxidized, reducing the amount of fly ash in the exhaust. Air is thoroughly mixed with the burning refuse to maintain a constant supply of oxygen. Multiple-hearth incinerator has several hearths, stacked vertically. The fluidized bed incinerator is full of hot sand and is suspended by air "*invection.*" The hot sand acts as "thermal flywheel" and the sludge is incinerated within the moving sand. Catalytic incineration process is a desired method to oxidize volatile organic compounds (VOCs) by using a catalyst to promote the combustion process. Catalytic incinerators require lower temperatures than conventional thermal incinerators, thus saving fuel and other costs.

## BIBLIOGRAPHY


Andrews, W. (ed.) (1972), Prentice-Hall: Englewood Cliffs, NJ. "*Environmental Pollution.*"

Baier, R. E., Mark, H. B. and Mattson, D. S. (eds.) (1981), Marcel Dekker: New York. "*Water Quality Measurement.*"

Ciaccio, L. L. (ed.) (1972), Marcel Dekker: New York. "*Water and Water Pollution Handbook.*"

Conway, R. A. and Ross, R. D. (1980), Van Nostrand-Reinhold: New York. "*Handbook of Industrial Waste Disposal.*"

Dart, R. K. and Stretton, R. J. (1977), Elsevier: New York. "*Microbial Aspects of Pollution Control.*"

Eaton, A. D. *et al.* (eds.) (1995), APHA: Washington, DC. "*Standard Methods for Examination of Water and Wastewater,*" (19th edn.).

Fry, J. C., *et al.* (eds.) (1992), Cambridge University Press: Cambridge, UK. "*Microbial Control of Pollution.*"

Hammer, M. J. (1975), John Wiley: New York. "*Water and Wastewater Technology.*"

Hodges, L. (1973), Holt-Reinhart & Wiley: New York. "*Environmental Pollution.*"

Hosetti, B. and Frost, S. (1998), *Crict. Rev. Envi. Sci. & Technol.*, **28**(2): 193. "A review of the control of biological waste treatment in stabilization ponds."

Koren, H. (1974), Pergamon Press: New York. "*Environmental Health and Safety.*"

Lamb, J. C. (1985), Wiley & Sons: New York. "*Water Quality and its Control.*"

McGauhey, P. H. (1968), McGraw-Hill: New York. "*Engineering Management of Water Quality.*"

McKay, G. (ed.) (1996), CRC Press: Boca Raton, FL. "*Use of Absorbents for the Removal of Pollutants from Wastewaters.*"

Manahan, S. E. (1993), Lewis Pubs: New York. "*Environmental Chemistry,*" (5th edn.).

Minear, R. A. & Keith, L. H. (eds.) (1982), Academic Press: London. "*Water Analysis.*"

Mizrahi, A. (ed.) (1989), Wiley-Liss: New York. "*Biological Waste Treatment.*"

Omenn, G. S. (ed.) (1988), Plenum Press: New York. "*Environmental Biotechnology.*"

Peirce, J. J., Weiner, R. F. & Vesiland, P. A. (1998), Butterworth-Heinemann: New Delhi. "*Environmental Pollution and Control,*" (4th edn.).

Pfafflin, J. R. & Ziegler, E. N. (1976), Gordon & Breach: New York. "*Encyclopedia of Environmental Engineering,*" (Vol. 2).

Purdom, P. W. (ed.) (1980), Academic Press: New York. "*Environmental Health,*" (2nd edn.).

Van Loon, J. C. (1982), CRC Press: Boca Raton, FL. "*Chemical Analysis of Inorganic Constituents of Water.*"

Verstraete, W. and Vandevivere, P. (1999), *Crict. Rev. Envi. Sci. Technol.*, **29**(2): 151. "New and broader applications of anaerobic digestion."

Wojtenko, I., Stinson, M. K. and Field, R. (2001), *Crict. Rev. Envi. Sci. & Technol.*, **31**(3): 241."Challenges of combined sewer overflow disinfection by ultraviolet irradiation."

Wojtenko, I., Stinson, M. K. and Field, R. (2001), *Crict. Rev. Envi. Sci. & Technol.*, **31**(3): 295."Performance of ozone as a disinfectant for combined sewer overflow."

... (1993), McGraw-Hill: New York. "*Environmental Science and Engineering.*"

... (1996), Water Pollution Control Federation (WPCF): Washington, DC. "*Standard Methods for the Examination of Water and Wastewater,*" (20th edn.).

# 29

# Land Pollution Control Methods

Land pollution is primarily due to mining, agriculture and animal husbandry and forest degradation. Strict control of land degradation operations is one of the primary steps needed to avoid or minimize atmospheric and hydrospheric pollution and degradation of ecosystems.

Solid waste consists of garbage, rubbish, trash, industrial and agriculture runoff, animal and human waste and sewage sludge. Waste disposal protocols involve collection, storage, processing, transport and dispersal. Processing includes separation, screening, size reduction, compaction and treatment and incineration/pyrolisis, etc.

The principal disposal operations are land filling, composting (biological degradation), incineration (volume reduction) and recycling methods. Incineration is not a total solution as it contributes to atmospheric as well as thermal pollution.

This chapter deals with some of the land pollution control methods.

Land pollution arises due to acid rain, mining, leachates, and oil shale recovery and from coal burning and other industrial operations. These operations contribute pollutants of different types. They include metallurgical waste, metal dust and fly ash. In addition, materials from building and road construction such as asphalt, cement, asbestos and dust and chemical pollutants from agriculture, animal husbandry and timber and paper mills find their way to the soil, unless properly disposed off. Control of land pollution is mandated not only for its own sake but is also an important step in avoiding or minimizing atmospheric and hydrological pollution.

## 29.1 LAND POLLUTANTS AND THEIR DISPOSAL

Some of the land pollutants and their control methods are given in Table 29.1.

**TABLE 29.1:** LAND POLLUTANTS AND THEIR CONTROL METHODS

| *Material* | *Process* | *Pollutants* | *Control Methods* |
|---|---|---|---|
| Asphalt | Mining, processing; land surfacing. | Particulate; hydrocarbons & CO | Baghouses, cyclones & electrostatic precipitators |
| Clay | Mining; dredging & tilling | Dust; $NO_x$ & fluorides | Precipitators; scrubbers & catalytic conversion |
| Cement | Lime + silica + Fe | Particulate matter & dust | Baghouses & multi-cyclones |
| Calcium oxide (Lime, CaO) | Calicination of $CaCO_3$ in kilns | Particulate matter | Cyclones; wet-scrubbers |
| Calcium carbide | CaO + C + lime in kilns | CO; $SO_x$ & Hydrocarbons (HCs) | Precipitators & wet-scrubbers |
| Glass | Soda lime + silica in kilns | Particulates | Cyclones and scrubbers |
| Gypsum | $CaSO_4$ (hydrated) | Dust & particulates | Precipitators; scrubbers & cyclones |
| Carbon (black) | Coal combustion | Soot; CO and hydrocarbons | Multi-cyclones |
| Coal & fly ash | Mining; process & pyrolysis of wood | Particulates; leachates; CO & hydrocarbons | Precipitators; Multi-cyclones |
| Timber & paper | Wood pulping by kraft process | Particulates; cellulose; alkalis & chemicals | Precipitators; neutralization; wet-scrubbers and recovery |
| Agriculture | Various processes | Particulates; organic and hazardous chemicals | Cyclones & wet-scrubbers |
| Food & beverage | Various processes | Particulates; organic wastes & bacterial and other pathogens | Precipitation; coagulation; filtration & biological treatment |

Solid waste consists of garbage, rubbish, trash, ash, street refuse, industrial, agricultural, animal and human wastes, sewage treatment sludge and other. Waste disposal management involves collection, storage, processing, transport and dispersion. Processing includes separation, size-reduction, compaction, treatment or incineration/pyrolysis. Shredding and grinding are common size-reduction methods of garbage disposal. Trommel is the most widely used screen in material separation and recovery, prior to disposal. The principal disposal options are (Table 29.2) land filling, compost, incineration and recycling methods. Proper planning and care should be taken in the waste disposal management as chemicals from landfills, dumps, and lagoons can leach into aquifers and streams.

**TABLE 29.2:** SOLID-WASTE DISPOSAL METHODS

| *Process* | *Purpose* | *Waste Disposal* |
|---|---|---|
| Filtration | Dewatering | Sludge |
| Chemical addition | Precipitation | Sludge |
| Landfill | Storage and disposal | Inert and radioactive materials |
| Submerged combustion | Dewatering | Liquids |
| Incineration | Volume reduction | Most organic materials |
| Biological degradation (compost) | Dilution | Biodegradable organics |
| Recycling | Reuse | Metals |

### 29.1.1 Landfill

Sanitary land filling method is the simplest and widely used waste disposal management method of solid, non-hazardous and non-radioactive wastes. The operation is basically a biological method of waste management. The solid wastes are spread in layers, compacted to the smallest practical volume, and covered by material applied at the end of each operating day.

Hazardous wastes are usually generated in chemical industries. They are toxic, reactive, ignitable, corrosive, infectious or radioactive. Wherever possible their disposal is taken up separately from ordinary waste material. The methods adopted for their treatment are more or less the same as those used for the other waste, except that more stringent precautions are taken in the following procedures, designed to minimize the chance of release of hazardous substances into the environment. Radioactive wastes emit ionizing radiations that can harm living organisms. However, the handling and disposal of such materials under specialized category and is not a responsibility of municipal authorities but is usually entrusted to specialist organizations in practically all countries.

The danger of improper waste storage lies, mainly, in its potential to contaminate surface and ground water supplies. Modern storage methods tend to eliminate or minimize such dangers. Some times, temporary storage of solid waste is done by forming new waste piles that are carefully constructed over an impervious base. To prevent dispersal, such piles have to be protected from wind and erosion. Long-term storage of waste is done in sanitary landfills. One of the important aspects of these modern landfills is that buried waste never comes in contact with ground water. An adequate distance is maintained between the bottom of the landfill and seasonally high ground water table. The landfill is built up in units called "cells" (Fig. 29.1). It is a disposal site that is carefully selected, designed, constructed, and operated to protect the environment and public health. The advantages are low cost, flexible operation and final disposal. Disadvantages are that the process is slow, requires large land area and there exists the possibility of leaching of pollutants and toxic metals from the site into the groundwater.

Daily solid waste is transported to a landfill site, spread in a layer and covered with 10-20 cm thick soil and a plastic liner. When refuse "cells" in a site are filled a layer of ~0.5-meter impervious soil, called final cover, is spread on top of the landfill. Area and trench methods are also in use. The geographic and topographic selection for a landfill is important. Monitoring and control systems are provided for detecting and, if necessary, eliminating contamination of the soil around the pits and under-ground water sources with offensive leachates from the piles.

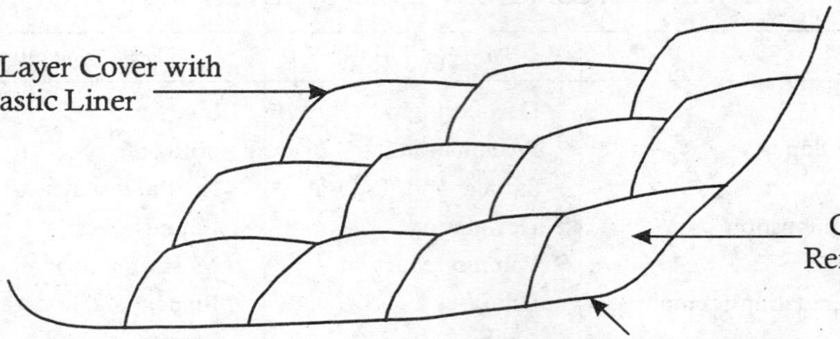

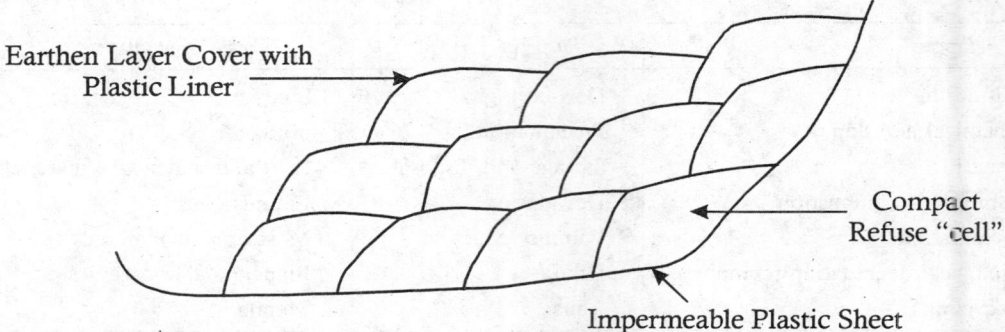

**Fig. 29.1:** Sanitary Landfill Scheme

### 29.1.2 Composting

Composting is the biochemical degradation of organic materials (biodegradable trash) under carefully controlled conditions to yield humus-like sanitary soil supplement. Composting offers a method of processing and recycling both garbage and sewage sludge in one operation. The volume of the waste can be reduced by as much as 50 per cent. Digested compost is processed and used as mulch. Processing includes drying and screening. The decomposable materials in refuse are isolated from inorganic materials through sorting and separating operations based on size, density and magnetic and other physical properties. Shredding reduces the size of the waste articles to uniform size.

Both aerobic and anaerobic decomposition can extract useful products. Aerobic system of decomposition is known as composting. It is the only process that provides for recycling of organic residue. In a common procedure of composting, the polluted material is mixed with a solid organic substance that is readily degraded (e.g. wood chips, sawdust straw, etc.). The pile is supplemented with nitrogen, phosphorous and other inorganic nutrients and is placed in a single heap Moisture and aeration are maintained by periodic mixing and turning and exposing the materials to air.

Composting method can be developed into a major process for handling sewage sludge to maintain an eco-friendly environment. As more stringent environmental rules and siting constraints limit the use of solid waste incineration and land filling options, the application of composting is likely to increase.

### 29.1.3 Incineration

Incineration is a controlled combustion process used for volume reduction and destruction of residual organic matter and pathogens. Advantages of incineration are:

1. Applicable to all combustible (organic) materials
2. Suitable to handle biohazard waste
3. Large land areas are not required.

The disadvantages are:

- Environmental nuisance
- Products may be hazardous to health
- It is not the ultimate disposal method. Combusted residues still have to be disposed.

By the early part of the twentieth century very few urban communities in the world were incinerating solid waste. Most cities were following the primitive method of dumping it on land or water. General recognition of the pollution and public health problems, resulting from the open dumping of waste as well as of improper incineration, resulted in change in attitudes in adoption of sanitary landfills, which were designed and operated in a manner that minimized risks to environment and public health. At the same time, the incinerators have been redesigned to recover heat energy from the waste and to provide extensive air pollution control devices to satisfy stringent standards of air quality.

Modern municipal incinerators are designed to operate on the basis of continuous feeding and burning of the solid waste. From storage pit the waste is lifted and deposited into a hopper and chute above the furnace and released on to a charging grate or stoker. The grate shakes and moves the waste through the furnace allowing air to circulate around the burning material. Rectangular as well as rotary kiln and vertical circular furnaces are in use. The furnaces are lined with refractory bricks to withstand high temperatures.

Complete burning in any incineration process depends on: (*i*) combustibility of the pollutants, (*ii*) residence time, (*iii*) flame temperature and (*iv*) turbulence. For burning carbonaceous waste without smoke, temperature >750° C has to be maintained. The degree of turbulence of the air in the incinerator affects oxidation and its overall performance. If the material to be disposed is in the form of waste gas containing organic materials that are combustible, incineration can be considered as the final method of waste disposal.

Pyrolysis is a thermochemical process, under anaerobic conditions, for conversion of organic solids to combustible gases, water vapor and a solid residue. The liberated gases can be used as fuel. 'Synpyrol' is a new technology, a combination of synthesis and pyrolysis, during which cellulose and water molecules react together to produce hydrolyzed cellulose, which is broken into hydrocarbons and other organic compounds. Conversion involves neutralization of acidic or alkali wastes and proper disposal. Special storage and handling facilities are required for radioactive waste.

## 29.2 RECYCLING

Recycling of part of the solid waste generated in the domestic and industrial sectors is an attractive way of conserving resources as well as reducing the burden on storing and final disposal. For example, converting wastepaper from offices into corrugated boxes or newsprint (post-consumer recycling).

Evidently, recycling involves separation; recovery and reuse of components of solid waste that may still have economic value. Two of the safe disposal methods mentioned above namely, composting and incineration with heat recovery, can be considered as recycling technologies. From the municipal waste paper, metals, glass, plastics and rubber are potentially reusable if not as they are, at least after some processing. Of the various steps involved in recycling separation poses greatest problems. After experimenting with various alternatives the general consensus arrived at is that separation of recyclable material from the garbage can be done at centralized mechanical processing plants.

In developed countries major cities have established Material Recycling Facilities (MRF). At a typical MRF the collected garbage is loaded on to a conveyor. Electromagnetic separators are employed to removes steel cans. The remaining material passes over a vibrating screen where broken glass is removed. The conveyor then passes through an air classifier, which separates plastic and aluminum containers from the heavier glass containers.

Methods of recycling solid-wastes are given in Table 29.3 and solid-waste management methods in Table 29.4.

**TABLE 29.3:** METHODS OF RECYCLING SOLID-WASTES

| *Recycling Process* | *Components* |
|---|---|
| Steam generation | Combustible components |
| Composting | Garbage, rags wood chips and paper |
| Waste recovery | All refuse |
| Metal recovery | All refuse |
| Gasification and pyrolysis | Organic fraction of wastes |

**TABLE 29.4:** SOLID-WASTE MANAGEMENT METHODS

| *Process* | *Waste Disposed* | *Cost* | *Remarks* |
|---|---|---|---|
| Compaction | Non-hazardous materials | – | Volume reduction for transport & refill |
| Landfill | Non-hazardous materials | Low | Simple operation |
| Composting | Biodegradable organic materials | Medium | Better suited for clean environment economical |
| Recycling | Selected materials | – | Depends on economics, usage, impact and social acceptance |
| Incineration | Combustible organics (except for explosives and special compounds) | High | Reduction in volume. End disposal for combustibles |
| Pyrolysis | Combustible organics (except for explosives and special compounds) | High | Reusable products |
| Special storage & handling | Radioactive waste | High | Very necessary |

## 29.3 RADIOACTIVE WASTE DISPOSAL

Nuclear industry, nuclear power plants, medical establishments, research facilities and some Government organizations are the sources of generating commercial low-level radioactive waste.

The International Atomic Energy Agency (IAEA) defines low- and intermediate-level waste as "radioactive wastes in which the concentration or quantity of radionuclides is above clearance levels established by the regulatory body, but with a radionuclide content and thermal power below those of high level waste (i.e. about 2 $kW/m^3$)." This type of waste is, usually, separated into short-lived and long-lived wastes, for purposes of storing and disposal. In the context of radioactive waste, 'disposal' is defined as "an emplacement of waste in an approved, specified facility without the intention of retrieval." Land disposal is the prevailing current common practice. In that case, the objective is to provide sufficient isolation of waste to protect humans and the environment and not to impose any undue burden on future generations.

### 29.3.1 Low-level Radioactive Waste Disposal

In most of the countries the handling and disposal of radioactive waste is managed through a system of licensing from the regulatory authorities, designated by the respective Governments. Because of the known hazards of radiations a number of issues and challenges have arisen in countries pursuing nuclear programs and radioactive waste disposal options. The exercise of selecting a suitable location for a radioactive waste disposal facility invokes a number of technical considerations like geology, geochemistry, tectonics, and seismicity, surface processes, meteorology, human induced events, transportation of waste from the point of its generation, land use, population distribution and environmental protection. Above all these lies the public acceptance of the selected site. Therefore, of all the waste disposal solutions the one that is most regulated and controlled is the case of radioactive waste.

The design of a disposal facility aims to limit the release of radionuclides to the biosphere, minimize exposure to the workers and the public, and minimizing maintenance during the "post-closure" phase. Most of the designs rely on a system of multiple engineered barriers to contain the waste. These systems include concrete vaults, back-filing materials, chemical barriers and measures for gas venting, drainage and buffer zones.

Most of the existing facilities for handling low-level waste are engineered within 10 m of the earth's surface; some are mined cavities located deeper in the earth (more than 50 m) and a few are geological repositories (hundreds of meters deep). Low-activity waste is packed in boxes, drums and casks, and is stacked in concrete trenches. High activity wastes are conditioned with concrete, bitumen or other low leachability materials, or placed in high integrity containers for structural stability. The space between the containers is filled with dry soil. A provision is made to monitor any liquid drainage within or outside the trenches and if there is any, it can be directed to an on-site water management facility.

Once a disposal facility is loaded to its capacity, processes known as 'closure' and 'post-closure'are set into operation. These consist of covering or sealing the disposal area, preparing documents, and performing safety assessments. Several hundreds of years are foreseen for 'post-closure' institutional control. The provisions of this process include access control, maintenance, site monitoring, record keeping and corrective actions, if required.

### 29.3.2 High-Level Radioactive Waste Disposal

The disposal of radioactive waste, high-level waste (HLW) in particular, is becoming a matter of concern. High-level radioactive waste generated in core fuel of a nuclear reactor, found at nuclear reactors or by nuclear fuel reprocessing; is a serious threat to anyone who comes near the waste without shielding. The nuclear power reactors that are approaching the end of their operational life and the radioactive waste from decommissioned nuclear missiles (after the cold-war) are some among the serious problems, the solutions for which are still being sought. Most of the high-level waste comes from the core of the reactors. This comprises uranium, plutonium and the highly radioactive fission products. Many of the isotopes in this waste have long half-lives (some longer than 100,000 y), which creates the need for taking care of it for millennia.

Invariably, a short-term storage is planned for materials like spent nuclear fuel before a more permanent placement is decided. A ten-year storage can bring approximately a 100 times decrease in radioactivity. Therefore, a ten-year period is envisaged for temporary storage of this material. At that stage that material becomes amenable for handling and shipment. Any 'long-term' storage covers a period of thousands of years. Virtually, security of this radioactive waste must be assured for geologic time periods. The factors to be guarded against are many: accidental uncovering, leaching of radio-nuclides into water sources, and standing against natural catastrophes like earthquakes and above all against stealth groups clandestinely removing it with criminal motives. The presence of elements like plutonium in this type of waste makes it all the more important to guard since they are not only highly radioactive but are also highly poisonous. Plutonium, perhaps, is the most toxic among all known elements.

At this stage it cannot be said with certainty that the world and even the countries holding large amounts of this potentially dangerous waste, have found a long-term, safe solution for ultimate disposal radioactive wastes.

## BIBLIOGRAPHY

Adamson, R. G. (1973), Bellhaven House: Ontario. "*Pollution: An Ecological Approach.*"
Andrews, W. (ed.) (1972), Prentice-Hall: Engelwood Cliffs, NJ. "*Environmental Pollution.*"
Conway, R. A. and Ross, R. D. (1980), Van Nostrand-Reinhold: New York. "*Handbook of Industrial Waste Disposal.*"
Cunningham, W. P. (ed.) (1994), Gale Pubs: Detroit, IL. "*Environmental Encyclopedia.*"
Eaton, A. D. *et al.* (eds.) (1995), APHA: Washington, DC. "*Standard Methods for Examination of Water and Wastewater,*" (19th ed.).
Edwards, C. E. (1974), CRC Press: Boca Raton, FL. "*Persisting Pesticides in the Environment.*"
Fry, J. C. *et al.* (eds.) (1992), Cambridge University Press: Cambridge, UK. "*Microbial Control of Pollution.*"
Hammer, M. J. (1975), John Wiley: New York. "*Water and Wastewater Technology.*"
Hutzinger, O. (ed.) (1990), Springerverlag: Berlin. "*Handbook of Environmental Chemistry.*"
Jorgensen, S. E. and Peterson, M. S. (eds.) (1981), Elsevier: Amsterdam. "*Principles of Environmental Science and Technology,*" (Vol. 14).
Lamb, J. C. (1985), Wiley & Sons: New York. "*Water Quality and its Control.*"
Manahan, S. E. (1993), Lewis Pubs: New York. "*Environmental Chemistry,*" (5th edn.).
Mizrahi, A. (ed.) (1989), Wiley-Liss: New York. "*Biological Waste Treatment.*"
Omenn, G. S. (ed.) (1988), Plenum Press: New York. "*Environmental Biotechnology.*"
Peirce, J. J., Weiner, R. F. & Vesiland, P. A. (1998), Butterworth-Heinemann: New Delhi. "*Environmental Pollution and Control,*" (4th edn.).
Pfafflin, J. R. & Ziegler, E. N. (1976), Gordon & Breach: New York. "*Encyclopedia of Environmental Engineering,*" (Vol. 2).
Sayler, G. S., Fox, R. & Blackburn, J. W. (eds.) (1991), Plenum Press: New York. "*Environmental Biotechnology for Waste Treatment.*"
Turk, A., *et al.* (1978), Saunders: Philadelphia, PA. "*Environmental Science,*" (2nd edn.).
Yaron, B., Calvet, R. and Prost, R. (1996), Springerverlag: Berlin. "*Soil Pollution, Processes and Dynamics.*"
—(1990), Natl. Inst. Standard and Technology: Gaithersburg, MD. "*Standard Materials for Environmental Research, Analysis and Control.*"
—(1993), McGraw-Hill: New York. "*Environmental Science and Engineering.*"

# 30

# Environmental Health

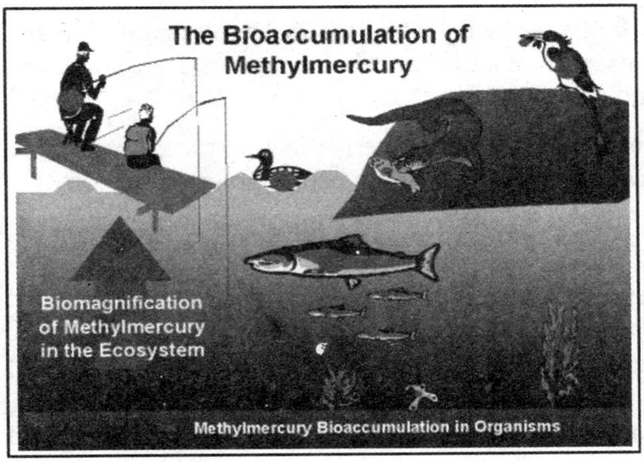


Environmental health refers to environmental conditions that determine the quality of living organisms that inhabit various ecosystems. Maintaining environmental health requires knowledge of environmental science and technology. It involves evaluation and optimization of various solutions to minimize the environmental pollution level.

There is an inverse correlation between environmental pollution and health and wellbeing of the populations. Urbanization and industrialization have ushered in a host of consumer products, synthetic hazardous chemicals. Many synthetic chemicals used in agriculture and food preservation are chronic poisons and carcinogens. Their production, usage and disposal norms have created complex environmental pollution problems. While it might be possible to find remedies to point-source pollutants, it is rather difficult to monitor and find suitable remedies for end-product disposal problems.

Biodegradation and bioremediation are part of the strategy to keep the concentration of environmental pollutants at very low levels. Microbial biodegradation is biologically catalyzed reduction of complex chemicals and some of them may be environmental toxicants or pollutants. Microbial biodegradation is carried out by various chemical transformations—hydrolysis, dehalogenation, dealkylation, etc. Controlled application of microorganisms for destruction of chemical pollutants is common in wastewater treatment, toxic waste disposal by bioreactors and biofilters.

Genetic engineering methods hold a great promise in providing increased and improved agronomical products with quality enhancement and shelf life. Genetic engineering also promises to provide organisms with novel and presently non-existing biodegrading abilities. A variety of problems can, perhaps, be tackled in future through the use of genetically engineered micro-organisms.

This chapter adresses some aspects concerned with environmental engineering and health.

Health of the environment is mandatory for the survival and wellbeing of all the living organisms on the planet. Health is a state of physical, physiological, mental and social wellbeing. Environmental health refers to characteristics of environmental conditions that affect the quality of health of organisms inhabiting various ecosystems. Quality of life of an organism is directly related to the quality of the environment surrounding it. Absence of disease itself is not a criterion for quality of life and clean environment.

## 30.1 ENVIRONMENTAL HEALTH AND ENGINEERING

"Environmental Engineering" encompasses knowledge of science and technology to enhance the quality of the environment. The objectives of environmental engineering are:

1. Protection of humans, animals and vegetation from the effects of adverse environmental factors.
2. Protection of the environment from the deleterious effects due to industrialization and urbanization.
3. Improvement of environmental quality for better living. This includes public sanitation and hygiene, occupational health, energy conservation and eco-friendly habitats, transportation, maintenance of clean air and water, and forest and land conservation, etc.

Environmental engineering involves recognition of the parameters affecting the environment, and evaluating and optimizing of solutions, wherever needed. Some of the polluting factors that affect the environment are:

1. *Ergonomic*: Social and urban, and local socioeconomic factors.
2. *Geographical*: Soil degradation due to excessive mining, stripping, deforestation, acid rain and other factors.
3. *Physical*: Radiation loss or excess (ozone depletion), noise, visibility (smog), thermal imbalance (greenhouse effect).
4. *Chemical*: Release of corrosive, hazardous gases, fumes and chemicals and their effects on the atmosphere.
5. *Biological*: Harm caused by insects, worms, bacteria, viruses and pathogens and other health-deteriorating factors.

### 30.1.1 Environmental Ecology

Energy is the basis of life support systems of all organisms. Sun is the main source of ths energy—the radiation energy transmitted through the atmosphere we live in. Any disturbance to the atmospheric "energy-balance" (e.g. photochemical smog, acid rain, ozone depletion, etc.) would disrupt the balance of all the ecological systems on this planet.

Nutritive energy is derived from the conversion of light energy into chemical energy by plants through photosynthesis. Plant kingdom provides food web upon which all other organisms depend. Therefore, plants are primary producers. Consumers are those who depend on the primary producers, and decomposers are degraders (micro-organisms). Micro-organisms mediate most of the oxidation-reduction processes in the hydrosphere. Aquatic micro-organisms (decomposers) play a key role in nutrient cycling. They mineralize organic chemicals and biomass to inorganic $H_2O$, $CO_2$ and mineral salts, and thus make nutrients available to the producers once again. Micro-organisms are also directly responsible for the degradation and detoxication of many pollutants in aquatic systems.

The primary productivity can be measured as the rate of photosynthesis, biomass (vegetation) yield, or by the nutrient loss or respiration of the plant kingdom. Various types and components of environmental pollution—the increased nutrients, decrease in sunlight penetration, thermal and chemical—can be evaluated by monitoring primary productivity as a function of various parameters.

Another approach towards describing an ecosystem is to trace the flow of nutritive energy through producers, consumers and decomposers and study the ecological effects of pollutants. Transport of nutritive compounds (food chain) is one of the via media for conveying the presence and quantification of micro-organisms and toxic substances. The geographical parameters include air, water and environment. Air and water may also be the media for transporting and dispersal of toxic gases and other hazardous compounds, micro-organisms and pathogens.

### 30.1.2 Industrial Ecology

Urbanization and industrialization have ushered in a host of consumer products, synthetic chemicals and hazardous compounds. Their usage, management and disposal have created problems, and with increased urbanization, the proposed remedies have become exceedingly complex and inadequate. Therefore, novel insights and solutions to the present day environmental woes have to be evolved. It might be possible to find a solution for the point-source pollutants (alternate methods or stoppage of production), but not for the end-product dispersal. For example, lead is dispersed by various means—from factories, waste pipes, agriculture drainage and scrap discards. While it is possible to control release of lead at the point of its production and lead batteries can be recycled, lead let out into the atmosphere cannot be recycled. In various instances, the consumer-related releases are more than production-related releases. For example, in a modern metallurgical setup, copper released to the environment from smelting operations is much less than the amount of copper released from the wiring in various small domestic and industrial appliances. Pollution due to end-product dispersal is going to increase with urbanization and demand for more appliances and greater comforts of life. Thus consumer society is increasingly becoming a pollution-creating system. Taking this trend into account, industrial ecology has to treat consumer society as an ecosystem. Assessment of the sources of raw materials, the effects of their extraction and the fate of products after their "useful lives" end, enables meaningful evaluation of resources and the pollutants generated at each stage of a product's existence.

## 30.2 ENVIRONMENTAL TOXICOLOGY TESTING

One of the most important properties of a molecule in conferring environmental toxicity is its propensity to concentrate up the food chain. Toxicology evaluation includes physicochemical, biochemical, immunological and animal and other screening tests.

### 30.2.1 Physicochemical Methods

The main abiotic factors that seem to enhance the degradation process are elevated temperature and aerobic activity. There are a number of tests, based on physicochemical properties such as solubility, adsorption, volatility, and structure-function relationships, which can be used to monitor the toxicity of a chemical in the environment.

#### *30.2.1.1 Solubility Tests*

If a chemical is highly soluble in water, it can be dispersed fast and is less likely to have a long-term environmental impact. The Hansch coefficient (the partition coefficient, $\log P_{OW}$) can be used as a physicochemical parameter in predicting the fate and behavior of a chemical in the environment. The lipophilicity of a molecule can be measured by determining its solubility in water and in n-octonol. In Hansch test, equal volumes of octonol and water are mixed in a separating funnel; the test chemical is added and the mixture is shaken vigorously. The concentration of the test chemical in both octonol and water phases is determined. The Hansch coefficient, $\log P_{OW}$, is then calculated from the retention times in reverse-phase HPLC.

$$\log P_{OW} = \log_{10} \frac{[X]_{octonoal}}{[X]_{water}} \quad \text{...(30.1)}$$

where, $[X]$ is the concentration of the test chemical.

If $\log P_{OW} > 3$ then the chemical is lipophilic, (e.g. the value for DDT is 6.2), and is likely to bioaccumulate, bioconcentrate, or biomagnify within the environment, due to slow metabolism or excretion.

Hydrolysis tests, as a function of pH, can be carried out to assess the stability of a chemical in aqueous ecosystems. The test chemical is added to water buffered to different pH values, so that an assessment of its solubility and stability under different environmental conditions can be made. For example, ester groups are readily hydrolyzed.

#### 30.2.1.2 Adsorption

The factors that affect whether a chemical is bound to or released from soil are particle size, soil type, presence of organic matter (humus) in the soil, ion-exchange capacity of soil, water solubility and $\log(P_{OW})$ coefficient of the substance. For testing, the chemical under consideration is added to different types of soil and the free and bound (adsorbed) chemical are determined. Chemicals with high $\log(P_{OW})$ tend to adsorb better.

#### 30.2.1.3 Mobility/Persistence

Mobility of a chemical in soil can be determined by mixing the chemical in water, adding a weighed amount of soil, equilibrating and determining the distribution coefficient, $K_d$,

$$K_d = \{[\text{chemical in aq. phase}]\}/\{[\text{total chemical}] - [\text{chemical in aq. phase}]\} \quad \text{...(30.2)}$$

The pH of the soil is important in assessing the environmental mobility of a chemical in the soils. Negative ions are mobile in soil and silt systems, because they are repelled by negatively charged clay particles; positively charged ions are attracted and, therefore, their passage is retarded. Acidic pH, generally, enhances the mobility of chemicals.

Persistence, along with mobility, is considered the most important factor that indicates the possibility of the chemical having significant effect on the environment. Persistent chemicals normally have half-lives > 20 days in a hydrolysis test.

#### 30.2.1.4 Volatility

If a chemical is volatile, it is unlikely to have a major impact upon the terrestrial environment. If the molecule has low water solubility, and is volatile, it is unlikely to have a significant impact on aqueous environment (e.g. ether, benzene).

### 30.2.2 Biomonitoring

Pollutants affect organisms, and the affected organisms can be used as indicators (ecological indicators) of the nature and degree of contamination. Ecological indicators are characteristic of the environment that, when measured, they quantify the magnitude of stress, habitat characteristics, degree of exposure to a stressor, or ecological response to exposure. The term is a collective term for response, exposure, habitat, and stressor indicators.

Biomonitors can act as warning beacons and as precursors to detailed chemical analysis. Bioindicator methods make use of the presence of certain taxonomic groups that are known to be sensitive in some detectable way to the presence of pollutants in their environment. The parameters to look for include changes in species diversity. Morphological, biochemical, histological, and physiological changes are also indicative. Above all increased mortality rates in populations give definite proof of pollution effects. Transgenic plants can also be used as biomarkers for environmental biomonitoring. This appoach allows rapid, cheap and precise assays of genotoxicity of soil.

Biomonitoring methods include screening of organisms and vegetation known to be sensitive to particular pollutants, as well as animal toxicity tests. Lichens are amongst the most sensitive of organisms to a variety of air pollutants, especially to $SO_2$ and fluorides. The presence of specialized plants that absorb specific contaminants from the soil through their roots or foliage (*phytomonitoring*) is an indication of the presence of the contaminants in soil and water. Morphological and histological changes, such as chlorosis (discoloration of normally green plant parts caused by disease, lack of nutrients, or various air pollutants), and necrosis (death of plant or animal cells or tissues) in higher plants, can also be used for monitoring water quality. Toxicity in marine water samples can be measured by decrease in light production of the luminescent marine bacteria, in which the productivity of light is intrinsically linked to cell respiration. Any changes in cell metabolism or damage to cell structure results in a change in cellular bioluminescence.

### *30.2.2.1 Biochemical Tests*

Biochemical indicators (enzymes involved in metabolism), such as cytochrome P450 can be used to monitor both acute and chronic effects from a wide range of pollutants in the natural environment. They can be used to: (*i*) assess potential effects of toxicant(s) on individuals and communities, (*ii*) determine thresholds of lethal and sub-lethal doses, (*iii*) give an early warning of potential problems well before signs of morbidity become evident and (*iv*) develop control mechanisms. These metabolic enzymes convert hydrophobic molecules (substrates) to polar products (Fig. 30.1) so that they could be excreted faster. Increase in polarity is very important in the overall fate of the molecule in the environment, because the greater the polarity of the molecule the lower is its persistence and therefore the molecule will have a lower impact in the long run. The excreted products need not have to be less toxic; in fact they could be more toxic, but they will be less persistent in the atmosphere.

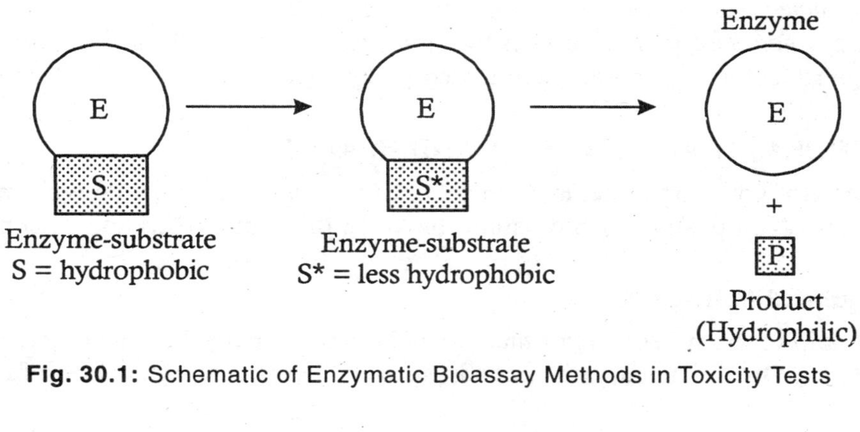


**Fig. 30.1:** Schematic of Enzymatic Bioassay Methods in Toxicity Tests

#### 30.2.2.2 Animal Toxicity Tests

Toxicity studies (acute toxicity, sub-acute toxicity and chronic toxicity) in animals are used to assess the potential harm of human and veterinary medicines, pesticides and industrial chemicals. Such toxicity tests assess the hazard (the intrinsic toxicity of a chemical). Risk is dependent on exposure and duration.

$$\text{Risk} = \text{Hazard} \times \text{Probability of exposure} \times \text{Duration}$$

Acute toxicity is the degree of a substance to cause severe biological harm or death soon after a single short-term (< day) exposure or dose. Acute toxicity test is aimed at determining the immediate effects of the chemical following a single dose. Acute toxicity tests look at the dose needed to reach an endpoint, and they do not address the mechanisms of action. Sub-acute toxicity test is aimed at determining the effects of multiple doses of the test chemical over a longer duration (~90 days). Chronic toxicity is the capacity of a substance to cause long-term poisonous health effects in humans, animals, fish, and other organisms, in which symptoms recur frequently or develop slowly over a long period of time. Chronic toxicity test, usually, extends over years (>2 years). Sub-chronic toxicity is the ability of the substance to cause effects for more than one year but less than the lifetime of the exposed organism.

#### 30.2.2.3 Mutagenicity Assays

All genotoxic carcinogens are mutagens, but not all mutagens are carcinogens. Genotoxic chemicals (chemicals that affect nucleic acids) impair the function of nucleic acids by disrupting the genetic code. They modify the nucleic acid bases (e.g. 8-hydroxylation of guanine; thymine dimerization) and/or disrupt the nucleic acid strands and structures (e.g. intercalation; strand scission). Whether a chemical is a potential mutagen or not can be assayed by bacterial assay (Ames test) or by mononucleus test.

The Ames test is a bacterial assay method, in which a mutant of a bacterium that is unable to synthesize, on its own, a particular amino acid is assayed. The *Salmonella* bacterium mutant, *Salmonella typhimurium*$^{His-}$, requires histidine in its culture medium. If a mutagen is applied to the sparse culture it results in the mutation of the bacterial DNA (S.t$^{His+}$) (reversion). An increase in the number of reversion colonies resulting after exposure to the test chemical constitutes a positive Ames test and means that the chemical is mutagenic. Ames S9 mix test is used to test for pro-mutagens.

The mononucleus test is an *in vivo* test in which the test chemical is administered to the animal. After a period, when the chemical reaches the target cells and takes effect, the animal is killed and histopathalogical features are observed microscopically.

## 30.3 ENVIRONMENTAL POLLUTION AND HEALTH

Of all the media of pollution, water is the major one for transmitting hazardous chemicals (Cu, Se, Zn, Cr, Hg, Mn, Ni, Pb, and Sn) and pathogenic organisms, and toxins and poisons.

### 30.3.1 Chemical Pollutants

For example, anthropogenic sources of mercury pollution are many; but micro-organisms in water alkylate mercury. Organomercury compounds, such as methylmercury are powerful neurological poisons (Minamata disease). Organomercury compounds consumed by and concentrated find their way into human food chain. Other metals, present in water, also have their ill effects on health. Copper-related diseases are Wilson's disease causing damage to liver (cirrhosis) and proximal renal tubules (Fanconi syndrome). Selenium pollution leads to cardiomyopathy and a deforming arthritic bone disease. Zinc pollution leads to delayed wound healing, diarrhea and dermatitis.

### 30.3.2 Biological Pollutants

Sewage normally contains various food-borne bacteria (*Salmonella*, *Staphylococcus*) and pathogenic organisms responsible for various types of diarrhea, fever and other infectious diseases. In addition, disease-causing vectors such as mosquitoes and flies require water habitat.

Mosquitoes require water for hatching eggs. Larva and pupa stages require water for development. Man-made mosquito breeding places develop due to water stagnation, poor sanitation and growth of slums. The female mosquito is the vector for a number of human diseases—malaria (anopheles mosquito), encephalitis, yellow fever and dengue fever.

Proliferation of flies is due to breakdown of sanitation. Flies are extremely obnoxious, and are mechanical and biological vectors of diseases, such as typhoid fever, enteric diseases, parasitic worms and anthrax. Cockroaches are carriers of filth and potential disease vectors and allergies. Ticks are vectors of fever and tularemia. Lice transmit typhus and fever (pediculosis).

Rodents (rats) are responsible for various types of plagues and other diseases. The rat flea (*Xenopsylla cheopis*) is primarily responsible for the transmitting of plague from rodent to rodent and rodent to man. As the rat dies, the fleas abandon the rodents and attack humans. They also transmit *Rickettsia typhi* (murine typhus) and *Rickettsia akari* (rickettsia pox, similar to the chicken pox) from rodents to humans.

### 30.3.3 Toxins and Poisons

A toxicant/poison is an agent that has harmful effect on a biological system at all levels, from the sub-cellular, through the whole organisms to communities and ecosystems. Effects of toxicants are damage to chromosomes and tissues, disruption of metabolic pathways and photosynthesis and eventual death of organisms. The effects of toxicants may be elicited immediately on exposure or after a delay, depending on the properties of toxicants, their mode of action and their susceptibility to metabolic breakdown (biotransformation). Toxicants that are bio-transformable tend to be rapidly excreted and not likely to have delayed effects.

Many plants and seeds contain toxins and poisons—lathyrus in raw beans, favism in raw fava beans, and ricin in castor beans and HCN in bitter almonds, sorghum and lima beans. Many alkaloids are toxic to humans. Microbial action on food produces toxic effects in humans.

1. *Staphylococcus*: Nausea, vomiting, diarrhea, gastrointestinal pain, dizziness, and headache.
2. *C. botulinum*: Respiratory failure.
3. *Mycotoxins*: *Aspergillus flavus* (aflatoxin), a mold virus, causes liver damage and hemorrhage.
4. *Salmonella*: Fever, nausea, abdominal cramps, diarrhea and dizziness.

Many chemicals used in the preservation of agricultural produce, additives used for the preservation of foodstuffs and beverages and other industrial contaminants are chronic poisons, carcinogens and mutagens (Table 30.1).

**TABLE 30.1:** SUBSTANCES ASSOCIATED WITH CHRONIC POISONS/CARCINOGENS

| *Substance* | *Industrial Production* | *Health Effects* |
|---|---|---|
| Acrylonitrile | Synthetic fibers; plastics; pesticides | Lung & bowl cancer |
| Arsenic | Ceramics, drugs and pesticides | Lung & skin cancer |
| Asbestos | Mining & building works | Asbestosis |
| Benzene | Industrial solvent | Leukemia |
| Beryllium | Chemical industry | Carcinogenic |
| Chloromethylether | Organic polymer | Lung cancer |
| Cadmium | Fertilizer, petrochemical and electrochemical industries | Kidney damage. |
| Coal dust | Coal operations | Black-lung disease |
| Chloroform | Industrial solvent | Kidney damage |
| Copper | Electrical goods | Toxic |
| Hydrogen sulfide | Petroleum & $H_2SO_4$ | Respiratory failure |
| Lead | Mining, paints and plumbing | CNS toxicity |
| Mercury | Fungicides | CNS toxicity |
| Ozone | Arc welding, bleaching | Mutant |
| Pesticides | Pesticides | Poisonous and toxic; cholinesterase crisis |
| Silica | Glass & ceramics | Silicosis |
| Sulfur dioxide | Mining & $H_2SO_4$ | Bronchitis |
| Trichloroethane | Industrial solvent | Carcinogen |
| Vinylchloride | Rubber and resin industries | Lung and brain damage |

## 30.4 FOOD PRESERVATION

Mankind depends on the environment for food. Wastage of food grains and prepared food either due to lack of storage and preservation or due to spoilage, spillage and discard has economic, social, hygienic and pollution implications. In addition to chemical (e.g. oxidation of fats and oils) and biochemical (e.g. enzyme degradation) reactions, micro-organisms such as bacteria (e.g. *Clostridium botulism* and *Salmonella*) and molds are major causes of food spoilage. Food preservation is an important factor in the prevention of wastage of food materials and control of nutritive pollutants. Food preservation methods destroy or reduce the toxic effects of microbial action in foods. There are several food-preservation techniques and technologies, from conventional drying, pickling, and canning to modern methods of vacuum and freeze-drying, high-temperature exposure and irradiation.

Microbial biosensor methods are used for assessing quality control of perishable food products, especially meat, poultry and seafood. Samples of food stored at 5° C are periodically removed and washed with water. The amount of amino acids, polyamines and viable counts (number of bacteria) in the washed water are measured using HPLC and colony counting method. At the same time, the wash water is charged into a flow injection analysis (FIA) system, coupled with a microbial sensor (yeast) for monitoring the freshness of food. Relationship between the sensor signals obtained from FIA system and the viable counts obtained by the colony counting method is correlated to the freshness/aging of the food products.

### 30.4.1 Moisture-Solids Balance

Basically moisture-solids control is osmotic pressure control. Dehydration and drying are the oldest food preservation techniques. Dehydration arrests the life processes of micro-organisms. Adding sugar or salt reduces water and provides high osmotic pressure in the food relative to the micro-organisms. High osmotic pressure differential has proven to be a major factor in the preservation of food by dehydration, salting and sugar pickling. Salting removes water from food tissues, thereby retarding or preventing microbial growth. Smoking, curing and pickling are preservation methods, involving addition of antioxidants. Smoking slows the growth of micro-organisms and reduces oxidation of fats. But, smoking is now used more for flavoring beef and ham. Pickling creates an acidic environment that is not conducive for growth of micro-organisms. Smoking, pickling and addition of sugar, fermentation and other conventional methods are still practiced not only for preservation but also for flavor, taste and nutritive considerations.

Latest methods such as vacuum drying reduce the heating step. Spray drying operates by conversion of the liquid product to an aerosol, which is allowed to flow through a hot-air blast. Freeze-drying, lyophilization or drying by sublimation helps reducing adverse effects of dehydration by heating. In puff drying partial vacuum is created and dehydration is carried out at reduced temperature.

### 30.4.2 High-temperature Treatment

Many food spoiling organisms have two stages of existence: (*i*) the vegetative or active growth stage and (*ii*) spore or inactive stage. Most thermal processes are designed to inactivate the vegetative form. They are "commercially" sterile, but not aseptically sterile. Pasteurization (at 100° C) is the most commonly used technique of this type for milk and milk products. At such temperatures most pathogens are killed but the food is not sterilized. Pasteurized milk still contains spoilage micro-organisms. The milk must be refrigerated until it is used to retard the growth of these micro-organisms.

All bacteria subject to moist heat at lethal temperature will reduce exponentially with time (will exhibit a logarithmic death rate). *Clostridium botulinum* is highly heat-resistant in spore form and is considered an obligate anaerobic (will grow only in the absence of oxygen). When the spores germinate, the metabolizing organism generates a highly potent toxin. Being an obligate anaerobic, the conditions inside hermetically sealed cans are ideal to *C. botulinum* growth and subsequent toxin formation. Heat-treatment should be at sufficiently high temperature and for longer period to kill its spores.

Canning is a high temperature treatment to destroy bacteria in food products contained in hermetically sealed vessels. It protects food products against environmental contamination. In this process food is filled in cans, and boiled for varying duration, depending on the type of food, and hermetically sealed immediately after cooking. Spores of some pathogens are not killed under these conditions. Vacuum canning is a low-temperature treatment, resulting in crystallization of water to make it nonavailable to microbial activity.

### 30.4.3 Freeze-drying

Freezing prevents spoilage of food by stopping or retarding micro-organisms, which cannot function at very low temperatures. Most of the food items (except some vegetables and tomatoes) are refrigerated to increase their shelf-life. Freeze-drying is another low-temperature preservation method.

The basic principle behind freeze-drying is—sublimation and lyophilization depend upon maintaining temperature and pressure low, so that water can exist as a solid or vapor by the triple-point effect (Fig. 30.2). If both temperature and pressure are maintained low, it is possible for water to change directly from solid to vapor, without passing though a liquid phase (sublimation).

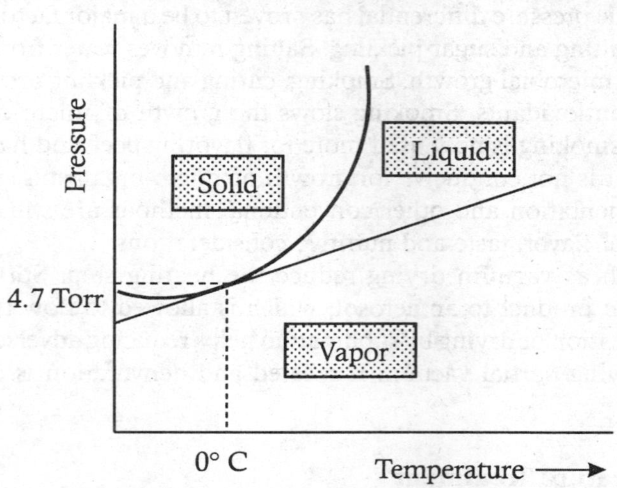


**Fig. 30.2:** Triple-point Effect in Water

Water vapors can be removed by sweeping with inert gases or maintaining high vacuum. Most systems operate at low pressure and high temperature so that

1. End products may be stored at room temperature
2. Avoid heat damage
3. Microbial activity during processing is halted
4. Rehydration occurs immediately.

### 30.4.4 Additives

Any substance introduced into food is called an additive. Some of the additives are necessary for preservation as well as for taste, flavor and aesthetic considerations. For example, salt, sugar, spices and vinegar are used for preservation as well as for taste. Smoking and curing of meat products is carried out for preservation, flavor and taste. Citric acid is an antioxidant that retards rancidity of oils and fats. More than 3000 additives are in use in food and beverage industries to bring out, replace, or mask the natural flavor and appearance of the food items or decrease the production costs. Some of these are meat tenderizers (e.g. monosodium glutamate), flavoring additives (e.g. maleic acid to give fruity-flavor), ice cream mixes, many dyes to give attractive appearance and emulsifiers to increase foaminess to creams and beverages. Many of these chemicals are not systematically tested or studied for their ill effects. For example, nitrate additives ($NaNO_3$) lead to nitrosamines, which are powerful cancer-inducing agents. Several of the food additives are aromatic, conjugated compounds that may be potentially carcinogenic. In addition, fertilizers, pesticides, herbicides, and veterinary drugs, artificially enriched animal feeds and remnants of farm and dairy operations are contaminating food products, causing outbreak of epidemics (e.g. mad-cow disease outbreak in Britain).

### 30.4.5 Ionizing Radiation Treatment

Ionizing radiation causes breakage of chemical bonds, leading to the formation of ions and free radicals. Those, in turn, react with food constituents to form stable compounds. The chemical changes that occur in food by irradiation are several orders of magnitude less than those caused by heat treatment for comparable effects. No nuclear changes that would make the food or pests radioactive are noticed, following irradiation. Still, there are some apprehensions among the public about the possible hazards involved in this treatment.

Regardless of how the food has been treated proper storage and handling, including proper temperature controls, after processing is necessary to insure that the food would remain safe and nutritious.

Two types of radiation sources are used—machines or radioactive nuclides. In the machine type electron accelerators and X-ray generators find practical application. Radionuclides are, mainly, $\gamma$-ray emitters and include Co-60 ($t_{1/2}$ = 5.3 y; 1.25 MeV).

Irradiation inactivates food spoilage organisms, including bacteria, mold and yeasts. The main purpose served by irradiation is lengthening of shelf-life of fresh fruits and vegetables by inhibiting the normal biological changes associated with growth and maturation processes, such as sprouting and ripening. In the context of food irradiation the dose is always reported as the amount of radiation absorbed by the food and is in units of Gray (Gy). The absorbed dose is determined from the intensity of the irradiating source and the exposure time. Food products, packaged and placed on a conveyor belt, are passed through a radiation chamber, where they receive a predetermined dose of irradiation. The dose required varies with the type of food and the desired effect, as given below.

#### *30.4.5.1 'Low'-dose (<1 kGy) Applications*

(*i*) Inhibition of sprouting of fresh fruits and vegetables, for long-term storage. This is achieved without the use of chemical sprout inhibitors
(*ii*) Delaying ripening of fruits
(*iii*) Prevention of losses in stored grains, fruits and vegetables, caused by insects. This technique avoids the use of fumigants
(*iv*) Destruction of parasites in food.

#### *30.4.5.2 'Medium'-Dose (1-10 kGy) Applications*

(*i*) Elimination of spoilage and pathogenic micro-organisms in various foods
(*ii*) Extending the shelf-life of many foods
(*iii*) Improving technological properties of foods, such as reducing cooking time of dehydrated vegetables.

At <10 kGy irradiation, many spoilage and pathogenic micro-organisms are greatly reduced but it is less effective in reducing the numbers of relatively radiant-resistant spores of certain pathogenic bacteria such as *Clostridium botulinum*. Hence, foods that are likely to harbor spore-forming bacteria call for special handling to ensure microbiological safety.

#### *30.4.5.3 'High'-dose (10-50 kGy) Applications*

(*i*) Sterilization of meat, poultry, seafoods, and other prepared foods in combination with mild heating to inactivate enzymes.
(*ii*) Disinfection of certain foods or ingredients such as dry spices and enzyme preparations.

### 30.4.6 Role of Genetic Engineering in Food Technology

Traditional plant breeding methods have paid major emphasis towards improving agronomic traits (resistance to biotic and abiotic stresses) of crops and plants to obtain better crop yields. Molecular genetics and biotechnological approaches, such as somaclonal variation or gene transformation, applied to food technology offer attractive alternatives to conventional genetic improvement methods, since they make possible a greater range of improvements to commercial varieties of crops and plants in a relatively short period of time with minimal or no change to other characteristics. Thus, genetic engineering methods are poised for a great leap forward in all aspects of food technology. Some of the advantages of genetic engineering over conventional methods are:

1. Increased productivity of transgenic crops/plants with increased resistance to various stresses, with incorporation of pesticide-, herbicide-, virus-, and other stress-resistant genes
2. Reduced environmental pollution due to reduced use of agrochemicals
3. Nutritional enhancement with addition of "essential" amino acids, enzymes, vitamins and dietary suplements
4. Better preservation and improved shelf life
5. Improved texture and processing, and flavour and taste, etc.
6. Production of biopharmaceutical antibodies and edible vaccines, with reduced production costs
7. Bio-based, renewable and environmental-friendly procedures, which can be easily scaled-up.

#### *30.4.6.1 Improvement in Agronomical Traits*

Improvements in agronomic traits include crops with improved resistance biotic and abiotic tolerance- e.g. resistance to bacterial, fungal diseases, resistance to pathogens, pests, and insects and tolerance to salt, water and thermal stresses.

Current methods prevailing in controlling insects, pests and other pathogens relay heavily on the use of chemical insecticides, pesticides and herbicides. Overuse of these many-made chemicals has become a major concern in environmental ecological pollution.

Genetic engineering methods offer new environmental-friendly strategies in the improvement of agronomical traits of food and cash crops, vegetables and fruits. Deployment of genes encoding pest and disease resistance will result in reduction and changes in pesticide usage. In terms of resistance to diseases and adaptation to harsh environments, genetic engineering methods have better potential. A general approach towards improvement in agronomical traits is:

1. Incorporation of high dose of insecticidal protein genes in order to kill resistant insects
2. Use of proteinase inhibitors
3. Use of ribosome inactivating proteins (RIPs)
4. Transferring alien genes by recombinant technologies
5. Somatic hybridization (fusion of cells belonging to different species) to create stable new variants.

##### *30.4.6.1.1 Anti-insecticidal Approach*

Lectins are a group of proteins with specific carbohydrate-binding activity, and some of them are known to have insecticidal properties. Incorporation of lectin genes into targeted crop seeds could be an interesting possibility for agronomic crop improvement.

#### 30.4.6.1.2 *Proreinase Inhibitors*

In plants, preteinase inhibitors that are highly specific for some classes of insect proteinases are induced in response to mechanical or other stresses, reversibly binding to the active sites of proteinases and forming inactive complexes, which reduce the availability of amino acids necessary for insect nutrition. For example, incorporation of α-amylase inhibitors seems to reduce the ability of insects to catabolize starch. Similarly, contact of polyphenolic oxidases (PPOs), and peroxidases with phenolic substrates produces quinines, which can irreversibly degrade nucleophilic amino acids that are essential to insects, thus limiting insect growth.

Ribosome inactivating proteins (RIPs) are another class of candidates to suppress insects and other plant pathogens.

### 30.4.6.2 Product Quality Enhancement

At present, many of the commercial applications of genetic engineering method in food technology are focused on developing new varieties of transgenic plants/crops with improved agronomic benefits, aimed at obtaining better crop yields. This trend will continue in developing countries to cater to the food requirements for teeming population. However, genetic engineering methods have the ability to create new varieties of plants and crops, focused on product quality and output traits (enhancement of "essential" amino acids and dietary supplements), in addition to improved agronomic traits. Future trends in plant biotechnology will be aimed more towards employing multiple gene technology (e.g. protoplast technology) to increase crop yield, and foods with enhanced nutritional value (proteins, carbohydrates and fat content), dietary supplements (e.g. vitamins, minerals and antioxidants) and production of biopharmaceutical, functional and medicinal foods.

#### 30.4.6.2.1 *Nutritional Enhancement of Plant Foods*

Specific intake levels of essential amino acids, proteins and fats and nutrients are required to ensure healthy growth and development.

(*a*) *Protein Modification*: Human beings cannot live on a protein-free diet, because they are incapable of synthesizing "essential" amino acids (e.g. Cys, Ile, Leu, Lys, Met, Phe, Thr, Trp and Val) present in proteins. The nutritional value of a protein-based diet is directly related to these essential amino acids. In general, cereals (rice, corn, wheat, etc.) have proteins with low content of three essential amino acids. Consequently, these essential amino acids must be provided in the diet. This can be achieved by protein modification methods, which include:

1. Increasing the essential amino acid content of the crop by transfer of genes encoding proteins rich in desirable amino acids. For example, incorporating maize gene encoding protein zein in soybean seeds increased methionine content. Transgenic plants with thioredoxin gene are a good source of sulfur amino acids. Introduction of "essential" amino acid-rich proteins, combined with manipulation of biosynthetic pathways, shows a great promise in creating foods that are able to supply a balanced level of all amino acids.
2. Improving the functionality of target crop proteins.
3. Reducing the content of those proteins with specific allergic properties (by the use antisense gene transformation protocols).

(*b*) *Carbohydrate Modification*: Carbohydrate modification is aimed at qualitative and quantitative change of existing carbohydrates (sucrose or starch and introducing novel non-caloric carbohydrates (e.g. fructans) in targeted crops. Increasing or modifying starch content is possible by modifying the enzymes responsible for its synthesis.

As an example, potato is one of the important staple foods. Improvement of potato by transgenic approaches are aimed not only at disease control strategies for late blight and bacterial wilt, to improve its quality and shelf-life, but also at improving low-temperature storage of potato tubers with minimal enzymatic degradation of starch and also at enhancing/altering starch content (amylopectin : amylose ratio) with higher amylopectin content. Starch with increased amylopectin has reduced calories, improved tuber texture, better flavor and less greasy taste.

(*c*) *Fatty Acid Modification*: Lipids and fats are also components of the human diet. The difference among the fatty acids in the seed oils occur: (*i*) in the length of the carbon skeleton (12 → 20), and (*ii*) in the presence of double bonds (unsaturation). Dietary studies have established that consuming unsaturated fatty acids and low-cholesterol foods is associated with low incidence of coronary artery diseases. Therefore, unsaturation of fatty acids is one of the objectives of genetic engineering methods. Fatty acid properties such as melting point, color flavor, spreadabililty, stability and shelf-life are related to the fatty acid composition of triacylglycerols. Strategies that can be used to modify oil composition and triacylglycerol content are:

1. Alteration of the major fatty acid level by antisense or co-suppression methods.
2. Over-expression of a specific key enzyme in lipid biosynthesis.

### 30.4.6.3 Enhancement of Dietary Supplements

Vitamins are essential compounds that must be obtained from diet. In addition, some vitamins are used as functional additives in food products (e.g. carotenoids, ascorbic acid). Large sections of human population in developing countries, and susceptible groups (e.g. people on medication, diabetics) are prone to inadequate vitamin uptake.

It is well known that many staple plant foods deficient in essential nutrients and, consequently, malnutrition is wide spread in many countries. For example, rice grain, a basic diet in many Asian countries, lacks an essential nutrient, vitamin A. It is estimated that rice with enriched vitamin A content would prevent death of at least several million children annually in these countries.

The worldwide deficiency of vitamin A in crops and vegetables is being tackled both through conventional plant breeding as well as by genetic engineering methods. However, the use of conventional plant breeding to deliver adequate intakes is dependent on the availability of caratenoid-rich staple foods. Therefore, in the countries where rice is a dietary staple food, the deficiency of vitamin A is likely to be eliminated only by the introduction of transgenic rice that has been genetically altered to produce β-carotene.

Similarly, enrichment of iron (e.g. by introduction of metallothionin in cysteine-rich proteins) in vegetables and fruits to reduce iron deficiency, and enrichment of lipid-soluble vitamin E (an antioxidant) to decrease incident of several disease are some of the areas where genetic engineering methods would play a prominent role.

### 30.4.6.4 Molecular Farming and Therapeutic Foods

The production of plant-derived biopharmaceutical products is known as molecular farming. The main categories of biopharmaceutical products are modified proteins, therapeutic antibodies, and antisense oligonucleotides and edible vaccines, which are used for therapeutic or *in vivo* diagnostics in humans. Transgenic plants have also been developed to produce expensive drugs, such as enkephalins, interferons, etc.

#### 30.4.6.5 Control of Flavor, Texture and Shelf-life

Changing light perception, by over- or down-expressing light receptors, can modify plant growth and reproduction. It is possible to modify some characters, specifically regulated by phytochromes, such as development of photosynthetic systems, hormone synthesis, circadian rhythm, growth and fruit ripening.

By regulating the activity of enzymes (by antisense techniques or by enzyme over-expression methods), it is possible to control or delay fruit softening, with better flavor, texture and increased shelf-life. Chloroplast-derived biopharmaceuticals are inexpensive to produce and store, and they are safer as they do not harbor human and animal pathogens. Edible vaccines are highly desirable as they can be delivered directly into food/feed, thus avoiding the need for shots and medical assistance. Color modification can be achieved with induction of carotenoids and flavonoids.

#### 30.4.6.6 Enhancement of Food-processing Properties

Food-processing properties depend on physicochemical properties such as viscoelasticity and texture, and biodegradation upon storage. As an example, wheat is a cereal commonly used to make bread and number of pastry products. The adequacy of wheat grains for these purposes rests on the grain hardness that affects the milling and backing properties of the grain, and composition of gluten that determines the viscoelastic properties of wheat dough. Cereals such as rice and corn lack the texture and softness of wheat grain. Transgenic rice and corn show marked softness of texture, with incorporation of genes that express modified wheat proteins that control grain softness.

Similarly, transgenic potatoes, with increased starch (amylopectin) content show improved texture (crispness), flavor, and less greasy taste. Regulation of autocatalytic production of ethylene, by introducing tissue-specific S-adenosylmethionine (SAM) hydrolase gene, would help in the control of ripening of fruits and improved shelf-life.

#### 30.4.6.7 Future Prospects

Undoubtedly, applications of genetic engineering methods in food technology have ushered in greater possibilities- increased food crops, with nutritional, dietary and therapeutic enrichments and concomitantly wider ramifications. Future trends in genetic engineering-based food technology look exciting and challenging. However, its acceptance by the consumers, at least in the coming years, depends on sustained and transparent risk evaluation, better channels of communication between industrial producers and end users about the importance and benefits of accepting such transgenic species over conventional plant breeding methods for the betterment of larger sections of human population. It also requires a paradigm change of mindset on the part of the consumers in accepting these modern ways of food production aimed at catering to the food and dietary needs of diverse groups human groups on this planet. Ultimately, the necessity for alleviating hunger and providing better quality of life of larger sections of population with environment-friendly methods will decide the efficacy of the modern trends.

Health of the environment is mandatory for the survival and wellbeing of all the living organisms on the planet. Health is a state of physical, physiological, mental and social wellbeing. Environmental health refers to characteristics of environmental conditions that affect the quality of health of organisms inhabiting various ecosystems. Quality of life of an organism is directly related to the quality of the environment surrounding it. Absence of disease itself is not a criterion for quality of life and clean environment.

## 30.5 BIODEGRADATION AND BIOREMEDIATION

Biodegradation, bioremediation and bioaugmentation are part of the strategy to keep the concentration of environmental pollutants at very low levels.

### 30.5.1 Biodegradation

"*Biodegradation*" refers to any biologically mediated structural alteration (by micro-organisms) of a compound. For example, many chemicals, food scraps, cotton, wool, and paper are biodegradable; plastics and polyester generally are not.

Microbial biodegradation is biologically catalyzed reduction of complex chemicals and some of them may be environmental toxicants or pollutants. Major agents causing biological transformation of organic chemicals to inorganic products (*mineralization*) in soil, sediment, surface, ground and wastewater are micro-organisms. Biodegradation by micro-organisms in sewage treatment systems is an example. As a rule, mineralization of organic compounds is characteristic of growth-linked biodegradation in which the organism converts the substrate (pollutant) to $CO_2$, cell components and other catabolic products. If organic pollutants are persisting, it is an indication that micro-organisms are not functioning or are not present.

Microbial biodegradation is carried out by various chemical transformations—hydrolysis, dehalogenation, dealkylation, etc. Hydrolysis is the common means by which micro-organisms inactivate toxicants—esters, anhydrides, amides and acyl derivatives.

#### *30.5.1.1 Acclimation*

Prior to the degradation of many organic chemicals, a period is noted in which no destruction of the chemical is evident. This period is designated 'acclimation' or 'adaptation' or 'drag period'. It may be defined as the time between the addition/entry of the chemical into an environment and evidence of its detectable loss. Many chemicals are rapidly degraded only after a certain acclimation period. The acclimation phase may be of considerable public health or ecological significance because if the chemical (pollutants such as insecticides, herbicides and fungicides) persists (see chapter 12), the possibility of an undesirable effect is increased. The length of acclimation is affected by several environmental factors (temperature, solubility, bioaccumulation, bioavailability, etc.).

#### *30.5.1.2 Activation and Detoxication*

One of the most undesirable aspects of microbial transformations in nature is the formation of toxicants. The process of forming toxic products from harmless precursors is known as 'activation' (Fig. 30.3). The action of micro-organisms on a particular chemical would be of considerable importance to environmental balance and human health.

Examples of microbial activation leading to biosynthesis of carcinogens, mutagens and neurotoxins are well known in the case of mercury, arsenic and tin. The concentration of methyl mercury in aquatic animals (fish) may be several orders of magnitude greater than that in the ambient water because of bioconcentration of the element. Though organic arsenic compounds are less toxic than inorganic forms, some of them are volatile and thus respiratory exposures may occur with harmful outcome. In the case of inorganic forms, arsenites, As(III), are more toxic than arsenates, As(V), and remediation of arsenic in drinking water involves oxidation of As(III) to As(V) followed by co-precipitation. Methylated tin compounds are highly toxic. They are highly lipophilic, have low aqueous solubility and are adsorbed strongly on soil sediment. Thus they are highly persistent environmental toxicants. Trimethyl-tin produces irreversible neuronal damage and neuronal necrosis in the brain. Other examples are microbial metabolism of trichloroethylene (TCE) leading to vinyl chloride ($ClHC = CH_2$) (a potent carcinogen) and transformation of phosphothionates to more toxic phosphates (insecticides) and thioethers to more toxic sulfoxides.

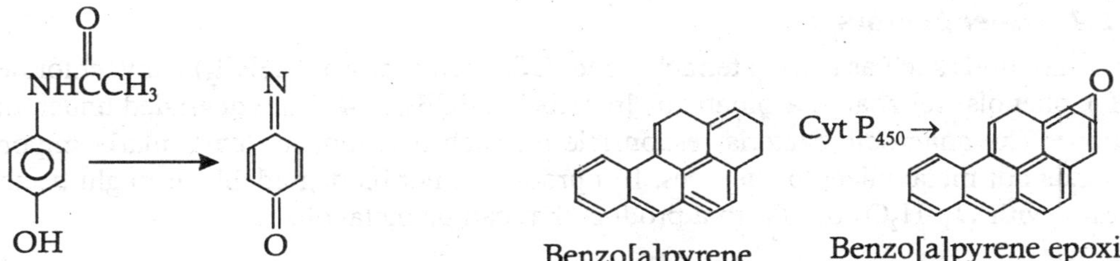


Paracetamol Quinoneimine (Toxic) Benzo[a]pyrene (Carcinogenic) Benzo[a]pyrene epoxide (Highly Carcinogenic)

**Fig. 30.3:** Examples of Bioactivation (from less toxic to more toxic Compounds)

Micro-organisms have also the ability to transform toxic pollutants to less harmful compounds. This is known as 'detoxication'. The processes of detoxication include hydrolysis, dehalogenation, methylation, demethylation, deamination and other chemical transformations.

### *30.5.1.3 Bioavailability*

*Bioavailabiliity* is the degree of ability of a substance to be absorbed and ready to interact in organism metabolism. The bioavailability for degradation of a molecule depends on its accessibility to degradative enzymes, microbial action, structural modification and other factors. If the molecule is not accessible to the degradative enzymes, the enzymes cannot act upon it. As an example, DDT is a lipophilic molecule and the cells take it up rapidly. Therefore, oxidative enzymes like Cyt $P_{450}$ do not get access to the molecule and so oxidation reactions do not readily occur. Thus, DDT is a persistent environmental pollutant.

The accessibility of a chemical to micro-organisms depends on the sorption characteristics of the soil. In the 1 : 1 ratio clay (Kaolinite), the Si and Al layers are tightly held together. Such non-expanding clays have lesser capacities for sorption and molecules are adsorbed only on the outer surfaces of the soil. As clay minerals and colloidal organic materials have a net negative charge only the molecules that are positively charged are retained by such surfaces. In a 2 : 1 ratio type clay the lattice structure of clay can expand, and such clays adsorb organic compounds both on external and internal surfaces. Many organic compounds, particularly the low molecular mass pollutants that get between the silicate sheets of the clay are generally unavailable to the microbes. But, the bioavaialbility would be different if sorption involves solely the weak retention associated with hydrogen bonds and Van der Waals forces. For example, cationic compounds such as the insecticide, paraquat that enters the clay lattice are resistant to biodegradation due to biononavailability. The poor availability or non-availability of a compound may be a major determinant of its persistence. Molecules that persist in nature (*recalcitrant molecules*) are undesirable for many reasons. Some are toxic and affect living organisms. The longer the recalcitrant molecule remains in nature, the greater is the risk of harmful effects. PCBs, insecticides and herbicides (DDT, dieldrin, heptachlor and chlordane) are some of the more prominent groups of persistent compounds. In addition, synthetic polymers that constitute the most commonly used plastics and fibers (Rayon, Nylon, and Teflon) are wholly resistant to microbial degradation and remain in environment for long periods.

Organic compounds that are mineralizable may be rendered partially or wholly resistant to mineralization by the addition of a single substituent (*xenophores*). The position of the xenophore on the molecule has a pronounced influence on the degradation.

#### *30.5.1.4 Other Factors*

Many compounds such as carbon tetrachloride ($CCl_4$), chloroform ($CHCl_3$), vinyl chloride ($ClHC = CH_2$), phenols, benzoates, aromatic hydrocarbons, PCBs, 2, 4-D are degraded under anaerobic conditions. The anaerobic bacteria responsible for such reactions are particularly important for compounds not metabolized by aerobes. If a product is not biodegradable, it might be converted abiotically with $O_3$, $H_2O_2$ or UV, to a product that can be metabolized.

### 30.5.2 Bioremediation

*Bioremediation* is the use of living organisms (e.g., bacteria) to clean up oil spills or remove other pollutants from soil, water, and wastewater; use of organisms such as non-harmful insects to remove agricultural pests or counteract diseases of trees, plants, and garden soil.

Controlled application of micro-organisms for destruction of chemical pollutants is common in wastewater treatment, toxic waste disposal by bioreactors and biofilters. For bioremediation to be effective, the following criteria have to be met: (*i*) micro-organisms can be identified that have the needed catabolic activity, (*ii*) they do not produce toxic products, (*iii*) the target compounds are available to the micro-organisms and (*iv*) the technology is cost effective.

Micro-organisms are being used to destroy a variety of volatile compounds (naphthalene, acetone, toluene, benzene, vinyl chloride, ammonia, amines and esters), by 'biofiltration' process. In this method, the micro-organisms are allowed to grow on some solid support (compost, peat or soil), and a stream of gas containing the unwanted molecules is passed through a bed made of such solid support (Fig. 30.4). For optimal operation the unit is spiked with adequate inorganic nutrients. The solid phase biofilter should be porous and should have a large surface area to maximize sorption of the compounds to be treated and to permit the development of biomass.

*Bioaugmentation* is the introduction of cultured micro-organisms into the subsurface environment for the purpose of enhancing bioremediation of organic contaminants. Generally the micro-organisms are selected for their ability to degrade the organic compounds present at the remediation site. Nutrients are usually also blended with the aqueous solution containing the microbes to serve as a carrier and dispersant. The liquid is introduced into the subsurface under natural conditions (gravity fed) or injected under pressure.

*Phytoremediation (phytotreatment)* is cultivation of specialized plants that absorb specific contaminants from the soil through their roots or foliage. Phytoremediation is the role of higher plants in the removal or degradation of pollutants in the rhizosphere. Plants and bacteria can form specific nonspecific associations in which the plant provides the bacteria with a particular carbon source that induces them to reduce the phytotoxicity of the contaminated soil. The specificity of the plant-bacteria interaction is dependent upon soil conditions, which can alter bioavailability. It appears that this role of the plant is related to the presence of allelopathic chemicals in the rhizosphere. Phytoremediation by transgenic plants may become an essential tool in cleaning the environment and reducing human and animal exposure to potential carcinogens and toxins.

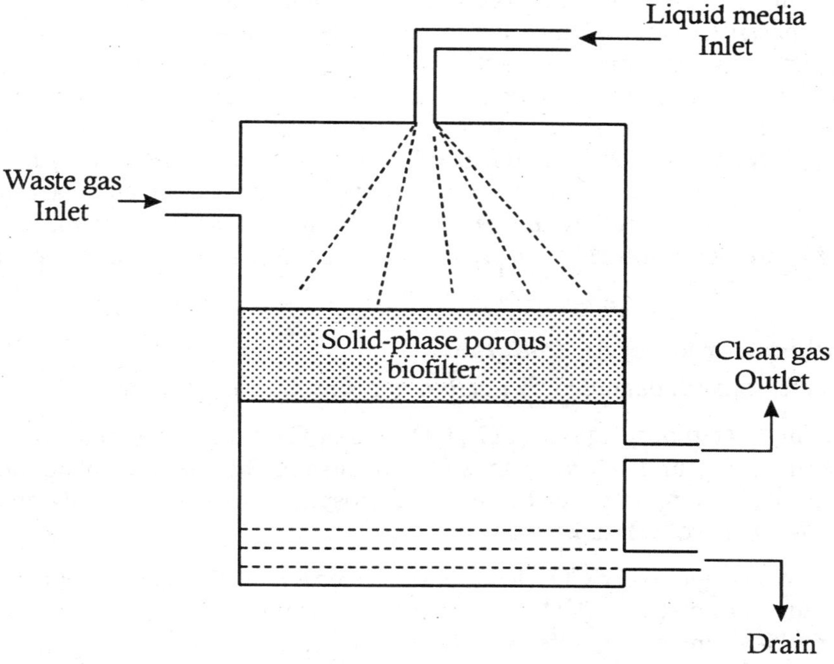


**Fig. 30.4:** Schematic of a Biofilter Setup

Efficient phytoremediation of elemental pollutants like Hg, and As can be achieved by controlling of their chemical species (e.g. conversion of toxic $MgHg^{+}$ to less toxic elemental Hg), electrochemical state (e.g. conversion of highly toxic As(III) to less toxic state), transport and above-ground binding. This will prevent toxic metals entering the food-chain or water supplies. A major advantage of phytoremediation is its low cost and environment friendly nature. The disadvantage is that though it reduces the concentration of contaminants in the soil, but incorporates them into biomasses that may be released back into the environment when the plant dies or is harvested.

Biological procedures such as land forming and composting tend to be inexpensive. A major advantage of land forming is its low cost, but the operation is slow and requires large area. However, organic pollutants and toxic metals might be leached from the site into the groundwater. Composting has been used as a means of treating soil contaminated with pollutants.

Genetic engineering promises to provide organisms with novel and presently non-existing biodegrading abilities. New organisms can be created that have the capacity to carry out catabolic processes that are not possible by existing organisms or under conditions not suitable for existing organisms. A variety of problems can, perhaps, be tackled in future through the use of genetically engineered micro-organisms. These include enhancing the rate of degradation, and having micro-organisms that can destroy pollutants as well as prove resistant to inhibitors.

#### 30.5.2.1 Bioremediation of Metals and Inorganic Pollutants

Metals, metalloids, and inorganic pollutants such as nitrates, sulfates, and phosphates frequently contaminate soils, aquatic sediments and wastewater discharge. These pollutants are derived from mining operations, metallurgy, and fly ash, paints and pesticides and sludge. Microbial processes do not destroy metals and metalloids. However, they can be modified or detoxified by bioremediation involving bio-sorption, reduction, solubilization, precipitation or other means. Bio-sorption is passive uptake of metals (no energy required) by microbial cells. Reasonable proportion of metals can be removed by activated sludge treatment process.

Micro-organisms can also reduce the concentration of a number of inorganic anions and cations. From the environmental viewpoint, the reduction processes are important, because they may change the toxicity, solubility and/or mobility of the ions. Wastewater from sewage-treatment plants contains a high concentration of nitrate. A widely used remedial method is to create anaerobic conditions, provide a carbon source, and allow the denitrifying bacteria to reduce the nitrate to $N_2$. Other approaches of bioremediation rely on precipitation, solubilization, oxidation/reduction and methylation.

#### 30.5.2.2 Bioremediation of Air Pollutants

The air pollutants, formed/degraded by micro-organisms, may contribute to:

1. Increase in "greenhouse" gases ($CO_2$, $CH_4$ and $N_2O$) leading to global warming.
2. Increase in $NO_x$, and hydrocarbons (HCs), responsible for photochemical smog and destruction of ozone layer in the upper atmosphere, exposing the living organisms to deleterious effects of UV radiation.

Anthropogenic contribution of $CO_2$ (fossil burning) represents the perturbation of great climatic importance. After $CO_2$, methane ($CH_4$) is the most important "greenhouse" gas. On a molecular basis, it is about 25-30 times more effective in absorbing IR radiation and thus more effective as a "greenhouse" gas. In addition, $CH_4$ breaks down in the atmosphere to CO, which itself is an important pollutant. Microbial processes (decomposition of plant and animal waste) in soils produce >2/3 of $N_2O$ that enters the atmosphere. $N_2O$ is of practical concern because (*i*) it is a good absorber of IR radiation and, therefore, is a "greenhouse" gas (on a molecular basis, it is 100 times as effective in this regard as $CO_2$), and (*ii*) in the stratosphere it is converted to nitric oxide (NO), which destroys $O_3$. In a series of steps involving $O_3$, $NO_x$, HCs, and sunlight photochemical smog is formed. Both photochemical smog and $O_3$ affect bronchial functions. $O_3$ is toxic to plants too. $NO_x$, combines with water to form $HNO_3$, a major component of acid rain.

Removal (adsorption) and oxidation of organic compounds, which include odorous chemicals, from contaminated air can be achieved by biofiltration methods. Microbes oxidize organic compounds to $CO_2$ and $H_2O$.

## BIBLIOGRAPHY

Alexander, M. (1999), Academic Press: New York. "*Biodegradation and Bioremediation.*"

Barkley, P. W. and Seckler, D. (1972), Harcourt & Brace: New York. "*Economic Growth and Environmental Decay.*"

Cember, H. (1996), Pergamon Press: New York. "*Introduction to Health Physics,*" (3rd edn.).

Cladwell, J. and Jakolay, W. B. (eds.) (1983), Academic Press: New York. "*Biological Basis of Detoxication.*"

Cunningham, W. P. (ed.) (1994), Gale: Detroit, IL. "*Environmental Encyclopedia.*"

Daniell, H., Wyroff, K. and Streatfield, S. J. (2001), *Trends Plant Sci.*, **6**: 219. "Medical molecular farming: Production of antibodies, biopharmaceuticals and edible vaccines."

Dart, R. K. and Stretton, R. J. (1977), Elsevier: New York. "*Microbial Aspects of Pollution Control.*"
Da silva, J. J., Frausto, R. and Williams, R. J. P. (1991), Clarendon Press: Oxford, UK. "*The Biological Chemistry of Elements.*"
Diehl, J. F. (1995), Marcel Dekker: New York. "*Safety of Irradiated Foods.*"
Eisenbud, M. and Gessell, T. (1997), Academic Press: San Diego, CA. "*Environmental Radioactivity from Natural, Industrial, and Military Sources,*" (4th edn.).
Fergusson, (1990), Pergamon Press: Oxford, UK. "*The Heavy Elements—Chemistry, Environmental Impact and Health Effects.*"
Finkel, A. J. Duel, W. C. (eds.) (1976), Pub. Sciences Group: Acton, MA. "*Clinical Implications of Air Pollution Research.*"
Folk, G. E., Jr. (1969), Lea & Febiger Pubs: Philadelphia, PA. "*Introduction to Environmental Physiology.*"
Fry, J. C., *et al.* (eds.) (1992), Cambridge University Press: Cambridge, UK. "*Microbial Control of Pollutants.*"
Giddings, G. *et al.* (2000), *Trends Biotechnol.*, **18**: 1151. "Transgenic plants as factories for biopharmaceuticals."
Hanlan, J. J. (1969), Mosby: St. Luis, MO. "*Health Effects of Environmental Pollutants.*"
Jablonkai, I. (2000), *Intl. J. Environ. Anal. Chem.* "Microbial and photolytic degradation of herbicide acetochlor."
Josephson, E. S. and Peterson, M. S. (eds.) (1983), CRC Press: Boca Raton, FL. "*Preservation of Food by Ionizing Radiation.*"
Kishore, G. M. & Schewmaker, C. (1999), Proc Natl Acad Sci., USA, 96; 5968. "*Biotechnology: Enhancing human nutrition in developing and developed countries.*"
Krishnamoorthy, T. M. and Nair, R. N. (1999), *IANCAS Bulletin*, **15**(3): 25. "Environmental radiation exposure: natural and man-made."
Landsberg, H. E. (1969), Doubleday: New York. "*Weather and Health.*"
Lee, S. D. (1977), Ann Arbor Pubs: Ann Arbor, MI. "*Biochemical Effects of Environmental Pollutants.*"
Markert, B. (ed.) (1993), VCH Pubs: New York. "*Plants as Biomonitors: Indicators for Heavy Metals in the Terrestrial Environment.*"
Meagher, R. B. (2000), *Curr. Opin. Plants Biol.*, **3**: 153. "Phytoremediation of toxic elemental and organic pollutants."
Mettler, F. A. and Upton, A. C. (1995), Saunders: Philadelphia, PA. "*Medical Effects of Ionizing Radiation,*" (2nd edn.).
Monteith, J. L. (ed.) (1977), Academic Press: New York. "*Vegetation and the Atmosphere.*"
Purdom, P. W. (ed.) (1980), Academic Press: New York. "*Environmental Health,*" (2nd edn.).
Shaw, I. C. and Chadwick, J. (1998), Taylor & Francis: London. "*Principles of Environmental Toxicology.*"
Siciliano, S. D. and Germida, J. J. (1998), *Environmental Reviews*, **6**(1): 65. "Mechanisms of phytoremediation: biochemical and ecological interactions between plants and bacteria."
Sayler, G. S., Fos, R. and Blackburn, J. W. (eds.) (1991), Plenum Press: New York. "*Environmental Biotechnology for Waste Treatment.*"
Thain, W. (1980), Pergamon Press: Oxford, UK. "*Monitoring Toxic Gases in the Atmosphere for Hygiene and Pollution Control.*"
United Nations Scientific Committee on the Effects of Atomic Radiation (UNSCEAR) (1993), UNSCEAR: Vienna. "*Sources and Effects of Ionizing Radiation.*"
Waldbott, G. L. (1978), Mosby: St. Luis, MO. "*Health Effects of Environmental Pollutants,*" (2nd edn.).
Yano, Y., *et al.* (2001), *Talanta*, **54**(2): 255. "Application of a microbial sensor to the quality of meat freshness."
... (1993), McGraw-Hill: New York. "*Environmental Science and Engineering.*"
Yuan, L. & Kraut, V. C. (1997), *Curr Opin Biotechnoi.*, **8**: 227. "Modification of plant components."

# APPENDICES

# APPENDICES

## APPENDIX I: SYSTEM INTERNATIONALE (SI) UNIT PREFIXES

| *Prefix* | *Symbol* | *Multiplier* | *Prefix* | *Symbol* | *Multiplier* |
|---|---|---|---|---|---|
| Deca | da | $10^{1}$ | Deci | d | $10^{-1}$ |
| Hecto | h | $10^{2}$ | Centi | c | $10^{-2}$ |
| Kilo | k | $10^{3}$ | Milli | m | $10^{-3}$ |
| Mega | M | $10^{6}$ | Micro | μ | $10^{-6}$ |
| Giga | G | $10^{9}$ | Nano | n | $10^{-9}$ |
| Tera | T | $10^{12}$ | Pico | p | $10^{-12}$ |
| Peta | P | $10^{15}$ | Femto | f | $10^{-15}$ |
| Exa | E | $10^{18}$ | Atto | a | $10^{-18}$ |

## APPENDIX II: FUNDAMENTAL AND DERIVED CONSTANTS IN SYSTEM INTERNATIONALE (SI) UNITS

| *Parameter* | *Unit* | *Symbol* | *Value* | *Dimensionality* |
|---|---|---|---|---|
| Length | Meter | m | – | m |
| Mass | Kilogram | kg | 1 dalton = $1.661 \times 10^{-27}$ kg | kg |
| Time | Second | s | – | s |
| Temperature | Kelvin | K | 0° C = 273.16 K | K |
| Frequency | Hertz | Hz | – | Cycles.$s^{-1}$ |
| Atomic mass unit | Amu or u | μ | $1.6606 \times 10^{-27}$ kg | kg |
| Velocity of light | – | c | $299.792 \times 10^{8}$ (in vacuum) | m.$s^{-1}$ |
| Amount of substance | Mole | mol | – | – |
| Concentration | Molarity | M | – | mol·$L^{-1}$ |
| | Molality | $\overline{M}$ | – | mcl·$kg^{-1}$ |
| Avogadro's number | – | $N_A$ | $6.00225 \times 10^{23}$ | $mol^{-1}$ |
| Force | Newton | N | – | Kg·m·$s^{-2}$ |
| Pressure | Pascal | Pa | –<br>= $10 \times 10^{-6}$ bar<br>= $7.5 \times 10^{-3}$ torr<br>= $9.869 \times 10^{-6}$ atm | N·$m^{-2}$ |
| Energy (work) | Joule | J | –<br>= $10^{7}$ ergs<br>= 0.239 cal<br>= $6.242 \times 10^{18}$ eV | N·m |
| Power | Watt | W | – | J·$s^{-1}$ |
| Radioactivity | Becquerel | Bq | = 27x$10^{-12}$ curie | $s^{-1}$ |
| Absorbed dose | Gray | Gy | = 100 rad | J·$kg^{-1}$ |
| Dose-equivalent | Sievert | Sv | = 100 rem | J·$kg^{-1}$ |
| Gravitational constant | – | G | $66.72 \times 10^{-12}$ | N·$m^{2}$·$kg^{-2}$ |
| Planck's constant | – | h | $0.6626 \times 10^{-33}$ | J·s |
| Boltzmann constant | – | $k_B$ | $13.807 \times 10^{-24}$ | J·$K^{-1}$ |
| Molar gas constant | – | R | 8.314 | J·$mol^{-1}$·$K^{-1}$ |

| *Parameter* | *Unit* | *Symbol* | *Value* | *Dimensionality* |
|---|---|---|---|---|
| Electric current | Ampere | A | – | A |
| Electric charge | Coulomb | C | – | A·s |
| Elementary charge | – | e | $0.1602 \times 10^{-18}$ C | C |
| Electric potential | Volt | V | – | $W{\cdot}A^{-1}$ |
| Electric resistance | Ohm | Ω | – | $V{\cdot}A^{-1}$ |
| Electric conductance | Siemens | S | – | $A{\cdot}V^{-1}$ |
| Electric inductance | Henry | H | – | $V{\cdot}s{\cdot}A^{-1}$ |
| Electric capacitance | Farad | $\mathfrak{J}$ | $96.485 \times 10^{3}$ | $C{\cdot}mol^{-1}$ |
| Magnetic flux | Weber | Wb | – | V·s |
| Magnetic flux density | Tesla | T | – | $V{\cdot}s{\cdot}m^{-2}$ |
| Bohr magneton | – | $\mu_B$ | $9.274 \times 10^{-24}$ | $J{\cdot}T^{-1}$ |
| Electron g-factor | – | $G_e$ | 2.00232 | – |
| Luminosity | Candella | Cd | – | Cd |

**APPENDIX III:** ELEMENTS AND THEIR ATOMIC MASS (BASED ON THE ATOMIC MASS $^{12}C$)

| *Element* | *Symbol* | *Atomic Number (Z)* | *Atomic Mass ($M_r$)* | *Element* | *Symbol* | *Atomic Number (Z)* | *Atomic Mass ($M_r$)* |
|---|---|---|---|---|---|---|---|
| Actinium | Ac | 89 | 227.03 | Aluminum | Al | 13 | 26.98 |
| Americium | Am | 95 | 243 | Antimony | Sb | 51 | 121.75 |
| Argon | Ar | 18 | 39.95 | Arsenic | As | 33 | 74.92 |
| Astatine | At | 85 | 210 | Barium | Ba | 56 | 137.33 |
| Berkelium | Bk | 97 | 247.07 | Beryllium | Be | 4 | 9.01 |
| Bismuth | Bi | 83 | 208.98 | Boron | B | 5 | 10.81 |
| Bromine | Br | 35 | 79.91 | Cadmium | Cd | 48 | 112.41 |
| Calcium | Ca | 20 | 40.08 | Californium | Cf | 98 | 242 |
| Carbon | C | 6 | 12.01 | Cerium | Ce | 58 | 140.12 |
| Cesium | Cs | 55 | 132.91 | Chlorine | Cl | 17 | 35.45 |
| Chromium | Cr | 24 | 52.00 | Cobalt | Co | 27 | 58.93 |
| Copper | Cu | 29 | 63.55 | Curium | Cm | 96 | 247.07 |
| Dysprosium | Dy | 66 | 162.50 | Einsteinium | Es | 99 | 252 |
| Erbium | Er | 68 | 167.26 | Europium | Eu | 63 | 151.97 |
| Fermium | Fm | 100 | 257 | Fluorine | F | 9 | 19.00 |
| Francium | Fr | 87 | 223 | Gadolinium | Gd | 64 | 157.25 |
| Gallium | Ga | 31 | 69.72 | Germanium | Ge | 32 | 72.60 |
| Gold | Au | 79 | 196.97 | Hafnium | Hf | 72 | 178.49 |
| Helium | He | 2 | 4.00 | Holmium | Ho | 67 | 164.93 |
| Hydrogen | H | 1 | 1.01 | Indium | In | 49 | 114.82 |
| Iodine | I | 53 | 126.90 | Iridium | Ir | 77 | 192.22 |
| Iron | Fe | 26 | 55.85 | Krypton | Kr | 36 | 83.80 |
| Kurchatavium | Ku | 104 | 257 | Lanthanum | La | 57 | 138.91 |
| Lawrencium | Lr | 103 | 260 | Lead | Pb | 82 | 207.21 |
| Lithium | Li | 3 | 6.94 | Lutetium | Lu | 71 | 174.97 |
| Magnesium | Mg | 12 | 24.31 | Manganese | Mn | 25 | 54.94 |

| *Element* | *Symbol* | *Atomic Number (Z)* | *Atomic Mass ($M_r$)* | *Element* | *Symbol* | *Atomic Number (Z)* | *Atomic Mass ($M_r$)* |
|---|---|---|---|---|---|---|---|
| Mendelevium | Md | 101 | 258 | Mercury | Hg | 80 | 200.59 |
| Molybdenum | Mo | 42 | 95.94 | Neodymium | Nd | 60 | 144.25 |
| Neon | Ne | 10 | 20.18 | Neptunium | Np | 93 | 237.05 |
| Nickel | Ni | 28 | 58.69 | Niobium | Nb | 41 | 92.91 |
| Nitrogen | N | 7 | 14.01 | Nobelium | No | 102 | 259 |
| Osmium | Os | 76 | 190.20 | Oxygen | O | 8 | 16.00 |
| Palladium | Pd | 46 | 106.42 | Phosphorous | P | 15 | 30.97 |
| Platinum | Pt | 78 | 195.08 | Plutonium | Pu | 94 | 244.06 |
| Polonium | Po | 84 | 209 | Potassium | K | 19 | 39.10 |
| Praseodymium | Pr | 59 | 140.91 | Promethium | Pm | 61 | 145 |
| Protactinium | Pa | 91 | 231.04 | Radium | Ra | 88 | 226.03 |
| Radon | Rn | 86 | 222 | Rhenium | Re | 75 | 186.21 |
| Rhodium | Rh | 45 | 102.91 | Rubidium | Rb | 37 | 85.47 |
| Ruthenium | Ru | 44 | 101.07 | Samarium | Sm | 62 | 150.36 |
| Scandium | Sc | 21 | 44.96 | Selenium | Se | 34 | 78.96 |
| Silicon | Si | 14 | 28.09 | Silver | Ag | 47 | 107.87 |
| Sodium | Na | 11 | 22.99 | Strontium | Sr | 38 | 87.62 |
| Sulfur | S | 16 | 32.07 | Tantalum | Ta | 73 | 180.95 |
| Technetium | Tc | 43 | 97.91 | Tellurium | Te | 52 | 127.60 |
| Terbium | Tb | 65 | 158.93 | Thallium | Tl | 81 | 204.38 |
| Thorium | Th | 90 | 232.04 | Thulium | Tm | 69 | 168.93 |
| Tin | Sn | 50 | 118.71 | Titanium | Ti | 22 | 47.88 |
| Tungsten | W | 74 | 183.85 | Uranium | U | 92 | 238.03 |
| Vanadium | V | 23 | 50.94 | Xenon | Xe | 54 | 131.29 |
| Ytterbium | Yb | 70 | 173.04 | Yttrium | Y | 39 | 88.91 |
| Zinc | Zn | 30 | 65.38 | Zirconium | Zr | 40 | 91.22 |

# Glossary

**A**

| | |
|---|---|
| *Abatement* | Reducing the degree or intensity of, or eliminating pollution. |
| *Abiotic* | Non-living factors or processes. |
| *Absorbance (A)* | A measure of the decrease in intensity of the transmitted light. |
| *Absorbed dose* | The amount of absorbed radiation per unit mass. |
| *Accelerator* | Device for imparting high kinetic energy to a charged particle. |
| *Acclimatization* | The physiological and behavioral adjustments of an organism to changes in its environment. |
| *Accuracy* | Closeness of experimental measurement to the true value. |
| *Acetylcholine (ACh)* | A neurotransmitter that binds to nicotinic (ligand-gated) and muscarinic (G-protein-coupled) receptors and regulates nerve impulse transmission. |
| *Acidosis* | A process that generates hydrogen ions. |
| *Acid mine drainage* | Acidic water that drains from mine tailings. |
| *Acid rain* | Precipitation of acidified aerosols ($pH < 5.5$) in the atmosphere, due to air pollutants (oxides of sulfur and nitrogen). |
| *Activated carbon* | A highly adsorbent form of carbon used to remove organics and toxic substances from liquid or gaseous emissions. |
| *Activated sludge* | Suspension of micro-organisms, taken from the bottom of the final settling tank from the primary effluent, to enhance the rate of microbial decomposition. |
| *Activated sludge system* | An aerated basin in which micro-organisms reduce organic compounds to $CO_2$ and $H_2O$, and more micro-organisms, and then a settling tank (final clarifier) in which the micro-organisms are separated out and recirculated into the aeration tank. |
| *Acute exposure* | A single exposure to a toxic substance which may result in severe biological harm or death. |
| *Acute toxicity* | The ability of a substance to cause severe biological harm or death soon after a single exposure or dose. |
| *Adhesion* | Force of attraction between unlike molecules. |
| *Adiabatic lapse rate* | Change in temperature with altitude, due to atmospheric pressure. It is the rate at which unsaturated air cools adiabatically (without heat exchange) as it rises. |
| *Advanced waste treatment* | Wastewater treatment beyond the secondary or biological stage (for removal of phosphorus and nitrogen). |
| *Advection* | Transportation. |
| *Aerated lagoon* | Air-circulated treatment pond for speedy biological decomposition of organic waste. |
| *Aeration* | A process to promote biological degradation of organic pollutants in water. |
| *Aerobic* | Oxygen-present environment. |

*Aerosol* : Dispersion of solid or liquid particles in a gaseous medium (e.g. fog, fumes, mist and smoke).
*Air quality assessment* : A prescribed level of atmospheric pollution allowed for a certain compound during a specific time in a specific geographic area.
*Air quality criteria* : The levels of pollution and lengths of exposure above which adverse health and welfare effects may occur.
*Air quality standards* : The level of atmospheric pollution prescribed by regulations that may not exceed during a given time and in a defined area.
*Aitken particle* : Condensation nuclei for rain drops.
*Albedo* : A fraction of the solar radiation reflected back by an object (e.g. by Earth's surface).
*Algal bloom* : Proliferation of living algae on the surface of lakes, often related to nutrient pollutants.
*Alpha particle* : High-energy helium nucleus ($^4He^{2+}$).
*Alum* : Aluminum sulfate.
*Amphiphilic* : Molecules containing both polar (hydrophilic) and apolar (hydrophobic) groups.
*Anaerobic* : Oxygen-absent environment.
*Anemometer* : Instrument to measure the air speed.
*Anion* : Negatively charged ion (attracted to the positive electrode, anode).
*Anisotropy* : Spatial non-uniformity of a physical property.
*Annihilation* : Conversion of a pair of particles of opposite charges to photon energy.
*Anode* : Positive electrode (toward which the negatively charged anions are attracted).
*Anthropogenic* : Resulting from human activity.
*Antibody* : A protein synthesized by an organism against a substance (antigen).
*Anticyclone* : High-pressure cell (clockwise in Northern Hemisphere).
*Antigen* : A substance that induces a specific immune response.
*Aquifer* : Porous water-bearing geologic stratum.
*Atom* : The smallest unit of an element that can still maintain the properties of that element.
*Atomic mass unit (amu)*: (1/12) the mass of carbon 12 ($1.661 \times 10^{-24}$ g).
*Atomic number (Z)* : Number of protons in an atom.
*Atomization* : Process of converting a sample into vapor/mist form.
*Auger electron* : Electron ejected from K- or L-shell as an alternative to characteristic X-ray emission, due to electron capture or internal conversion.
*Autotrophs* : Primary producers of organic compounds from inorganic raw materials. (1) Photoautotrophs (plants, algae, and some bacteria) use light as the source of the needed energy (e.g. photosynthesis). (2) Chemoautotrophs use the energy secured by oxidizing some inorganic substances in their surroundings.
*Avogadro's number* : Number of atoms in the gram atomic weight of a given element ($6.023 \times 10^{23} g^{-1} mole^{-1}$).

**B**

*β-Particle* : Electron or positron.
*Bar screen* : A device used in wastewater treatment to remove large floating and suspended solids.
*Bathochromic shift* : Displacement of absorption profile towards longer wavelength region.
*Benthic region* : Bottom of a body of water.
*Bioaccumulation* : Increase in concentration of substances in living organisms due to slow metabolism or excretion.
*Bioaugmentation* : The introduction of cultured micro-organisms into the subsurface environment for the purpose of enhancing bioremediation of organic contaminants.
*Bioavailability* : Degree of availability (ability) of a substance to take part in metabolic processes.
*Biodegradation* : Metabolic process by which high-energy organic compounds are converted to low-energy $CO_2$ and $H_2O$.
*Biomass* : The total amount of living matter (vegetation) within each trophic level.
*Biome* : Entire community of living organisms in a single major ecological area.

| | | |
|---|---|---|
| *Bioremediation* | : | Use of living organisms to remove pollutants from soil, water, and agricultural pests. |
| *Biosensor* | : | Analytical device comprising a biological recognition element (e.g. enzyme, antibody, receptor, DNA). |
| *Biosphere* | : | Areas of the Earth occupied by living organisms (the sum of biological habitats). |
| *Biota* | : | The flora and fauna of a region. |
| *Blackbody* | : | Ideal absorber and emitter of electromagnetic radiation (light, heat, etc). |
| *Breeder reactor* | : | A nuclear reactor that produces more fuel than it consumes. |
| *Bremsstrahlung* | : | X-ray emissions caused by Coulombic attenuation of electrons in matter. |
| *Brown-lung disease* | : | Lung disease due to inhalation of coal dust and cotton dust. |
| *Bubbler* | : | Device for measuring gaseous air pollutants. |
| *Buffer* | : | Medium incorporating resistance to change (e.g. pH buffer). |
| *Byssinosis* | : | Brown-lung disease due to inhalation of cotton dust. |

## C

| | | |
|---|---|---|
| *Carbon cycle* | : | Combined processes, which include photosynthesis, decomposition and respiration. |
| *Carcinogen* | : | Agent or substance that can induce cancer in a cell. |
| Catalyst | : | Substance that accelerates the rate of chemical reaction without being used up in the process. |
| *Catalytic converter* | : | Air pollution abatement device that removes pollutants from motor vehicle exhaust, either by oxidizing them into carbon dioxide and water or reducing them to nitrogen. |
| *Cathode* | : | The negative electrode (toward which the positively charged cations are attracted). |
| *Cation* | : | Positively charged ion (attracted towards the negative anode, cathode). |
| *Chelation* | : | Complexion of metal cations with organic compounds. |
| *Chemical potential* | : | Molecular Gibbs "free energy". |
| *Chirality* | : | Handedness (property of a molecule, non-superimposable on its mirror image). |
| *Chlorosis* | : | Loss of green color of plants due to action of pollutants. |
| *Cholinergic* | : | Synaptic transmission mediated by acetylcholine (ACh). |
| *Chromatography* | : | Technique of molecular separation of compounds in a mixture. In the normal phase technique, less polar compounds are eluted first; and in the reverse phase technique, the more polar compounds are eluted first. |
| *Chronic toxicity* | : | The capacity of a substance to cause long-term poisonous health effects in organisms. |
| *Cis* | : | Configuration as on the same side. |
| *Clarifier* | : | Settling tank in wastewater treatment. |
| *Clathrate* | : | Cage-like structure. |
| *Climate* | : | Description of atmospheric pattern in a particular region averaged over an extended period (three or four decades). |
| *Closed system* | : | A thermodynamic system where only energy (and not matter) is exchanged with the environment. |
| *Cohesion* | : | Force of attraction between like molecules. |
| *Coliform* | : | Micro-organisms found in the intestinal tracks of humans and animals, which produce gas and ferment lactose. |
| *Coliform index* | : | A rating of the purity of water based on a count of fecal bacteria. |
| *Colloid* | : | Substance whose particles range from 1-1,000 nm in size. |
| *Community* | : | The population of plants, animals, and microbes found in a particular area and often interacting with one another. |
| *Compensation level* | : | The depth at which photosynthesis is equal to respiration in a water body. |
| *Composting* | : | Controlled biological decomposition of organic material in the presence of air to form a humus-like material. |
| *Compton scattering* | : | Transformation of photon (X-ray) energy through collisions with "free" electrons (by inelastic scattering). |
| *Conduction* | : | Transfer of energy through physical contact, without mass transfer. |
| *Convection* | : | Transfer of heat energy through mass movement. |
| *Corioli's effect* | : | Trajectory of wind circulation arising due to rotation of the Earth around its axis. |

| | | |
|---|---|---|
| *Correlation* | : | A statistical measure that indicates the extent to which two factors vary together and thus how well either factor predicts the other. |
| *Covalent bond* | : | A chemical bond where an electron pair is shared by two atoms. |
| *Critical temperature* | : | Temperature above which a gas cannot be liquefied by increasing the pressure. |
| *Cyclone* | : | (1) Low-pressure system of circulating winds (Counter clockwise in the Northern Hemisphere). (2) A device for removing large dust particles. |

**D**

| | | |
|---|---|---|
| *Dalton* | : | Unit of mass that is equivalent to one-twelfth the mass of an atom of carbon-12 (~ mass of hydrogen atom). |
| *Decomposers* | : | Organisms that depend on primary producers for living. |
| *Decomposition* | : | The conversion of chemically unstable materials to more stable forms by chemical or biological action. |
| *Defoliant* | : | Herbicide that removes leaves from trees and plants. |
| *Denitrification* | : | Bacterial reduction of nitrates to nitrites and molecular $N_2$ under anaerobic conditions. |
| *Detection limit* | : | The lowest concentration of a chemical that can be detected. |
| *Detritivores* | : | A feeding group (earthworms, beetles, ants, termites, bacteria, fungus, etc.) that feed on detritus. |
| *Dew point* | : | Temperature (with 100% relative humidity) below which moisture condenses. |
| *Diffusion* | : | Transport of matter down the concentration gradient. |
| *Dipole* | : | A substance with regions of net negative and positive charge. |
| *Dissociation* | : | Separation of ions from a molecule or crystal lattice. |
| *Dose/Dosage* | : | The actual quantity of a substance (radiation) received by an organism. |
| *Dose equivalent (H)* | : | A unit of biologically effective dose. |
| *Dose-response analysis* | : | Environmental risk analysis procedure to correlate relationship between dose and toxicological responses. |
| *Dosimetry* | : | Measurement of high-energy radiation dosage. |
| *Dystrophic lakes* | : | Lakes between eutrophic and swamp stages. |

**E**

| | | |
|---|---|---|
| *Ecological impact* | : | Effect of a factor on living organisms and their non-living (abiotic) environment. |
| *Ecology* | : | The relationship of living organisms to one another and their environment. |
| *Ecosystem* | : | The smallest unit of the biotic system that adequately sustains life. |
| *Edema* | : | Swelling of tissues and accumulation of fluid. |
| *Electrode* | : | A device used to interface ionic potentials and currents. |
| *Electrolyte* | : | A substance that dissociates into ions when dissolved in a solvent. |
| *Electron affinity* | : | The energy released in the formation of an ion from a free atom. |
| *Electron capture* | : | Radioactive decay process in which the nucleus captures an orbital electron, transmuting the nucleus to that of another element. |
| *Electrophoresis* | : | Molecular separation method based on the differential migration of charged particles under the influence of applied electric field. |
| *Emphysema* | : | Loss of air exchange in lungs. |
| *Endergonic* | : | Characterized by concomitant absorption of energy. |
| *Enthalpy (H)* | : | Heat produced/absorbed in a chemical reaction (at constant pressure). |
| *Entropy (S)* | : | A thermodynamical measure of molecular disorder. |
| *Epilimnion* | : | Top layer of a thermally stratified water body (lake). |
| *Equilibrium potential* | : | The membrane potential at which there is an electrochemical potential equilibrium for a particular ion. |
| *Eutectic* | : | A compound that melts at lower temperature than its components. |
| *Eutrophication* | : | Aging of aquatic environment with excessive aquatic plant growth, as a result of introducing abundant nutrients into it. |
| *Evapotranspiration* | : | The sum of water loss due to evaporation and plant transpiration. |

| | | |
|---|---|---|
| *Excitation* | : | The process of raising an electron from its ground energy state to a higher energy state. |
| *Exergonic* | : | Characterized by a concomitant release of energy. |

**F**

| | | |
|---|---|---|
| *Fermentation* | : | Anaerobic decomposition of an organic compound (e.g., glucose) by a living organism. |
| *Fingerprint region* | : | The region between 400-1100 $cm^{-1}$, characteristic of specific molecules/moieties. |
| *Flue gas* | : | Air coming out of burner after combustion. |
| *Flurosis (bone)* | : | Bone disease. |
| *Flux* | : | The rate of flow across a unit area. |
| *Fly ash* | : | Non-combustible fine particles released in coal-burning operations. |
| *Food chain* | : | Pathway outlining how members of an ecosystem obtain energy and nutrients. |
| *Free energy* | : | The energy available to do work at a given temperature. |
| *Free radical* | : | Highly reactive (electrophilic) atoms or molecules with incomplete electronic configuration. |
| *Fuel cell* | : | An electrochemical device that converts chemical energy of a fuel (e.g. hydrogen + oxygen) directly into electricity. |
| *Fume(s)* | : | Dust (metal) vapor(s). |

**G**

| | | |
|---|---|---|
| *Gamma ray* | : | Electromagnetic radiation at very high-energy range. |
| *Gaussian profile* | : | Normal distribution profile. |
| *Garbage* | : | Biological domestic solid waste. |
| *Geiger counter* | : | An instrument to detect ionizing radiation. |
| *Greenhouse effect* | : | Global warming due to trapping of infrared (heat) radiation, caused by the presence of IR-absorbing constituents in the atmosphere ($CO_2$, $NH_4$, and water vapor) that allow incoming solar radiation to pass through but absorb the heat radiated back from the Earth's surface. |

**H**

| | | |
|---|---|---|
| *Habitat* | : | The place where a population can successfully live and its surroundings, (both biotic and abiotic). |
| *Half-life* | : | Time required for a radioactive material, or a pollutant to decrease by half the initial amount. |
| *Heat capacity (C)* | : | The amount of heat required to raise the temperature of a body by one degree ($C = \Delta Q/\Delta T$). |
| *Heat exchange* | : | Transfer of energy from one body to another without doing work. |
| *Hertz (Hz)* | : | Frequency (cycles per second). |
| *Heterotrophs* | : | Consumers of primary producers (autotrophs) for their food requirement. |
| *Homeosatis* | : | Tendency to maintain a balanced or constant internal state that is optimal for functioning. |
| *Humus* | : | Organic portion of the soil remaining after prolonged microbial decomposition. |
| *Hydrogen bond* | : | A polar non-covalent interaction between electronegative atoms, involving a hydrogen atom. |
| *Hydrolysis* | : | Decomposition of a substance by the insertion of water molecules between certain of its bonds. |
| *Hydrophilic* | : | Water-loving. |
| *Hydrophobic* | : | Water-repelling. |
| *Hypolimnion* | : | Bottom region of a thermally stratified water body (lake). |

**I**

| | | |
|---|---|---|
| *Impedance* | : | The dynamic resistance to flow met by fluids. |

*Inertia* : Resistance to change.
*Influx* : The flow of matter into a cell.
*Internal conversion* : The transition between two energy states of a nucleus where the energy difference is absorbed by an orbital electron and ejected.
*Inversion (atmospheric)* : Atmospheric condition where a layer of air afloat is warmer than the underlying air mass (and thus can trap air pollutants beneath it).
*Ion* : Atom or group of atoms that has an electrical charge arising from the gain or loss of electrons.
*Ionization* : The process of removal of electron(s) from a neutral atom and thus making it an ionic species.
*Ionization potential* : The energy required to remove an electron from an isolated atom or a molecule.
*Ionizing radiation* : Radiation that produces ion pairs (e.g. $\alpha$– and $\gamma$-rays).
*Isobaric* : Change that takes place at constant pressure.
*Isoelectric* : State of zero potential difference.
*Isoelectric point* : The pH of a solution at which an amphoteric molecule exists as a neutral species (net charge is zero).
*Isothermal* : Change that takes place at constant temperature.
*Isotonic* : Having the same concentration of water as the solution under comparison.
*Isotopes* : Nuclides having the same number of protons (Z equal) but different number of neutrons (e.g. C-12; C-13; C-14).
*Isotropic* : Uniformity of a physical property in all directions.

**K**

*Kerma* : Kinetic energy released in a material.

**L**

*Landfills* : Disposal sites for solid wastes.
*Lapse rate* : Rate at which temperature varies with altitude in the atmosphere.
*Latent heat* : The energy required to change the state of a unit mass of a material from solid to liquid (fusion) or liquid to gas (evaporation), without a change in temperature.
*LC50* : Lethal concentration-50 is the median level concentration of a substance through respiratory route that is expected to kill 50% test animals within a specified period of time.
*LD50* : Lethal dose-50 is the dose of a substance that is expected to kill 50% test animals within a specified period of time, exposed through a route other than respiratory system.
*Leachate* : Water that collects contaminants as it trickles through wastes.
*Ligand* : A molecule that forms a non-covalent complex with another molecule.
*Lipophilic* : Affinity to lipids.
*Lipoprotein* : A protein with covalently bound lipid molecule(s).
*Litter* : Street and highway refuse.
*Luminescence* : Measure of brightness per unit area of light-emitting surface ($cd/m^2$).

**M**

*Magnification* : Ratio of apparent image size and object size in visual observation.
*Masking* : Treatment of a sample with a reagent to prevent interference from other constituents.
*Mass analyzer* : Mass spectrometer device to separate ionic components according to their mass/charge (m/ze) ratios.
*Metalimnion* : Region between surface (epilimnion) and benthal (hypolimninon) regions in a thermally stratified water body.
*Meteorology* : Study of the atmospheric phenomena.
*Mist* : Small liquid droplets (50-500 $\mu$m) in air.

*Molality* : The number of moles of a solute in a kilogram of solvent.
*Molarity* : The number of moles of a solute in a litre of solvent.
*Mole* : Avogadro's number ($6.023 \times 10^{23}$) of molecules of an element (or compound); equal to molecular mass in grams.
*Monochromator* : An optical device to select radiation of particular wavelength (energy).
*Mutagen* : Agent (chemical or radiation) that can cause damage to genetic material in cells.

**N**

*Necrosis* : Death of living cells or tissues.
*Niche* : The place occupied by a particular species and its relation to other species in an ecological community.
*Nitrification* : The process of converting ammonia to nitrate.
*Nitrogen cycle* : Processes of circulation of nitrogen (nitrification and de-nitrification) in nature.
*Nitrogen fixation* : The process of chemically converting nitrogen gas from the atmosphere into nitrogen-containing compounds that can be used by plants and animals.
*Numerical Aperture (NA)* : The ratio of half the aperture angle of an optical system and the refractive index.
*Neutron* : One of the fundamental constituents of an atomic nucleus with no charge.

**O**

*Oligotrophic (lakes)* : Deep lakes that have a low supply of nutrients and contain little organic matter and high dissolved oxygen level.
*Orbitals* : Probability surfaces associated with an atom, within which an electron is likely to found.
*Osmolarity* : The effective osmotic pressure.
*Osmosis* : Diffusion of solvent across a semi-permeable membrane.
*Osmotic pressure* : The hydrostatic pressure necessary to prevent osmotic flow.
*Oxidation* : Process of removing electrons from a substance

**P**

*Pair-production* : Conservation of photon energy (>1.02 MeV) into matter ($\beta^+\beta^-$).
*Pathogen* : Micro-organism that can cause disease in host organisms.
*Permeability* : Ability to permeate (diffuse).
*pH* : Measure of hydrogen ion [$H^+$] concentration; represents acidity or alkalinity of a solution.
*Phonon* : A quantum of sound energy.
*Phosphorus cycle* : The movement of phosphorus from lithosphere through biosphere and back to lithosphere.
*Photoelectric absorption* : Absorption of a photon by interaction with a "bound" electron.
*Photoelectron* : Electron ejected from a K– or L–shell as a result of photoelectric reaction.
*Photon* : A quantum of light energy.
*Photoreceptor* : Sensory cell that is tuned to receive light energy.
*Photosynthesis* : The process by which green plants produce carbohydrates using simple inorganic components such as water and carbon dioxide, in the presence of light energy.
*Phytotreatment* : Cultivation of specialized plants that absorb specific contaminants from the soil through their roots or foliage.
*Polarizer* : Device for producing polarized light.
*Pollution* : Introduction of products in such quantities and at rates that are far in excess of nature's capacity to dilute and disperse them in order to render them harmless to the environment.
*Pollutants (primary)* : Pollutants emitted directly from the sources.
*Pollutants (secondary)* : Pollutants that are produced due to reactions of primary pollutants.

*Potentiation* : The ability of one chemical to increase the effect of another chemical.

*Precision* : Random error associated with the result.

*Primary standard* : (1) A substance whose purity and quality are well established and which is suitable for making calibrations.
(2) National ambient air quality standards designed to protect human health with an adequate margin for safety.

*Primary waste treatment* : Removal of solids in wastewater (physical treatment process) before it is discharged or treated further.

*Proton* : One of the fundamental constituents of an atomic nucleus with positive charge (e.g. $^1H^+$).

*Psychrometry* : The study of the interactions between temperature and water vapor in air.

## Q

*Quantum efficiency* : A measure of radiation absorption/conversion.

## R

*Radiation* : Transfer of energy by direct process (no medium is required).

*Radioactivity* : Spontaneous disintegration of nuclei of higher atomic number elements.

*Radioimmunoassay (RIA)* : An immunological technique for quantitative measurement of substance by radio-labeling.

*Radioisotope (radionuclide)* : A radioactive nuclide that transmutes by way of nuclear decay.

*Random sampling* : Sampling that fairly represents a population because each member has an equal chance of inclusion.

*Rayleigh (& Mie) scattering* : Elastic molecular scattering of light of different wavelengths by molecules (particles).

*Reaction kinetics* : Phenomenological approach towards the dynamics of reaction systems.

*Redox reaction* : A chemical reaction in which electrons are transferred from one atom (which is thereby oxidized) to another (which is thereby reduced).

*Reduction* : Process of adding electrons to a substance.

*Refraction* : Bending of a wave as it enters from one medium to another.

*Refractive index* : The ratio of the velocity of light in vacuum and in a medium.

*Refuse* : Urban solid waste.

*Regression* : A statistical technique by which the variation of a dependent variable can be predicted from the value of an independent variable (or variables) through a regression coefficient (or coefficients) plus a constant term. ($y = ax + b$)

*Relative risk* : Ratio of the probabilities that an adverse effect will occur in two different populations (exposed/non-exposed).

*Resolving power* : The smallest lateral distance that can be resolved in an image.

*Resolution* : The degree of detail realizable.

*Reynolds number (Re)* : Ratio between inertial and viscous forces.

*Risk assessment* : Estimating the risks associated with different stressors or management actions.

*Rubbish* : Non-degradable solid waste.

*Runoff* : Excess of water that cannot be absorbed by the soil.

## S

*Scintillation* : Flash of light produced in a phosphor by radiation.

*Scrubber* : Device for removing air pollutants by bringing them into contact with water.

*Secondary waste treatment* : Removal of oxygen-demanding substances in wastewater (biological treatment process).

*Sensitivity* : Ability to detect a low level of something. The more sensitive the test, the less likely that it will cause "false negatives" (a failure to detect something that is actually present).

| | | |
|---|---|---|
| *Sewage* | : | Domestic and commercial wastewater and products. |
| *Sludge* | : | Solid suspensions in wastewater. |
| *Smog* | : | Polluted air (smoke + fog). |
| *Smog (photochemical)* | : | Chemical reaction of $NO_x$ and hydrocarbons with UV radiation as catalyst. |
| *Solar constant* | : | The rate at which solar radiation reaches the earth's atmosphere. |
| *Specific heat* | : | Molar heat capacity. |
| *Specificity* | : | Capability of discriminating between two things. The more specific a test is, the fewer the "false positives." |
| *Spirometer* | : | Instrument for measuring respiratory volumes. |
| *Steady state* | : | Dynamic equilibrium. |
| *Steric* | : | Pertaining to the spatial arrangement. |
| *Standard* | : | A pure substance whose quantitative characteristics are known. |
| *Stochastic effect* | : | An effect (random) whose probability of occurrence in an irradiated species is a function of dose. |
| *Stressor* | : | Physical, chemical, or biological entity that can induce adverse effects on ecosystems or human health. |
| *Substrate* | : | A substance that is acted upon by an enzyme. |
| *Suspension* | : | Mixture containing solid particles larger than 100 micrometers distributed throughout a liquid. |
| *Synergism* | : | An interaction of two or more chemicals that results in an effect greater than the sum of their separate effects. |

**T**

| | | |
|---|---|---|
| *Tertiary waste treatment* | : | Removal of algal (nitrogen- and phosphorus-containing) nutrients in wastewater (chemical treatment process). |
| *Thermistor* | : | Thermal resistor (a measuring device of temperature). |
| *Thermocline* | : | Inflection point (middle layer) in a thermally stratified water body. |
| *Thermophilic* | : | Heat-seeking. |
| *Thermophobic* | : | Heat-avoiding. |
| *Threshold dose* | : | The minimum dose of a radiation, or an agent that will produce a detectable biological effect. |
| *Tidal volume* | : | The volume of air moved in or out of the lungs with each breath. |
| *Tone (sound)* | : | Pure sound of uniform frequency. |
| *Toxicity* | : | The degree to which a substance or mixture of substances can harm humans or animals.<br>(1) Acute toxicity involves harmful effects in an organism through a single or short-term exposure.<br>(2) Chronic toxicity is the ability of a substance or mixture of substances to cause harmful effects over an extended period. |
| *Toxin* | : | Metabolic product of an organism that is poisonous to another organism. |
| *Transducer* | : | A device that transforms energy from one form to another. |
| *Transduction* | : | Process of converting one form of energy to another form. |
| *Transpiration* | : | Evaporation of water through (leaves of) plants. |
| *Triple point* | : | Critical point in the pressure-temperature diagram, where gaseous, liquid and solid phases are in equilibrium. |
| *Tropogenic zone* | : | The region above the compensation level in a thermally stratified water body where photosynthesis > respiration. |
| *Tropolytic zone* | : | The region below the compensation level in a thermally stratified water body where photosynthesis < respiration. |
| *Troposphere* | : | The lowest strata of the atmosphere. |
| *Turbulence flow* | : | Flow pattern with sharp gradients and discontinuities in the direction of flow. |

## U

*Unit risk* : Risk to an individual from exposure to a standard concentration of airborne or water-borne pollutant.

## V

*Valence* : Number of electrons gained, lost, or shared by an atom in bonding to one or more other atoms.

*Van der Waals forces* : Weak non-bonded interactions between atoms and molecules.

## W

*Washout* : Removal of air pollutants by rain (water).

*Wastewater* : The spent or used water from a home, community, farm, or industry that contains dissolved or suspended matter.

*Water hardness* : Quality of water with dissolved carbonates and sulfates of calcium and magnesium.

*Water potential* : Gibbs "free energy" per mole of water in a system.

*Water quality criteria* : Levels of water quality expected to render a body of water suitable for its designated use.

*Waveguide* : Channeled wave propagation (due to total internal reflection).

*Weather* : The behavior of the wind patterns during short time period.

*Wind rose* : Graphic representation of wind rate.

## X

*Xenobiota* : Biota or materials alien to the environment.

*X-rays* : High-energy electromagnetic radiation (~0.1-10 Å photons) by the bombardment of heavy element targets with high-energy electrons.

## Z

*Zooplankton*(s) : Tiny aquatic animal(s).

# Index


A

Abbe's criterion 566
Agrochemicals 205-11
  botanicals 209-10
  carbamates 209
  organochlorines 207
  organophosphates 208
  pesticides 206-11
  phenolics 209
Air pollution. *See* under Pollutants
Air quality monitoring
  ATMOS 291
  chemical methods 283-86
  colorimetric methods 281, 286
  combustion methods 288
  continuous monitoring 284
  electrical conductivity method 288-89
  electrochemical methods 286, 288
  gravimetric methods 283-84
  indicator methods 280-81
  interferometric method 290-91
  LIDAR 289, 291
  LIF 342
  long-path monitoring 291-92
  LPAS 342
  manometric methods 285
  nuclear analytical methods 293, 531-59
  odor 283
  physical methods 286
  qualitative methods 280-83
  quantitative methods 283-88
  radiometers 289-91
  radiosonde 42
  remote heterodyne 292-93
  Ringlemann charts 282
  SBUV 341
  thermal methods 287-88
  titrimetric methods 285
  TOMS 341
  UV-DOAS 496
  visibility 282
  volumetric methods 284-85
Airway resistance 110, 119
Albedo 26, 294, 652
Algal bloom 140, 164, 652
Analysis of environmental pollutants
  chemical methods 333-414
  electroanalytical methods 465-84
  electron diffraction 567
  microscopic methods 566-67
  nuclear analytical methods 531-59
  particulate matter of 295-300
  physical methods 415-574
  separation techniques 421-63
  spectroscopic methods 485-529
  statistical methods 268-70
  X-ray diffraction 562-66
Analysis of pollutants (Air)
  ammonia 354-55, 477
  asbestos 398-99
  bromide 363
  carbon dioxide 346
  carbon monoxide 345-46
  chloride 361-62
  chlorine dioxide 362
  combustible gases 347
  cyanide 355, 364
  detection methods 294
  elemental carbon 343
  fluoride 361

gases and aerosols 337-65, 508-09
halogens 360-64
hazardous chemicals (HAPs) 301-03
hydrogen sulfide 356
inorganic carbon 343
iodide 363
Kjeldahl method 349
methane 347
nitrate 353, 478
nitric acid 352
nitric oxide 349-50
nitrite 350-52
nitrogen 347-49
nitrogen dioxide 350-52
nitrogenous compounds 347-55
nitrous acid 352, 496
nitrous oxide 349
oxygen 340, 478
ozone 341
particulate matter 294-301
peroxides 342
phosphorus 399
photochemical smog 355, 659
radicals 342
silicon 398
sulfate 360
sulfide 356
sulfite 360
sulfur compounds 356-60
sulfur dioxide 357-59
sulfuric acid 360
total organic carbon 343-45

Analysis of pollutants (Metals) 367-96
alkali metals 371-73
alkaline-earth metals 375-78
arsenic 386-88
boron subgroup 382-83
cadmium 379
chromium 389-90
copper subgroup 373-75
iron 391-92
lead 384-85
mercury 380-81, 518
speciation 368-71
trace metals 482
wet-chemical methods 394
zinc 378-79

Analysis of pollutants (Organic) 401-14
agrochemicals 413-14
aliphatic hydrocarbons 408
alkyl nitrates 405-06
aromatic hydrocarbons 408-10
classification of 402-03
halogenated hydrocarbons 410-13
HAPs 301-03
hydrocarbons 408-13
hydrohalocarbons 413
mononuclear aromatic HCs 408
oils and greases 406
oxygen-containing 403-05
pesticides 413, 482
phenols 406-08
polyaromatic (PAH) 409-10
synthetic compounds 410-14
total organic halogen (TOX) 412
volatile organic (VOC) 402-05
VOST procedure 301

Analysis of pollutants (Radioactive) 303-04, 531-59
molecular activation analysis 552-53
neutron activation analysis 551-55
nuclear analytical methods 531-59
nuclear reaction analysis 550-51

Analysis of pollutants (Water) 307-31
bacterial estimation of 329-30
biochemical oxygen demand (BOD) 321-23
biological indicators 325-30
chemical oxygen demand (COD) 323-24
chemical parameters 315-25
color 310
conductimetric methods 473
dew point temperature 314, 654
dissolved and suspended matter 315
dissolved oxygen (DO) 321
electrical conductivity 314-15
factors affecting 308
hardness 316
humidity 312
inorganic pollutants 319-21
IR method 509
monitoring 473
odor 310
organic pollutants 321-25
organoleptic properties 309-11
oxygen demand (OD) 325
parameters influencing 309
pathogens 320
pH 316, 657
physical parameters 308-15
plant water potential 326-27
potentiometric methods 475
salinity 316
soil water 325-27
taste 310
total oxygen demand (TOD) 324-25
turbidity 310-12

AQIA 49

Arrhenius equation 65
Assays
  bioassay 629
  bioluminescence 505
  chemiluminescence immuno 505-06
  EMIA 506
  enzyme-mediated 504-05
  IRMA 556-58
  luminescence-based 503-06
  mutagenecity 630
  radioimmunnoassay (RIA) 554-56, 658
Atmosphere
  biosphere 2-3, 156-57
  chemosphere 80
  composition of 22-25
  exosphere 23
  heterosphere 22
  homosphere 22
  ionosphere 23
  mesosphere 23-24
  stratopause 24
  stratosphere 23-24
  thermosphere 23-24
  troposphere 23-24
Atmosphere-earth balance 96
Atmospheric measurements
  anemometer 41-42
  barograph 44
  barometer 43
  Beaufort's classification 41
  Doppler sodar 42
  interferometric methods 280
  LIDAR 42, 289, 291
  long-path monitoring 291
  manometric 45
  of humidity 138, 312
  of pressure 42
  of temperature 45
  of wind direction 41
  of wind speed 41
  pitot tube anemometer 41
  radiosonde 42
  sonic anemometer 41-42, 289
Atmospheric stability
  adiabatic 38-40
  dispersion 32, 50
  lapse rates 35-40, 656
  Pasquill stability class 32, 34
  sub-adiabatic 37, 39
  super-adiabatic 39
  temperature inversion 35, 53
  turbulence 33-34
ATP determination 329
Autotrophs 156-57, 652

**B**

Beer-Lambert law 488-89, 500
Bernoulli equation 135
Bioluminescence 329, 369, 505-05
Biome 156, 652
Biomonitoring 628-30
  biochemical tests 629-30
  mutagenecity tests 630
  persistence tests 628
  phytomonitoring 629
  solubility tests 627-28
  toxicology tests 627-28
Bioremediation 640-45, 653
  acclimation 641
  activation 641
  bioavailability 642, 652
  biodegradation 640-43, 652
  biofilter 643
  detoxication 641
  of air pollutants 644
  of elemental pollutants 644
  phytoremediation 644
Blackbody 26, 653
  temperature 95
Boyle Mariotte law 44
Bremsstrahlung 540, 653
Brown lung disease 112
Brownian motion 109, 234, 236
Byssinosis 112, 653

**C**

Carbon
  cycle 3, 80, 653
  elemental 343
  inorganic 343-45
  microbial transformation of 160
  TOC 343-45
Carbon dioxide 90-92
  cycle 92
  determination of 345-46, 510-11
Carbon monoxide 90-92
  control of 586-87
  determination of 345-46, 510-11
  toxicity 119-20
Carnot cycle 62-63
Chelation 198, 374-75, 517-18, 653
Chemical methods in pollution analysis 333-414
  agrochemicals 413-14
  gases and aerosols 337-65
  hazardous chemicals 301
  inorganics 343

metals 367-96
organics 401-14
Chemiluminescence 350-54
Chromatography 427-45, 453-54
adsorption 434-36
affinity 438-39
chiral 439
gas (GC) 442-45
HPLC 432-34
ion-exchange 439-42
liquid (LC) 432-42
paper 429-31
size-exclusion 436-38
supercritical fluid (SFC) 453-54
thermal diffusion 442
thermal field flow fractionation 442
thin layer (TLC) 431-32
Clausius-Clapeyron equation 134, 140
Climate 29-31, 653
Köppen's classification of 30-31
zones 30
Colorimetry 280-81, 406-07
Control of pollutants (Air)
absorption 582-83
adsorption 583
Anderson multistage sampler 239-40
bioremediation 644
catalytic oxidation 584
centrifugal separators 244
chemical conversion 583
collection methods 234-49, 582-84
electrostatic precipitation 241
filters 242-43
Hi-vol sampler 237-38
in chemical industry 590-91
in metallurgy 589-90
in power generation plants 587
incineration 582, 591
inertial impactors 236-40
lean $NO_x$ technology 585
Low-vol sampler 238-39
mineralization 264
of carbon monoxide 586-87
of gases 584-91
of hydrocarbons 586
of $NO_x$ 584-86
of organics 591
of particulate matter 253
of $SO_x$ 584
precipitation 262, 582
pyrolysis 588
sampling 253-56
scrubbers 245-247
selective catalytic reduction 585
settling chambers 242
sorbent-trap 256
sorption 263
strategy 582
sublimation 261-62
thermal oxidation 583-84
zero ammonia technology (ZAT) 586
Control of pollutants (Soil) 617-24
composting 620
incineration 614-15, 620-21
landfill 614, 619-20
radioactive waste 622-24
recycling 621-22
sampling 256-58
Control of pollutants (Water) 595-615
activated-sludge process 602-04, 651
advanced treatment 606-10
aerated lagoons 604-05, 651
aerobic microbial degradation 601-04
anaerobic microbial degradation 604-05
coagulation 234-35, 598-600
denitrification 160-61, 604, 607
disinfection 611
filtration 262, 602
incineration 614
physicochemical treatment 611
point-of-use treatment 612
pre-treatment 599-600
primary treatment 598-600, 658
removal of inorganics 608-10
removal of metals 612-13
removal of organics 608
removal of sludge 613-15
reverse osmosis 608
sampling 259-63
sand filtration 602
screening 599
secondary treatment 600-05
secondary-tertiary treatment 610-12, 659
stabilization ponds 604
tertiary treatment 605-12
wastewater treatment 597-615
Convection 26-27, 31-35, 52-53
Criteria air pollutants 72, 81


## D


Detectors
atomic emission (AED) 446-47
chemiluminescence (CLD) 450
electrical conductivity (ELCD) 447-49
electron capture (ECD) 447
flame ionization (FID) 449-52
flame photometric (FPD) 450-51
for VOCs 446
gas-filled 532-33

Geiger-Müller counters 534-35
ionization chambers 533-34
photocathode 492
photodiode 491-92
photoemissive 491
photoionization (PID) 451-52
photomultiplier tube (PMT) 491-93
photon 490, 542
photovoltaic 490-91
proportional counters 532-34
radiation 303-04
radon progeny 240-41
scintillation counters 535-37
semiconductor 491-92, 535
thermal conductivity (TCD) 452-53
thermoionic (TID) 450-51
thermoluminescence 303
X-ray 542
Dew point 35, 42, 139
temperature 314, 654
Dialysis
Donnan 424
electro 424-25, 608-10
equilibrium 259
reverse osmosis 425
ultra-filtration 425
Diffusion 51-52, 109
Dispersion
atmospheric 32, 50
of air pollutants 52
plume 53
Dissolved oxygen (DO) 162, 168-70
measurement of 480-81

E

Ecology
environmental 156, 626-27
food chain 157, 655
industrial 627
microbial 159-60
of water bodies 162, 174
plant 158-59
standing crop 158
Electroanalytical methods 465-83
AdSV 470
amperometry 471-72
ASV 468-70
conductimetry 472-74
coulometry 466, 471-72
CSV 470
gasometry 472-73
in pollution monitoring 481-83
oscillometry 472
polarography 470-71
potentiometry 474-78
voltammetry 467-71
Electrophoresis 459-61
capillary zone (CZE) 461
electrophoretic mobility 459
gel 460
immuno 460
PAGE 460
Energy sources
biomass 11, 652
carbon-based 95-96
fuel cells 11-12
geothermal 10, 588
hydro power 11
MHD 13-14
non-conventional 588-89
nuclear 12
problem 9
solar 10, 156-58
wind power 10, 588-89
Environmental cycles
carbon 3, 80, 653
carbon dioxide 92
fluoride 80
hydrologic 6, 128-29
nitrogen 4, 84
nutrient 6
sulfur 5
Environmental health 625-44
bioaccumulation 640, 652
bioaugmentation 642, 652
bioavailability 641-42, 652
biochemical tests 629
biodegradation 640-42, 652
biomonitoring 628-30
bioremediation 642-44
environmental ecology 626-27
environmental engineering 66, 626
industrial ecology 627
phytomonitoring 629, 642-48
toxicity tests 627-30
Environmental systems
atmosphere 17-46
biosphere 2-3, 102, 156-57, 653
ecosphere 156
hydrosphere 123-83
lithosphere 185-200
Eutrophication 4, 159, 176, 654
Evapotranspiration 129, 654

F

FACS 461
Fick's law 33, 51
Flow injection analysis (FIA) 320, 351

Food preservation 632-40
  additives 609, 634
  dietary supplements 638
  freeze-drying 633-34
  genetic engineering methods 636-40
  high-temperature 633
  moisture-solid balance 633
  molecular farming 638-39
  nutritional enhancement 637-38
  quality enhancement 637-40
  radiation treatment 635
  therapeutic foods 638-40
Force
  buoyant 35, 51
  coriolis 31-32, 40-41
  frictional 31
  gravitational 51
Fugacity 137-38

G

Gas chromatography (GC) 442-45
  gas-liquid (GLC) 442
  gas-solid (GSC) 442
  principles of operation 443
Greenhouse effect 93, 95-97, 655
Griess-Salzman method 351

H

Hagen-Poiseulli equation 173
Haldane 119
Hazardous pollution. *See* under Pollutants
Heat-island effect 54
Henderson-Hasselbalch equation 317
Heterotrophs 157, 655
Humidity 138-39, 312
  absolute 138, 312
  relative 139, 312
  specific 138
Hydrogen bond 130-31, 655
Hydrosphere 123-83
Hygrometer 139, 312-13

I

Integrated waste-management 13
Interferometry 290-91, 511-12
Ion-sensitive electrodes 316-19, 475-77
  Clark's cell 319
  enzyme electrodes 477
  gas-sensing electrode 319, 475-77
  ISFET 476
  pH electrode 318, 475
  Severinghaus electrode 319, 345

J

Jablonski diagram 501
Jacobs-Hochmeister method 351, 364

K

Kerma 218, 656

L

Lapse rate
  adiabatic 38-40, 651
  ambient 39
  neutral 39
  sub-adiabatic 39
  super-adiabatic 39
Lithosphere 185-200

M

Mass spectrometry (MS) 455-59
  hyphenated techniques 454-55
  ionization chamber 456-57
  LMMS 294
  mass analyzer 457-59
  quadruple mass analyzer 458-59
  secondary ion (SIMS) 456
Metal toxicity 112-119
  aluminium 115
  anthropogenic effects 8
  arsenic 118-19
  cadmium 115
  encephalopathy 115-16
  itai-itai disease 115, 204
  lead 115-17
  lead palsy 116
  mercury 117-18
  mesothelioma 112, 119
  minamata disease 118
Meteorology 31, 49-50
Methemoglobinemia 350

N

Navier-Stokes equation 135
Nephelometry 295, 492-93
Nernst equation 475
Nernst-Planck equation 466
Nessler's reagent 354
Nitric oxide 85, 349-50
Nitrogen
  cycle 4, 84
  denitrification 160
  determination 349
  fixation 160, 657
  microbial transformation of 160-61

nitrification 160-61, 657
Nitrogen dioxide 85, 350-52
physiological effects 119
Nitrous oxide 84-85, 349
Non-dispersive IR (NDIR) 345, 509-11

**O**

Osmotic pressure 138
Oxygen 4, 88
determination of 340, 478-81
dissolved (DO) 162, 480-81
magnetic detection of 479-80
Ozone 88
depletion 88
determination 341
ground level 80-81, 88
hole 4, 90
layer 88-90
SUBV 341
TOMS 341
toxicity 119

**P**

Pascal 43
Peroxyacetyl nitrate (PAN) 78, 89, 203
Pesticides 205-11, 483
carbamates 209-10
DDT 206-07
mode of action 206
organochlorine 207
organophosphorus 208
persistent 210-11
phenolics 209
pH 316-18
Phagotrophs 157
Photolytic cycle 86
Photosynthesis 3, 105, 159, 657
chlorophyll determination 328
leaf 328
measurement 328
Physical methods in pollution analysis 415-574
barrier separation 424
chromatographic techniques 427-55
diffraction methods 562-64
electroanalytical methods 465-83
electrophoresis 459-61
mass spectrometry (MS) 455-59
microscopic methods 566-67
nuclear analytical techniques 531-59
osmotic methods 424-25
sedimentation 422-23
sorption methods 427-55
spectroscopic methods 485-529, 537-47
Phytoremediation 644
Plume behavior
coning 56
dispersion 53
fanning 57
lofting 57
looping 56
trapping 57
Poiseuille's equation 135
Poisson's equation 50
Pollutants (Air)
acid rain 4, 6, 104-05, 651
aerosols 74-75, 82-83, 652
asbestos 120
chemical 156, 163
chlorofluorocarbons (CFCs) 89-90
classification of 73-75, 234
clinical implications 120
criteria pollutants 72, 81
determinants 48
dispersion modeling 52
diurnal variation 55
effects on animal kingdom 106, 120
effects on breathing 110
effects on ecosystems 103
effects on matter 102
effects on plants 104, 106
episodes 54-55
gaseous 76, 82-93
global scale 80, 103
greenhouse gases 80, 347, 349
hazardous chemicals 203-05
hydrocarbons 76-77
impact of 79-99
infective agents 7
inorganic 74
local scale 80
nature of 72
noise 7, 567
organic 74, 77
origin of 72, 77-78
oxides of carbon 90-93
oxides of nitrogen 84-88
oxides of sulfur 83
oxygen demanding wastes 7
ozone 88-89
particulate 75, 81-82
pesticides 206-10
photochemical smog 85-88
physical forms 72
physicochemical effects 107, 119
physiological effects 107, 111, 119
plant nutrients 7
primary 73, 274, 280, 657

primary standard 121, 658
radioactive 7, 77
reacting constituents 58
regional scale 80
sampling 233, 253-56
secondary 73, 274, 280, 657
secondary standard 121
sulfur dioxide 83
synergic effect 120, 659
thermal stress 93-95, 162-63
toxic effects of 112
toxic substances 202
trace metals 7
Pollutants (Hazardous) 201-12
agrochemicals 205-11
CFCs 93
chemicals 203-05
industrial chemicals 204
organochlorines 207
organophosphates 208
pesticides 206-10
physical classification 202
physiological classification 203
PPCPs 211
toxic substances 202
Pollutants (Particulate)
aerosol 75, 82-83, 297-300
Aitken particles 74, 103
dust 74-75
fly ash 75
fumes 75
mass concentration 295, 300
number concentration 299-300
PM-10 76, 82
PM2.5 75, 82
size concentration 296, 299
smog 75
smoke 75
surface composition 300
toxic effects 112
Pollutants (Radioactive)
artificially produced 217, 221
biological effects 225-26
exposure to 219-24
natural 215-17, 220
radionuclides 214-17, 658
radon and progeny 77, 221
sources of 214-17
Pollutants (Soil) 192-227
agriculture 196
chemical 196-97
domestic waste 195
erosion 193-94
garbage 194
industrial waste 199-200
litter 194
mining waste 195-96
municipal waste 195
radioactive waste 200
refuse 194
rubbish 194
sampling 256-58
solid waste 194-96
toxic waste 202
Pollutants (Water)
acid rain 163-64, 651
agricultural sources 151-52
anthropogenic sources 149-50
atmospheric deposition 152-53
biological factors 164-65
BOD 169-71
chemical factors 163
classification 147
COD 171
domestic sewage 151
ecological effects of 153, 155-63
environmental risks 179
factors affecting 147
in lakes 174
in streams 172
industrial sewage 151
mining wastes 152
monitoring 308, 473
N-BOD 171
non-point sources 150
oxygen demand (OD) 168-71
physical characteristics of 149
point sources 150
pollution tree 150
sources of 149-53, 308
storm sewage 151
thermal stress 162-63
trace metals 153
turbidity 139, 162
Pollution (Noise) 567-73
control of 573
intensity of sound 567-68
measurement of 568-72
physical effects of 572-73
sound levels 568
Pollution (Thermal) 93-95
$CO_2$-balance 95
effects on flora 94
effects on organisms 93-94
global impact 94-95
PPCPs 211
Psychrometer 139, 313
Pulse-height analyser 543

## Q

Quality assurance 245, 266-68
Quantum efficiency 500, 658

## R

Radiation
 biological effects of 225-26
 blackbody 26, 653
 detectors 303-04, 532-37
 electromagnetic 26
 exposure 219, 226
 hazards 220-25
 infrared (IR) 26
 ionizing 217, 486
 solar 24-28
 sources of 490, 514-16
 terrestrial 26
 thermal 28
Radiation-energy balance 26-28
Radioactive pollution. *See* under Pollutants
Radioactivity
 absorbed dose 218, 651
 biological effects of 219-226
 dose equivalent 218, 654
 half-life 218, 655
 hazards 220-25
 measurement of 217-19, 551
 series 215
 sources of 214
Radionuclides 214-15, 224, 658
Raoult's law 136
Reaction kinetics 64-69, 658
 1st-order 65-66
 2nd-order 65-66
 3rd-order 65, 68
 rate constant 65
 zero-order 65-66
Respiration 3, 106, 159
Reverse osmosis 263, 612-13
Reynolds number 107, 658
Richardson number 34, 53
Ring-oven technique 338-39

## S

Sampling methods
 ashing 264-65
 derivatization 258
 dialysis 259, 424-25
 electrochemical 264
 evaporation 261
 extraction 263
 filtration 262-63
 floatation 263
 freeze-out 242
 of gaseous pollutants 253-56
 of liquid-phase pollutants 259
 of organic pollutants 256
 of particulate matter 253
 of soil pollutants 256-58
 sample collection 234-48
 sample preparation 252, 260
 sorbent-trap method 256
 sorption 235, 263
 Soxhlet extraction 258
 statistical methods 268-70
 sublimation 261
 zone melting 265-66
Scattering
 light 51, 297
 Mie 51, 297, 658
 Raman 512
 Rayleigh 51, 297, 512, 658
 Tyndall 310, 492
Separation methods 421-61
 barrier separation 424-25
 chromatography 427-45, 453-55
 classification 422
 dialysis 424-25
 electrodialysis 425
 electrophoresis 459-61
 mass spectrometry (MS) 455-58
 reverse osmosis 425
 sedimentation 108, 422-23
 solid-phase extraction 461
 solvent extraction 425-27
 sorption 427-45, 453
 ultra-filtration 425
Sievert 218
Soil
 chemistry of 191, 97
 clay 191
 components 190-91
 humus 191
 physical characteristics 190
 role of 191
Soil pollution. *See* under Pollutants
Solar constant 26, 659
Spectrophotometry 488-90
Spectroscopy 485-529
 absorption 493-99
 action 493-94
 atomic 514-28
 atomic absorption (AAS) 514-24
 atomic emission (AES) 524-28
 Auger electron. *See* X-ray spectroscopy
 CARS 513
 correlation 495-96

diffusion reflectance (DRS) 494
DOAS 350, 496
Doppler-limited 497-98
electron spin resonance (ESR) 513-14
electrothermal AAS (ET-AAS) 519-20
flame resonance (FR-AAS) 521-22
fluorescence 499-503, 528-29
ICP-AES 526
infrared (IR) 506-12
molecular 487
nondispersive IR (NDIR) 345, 509-12
photoacoustic (PAS) 498-99
Raman 512-13
resonance-ionization (RIS) 498
second-derivative 497
TDLAS 350
thermal lens 498-99
UV-visible 486-98
X-ray. *See* X-ray spectroscopy
Stefan-Boltzmann law 26, 289
Stokes
equation 135, 459
number 108
shift 505-06
Sulfur cycle 5
Sulfur dioxide 83
determination of 357-59
physiological effects 119

**T**

Thermodynamics 59-64
1st law of 59
2nd law of 60
3rd law of 59
endothermic 58, 60
enthalpy 59, 654
entropy 61, 654
exothermic 60
free energy 61-62
heat capacity 60, 655
internal energy 59
principles of 59
specific heat 60
Thermometers
glass bulb 45
liquid-filled 287
ohmic resistance 46, 287
thermistors 287
thermocouples 288
thermopiles 288
Transpiration 105, 129, 659
Turbidimetry 311, 492-93
Tyndall effect 311-12
Tyndallometer 295

**V**

Van der Waals equation 43
Van't Hoff equation 138

**W**

Water
acidity 142, 317
alkalinity 142, 317
biological characteristics 142
characteristics of 130, 596
chemical properties of 139-42
clathrate structure of 131-32
colligative properties of 135
color 139
cycle 129
density of 132
dipole nature of 134
fugacity of 137
hardness 141, 316
hydrogen bonding 130-31
physical properties of 132-35
physicochemical properties of 135-39
potential 133, 325
quality 149
salinity 140-41
solubility 139
spatial structure of 130-31
standards 596
surface tension 133
thermal properties of 133-34
vapour pressure 137
viscosity of 135
Water bodies 167-183
(P-R) cycle 177-79
compensation level 174-75, 653
dissolved oxygen (DO) 168, 172-73
dystrophic 173-74, 654
ecology of 162, 174-75
epilimnion region 176, 654
eutrophic phase 176, 654
hypolimnion region 176, 655
mesotrophic phase 177
metalimnion region 176, 656
mobile 168
oligotrophic phase 177, 657
stationary 168, 173-77
stratification of 175
thermal stratification 176-77
thermocline 176-77, 659
trophogenic zone 174-75, 659
tropholytic zone 174-75, 659
Water pollution. *See* under Pollutants

Water potential 133, 325
Water quality index (WQI) 148
Weather 29
Weber-Fechner law 102
West-Gaeke method 357
Wien's law 26
Winkler method 340

X

X-ray diffraction 562-67
  Bragg's law 545, 564-65
  Debye-Sherrer method 564-65
  phase problem 565
  principles of 563-64
  space group 564
  structure determination by 564-66
  unit cell 564
X-ray spectrometer
  energy dispersion 545-47
  wavelength dispersion 545-46
X-ray spectroscopy 537-47
  Auger electron 548-50, 652
  ion beam analysis 550-55
  ion scattering (ISS) 547-48
  photoelectron absorption 547-48, 657
  Rutherford backscattering (RBS) 550
  TRXF 544-45
X-ray absorption 543-44
X-ray fluorescence (XRF) 544-45
X-ray line spectra 538-40
X-ray sources 540

Z

Zwaarder number 283